AF616045

ASTÉRISQUE 316

CRYSTALLINE COHOMOLOGY OF ALGEBRAIC STACKS AND HYODO-KATO COHOMOLOGY

Martin C. Olsson

Société Mathématique de France 2007
Publié avec le concours du Centre National de la Recherche Scientifique

Martin C. Olsson

Department of Mathematics, University of California,
Berkeley, CA 94720-3840, USA.

E-mail : molsson@math.berkeley.edu

2000 ***Mathematics Subject Classification.*** — 14F20, 14F30, 14G20, 11G25.

Key words and phrases. — Algebraic stacks, crystalline cohomology, C_{st}-conjecture, Hyodo-Kato isomorphism, de Rham-Witt complex.

CRYSTALLINE COHOMOLOGY OF ALGEBRAIC STACKS AND HYODO-KATO COHOMOLOGY

Martin C. Olsson

Abstract. — In this text we study using stack-theoretic techniques the crystalline structure on the de Rham cohomology of a proper smooth scheme over a p-adic field and applications to p-adic Hodge theory. We develop a general theory of crystalline cohomology and de Rham-Witt complexes for algebraic stacks, and apply it to the construction and study of the (φ, N, G)-structure on de Rham cohomology. Using the stack-theoretic point of view instead of log geometry, we develop the ingredients needed to prove the C_{st}-conjecture using the method of Fontaine, Messing, Hyodo, Kato, and Tsuji, except for the key computation of p-adic vanishing cycles. We also generalize the construction of the monodromy operator to schemes with more general types of reduction than semistable, and prove new results about tameness of the action of Galois on cohomology.

Résumé **(Cohomologie cristalline des champs algébriques et isomorphisme de Hyodo-Kato)**

Dans ce texte, nous étudions, par des techniques « champêtres », la structure cristalline sur la cohomologie de de Rham d'un schéma propre et lisse sur un corps p-adique et ses applications à la théorie de Hodge p-adique. Nous développons une théorie générale de la cohomologie cristalline et des complexes de de Rham-Witt associés aux champs algébriques, et l'appliquons à la construction et à l'étude de la (φ, N, G)-structure sur la cohomologie de de Rham. Nous plaçant du point de vue des champs plutôt que de la géométrie logarithmique, nous développons les ingrédients nécessaires à la démonstration de la conjecture C_{st} suivant la voie de Fontaine, Messing, Hyodo, Kato et Tsuji (en laissant de côté le calcul-clé des cycles évanescents p-adiques). Nous généralisons aussi la construction des opérateurs de monodromie aux schémas au-delà du cas semi-stable, et obtenons de nouveaux résultats sur le caractère modéré de l'action galoisienne sur la cohomologie.

CONTENTS

INTRODUCTION

This text grew out of an attempt to understand the theory of log crystalline cohomology and its application to the C_{st} conjecture of Fontaine and Jannsen using the stack-theoretic techniques introduced in [**62**].

Before explaining the contents of the paper let us briefly review the statements of these conjectures, now proven in different ways by Tsuji [**73**], Faltings [**20, 21**], and Niziol [**53, 54**].

Let K be a complete discrete valuation field of mixed characteristic $(0, p)$ with ring of integers V and perfect residue field k. Let $K \hookrightarrow \overline{K}$ be an algebraic closure of K, let $K_0 \subset K$ be the field of fractions of the ring of Witt-vectors of the residue field of V, and let $K_0^{ur} \subset \overline{K}$ denote the maximal unramified extension of K_0 in $\overline{K}$. There is a canonical automorphism $\sigma : K_0^{ur} \to K_0^{ur}$ induced by the Frobenius on the residue fields. Let G denote the Galois group $\mathrm{Gal}(\overline{K}/K)$. The group G acts by restriction also on K_0^{ur}.

Let X/K be a smooth proper scheme. Associated to X are the de Rham cohomology groups $H^*_{\mathrm{dR}}(X/K)$ and the p-adic étale cohomology groups $H^*(\overline{X}, \mathbb{Q}_p)$, where $\overline{X}$ denotes the base change of X to $\overline{K}$. The space $H^*_{\mathrm{dR}}(X/K)$ comes equipped with the Hodge filtration Fil_H, and the space $H^*(\overline{X}, \mathbb{Q}_p)$ has a continuous action of the absolute Galois group $G_K := \mathrm{Gal}(\overline{K}/K)$. The conjectures of Fontaine concern the relationship between $H^*_{\mathrm{dR}}(X/K)$ and $H^*(\overline{X}, \mathbb{Q}_p)$.

One of the key ingredients in the C_{dR}-conjecture relating these two cohomology theories is the construction of a so-called (φ, N, G)-module structure on $H^*_{\mathrm{dR}}(X/K)$ in the following sense.

Definition 0.1.1. — A *(φ, N, G)-module* is a collection of data (D, φ, N) as follows:

(0.1.1.1) A finite-dimensional K_0^{ur}-vector space D with a σ-linear automorphism φ.

(0.1.1.2) A K_0^{ur}-linear nilpotent endomorphism N of D such that $N\varphi = p\varphi N$.

(0.1.1.3) A continuous semilinear (with respect to the natural action of G on K_0^{ur}) action of G on D such that for all $g \in G$, $\varphi \circ g = g \circ \varphi$ and $N \circ g = g \circ N$.

A *filtered* (φ, N, G)*-module* is a collection of data $(D, \varphi, N, \mathrm{Fil})$, where (D, φ, N) is a (φ, N, G)-module and in addition there is the following structure.

(0.1.1.4) A decreasing, separated, and exhaustive filtration Fil on $D_{\overline{K}} := D \otimes_{K_0^{ur}} \overline{K}$ stable under the diagonal action of G.

The category of filtered (φ, N, G)-modules is denoted $\underline{MF}_{\overline{K}/K}(\varphi, N)$.

If T is a finite dimensional K-vector space, then a *(filtered)* (φ, N, G)*-module structure on* T is a (filtered) (φ, N, G)-module (D, φ, N) together with isomorphisms $\rho_\pi : T \simeq (D \otimes_{K_0^{ur}} \overline{K})^G$ for each choice of uniformizer π in K such that if $\pi' = u\pi$ then

$$\rho_\pi = \rho_{\pi'} \exp(\log(u)N), \tag{0.1.1.5}$$

where log denotes the p-adic logarithm.

It follows from the proof of the C_{st}-conjecture that the de Rham cohomology $H^*_{\mathrm{dR}}(X/K)$ has a natural filtered (φ, N, G)-structure such that the filtration on $H^*_{\mathrm{dR}}(X/K)$ is the Hodge filtration. More precisely, consider the ring B_{st} of Fontaine [**23**, §3]. Let $\mathrm{Rep}(G)$ denote the category of finite dimensional $\mathbb{Q}_p$-vector spaces with continuous action of the Galois group G. To any Galois representation $V \in \mathrm{Rep}(G)$ one can associate a (φ, N, G)-module as follows. The ring B_{st} comes equipped with a semi-linear Frobenius endomorphism φ, an operator N, and an action of G satisfying certain compatibilities. Furthermore, the choice of a uniformizer $\pi \in K$ defines an inclusion $B_{\mathrm{st}} \otimes_{K_0} K \hookrightarrow B_{\mathrm{dR}}$, where B_{dR} is as in [**23**, §1]. In particular, $B_{\mathrm{st}} \otimes_{K_0} K$ inherits a filtration from B_{dR}. For any finite extension $K \subset L \subset \overline{K}$ let $G_L \subset G$ denote the subgroup $\mathrm{Gal}(\overline{K}/L)$. Since $B_{\mathrm{st}}^{G_L} = L_0$ (the ring of Witt vectors of the residue field of L), we can define a K_0^{ur}-space

$$D_{\mathrm{pst}}(V) := \varinjlim_{K \subset L \subset \overline{K}} (B_{\mathrm{st}} \otimes_{\mathbb{Q}_p} V)^{G_L}. \tag{0.1.1.6}$$

The operators φ and N on B_{st} induce a (φ, N, G)-module structure on $D_{\mathrm{pst}}(V)$, and the inclusion $D_{\mathrm{pst}}(V) \otimes_{K_0^{ur}} \overline{K} \subset B_{\mathrm{dR}} \otimes_{\mathbb{Q}_p} V$ obtained by passing to the limit from the inclusions $B_{\mathrm{st}} \otimes_{L_0} L \subset B_{\mathrm{dR}}$ induces a filtration on $D_{\mathrm{pst}}(V) \otimes_{K_0^{ur}} \overline{K}$. We therefore obtain a functor

$$D_{\mathrm{pst}} : \mathrm{Rep}(G) \longrightarrow \underline{MF}_{\overline{K}/K}(\varphi, N). \tag{0.1.1.7}$$

There is also a functor

$$V_{\mathrm{pst}} : \underline{MF}_{\overline{K}/K}(\varphi, N) \longrightarrow \mathrm{Rep}(G) \tag{0.1.1.8}$$

sending (D, φ, N) to the G-representation

(0.1.1.9)

$$V_{\mathrm{pst}}(D, \varphi, N) := \{v \in B_{\mathrm{st}} \otimes_{K_0^{ur}} D | Nv = 0, \varphi(v) = v, \text{ and } v \otimes 1 \in \mathrm{Fil}^0(D \otimes_{K_0^{ur}} \overline{K})\}.$$

For $(D, \varphi, N) \in \underline{MF}_{\overline{K}/K}(\varphi, N)$ one can also define $(D \otimes_{K_0^{ur}} \overline{K})^G$. This is a filtered K-vector space.

For $V \in \mathrm{Rep}(G)$, there is a natural map

$$\alpha : B_{\mathrm{st}} \otimes_{K_0^{ur}} D_{\mathrm{pst}}(V) \longrightarrow B_{\mathrm{st}} \otimes_{\mathbb{Q}_p} V. \tag{0.1.1.10}$$

The representation V is called *potentially semistable* if this map α is an isomorphism [**24**, 5.6.1]. By [**24**, 5.6.7] the functor D_{pst} is fully faithful when restricted to the subcategory $\mathrm{Rep}^{\mathrm{pst}}(G) \subset \mathrm{Rep}(G)$ of potentially semistable representations, and if $\underline{MF}^{\mathrm{adm}}_{\overline{K}/K}(\varphi, N) \subset \underline{MF}_{\overline{K}/K}(\varphi, N)$ denotes its essential image, then a quasi-inverse

$$\underline{MF}^{\mathrm{adm}}_{\overline{K}/K}(\varphi, N) \longrightarrow \mathrm{Rep}^{\mathrm{pst}}(G) \tag{0.1.1.11}$$

is provided by V_{pst}.

We also consider the subcategory $\mathrm{Rep}^{\mathrm{st}}(G) \subset \mathrm{Rep}^{\mathrm{pst}}(G)$ consisting of representations V for which the natural map

$$B_{\mathrm{st}} \otimes_{K_0} (B_{\mathrm{st}} \otimes_{\mathbb{Q}_p} V)^G \longrightarrow B_{\mathrm{st}} \otimes_{\mathbb{Q}_p} V \tag{0.1.1.12}$$

is an isomorphism. Such representations are called *semistable* and the essential image of $\mathrm{Rep}^{\mathrm{st}}(G)$ in $\underline{MF}^{\mathrm{adm}}(\varphi, N)$ is precisely those (φ, N, G)-modules (D, φ, N) for which the action of G on D is trivial (that is, the natural map $K_0^{ur} \otimes_{K_0} D^G \to D$ is an isomorphism). Let D_0 denote the space D^G. The operators φ and N induce operators, denoted by the same letters, on D_0 and the filtration descends to $D_0 \otimes_{K_0} K$.

We can now state the C_{st}-conjecture as follows:

Theorem 0.1.2 **([73, 20, 21, 53])**. — *Let X/K be a smooth proper scheme of dimension d with semistable reduction, and let m be an integer. Then the Galois representation $V = H^m(\overline{X}, \mathbb{Q}_p)$ is semistable and for any choice of uniformizer in K there is an isomorphism $K \otimes_{K_0} D_0 \simeq H^m_{\mathrm{dR}}(X/K)$ compatible with the filtrations (where $H^m_{\mathrm{dR}}(X/K)$ is filtered by the Hodge filtration).*

Using de Jong's alterations theorem [**37**] one can deduce from this the following so-called C_{pst}-conjecture:

Theorem 0.1.3 **([74])**. — *Let X/K be a smooth proper scheme and m an integer. Then the Galois representation $V = H^m(\overline{X}, \mathbb{Q}_p)$ is potentially semistable, and for any choice of uniformizer in K there is an isomorphism $(\overline{K} \otimes_{K_0^{ur}} D_{\mathrm{pst}}(V))^G \simeq H^m_{\mathrm{dR}}(X/K)$ compatible with the filtrations.*

Since one can recover the Galois representation $H^m(\overline{X}, \mathbb{Q}_p)$ from $D_{\mathrm{pst}}(V)$, it is of great interest to understand in more detail the (φ, N, G)-module $D_{\mathrm{pst}}(V)$. In the case of semistable reduction, the module $D_{\mathrm{pst}}(V)$ can be constructed using the theory of log geometry and log crystalline cohomology developed by Fontaine, Illusie, and Kato [**40**]. Unfortunately, the construction in general is based on an abstract "independence of model argument" and de Jong's alterations and so does not directly yield information about the Galois representation.

The starting point for this work is the paper [**62**] which gives a dictionary between logarithmic geometry and "ordinary" geometry of schemes over certain algebraic stacks. This suggests the possibility of giving a construction of the (φ, N, G)-structure using a theory of crystalline cohomology for algebraic stacks rather than logarithmic geometry. One of the principal aims of this text is to work out this idea and give such a new interpretation of the (φ, N, G)-structure. Of course since there truly is a dictionary between the stack-theoretic approach to the (φ, N, G)-structure and the approach using log geometry (as explained in Chapter 9), in most situations one can use either approach depending on one's taste. Nonetheless, we feel that the stack-theoretic approach is more concrete in the sense that it essentially relies only on the classical theory of crystalline cohomology and descent theory, whereas the logarithmic approach requires the development of a new theory mimicking the classical situation in the logarithmic world. The stack-theoretic approach also helps "explain" why the logarithmic theory works out so smoothly, and of course the general theory of crystalline cohomology developed in this text is more general than the logarithmic theory.

Aside from giving a new perspective on the (φ, N, G)-structure in the semistable reduction case, this stack-theoretic approach also yields more information in some cases than the logarithmic theory. In particular, in the case when the scheme X has so-called "log smooth reduction" but not necessarily semistable reduction (or not "Cartier type" reduction in the logarithmic language), the stack-theoretic approach gives a direct construction of the (φ, N, G)-structure which does not involve the use of alterations in the sense of de Jong [**37**]. This construction yields new theorems about tameness of the Galois action on the module $D_{\mathrm{pst}}(V)$ (see the discussion of Chapter 7 below).

Before studying the (φ, N, G)-structure, however, we must first develop a theory of crystalline cohomology for algebraic stacks. This is done in Chapters 1–4. Because of the importance that stacks and their cohomology have played in recent years, we develop the theory in greater generality than strictly necessary for the applications to the conjecture of Fontaine and Jannsen. Aside from being more aesthetically pleasing, we are hopeful that this general theory will find applications in the future.

In Chapter 1 we develop the basic definitions and results about the crystalline topos of a stack. We consider both the lisse-étale crystalline site defined for an Artin stack and the étale crystalline site defined for Deligne-Mumford stacks. Because of the non-functoriality of the lisse-étale topos of an algebraic stack (see for example [**68**]), there are a number of technical issues which are worked out in this chapter.

In Chapter 2 we study the notion of crystal and differential calculus for algebraic stacks. Most of the results are developed for representable morphisms from Deligne-Mumford stacks to Artin stacks which is a somewhat restrictive setting. However, since computations using the relationship between crystalline and de Rham cohomology are usually performed locally on the source this is sufficient for most purposes.

In Chapter 3 we turn to the main technical difficulty in trying to extend the crystalline theory to algebraic stacks. The crystalline cohomology of stacks is in general a mixture of crystalline cohomology of schemes, Lie algebra cohomology, and group cohomology. It is therefore not surprising that one must impose some conditions in order to carry over some of the classical results to stacks. The main difficulty arises when considering the Frobenius morphism $F : \mathcal{S} \to \mathcal{S}$ of an algebraic stack in characteristic $p > 0$. Not only is the Frobenius morphism F usually not representable, the cohomology sheaves $R^iF_*\mathcal{O}_{\mathcal{S}}$ may be not be zero for $i > 0$. Thus the Frobenius morphism of a stack behaves wildly when compared to the Frobenius morphism of a scheme. This difficulty becomes especially problematic when trying to generalize the Cartier isomorphism to stacks. Nonetheless, with some assumptions which can easily be checked in practice (in particular for the stacks we consider in subsequent chapters), the Cartier isomorphism and its consequences (in particular Ogus' generalization of Mazur's theorem) can be generalized to the stack-theoretic context.

In Chapter 4 we generalize the theory of the de Rham-Witt complex [**34**] to the stack-theoretic setting. The main case of interest is the situation of a smooth representable morphism $\mathcal{X} \to \mathcal{S}$ over a perfect field of characteristic p, with $\mathcal{X}$ a Deligne-Mumford stack and $\mathcal{S}$ an Artin stack such that the cohomology of the Frobenius morphism behaves as if $\mathcal{S}$ was a perfect scheme (we call such a stack a "perfect stack"). Because we wish to work locally in the lisse-étale topology on $\mathcal{S}$, in order to develop a de Rham-Witt theory with an algebraic stack as a base we first study de Rham-Witt theory for schemes over a non-perfect base.

There are at least two different descriptions of the classical de Rham-Witt procomplex $W.\Omega^\bullet_{X/S}$ of a smooth morphism $X \to S$ with S perfect. In the first approach, one defines $W.\Omega^\bullet_{X/S}$ as the initial object in a certain category of differential graded algebras with operators F and V. The second description is based on the comparison with crystalline cohomology which shows that one can also define the de Rham-Witt pro-complex by taking $W_n\Omega^q_{X/S} := \sigma^{-n*}R^qu_{X/W_n(S)*}\mathcal{O}_{X/W_n(S)}$, where $\sigma : W_n(S) \to W_n(S)$ denotes the canonical lift of Frobenius and $u_{X/S} : (X/W_n(S))_{\text{cris}} \to X_{\text{et}}$ is the projection from the crystalline topos to the étale topos.

When S is not perfect, one can try to generalize both approaches. A generalization of the first one has been developed by Langer and Zink [**47**] and gives rise to a pro-differential graded algebra we denote by $W_n^{\text{LZ}}\Omega^\bullet_{X/S}$. The second approach via crystalline cohomology gives a pro-differential graded algebra which we denote by $\mathcal{A}^\bullet_{n,X/S}$. There is a canonical map $W_n^{\text{LZ}}\Omega^\bullet_{X/S} \to \mathcal{A}^\bullet_{n,X/S}$, but this map is usually not an isomorphism.

Let k be a perfect field of characteristic $p > 0$ and let W be the ring of Witt vectors of k. Let $\mathcal{S}/W$ be a flat algebraic stack with reduction $\mathcal{S}_0$ a perfect stack, and let $\mathcal{X} \to \mathcal{S}_0$ be a locally separated smooth representable morphism of algebraic stacks with $\mathcal{X}$ a Deligne-Mumford stacks. The second approach to the de Rham-Witt complex via crystalline cohomology gives rise to a pro-differential graded algebra $\mathcal{A}^\bullet_{n,\mathcal{X}/\mathcal{S}}$ with operators F and V, and we define the de Rham-Witt complex of $\mathcal{X}/\mathcal{S}$ by $W_n\Omega^\bullet_{\mathcal{X}/\mathcal{S}} := \sigma^{-n*}\mathcal{A}^\bullet_{n,\mathcal{X}/\mathcal{S}}$. In Chapter 4 we prove that this enjoys many of the usual properties of the de Rham-Witt complex. In particular, there is a comparison theorem relating it to crystalline cohomology which gives rise to a slope spectral sequence which can be studied using the classical methods of [**34**]. On the other hand, we show in Chapter 4 that the de Rham-Witt complex $W_n\Omega^\bullet_{\mathcal{X}/\mathcal{S}}$ can also be described using descent theory by choosing a smooth cover $S \to \mathcal{S}$ of $\mathcal{S}$ by a scheme and considering the Langer-Zink de Rham-Witt complex $W_n^{\text{LZ}}\Omega^\bullet_{X_\bullet/S_{0,\bullet}}$, where $S_\bullet$ denotes the 0-coskeleton of $S \to \mathcal{S}$ and $X_\bullet := \mathcal{X}\times_{\mathcal{S}}S_\bullet$ (cf. 4.6.7). This is important for technical reasons as it enables one to construct maps from $W_n\Omega^\bullet_{\mathcal{X}/\mathcal{S}}$ to other differential graded algebras using the universal property of the Langer-Zink de Rham-Witt complex. In particular, this gives a key technical result (4.6.9) needed for the study of the C_{st}-conjecture which also "explains" the key technical lemma of Hyodo-Kato [**31**, 4.8] in the logarithmic approach.

In Chapters 5–7 we turn to the study of the Hyodo-Kato isomorphism.

Let k denote a perfect field of characteristic p, W the ring of Witt vectors of k, and $W\langle t\rangle$ the p-adic completion of the divided power envelope of the surjection $W[t] \to k$ sending t to 0. The ring $W\langle t\rangle$ has a canonical lift of Frobenius defined to by the canonical lift of Frobenius σ to W and the map $t \mapsto t^p$.

In Chapter 5 we study some abstract semi-linear algebra over the ring $W\langle t\rangle$, as well as projective systems of $W_n\langle t\rangle := W\langle t\rangle \otimes \mathbb{Z}/p^{n+1}$-modules. Our interest in this semi-linear algebra arises as follows. Let $\mathcal{X}/W$ be a semistable proper scheme (more generally we consider also $\mathcal{X}$ defined over a ramified extension of W, but in this introduction we restrict for simplicity to the case when $\mathcal{X}$ is defined over W). Then we construct in Chapter 6 an algebraic stack $\mathcal{S}_{W\langle t\rangle}$ flat over $W\langle t\rangle$ and a smooth map $\mathcal{X}_0 \to \mathcal{S}_{W\langle t\rangle} \otimes_{W\langle t\rangle} k$. The stack $\mathcal{S}_{W\langle t\rangle}$ is obtained from a modification of the following construction. Let $\alpha_1, \dots, \alpha_r$ be positive integers and consider the scheme

$$\mathrm{Spec}\big(W\langle t\rangle[X_1, \dots, X_r, V^{\pm}]/(X_1^{\alpha_1} \cdots X_r^{\alpha_r} V - t)\big). \tag{0.1.3.1}$$

There is an action of $\mathbb{G}_m^r$ on this scheme for which $(u_1, \dots, u_r) \in \mathbb{G}_m^r$ acts by

$$X_i \longmapsto u_i X_i, \quad V \longmapsto u_1^{-\alpha_1} \cdots u_r^{-\alpha_r} V. \tag{0.1.3.2}$$

The quotient

$$\big[\mathrm{Spec}\big(W\langle t\rangle[X_1, \dots, X_r, V^{\pm}]/(X_1^{\alpha_1} \cdots X_r^{\alpha_r} V - t)\big)/\mathbb{G}_m^r\big] \tag{0.1.3.3}$$

is then étale over $\mathcal{S}_{W\langle t\rangle}$.

Remark 0.1.4. — In fact it is possible to prove the C_{st}-conjecture using only the stacks (0.1.3.3). However working with the stack $\mathcal{S}_{W\langle t\rangle}$ makes certain constructions more canonical so we choose to use $\mathcal{S}_{W\langle t\rangle}$ instead of (0.1.3.3).

Let $\mathrm{ps}(W\langle t\rangle)$ denote the category of projective systems $M.$ of $W\langle t\rangle$-modules with M_n annihilated by p^{n+1} (so M_n is a $W_n\langle t\rangle$-module), and let $\mathrm{ps}(W\langle t\rangle)_{\mathbb{Q}}$ denote the category whose objects are the same as the objects of $\mathrm{ps}(W\langle t\rangle)$ but whose morphisms are given by

$$\mathrm{Hom}_{\mathrm{ps}(W\langle t\rangle)_{\mathbb{Q}}}(M., N.) := \mathrm{Hom}_{\mathrm{ps}(W\langle t\rangle)}(M., N.) \otimes \mathbb{Q}. \tag{0.1.4.1}$$

One of the key technical results (a variant of [**31**, 5.2]) used in the proof of the C_{st}-conjecture is then to show that there is a canonical isomorphism in the category $\mathrm{ps}(W\langle t\rangle)_{\mathbb{Q}}$

$$\big\{H^*((\mathcal{X}_0/\mathcal{S}_{W_n})_{\mathrm{cris}}, \mathcal{O}_{\mathcal{X}_0/\mathcal{S}_{W_n}}) \otimes_W W\langle t\rangle\big\} \simeq \big\{H^*((\mathcal{X}_0/\mathcal{S}_{W_n\langle t\rangle})_{\mathrm{cris}}, \mathcal{O}_{\mathcal{X}_0/\mathcal{S}_{W_n\langle t\rangle}})\big\} \tag{0.1.4.2}$$

compatible with the Frobenius endomorphisms and the projections to the projective system $\{H^*((\mathcal{X}_0/\mathcal{S}_{W_n})_{\mathrm{cris}}, \mathcal{O}_{\mathcal{X}_0/\mathcal{S}_{W_n}})\}$.

Following Hyodo and Kato's approach in [**31, 41**], we construct the isomorphism (0.1.4.2) by studying the slope spectral sequence
(0.1.4.3)

$$E_1^{pq} = H^q(\mathfrak{X}_{0,\mathrm{et}}, R^p u_{\mathfrak{X}_0/\mathcal{S}_{W_n\langle t\rangle}*}\mathcal{O}_{\mathfrak{X}_0/\mathcal{S}_{W_n\langle t\rangle}}) \Longrightarrow H^{p+q}((\mathfrak{X}_0/\mathcal{S}_{W\langle t\rangle})_{\mathrm{cris}}, \mathcal{O}_{\mathfrak{X}_0/\mathcal{S}_{W\langle t\rangle}}).$$

The construction has two basic components.

(1) In Chapter 5 we prove an "abstract Hyodo-Kato isomorphism" (see 5.3.33 for the precise statement) in a general setting of a spectral sequence in the category $\mathrm{ps}(W\langle t\rangle)_{\mathbb{Q}}$. This result is a corollary of our study of F-isocrystals over $W\langle t\rangle$, based on Ogus' notion of "twisted inverse limits" [**59**].

(2) In order to apply this abstract Hyodo-Kato isomorphism to the slope spectral sequence (0.1.4.3) we must show that the E_1-terms

$$H^q(\mathfrak{X}_{0,\mathrm{et}}, R^p u_{\mathfrak{X}_0/\mathcal{S}_{W_n\langle t\rangle}*}\mathcal{O}_{\mathfrak{X}_0/\mathcal{S}_{W_n\langle t\rangle}}) \qquad (0.1.4.4)$$

satisfy a certain finiteness condition (they are "free of finite type modulo torsion" in the sense of 5.1.14). This finiteness property depends on some subtle geometric properties of the stacks $\mathcal{S}_{W\langle t\rangle}$ and $\mathcal{S}_W$ (and variants of these stacks), and the comparison of the de Rham-Witt complex $W_n\Omega^\bullet_{\mathfrak{X}/\mathcal{S}_W}$ to the Langer-Zink de Rham-Witt complex proven in Chapter 4.

In Chapter 6, we describe the algebraic stacks $\mathcal{S}_H(\alpha)$ which serve as the base stacks in the construction of the (φ, N, G)-structure on de Rham cohomology when the scheme X/K in the above has "log smooth reduction" in a suitable sense. We verify in particular the necessary finiteness conditions on the E_1-terms (0.1.4.4) for the results of Chapter 5 to be applicable. The stacks $\mathcal{S}_H(\alpha)$ are defined over $\mathbb{Z}[t]$, and for every integer $e \geq 1$ there is a canonical map

$$\Lambda_e : \mathcal{S}_H(\alpha) \longrightarrow \mathcal{S}_H(\alpha) \qquad (0.1.4.5)$$

covering the map $\mathbb{Z}[t] \to \mathbb{Z}[t]$ sending t to t^e. A case of particular interest is when e is equal to a prime p, in which case the map Λ_p defines a lifting of the relative Frobenius morphism of the reduction mod p of $\mathcal{S}_H(\alpha)$. The key technical result is the calculation 6.3.18 computing $\Lambda_{p^n*}\mathcal{O}_{\mathcal{S}_H(\alpha)}$ in certain cases. This computation then yields 6.3.26 which in turn enables us to apply the stack-theoretic version of Ogus generalization of Mazur's theorem 3.4.38 which is the heart of the Hyodo-Kato isomorphism.

At the end of Chapter 6 and in Chapter 7 we then construct the Hyodo-Kato isomorphism and the (φ, N, G)-structure on de Rham cohomology, first in the case of semistable reduction (6.4 and 6.5) and then in Chapter 7 for schemes with more general kinds of reduction. This direct construction of the (φ, N, G)-structure yields information not readily available from the approach using alterations.

In Chapter 8 we then turn to the C_{st}-conjecture. In this chapter, we rework most of the paper [**41**] from the stack-theoretic point of view. The reader familiar with the approach using log geometry may wish to read this section with [**41**] nearby to compare the two approaches. We develop in detail using stacks instead of log geometry the aspects of Tsuji's proof of the C_{st}-conjecture (at least when the dimension d of X satisfies $d < p - 1$) which previously used the logarithmic theory. This includes a crystalline description of the ring B_{st}, a crystalline description of $(B_{\mathrm{st}} \otimes D^m)^{N=0}$, as well as the definition of syntomic complexes. For the convenience of the reader we also give an outline of the remaining aspects of the proof (essentially Tsuji's key computation of p-adic vanishing cycles). We hope that the reader not familiar with Tsuji's proof can read this chapter, understand the outline of the proof, and fill the remaining parts using Tsuji's paper [**73**].

Let V, K, K_0, G etc. be as in the beginning of the introduction, and let X/K be a smooth proper scheme. We say that X has *log smooth reduction* if there exist a flat proper integral scheme $\mathcal{X}/V$ with generic fiber X and a dense open set $\mathcal{U} \subset \mathcal{X}$ such that étale locally on $\mathcal{X}$ there exists a finitely generated integral monoid P, a map $\theta : \mathbb{N} \to P$ such that the following hold:

0.1.4 (i) The induced map $\mathbb{Z} \to P^{\mathrm{gp}}$ is injective with cokernel p-torsion free (where $P \to P^{\mathrm{gp}}$ denotes universal map from P to a group)

0.1.4 (ii) There exists an étale V-morphism

$$\mathcal{X} \longrightarrow \mathrm{Spec}(V \otimes_{\mathbb{Z}[\mathbb{N}]} \mathbb{Z}[P]), \tag{0.1.4.6}$$

where $\mathbb{Z}[P]$ and $\mathbb{Z}[\mathbb{N}]$ denote the monoid algebras, the map $\mathbb{Z}[\mathbb{N}] \to \mathbb{Z}[P]$ is the map induced by θ, and the map $\mathbb{Z}[\mathbb{N}] \to V$ sends $1 \in \mathbb{N}$ to π.

0.1.4 (iii) The inverse image under (0.1.4.6) of the open set

$$\mathrm{Spec}(V \otimes_{\mathbb{Z}[\mathbb{N}]} \mathbb{Z}[P^{\mathrm{gp}}]) \subset \mathrm{Spec}(V \otimes_{\mathbb{Z}[\mathbb{N}]} \mathbb{Z}[P]) \tag{0.1.4.7}$$

is equal to $\mathcal{U}$.

Remark 0.1.5. — As explained in 9.7, if in addition we can choose the monoid P to be saturated, then a result of Kato (see [**42**, 11.6] and [**55**, 2.6]) implies that the open set $\mathcal{U} \subset \mathcal{X}$ defines a natural log structure $M_{\mathcal{X}}$ on $\mathcal{X}$ such that if M_V denotes the natural log structure on M_V then the morphism $(\mathcal{X}, M_{\mathcal{X}}) \to (\mathrm{Spec}(V), M_V)$ is log smooth. In 9.7.3 we show that if $(\mathcal{X}, \mathcal{U})$ is as above (with P not necessarily saturated), then the normalization $\mathcal{X}'$ of $\mathcal{X}$ with the inverse image $\mathcal{U}' \subset \mathcal{X}'$ of $\mathcal{U}$ also satisfies the above conditions. Hence a smooth proper scheme X/K has log smooth reduction if and only if there exists a log smooth proper morphism $(\mathcal{X}, M_{\mathcal{X}}) \to (\mathrm{Spec}(V), M_V)$ whose underlying morphism of schemes on the generic fiber is equal to $X \to \mathrm{Spec}(K)$.

Remark 0.1.6. — The open set $\mathcal{U} \cap X \subset X$ may be strictly smaller than X. For example, if $\mathcal{U} \subset \mathcal{X}$ is étale locally isomorphic to

$$\mathrm{Spec}\big(K[x^{\pm}, y^{\pm}, z^{\pm}]/(z = \pi^{-1}xy)\big) \subset \mathrm{Spec}\big(V[x, y, z]/(xy = \pi z)\big). \tag{0.1.6.1}$$

Hence the above terminology is perhaps misleading since there may not exist a log smooth proper morphism $(\mathcal{X}, M_{\mathcal{X}}) \to (\mathrm{Spec}(V), M_V)$ whose generic fiber has trivial log structure and is equal to X. We hope this does not cause confusion.

The following is a p-adic analogue of results of Rapoport and Zink [**70**] and C. Nakayama [**52**, 0.1.1] for ℓ-adic cohomology.

Theorem 0.1.7. — *If X/K has log smooth reduction, then the action of G on D^m factors through a tame quotient.*

Theorem 0.1.8. — *Let X/K be as in 0.1.7, and assume X/K admits a log smooth model $(\mathfrak{X}, \mathfrak{U})/V$ with $\mathfrak{X}$ regular and such that the reduced closed fiber of $\mathfrak{X}$ is a divisor with simple normal crossings (note in this case we can just take $\mathfrak{U}$ equal to $X \subset \mathfrak{X}$). Let $\alpha_1, \ldots, \alpha_r$ be the multiplicities of the components of the closed fiber. If N denotes the product of the integers $\alpha_i/p^{\nu_p(\alpha_i)}$, then the action of G on D^m factors through the Galois group of the Galois closure of the field extension $K \subset K(\pi^{1/N})$.*

Theorem 0.1.8 is proven in Chapter 7. As explained in 9.7, theorem 0.1.7 follows from 0.1.8. In certain non log smooth cases we are also able to apply our techniques to obtain control over the amount of wildness involved in the action of G on D^m (see 7.2.14).

In the final Chapter 9, we explain how the theory developed in this text relates to the logarithmic theory. We explain how the stack-theoretic approach enables one to recover the theory of log crystalline cohomology, the log Cartier isomorphism, and also explain the equivalence between the approach to the (φ, N, G)-structure taken in this paper and the logarithmic approach.

Acknowledgements. — It should be clear that our construction of the (φ, N, G)-structure is based to a large extent on the ideas in [**31, 41, 59, 73, 74**]. In particular, Chapter 8 is essentially a translation into the stack-theoretic language of [**41**] and section 8.5 is essentially contained in [**74**, Appendix]. Our aim in the discussion on the C_{st}-conjecture is largely to provide a new point of view, though of course there are new results as well.

The author is grateful to L. Illusie whose questions and comments have been very helpful and inspiring, A. Abbes who among other things provided very helpful feedback on the preprint "Crystalline cohomology of schemes with tame reduction", as well as to two referees of this preprint whose helpful comments and suggestions encouraged us to pursue the more ambitious project found in this text. The referee of the present monograph did a truly remarkable job making an extraordinary number of corrections and helpful suggestions, both mathematical and stylistic. We also benefited from conversations with A. J. de Jong. Finally the author is grateful to A. Ogus for his support and encouragement. This paper grew out of the author's thesis written under professor Ogus, and it would not be in existence without his help and advice.

At various times during this research the author has received financial support from the Clay mathematics Institute, the Mathematical Sciences Research Institute in Berkeley, the Institute for Advanced Study in Princeton, and an NSF post-doctoral research fellowship.

0.2. Preliminaries and conventions

0.2.1. — We generally follow the conventions about algebraic stacks used in [**49**] except we need more relaxed hypotheses on the diagonal than in [**49**, 4.1]. Precisely, an algebraic stack $\mathfrak{X}$ over some scheme S will mean a stack over the category of S-schemes with the étale topology such that the following two conditions hold:

(0.2.1.1) The diagonal $\Delta : \mathfrak{X} \to \mathfrak{X} \times_S \mathfrak{X}$ is representable, of finite type, and locally separated.

(Recall that a representable morphism of algebraic stacks $f : \mathcal{Z} \to \mathcal{W}$ is locally separated if the diagonal $\mathcal{Z} \to \mathcal{Z} \times_{\mathcal{W}} \mathcal{Z}$ is a quasi-compact immersion.) This implies in particular that the diagonal Δ is quasi-compact and quasi-separated.

(0.2.1.2) There exists a smooth surjection $X \to \mathcal{X}$ with X a scheme.

We call an algebraic stack $\mathcal{X}$ for which there exists an étale surjection $X \to \mathcal{X}$ with X a scheme a *Deligne-Mumford stack.*

A morphism of algebraic S-stacks $f : \mathcal{X} \to \mathcal{Y}$ is called *representable by Deligne-Mumford stacks* if for any scheme Y and smooth morphism $Y \to \mathcal{Y}$ the fiber product $\mathcal{X} \times_{\mathcal{Y}} Y$ is a Deligne-Mumford stack over Y.

Lemma 0.2.2. — *Let S be a scheme, $\mathcal{X} \to S$ an algebraic stack, and $X \to \mathcal{X}$ a morphism from a locally separated S-space X. Then the diagonal $X \to X \times_{\mathcal{X}} X$ is a quasi-compact immersion (*i.e., *the morphism $X \to \mathcal{X}$ is locally separated).*

Proof. — Let P denote the fiber product $(X \times_{\mathcal{X}} X) \times_{X \times_S X} X$, and let $\sigma : X \to P$ be the section defined by the diagonal map. Since the map $\mathcal{X} \to \mathcal{X} \times_S \mathcal{X}$ is locally separated, the space P is locally separated over X. To prove that the map $X \to X \times_{\mathcal{X}} X$ is a quasi-compact immersion, it suffices to show that σ is a quasi-compact immersion since the map $P \to X \times_{\mathcal{X}} X$ is a quasi-compact immersion being the base change of $X \subset X \times_S X$. The result therefore follows from [**46**, I.1.21] applied to the category of quasi-compact immersions. □

0.2.3. — If $\mathcal{S}$ is an algebraic stack over some base scheme B, we define the 2-category of $\mathcal{S}$-*stacks* to be the 2-category whose objects are pairs $(\mathcal{X}, x)$, where $\mathcal{X}/B$ is an algebraic stack and $x : \mathcal{X} \to \mathcal{S}$ is a morphism of algebraic stacks. A 1-morphism $(\mathcal{X}', x') \to (\mathcal{X}, x)$ is a pair (f, f^b), where $f : \mathcal{X}' \to \mathcal{X}$ is a morphism of stacks and $f^b : x' \simeq x \circ f$ is an isomorphism of functors. A 2-isomorphism $(f, f^b) \to (g, g^b)$ between two 1-morphisms $(\mathcal{X}', x') \to (\mathcal{X}, x)$ is an isomorphism of functors $\iota : f \to g$ such that the two isomorphisms

$$g^b, x(\iota) \circ f^b : x' \longrightarrow x \circ g \tag{0.2.3.1}$$

are equal. An $\mathcal{S}$-*algebraic space* (resp. $\mathcal{S}$-*scheme*) is an $\mathcal{S}$-stack $(\mathcal{X}, x)$ with $\mathcal{X}$ an algebraic space (resp. scheme). Observe that the 2-subcategory consisting of $\mathcal{S}$-stacks $(\mathcal{X}, x)$ for which the morphism $x : \mathcal{X} \to \mathcal{S}$ is representable is equivalent to a 1-category. This is because if $(f, f^b) : (\mathcal{X}', x') \to (\mathcal{X}, x)$ is a morphism of such $\mathcal{S}$-stacks and $\iota : f \to f$ is an isomorphism of f such that $x(\iota)$ is the identity, then ι is the identity since the functor x is faithful. Thus it makes sense to talk about the category of $\mathcal{S}$-spaces or $\mathcal{S}$-schemes.

0.2.4. — When working with general Artin stacks we usually consider the lisse-étale topology as defined in [**49**, §12]. The reader should be aware that there are some problems with *loc. cit.* as the lisse-étale topology is not functorial as asserted there. However, in [**68**] the necessary aspects of the theory needed for this paper have been worked out. In particular, there is a good theory of quasi-coherent sheaves on Artin stacks. For an Artin stack $\mathcal{X}$ we write Lis-Et($\mathcal{X}$) for the site whose underlying category is the category of smooth $\mathcal{X}$-schemes, and whose topology is generated by étale surjective morphisms. We write $\mathcal{X}_{\text{lis-et}}$ for the associated topos.

0.2.5. — We write Δ for the standard simplicial category whose objects are the ordered sets $[n] = \{0, 1, \ldots, n\}$ ($n \in \mathbb{N}$) and whose morphisms are order preserving maps. We write Δ^+ for the full subcategory of Δ with the same objects but morphisms only the injective order preserving maps. For a category $\mathcal{C}$, a simplicial (resp. strictly simplicial) object in $\mathcal{C}$ is a functor $X_\bullet : \Delta^{\text{op}} \to \mathcal{C}$ (resp. $X_\bullet^+ : \Delta^{+\text{op}} \to \mathcal{C}$). For $n \in \mathbb{N}$ we usually write X_n for $X_\bullet([n])$ (resp. $X_\bullet^+([n])$). A simplicial object $X_\bullet$ in $\mathcal{C}$ induces a strictly simplicial object $X_\bullet^+$ defined to be the composite

$$\Delta^{+\text{op}} \subset \Delta^{\text{op}} \xrightarrow{X_\bullet} \mathcal{C}. \tag{0.2.5.1}$$

0.2.6. — We sometimes consider simplicial and strictly simplicial topoi which are defined for example in [**5**, V^{bis}.1.2.1].

In particular, for a simplicial (resp. strictly simplicial) algebraic space $X_\bullet$ (resp. $X_\bullet^+$), we denote by $X_{\bullet\text{et}}$ (resp. $X_{\bullet\text{et}}^+$) the étale topos. Recall that a sheaf $F_\bullet$ in $X_{\bullet\text{et}}$ (resp. $X_{\bullet\text{et}}^+$) consists of a sheaf F_n in $X_{n,\text{et}}$ for every $n \in \mathbb{N}$ together with a map $X_\bullet(\delta)^{-1}F_n \to F_{n'}$ (resp. $X_\bullet^+(\delta)^{-1}F_n \to F_{n'}$) for every morphism $\delta : [n'] \to [n]$ in Δ (resp. Δ^+). Furthermore, these transition morphisms are require to be compatible with compositions in Δ (resp. Δ^+).

In particular there is a natural structure sheaf $\mathcal{O}_{X_\bullet}$ (resp. $\mathcal{O}_{X_\bullet^+}$) defined by the structure sheaf on each X_n. A sheaf $F_\bullet$ of $\mathcal{O}_{X_\bullet}$-modules (resp. $\mathcal{O}_{X_\bullet^+}$-modules) is *quasi-coherent* if each sheaf F_n is a quasi-coherent sheaf on X_n and for every morphism $\delta : [n] \to [n']$ the induced map $X_\bullet(\delta)^*F_n \to F_{n'}$ (resp. $X_\bullet^+(\delta)^*F_n \to F_{n'}$) is an isomorphism.

0.2.7. — Recall [**34**, I.1.5] that for a scheme X over $\mathbb{F}_p$, one can define for any integer n the Witt scheme $W_n(X)$. The underlying topological space of $W_n(X)$ is equal to that of X, and the structure sheaf associates to any open set $U \subset X$ the ring of Witt vectors $W_n(\Gamma(U, \mathcal{O}_U))$. This can be generalized to Deligne-Mumford stacks as follows.

Let $\mathcal{X}$ be a Deligne-Mumford stack over some affine $\mathbb{F}_p$-scheme Spec(A). Let $W_n(\mathcal{X})$ be the fibered category over Spec($W_n(A)$) which to any $W_n(A)$-algebra R associates the groupoid of pairs (f, ρ), where $f : \text{Spec}(R \otimes_{W_n(A)} A) \to \mathcal{X}$ is a 1-morphism and $\rho : f^{-1}W_n(\mathcal{O}_{\mathcal{X}_{\text{et}}}) \to \mathcal{O}_{\text{Spec}(R)}$ is a ring homomorphism such that the diagram

$$\begin{array}{ccc} f^{-1}W_n(\mathcal{O}_{\mathcal{X}_{\text{et}}}) & \longrightarrow & f^{-1}\mathcal{O}_{\mathcal{X}_{\text{et}}} \\ {\scriptstyle\rho}\downarrow & & \downarrow \\ \mathcal{O}_{\text{Spec}(R)} & \longrightarrow & \mathcal{O}_{\text{Spec}(R \otimes_{W_n(A)} A)}. \end{array} \tag{0.2.7.1}$$

Here f^{-1} is the inverse image functor for the morphism of topoi $\text{Spec}(R \otimes_{W_n(A)} A)_{\text{et}} \to \mathcal{X}_{\text{et}}$ and we have identified the étale topoi of Spec(R) and $\text{Spec}(R \otimes_{W_n(A)} A)$ using the invariant of the étale site under infinitesimal thickenings.

Proposition 0.2.8. — *The stack $W_n(\mathcal{X})$ is a Deligne-Mumford stack. The étale topos $W_n(\mathcal{X})_{\mathrm{et}}$ is canonically equivalent to the ringed topos $(\mathcal{X}_{\mathrm{et}}, W_n(\mathcal{O}_{\mathcal{X}_{\mathrm{et}}}))$, where $W_n(\mathcal{O}_{\mathcal{X}_{\mathrm{et}}})$ is the sheaf which to any étale $\mathcal{X}$-scheme U associates $W_n(\Gamma(U, \mathcal{O}_U))$.*

Proof. — Consider first the case when the diagonal of $\mathcal{X}$ is representable by schemes.

Let $U \to \mathcal{X}$ be an étale surjection with U a scheme. Set $U' := U \times_{\mathcal{X}} U$ and let $(U' \rightrightarrows U)$ denote the resulting étale groupoid in schemes. Applying the functor $W_n(-)$ to this groupoid we obtain a groupoid in $W_n(A)$-schemes $(W_n(U') \rightrightarrows W_n(U))$. Since the functor $W_n(-)$ takes étale morphisms to étale morphisms [**34**, I.1.5.8], it follows that in fact $(W_n(U') \rightrightarrows W_n(U))$ is an étale groupoid. We claim that the resulting Deligne-Mumford stack is equal to $W_n(\mathcal{X})$. Temporarily denote by $W_n(\widetilde{\mathcal{X}})$ the algebraic space defined by $(W_n(U') \rightrightarrows W_n(U))$. There is a natural map $W_n(\widetilde{\mathcal{X}}) \to W_n(\mathcal{X})$. It is clear that any morphism $T \to \mathcal{X}$ étale locally lifts to a map to $W_n(U)$ and hence also $W_n(\widetilde{\mathcal{X}})$ since this is true over $\mathrm{Spec}(A)$. Thus it suffices to show that two maps $f, f' : \mathrm{Spec}(R) \to W_n(\widetilde{\mathcal{X}})$ are equal if the induced maps $(\bar{f}, \rho)$ and $(\bar{f}', \rho')$ to $W_n(\mathcal{X})$ are isomorphic by an isomorphism ι. This is an étale local assertion, and hence we may assume that f and f' factor through maps $\tilde{f}$ and $\tilde{f}'$ to $W_n(U)$. We then need to show that the resulting map $\tilde{f} \times \tilde{f}' : \mathrm{Spec}(R) \to W_n(U) \times_{W_n(A)} W_n(U)$ factors through $W_n(U')$. Let $g : \mathrm{Spec}(R \otimes_{W_n(A)} A) \to W_n(U')$ be the map induced by the reduction of the isomorphism ι. Since the map $\mathrm{pr}_i : W_n(U') \to W_n(U)$ $(i = 1, 2)$ is étale, there exists a unique lifting $g_i : \mathrm{Spec}(R) \to W_n(U')$ of g such that $\mathrm{pr}_1 \circ g_1 = \tilde{f}$ and $\mathrm{pr}_2 \circ g_2 = \tilde{f}'$. The assumption that $\rho = \rho'$ implies that the two induced maps

$$g_i^* : g^{-1} W_n(U') \longrightarrow R \qquad (0.2.8.1)$$

are equal and hence $g_1 = g_2$. Thus $\tilde{f} \times \tilde{f}'$ factors through $W_n(U')$. This proves the case when $\mathcal{X}$ is a Deligne-Mumford stack with diagonal representable by schemes. In particular, it proves the result for algebraic spaces.

But once we know that the result holds for algebraic spaces, we can repeat the above argument to get the result for an arbitrary Deligne-Mumford stack using a presentation in algebraic spaces $(U' \rightrightarrows U)$ of $\mathcal{X}$. □

CHAPTER 1

DIVIDED POWER STRUCTURES ON STACKS AND THE CRYSTALLINE TOPOS

1.1. PD-stacks

Definition 1.1.1. — A *PD-stack* is a triple $(\mathcal{S}, I, \gamma)$, where $\mathcal{S}$ is an algebraic stack over $\mathbb{Z}/p^n\mathbb{Z}$ for some integer n and (I, γ) is a quasi-coherent sheaf of ideals with divided power structure in the ringed topos $(\mathcal{S}_{\text{lis-et}}, \mathcal{O}_{\mathcal{S}_{\text{lis-et}}})$ (see [**7**, I.1.9.1]) for the definition of a divided power ideal in a ringed topos). The pair (I, γ) will be referred to as a *quasi-coherent PD-ideal in* $\mathcal{O}_{\mathcal{S}_{\text{lis-et}}}$.

Let $(\mathcal{S}', I', \gamma')$ and $(\mathcal{S}, I, \gamma)$ be PD-stacks, and let $f : \mathcal{S}' \to \mathcal{S}$ be a morphism of stacks. We say that f is a *PD-morphism* if the image of I under $\mathcal{O}_{\mathcal{S}_{\text{lis-et}}} \to f_*\mathcal{O}_{\mathcal{S}'_{\text{lis-et}}}$ is contained in f_*I and if for every integer $i \geq 0$ the diagram

$$\begin{array}{ccc} I & \longrightarrow & f_*I' \\ {\scriptstyle \gamma_i}\downarrow & & \downarrow{\scriptstyle f_*(\gamma'_i)} \\ \mathcal{O}_{\mathcal{S}_{\text{lis-et}}} & \longrightarrow & f_*\mathcal{O}_{\mathcal{S}'_{\text{lis-et}}} \end{array} \tag{1.1.1.1}$$

commutes.

1.1.2. — Let $(\mathcal{S}, I, \gamma)$ be a PD-stack and let $f : \mathcal{X} \to \mathcal{S}$ be a morphism of algebraic stacks. Recall [**68**, 6.5 (ii)] that even though the lisse-étale site of an algebraic stack is not functorial, it still makes sense to pullback a quasi-coherent sheaf $\mathcal{F}$ on $\mathcal{S}$ to a quasi-coherent sheaf $f^*\mathcal{F}$ on $\mathcal{X}$. If

$$\begin{array}{ccc} U & \xrightarrow{\tilde{f}} & V \\ {\scriptstyle p}\downarrow & & \downarrow{\scriptstyle q} \\ \mathcal{X} & \longrightarrow & \mathcal{S} \end{array} \tag{1.1.2.1}$$

is a commutative diagram with p and q smooth and U and V algebraic spaces, then the restriction $(f^*\mathcal{F})_U$ of $f^*\mathcal{F}$ to the étale site of U is equal to the pullback via the morphism of ringed topoi $U_{\text{et}} \to V_{\text{et}}$ induced by $\tilde{f}$ of $\mathcal{F}_V$. In particular, f^*I is a quasi-coherent sheaf on $\mathcal{X}$ and we write $f^*I \cdot \mathcal{O}_{\mathcal{X}}$ for the quasi-coherent sheaf of ideals which is the image of the map $f^*I \to \mathcal{O}_{\mathcal{X}_{\text{lis-et}}}$.

We say that γ *extends to* $\mathcal{X}$ if there exists a divided power structure γ' on $f^*I \cdot \mathcal{O}_{\mathcal{X}}$ such that the morphism f induces a PD-morphism

$$(\mathcal{X}, f^*I \cdot \mathcal{O}_{\mathcal{X}}, \gamma') \longrightarrow (\mathcal{S}, I, \gamma). \tag{1.1.2.2}$$

We will see in 1.1.10 below that such an extension is unique if it exists.

Definition 1.1.3. — Let $(\mathcal{S}', I', \gamma')$ and $(\mathcal{S}, I, \gamma)$ be PD-stacks, and $f : \mathcal{S}' \to \mathcal{S}$ a morphism. We say that f is *compatible with* γ *and* γ' if γ extends to $\mathcal{S}'$ and if there exists a PD-structure δ on $I' + f^*I \cdot \mathcal{O}_{\mathcal{S}'_{\text{lis-et}}}$ such that $\delta|_{I'} = \gamma'$ and $\delta|_{f^*I \cdot \mathcal{O}_{\mathcal{S}'_{\text{lis-et}}}}$ is equal to the given extension of γ.

1.1.4. — In the setting of algebraic spaces, the above definitions are equivalent to the usual notions using the étale topology instead of the lisse-étale topology. Precisely, let S be an algebraic space and

$$s : S_{\text{lis-et}} \longrightarrow S_{\text{et}} \tag{1.1.4.1}$$

the morphism of ringed topoi for which s_* is the functor which restricts a sheaf in $S_{\text{lis-et}}$ to the étale site of S and s^{-1} is the functor which sends an étale sheaf F to the sheaf which to any smooth $g : Z \to S$ associates $\Gamma(Z_{\text{et}}, g^{-1}F)$.

For any quasi-coherent sheaf of ideals $I \subset \mathcal{O}_{S_{\text{et}}}$ the pullback s^*I is a quasi-coherent sheaf of ideals in $\mathcal{O}_{S_{\text{lis-et}}}$. Furthermore, if γ is a divided power structure on I, then for any smooth $g : Z \to S$ the pullback $g^*I \subset \mathcal{O}_{Z_{\text{et}}}$ has by [**7**, I.2.7.4] a unique divided power structure γ^Z such that the map g defines a PD-morphism

$$(Z, g^*I, \gamma^Z) \longrightarrow (S, I, \gamma). \tag{1.1.4.2}$$

It follows that the divided power structure γ on I induces a unique divided power structure $s^*\gamma$ on s^*I such that for any smooth $g : Z \to S$ the map g defines a PD-morphism

$$(Z, s^*I|_{Z_{\text{et}}}, s^*\gamma|_{Z_{\text{et}}}) \longrightarrow (S, I, \gamma). \tag{1.1.4.3}$$

Lemma 1.1.5. — *Let S be an algebraic space. Then the map $(I, \gamma) \mapsto (s^*I, s^*\gamma)$ defines a bijection between the set of quasi-coherent PD-ideals (I, γ) in $\mathcal{O}_{S_{\text{et}}}$, and the set of quasi-coherent PD-ideals $J \subset \mathcal{O}_{S_{\text{lis-et}}}$.*

Proof. — By descent theory, the functor s^* induces a bijection between the set of quasi-coherent ideals $I \subset \mathcal{O}_{S_{\text{et}}}$ and the set of quasi-coherent ideals $J \subset \mathcal{O}_{S_{\text{lis-et}}}$. Thus it suffices to show that if $I \subset \mathcal{O}_{S_{\text{et}}}$ is a quasi-coherent ideal, then any divided power structure γ on s^*I is equal to $s^*(\gamma|_{S_{\text{et}}})$, where $\gamma|_{S_{\text{et}}}$ denotes the divided power structure on I obtained by restriction to S_{et}. If $g : Z \to S$ is any smooth morphism, then $\gamma|_{Z_{\text{et}}}$ and $s^*(\gamma|_{S_{\text{et}}})|_{Z_{\text{et}}}$ both define extensions of $\gamma|_{S_{\text{et}}}$ to the ideal g^*I, and hence by [**7**, comment after I.2.1] we have $\gamma|_{Z_{\text{et}}} = s^*(\gamma|_{S_{\text{et}}})|_{Z_{\text{et}}}$. □

Lemma 1.1.6. — *Let (S', I', γ') and (S, I, γ) be algebraic spaces with PD-ideals $I' \subset \mathcal{O}_{S'_{\text{et}}}$ and $I \subset \mathcal{O}_{S_{\text{et}}}$. Then a quasi-compact and quasi-separated morphism $f : S' \to S$ induces a PD-morphism $(S'_{\text{et}}, I', \gamma') \to (S_{\text{et}}, I, \gamma)$ if and only if f induces a PD-morphism in the sense of 1.1.1 between $(\mathcal{S}', s^*I', s^*\gamma')$ and $(\mathcal{S}, s^*I, s^*\gamma)$.*

Proof. — Let f^{et} denote the morphism of topoi $S'_{\text{et}} \to S_{\text{et}}$ induced by f, and write f_* for the pushforward of lisse-étale sheaves. Then by [**68**, proof of 6.20] for any quasi-coherent sheaf $\mathcal{M}$ on S'_{et} we have $s^*(f^{\text{et}}_*\mathcal{M}) = f_*(s^*\mathcal{M})$. It follows that the map $I \to f^{\text{et}}_*\mathcal{O}_{S'_{\text{et}}}$ factors through f^{et}_*I' if and only if $s^*I \to f_*\mathcal{O}_{\mathcal{S}_{\text{lis-et}}}$ factors through $f_*(s^*I')$. It is also clear that if f induces a PD-morphism in the sense of 1.1.1, then f defines a PD-morphism $(S'_{\text{et}}, I', \gamma') \to (S_{\text{et}}, I, \gamma)$ since for any $i \geq 0$ the diagram

$$\begin{array}{ccc} I & \longrightarrow & f^{\text{et}}_*I' \\ {\scriptstyle \gamma_i}\downarrow & & \downarrow{\scriptstyle f_*(\gamma'_i)} \\ \mathcal{O}_{\mathcal{S}_{\text{et}}} & \longrightarrow & f_*\mathcal{O}_{\mathcal{S}'_{\text{et}}} \end{array} \tag{1.1.6.1}$$

can be obtained from the diagram (1.1.1.1) by restricting to the étale topology of S.

Conversely, to prove that if f defines a PD-morphism $(S'_{\text{et}}, I', \gamma') \to (S_{\text{et}}, I, \gamma)$ then the diagram (1.1.1.1) commutes for every $i \geq 0$, note first that we may work étale locally on S and S' and hence may assume that S and S' are affine schemes. Write $S = \text{Spec}(R)$, $S' = \text{Spec}(R')$, $M = \Gamma(S, I)$, and $M' = \Gamma(S', I')$. For any smooth affine $\text{Spec}(B) \to \text{Spec}(R)$ the evaluation of the diagram (1.1.1.1) on $\text{Spec}(B)$ is identified with the diagram

$$\begin{array}{ccc} M \otimes_R B & \longrightarrow & M' \otimes_R B \\ {\scriptstyle \gamma_i}\downarrow & & \downarrow{\scriptstyle \gamma'_i} \\ B & \longrightarrow & R' \otimes_R B. \end{array} \tag{1.1.6.2}$$

Here $\gamma_i(m \otimes b) = b^i\gamma_i(m \otimes 1)$ and $\gamma'_i(m' \otimes b) = b^i\gamma'_i(m' \otimes 1)$. To prove that (1.1.6.2) commutes it therefore suffices to show that the diagram

$$\begin{array}{ccc} M & \longrightarrow & M' \\ {\scriptstyle \gamma_i}\downarrow & & \downarrow{\scriptstyle \gamma'_i} \\ R & \longrightarrow & R' \end{array} \tag{1.1.6.3}$$

commutes which follows from the fact that $(S'_{\text{et}}, I', \gamma') \to (S_{\text{et}}, I, \gamma)$ is a PD-morphism. □

1.1.7. — Let $\mathcal{S}$ be a quasi-compact algebraic stack and $I \subset \mathcal{O}_{\mathcal{S}_{\text{lis-et}}}$ a quasi-coherent sheaf of ideals. Let $f : U \to \mathcal{S}$ be a quasi-compact smooth surjection with U an affine scheme. Set $P = U \times_{\mathcal{S}} U$, and let $g : P \to \mathcal{S}$ be the projection. Since the diagonal of $\mathcal{S}$ is quasi-compact and quasi-separated by our conventions 0.2.1, the map g is also quasi-compact and quasi-separated. Furthermore, both of the projections $p_i : P \to U$ $(i = 1, 2)$ are smooth. Denote by $I|_{U_{\text{et}}}$ (resp. $I|_{P_{\text{et}}}$) the restriction of I to the étale site of U (resp. P). Since I is quasi-coherent, for each i there is a natural isomorphism $\sigma_i : p_i^* I|_{U_{\text{et}}} \to I|_{P_{\text{et}}}$. If γ_U is a divided power structure on $I|_{U_{\text{et}}}$, then since p_i is flat (in fact smooth) we can by [**7**, I.2.7.4] pullback the PD-structure to a divided power structure on $p_i^* I|_{U_{\text{et}}}$. We write $p_i^*(\gamma_U)$ for the divided power structure on $I|_{P_{\text{et}}}$ obtained from this pullback and the canonical isomorphism σ_i.

Lemma 1.1.8. — *Restriction defines a bijection between the set of divided power structures on I in the topos $\mathcal{S}_{\text{lis-et}}$ and the set of divided power structures γ_U on $I|_{U_{\text{et}}}$ such that the PD-structures $p_1^*(\gamma_U)$ and $p_2^*(\gamma_U)$ on $I|_{P_{\text{et}}}$ are equal.*

Proof. — Let γ_U be a divided power structure on $I|_{U_{\text{et}}}$ as in the lemma, and let γ_P denote $p_1^*(\gamma_U) = p_2^*(\gamma_U)$. Write simply s for the morphisms of topoi $U_{\text{lis-et}} \to U_{\text{et}}$ and $P_{\text{lis-et}} \to P_{\text{et}}$ defined in 1.1.4. For any $n \geq 0$, there is a morphism of exact sequences

$$\begin{array}{ccccccc} 0 & \longrightarrow & I & \xrightarrow{f^*} & f_* s^*(I|_{U_{\text{et}}}) & \rightrightarrows & g_* s^*(I|_{P_{\text{et}}}) \\ & & & & \downarrow{\scriptstyle s^*\gamma_{U,n}} & & \\ 0 & \longrightarrow & \mathcal{O}_{\mathcal{S}_{\text{lis-et}}} & \xrightarrow{f^*} & f_* \mathcal{O}_{U_{\text{lis-et}}} & \rightrightarrows & g_* \mathcal{O}_{P_{\text{lis-et}}}. \end{array} \tag{1.1.8.1}$$

The map $\gamma_{U,n}$ therefore defines a map $\gamma_n : I \to \mathcal{O}_{\mathcal{S}_{\text{lis-et}}}$. If we restrict this diagram to U we obtain the diagram

$$\begin{array}{ccccccc} 0 & \longrightarrow & I|_{U_{\text{et}}} & \xrightarrow{p_2^*} & p_{2,*} I|_{P_{\text{et}}} & \rightrightarrows & p_{3,*} I|_{(U \times_{\mathcal{S}} U \times_{\mathcal{S}} U)_{\text{et}}} \\ & & & & \downarrow{\scriptstyle p_1^*\gamma_{U,n}} & & \downarrow{\scriptstyle p_1^*\gamma_{U,n}} \\ 0 & \longrightarrow & \mathcal{O}_{U_{\text{et}}} & \xrightarrow{p_2^*} & p_{2,*} \mathcal{O}_{P_{\text{et}}} & \rightrightarrows & p_{3,*} \mathcal{O}_{(U \times_{\mathcal{S}} U \times_{\mathcal{S}} U)_{\text{et}}}, \end{array} \tag{1.1.8.2}$$

where the right horizontal arrows are p_{13}^* and p_{23}^*. It follows that the restriction of γ_n to U_{et} is equal to $\gamma_{U,n}$. In particular, the maps γ_n locally on $\mathcal{S}$ define a divided power structure, and hence also globally, and we can recover γ_U from γ_n by restriction. This implies the lemma. □

Example 1.1.9. — The condition that $p_1^*(\gamma_U) = p_2^*(\gamma_U)$ is not vacuous. For example, let S be a scheme and G a finite group acting on S and let $\mathcal{S}$ be the stack quotient $[S/G]$. For any quasi-coherent sheaf of ideals $\mathcal{I}$ on $\mathcal{S}$ corresponding to a G-linearized quasi-coherent sheaf of ideals I on S and divided power structure γ on I, the condition that $p_1^*(\gamma) = p_2^*(\gamma)$ then means that for element $g \in G$ the divided power structure $g^*(\gamma)$ on $g^* I \simeq I$ (isomorphism given by the G-linearization) is equal to γ. This can fail. For example, let k be an algebraically closed field of characteristic 2, $S = \operatorname{Spec}(k[\epsilon]/\epsilon^2)$ and let $G = \mu_l$ for some odd prime l. Define the action of G on S by having $\zeta \in \mu_l$ act by $\epsilon \mapsto \zeta\epsilon$. Define a divided power structure γ on $I = (\epsilon)$ by setting $\gamma_i(\epsilon) = \epsilon$ if $i = 2^k$ for some $k \geq 0$ and zero otherwise. Using the fact that $\operatorname{char}(k) = 2$ one sees that this defines a divided power structure on (ϵ) (note that since for any $a \in k$ we must have $\gamma_i(a\epsilon) = a^i \gamma_i(\epsilon)$ this gives the formula for γ_i evaluated on any element of (ϵ)). For any $\zeta \in \mu_l$, the pullback $\zeta^*(\gamma)$ is a divided power structure with

$\gamma_2(\epsilon) = \zeta^{-1}\epsilon$. Hence the divided power structure γ does not descend to the stack $[\mathrm{Spec}(k[\epsilon]/(\epsilon^2))/\mu_l]$.

Lemma 1.1.10. — *Let $(\mathcal{S}, I, \gamma)$ be a PD-stack, and $\mathcal{X} \to \mathcal{S}$ a morphism of algebraic stacks. If γ extends to $\mathcal{X}$, then the extension γ' is unique.*

Proof. — Assume γ' and $\tilde{\gamma}'$ are two extensions, and fix a commutative square as in (1.1.2.1) with p surjective. Then the restrictions of γ' and $\tilde{\gamma}'$ to U_{et} define two extensions of $\gamma|_{V_{\mathrm{et}}}$ to U_{et} which by [**7**, comment after I.2.1] must be equal. It follows that γ' and $\tilde{\gamma}'$ are equal when restricted to U_{et}. By 1.1.8 we conclude that $\gamma' = \tilde{\gamma}'$. □

Lemma 1.1.11. — *Let $(\mathcal{S}, I, \gamma)$ be a PD-stack and $f : \mathcal{X} \to \mathcal{S}$ a morphism of algebraic stacks. Fix a diagram as in (1.1.2.1) with p surjective. Then γ extends to $\mathcal{X}$ if and only if the PD-ideal $(I, \gamma)|_{V_{\mathrm{et}}}$ in the ringed topos $(V_{\mathrm{et}}, \mathcal{O}_{V_{\mathrm{et}}})$ extends to the ringed topos $(U_{\mathrm{et}}, \mathcal{O}_{U_{\mathrm{et}}})$ in the sense of* [**7**, I.2.1].

Proof. — The "only if" direction is immediate. For the "if" direction, let $J \subset \mathcal{O}_{\mathcal{X}_{\text{lis-et}}}$ denote the ideal $f^*I \cdot \mathcal{O}_{\mathcal{X}_{\text{lis-et}}}$. By assumption the restriction $J|_{U_{\mathrm{et}}}$ has a PD-structure γ'_U extending that on $I|_{V_{\mathrm{et}}}$. For $i = 1, 2$, we have a commutative diagram

(1.1.11.1)
$$\begin{array}{ccc} U \times_{\mathcal{X}} U & \longrightarrow & V \times_{\mathcal{S}} V \\ {\scriptstyle p_i}\downarrow & & \downarrow{\scriptstyle q_i} \\ U & \longrightarrow & V, \end{array}$$

which shows that $p_i^*(\gamma'_U)$ is a PD-structure extending the PD-structure $q_i^*(\gamma|_V)$ on $I|_{(V\times_{\mathcal{S}}V)_{\mathrm{et}}}$. By 1.1.8, the pullback $q_i^*(\gamma|_V)$ is equal to the restriction of γ to $(V \times_{\mathcal{S}} V)_{\mathrm{et}}$. It follows that $q_1^*(\gamma|_V) = q_2^*(\gamma|_V)$ and hence $p_1^*(\gamma'_U) = p_2^*(\gamma'_U)$ by [**7**, comment after I.2.1]. By 1.1.8 the PD-structure γ'_U therefore descends to a PD-structure γ' on J.

It remains to see that for any integer $i \geq 0$ the diagram

(1.1.11.2)
$$\begin{array}{ccc} I & \longrightarrow & f_*J \\ {\scriptstyle \gamma_i}\downarrow & & \downarrow{\scriptstyle f_*(\gamma'_i)} \\ \mathcal{O}_{\mathcal{S}_{\text{lis-et}}} & \longrightarrow & f_*\mathcal{O}_{\mathcal{X}_{\text{lis-et}}} \end{array}$$

commutes. Let $g : U \to \mathcal{S}$ denote the composite $U \to \mathcal{X} \to \mathcal{S}$. We then have a commutative diagram

(1.1.11.3)
$$\begin{array}{ccccc} I & \longrightarrow & f_*J & \xrightarrow{\ j\ } & g_*(I|_U) \\ {\scriptstyle \gamma_i}\downarrow & & \downarrow{\scriptstyle f_*(\gamma'_i)} & & \downarrow{\scriptstyle g_*(\gamma'_i|_U)} \\ \mathcal{O}_{\mathcal{S}_{\text{lis-et}}} & \longrightarrow & f_*\mathcal{O}_{\mathcal{X}_{\text{lis-et}}} & \xrightarrow{\ k\ } & g_*\mathcal{O}_{U_{\text{lis-et}}}, \end{array}$$

where the right hand square commutes and the maps j and k are injective since $U \to \mathcal{X}$ is smooth and surjective. It follows that it suffices to verify the commutativity of (1.1.11.2) in the case when $U = \mathcal{X}$.

This special case can be seen by observing that there is also a diagram

$$\begin{array}{ccccc} I & \longrightarrow & q_*(I|_V) & \longrightarrow & q_*\tilde{f}_*J = f_*J \\ {\scriptstyle \gamma_i}\downarrow & & {\scriptstyle q_*(\gamma_i|_V)}\downarrow & & \downarrow{\scriptstyle f_*\gamma_i'} \\ \mathcal{O}_{\mathcal{S}} & \longrightarrow & q_*\mathcal{O}_{V_{\text{lis-et}}} & \longrightarrow & f_*\mathcal{O}_{U_{\text{lis-et}}}, \end{array} \tag{1.1.11.4}$$

where the left square commutes since γ_i is a morphism of sheaves, and the right square commutes since γ_U' is an extension of $\gamma|_V$ to U. □

Corollary 1.1.12. — *Let $(\mathcal{S}, I, \gamma)$ be a PD-stack and $f : \mathcal{X} \to \mathcal{S}$ a morphism which is flat (resp. a closed immersion defined by a sub-PD-ideal $J \subset I$). Then γ extends to $\mathcal{X}$.*

Proof. — Let $V \to \mathcal{S}$ be a smooth cover by a scheme V, and let $U \to V \times_{\mathcal{X}} \mathcal{S}$ be a smooth surjection with U a scheme (resp. set $U = V \times_{\mathcal{X}} \mathcal{S}$). Then $U \to V$ is flat (resp. a closed immersion defined by a sub-PD-ideal in $I|_{V_{\text{et}}}$), and hence the result follows from [**7**, I.2.7.4] (resp. [**7**, I.1.6.2]). □

1.2. Divided power envelopes

1.2.1. — Fix a PD-stack $(\mathcal{S}, I, \gamma)$, and define 2-categories $\mathcal{C}$ and $\mathcal{C}'$ as follows.

The objects of $\mathcal{C}$ are closed immersions $j : \mathcal{X} \hookrightarrow \mathcal{Y}$ of algebraic stacks over $\mathcal{S}$ defined by a quasi-coherent sheaf of ideals such that γ extends to $\mathcal{X}$, together with an isomorphism of functors $\sigma_j : s_{\mathcal{X}} \simeq s_{\mathcal{Y}} \circ j$, where $s_{\mathcal{X}} : \mathcal{X} \to \mathcal{S}$ and $s_{\mathcal{Y}} : \mathcal{Y} \to \mathcal{S}$ are the structure morphisms. We usually omit σ_j from the notation and write simply $j : \mathcal{X} \hookrightarrow \mathcal{Y}$ for an object of $\mathcal{C}$, but the isomorphism σ_j is always assumed part of the data.

For any two objects $j_i : \mathcal{X}_i \hookrightarrow \mathcal{Y}_i$ $(i = 1, 2)$ of $\mathcal{C}$ the category

$$\mathrm{HOM}_{\mathcal{C}}(\mathcal{X}_1 \hookrightarrow \mathcal{Y}_1, \mathcal{X}_2 \hookrightarrow \mathcal{Y}_2) \tag{1.2.1.1}$$

is the category whose objects are two-commutative diagrams of $\mathcal{S}$-morphisms

$$\begin{array}{ccc} \mathcal{X}_1 & \xrightarrow{j_1} & \mathcal{Y}_1 \\ {\scriptstyle f}\downarrow & & \downarrow{\scriptstyle g} \\ \mathcal{X}_2 & \xrightarrow{j_2} & \mathcal{Y}_2. \end{array} \tag{1.2.1.2}$$

In other words, an object of (1.2.1.1) consists of a pair of morphisms $f : \mathcal{X}_1 \to \mathcal{X}_2$ and $g : \mathcal{Y}_1 \to \mathcal{Y}_2$, isomorphisms $\iota_f : s_{\mathcal{X}_1} \simeq s_{\mathcal{X}_2} \circ f$ and $\iota_g : s_{\mathcal{Y}_1} \simeq s_{\mathcal{Y}_2} \circ g$, and an isomorphism $\rho : g \circ j_1 \simeq j_2 \circ f$ of functors such that the diagram of isomorphisms

$$\begin{array}{ccccc} s_{\mathcal{X}_2} \circ f & \xleftarrow{\iota_f} & s_{\mathcal{X}_1} & \xrightarrow{\sigma_{j_2}} & s_{\mathcal{Y}_1} \circ j_1 \\ {\scriptstyle \sigma_{j_2}}\downarrow & & & & \downarrow{\scriptstyle \iota_g} \\ s_{\mathcal{Y}_2} \circ j_2 \circ f & \xrightarrow{\rho} & s_{\mathcal{Y}_2} \circ g \circ j_1 & \xrightarrow{\text{id}} & s_{\mathcal{Y}_2} \circ g \circ j_1 \end{array} \tag{1.2.1.3}$$

commutes. If $(f, g, \iota_f, \iota_g, \rho)$ and $(f', g', \sigma_{f'}, \sigma_{g'}, \rho')$ are two 1-morphisms in $\mathcal{C}$, then a 2-morphism in $\mathcal{C}$ is defined to be a pair (α, β), where $\alpha : f \to f'$ and $\beta : g \to g'$ are isomorphisms of functors such that

$$\begin{array}{ccc} s_{\mathcal{X}_1} & \xrightarrow{\iota_f} & s_{\mathcal{X}_2} \circ f \\ {\scriptstyle \mathrm{id}}\downarrow & & \downarrow{\scriptstyle \alpha} \\ \mathcal{X}_1 & \xrightarrow{\sigma_{f'}} & s_{\mathcal{X}_2} \circ f', \end{array} \tag{1.2.1.4}$$

$$\begin{array}{ccc} s_{\mathcal{Y}_1} & \xrightarrow{\iota_g} & s_{\mathcal{Y}_2} \circ g \\ {\scriptstyle \mathrm{id}}\downarrow & & \downarrow{\scriptstyle \beta} \\ \mathcal{Y}_1 & \xrightarrow{\sigma_{g'}} & s_{\mathcal{Y}_2} \circ g', \end{array} \tag{1.2.1.5}$$

and

$$\begin{array}{ccc} j_2 \circ f & \xrightarrow{\rho} & g \circ j_1 \\ {\scriptstyle \alpha}\downarrow & & \downarrow{\scriptstyle \beta} \\ j_2 \circ f' & \xrightarrow{\rho'} & g' \circ j_1 \end{array} \tag{1.2.1.6}$$

commute.

The 2-category $\mathcal{C}'$ is defined as follows. The objects of $\mathcal{C}'$ are objects $j : \mathcal{X} \hookrightarrow \mathcal{Y}$ of $\mathcal{C}$ together with a divided power structure δ on the ideal of $\mathcal{X}$ in $\mathcal{Y}$ such that δ is compatible with γ. We usually omit the PD-structure δ from the notation and write $j : \mathcal{X} \hookrightarrow \mathcal{Y}$ also for objects of $\mathcal{C}'$. For two objects $j_i : \mathcal{X}_i \hookrightarrow \mathcal{Y}_i$ of $\mathcal{C}'$ $(i = 1, 2)$, the category

$$\mathrm{HOM}_{\mathcal{C}'}(j_1 : \mathcal{X}_1 \hookrightarrow \mathcal{Y}_1, j_2 : \mathcal{X}_2 \hookrightarrow \mathcal{Y}_2) \tag{1.2.1.7}$$

is defined to be the full subcategory of (1.2.1.1) consisting of morphisms as above for which g is compatible with the divided power structures on $\mathcal{Y}_1$ and $\mathcal{Y}_2$.

There is a natural 2-functor $F : \mathcal{C}' \to \mathcal{C}$ which forgets the divided power structure.

Remark 1.2.2. — If $j_i : \mathcal{X}_i \hookrightarrow \mathcal{Y}_i$ are two objects of $\mathcal{C}$ with $s_{\mathcal{Y}_i} : \mathcal{Y}_i \to \mathcal{S}$ representable, then the category

$$\mathrm{HOM}_{\mathcal{C}}(\mathcal{X}_1 \hookrightarrow \mathcal{Y}_1, j_2 : \mathcal{X}_2 \hookrightarrow \mathcal{Y}_2) \tag{1.2.2.1}$$

is equivalent to a set. That is, the objects admit no automorphisms. This can be seen by observing that if (α, β) is an automorphism of an object (f, g, ι_f, ι_g) then $s_{\mathcal{X}_2}(\alpha) = \mathrm{id}$ and $s_{\mathcal{Y}_2}(\beta) = \mathrm{id}$ by (1.2.1.4) and (1.2.1.5). On the other hand, $s_{\mathcal{X}_2}$ and $s_{\mathcal{Y}_2}$ are faithful functors by [**49**, 8.1.2] and hence $(\alpha, \beta) = \mathrm{id}$.

A similar remark applies to the 2-category $\mathcal{C}'$.

Theorem 1.2.3. — *For any object $j : \mathcal{X} \hookrightarrow \mathcal{Y}$ of $\mathcal{C}$ there is a canonically associated object $j_D : \mathcal{X} \hookrightarrow D_{\mathcal{X},\gamma}(\mathcal{Y})$ of $\mathcal{C}'$ with a morphism $F(\mathcal{X} \hookrightarrow D_{\mathcal{X},\gamma}(\mathcal{Y})) \to (j : \mathcal{X} \hookrightarrow \mathcal{Y})$ in $\mathcal{C}$ such that for any other object $j' : \mathcal{X}' \hookrightarrow \mathcal{Y}'$ of $\mathcal{C}'$ the induced functor*

$$(1.2.3.1) \qquad \mathrm{HOM}_{\mathcal{C}'}(\mathcal{X}' \hookrightarrow \mathcal{Y}', \mathcal{X} \hookrightarrow D_{\mathcal{X},\gamma}(\mathcal{Y})) \longrightarrow \mathrm{HOM}_{\mathcal{C}}(F(\mathcal{X}' \hookrightarrow \mathcal{Y}'), \mathcal{X} \hookrightarrow \mathcal{Y})$$

is an equivalence of categories. Moreover, if $\mathcal{Z} \to \mathcal{Y}$ if a flat morphism of algebraic stacks, then the natural map $D_{\mathcal{X}\times_{\mathcal{Y}}\mathcal{Z},\gamma}(\mathcal{Z}) \to D_{\mathcal{X},\gamma}(\mathcal{Y}) \times_{\mathcal{Y}} \mathcal{Z}$ is an isomorphism.

The proof of 1.2.3 will be in several steps 1.2.4–1.2.9.

1.2.4. — Let us first consider the case when $j : \mathcal{X} \hookrightarrow \mathcal{Y}$ is a closed immersion of algebraic spaces. Let $P : S \to \mathcal{S}$ be a smooth cover by a scheme, and let $S_\bullet \to \mathcal{S}$ be the 0-coskeleton of P. Define $j_\bullet : X_\bullet \hookrightarrow Y_\bullet$ to be the base change to $S_\bullet$ of j. Denote by $I_\bullet$ the quasi-coherent sheaf on $S_{\bullet,\mathrm{et}}$ obtained by restricting the quasi-coherent sheaf I to $S_{\bullet,\mathrm{et}}$. The divided power structure on I defines a divided power structure $\gamma^\bullet$ on $I_\bullet$. Let $\mathcal{O}_{S_{\bullet,\mathrm{et}}}|_{Y_{\bullet,\mathrm{et}}}$ denote the inverse image via the morphism of topoi $Y_{\bullet,\mathrm{et}} \to S_{\bullet,\mathrm{et}}$ of the structure sheaf on $S_{\bullet,\mathrm{et}}$ so that there is a diagram of sheaves of algebras on $Y_{\bullet,\mathrm{et}}$

$$(1.2.4.1) \qquad \mathcal{O}_{S_{\bullet,\mathrm{et}}}|_{Y_{\bullet,\mathrm{et}}} \longrightarrow \mathcal{O}_{Y_{\bullet,\mathrm{et}}} \to j_{\bullet *}\mathcal{O}_{X_{\bullet,\mathrm{et}}}.$$

Furthermore, the restriction $(I_\bullet, \gamma^\bullet)|_{Y_{\bullet,\mathrm{et}}}$ defines a divided power ideal in $\mathcal{O}_{S_{\bullet,\mathrm{et}}}|_{Y_{\bullet,\mathrm{et}}}$. Let $\mathcal{D}_{X_\bullet,\gamma^\bullet}(Y_\bullet)$ be the divided power envelope in the sense of [**7**, I.2.3.1] of the morphism of algebras with ideals in the topos $Y_{\bullet,\mathrm{et}}$.

$$(1.2.4.2) \qquad (\mathcal{O}_{S_{\bullet,\mathrm{et}}}|_{Y_{\bullet,\mathrm{et}}}, I_\bullet) \longrightarrow (\mathcal{O}_{Y_{\bullet,\mathrm{et}}}, \mathrm{Ker}(\mathcal{O}_{Y_{\bullet,\mathrm{et}}} \longrightarrow j_{\bullet *}\mathcal{O}_{X_{\bullet,\mathrm{et}}})).$$

Since γ extends to $\mathcal{X}$ by assumption, the divided powers $\gamma^\bullet$ induces a divided power structure on the image of $I_\bullet$ in $j_{\bullet *}\mathcal{O}_{X_{\bullet,\mathrm{et}}}$. By the universal property of $\mathcal{D}_{X_\bullet,\gamma^\bullet}(Y_\bullet)$ we therefore have a factorization

$$(1.2.4.3) \qquad \mathcal{O}_{S_{\bullet,\mathrm{et}}}|_{Y_{\bullet,\mathrm{et}}} \longrightarrow \mathcal{O}_{Y_{\bullet,\mathrm{et}}} \longrightarrow \mathcal{D}_{X_\bullet,\gamma^\bullet}(Y_\bullet) \longrightarrow j_{\bullet *}\mathcal{O}_{X_{\bullet,\mathrm{et}}},$$

and the kernel of $\mathcal{D}_{X_\bullet,\gamma^\bullet}(Y_\bullet) \to j_{\bullet *}\mathcal{O}_{X_{\bullet,\mathrm{et}}}$ is equipped with a divided power structure compatible with $\gamma^\bullet$.

Lemma 1.2.5. — *The sheaf $\mathcal{D}_{X_\bullet,\gamma^\bullet}(Y_\bullet)$ is a quasi-coherent sheaf on $Y_{\bullet,\mathrm{et}}$.*

Proof. — The sheaf $\mathcal{D}_{X_\bullet,\gamma^\bullet}(Y_\bullet)|_{Y_{n,\mathrm{et}}}$ is the divided power envelope $\mathcal{D}_{X_n,\gamma^n}(Y_n)$ of the closed immersion of S_n-spaces $X_n \hookrightarrow Y_n$. That $\mathcal{D}_{X_n,\gamma^n}(Y_n)$ is quasi-coherent therefore follows from [**7**, I.2.7.1]. To see that for any morphism $\delta : [n] \to [n']$ in Δ the induced map

$$(1.2.5.1) \qquad \mathcal{O}_{Y_{n'},\mathrm{et}} \otimes_{\mathcal{O}_{Y_{n,\mathrm{et}}}} \mathcal{D}_{X_\bullet,\gamma^\bullet}(Y_\bullet)|_{Y_{n,\mathrm{et}}} \longrightarrow \mathcal{D}_{X_\bullet,\gamma^\bullet}(Y_\bullet)|_{Y_{n',\mathrm{et}}}$$

is an isomorphism, note that as in [**49**, 12.8.1] it suffices to consider injective maps $\delta : [n] \to [n']$ in Δ. Therefore this also follows from [**7**, I.2.7.1] since for an injective map δ the map $Y_\bullet(\delta) : Y_{n'} \to Y_n$ is smooth by construction of $Y_\bullet$. □

1.2.6. — It follows that $\mathcal{D}_{X_\bullet,\gamma^\bullet}(Y_\bullet)$ descends to a quasi-coherent sheaf $\mathcal{D}_{\mathcal{X},\gamma}(\mathcal{Y})$ of $\mathcal{O}_{\mathcal{Y}_{\text{lis-et}}}$-algebras on $\mathcal{Y}$ such that the surjection $\mathcal{O}_{\mathcal{Y}_{\text{lis-et}}} \to j_*\mathcal{O}_{\mathcal{X}_{\text{lis-et}}}$ factors canonically through $\mathcal{D}_{\mathcal{X},\gamma}(\mathcal{Y})$. Moreover, the kernel of the map $\mathcal{D}_{\mathcal{X},\gamma}(\mathcal{Y}) \to j_*\mathcal{O}_{\mathcal{X}_{\text{lis-et}}}$ has by 1.1.8 a PD-structure δ induced by the PD-structure on $\mathcal{D}_{X_\bullet,\gamma^\bullet}(Y_\bullet)$, and by 1.1.11 this PD-structure δ is compatible with γ. Define $D_{\mathcal{X},\gamma}(\mathcal{Y})$ to be the relative spectrum

$$\underline{\text{Spec}}_{\mathcal{Y}}(\mathcal{D}_{\mathcal{X},\gamma}(\mathcal{Y})). \tag{1.2.6.1}$$

Then by construction we have an object

$$\mathcal{X} \hookrightarrow D_{\mathcal{X},\gamma}(\mathcal{Y}) \tag{1.2.6.2}$$

of $\mathcal{C}'$ with a morphism

$$\begin{array}{ccc} \mathcal{X} & \longrightarrow & D_{\mathcal{X},\gamma}(\mathcal{Y}) \\ {\scriptstyle \text{id}}\downarrow & & \downarrow \\ \mathcal{X} & \longrightarrow & \mathcal{Y} \end{array} \tag{1.2.6.3}$$

in $\mathcal{C}$. Furthermore, it follows from the construction and [**7**, I.2.7.1] that the formation of $D_{\mathcal{X},\gamma}(\mathcal{Y})$ is compatible with flat base change $\mathcal{Y}' \to \mathcal{Y}$, with $\mathcal{Y}'$ an algebraic space.

Here it should be noted that *a priori* the space $D_{\mathcal{X},\gamma}(\mathcal{Y})$ depends on the choice of the smooth cover $S \to \mathcal{S}$. That it does not will follow once we verify the universal property.

This base change property enables us to define $D_{\mathcal{X},\gamma}(\mathcal{Y})$ for an arbitrary object $j : \mathcal{X} \hookrightarrow \mathcal{Y}$ of $\mathcal{C}$. Choose a smooth cover $q : Y \to \mathcal{Y}$ with Y an algebraic space, and let $Y_\bullet$ be the 0-coskeleton of q. Denote by $j_\bullet : X_\bullet \hookrightarrow Y_\bullet$ the closed immersion obtained by base change from j. Then for each $[n] \in \Delta$, the closed immersion $j_n : X_n \hookrightarrow Y_n$ is an object of $\mathcal{C}$ and hence we can form $\mathcal{D}_{X_n,\gamma}(Y_n)$ (defined using some fixed cover $S \to \mathcal{S}$), which is a quasi-coherent sheaf on $Y_{n,\text{et}}$. For any injective morphism $\delta : [n] \to [n']$ the natural map $Y(\delta)^*\mathcal{D}_{X_n,\gamma}(Y_n) \to \mathcal{D}_{X_{n'},\gamma}(Y_{n'})$ is an isomorphism since the map $Y_\bullet(\delta) : Y_{n'} \to Y_n$ is smooth. By descent theory for quasi-coherent sheaves, the simplicial sheaf $[n] \mapsto \mathcal{D}_{X_n,\gamma}(Y_n)$ is obtained from a quasi-coherent sheaf of $\mathcal{O}_{\mathcal{Y}}$-algebras $\mathcal{D}_{\mathcal{X},\gamma}(\mathcal{Y})$ on $\mathcal{Y}_{\text{lis-et}}$.

As above the kernel of the map $\mathcal{D}_{\mathcal{X},\gamma}(\mathcal{Y}) \to j_*\mathcal{O}_{\mathcal{X}_{\text{lis-et}}}$ has by 1.1.8 a PD-structure δ induced by the PD-structure on $[n] \mapsto \mathcal{D}_{X_n,\gamma}(Y_n)$, and by 1.1.11 this PD-structure δ is compatible with γ. We therefore obtain an object

$$\mathcal{X} \hookrightarrow D_{\mathcal{X},\gamma}(\mathcal{Y}) := \underline{\text{Spec}}_{\mathcal{Y}}(\mathcal{D}_{\mathcal{X},\gamma}(\mathcal{Y})) \tag{1.2.6.4}$$

of $\mathcal{C}'$ with a morphism to $(\mathcal{X} \hookrightarrow \mathcal{Y}) \in \mathcal{C}$ which is the identity on $\mathcal{X}$.

1.2.7. — For any object $\mathcal{X} \hookrightarrow \mathcal{Y}$ of $\mathcal{C}$ we have now constructed an object $\mathcal{X} \hookrightarrow D_{\mathcal{X},\gamma}(\mathcal{Y})$ of $\mathcal{C}'$ with a morphism $F(\mathcal{X} \hookrightarrow D_{\mathcal{X},\gamma}(\mathcal{Y})) \to (\mathcal{X} \hookrightarrow \mathcal{Y})$ in $\mathcal{C}$. We now verify that the functor (1.2.3.1) is an equivalence of categories. This will also imply that $D_{\mathcal{X},\gamma}(\mathcal{Y})$ is independent of all the choices (up to canonical isomorphism).

For this we reduce to the case of algebraic spaces as follows. For a smooth morphism $h_U : U \to \mathcal{S}$, let $\mathcal{C}_U$ (resp. $\mathcal{C}'_U$) be the category $\mathcal{C}$ (resp. $\mathcal{C}'$) obtained using $(U, I|_U, \gamma)$ instead of $(\mathcal{S}, I, \gamma)$. There are natural base change functors

$$h_U^* : \mathcal{C} \longrightarrow \mathcal{C}_U, \quad h_U^* : \mathcal{C}' \longrightarrow \mathcal{C}'_U \tag{1.2.7.1}$$

sending an object $\mathcal{X} \hookrightarrow \mathcal{Y}$ to $(\mathcal{X} \times_{\mathcal{S}} U) \hookrightarrow (\mathcal{Y} \times_{\mathcal{S}} U)$. The functor (1.2.3.1) then extends to a morphism of fibered categories over $\mathcal{S}_{\text{lis-et}}$

$$\begin{array}{c} \big(U \longmapsto \mathrm{HOM}_{\mathcal{C}'_U}\big(h_U^*(\mathcal{X}' \hookrightarrow \mathcal{Y}'), h_U^*(\mathcal{X} \hookrightarrow D_{\mathcal{X},\gamma}(\mathcal{Y}))\big)\big) \\ \downarrow \\ \big(U \longmapsto \mathrm{HOM}_{\mathcal{C}_U}\big(h_U^* F(\mathcal{X}' \hookrightarrow \mathcal{Y}'), h_U^*(\mathcal{X} \longhookrightarrow \mathcal{Y})\big)\big). \end{array} \tag{1.2.7.2}$$

On the other hand, for any smooth $h : U \to \mathcal{S}$ we have

$$h_U^* F(\mathcal{X}' \hookrightarrow \mathcal{Y}') \simeq F_U(h_U^*(\mathcal{X}' \hookrightarrow \mathcal{Y}')), \tag{1.2.7.3}$$

and by the construction of $D_{\mathcal{X},\gamma}(\mathcal{Y})$ we have

$$h_U^*(\mathcal{X} \hookrightarrow D_{\mathcal{X},\gamma}(\mathcal{Y})) \simeq (\mathcal{X} \times_{\mathcal{S}} U \hookrightarrow D_{\mathcal{X} \times_{\mathcal{S}} U, \gamma|_U}(\mathcal{Y} \times_{\mathcal{S}} U)), \tag{1.2.7.4}$$

where the divided power envelope is formed with respect to the cover $S \times_{\mathcal{S}} U \to U$ and the cover $Y \times_{\mathcal{S}} U \to \mathcal{Y} \times_{\mathcal{S}} U$ (where $S \to \mathcal{S}$ and $Y \to \mathcal{Y}$ are the covers used in the construction above). Moreover, it follows from the definitions that both source and target of (1.2.7.2) are stacks (though of course not stacks in groupoids). Consequently to prove that (1.2.7.2), and hence also (1.2.3.1), is an equivalence it suffices to prove that (1.2.7.2) is locally an equivalence. We may therefore replace $\mathcal{S}$ by a smooth cover, and hence for the remainder of the proof we assume that $\mathcal{S}$ is an algebraic space.

Next observe that if $\mathcal{X} \hookrightarrow \mathcal{Y}$ is a closed immersion of algebraic spaces over some smooth $\mathcal{S}$-space $U \to \mathcal{S}$ such that γ extends to $\mathcal{X}$, then the divided power envelope of $\mathcal{X}$ in $\mathcal{Y}$ is the same whether constructed with respect to the base $(\mathcal{S}, I, \gamma)$ or the base $(U, I|_U, \gamma|_U)$. Indeed by the construction this amounts to the following statement:

Lemma 1.2.8. — *Let T be a topos and $A \to A' \to B$ be a diagram of algebras in T with $A \to A'$ flat. Let (I, γ) be a PD-ideal in A and $J \subset B$ an ideal containing the image of I. Denote by (I', γ') the PD-ideal $A' \otimes_A I$ in A'. Then*

$$D_{A,\gamma}(B, J) = D_{A',\gamma'}(B, J). \tag{1.2.8.1}$$

Proof. — This follows from [**7**, I.2.3.2 (i)]. □

1.2.9. — It follows that we may further assume that the cover $S \to \mathcal{S}$ is simply $\mathrm{id} : \mathcal{S} \to \mathcal{S}$.

Observe that in the case when $\mathcal{S}$ is an algebraic space, the data of the isomorphisms ι_f and ι_g in the definition of morphisms in $\mathcal{C}$ and $\mathcal{C}'$ can be omitted in the definitions. Therefore, if $j' : \mathcal{X}' \hookrightarrow \mathcal{Y}'$ is an object of $\mathcal{C}'$, the category

$$\mathrm{HOM}_{\mathcal{C}}(F(j' : \mathcal{X}' \hookrightarrow \mathcal{Y}'), j : \mathcal{X} \hookrightarrow \mathcal{Y}) \tag{1.2.9.1}$$

can be described as the category of triples (f, g, ρ), where $f : \mathcal{X}' \to \mathcal{X}$ and $g : \mathcal{Y}' \to \mathcal{Y}$ are morphisms of $\mathcal{S}$-stacks, and $\rho : j \circ f \to g \circ j'$ is an isomorphism of functors. Similarly the category

$$\mathrm{HOM}_{C'}(j' : \mathcal{X}' \hookrightarrow \mathcal{Y}', \tilde{j} : \mathcal{X} \hookrightarrow D_{\mathcal{X},\gamma}(\mathcal{Y})) \tag{1.2.9.2}$$

is equivalent to the category of triples (f, g, ρ), where $f : \mathcal{X}' \to \mathcal{X}$ and $g : \mathcal{Y}' \to D_{\mathcal{X},\gamma}(\mathcal{Y})$ are morphisms of $\mathcal{S}$-stacks with g compatible with the PD-structures, and $\rho : \tilde{j} \circ f \to g \circ j'$ is an isomorphism of functors.

Since the functor $D_{\mathcal{X},\gamma}(\mathcal{Y}) \to \mathcal{Y}$ is faithful being representable [**49**, 8.1.2], it follows immediately that (1.2.3.1) is faithful.

To show that it is an equivalence of categories, fix a morphism (f, g, ρ) in (1.2.9.1), and define $\mathcal{P}$ to be the category of data $((f', g', \rho'), \sigma)$ consisting of a morphism (f', g', ρ') in (1.2.9.2) together with an isomorphism $\sigma : F(f', g', \rho') \simeq (f, g, \rho)$ in (1.2.9.1). To complete the proof of 1.2.3 it suffices to show that $\mathcal{P}$ is equivalent to the punctual category (the category with 1 object and 1 morphism).

By the universal property of the relative spectrum of a quasi-coherent sheaf of algebras, the category $\mathcal{P}$ is equivalent to the set of augmentations $\lambda : f^*\mathcal{D}_{\mathcal{X},\gamma}(\mathcal{Y}) \to \mathcal{O}_{\mathcal{Y}'_{\text{lis-et}}}$ of the quasi-coherent $\mathcal{O}_{\mathcal{Y}_{\text{lis-et}}}$-algebra $f^*\mathcal{D}_{\mathcal{X},\gamma}(\mathcal{Y})$ obtained by pullback such that the composite

$$f^*\mathcal{D}_{\mathcal{X},\gamma}(\mathcal{Y}) \xrightarrow{\ \lambda\ } \mathcal{O}_{\mathcal{Y}'_{\text{lis-et}}} \longrightarrow j'_*\mathcal{O}_{\mathcal{X}'_{\text{lis-et}}} \tag{1.2.9.3}$$

is the map induced by $\tilde{j} \circ g$, and such that the resulting morphism of stacks $\mathcal{Y}' \to D_{X,\gamma}(\mathcal{Y})$ is compatible with the divided power structure. Since morphisms of sheaves may be constructed locally and the compatibility with divided power structures can also be checked locally by 1.1.11 we see that to prove that $\mathcal{P}$ is equivalent to the punctual category we may replace $\mathcal{Y}'$ by a smooth cover and hence may assume that $\mathcal{Y}'$ is an affine scheme and that we are given a factorization of the map $g : \mathcal{Y}' \to \mathcal{Y}$ through the smooth cover $Y \to \mathcal{Y}$. Since $D_{\mathcal{X},\gamma}(\mathcal{Y}) \times_{\mathcal{Y}} Y \simeq D_{\mathcal{X}\times_{\mathcal{Y}} Y,\gamma}(Y)$ by construction, we see from this and 1.1.11 that we may further replace $\mathcal{Y}$ by Y and hence may assume that $\mathcal{Y}$ is an algebraic space. In this case it follows from the universal property of the divided power envelope [**7**, I.2.3.1] that $\mathcal{P}$ is equivalent to the punctual category.

This completes the proof of 1.2.3. □

Let us also record the following corollary which follows from the proof of 1.2.3 and the corresponding result for schemes [**7**, I.2.7.1].

Corollary 1.2.10. — *With notation as in 1.2.3, if $h : \mathcal{S}' \to \mathcal{S}$ is a smooth morphism of stacks, and we view $\mathcal{S}'$ as a PD-stack with ideal $I' := h^*I$ with divided structure γ' defined by γ (1.1.12), then for any object $\mathcal{X} \hookrightarrow \mathcal{Y}$ of C there is a canonical isomorphism*

$$(\mathcal{X} \times_{\mathcal{S}} \mathcal{S}' \longhookrightarrow D_{\mathcal{X},\gamma}(\mathcal{Y}) \times_{\mathcal{S}} \mathcal{S}') \simeq (\mathcal{X} \times_{\mathcal{S}} \mathcal{S}' \longhookrightarrow D_{\mathcal{X}\times_{\mathcal{S}}\mathcal{S}',\gamma'}(\mathcal{Y} \times_{\mathcal{S}} \mathcal{S}')). \tag{1.2.10.1}$$

1.2.11. — As in [**7**, proof of III.2.1.3], theorem 1.2.3 can be generalized as follows. Consider two PD-stacks $(\mathcal{S}_i, I_i, \gamma^i)$ $(i = 1, 2)$, and let $\mathcal{E}$ denote the category of closed immersions $\mathcal{X} \hookrightarrow \mathcal{Y}$ defined by a quasi-coherent sheaf of ideals in $\mathcal{O}_{\mathcal{Y}_{\text{lis-et}}}$ over $\mathcal{S}_1 \times \mathcal{S}_2$

for which both γ^1 and γ^2 extend to $\mathcal{X}$. Let $\mathcal{E}'$ be the category of objects $(\mathcal{X} \hookrightarrow \mathcal{Y}) \in \mathcal{E}$ together with divided power structure on the ideal of $\mathcal{X}$ in $\mathcal{Y}$ compatible with both γ_1 and γ_2. There is a natural forgetful functor $F : \mathcal{E}' \to \mathcal{E}$. Then using the same arguments used to prove 1.2.3 and [**7**, proof of III.2.1.3] one obtains the following proposition:

Proposition 1.2.12. — *For any object $j : \mathcal{X} \hookrightarrow \mathcal{Y}$ of $\mathcal{E}$ there is a canonically associated object $j_D : \mathcal{X} \hookrightarrow D_{\mathcal{X},\gamma_1,\gamma_2}(\mathcal{Y})$ of $\mathcal{E}'$ with a morphism $F(\mathcal{X} \hookrightarrow D_{\mathcal{X},\gamma_1,\gamma_2}(\mathcal{Y})) \to (j : \mathcal{X} \hookrightarrow \mathcal{Y})$ in $\mathcal{E}$ such that for any other object $j' : \mathcal{X}' \hookrightarrow \mathcal{Y}'$ of $\mathcal{E}'$ the induced functor*

$$\mathrm{HOM}_{\mathcal{E}'}(\mathcal{X}' \hookrightarrow \mathcal{Y}', \mathcal{X} \hookrightarrow D_{\mathcal{X},\gamma_1,\gamma_2}(\mathcal{Y})) \longrightarrow \mathrm{HOM}_{\mathcal{E}}(F(\mathcal{X}' \hookrightarrow \mathcal{Y}'), \mathcal{X} \hookrightarrow \mathcal{Y}) \tag{1.2.12.1}$$

is an equivalence of categories.

1.3. The crystalline topos

1.3.1. — Let $(\mathcal{S}, I, \gamma)$ be a PD-stack and $f : \mathcal{X} \to \mathcal{S}$ a morphism of algebraic stacks such that γ extends to $\mathcal{X}$. We define the *lisse-étale crystalline site of* $\mathcal{X}/\mathcal{S}$, denoted $\mathrm{Cris}(\mathcal{X}_{\text{lis-et}}/(\mathcal{S}, I, \gamma))$ (or just $\mathrm{Cris}(\mathcal{X}_{\text{lis-et}}/\mathcal{S})$ if there is no chance of confusion), as follows.

The objects of $\mathrm{Cris}(\mathcal{X}_{\text{lis-et}}/\mathcal{S})$ are triples $(U, j : U \hookrightarrow T, \delta)$, where $U \to \mathcal{X}$ is a smooth morphism with U a scheme, j is a closed immersion of $\mathcal{S}$-schemes, and δ is a divided power structure on the ideal of U in T compatible with γ. We usually write just $U \hookrightarrow T$, or even just T, for an object of $\mathrm{Cris}(\mathcal{X}_{\text{lis-et}}/\mathcal{S})$. A morphism $(U', j' : U' \to T', \delta') \to (U, j : U \hookrightarrow T, \delta)$ is a pair (f, g), where $f : U' \to U$ is a morphism of $\mathcal{X}$-schemes, and $g : T' \to T$ is a morphism of $\mathcal{S}$-schemes compatible with δ and δ' such that the diagram of $\mathcal{S}$-schemes

$$\begin{array}{ccc} U' & \xrightarrow{j'} & T' \\ {\scriptstyle f}\downarrow & & \downarrow{\scriptstyle g} \\ U & \xrightarrow{j} & T \end{array} \tag{1.3.1.1}$$

commutes. The topology on $\mathrm{Cris}(\mathcal{X}_{\text{lis-et}}/\mathcal{S})$ is the topology defined by the pre-topology for which the covering families are morphisms $\{(U_i, T_i, \delta_i) \to (U, T, \delta)\}$ for which the morphisms of schemes $\{T_i \to T\}$ form an étale cover.

The *lisse-étale crystalline topos of* $\mathcal{X}/\mathcal{S}$, denoted $(\mathcal{X}_{\text{lis-et}}/(\mathcal{S}, I, \gamma))_{\text{cris}}$ (or sometimes simply $(\mathcal{X}_{\text{lis-et}}/\mathcal{S})_{\text{cris}}$), is the topos associated to the site $\mathrm{Cris}(\mathcal{X}_{\text{lis-et}}/\mathcal{S})$. The topos $(\mathcal{X}_{\text{lis-et}}/\mathcal{S})_{\text{cris}}$ is ringed with structure sheaf given by

$$\mathcal{O}_{\mathcal{X}_{\text{lis-et}}/\mathcal{S}}(U \hookrightarrow T) := \Gamma(T_{\text{et}}, \mathcal{O}_{T_{\text{et}}}). \tag{1.3.1.2}$$

1.3.2. — When $\mathcal{X}$ in the above is a Deligne-Mumford stack, we will also consider the full subcategory $\mathrm{Cris}(\mathcal{X}_{\text{et}}/\mathcal{S}) \subset \mathrm{Cris}(\mathcal{X}_{\text{lis-et}}/\mathcal{S})$ whose objects are the objects $U \hookrightarrow T$ with U an étale $\mathcal{X}$-scheme. The topology on $\mathrm{Cris}(\mathcal{X}_{\text{lis-et}}/\mathcal{S})$ induces a topology on $\mathrm{Cris}(\mathcal{X}_{\text{et}}/\mathcal{S})$ and hence $\mathrm{Cris}(\mathcal{X}_{\text{et}}/\mathcal{S})$ is a site which we refer to as the *étale crystalline site of* $\mathcal{X}/\mathcal{S}$. We write $(\mathcal{X}_{\text{et}}/\mathcal{S})_{\text{cris}}$ for the associated topos which we call the *étale* *crystalline topos of* $\mathcal{X}/\mathcal{S}$. The structure sheaf in $(\mathcal{X}_{\text{lis-et}}/\mathcal{S})_{\text{cris}}$ induces by restriction a sheaf of rings in $(\mathcal{X}_{\text{et}}/\mathcal{S})_{\text{cris}}$. In what follows we view $(\mathcal{X}_{\text{et}}/\mathcal{S})_{\text{cris}}$ as a ringed topos with this structure sheaf.

1.3.3. — As in the classical case, a sheaf $F \in (\mathcal{X}_{\text{lis-et}}/\mathcal{S})_{\text{cris}}$ is equivalent to the data of a sheaf F_T on T_{et} for each object $(U,T,\delta) \in \text{Cris}(\mathcal{X}_{\text{lis-et}}/\mathcal{S})$ together with a morphism $q^{-1}F_T \to F_{T'}$ for every morphism $T' \to T$ in $\text{Cris}(\mathcal{X}_{\text{lis-et}}/\mathcal{S})$. Furthermore, these morphisms are required to be compatible with compositions of morphisms in $\text{Cris}(\mathcal{X}_{\text{lis-et}}/\mathcal{S})$.

Similarly, a $\mathcal{O}_{\mathcal{X}_{\text{lis-et}}/\mathcal{S}}$-module M can be described by a $\mathcal{O}_{T_{\text{et}}}$-module M_T for every object $T \in \text{Cris}(\mathcal{X}_{\text{lis-et}}/\mathcal{S})$ together with transition morphisms $q^*M_T \to M_{T'}$ in $\text{Cris}(\mathcal{X}_{\text{lis-et}}/\mathcal{S})$ compatible with compositions.

Remark 1.3.4. — In the definition of the crystalline topos, we could also have used the bigger site consisting of PD-immersions $U \hookrightarrow T$ with U and T algebraic spaces. Since any such closed immersion admits an étale cover by an object of $\text{Cris}(\mathcal{X}_{\text{lis-et}}/\mathcal{S})$ the resulting topoi are equivalent. We will therefore usually just work with schemes, but occasionally it is useful to note that a sheaf $F \in (\mathcal{X}_{\text{lis-et}}/\mathcal{S})_{\text{cris}}$ can be evaluated on a PD-immersion $U \hookrightarrow T$ with U and T only algebraic spaces.

1.4. Three basic lemmas and functoriality

1.4.1. — Let $f : (\mathcal{S}', I', \gamma') \to (\mathcal{S}, I, \gamma)$ be a morphism of PD-stacks, and consider a 2-commutative diagram of algebraic stacks

$$\begin{array}{ccc} \mathcal{X}' & \xrightarrow{g} & \mathcal{X} \\ \downarrow & & \downarrow \\ \mathcal{S}' & \xrightarrow{f} & \mathcal{S} \end{array} \tag{1.4.1.1}$$

such that γ (resp. γ') extends to $\mathcal{X}$ (resp. $\mathcal{X}'$).

Definition 1.4.2. — Fix $(U',T',\delta') \in \text{Cris}(\mathcal{X}'_{\text{lis-et}}/\mathcal{S}')$ and $(U,T,\delta) \in \text{Cris}(\mathcal{X}_{\text{lis-et}}/\mathcal{S})$. A *g-PD morphism* $h : T' \to T$ is a 2-commutative diagram of $\mathcal{S}$-stacks

$$\begin{array}{ccc} T' & \xrightarrow{h_T} & T \\ \uparrow & & \uparrow \\ U' & \xrightarrow{h_U} & U \\ \downarrow & & \downarrow \\ \mathcal{X}' & \xrightarrow{g} & \mathcal{X}, \end{array} \tag{1.4.2.1}$$

where h_T is a PD-morphism. We denote the set of such g-PD morphisms $T' \to T$ by $\text{Hom}_{g-PD}(T',T)$.

Lemma 1.4.3. — *For any object* $(U', T', \delta') \in \mathrm{Cris}(\mathcal{X}'_{\text{lis-et}}/\mathcal{S}')$ *there exist a covering*

$$(\tilde{U}', \tilde{T}', \tilde{\delta}') \longrightarrow (U', T', \delta') \tag{1.4.3.1}$$

and a g-PD morphism $h' : \tilde{T}' \to T$ *for some* $(U, T, \delta) \in \mathrm{Cris}(\mathcal{X}_{\text{lis-et}}/\mathcal{S})$.

Proof. — Let $U \to \mathcal{X}$ be a smooth surjection with U a scheme. Then the product $U \times_{\mathcal{X}} \mathcal{X}'$ is a smooth algebraic stack which surjects onto $\mathcal{X}'$. By the existence of quasi-sections for smooth morphisms [**15**, IV.17.16.3], we can find an étale surjection $V' \to U'$ and a lifting $V' \to U \times_{\mathcal{X}} \mathcal{X}'$ of the composite $V' \to U' \to \mathcal{X}'$. Let $\tilde{V}'$ be the unique lift of V' to an étale T'-scheme. Then replacing $U' \hookrightarrow T'$ by $(V' \hookrightarrow \tilde{V}')$ we see that we may assume that there exists a morphism $U' \to U$ for some smooth $\mathcal{X}$-scheme U. Replacing $\mathcal{X}$ by U and $\mathcal{X}'$ by U' we see that it suffices to consider the case when $\mathcal{X}$ and $\mathcal{X}'$ are schemes and $U' = \mathcal{X}'$. Furthermore, we can without loss of generality replace U' and U by Zariski covers and hence may even assume that $U' = \mathcal{X}'$ and $\mathcal{X}$ are affine schemes.

Set $X = \mathrm{Spec}(B)$, $X' = \mathrm{Spec}(B')$, and $T' = \mathrm{Spec}(C')$. Let $C = C' \times_{B'} B$ be the product in the category of rings and let $T = \mathrm{Spec}(C)$. We claim that T has a unique structure of an $\mathcal{S}$-scheme making $T' \to T$ and $X \hookrightarrow T$ morphisms of $\mathcal{S}$-schemes. Let $x' : X' \to \mathcal{S}$ be the given map. For any map $f : X' \to Y$, let $\mathcal{S}_{x'}(Y)$ denote the category of pairs (y, ϵ) where $y : Y \to \mathcal{S}$ is a 1-morphism and $\epsilon : f^*y \simeq x'$ is an isomorphism in $\mathcal{S}_{X'}$. Then what is needed is that the natural functor

$$\mathcal{S}_{x'}(C' \times_{B'} B) \longrightarrow \mathcal{S}_{x'}(C') \times \mathcal{S}_{x'}(B) \tag{1.4.3.2}$$

is an equivalence of categories. This follows from the following lemma. The closed immersion $X \hookrightarrow T$ is given the structure of a PD-thickening as in [**7**, III.2.1.2]. □

Lemma 1.4.4. — *Let* $C' \to B'$ *be a surjection of rings with nilpotent kernel, and* $B \to B'$ *any morphism of rings. Then for any algebraic stack* $\mathcal{S}$ *and morphism* $x' : \mathrm{Spec}(B') \to \mathcal{S}$, *the natural functor*

$$\mathcal{S}_{x'}(C' \times_{B'} B) \longrightarrow \mathcal{S}_{x'}(C') \times \mathcal{S}_{x'}(B) \tag{1.4.4.1}$$

is an equivalence of categories.

Proof. — For ease of notation, write $X' = \mathrm{Spec}(C' \times_{B'} B)$, $V = \mathrm{Spec}(B')$, $V' = \mathrm{Spec}(C')$, and $X = \mathrm{Spec}(B)$.

If $U \to X$ is an étale morphism, denote by $U' \to X'$ (resp. $U_V \to V$, $U'_V \to V'$) the étale morphism defined by the identification of the étale sites [**28**, I.8.3] of X' and X (resp. defined by pullback of U, defined by pullback of U'). Let $\mathcal{Z}_1$ (resp. $\mathcal{Z}_2$) denote the fibered category over the small étale site of X which to any U associates $\mathcal{S}_{x'|_{U_V}}(U')$ (resp. $\mathcal{S}_{x'|_{U_V}}(U) \times \mathcal{S}_{x'|_{U_V}}(U'_V)$). There is a natural functor $F : \mathcal{Z}_1 \to \mathcal{Z}_2$ whose value on X is (1.4.4.1). Since $\mathcal{Z}_1$ and $\mathcal{Z}_2$ are both stacks over X_{et}, F is an equivalence if and only if F is étale locally on X an equivalence, and hence to prove the lemma it suffices to show that $F(X) : \mathcal{Z}_1(X) \to \mathcal{Z}_2(X)$ is fully faithful and that every object of $\mathcal{Z}_2$ is étale locally on X in the image of F.

Note first that the result is immediate in the case when $\mathcal{S}$ is a scheme since the diagram

$$\begin{array}{ccc} V & \longrightarrow & X \\ \downarrow & & \downarrow \\ V' & \longrightarrow & X' \end{array} \tag{1.4.4.2}$$

is cocartesian in the category of schemes.

In fact it is also cocartesian in the category of algebraic spaces. To see this, let T be an algebraic space, and let

$$\mathrm{Hom}(X', T) \longrightarrow \mathrm{Hom}(X, T) \times_{\mathrm{Hom}(V,T)} \mathrm{Hom}(V', T) \tag{1.4.4.3}$$

be the natural map. Fix an object (f_1, f_2) of the target set of (1.4.4.3), and for any étale T-scheme U/T define X'_U to be the unique lift of $X \times_{f_1,T} U$ to an étale X' scheme. Note that X'_U is the scheme $(|X_U|, \mathcal{O}_{X_U} \times_{g_{U*}\mathcal{O}_{V_U}} g_{U*}\mathcal{O}_{V'_U})$, where V_U (resp. V'_U) denotes $V \times_T U$ (resp. $V' \times_T U$). Let F be the sheaf on T_{et} which to any étale T-scheme $U \to T$ associates the set of maps $X'_U \to U$ inducing the pair

$$(f_{1,U}, f_{2,U}) \in \mathrm{Hom}(X \times_T U, U) \times_{\mathrm{Hom}(V \times_T U, U)} \mathrm{Hom}(V' \times_T U, U) \tag{1.4.4.4}$$

obtained from (f_1, f_2) by base change. To prove that (1.4.4.3) is bijective it suffices to show that $F \simeq \{*\}$. Since F is a sheaf this can be verified étale locally on T, and hence follows from the case when T is a scheme.

To see that $F(X)$ is fully faithful, let (t_i, ϵ_i) $(i = 1, 2)$ denote two objects of $\mathcal{Z}_1(X)$. To give an isomorphism $(t_1, \epsilon_1) \to (t_2, \epsilon_2)$ (resp. $F(t_1, \epsilon_1) \to F(t_2, \epsilon_2)$) in $\mathcal{Z}_1$ (resp. $\mathcal{Z}_2$) is equivalent to giving a morphism i (resp. two morphisms i_1 and i_2)

$$\begin{gathered} i : X' \longrightarrow X' \times_{t_1 \times t_2, \mathcal{S}\times\mathcal{S}, \Delta} \mathcal{S} \\ (\text{resp. } i_1 : V' \longrightarrow X' \times_{t_1 \times t_2, \mathcal{S}\times\mathcal{S}, \Delta} \mathcal{S}, \quad i_2 : X \longrightarrow X' \times_{t_1 \times t_2, \mathcal{S}\times\mathcal{S}, \Delta} \mathcal{S}) \end{gathered} \tag{1.4.4.5}$$

over X' such that $g^*j^*(i) = \epsilon_2^{-1} \circ \epsilon_1$ (resp. $j_V^*(i_1) = \epsilon_2^{-1} \circ \epsilon_1 = g^*(i_2)$). The full faithfulness therefore follows from the case when $\mathcal{S}$ is an algebraic space.

To see that every object $O := ((t_X, \epsilon_X), (t_{V'}, \epsilon_{V'})) \in \mathcal{Z}_2(X)$ is étale locally in the image of F, let $Y \to \mathcal{S}$ be a smooth cover. After replacing X by an étale cover, we may by the existence of quasi-sections for smooth morphisms [**15**, IV.17.16.3] assume that $t_X : X \to \mathcal{S}$ factors through a map $\tilde{t}_X : X \to Y$. Moreover, in this case, we have a 2-commutative diagram

$$\begin{array}{ccc} V & \xrightarrow{\tilde{t}_X \circ g} & Y \\ {\scriptstyle j_V}\downarrow & & \downarrow \\ V' & \xrightarrow{t_{V'}} & \mathcal{S}, \end{array} \tag{1.4.4.6}$$

and since V' is affine it follows that O is induced by a commutative diagram

$$\begin{array}{ccc} V & \xrightarrow{j_V} & V' \\ {\scriptstyle g}\downarrow & & \downarrow{\scriptstyle \tilde{t}_{V'}} \\ X & \xrightarrow{\tilde{t}_X} & Y. \end{array} \tag{1.4.4.7}$$

Hence by the case of an algebraic space the pair $(\tilde{t}_X, \tilde{t}'_V)$ is induced by a unique morphism $X' \to Y$ and the image under F of the resulting object $O_1 \in \mathcal{Z}_1(X)$ is isomorphic to O. □

The proof of 1.4.3 also yields the following:

Corollary 1.4.5. — *Assume in addition that $\mathcal{X}$ is a Deligne-Mumford stack. Then any object $(U, T, \delta) \in \mathrm{Cris}(\mathcal{X}_{\text{lis-et}}/\mathcal{S})$ admits étale locally a morphism to an object of $\mathrm{Cris}(\mathcal{X}_{\text{et}}/\mathcal{S})$.*

Lemma 1.4.6. — *Let $U \to \mathcal{X}$ be a smooth morphism from a scheme U to $\mathcal{X}$ and let $U \hookrightarrow T_1$ and $U \hookrightarrow T_2$ be two PD-immersions over $\mathcal{S}$ defining two objects of $\mathrm{Cris}(\mathcal{X}_{\text{lis-et}}/\mathcal{S})$. Suppose further that $\mathcal{Y} \to \mathcal{S}$ is a morphism of algebraic stacks, and that we are given a 2-commutative diagram of $\mathcal{S}$-stacks*

$$\begin{array}{ccc} U & \longrightarrow & T_1 \\ \downarrow & & \downarrow{\scriptstyle q_1} \\ T_2 & \xrightarrow{q_2} & \mathcal{Y}. \end{array} \tag{1.4.6.1}$$

Then there exist a PD-immersion $U \hookrightarrow T$ over $\mathcal{S}$, morphisms in $\mathrm{Cris}(\mathcal{X}_{\text{lis-et}}/\mathcal{S})$

$$p_1 : T \longrightarrow T_1, \qquad p_2 : T \longrightarrow T_2, \tag{1.4.6.2}$$

and an isomorphism of functors $\epsilon : q_1 \circ p_1 \simeq q_2 \circ p_2$ inducing the given one over U which is universal in the following sense: for any object $(U', T', \delta') \in \mathrm{Cris}(\mathcal{X}'_{\text{lis-et}}/\mathcal{S}')$ and g-PD morphisms $h_1 : T' \to T_1$ and $h_2 : T' \to T_2$ together with an isomorphism $\epsilon' : q_1 \circ h_1 \simeq q_2 \circ h_2$ inducing the pullback of that on U to U', there exists a unique g-PD morphism $h : T' \to T$ such that $h^*\epsilon = \epsilon'$.

Proof. — Set T equal to the divided power envelope $D_{U,\delta_1,\delta_2}(T_1 \times_{\mathcal{Y}} T_2)$ with compatibility with δ_1 and δ_2 as defined in 1.2.12. □

Corollary 1.4.7. — *If $\mathcal{X}$ is a Deligne-Mumford stack then finite fiber products are representable in $\mathrm{Cris}(\mathcal{X}_{\text{et}}/\mathcal{S})$.*

Proof. — Consider a diagram

$$\begin{array}{ccc} & & (U_1, T_1, \delta_1) \\ & & \downarrow \\ (U_2, T_2, \delta_2) & \longrightarrow & (U_3, T_3, \delta_3) \end{array} \tag{1.4.7.1}$$

in $\mathrm{Cris}(\mathcal{X}_{\mathrm{et}}/\mathcal{S})$. Set $U = U_1 \times_{U_3} U_2$, and let $z : U \hookrightarrow T_1 \times_{T_3} T_2$ be the induced closed immersion. Let $U \hookrightarrow T$ be the divided power envelope $D_{U,\delta_1,\delta_2}(T_1 \times_{T_3} T_2)$ with compatibility with respect to δ_1 and δ_2. Then $U \hookrightarrow T$ represents the fiber product of the diagram (1.4.7.1).

The representability of a general finite fiber product follows from this special case by induction. □

Corollary 1.4.8. — *Let (U', T', δ') be an object of $\mathrm{Cris}(\mathcal{X}'_{\text{lis-et}}/\mathcal{S}')$ and let*

$$h_i : (U', T', \delta') \longrightarrow (U_i, T_i, \delta_i) \quad (i = 1, 2) \tag{1.4.8.1}$$

be two g-PD-morphisms to objects of $\mathrm{Cris}(\mathcal{X}_{\text{lis-et}}/\mathcal{S})$. Then there exist an object $(U, T, \delta) \in \mathrm{Cris}(\mathcal{X}_{\text{lis-et}}/\mathcal{S})$ and morphisms $p_i : (U, T, \delta) \to (U_i, T_i, \delta_i)$ in $\mathrm{Cris}(\mathcal{X}_{\text{lis-et}}/\mathcal{S})$ such that after replacing (U', T', δ') by a covering there exists a g-PD-morphism $h : T' \to T$ such that $h_i = p_i \circ h$ $(i = 1, 2)$.

Proof. — By replacing (U', T', δ') by an étale covering, we may without loss of generality assume that T' is an affine scheme. Consider $U_1 \times_{\mathcal{X}} U_2$ and let $\lambda : U' \to U_1 \times_{\mathcal{X}} U_2$ be the morphism $h_1 \times h_2$. Since U' is quasi-compact, there exist an affine scheme U and an étale morphism $U \to U_1 \times_{\mathcal{X}} U_2$ whose image contains the image of U'. After replacing U' by another étale cover we may therefore assume that we have a morphism $h_U : U' \to U$ such that $p_i \circ h_U = h_i$ $(i = 1, 2)$, where $p_i : U \to U_i$ denotes the projections. Since U is affine and smooth over U_i and $U_i \hookrightarrow T_i$ is defined by a nil-ideal, there exists a smooth lifting $\widetilde{T}_i$ of U to T_i. Since $\widetilde{T}_i \to T_i$ is smooth the PD-structure on T_i extends to $\widetilde{T}_i$ so we have morphisms

$$\begin{array}{ccc} U & \longrightarrow & \widetilde{T}_i \\ {\scriptstyle p_i}\downarrow & & \downarrow \\ U_i & \longrightarrow & T_i \end{array} \tag{1.4.8.2}$$

in $\mathrm{Cris}(\mathcal{X}_{\text{lis-et}}/\mathcal{S})$. Let $(U, T, \delta) \in \mathrm{Cris}(\mathcal{X}_{\text{lis-et}}/\mathcal{S})$ be the object obtained from 1.4.6 applied to the thickenings $U \hookrightarrow \widetilde{T}_i$ and $\mathcal{Y} = \mathcal{S}$. Since the $\widetilde{T}_i \to T_i$ are smooth, the existence of quasi-sections for smooth morphisms [**15**, IV.17.16.3] implies that after replacing T' by an étale cover there exist liftings $\tilde{h}_i : T' \to \widetilde{T}_i$ of the maps $h_i : T' \to T_i$. From 1.4.6 we therefore obtain a map $h : T' \to T$ with the desired properties. □

Corollary 1.4.9. — *Assume $\mathcal{X}$ is a Deligne-Mumford stack. Then finite and nonempty products in $\mathrm{Cris}(\mathcal{X}_{\mathrm{et}}/\mathcal{S})$ are representable.*

Proof. — Let (U_1, T_1, δ_1) and (U_2, T_2, δ_2) be two objects of $\mathrm{Cris}(\mathcal{X}_{\mathrm{et}}/\mathcal{S})$ and set $U := U_1 \times_{\mathcal{X}} U_2$. The projections to U_1 and U_2 are étale, and hence for each $i = 1, 2$ there exists a unique étale morphism $\widetilde{T}_i \to T_i$ such that the diagram

(1.4.9.1)
$$\begin{array}{ccc} U & \longrightarrow & \widetilde{T}_i \\ \downarrow & & \downarrow \\ U_i & \longrightarrow & T_i \end{array}$$

is cartesian. Replacing (U_i, T_i, δ_i) by $U \hookrightarrow \widetilde{T}_i$ with the PD-structure induced by that on T_i, we may assume that $U_1 = U_2 = U$. Applying 1.4.6 with $\mathcal{Y} = \mathcal{S}$ it follows that products of two objects are representable. The case of a general finite product follows from this by induction. □

Example 1.4.10. — In general products are not representable in $\mathrm{Cris}(\mathcal{X}_{\text{lis-et}}/\mathcal{S})$. An explicit example can be constructed as follows. Let k be an algebraically closed field of positive characteristic and G/k a smooth affine group scheme. Set $U = U' = \mathrm{Spec}(k)$ with maps to BG given by the trivial torsor, $T = \mathrm{Spec}(k[\epsilon]/\epsilon^2)$ with the unique divided power structure δ with $\delta_n = 0$ for $n \geq 2$, and $T' = \mathrm{Spec}(k)$. Then we claim that the product of (U, T, δ) and (U', T', δ') (where δ' is the unique divided power structure on the zero ideal in k) is not representable. Since $U \times_{BG} U'$ is canonically isomorphic to G, such a product would be a PD-immersion $j : G \hookrightarrow P$ sitting in a commutative diagram

(1.4.10.1)
$$\begin{array}{ccc} G & \xrightarrow{j} & P \\ \downarrow & & \downarrow \\ \mathrm{Spec}(k) & \longrightarrow & T. \end{array}$$

Let $\widetilde{G}/T$ be the smooth lifting of G given by $T \times_{\mathrm{Spec}(k)} G$ so that $G \hookrightarrow \widetilde{G}$ is a PD-immersion. Since $j : G \hookrightarrow P$ represents the product in $\mathrm{Cris}(BG_{\text{lis-et}}/k)$, we obtain a unique morphism $\rho : \widetilde{G} \to P$ over T restricting to $j : G \hookrightarrow P$. On the other hand, for any automorphism σ of $\widetilde{G}$ (as a scheme, not necessarily respecting the group structure) restricting to the identity on G the composite

(1.4.10.2)
$$\widetilde{G} \xrightarrow{\sigma} \widetilde{G} \xrightarrow{\rho} P$$

must also equal ρ. It follows that the image of $\mathcal{O}_P$ in $\mathcal{O}_{\widetilde{G}} = \mathcal{O}_G[\epsilon]$ must be contained in the elements invariant under the group of derivations $\mathcal{O}_G \to \mathcal{O}_G$. On the other hand, if G is positive dimension there exists a nontrivial derivation ∂ of $\mathcal{O}_G$. Let $f \in \mathcal{O}_G$ be an element with $\partial(f) \neq 0$. Since $\mathcal{O}_P \to \mathcal{O}_G$ is surjective, there exists an element $s \in \mathcal{O}_P$ mapping to $f \in \mathcal{O}_G$. Then $\rho^*(s) = f + g \cdot \epsilon$ for some $g \in \mathcal{O}_G$. If σ is the infinitesimal automorphism corresponding to ∂ we have $\sigma^*(f + g\epsilon) = f + (g + \partial(f)\epsilon)$. In particular $\sigma^*(\rho^*(s)) \neq \rho^*(s)$ which is a contradiction. It follows that the product is not representable.

Lemma 1.4.11. — *Let $\mathcal{X}$ and $\mathcal{X}'$ be algebraic stacks with $\mathcal{X}$ Deligne-Mumford. Let $\mathcal{Z}$ be an algebraic stack with a morphism $\mathcal{S} \to \mathcal{Z}$, $(U, T, \delta) \in \mathrm{Cris}(\mathcal{X}_{\text{et}}/\mathcal{S})$ an object and $q_1, q_2 : T \to Y$ two morphisms of $\mathcal{Z}$-schemes such that $q_{1,U} = q_{2,U}$. Then there exists* *an object* $(U, T_0, \delta) \in \mathrm{Cris}(\mathcal{X}_{\mathrm{et}}/\mathcal{S})$ *together with a morphism* $p : T_0 \to T$ *inducing the identity on* U*, such that* $q_1 \circ p = q_2 \circ p$ *with the following universal property: for any* $(U', T', \delta') \in \mathrm{Cris}(\mathcal{X}'_{\text{lis-et}}/\mathcal{S}')$ *and* g*-PD-morphism* $h : T' \to T$ *such that* $q_1 \circ h = q_2 \circ h$ *there exists a unique* g*-PD-morphism* $T' \to T_0$ *filling in the diagram:*

(1.4.11.1)

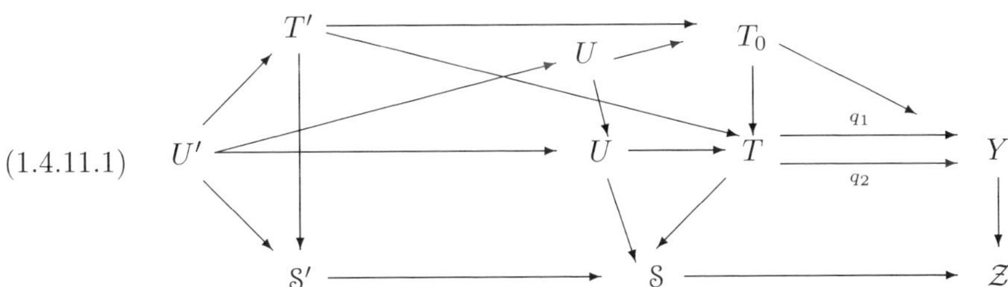

Proof. — Let T'_0 be the equalizer of the two maps of schemes $q_i : T \to Y$ and let $q : T'_0 \to Y$ be the resulting morphism of schemes. Over T'_0 we have an automorphism α of $q^*(y)$ (where $y \in \mathcal{Z}(Y)$ is the given 1-morphism) which reduces to the identity over U. Since $\mathcal{Z}$ is locally separated (*i.e.*, the diagonal of $\mathcal{Z}$ is of finite type by assumption 0.2.1), the condition $\alpha = 1$ is representable by a locally closed subscheme of T'_0, and since $U \hookrightarrow T'_0$ is a homeomorphism on the underlying spaces the condition $\alpha = 1$ is in fact represented by a closed subscheme T''_0 of T'_0. Let $J \subset \mathcal{O}_T$ be the PD-ideal generated by the ideal defining T''_0 in T, and let $U \hookrightarrow T_0$ be the resulting PD-immersion. This has the required properties. □

Corollary 1.4.12. — *If* $\mathcal{X}$ *is a Deligne-Mumford stack and* $q_1, q_2 : (U_1, T_1, \delta_1) \to (U_2, T_2, \delta_2)$ *are two morphisms in* $\mathrm{Cris}(\mathcal{X}_{\mathrm{et}}/\mathcal{S})$*, then the equalizer of* q_1 *and* q_2 *is representable in* $\mathrm{Cris}(\mathcal{X}_{\mathrm{et}}/\mathcal{S})$.

Proof. — Let $U \subset U_1$ denote the equalizer of the two $\mathcal{X}$-maps $q_{U,i} : U_1 \to U_2$. This equalizer in the category of $\mathcal{X}$-spaces can be constructed as follows. First let $U' \subset U$ be the equalizer of the two morphisms of schemes $U_1 \to U_2$ and let $q_U : U' \to U_2$ be the induced map. Let $u_i : U_i \to \mathcal{X}$ be the two structure maps and $u' : U' \to \mathcal{X}$ the restriction of u_1. The two isomorphisms $\iota_i : q^*_{U,i}(u_2) \to u_1$ restrict to two isomorphisms $q^*_U(u_2) \to u'$ over U'. Taking their difference we obtain an automorphism α of u'. The condition that α is equal to the identity is then represented by a locally closed subscheme of U' since the diagonal of $\mathcal{X}$ is assume locally separated. The equalizer U of the two morphisms $q_{U,i} : U_1 \to U_2$ in the category of $\mathcal{X}$-schemes is this subscheme of U' representing the condition that $\alpha = \mathrm{id}$.

The corollary now follows from 1.4.11 by taking T to be the divided power envelope of U in T_1 with the maps to (U_2, T_2, δ_2) induced by the maps q_i, $\mathcal{S}' = \mathcal{Z} = \mathcal{S}$, $\mathcal{X} = \mathcal{X}'$, and $Y = T_2$. □

Corollary 1.4.13. — *If* $\mathcal{X}$ *is a Deligne-Mumford stack, then finite nonempty projective limits in* $\mathrm{Cris}(\mathcal{X}_{\mathrm{et}}/\mathcal{S})$ *are representable.*

Proof. — As in [**7**, III.2.1.8] it suffices to show that finite nonempty products and equalizers are representable which follows from 1.4.9 and 1.4.12. □

Using the same argument as in the classical case we now obtain the following:

Theorem 1.4.14. — *Assume $\mathcal{X}$ and $\mathcal{X}'$ in (1.4.1.1) are Deligne-Mumford stacks. Then there exists a unique morphism of topoi*

(1.4.14.1) $$g_{\mathrm{cris}} : (\mathcal{X}'_{\mathrm{et}}/\mathcal{S}')_{\mathrm{cris}} \longrightarrow (\mathcal{X}_{\mathrm{et}}/\mathcal{S})_{\mathrm{cris}}$$

with the property that for any $T \in \mathrm{Cris}(\mathcal{X}_{\mathrm{et}}/\mathcal{S})$,

(1.4.14.2) $$g^*_{\mathrm{cris}}(\tilde{T})(U', T', \delta') = \mathrm{Hom}_{g-PD}(T', T),$$

where $\tilde{T}$ denotes the sheaf associated to T. Moreover, g_{cris} is naturally a morphism of ringed topoi.

Proof. — This follows from the lemmas and the argument used in [**7**, III.2.2]. □

Remark 1.4.15. — More generally, if only $\mathcal{X}$ is a Deligne-Mumford stack then there is a morphism of topoi

(1.4.15.1) $$g : (\mathcal{X}'_{\text{lis-et}}/\mathcal{S}')_{\mathrm{cris}} \longrightarrow (\mathcal{X}_{\mathrm{et}}/\mathcal{S})_{\mathrm{cris}}$$

with pullback of representable objects defined by the formula (1.4.14.2).

Proposition 1.4.16. — *Consider a diagram of PD-stacks*

(1.4.16.1) $$(\mathcal{S}'', I'', \gamma'') \xrightarrow{u'} (\mathcal{S}', I', \gamma') \xrightarrow{u} (\mathcal{S}, I, \gamma),$$

and a 2-commutative diagram of algebraic stacks

(1.4.16.2) $$\begin{array}{ccccc} \mathcal{X}'' & \xrightarrow{g'} & \mathcal{X}' & \xrightarrow{g} & \mathcal{X} \\ \downarrow & & \downarrow & & \downarrow \\ \mathcal{S}'' & \xrightarrow{u'} & \mathcal{S}' & \xrightarrow{u} & \mathcal{S}, \end{array}$$

where $\mathcal{X}''$, $\mathcal{X}'$, and $\mathcal{X}$ are Deligne-Mumford stacks such that γ (resp. γ', γ'') extends to $\mathcal{X}$ (resp. $\mathcal{X}'$, $\mathcal{X}''$). Then there is a natural isomorphism of morphims of topoi

(1.4.16.3) $$(g \circ g')_{\mathrm{cris}} \simeq g_{\mathrm{cris}} \circ g'_{\mathrm{cris}}.$$

Proof. — This follows from the argument used in the proof of [**7**, III.2.2.6]. □

1.4.17. — Just as the ordinary lisse-étale topos of an algebraic stack is not functorial [**68**], there does not in general exist a morphism of topoi between $(\mathcal{X}'_{\text{lis-et}}/\mathcal{S})_{\mathrm{cris}}$ and $(\mathcal{X}_{\text{lis-et}}/\mathcal{S})$. The method of [**7**, III.2.2] still yields adjoint functors (g^{-1}, g_*), but the functor g^{-1} need not commute with finite inverse limits. Explicitly, the pullback functor g^{-1} can be described on stalks as follows. Let $F \in (\mathcal{X}_{\text{lis-et}}/\mathcal{S})_{\mathrm{cris}}$ be a sheaf and $(U', T', \delta') \in \mathrm{Cris}(\mathcal{X}'_{\text{lis-et}}/\mathcal{S})$ an object. Choose a geometric point $\bar{t}' \to T'$. Then the stalk $(g^{-1}F)_{T', \bar{t}'}$ of the restriction of $g^{-1}F$ to T'_{et} can be computed as follows. Let

I be the category with objects pairs (V', T, h), where $V' \to T'$ is an étale neighborhood of $\bar{t}'$, T is an object of $\mathrm{Cris}(\mathcal{X}_{\text{lis-et}}/\mathcal{S})$ and $h : V' \to T$ is a g-PD-morphism. Then the stalk $(g^{-1}F)_{T',\bar{t}'}$ is equal to

$$\varinjlim_{(V',T,h)\in I} F(T). \tag{1.4.17.1}$$

The category I is nonempty by 1.4.3 and connected by 1.4.8. However, equalizers do not in general exist as the equalizer of two maps between smooth $\mathcal{X}$-schemes need not be smooth. Therefore, the category I is not co-filtering and the limit $\varinjlim_I$ is not an exact functor. As we will see in 2.1.3, however, when $\mathcal{X} \to \mathcal{S}$ is representable the functor g^{-1} still defines a reasonable pullback for crystals in $(\mathcal{X}_{\text{lis-et}}/\mathcal{S})_{\text{cris}}$.

Warning 1.4.18. — Consider a diagram as in (1.4.16.2) except assume only that $\mathcal{X}$, $\mathcal{X}'$, and $\mathcal{X}''$ are Artin stacks. Then there are natural morphisms of functors

$$(g \circ g')_* \longrightarrow g_* \circ g'_*, \quad g'^{-1} \circ g^{-1} \longrightarrow (g \circ g')^{-1}, \tag{1.4.18.1}$$

but we are unable to show that these maps are isomorphisms in general. Nonetheless, for any sheaf $F \in (\mathcal{X}_{\text{lis-et}}/\mathcal{S})_{\text{cris}}$ there is a natural commutative diagram

$$\begin{array}{ccc} \Gamma((\mathcal{X}_{\text{lis-et}}/\mathcal{S})_{\text{cris}}, F) & \longrightarrow & \Gamma((\mathcal{X}'_{\text{lis-et}}/\mathcal{S})_{\text{cris}}, g^{-1}F) \\ \downarrow & & \downarrow \\ \Gamma((\mathcal{X}''_{\text{lis-et}}/\mathcal{S})_{\text{cris}}, (g \circ g')^{-1}F) & \longleftarrow & \Gamma((\mathcal{X}''_{\text{lis-et}}/\mathcal{S})_{\text{cris}}, g'^{-1} \circ g^{-1}F). \end{array} \tag{1.4.18.2}$$

Corollary 1.4.19. — *Assume the morphism g in (1.4.1.1) is smooth. Then there is a unique morphism of topoi*

$$g_{\text{cris}} : (\mathcal{X}'_{\text{lis-et}}/\mathcal{S}')_{\text{cris}} \longrightarrow (\mathcal{X}_{\text{lis-et}}/\mathcal{S})_{\text{cris}} \tag{1.4.19.1}$$

with pullback of representable objects given by (1.4.14.2).

Proof. — In this case, the functor g^{-1} is just restriction which is exact. □

1.4.20. — Let $\mathcal{X} \to \mathcal{S}$ be a morphism of algebraic stacks such that γ extends to $\mathcal{X}$, and let $X \to \mathcal{X}$ be a smooth representable morphism of algebraic stacks (in what follows X will usually be an algebraic space but this is not necessary here). Define X^s to be the sheaf on $(\mathcal{X}_{\text{lis-et}}/\mathcal{S})_{\text{cris}}$ which to any object (U, T, δ) associates the set of $\mathcal{X}$-morphisms $U \to X$. If e denote the initial object of $(\mathcal{X}_{\text{lis-et}}/\mathcal{S})_{\text{cris}}$ then the map $X^s \to e$ is a covering. Furthermore, if $X' \to \mathcal{X}$ is a second smooth morphism then there is a natural isomorphism

$$X^s \times_e X'^s \simeq (X \times_{\mathcal{X}} X')^s. \tag{1.4.20.1}$$

Denote by $(\mathcal{X}_{\text{lis-et}}/\mathcal{S})_{\text{cris}}|_{X^s}$ the localized topos of sheaves over X^s. If $(U, T, \delta) \in \mathrm{Cris}(X_{\text{lis-et}}/\mathcal{S})$ is an object, then the associated sheaf $\widetilde{T} \in (\mathcal{X}_{\text{lis-et}}/\mathcal{S})_{\text{cris}}$ admits a canonical morphism to X^s induced by the map $U \to X$. We therefore obtain a map

$$\varphi : (\mathcal{X}_{\text{lis-et}}/\mathcal{S})_{\text{cris}}|_{X^s} \longrightarrow (X_{\text{lis-et}}/\mathcal{S})_{\text{cris}} \tag{1.4.20.2}$$

by sending a sheaf $F \in (\mathcal{X}_{\text{lis-et}}/\mathcal{S})_{\text{cris}}|_{X^s}$ to the sheaf $\varphi(F)$ which to any $(U, T, \delta) \in \text{Cris}\,(X_{\text{lis-et}}/\mathcal{S})$ associates

$$\text{Hom}_{(\mathcal{X}_{\text{lis-et}}/\mathcal{S})_{\text{cris}}|_{X^s}}(\widetilde{T}, F). \tag{1.4.20.3}$$

Warning 1.4.21. — The functor φ is not a morphism of topoi. Again it has a left adjoint which need not be exact.

1.4.22. — By [**5**, III.5.4] the category $(\mathcal{X}_{\text{lis-et}}/\mathcal{S})_{\text{cris}}|_{X^s}$ can also be described as the topos associated to the site $\text{Cris}\,(\mathcal{X}_{\text{lis-et}}/\mathcal{S})|_X$ whose objects are pairs $((U, T, \delta), s)$, where $(U, T, \delta) \in \text{Cris}\,(\mathcal{X}_{\text{lis-et}}/\mathcal{S})$ and $s : U \to X$ is an $\mathcal{X}$-morphism. There is a natural inclusion

$$\text{Cris}\,(X_{\text{lis-et}}/\mathcal{S}) \subset \text{Cris}\,(\mathcal{X}_{\text{lis-et}}/\mathcal{S})|_X. \tag{1.4.22.1}$$

The functor (1.4.20.2) is then identified with the functor which restricts a sheaf to $\text{Cris}\,(X_{\text{lis-et}}/\mathcal{S})$. In particular, φ is an exact functor. The cohomology functors

$$\{H^i((X_{\text{lis-et}}/\mathcal{S})_{\text{cris}}, \varphi(-))\} \tag{1.4.22.2}$$

define a cohomological δ-functor on the category of abelian sheaves in $(\mathcal{X}_{\text{lis-et}}/\mathcal{S})_{\text{cris}}$, and hence there is a unique δ-functorial map

$$\alpha : H^i((\mathcal{X}_{\text{lis-et}}/\mathcal{S})_{\text{cris}}|_{X^s}, -) \longrightarrow H^i((X_{\text{lis-et}}/\mathcal{S})_{\text{cris}}, \varphi(-)). \tag{1.4.22.3}$$

Proposition 1.4.23. — *The map (1.4.22.3) is an isomorphism of functors.*

Proof. — Consider first the case when X is an algebraic space. In this case we also have inclusions

$$\text{Cris}\,(X_{\text{et}}/\mathcal{S}) \subset \text{Cris}\,(X_{\text{lis-et}}/\mathcal{S}) \subset \text{Cris}\,(\mathcal{X}_{\text{lis-et}}/\mathcal{S})|_X \tag{1.4.23.1}$$

which induce morphisms of δ-functors
(1.4.23.2)

$$H^i((\mathcal{X}_{\text{lis-et}}/\mathcal{S})_{\text{cris}}|_{X^s}, -) \xrightarrow{\alpha} H^i((X_{\text{lis-et}}/\mathcal{S})_{\text{cris}}, \varphi(-)) \xrightarrow{\beta} H^i((X_{\text{et}}/\mathcal{S})_{\text{cris}}, \varphi'(-)),$$

where φ' denotes the functor which restricts a sheaf on $\text{Cris}\,(\mathcal{X}_{\text{lis-et}}/\mathcal{S})|_X$ to $\text{Cris}\,(X_{\text{et}}/\mathcal{S})$. To prove that α is an isomorphism it suffices to show that the maps β and $\beta \circ \alpha$ are isomorphisms. Since finite nonempty projective limits in $\text{Cris}\,(X_{\text{et}}/\mathcal{S})$ are representable by 1.4.13, the maps α and $\beta \circ \alpha$ are induced by morphisms of topoi

$$s : (\mathcal{X}_{\text{lis-et}}/\mathcal{S})_{\text{cris}}|_{X^s} \longrightarrow (X_{\text{et}}/\mathcal{S})_{\text{cris}}, \quad t : (X_{\text{lis-et}}/\mathcal{S})_{\text{cris}} \longrightarrow (X_{\text{et}}/\mathcal{S})_{\text{cris}}. \tag{1.4.23.3}$$

The functors s_* and t_* therefore take injective abelian sheaves to injective abelian sheaves, and are also exact. From this the case when X is an algebraic space follows.

For the general case, let $P : U \to X$ be a smooth surjection with U an algebraic space, and let $U_\bullet$ be the 0-coskeleton of P. For any $[n] \in \Delta$, there is a natural isomorphism

$$((\mathcal{X}_{\text{lis-et}}/\mathcal{S})_{\text{cris}}|_{X^s})|_{U_n^s} \simeq (\mathcal{X}_{\text{lis-et}}/\mathcal{S})_{\text{cris}}|_{U_n^s}. \tag{1.4.23.4}$$

Let $(\mathcal{X}_{\text{lis-et}}/\mathcal{S})_{\text{cris}}|_{U^s_\bullet}$ (resp. $(X_{\text{lis-et}}/\mathcal{S})_{\text{cris}}|_{U^s_\bullet}$) denote the simplicial topos

$$[n] \longmapsto (\mathcal{X}_{\text{lis-et}}/\mathcal{S})_{\text{cris}}|_{U^s_n} \quad (\text{resp. } [n] \longmapsto (X_{\text{lis-et}}/\mathcal{S})_{\text{cris}}|_{U^s_n}). \tag{1.4.23.5}$$

Let

(1.4.23.6)

$$\pi_1 : (\mathcal{X}_{\text{lis-et}}/\mathcal{S})_{\text{cris}}|_{U^s_\bullet} \longrightarrow (\mathcal{X}_{\text{lis-et}}/\mathcal{S})_{\text{cris}}|_{X^s}, \quad \pi_2 : (X_{\text{lis-et}}/\mathcal{S})_{\text{cris}}|_{U^s_\bullet} \longrightarrow (X_{\text{lis-et}}/\mathcal{S})_{\text{cris}}$$

be the projections.

Lemma 1.4.24. — *Let $\mathcal{T}$ be a topos and $X \to e$ a covering of the initial object. Denote by $X_\bullet$ the 0-coskeleton of $X \to e$ and let $\mathcal{T}|_{X_\bullet}$ be the simplicial topos $[n] \mapsto \mathcal{T}|_{X_n}$. Then for any abelian sheaf $F \in \mathcal{T}$ the adjunction map $F \to R\pi_*\pi^*F$ is an isomorphism, where $\pi : \mathcal{T}|_{X_\bullet} \to \mathcal{T}$ denotes the projection.*

Proof. — For each $[n] \in \Delta$ the restriction functor $F \mapsto F \times X_n$ from $\mathcal{T} \to \mathcal{T}|_{X_n}$ takes injective abelian sheaves to injective abelian sheaves (see for example [**5**, V.2.2]). Therefore, if F is an injective sheaf then $R\pi_*\pi^*F$ is isomorphic to the complex of sheaves

$$\cdots \pi_{n*}\pi_n^*F \longrightarrow \pi_{n+1*}\pi_{n+1}^*F \longrightarrow \cdots. \tag{1.4.24.1}$$

By [**5**, V.4.5] it follows that $R^i\pi_*\pi^*F = 0$ for $i > 0$, and since $X \to e$ is a covering the sequence

$$F \longrightarrow \pi_{0*}\pi_0^*F \longrightarrow \pi_{1*}\pi_1^*F \tag{1.4.24.2}$$

is exact. □

There is a diagram of δ-functors

(1.4.24.3)

$$\begin{array}{ccc} H^*((\mathcal{X}_{\text{lis-et}}/\mathcal{S})_{\text{cris}}|_{X^s}, (-)) & \xrightarrow{\alpha} & H^*((X_{\text{lis-et}}/\mathcal{S})_{\text{cris}}, \varphi(-)) \\ \downarrow \pi_1^* & & \downarrow \pi_2^* \\ H^*((\mathcal{X}_{\text{lis-et}}/\mathcal{S})_{\text{cris}}|_{U^s_\bullet}, (-)|_{(\mathcal{X}_{\text{lis-et}}/\mathcal{S})_{\text{cris}}|_{U^s_\bullet}}) & \xrightarrow{\tilde{\alpha}} & H^*((X_{\text{lis-et}}/\mathcal{S})_{\text{cris}}|_{U^s_\bullet}, (-)|_{(X_{\text{lis-et}}/\mathcal{S})_{\text{cris}}|_{U^s_\bullet}}) \end{array}$$

which commutes since it commutes for $i = 0$ and $H^*((\mathcal{X}_{\text{lis-et}}/\mathcal{S})_{\text{cris}}|_{X^s}, (-))$ is universal. Since π_1^* and π_2^* are isomorphisms by 1.4.24, to prove that α is an isomorphism it suffices to show that $\tilde{\alpha}$ is an isomorphism. Furthermore, consideration of the Leray spectral sequence [**5**, V.5.3] shows that to prove that $\tilde{\alpha}$ is an isomorphism it suffices to show that each of the maps

$$H^*((\mathcal{X}_{\text{lis-et}}/\mathcal{S})_{\text{cris}}|_{U^s_n}, (-)) \longrightarrow H^*((X_{\text{lis-et}}/\mathcal{S})_{\text{cris}}|_{U^s_n}, (-)|_{(X_{\text{lis-et}}/\mathcal{S})_{\text{cris}}|_{U^s_n}}) \tag{1.4.24.4}$$

are isomorphisms. But by the representable case already considered both sides of this morphism are isomorphic to

$$H^*((U_{\text{lis-et}}/\mathcal{S})_{\text{cris}}, (-)|_{(U_{\text{lis-et}}/\mathcal{S})_{\text{cris}}}). \tag{1.4.24.5}$$

Furthermore, with these identifications the map (1.4.24.4) becomes identified with the identity map since this is true for H^0 and the left hand side of (1.4.24.4) is universal. □

Proposition 1.4.25. — *Consider a diagram as in (1.4.1.1). Then for any abelian sheaf $F \in (\mathcal{X}'_{\text{lis-et}}/\mathcal{S}')_{\text{cris}}$ and $i \geq 0$ the sheaf $R^i g_*(F)$ is isomorphic to the sheaf associated to the presheaf which to any $(U, T, \delta) \in \text{Cris}(\mathcal{X}_{\text{lis-et}}/\mathcal{S})$ associates $H^i(((\mathcal{X}' \times_{\mathcal{X}} U)_{\text{lis-et}}/\mathcal{S}')_{\text{cris}}, F|_{((\mathcal{X}' \times_{\mathcal{X}} U)_{\text{lis-et}}/\mathcal{S}')_{\text{cris}}})$.*

Proof. — The functor which restricts a sheaf in $(\mathcal{X}_{\text{lis-et}}/\mathcal{S})_{\text{cris}}$ to T_{et} is exact, and hence for any $F \in (\mathcal{X}'_{\text{lis-et}}/\mathcal{S}')_{\text{cris}}$ and object $(U, T, \delta) \in \text{Cris}(\mathcal{X}_{\text{lis-et}}/\mathcal{S})$ the restriction $R^i g_*(F)_T$ is isomorphic to the sheaf associated to the presheaf which to an étale morphism $V \to T$ associates the group $H^i(R\text{Hom}_{(\mathcal{X}'_{\text{lis-et}}/\mathcal{S}')_{\text{cris}}}(g^*\widetilde{V}, F))$, where $\widetilde{V}$ denotes the free abelian sheaf defined by the object $(V \times_T U, V, \delta) \in \text{Cris}(\mathcal{X}_{\text{lis-et}}/\mathcal{S})$. There is a natural map of sheaves $\widetilde{V} \to U^s$ which induces a map of sheaves $g^*\widetilde{V} \to g^*U^s \simeq \mathcal{X}'^s_U$, where we write $\mathcal{X}'_U$ for $\mathcal{X}' \times_{\mathcal{X}} U$. Since the forgetful functor $(G \to \mathcal{X}'^s_U) \mapsto G$ is left adjoint to the restriction functor $(\mathcal{X}'_{\text{lis-et}}/\mathcal{S}')_{\text{cris}} \to (\mathcal{X}'_{\text{lis-et}}/\mathcal{S}')_{\text{cris}}|_{\mathcal{X}'^s_U}$ sending a sheaf F to $F \times \mathcal{X}'^s_U$ it follows that

(1.4.25.1)
$$R\text{Hom}_{(\mathcal{X}'_{\text{lis-et}}/\mathcal{S}')_{\text{cris}}}(g^*\widetilde{V}, F) \simeq R\text{Hom}_{(\mathcal{X}'_{\text{lis-et}}/\mathcal{S}')_{\text{cris}}|_{\mathcal{X}'^s_U}}(g^*\widetilde{V}, F|_{(\mathcal{X}'_{\text{lis-et}}/\mathcal{S}')_{\text{cris}}|_{\mathcal{X}'^s_U}}).$$

Let

$$\underline{\text{Hom}}_{(\mathcal{X}'_{\text{lis-et}}/\mathcal{S}')_{\text{cris}}|_{\mathcal{X}'^s_U}}(g^*\widetilde{V}, F|_{(\mathcal{X}'_{\text{lis-et}}/\mathcal{S}')_{\text{cris}}|_{\mathcal{X}'^s_U}}) \tag{1.4.25.2}$$

denote the internal Hom-sheaf. Then

$$H^i(R\text{Hom}_{(\mathcal{X}'_{\text{lis-et}}/\mathcal{S}')_{\text{cris}}|_{\mathcal{X}'^s_U}}(g^*\widetilde{V}, F|_{(\mathcal{X}'_{\text{lis-et}}/\mathcal{S}')_{\text{cris}}|_{\mathcal{X}'^s_U}})) \tag{1.4.25.3}$$

is isomorphic to

$$H^i((\mathcal{X}'_{\text{lis-et}}/\mathcal{S}')_{\text{cris}}|_{\mathcal{X}'^s_U}, R\underline{\text{Hom}}_{(\mathcal{X}'_{\text{lis-et}}/\mathcal{S}')_{\text{cris}}|_{\mathcal{X}'^s_U}}(g^*\widetilde{V}, F|_{(\mathcal{X}'_{\text{lis-et}}/\mathcal{S}')_{\text{cris}}|_{\mathcal{X}'^s_U}})), \tag{1.4.25.4}$$

which by 1.4.23 is isomorphic to

$$H^i((\mathcal{X}'_{U,\text{lis-et}}/\mathcal{S}')_{\text{cris}}, \varphi(R\underline{\text{Hom}}_{(\mathcal{X}'_{\text{lis-et}}/\mathcal{S}')_{\text{cris}}|_{\mathcal{X}'^s_U}}(g^*\widetilde{V}, F|_{(\mathcal{X}'_{\text{lis-et}}/\mathcal{S}')_{\text{cris}}|_{\mathcal{X}'^s_U}}))). \tag{1.4.25.5}$$

On the other hand $\varphi(R\underline{\text{Hom}}_{(\mathcal{X}'_{\text{lis-et}}/\mathcal{S}')_{\text{cris}}|_{\mathcal{X}'^s_U}}(g^*\widetilde{V}, F|_{(\mathcal{X}'_{\text{lis-et}}/\mathcal{S}')_{\text{cris}}|_{\mathcal{X}'^s_U}}))$ is isomorphic to $R\underline{\text{Hom}}_{(\mathcal{X}'_{U,\text{lis-et}}/\mathcal{S}')_{\text{cris}}}((g^*\tilde{V}), \varphi(F))$, and hence (1.4.25.5) is isomorphic to

$$H^i(((\mathcal{X}' \times_{\mathcal{X}} U)_{\text{lis-et}}/\mathcal{S}')_{\text{cris}}, F|_{((\mathcal{X}' \times_{\mathcal{X}} U)_{\text{lis-et}}/\mathcal{S}')_{\text{cris}}}). \tag{1.4.25.6}$$ □

1.5. Comparison of $(\mathcal{X}_{\text{lis-et}}/\mathcal{S})_{\text{cris}}$ and $(\mathcal{X}_{\text{et}}/\mathcal{S})_{\text{cris}}$ when $\mathcal{X}$ is Deligne-Mumford

1.5.1. — Let $(\mathcal{S}, I, \gamma)$ be a PD-stack, and let $\mathcal{X} \to \mathcal{S}$ be a morphism of algebraic stacks such that γ extends to $\mathcal{X}$. Assume further that $\mathcal{X}$ is a Deligne-Mumford stack. By [**5**, IV.4.9.2] the fact that finite nonempty projective limits are representable in $\text{Cris}(\mathcal{X}_{\text{et}}/\mathcal{S})$ (1.4.13) implies that the inclusion

$$\text{Cris}(\mathcal{X}_{\text{et}}/\mathcal{S}) \subset \text{Cris}(\mathcal{X}_{\text{lis-et}}/\mathcal{S}) \tag{1.5.1.1}$$

induces a morphism of ringed topoi

$$r_{\mathcal{X}} : (\mathcal{X}_{\text{lis-et}}/\mathcal{S})_{\text{cris}} \longrightarrow (\mathcal{X}_{\text{et}}/\mathcal{S})_{\text{cris}}. \tag{1.5.1.2}$$

The functor $r_{\mathcal{X}*}$ simply restrict a sheaf to $\text{Cris}(\mathcal{X}_{\text{et}}/\mathcal{S})$, and in particular is exact. The inverse image $r_{\mathcal{X}}^{-1}$ is given by the usual formula: For a sheaf $F \in (\mathcal{X}_{\text{et}}/\mathcal{S})_{\text{cris}}$ the inverse image $r_{\mathcal{X}}^{-1}$ is the sheaf associated to the presheaf whose value on an object $T \in \text{Cris}(\mathcal{X}_{\text{lis-et}}/\mathcal{S})$ is equal to

$$\varinjlim_{T \to T'} F(T'), \tag{1.5.1.3}$$

where the limit is taken over the category of morphisms $\text{Cris}(\mathcal{X}_{\text{lis-et}}/\mathcal{S})$ with $T' \in \text{Cris}(\mathcal{X}_{\text{et}}/\mathcal{S})$. Because finite nonempty projective limits in $\text{Cris}(\mathcal{X}_{\text{et}}/\mathcal{S})$ are representable this indexing category is filtering and hence the pullback functor $r_{\mathcal{X}}^{-1}$ is exact.

Since the exact functor $r_{\mathcal{X}*}$ has an exact left adjoint $r_{\mathcal{X}}^{-1}$ it follows that $r_{\mathcal{X}*}$ takes injectives to injectives.

Proposition 1.5.2. — *Consider a diagram as in (1.4.1.1) with $\mathcal{X}$ and $\mathcal{X}'$ Deligne-Mumford stacks, and let $g_{\text{cris}} : (\mathcal{X}'_{\text{et}}/\mathcal{S}')_{\text{cris}} \to (\mathcal{X}_{\text{et}}/\mathcal{S})_{\text{cris}}$ be the morphism of topoi defined in 1.4.14. Denote by $g^{\text{lis-et}}_{\text{cris}*} : (\mathcal{X}'_{\text{lis-et}}/\mathcal{S}')_{\text{cris}} \to (\mathcal{X}_{\text{lis-et}}/\mathcal{S})_{\text{cris}}$ the pushforward functor of lisse-étale sheaves defined in 1.4.17. Then for any $F \in (\mathcal{X}'_{\text{lis-et}}/\mathcal{S}')_{\text{cris}}$ there is a natural isomorphism*

$$Rr_{\mathcal{X}*}Rg^{\text{lis-et}}_{\text{cris}*}(F) \simeq Rg_{\text{cris}*}(r_{\mathcal{X}'*}F). \tag{1.5.2.1}$$

Proof. — Since $r_{\mathcal{X}*}$ is exact, there is a natural isomorphism of functors

$$Rr_{\mathcal{X}*}Rg^{\text{lis-et}}_{\text{cris}*}(-) \simeq R(r_{\mathcal{X}*} \circ g^{\text{lis-et}}_{\text{cris}*})(-). \tag{1.5.2.2}$$

Similarly, since $r_{\mathcal{X}'*}$ is exact and takes injectives to injectives, we have an isomorphism of functors

$$Rg_{\text{cris}*}(r_{\mathcal{X}'*}(-)) \simeq R(g_{\text{cris}*} \circ r_{\mathcal{X}'})(-). \tag{1.5.2.3}$$

From this the result follows since by definition

$$r_{\mathcal{X}*} \circ g^{\text{lis-et}}_{\text{cris}*} \simeq g_{\text{cris}*} \circ r_{\mathcal{X}'}. \tag{1.5.2.4}$$

□

1.5.3. — Fix a diagram (1.4.1.1) and a smooth cover $P : X' \to \mathcal{X}'$ with X' an algebraic space, and let $X'_\bullet$ be the 0-coskeleton of P. Denote by $X'^+_\bullet$ the strictly simplicial space obtained from $X'_\bullet$. Since each morphism in $X'^+_\bullet$ is smooth we obtain by 1.4.19 a strictly simplicial topos $(X'^+_{\bullet,\text{lis-et}}/\mathcal{S}')_{\text{cris}}$ with an augmentation

$$\pi : (X'^+_{\bullet,\text{lis-et}}/\mathcal{S}')_{\text{cris}} \longrightarrow (\mathcal{X}'_{\text{lis-et}}/\mathcal{S}')_{\text{cris}}. \tag{1.5.3.1}$$

Proposition 1.5.4. — *For any abelian sheaf $F \in (\mathcal{X}'_{\text{lis-et}}/\mathcal{S}')_{\text{cris}}$, the adjunction map $F \to R\pi_*\pi^*F$ is an isomorphism.*

Proof. — Fix an integer $s \geq 0$. By the same argument used in the proof of 1.4.25, the sheaf $R^s\pi_*\pi^*F$ is isomorphic to the sheaf associated to the presheaf which to any object $U \hookrightarrow T$ of $\text{Cris}(\mathcal{X}'_{\text{lis-et}}/\mathcal{S}')$ associates the group

$$H^s((X'^+_{U\bullet,\text{lis-et}}/T)_{\text{cris}}, F), \tag{1.5.4.1}$$

where $X'^+_{U\bullet}$ denotes $X'^+_\bullet \times_{\mathcal{X}'} U$. It follows that it suffices to consider the case when $\mathcal{X}'$ is a scheme and $T = \mathcal{S}'$. Furthermore, by replacing $\mathcal{X}'$ by an étale cover we may assume that $P : X' \to \mathcal{X}'$ admits a section $\sigma : \mathcal{X}' \to X'$.

Lemma 1.5.5. — *Let F be an injective abelian sheaf in $(\mathcal{X}'_{\text{lis-et}}/\mathcal{S}')_{\text{cris}}$ and $U \to \mathcal{X}'$ a smooth representable morphism of algebraic stacks. Then for any $i > 0$ the group $H^i((U'_{\text{lis-et}}/\mathcal{S}')_{\text{cris}}, F)$ is zero.*

Proof. — By 1.4.23 the group

$$H^i((U'_{\text{lis-et}}/\mathcal{S}')_{\text{cris}}, F) \tag{1.5.5.1}$$

is isomorphic to the group $H^i((\mathcal{X}'_{\text{lis-et}}/\mathcal{S}')_{\text{cris}}|_{U'^s}, F)$, where $(\mathcal{X}'_{\text{lis-et}}/\mathcal{S}')_{\text{cris}}|_{U'^s}$ is as in 1.4.20. On the other hand, as explained in [**5**, V.2.2] the restriction functor

$$(\mathcal{X}'_{\text{lis-et}}/\mathcal{S}')_{\text{cris}} \longrightarrow (\mathcal{X}'_{\text{lis-et}}/\mathcal{S}')_{\text{cris}}|_{U'^s} \tag{1.5.5.2}$$

takes injective abelian sheaves to injective abelian sheaves. □

By considering an injective resolution of F, we see that in order to prove 1.5.4 it suffices to consider the case when F is an injective sheaf. In this case (1.5.4.1) is equal by 1.5.5 to the s-th cohomology group of the complex $L^\bullet$

$$\cdots \longrightarrow \Gamma((X'_{n,\text{lis-et}}/T)_{\text{cris}}, F) \longrightarrow \cdots, \tag{1.5.5.3}$$

where the map $d : L^n \to L^{n+1}$ is given by the formula

$$\sum_{i=0}^{n+1}(-1)^i \text{pr}^*_{0\cdots\hat{i}\cdots(n+1)}. \tag{1.5.5.4}$$

For every n let $\kappa_n : X'_{n-1} \to X'_n$ be the map

$$\text{id} \times \sigma : X'_{n-1} \simeq X'_{n-1} \times_{\mathcal{X}'} \mathcal{X}' \longrightarrow X'_{n-1} \times_{\mathcal{X}'} X' \simeq X'_n, \tag{1.5.5.5}$$

and let

$$\kappa_n^* : \Gamma((X'_{n,\text{lis-et}}/T)_{\text{cris}}, F) \longrightarrow \Gamma((X'_{n-1,\text{lis-et}}/T)_{\text{cris}}, F) \tag{1.5.5.6}$$

be the map induced by pullback (note that as discussed in 1.4.17 even though κ_n does not induce a morphism of topoi this map is still defined). For $i < n$ there is a commutative diagram

$$\begin{array}{ccc} X_{n-1} & \xrightarrow{\kappa_n} & X_n \\ {\scriptstyle \text{pr}_{0\ldots\hat{\imath}\ldots(n-1)}}\downarrow & & \downarrow{\scriptstyle \text{pr}_{0\ldots\hat{\imath}\ldots n}} \\ X_{n-2} & \xrightarrow{\kappa_{n-1}} & X_{n-1}, \end{array} \tag{1.5.5.7}$$

and for $i = n$ we have $\text{pr}_{0\ldots\hat{n}} \circ \kappa_n = \text{id}$. From this and the commutativity of the diagram (1.4.18.2) it follows that for any section $m \in L^n$ we have

$$m = (-1)^{n+1}\Big(\kappa_{n+1}^*\Big(\sum_{i=0}^{n+1} \text{pr}^*_{0\ldots\hat{\imath}\ldots(n+1)}(m)\Big) - \sum_{i=0}^{n} \text{pr}^*_{0\ldots\hat{\imath}\ldots n}(\kappa_n^*(m))\Big). \tag{1.5.5.8}$$

From this it follows that $L^\bullet$ has no higher cohomology groups and that

$$H^0(L^\bullet) \simeq \Gamma((\mathcal{X}'_{\text{lis-et}}/\mathcal{S}')_{\text{cris}}, F). \tag{1.5.5.9}$$

□

Corollary 1.5.6. — *For any abelian sheaf $F \in (\mathcal{X}'_{\text{lis-et}}/\mathcal{S}')_{\text{cris}}$, there is a natural isomorphism in the derived category of abelian sheaves on $(\mathcal{X}_{\text{et}}/\mathcal{S})_{\text{cris}}$*

$$Rr_{\mathcal{X}*}Rg^{\text{lis-et}}_{\text{cris}*}(F) \simeq Rg_{\bullet*}(r_{X'^+_\bullet} \circ \pi^*(F)), \tag{1.5.6.1}$$

where $g_{\bullet*}$ is the pushforward functor for the morphism of topoi $g_\bullet : (X'^+_{\bullet,\text{et}}/\mathcal{S}')_{\text{cris}} \to (\mathcal{X}_{\text{et}}/\mathcal{S})_{\text{cris}}$.

Proof. — By 1.5.4 there is a natural isomorphism

$$Rr_{\mathcal{X}*}Rg^{\text{lis-et}}_{\text{cris}*}(F) \simeq Rr_{\mathcal{X}*}Rg^{\text{lis-et}}_{\bullet*}(\pi^*F), \tag{1.5.6.2}$$

where $g^{\text{lis-et}}_{\bullet*}$ denotes the direct image functor $(X'^+_{\bullet\text{lis-et}}/\mathcal{S}')_{\text{cris}} \to (\mathcal{X}_{\text{lis-et}}/\mathcal{S})_{\text{cris}}$. On the other hand from 1.5.2 it follows that

$$Rr_{\mathcal{X}*}Rg^{\text{lis-et}}_{\bullet*}(\pi^*F) \simeq Rg_{\bullet*}(r_{X'^+_\bullet} \circ \pi^*(F)). \tag{1.5.6.3}$$

□

Corollary 1.5.7. — *Let $\mathcal{X} \to \mathcal{S}$ be a morphism of algebraic stacks such that γ extends to $\mathcal{X}$, and let $X \to \mathcal{X}$ be a smooth cover with 0-coskeleton $X_\bullet$. Then for any abelian sheaf $F \in (\mathcal{X}_{\text{lis-et}}/\mathcal{S})_{\text{cris}}$, there is a natural spectral sequence*

$$E_1^{st} = H^t((X_{s,\text{et}}/\mathcal{S})_{\text{cris}}, F|_{(X_{s,\text{et}}/\mathcal{S})_{\text{cris}}}) \implies H^{s+t}((\mathcal{X}_{\text{lis-et}}/\mathcal{S})_{\text{cris}}, F). \tag{1.5.7.1}$$

Proof. — By 1.5.4, the right hand side of (1.5.7.1) is isomorphic to

$$H^{s+t}((X^+_{\bullet,\text{lis-et}}/\mathcal{S})_{\text{cris}}, F|_{X_\bullet}), \tag{1.5.7.2}$$

which by 1.5.2 is isomorphic to

$$H^{s+t}((X^+_{\bullet,\text{et}}/\mathcal{S})_{\text{cris}}, F|_{X_\bullet}). \tag{1.5.7.3}$$

The corollary therefore follows from [**68**, 2.7]. □

1.6. Projection to the lisse-étale topos

1.6.1. — Let $(\mathcal{S}, I, \gamma)$ be a PD-stack and let $\mathcal{X} \to \mathcal{S}$ be a morphism of algebraic stacks such that γ extends to $\mathcal{X}$. We define a morphism of topoi

$$u_{\mathcal{X}_{\text{lis-et}}/\mathcal{S}} : (\mathcal{X}_{\text{lis-et}}/\mathcal{S})_{\text{cris}} \longrightarrow \mathcal{X}_{\text{lis-et}} \tag{1.6.1.1}$$

as follows. The functor $u^*_{\mathcal{X}_{\text{lis-et}}/\mathcal{S}}$ sends a sheaf $F \in \mathcal{X}_{\text{lis-et}}$ to the sheaf which to any $(U, T, \delta) \in \text{Cris}(\mathcal{X}_{\text{lis-et}}/\mathcal{S})$ associates $F(U)$. Observe that $u^*_{\mathcal{X}_{\text{lis-et}}/\mathcal{S}}$ clearly commutes with finite projective limits. The functor $u_{\mathcal{X}_{\text{lis-et}}/\mathcal{S}*}$ sends a sheaf $G \in (\mathcal{X}_{\text{lis-et}}/\mathcal{S})_{\text{cris}}$ to the sheaf which to any smooth $U \to \mathcal{S}$ associates $\Gamma((U_{\text{lis-et}}/\mathcal{S})_{\text{cris}}, G)$. The functor $u^*_{\mathcal{X}_{\text{lis-et}}/\mathcal{S}}$ also has a left adjoint $u_{\mathcal{X}_{\text{lis-et}}/\mathcal{S}!}$ given by the formula

$$u_{\mathcal{X}_{\text{lis-et}}/\mathcal{S}!}F(U) = F(U \xrightarrow{\text{id}} U). \tag{1.6.1.2}$$

If $\mathcal{X}$ is a Deligne-Mumford stack then there is also a projection

$$u_{\mathcal{X}_{\text{et}}/\mathcal{S}} : (\mathcal{X}_{\text{et}}/\mathcal{S})_{\text{cris}} \longrightarrow \mathcal{X}_{\text{et}} \tag{1.6.1.3}$$

defined analogously to $u_{\mathcal{X}_{\text{lis-et}}/\mathcal{S}}$. It follows from the definitions that the diagram of topoi

$$\begin{array}{ccc} (\mathcal{X}_{\text{lis-et}}/\mathcal{S})_{\text{cris}} & \xrightarrow{u_{\mathcal{X}_{\text{lis-et}}/\mathcal{S}}} & \mathcal{X}_{\text{lis-et}} \\ {\scriptstyle r_{\mathcal{X}}}\downarrow & & \downarrow{\scriptstyle \tilde{r}} \\ (\mathcal{X}_{\text{et}}/\mathcal{S})_{\text{cris}} & \xrightarrow{u_{\mathcal{X}_{\text{et}}/\mathcal{S}}} & \mathcal{X}_{\text{et}} \end{array} \tag{1.6.1.4}$$

commutes, where $\tilde{r} : \mathcal{X}_{\text{lis-et}} \to \mathcal{X}_{\text{et}}$ is the natural morphism of topoi.

Proposition 1.6.2. — *Let $\mathcal{X} \to \mathcal{S}$ be a morphism of algebraic stacks such that γ extends to $\mathcal{X}$, and let $U \to \mathcal{X}$ be a smooth morphism with U an algebraic space. Then for any abelian sheaf $F \in (\mathcal{X}_{\text{lis-et}}/\mathcal{S})_{\text{cris}}$ the restriction of $Ru_{\mathcal{X}_{\text{lis-et}}/\mathcal{S}*}(F)$ to U_{et} is isomorphic to $Ru_{U_{\text{et}}/\mathcal{S}*}(F|_{\text{Cris}(U_{\text{et}}/\mathcal{S})})$.*

Proof. — Since the restriction functor $\mathcal{X}_{\text{lis-et}} \to U_{\text{et}}$ is exact, $Ru_{\mathcal{X}_{\text{lis-et}}/\mathcal{S}*}(F)|_{U_{\text{et}}}$ is isomorphic to $R\Lambda(F)$, where Λ is the composite of the functor $u_{\mathcal{X}_{\text{lis-et}}/\mathcal{S}*}$ and the restriction functor. On the other hand, the functor Λ can also be described as the composite of the restriction functor $(\mathcal{X}_{\text{lis-et}}/\mathcal{S})_{\text{cris}} \to (U_{\text{et}}/\mathcal{S})_{\text{cris}}$ with the functor $u_{U_{\text{et}}/\mathcal{S}*}$. It follows that there is a canonical map

$$R\Lambda(F) \longrightarrow Ru_{U_{\text{et}}/\mathcal{S}*}(F|_{(U_{\text{et}}/\mathcal{S})_{\text{cris}}}) \tag{1.6.2.1}$$

which we claim is an isomorphism.

Let $s \geq 0$ be an integer. Then $R^s\Lambda(F)$ is equal to the sheaf associated to the presheaf on U_{et} which associates to any étale $V \to U$ the group

$$H^s((\mathcal{X}_{\text{lis-et}}/\mathcal{S})|_{V^s}, F), \tag{1.6.2.2}$$

and $R^s u_{U_{\mathrm{et}}/\mathcal{S}*}(F|_{(U_{\mathrm{et}}/\mathcal{S})_{\mathrm{cris}}})$ is the sheaf associated to the presheaf which to $V \to U$ associates

$$H^s((U_{\mathrm{et}}/\mathcal{S})_{\mathrm{cris}}|_{V^s}, F|_{(U_{\mathrm{et}}/\mathcal{S})_{\mathrm{cris}}}). \tag{1.6.2.3}$$

By the isomorphisms in (1.4.23.2) both of these groups are isomorphic to

$$H^s((V_{\mathrm{et}}/\mathcal{S})_{\mathrm{cris}}, F|_{(V_{\mathrm{et}}/\mathcal{S})_{\mathrm{cris}}}), \tag{1.6.2.4}$$

and with these identifications the map $R^s\Lambda(F) \to R^s u_{U_{\mathrm{et}}/\mathcal{S}*}(F|_{(U_{\mathrm{et}}/\mathcal{S})_{\mathrm{cris}}})$ becomes identified with the identity map (to see this last compatibility note that it follows from the definitions that it holds for $s = 0$ and hence holds in general since both arrows are maps of universal cohomological δ-functors). □

CHAPTER 2

CRYSTALS AND DIFFERENTIAL CALCULUS ON STACKS

2.1. Crystals

Let $(\mathcal{S}, I, \gamma)$ be a PD-stack and $\mathcal{X} \to \mathcal{S}$ a morphism of algebraic stacks such that γ extends to $\mathcal{X}$.

Definition 2.1.1. — A *crystal* of $\mathcal{O}_{\mathcal{X}_{\text{lis-et}}/\mathcal{S}}$-modules is a sheaf E of $\mathcal{O}_{\mathcal{X}_{\text{lis-et}}/\mathcal{S}}$-modules on $\text{Cris}(\mathcal{X}_{\text{lis-et}}/\mathcal{S})$ such that for any morphism $u : T' \to T$ in $\text{Cris}(\mathcal{X}_{\text{lis-et}}/\mathcal{S})$ the induced map (see 1.3.3)

$$u^* E_T \longrightarrow E_{T'} \tag{2.1.1.1}$$

is an isomorphism. A crystal E is *quasi-coherent* if each E_T is a quasi-coherent sheaf on T_{et}.

If $\mathcal{X}$ is a Deligne-Mumford stack we also obtain a notion of (quasi-coherent) crystal in $(\mathcal{X}_{\text{et}}/\mathcal{S})_{\text{cris}}$ by replacing the site $\text{Cris}(\mathcal{X}_{\text{lis-et}}/\mathcal{S})$ in the preceding definition by the site $\text{Cris}(\mathcal{X}_{\text{et}}/\mathcal{S})$.

Proposition 2.1.2. — *Consider a 2-commutative diagram*

$$\begin{array}{ccc} \mathcal{X}' & \xrightarrow{g} & \mathcal{X} \\ \downarrow & & \downarrow \\ \mathcal{S}' & \longrightarrow & \mathcal{S} \end{array} \tag{2.1.2.1}$$

as in (1.4.1.1), with $\mathcal{X}'$ and $\mathcal{X}$ Deligne-Mumford stacks. If E is a (quasi-coherent) crystal in $(\mathcal{X}_{\text{et}}/\mathcal{S})_{\text{cris}}$ then the pullback $g^*_{\text{cris}}E$ to $(\mathcal{X}'_{\text{et}}/\mathcal{S}')_{\text{cris}}$ is a (quasi-coherent) crystal.

Proof. — This follows from the same argument used in [**7**, IV.1.2.4]. □

Concerning pullback for more general Artin stacks we have the following partial result.

Proposition 2.1.3. — *Consider a 2-commutative diagram*

$$\begin{array}{ccc} \mathcal{X}' & \xrightarrow{g} & \mathcal{X} \\ \downarrow & & \downarrow \\ \mathcal{S}' & \longrightarrow & \mathcal{S} \end{array} \tag{2.1.3.1}$$

as in (1.4.1.1) and assume $\mathcal{X} \to \mathcal{S}$ is representable and locally separated. Then for any crystal E on $\mathrm{Cris}(\mathcal{X}_{\text{lis-et}}/\mathcal{S})$ *the pullback g^*E on* $\mathrm{Cris}(\mathcal{X}'_{\text{lis-et}}/\mathcal{S}')$ *is a crystal. If E is quasi-coherent then the pullback is also quasi-coherent.*

Proof. — For an object $T' \in \mathrm{Cris}(\mathcal{X}'_{\text{lis-et}}/\mathcal{S}')$, let $I^{T'}$ denote the category of g-PD-morphisms $h : T' \to T$ to objects $T \in \mathrm{Cris}(\mathcal{X}_{\text{lis-et}}/\mathcal{S})$. By the definition of g^* the sheaf g^*E is isomorphic to the sheaf associated to the presheaf

$$T' \longmapsto \Gamma(T', \mathcal{O}_{T'}) \otimes_{\varinjlim_{(h:T'\to T)\in I^{T'}} \Gamma(T,\mathcal{O}_T)} \varinjlim_{(h:T'\to T)\in I^{T'}} E(T). \tag{2.1.3.2}$$

On the other hand, by the universal property of $\varinjlim$ and $\otimes$, for any $\mathcal{O}_{T'}(T')$-module N we have

(2.1.3.3)

$$\begin{aligned} \mathrm{Hom}_{\mathcal{O}_{T'}(T')}\big((\varinjlim_{h\in I^{T'}} E(T)) \otimes_{\varinjlim_{h\in I^{T'}} \mathcal{O}_T(T)} \mathcal{O}_{T'}(T'), N\big) & \\ \simeq \mathrm{Hom}_{\varinjlim_{h\in I^{T'}} \mathcal{O}_T(T)}\big(\varinjlim_{h\in I^{T'}} E(T), N\big) & \\ \simeq \varprojlim_{h\in I^{T'}} \mathrm{Hom}_{\mathcal{O}_T(T)}(E(T), N) & \\ \simeq \varprojlim_{h\in I^{T'}} \mathrm{Hom}_{\mathcal{O}_{T'}(T')}(E(T) \otimes_{\mathcal{O}_T(T)} \mathcal{O}_{T'}(T'), N) & \\ \simeq \mathrm{Hom}_{\mathcal{O}_{T'}(T')}\big(\varinjlim_{h\in I^{T'}} (E(T) \otimes_{\mathcal{O}_T(T)} \mathcal{O}_{T'}(T')), N\big). & \end{aligned}$$

By Yoneda's lemma it follows that g^*E is isomorphic to the sheaf associated to the presheaf which to any $T' \in \mathrm{Cris}(\mathcal{X}'_{\text{lis-et}}/\mathcal{S})$ associates the limit

$$\varinjlim_{(h:T'\to T)\in I^{T'}} \Gamma(T', h^*E_T). \tag{2.1.3.4}$$

We claim that for any fixed $h_0 : T' \to T_0$ in $I^{T'}$ the induced map $h_0^*E_{T_0} \to (g^*E)_T$ is an isomorphism.

To see this it suffices to check on stalks. Let $\bar{t}' \to T'$ be a geometric point and let $I^{T',\bar{t}'}$ denote the category of pairs (V', h), where V' is an étale neighborhood of $\bar{t}'$ in T' and $h : V' \to T$ is an object of $I^{V'}$. Then we want to show that the natural map

$$(h_0^*E_{T_0})_{\bar{t}'} \longrightarrow \varinjlim_{h\in I^{T',\bar{t}'}} (h^*E_T)_{\bar{t}'} \tag{2.1.3.5}$$

is an isomorphism. To prove this, note first that it would be immediate if the category $I^{T',\bar{t}'}$ were filtering, for by definition of crystal all the transition morphisms in the limit on the right hand side of (2.1.3.5) are isomorphisms. Unfortunately, the category $I^{T',\bar{t}'}$ is not filtering.

Let $\overline{I}^{T',\bar{t}'}$ denote the category whose objects are the same as those of $I^{T',\bar{t}'}$ but for which $\mathrm{Hom}_{\overline{I}^{T',\bar{t}'}}(h,h')$ is the empty set if there does not exist a morphism $h \to h'$ in $I^{T',\bar{t}'}$ and the unital set if $\mathrm{Hom}_{I^{T',\bar{t}'}}(h,h')$ is nonempty. Then the category $\overline{I}^{T',\bar{t}'}$ is filtering. To prove the proposition it suffices to show that the functor

$$(h \in I^{T',\bar{t}'}) \longmapsto (h^* E_T)_{\bar{t}'} \tag{2.1.3.6}$$

factors through $\overline{I}^{T',\bar{t}'}$. For then the right hand side (2.1.3.5) can be replaced by the limit

$$\varinjlim_{h\in \overline{I}^{T',\bar{t}'}} (h^* E_T)_{\bar{t}'} \tag{2.1.3.7}$$

which is a filtering limit.

The statement that the functor (2.1.3.6) factors through $\overline{I}^{T',\bar{t}'}$ amounts to the statement that if $h : T' \to T$ and $\tilde{h} : T' \to \tilde{T}$ are two g-PD-morphisms, and if $f, g : \tilde{T} \to T$ are two morphisms in $\mathrm{Cris}(\mathcal{X}_{\text{lis-et}}/\mathcal{S})$ such that $f \circ \tilde{h} = g \circ \tilde{h} = h$, then the two maps

$$f^*, g^* : h^* E_T \longrightarrow \tilde{h}^* E_{\tilde{T}} \tag{2.1.3.8}$$

are equal. For this note that since $\mathcal{X} \to \mathcal{S}$ is representable and hence faithful [**49**, 8.1.2], the map $U \times_{\mathcal{X}} U \to T \times_{\mathcal{S}} T$ is a monomorphism, where $U \to \mathcal{X}$ denotes the subscheme of T defined by the PD-ideal. Since $\mathcal{X} \to \mathcal{S}$ is locally separated by assumption and the diagram

$$\begin{array}{ccc} U \times_{\mathcal{X}} U & \longrightarrow & U \times_{\mathcal{S}} U \\ \downarrow & & \downarrow \\ \mathcal{X} & \xrightarrow{\Delta} & \mathcal{X} \times_{\mathcal{S}} \mathcal{X} \end{array} \tag{2.1.3.9}$$

is cartesian, the map $U \to U \times_{\mathcal{X}} U \to T \times_{\mathcal{S}} T$ is an immersion. Set

$$D := D_{U\times_{\mathcal{X}} U, \gamma}(T \times_{\mathcal{S}} T). \tag{2.1.3.10}$$

By the universal property of D the map $f \times g : \tilde{T} \to T \times_{\mathcal{S}} T$ factors through D, and hence there is a commutative diagram of $\mathcal{S}$-schemes

$$\begin{array}{ccccc} T' & \xrightarrow{\tilde{h}} & \tilde{T} & & \\ {\scriptstyle h}\downarrow & & \downarrow{\scriptstyle f\times g} & & \\ T & \xrightarrow{\Delta} & D & \xrightarrow{\mathrm{pr}_i} & T. \end{array} \tag{2.1.3.11}$$

The maps $f^*, g^* : h^*E_T \to \tilde{h}^*E_{\tilde{T}}$ are then identified with the pullbacks to T' of the two maps $\mathrm{pr}_i^* : \mathrm{pr}_i^*E_T \to E_D$ $(i = 1, 2)$. But these two maps both pullback to the same map $E_T \to \Delta^*E_D$ via the map Δ, and hence their pullbacks to T' are also equal. This completes the proof of 2.1.3. □

The proof of the proposition in fact gives a way to calculate g^*E:

Corollary 2.1.4. — *Let $g : \mathcal{X}' \to \mathcal{X}$ be as in 2.1.3. Let $(U', T', \delta') \in \mathrm{Cris}(\mathcal{X}'_{\text{lis-et}}/\mathcal{S})$ be an object and $\tilde{h} : U \to X$ a factorization of the map $U \to \mathcal{X}' \to \mathcal{X}$ through a smooth $\mathcal{X}$-space X. Then $g^*E(U, T, \delta)$ is equal to the value of the pullback $\tilde{h}^*_{\mathrm{cris}}(E|_{(X_{\mathrm{et}}/\mathcal{S})_{\mathrm{cris}}})$ on $(U, T, \delta) \in \mathrm{Cris}(U_{\mathrm{et}}/\mathcal{S})$.*

Proposition 2.1.5. — *Let $\mathcal{X} \to \mathcal{S}$ be a morphism of algebraic stacks such that γ extends to $\mathcal{X}$, and assume that $\mathcal{X}$ is a Deligne-Mumford stack. Then restriction induces an equivalence of categories between the category of (quasi-coherent) crystals in $(\mathcal{X}_{\text{lis-et}}/\mathcal{S})_{\mathrm{cris}}$ and the category of (quasi-coherent) crystals in $(\mathcal{X}_{\mathrm{et}}/\mathcal{S})_{\mathrm{cris}}$.*

Proof. — Let $r_{\mathcal{X}} : (\mathcal{X}_{\text{lis-et}}/\mathcal{S})_{\mathrm{cris}} \to (\mathcal{X}_{\mathrm{et}}/\mathcal{S})_{\mathrm{cris}}$ be the morphism of topoi defined in 1.5.1. The proposition follows from the following lemma which shows that $r_{\mathcal{X}}^*$ defines a quasi-inverse to $r_{\mathcal{X}*}$. □

Lemma 2.1.6

(i) *If E is a (quasi-coherent) crystal in $(\mathcal{X}_{\mathrm{et}}/\mathcal{S})_{\mathrm{cris}}$ then the pullback $r_{\mathcal{X}}^*E$ is a (quasi-coherent) crystal on $\mathrm{Cris}(\mathcal{X}_{\text{lis-et}}/\mathcal{S})$.*

(ii) *If M is a crystal in $(\mathcal{X}_{\text{lis-et}}/\mathcal{S})_{\mathrm{cris}}$ then the natural map $r_{\mathcal{X}}^*r_{\mathcal{X}*}M \to M$ is an isomorphism.*

Proof. — Let $(U, T, \delta) \in \mathrm{Cris}(\mathcal{X}_{\text{lis-et}}/\mathcal{S})$ be an object. By 1.4.5 there exists after replacing T by an étale cover a morphism $h : (U, T, \delta) \to (V, Z, \epsilon)$ with $(V, Z, \epsilon) \in \mathrm{Cris}(\mathcal{X}_{\mathrm{et}}/\mathcal{S})$. In this case the sheaf $(r_{\mathcal{X}}^*E)_T$ on T_{et} is equal to $h_T^*E_Z$, where $h_T : T_{\mathrm{et}} \to Z_{\mathrm{et}}$ denotes the morphism induced by h. Indeed the sheaf $(r_{\mathcal{X}}^*E)_T$ is equal to the sheaf associated to the presheaf on $\mathrm{Et}(T)$ which to any $T' \to T$ associates the limit

$$\varinjlim_{(h:T' \to Z) \in I^{T'}} h^*E_Z(T'), \tag{2.1.6.1}$$

where $I^{T'}$ denotes the category of morphisms $h : (U \times_T T' \hookrightarrow T') \to (V, Z, \delta)$ in $\mathrm{Cris}(\mathcal{X}_{\text{lis-et}}/\mathcal{S})$ with $(V, Z, \delta) \in \mathrm{Cris}(\mathcal{X}_{\mathrm{et}}/\mathcal{S})$. The category $I^{T'}$ is filtering by 1.4.13 and since E is a crystal it follows that for any given $h_0 : T' \to Z$ in $I^{T'}$ the map from $h_0^*E_Z$ to $(r_{\mathcal{X}}^*E)_T$ is an isomorphism.

From this description it follows that if E is a (quasi-coherent) crystal then the pullback $r_{\mathcal{X}}^*E$ is also a (quasi-coherent) crystal and (i) follows.

To see (ii), note that by (i) the map $r_{\mathcal{X}}^*r_{\mathcal{X}*}M \to M$ is a map of crystals which becomes an isomorphism when restricted to $\mathrm{Cris}(\mathcal{X}_{\mathrm{et}}/\mathcal{S})$. Since any object of the site $\mathrm{Cris}(\mathcal{X}_{\text{lis-et}}/\mathcal{S})$ admits étale locally a morphism to an object of $\mathrm{Cris}(\mathcal{X}_{\mathrm{et}}/\mathcal{S})$ by 1.4.5 this implies (ii). □

2.1.7. — It will also be useful to have a simplicial description of quasi-coherent crystals. Let $\mathcal{X} \to \mathcal{S}$ be a morphism of algebraic stacks such that γ extends to $\mathcal{X}$. Fix a smooth cover $P : S \to \mathcal{S}$ with S an algebraic space, and let $Q : X \to S \times_{\mathcal{S}} \mathcal{X}$ be a smooth representable surjection with X a Deligne-Mumford stack. Denote by $S_\bullet$ (resp. $X_\bullet$) the simplicial Deligne-Mumford stack which is the 0-coskeleton of the morphism $S \to \mathcal{S}$ (resp. $X \to \mathcal{X}$). We then have a commutative diagram

$$\begin{array}{ccc} X_\bullet & \longrightarrow & \mathcal{X} \\ \downarrow & & \downarrow \\ S_\bullet & \longrightarrow & \mathcal{S}. \end{array} \tag{2.1.7.1}$$

Let $(X_{\bullet,\mathrm{et}}/S_\bullet)_{\mathrm{cris}}$ denote the simplicial topos $[n] \mapsto (X_{n,\mathrm{et}}/S_n)_{\mathrm{cris}}$.

Definition 2.1.8. — A *(quasi-coherent) crystal* $E_\bullet$ in $(X_{\bullet,\mathrm{et}}/S_\bullet)_{\mathrm{cris}}$ is a sheaf of $\mathcal{O}_{X_{\bullet,\mathrm{et}}/S_\bullet}$-modules $E_\bullet$ such that for each n the sheaf $E_n \in (X_{n,\mathrm{et}}/S_n)_{\mathrm{cris}}$ is a (quasi-coherent) crystal, and such that for every morphism $\delta : [n] \to [n']$ in Δ the induced map

$$\delta^* E_n \longrightarrow E_{n'} \tag{2.1.8.1}$$

is an isomorphism of sheaves in $(X_{n',\mathrm{et}}/S_{n'})_{\mathrm{cris}}$.

2.1.9. — If E is a crystal in $(\mathcal{X}_{\text{lis-et}}/\mathcal{S})_{\mathrm{cris}}$ then for any $[n] \in \Delta$ the restriction of E to $\mathrm{Cris}(X_{n,\mathrm{et}}/S_n)$ is a crystal and hence we obtain by restriction a crystal $E_\bullet$ in $(X_{\bullet,\mathrm{et}}/S_\bullet)_{\mathrm{cris}}$.

Proposition 2.1.10. — *The functor $E \mapsto E_\bullet$ induces an equivalence between the category of quasi-coherent crystals in $(\mathcal{X}_{\text{lis-et}}/\mathcal{S})_{\mathrm{cris}}$ and the category of quasi-coherent crystals in $(X_{\bullet,\mathrm{et}}/S_\bullet)_{\mathrm{cris}}$.*

Proof. — We construct a quasi-inverse as follows. Consider first the case when $\mathcal{S}$ is an algebraic space and $S = \mathcal{S}$. Let $E_\bullet$ be a quasi-coherent crystal in $(X_{\bullet,\mathrm{et}}/\mathcal{S})_{\mathrm{cris}}$ and let $(U, T, \delta) \in \mathrm{Cris}(\mathcal{X}_{\text{lis-et}}/\mathcal{S})$ be an object. We construct a sheaf E_T on T_{et} as follows. After replacing T by an étale cover, there exists a lifting $\tilde{s} : U \to X$ of the structure morphism $s : U \to \mathcal{X}$. Define E_T to be the sheaf $(\tilde{s}^* E_0)_T$ on T_{et}. If $\tilde{s}'$ is a second lifting of s then there is a canonical isomorphism $\iota : \tilde{s}^* E_0 \simeq \tilde{s}'^* E_0$ obtained by setting $h : U \to X_1$ equal to $\tilde{s} \times \tilde{s}'$ and defining ι to be the isomorphism

$$\tilde{s}^* E_0 \simeq h^* \mathrm{pr}_1^* E_0 \xrightarrow{\mathrm{pr}_1^*} h^* E_1 \xleftarrow{\mathrm{pr}_2^*} h^* \mathrm{pr}_2^* E_0 \simeq \tilde{s}'^* E_0, \tag{2.1.10.1}$$

where the maps pr_1^* and pr_2^* are isomorphisms by the definition of a crystal in $(X_{\bullet,\mathrm{et}}/\mathcal{S})_{\mathrm{cris}}$. Furthermore, if $\tilde{s}'' : U \to X_0$ is a third factorization and if $\iota' : \tilde{s}'^* E_0 \to \tilde{s}''^* E_0$ and $\iota'' : \tilde{s}^* E_0 \to \tilde{s}''^* E_0$ are the resulting isomorphisms, then $\iota'' = \iota' \circ \iota$. This follows by setting $k : U \to X_2$ equal to the map $\tilde{s} \times \tilde{s}' \times \tilde{s}''$ and noting that if $\sigma : \mathrm{pr}_1^* E_0 \to \mathrm{pr}_2^* E_0$ denotes the isomorphism in $(X_{1,\mathrm{et}}/\mathcal{S})_{\mathrm{cris}}$ obtained from the isomorphisms

$$\mathrm{pr}_1^* E_0 \xrightarrow{\mathrm{pr}_1^*} E_1 \xleftarrow{\mathrm{pr}_2^*} \mathrm{pr}_2^* E_0, \tag{2.1.10.2}$$

then $\iota = k^*(\mathrm{pr}_{12}^*(\sigma))$, $\iota' = k^*(\mathrm{pr}_{23}^*(\sigma))$, and $\iota'' = k^*(\mathrm{pr}_{13}^*(\sigma))$. That $\iota'' = \iota' \circ \iota$ then follows from the fact that $\mathrm{pr}_{13}^*(\sigma) = \mathrm{pr}_{23}^*(\sigma) \circ \mathrm{pr}_{12}^*(\sigma)$ since $E_\bullet$ is a simplicial sheaf. It follows that E_T is defined globally on T_{et}. Furthermore, by the construction if $h : T' \to T$ is a morphism in $\mathrm{Cris}(\mathcal{X}_{\text{lis-et}}/\mathcal{S})$ then there is a natural isomorphism $h^*E_T \to E_{T'}$ of sheaves in T'_{et}. In this way we obtain a functor from the category of quasi-coherent crystals in $(X_{\bullet,\mathrm{et}}/\mathcal{S})_{\mathrm{cris}}$ to the category of quasi-coherent crystals in $(\mathcal{X}_{\text{lis-et}}/\mathcal{S})_{\mathrm{cris}}$.

To treat the general case, note first that if $E_\bullet$ is a quasi-coherent crystal in $(X_{\bullet,\mathrm{et}}/S_\bullet)_{\mathrm{cris}}$, then $E_\bullet$ defines a quasi-coherent crystal in $(\mathcal{X}_{n,\text{lis-et}}/S_n)_{\mathrm{cris}}$ for every $[n] \in \Delta$, where $\mathcal{X}_n := \mathcal{X} \times_{\mathcal{S}} S_n$. Indeed, let $X_{S_n,\bullet}$ denote the base change

$$X_\bullet \times_{\mathcal{S}} S_n. \tag{2.1.10.3}$$

Then $X_{S_n,\bullet}$ is isomorphic to the 0-coskeleton of the natural smooth surjective morphism $X \times_{\mathcal{S}} S_n \to \mathcal{X}_n$. Now the crystal $E_\bullet$ defines by restriction a quasi-coherent crystal in $(X_{S_n,\bullet}/S_n)_{\mathrm{cris}}$ which by the case already considered is obtained by restriction from a quasi-coherent crystal $\widetilde{E}_n$ in $(\mathcal{X}_{S_n,\text{lis-et}}/S_n)_{\mathrm{cris}}$. It follows that $E_\bullet$ is obtained by restriction from a collection of crystals $\{\widetilde{E}_n\}$ equipped with compatible isomorphisms $\delta^*\widetilde{E}_n \to \widetilde{E}_{n'}$ for every morphism $\delta : [n] \to [n']$ in Δ.

Let $(U, T, \delta) \in \mathrm{Cris}(\mathcal{X}_{\text{lis-et}}/\mathcal{S})$ be an object, and let $U_\bullet \hookrightarrow T_\bullet$ be the closed immersion of simplicial stacks defined by base change to $S_\bullet$. For each $[n] \in \Delta$, $U_n \hookrightarrow T_n$ is an object $\mathrm{Cris}(\mathcal{X}_{S_n,\text{lis-et}}/S_n)$ (this is not quite correct since U and T are only algebraic spaces but see 1.3.4), and hence we can evaluate $\widetilde{E}_n$ on $U_n \hookrightarrow T_n$ to obtain a quasi-coherent sheaf $\mathcal{E}_{T_n}$ on $T_{n,\mathrm{et}}$. The simplicial structure on $\{\widetilde{E}_n\}$ gives the sheaf $\mathcal{E}_{T_n}$ the structure of a simplicial sheaf on $T_{\bullet,\mathrm{et}}$, and since the pullback $\delta^*\widetilde{E}_n \to \widetilde{E}_{n'}$ is an isomorphism for every morphism $\delta : [n] \to [n']$ the map $\delta^*\mathcal{E}_{T_n} \to \mathcal{E}_{T_{n'}}$ is also an isomorphism. By descent theory for quasi-coherent sheaves, the simplicial sheaf $\mathcal{E}_{T_\bullet}$ is obtained from a quasi-coherent sheaf E_T on T_{et}.

It follows from the construction that for any morphism $h : T' \to T$ in $\mathrm{Cris}(\mathcal{X}_{\text{lis-et}}/\mathcal{S})$ there is a natural isomorphism $h^*E_T \to E_{T'}$ in T'_{et}. In this way we obtain a quasi-coherent crystal $\{E_T\}$ in $(\mathcal{X}_{\text{lis-et}}/\mathcal{S})_{\mathrm{cris}}$ from $E_\bullet$. It follows from the construction that $E_\bullet \mapsto \{E_T\}$ defines a quasi-inverse to the restriction functor $E \mapsto E_\bullet$. □

Corollary 2.1.11. — *Let $\mathcal{X} \to \mathcal{S}$ be a morphism of algebraic stacks such that γ extends to $\mathcal{X}$ and $\mathcal{X}$ is a Deligne-Mumford stack. Then pullback defines an equivalence of categories between the category of quasi-coherent crystals in $(\mathcal{X}_{\mathrm{et}}/\mathcal{S})_{\mathrm{cris}}$ and the category of quasi-coherent crystals in $(\mathcal{X}_{\bullet,\mathrm{et}}/S_\bullet)_{\mathrm{cris}}$, where $\mathcal{X}_\bullet := \mathcal{X} \times_{\mathcal{S}} S_\bullet$.*

Proof. — This follows from 2.1.5 and 2.1.10 taking $Q : X \to \mathcal{X} \times_{\mathcal{S}} S$ equal to the identity map $\mathcal{X} \times_{\mathcal{S}} S \to \mathcal{X} \times_{\mathcal{S}} S$. □

2.2. Modules with connection and the de Rham complex

2.2.1. — Let $\mathcal{X} \to \mathcal{S}$ be a morphism of algebraic stacks which we assume to be representable and locally separated (this assumption is satisfied for example if $\mathcal{X}$ is a locally separated algebraic space (0.2.2)). Then the diagonal map

$$\Delta : \mathcal{X} \longrightarrow \mathcal{X} \times_{\mathcal{S}} \mathcal{X} \tag{2.2.1.1}$$

is an immersion, and we define the sheaf of differentials $\Omega^1_{\mathcal{X}/\mathcal{S}}$ [**49**, p. 163] to be the conormal bundle of this immersion. In other words, let $\mathcal{I} \subset \mathcal{O}_{\mathcal{X}\times_{\mathcal{S}}\mathcal{X},\text{lis-et}}$ be the ideal of $\mathcal{X}$ and set $\Omega^1_{\mathcal{X}/\mathcal{S}} := \Delta^*\mathcal{I}$ (note that though there is not a morphism of topoi $\mathcal{X}_{\text{lis-et}} \to (\mathcal{X} \times_{\mathcal{S}} \mathcal{X})_{\text{lis-et}}$ the pullback of a quasi-coherent sheaf $\Delta^*\mathcal{I}$ is still well-defined by [**68**, 6.5]).

If $\mathcal{X}$ is a Deligne-Mumford stack we write $\Omega^1_{\mathcal{X}_{\text{et}}/\mathcal{S}}$ for the restriction of $\Omega^1_{\mathcal{X}/\mathcal{S}}$ to $\mathcal{X}_{\text{et}}$ if we wish to make clear that we are working with the étale topology. When no confusion seems likely to arise we sometimes also write simply $\Omega^1_{\mathcal{X}/\mathcal{S}}$ for $\Omega^1_{\mathcal{X}_{\text{et}}/\mathcal{S}}$.

Let

$$\begin{array}{ccc} \mathcal{X}' & \xrightarrow{g} & \mathcal{X} \\ {\scriptstyle f'}\downarrow & & \downarrow{\scriptstyle f} \\ \mathcal{S}' & \longrightarrow & \mathcal{S} \end{array} \tag{2.2.1.2}$$

be a 2-commutative diagram of algebraic stacks with f' and f representable. Then there is a natural commutative diagram

$$\begin{array}{ccc} \mathcal{X}' & \xrightarrow{g} & \mathcal{X} \\ {\scriptstyle \Delta}\downarrow & & \downarrow{\scriptstyle \Delta} \\ \mathcal{X}' \times_{\mathcal{S}'} \mathcal{X}' & \xrightarrow{f\times f} & \mathcal{X} \times_{\mathcal{S}} \mathcal{S}, \end{array} \tag{2.2.1.3}$$

and hence an induced morphism $f^*\Omega^1_{\mathcal{X}/\mathcal{S}} \to \Omega^1_{\mathcal{X}'/\mathcal{S}'}$.

If $S \to \mathcal{S}$ is a smooth morphism with S an algebraic space and $\mathcal{X}_S$ denotes the algebraic space (since $\mathcal{X} \to \mathcal{S}$ is representable) obtained by base change, then $\Omega^1_{\mathcal{X}/\mathcal{S}}|_{\mathcal{X}_{S,\text{et}}}$ is naturally isomorphic to the usual sheaf of differentials $\Omega^1_{\mathcal{X}_{S,\text{et}}/S_{\text{et}}}$. From this and the corresponding fact for algebraic spaces it follows that for any representable morphism of algebraic stacks $g : \mathcal{X}' \to \mathcal{X}$ the sequence

$$g^*\Omega^1_{\mathcal{X}/\mathcal{S}} \longrightarrow \Omega^1_{\mathcal{X}'/\mathcal{S}} \longrightarrow \Omega^1_{\mathcal{X}'/\mathcal{X}} \longrightarrow 0 \tag{2.2.1.4}$$

is exact.

Lemma 2.2.2. — *If $\mathcal{X} \to \mathcal{S}$ is smooth, then $\Omega^1_{\mathcal{X}/\mathcal{S}}$ is locally free.*

Proof. — It suffices to prove this after making a smooth base change $S \to \mathcal{S}$ with S an algebraic space. In this case the result follows from the corresponding result for schemes. □

2.2.3. — When $\mathcal{X}$ is a Deligne-Mumford stack (we discuss the case of a general Artin stack in 2.2.19 below) there is a derivation $d : \mathcal{O}_{\mathcal{X}_{\text{et}}} \to \Omega^1_{\mathcal{X}_{\text{et}}/\mathcal{S}}$ functorial in $\mathcal{X}$ defined as follows. Let $S \to \mathcal{S}$ be a smooth surjection with S an algebraic space and set $S' := S \times_{\mathcal{S}} S$. Let $\mathcal{X}_S$ and $\mathcal{X}_{S'}$ denote the spaces obtained by base change and let $g : \mathcal{X}_S \to \mathcal{X}$ and $g' : \mathcal{X}_{S'} \to \mathcal{X}$ be the projections. By descent theory, for any quasi-coherent sheaf M on $\mathcal{X}_{\text{et}}$ there is a natural exact sequence

$$M \longrightarrow g_* g^* M \rightrightarrows g'_* g'^* M. \tag{2.2.3.1}$$

In particular, we have a commutative diagram

$$\begin{array}{ccccc} \mathcal{O}_{\mathcal{X}_{\text{et}}} & \longrightarrow & g_* \mathcal{O}_{\mathcal{X}_{S,\text{et}}} & \rightrightarrows & g'_* \mathcal{O}_{\mathcal{X}_{S',\text{et}}} \\ & & \downarrow d_{\text{et}} & & \downarrow d_{\text{et}} \\ \Omega^1_{\mathcal{X}_{\text{et}}/\mathcal{S}} & \longrightarrow & g_* \Omega^1_{\mathcal{X}_{S,\text{et}}/S_{\text{et}}} & \rightrightarrows & \Omega^1_{\mathcal{X}_{S',\text{et}}/S'_{\text{et}}}, \end{array} \tag{2.2.3.2}$$

where the maps d_{et} are the derivations obtained from the theory for algebraic spaces. From the exactness of the horizontal rows it follows that there is a unique derivation $d : \mathcal{O}_{\mathcal{X}_{\text{et}}} \to \Omega^1_{\mathcal{X}_{\text{et}}/\mathcal{S}}$ such that the square

$$\begin{array}{ccc} \mathcal{O}_{\mathcal{X}_{\text{et}}} & \xrightarrow{d} & \Omega^1_{\mathcal{X}_{\text{et}}/\mathcal{S}} \\ \downarrow & & \downarrow \\ g_* \mathcal{O}_{\mathcal{X}_{S,\text{et}}} & \xrightarrow{d_{\text{et}}} & \Omega^1_{\mathcal{X}_{S,\text{et}}/S_{\text{et}}} \end{array} \tag{2.2.3.3}$$

commutes.

This derivation d is independent of the choice of $S \to \mathcal{S}$. For this note first that if $T \to \mathcal{S}$ is a second smooth cover and $h : T \to S$ an $\mathcal{S}$-morphism, then there is a commutative diagram

$$\begin{array}{ccccc} \mathcal{O}_{\mathcal{X}_{\text{et}}} & \longrightarrow & g_* \mathcal{O}_{\mathcal{X}_{S,\text{et}}} & \longrightarrow & g'_* \mathcal{O}_{\mathcal{X}_{T,\text{et}}} \\ & & d_{\text{et}} \downarrow & & \downarrow d_{\text{et}} \\ \Omega^1_{\mathcal{X}_{\text{et}}/\mathcal{S}} & \longrightarrow & g_* \Omega^1_{\mathcal{X}_{S,\text{et}}/S_{\text{et}}} & \longrightarrow & g'_* \Omega^1_{\mathcal{X}'_{\text{et}}/T_{\text{et}}}, \end{array} \tag{2.2.3.4}$$

where $g' : \mathcal{X}_T \to T$ denotes the projection. From this it follows that T and S define the same map $d : \mathcal{O}_{\mathcal{X}_{\text{et}}} \to \Omega^1_{\mathcal{X}_{\text{et}}/\mathcal{S}}$. If $T \to \mathcal{S}$ is an arbitrary second smooth cover, then by the preceding paragraph S and $S \times_{\mathcal{S}} T$ define the same derivation, as do T and $T \times_{\mathcal{S}} S$. It follows that the derivation d is independent of the choice of S.

2.2.4. — If $\mathcal{X} \to \mathcal{S}$ is smooth, then as noted in 2.2.2 above the sheaf $\Omega^1_{\mathcal{X}/\mathcal{S}}$ is locally free. For $i \geq 0$ let $\Omega^i_{\mathcal{X}/\mathcal{S}}$ denote $\Lambda^i \Omega^1_{\mathcal{X}/\mathcal{S}}$. An argument similar to the one defining the derivation d above shows that there exist unique morphisms $d_i : \Omega^i_{\mathcal{X}/\mathcal{S}} \to \Omega^{i+1}_{\mathcal{X}/\mathcal{S}}$ of abelian sheaves such that for any smooth morphism $S \to \mathcal{S}$ with S a scheme the diagram

$$\begin{array}{ccc} \Omega^i_{\mathcal{X}_S/S} & \xrightarrow{\tilde{d}_i} & \Omega^{i+1}_{\mathcal{X}_S/S} \\ \uparrow & & \uparrow \\ g^{-1}\Omega^i_{\mathcal{X}/\mathcal{S}} & \xrightarrow{d_i} & g^{-1}\Omega^{i+1}_{\mathcal{X}/\mathcal{S}} \end{array} \tag{2.2.4.1}$$

commutes, where $\tilde{d}_i$ denotes the usual differentiation of forms.

For any i the composite

$$\Omega^i_{\mathcal{X}/\mathcal{S}} \xrightarrow{d_i} \Omega^{i+1}_{\mathcal{X}/\mathcal{S}} \xrightarrow{d_{i+1}} \Omega^{i+2}_{\mathcal{X}/\mathcal{S}} \tag{2.2.4.2}$$

is zero since this can be verified after making a base change $S \to \mathcal{S}$. We therefore obtain a complex $\Omega^\bullet_{\mathcal{X}/\mathcal{S}}$ on $\mathcal{X}_{\text{et}}$ called the *de Rham complex* of $\mathcal{X}/\mathcal{S}$.

Remark 2.2.5. — The de Rham complex defined above is a special instance of the general construction given in [**33**, VIII.1.2] of the de Rham complex of a "catégorie formelle à PD" (note that [**33**, VIII.1.2.8] applies by 2.2.2). Since in our case all properties of the de Rham complex can be deduced by descent and we also need variants with coefficients, we do not appeal to *loc. cit.*.

Note also that $\Omega^1_{\mathcal{X}_{\text{et}}/\mathcal{S}}$ is usually not locally generated by the image of d (see for example 2.2.8 below). This means that we cannot quote [**7**] directly in the proofs that follow since the axioms of a so-called de Rham category fail [**7**, II.3.1.3].

2.2.6. — Let $\mathcal{X} \to \mathcal{S}$ be a smooth morphism of algebraic stacks with $\mathcal{X}$ Deligne-Mumford, and define $T_{\mathcal{X}_{\text{et}}/\mathcal{S}}$ to be the dual of $\Omega^1_{\mathcal{X}_{\text{et}}/\mathcal{S}}$. There is a natural Lie algebra structure

$$[\cdot,\cdot] : T_{\mathcal{X}_{\text{et}}/\mathcal{S}} \times T_{\mathcal{X}_{\text{et}}/\mathcal{S}} \longrightarrow T_{\mathcal{X}_{\text{et}}/\mathcal{S}} \tag{2.2.6.1}$$

defined as follows. Let $S \to \mathcal{S}$ be a smooth surjection with S an algebraic space and set $S' := S \times_{\mathcal{S}} S$. Let $\mathcal{X}_S$ and $\mathcal{X}_{S'}$ denote the spaces obtained by base change and let $g : \mathcal{X}_S \to \mathcal{X}$ and $g' : \mathcal{X}_{S'} \to \mathcal{X}$ be the projections. Then we have a commutative diagram of sheaves

$$\begin{array}{ccccc} T_{\mathcal{X}_{\text{et}}/\mathcal{S}} \oplus T_{\mathcal{X}_{\text{et}}/\mathcal{S}} & \longrightarrow & g_*T_{\mathcal{X}_S/S} \oplus g_*T_{\mathcal{X}_S/S} & \longrightarrow & g'_*T_{\mathcal{X}_{S'}/S'} \oplus g'_*T_{\mathcal{X}_{S'}/S'} \\ & & \big\downarrow [\cdot,\cdot]_{\text{et}} & & \big\downarrow [\cdot,\cdot]_{\text{et}} \\ T_{\mathcal{X}_{\text{et}}/\mathcal{S}} & \longrightarrow & g_*T_{\mathcal{X}_S/S} & \longrightarrow & g'_*T_{\mathcal{X}_{S'}/S'}, \end{array} \tag{2.2.6.2}$$

where $[\cdot,\cdot]_{\text{et}}$ denotes the Lie algebra structures obtained from the theory for algebraic spaces. Since the rows are exact we obtain a pairing $[\cdot,\cdot]$ on $T_{\mathcal{X}_{\text{et}}/\mathcal{S}}$. As in 2.2.3 this pairing is independent of the choice of $S \to \mathcal{S}$.

Proposition 2.2.7. — *For a local section $\xi \in T_{\mathcal{X}_{\text{et}}/\mathcal{S}}$ let d_ξ be the composite*

$$\mathcal{O}_{\mathcal{X}_{\text{et}}} \xrightarrow{d} \Omega^1_{\mathcal{X}_{\text{et}}/\mathcal{S}} \xrightarrow{\xi} \mathcal{O}_{\mathcal{X}_{\text{et}}}. \tag{2.2.7.1}$$

Then the induced map

$$T_{\mathcal{X}_{\text{et}}/\mathcal{S}} \longrightarrow \mathcal{D}er_{\mathcal{S}}(\mathcal{O}_{\mathcal{X}_{\text{et}}}), \quad \xi \longmapsto d_\xi \tag{2.2.7.2}$$

is a map of sheaves of Lie algebras.

Proof. — Let $S \to \mathcal{S}$ be a smooth cover and let $\mathcal{X}_S = \mathcal{X} \times_{\mathcal{S}} S$. Then if $\pi : \mathcal{X}_S \to \mathcal{X}$ denotes the projection, there an isomorphism $\pi^* T_{\mathcal{X}_{\mathrm{et}}/\mathcal{S}} \simeq T_{\mathcal{X}_{S,\mathrm{et}}/S}$ which is compatible with the maps d. Hence for any $\xi \in T_{\mathcal{X}_{\mathrm{et}}/\mathcal{S}}$ there is a commutative diagram

$$
\begin{array}{ccc}
\mathcal{O}_{\mathcal{X}_{S,\mathrm{et}}} & \xrightarrow{d_{\pi^*\xi}} & \mathcal{O}_{\mathcal{X}_{S_{\mathrm{et}}}} \\
\uparrow & & \uparrow \\
\pi^{-1}\mathcal{O}_{\mathcal{X}_{\mathrm{et}}} & \xrightarrow{\pi^{-1}d_\xi} & \pi^{-1}\mathcal{O}_{\mathcal{X}_{\mathrm{et}}}.
\end{array}
\tag{2.2.7.3}
$$

Since the map $\pi^{-1}T_{\mathcal{X}_{\mathrm{et}}/\mathcal{S}} \to T_{\mathcal{X}_{S,\mathrm{et}}/S}$ is a morphism of Lie-algebras by construction of the Lie algebra structure on $T_{\mathcal{X}_{\mathrm{et}}/\mathcal{S}}$ and the vertical arrows are injective, the lemma follows from the corresponding result for $\mathcal{X}_S/S$. □

Example 2.2.8. — Let G be a smooth affine group scheme over a scheme B and let $\mathcal{S} = BG$, $\mathcal{X} = B$. Let $\mathcal{X} = B \to \mathcal{S} = BG$ be the morphism defined by the trivial torsor. Then $\mathcal{X} \times_{\mathcal{S}} \mathcal{X} \simeq G$ and the diagonal map $B \hookrightarrow G$ is the identity section. Therefore in this case $\Omega^1_{\mathcal{X}/\mathcal{S}}$ is isomorphic to $\mathrm{Lie}(G)^*$ (the dual of the Lie algebra of G). In this case the differential $d : \mathcal{O}_B \to \mathrm{Lie}(G)^*$ is the zero map. The sheaf $T_{\mathcal{X}/\mathcal{S}}$ is equal to $\mathrm{Lie}(G)$. We leave to the reader the verification that the pairing (2.2.6.1) agrees with the usual Lie algebra structure on $\mathrm{Lie}(G)$. Furthermore the maps $\Omega^i_{\mathcal{X}/\mathcal{S}} \to \Omega^{i+1}_{\mathcal{X}/\mathcal{S}}$ occurring in the de Rham complex are equal to the maps

$$
d_i : \Lambda^i \mathrm{Lie}(G)^* \longrightarrow \Lambda^{i+1}\mathrm{Lie}(G)^* \tag{2.2.8.1}
$$

sending $\omega \in \Lambda^i\mathrm{Lie}(G)^*$ to the element of $\Lambda^{i+1}\mathrm{Lie}(G)^*$ characterized by the condition that for $\xi_1, \dots, \xi_{i+1} \in \mathrm{Lie}(G)$ we have

$$
d_i(\omega)(\xi_1 \wedge \cdots \wedge \xi_{i+1}) = \sum_{l<k} (-1)^{l+k} \omega([\xi_l, \xi_k] \wedge \cdots \hat{\xi}_l \wedge \cdots \wedge \hat{\xi}_k \wedge \cdots \wedge \xi_{i+1}). \tag{2.2.8.2}
$$

This is the usual complex computing Lie algebra cohomology [**38**, 4.27].

Example 2.2.9. — A second example which hints at the connection with logarithmic geometry is the following. Let k be a field and $\mathcal{S} = [\mathbb{A}^1/\mathbb{G}_m]$, where $\mathbb{A}^1 = \mathrm{Spec}(k[t])$ is the affine line over k and $\mathbb{G}_m$ acts on $\mathbb{A}^1$ via the usual multiplicative action. Let $\mathcal{X} = \mathbb{A}^1$ and $\mathcal{X} \to \mathcal{S}$ the projection. In this case $\mathcal{X} \times_{\mathcal{S}} \mathcal{X}$ is the scheme $\mathbb{A}^1 \times \mathbb{G}_m$ and the diagonal map is the morphism $\mathrm{id} \times e : \mathbb{A}^1 \to \mathbb{A}^1 \times \mathbb{G}_m$, where e denotes the identity section. If we write $\mathbb{G}_m = \mathrm{Spec}(k[u^{\pm}])$, then it follows that $\Omega^1_{\mathcal{X}/\mathcal{S}}$ in this case is free of rank 1 over $k[t]$ with basis $(u-1)$. Let us compute the differential $d : k[t] \to k[t] \cdot (u-1)$. The two projections $p_i : \mathbb{A}^1 \times \mathbb{G}_m \to \mathbb{A}^1$ $(i = 1, 2)$ are given by the projection to the first factor for $i = 1$ and the action for $i = 2$. From this one sees that the differential which sends $f \in k[t]$ to $p_2^*(f) - p_1^*(f)$ modulo $(u-1)^2$ sends t to $t \cdot (u-1)$. It follows that the map

$$
d : k[t] \longrightarrow k[t] \cdot (u-1) \tag{2.2.9.1}
$$

sends t^i to $it^i \cdot (u-1)$. In more familiar notation, the basis $(u-1)$ should be viewed as $\mathrm{dlog}(t)$.

This connection with log geometry is discussed in detail in Chapter 9. Briefly, the stack $[\mathbb{A}^1/\mathbb{G}_m]$ can be viewed as the stack associating to any scheme S the groupoid of pairs $(\mathcal{L}, \alpha)$, where $\mathcal{L}$ is a line bundle on S and $\alpha : \mathcal{L} \to \mathcal{O}_S$ is a morphism of $\mathcal{O}_S$-modules. As discussed in [**40**, Complement 1] such a pair $(\mathcal{L}, \alpha)$ is equivalent to the data of a fine log structure M on S together with a morphism $\beta : \mathbb{N} \to \overline{M}$ which étale locally on S lifts to a chart. With this interpretation of $[\mathbb{A}^1/\mathbb{G}_m]$, the quotient map $\mathbb{A}^1 \to [\mathbb{A}^1/\mathbb{G}_m]$ is the morphism corresponding to the log structure defined by the origin in $\mathbb{A}^1$. As we discuss in Chapter 9, using the characterizing infinitesimal lifting property of differentials (in either "ordinary" geometry or logarithmic geometry) one can then deduce a canonical isomorphism between $\Omega^1_{\mathbb{A}^1/[\mathbb{A}^1/\mathbb{G}_m]}$ and the logarithmic differentials.

Definition 2.2.10. — Let $\mathcal{X} \to \mathcal{S}$ be a smooth representable morphism of algebraic stacks with $\mathcal{X}$ Deligne-Mumford. A *module with connection on* $\mathcal{X}_{\mathrm{et}}/\mathcal{S}$ is a quasi-coherent sheaf of $\mathcal{O}_{\mathcal{X}_{\mathrm{et}}}$-modules $\mathcal{E}$ on $\mathcal{X}_{\mathrm{et}}$ together with a morphism

$$\nabla : \mathcal{E} \longrightarrow \mathcal{E} \otimes_{\mathcal{O}_{\mathcal{X}_{\mathrm{et}}}} \Omega^1_{\mathcal{X}_{\mathrm{et}}/\mathcal{S}} \tag{2.2.10.1}$$

such that for any local sections $f \in \mathcal{O}_{\mathcal{X}_{\mathrm{et}}}$ and $e \in \mathcal{E}$ we have

$$\nabla(fe) = f\nabla(e) + e \otimes d(f). \tag{2.2.10.2}$$

The pair $(\mathcal{E}, \nabla)$ is *integrable* if the induced map

$$T_{\mathcal{X}_{\mathrm{et}}/\mathcal{S}} \longrightarrow \mathcal{E}nd(\mathcal{E}) \tag{2.2.10.3}$$

is a morphism of sheaves of Lie-algebras. If $\xi \in T_{\mathcal{X}_{\mathrm{et}}/\mathcal{S}}$ is a local section, we denote the induced endomorphism of $\mathcal{E}$ by ∇_ξ. We denote by $MC(\mathcal{X}_{\mathrm{et}}/\mathcal{S})$ (resp. $MIC(\mathcal{X}_{\mathrm{et}}/\mathcal{S})$) the category of quasi-coherent modules with connection (resp. integrable connection).

Remark 2.2.11. — As in the classical case, a connection ∇ on a quasi-coherent $\mathcal{O}_{\mathcal{X}_{\mathrm{et}}}$-module $\mathcal{E}$ defines for every $i \geq 0$ a map of abelian sheaves

$$\nabla_i : \mathcal{E} \otimes_{\mathcal{O}_{\mathcal{X}}} \Omega^i_{\mathcal{X}/\mathcal{S}} \longrightarrow \mathcal{E} \otimes_{\mathcal{O}_{\mathcal{X}}} \Omega^{i+1}_{\mathcal{X}/\mathcal{S}}, \quad e \otimes \omega \longmapsto e \otimes d\omega + \nabla(e) \otimes \omega. \tag{2.2.11.1}$$

The condition that $(\mathcal{E}, \nabla)$ is integrable is then equivalent to the condition that the *curvature* $\nabla^2 := \nabla_1 \circ \nabla$ is zero.

2.2.12. — We will also need a simplicial description of modules with integrable connection. Let $\mathcal{X} \to \mathcal{S}$ be a smooth representable morphism of algebraic stacks with $\mathcal{X}$ a Deligne-Mumford stack. Fix a smooth cover $S \to \mathcal{S}$ and let $S_\bullet$ be the 0-coskeleton. Denote by $X_\bullet$ the simplicial algebraic space $\mathcal{X} \times_{\mathcal{S}} S_\bullet$.

Definition 2.2.13. — A *module with connection* $(\mathcal{E}, \nabla)$ on $X_\bullet/S_\bullet$ is a quasi-coherent sheaf of $\mathcal{O}_{X_{\bullet,\mathrm{et}}}$-modules $\mathcal{E}$ on $X_{\bullet,\mathrm{et}}$ together with a map

$$\nabla : \mathcal{E} \longrightarrow \mathcal{E} \otimes \Omega^1_{X_{\bullet,\mathrm{et}}/S_{\bullet,\mathrm{et}}} \tag{2.2.13.1}$$

such that for any local sections $f \in \mathcal{O}_{X_{\bullet\mathrm{et}}}$ and $e \in \mathcal{E}$ we have

$$\nabla(fe) = e \otimes df + f\nabla(e). \tag{2.2.13.2}$$

The pair $(\mathcal{E}, \nabla)$ is *integrable* if the induced map

$$T_{X_\bullet/S_\bullet} \longrightarrow \mathcal{E}nd(\mathcal{E}) \tag{2.2.13.3}$$

is a map of sheaves of Lie algebras. We denote by $MC(X_\bullet/S_\bullet)$ (resp. $MIC(X_\bullet/S_\bullet)$) the category of quasi-coherent modules with connection (resp. integrable connection).

Lemma 2.2.14. — *Let $\mathcal{X}/\mathcal{S}$ be smooth and let $X_\bullet \to S_\bullet$ be as above. Then there is a natural equivalence of categories between* $\mathrm{MC}(\mathcal{X}_{\mathrm{et}}/\mathcal{S})$ *(resp.* $\mathrm{MIC}(\mathcal{X}_{\mathrm{et}}/\mathcal{S})$*) and* $\mathrm{MC}(X_\bullet/S_\bullet)$ *(resp.* $\mathrm{MIC}(X_\bullet/S_\bullet)$*).*

Proof. — Let $P^1_{\mathcal{X}/\mathcal{S}}$ denote the first infinitesimal neighborhood of $\mathcal{X}$ in $\mathcal{X} \times_{\mathcal{S}} \mathcal{X}$, and let

$$\mathrm{pr}_1, \mathrm{pr}_2 : P^1_{\mathcal{X}/\mathcal{S}} \rightrightarrows \mathcal{X} \tag{2.2.14.1}$$

denote the two projections. As in the classical case, a connection on a quasi-coherent sheaf $\mathcal{E}$ is equivalent to an isomorphism of quasi-coherent sheaves $\mathrm{pr}_1^*\mathcal{E} \simeq \mathrm{pr}_2^*\mathcal{E}$ on $P^1_{\mathcal{X}/\mathcal{S}}$ which reduces to the identity when pulled back to $\mathcal{X}$ via the diagonal. Now if $U \to \mathcal{S}$ is any smooth morphism then the base change of the following diagram of stacks to U

$$\begin{array}{ccc} X & \hookrightarrow & P^1_{\mathcal{X}/\mathcal{S}} \\ \downarrow & & \downarrow\downarrow \\ \mathcal{S} & \longleftarrow & \mathcal{X} \end{array} \tag{2.2.14.2}$$

is canonically isomorphic to the diagram

$$\begin{array}{ccc} \mathcal{X} \times_{\mathcal{S}} U & \hookrightarrow & P^1_{\mathcal{X}\times_{\mathcal{S}} U/U} \\ \downarrow & & \downarrow\downarrow \\ U & \longleftarrow & \mathcal{X} \times_{\mathcal{S}} U \end{array} \tag{2.2.14.3}$$

Thus if $(\mathcal{E}, \nabla)$ is a module with connection on $\mathcal{X}/\mathcal{S}$, then the pullback to $\mathcal{X} \times_{\mathcal{S}} U$ has a canonical connection obtained by pulling back the isomorphism $\mathrm{pr}_1^*\mathcal{E} \simeq \mathrm{pr}_2^*\mathcal{E}$ to $P^1_{\mathcal{X}\times_{\mathcal{S}} U/U} \simeq U \times_{\mathcal{S}} P^1_{\mathcal{X}/\mathcal{S}}$. Thus for any $U \to \mathcal{S}$, there is a natural functor

$$\rho_U^* : MC(\mathcal{X}/\mathcal{S}) \longrightarrow MC(\mathcal{X} \times_{\mathcal{S}} U/U). \tag{2.2.14.4}$$

Moreover, if $g : U' \to U$ is a morphism of algebraic spaces, then the functor $\rho_{U'}^*$ is simply the composite of ρ_U^* with the natural pullback functor $MC((\mathcal{X} \times_{\mathcal{S}} U)_{\mathrm{et}}/U) \to MC((\mathcal{X} \times_{\mathcal{S}} U')_{\mathrm{et}}/U')$ [**44**, 1.1.4]. We thus obtain a functor

$$\rho^* : MC(\mathcal{X}/\mathcal{S}) \longrightarrow MC(X_\bullet/S_\bullet) \tag{2.2.14.5}$$

by sending $(\mathcal{E}, \nabla)$ to the module with connection on $X_\bullet/S_\bullet$ whose restriction to X_i is $\rho_{S^i}^*(\mathcal{E}, \nabla)$ and whose transition maps are the natural ones. The fact that ρ^* is an equivalence follows from [**49**, 13.5.4] applied to $P^1_{\mathcal{X}/\mathcal{S}}$.

More concretely, for any $U \to \mathcal{S}$, the pullback of $(\mathcal{E}, \nabla)$ to $\mathcal{X} \times_{\mathcal{S}} U$ is the sheaf $\pi^*\mathcal{E}$ (where $\pi : \mathcal{X} \times_{\mathcal{S}} U \to \mathcal{X}$ is the projection) with connection the unique connection ∇_U such that for any local section $\xi \in T_{\mathcal{X}/\mathcal{S}}$ the diagram

$$\begin{array}{ccc} \pi^*\mathcal{E} & \xrightarrow{\nabla_{U,\pi^*\xi}} & \pi^*\mathcal{E} \\ \uparrow & & \uparrow \\ \pi^{-1}\mathcal{E} & \xrightarrow{\nabla_\xi} & \pi^{-1}\mathcal{E} \end{array} \tag{2.2.14.6}$$

commutes. From this it follows that the functor ρ^* preserves the notion of "integrable connection". □

We also note the following which follows from the same argument.

Corollary 2.2.15. — *Let $\mathcal{X} \to \mathcal{S}$ be a smooth representable morphism of algebraic stacks with $\mathcal{X}$ Deligne-Mumford, and let $S \to \mathcal{S}$ be a smooth cover by an algebraic space. Then pullback defines an equivalence between the category $MIC(\mathcal{X}_{\mathrm{et}}/\mathcal{S})$ and the category of pairs $((\mathcal{E}, \nabla), \sigma)$, where $(\mathcal{E}, \nabla) \in MIC(\mathcal{X}_{S,\mathrm{et}}/S)$ and*

$$\sigma : \mathrm{pr}_1^*(\mathcal{E}, \nabla) \simeq \mathrm{pr}_2^*(\mathcal{E}, \nabla) \tag{2.2.15.1}$$

on $\mathcal{X}_{S\times_{\mathcal{S}}S}$ such that $\mathrm{pr}_{13}^*(\sigma) = \mathrm{pr}_{23}^*(\sigma) \circ \mathrm{pr}_{12}^*(\sigma)$ on $\mathcal{X}_{S\times_{\mathcal{S}}S\times_{\mathcal{S}}S}$.

2.2.16. — Let $\mathcal{X} \to \mathcal{S}$ be a smooth morphism of algebraic stacks with $\mathcal{X}$ a Deligne-Mumford stack, and let $(\mathcal{E}, \nabla) \in MIC(\mathcal{X}_{\mathrm{et}}/\mathcal{S})$ be a module with integrable connection. Define the *de Rham complex of* (E, ∇), denoted $\mathcal{E} \otimes \Omega^\bullet_{\mathcal{X}_{\mathrm{et}}/\mathcal{S}}$, by defining

$$\nabla_i : \mathcal{E} \otimes \Omega^i_{\mathcal{X}_{\mathrm{et}}/\mathcal{S}} \longrightarrow \mathcal{E} \otimes \Omega^{i+1}_{\mathcal{X}_{\mathrm{et}}/\mathcal{S}} \tag{2.2.16.1}$$

by the formula

$$\begin{aligned} &\nabla_i(e \otimes \omega)(\xi_1 \wedge \cdots \wedge \xi_{i+1}) = \\ &\quad \sum_{l=1}^{i+1} (-1)^{l+1} \nabla_{\xi_l}\big(e \cdot (\omega(\xi_1 \wedge \cdots \wedge \hat{\xi}_l \wedge \cdots \wedge \xi_{i+1}))\big) \\ &\quad + \sum_{l<k} (-1)^{k+l} e \cdot \omega([\xi_l, \xi_k] \wedge \cdots \wedge \hat{\xi}_l \wedge \cdots \wedge \cdots \wedge \hat{\xi}_k \wedge \cdots \xi_{i+1}), \end{aligned} \tag{2.2.16.2}$$

where $\Omega^i_{\mathcal{X}_{\mathrm{et}}/\mathcal{S}} := (\bigwedge^i T_{\mathcal{X}_{\mathrm{et}}/\mathcal{S}})^*$ and ∇_ξ denotes the endomorphism of $\mathcal{E}$ defined by $\nabla(\xi)$.

Observe that if $e \in \mathcal{E}$, $f \in \mathcal{O}_{\mathcal{X}_{\mathrm{et}}}$, $\omega \in \Omega^i_{\mathcal{X}_{\mathrm{et}}/\mathcal{S}}$ then

$$\nabla_i(fe \otimes \omega) = f\nabla_i(e \otimes \omega) + (-1)^i e \otimes \omega \wedge df. \tag{2.2.16.3}$$

Lemma 2.2.17. — *If $\mathcal{X}$ and $\mathcal{S}$ are schemes, then the above defined de Rham complex $\mathcal{E} \otimes \Omega^\bullet_{\mathcal{X}_{\mathrm{et}}/\mathcal{S}}$ agrees with the usual de Rham complex of a module with integrable connection* [**44**, 1.0].

Proof. — To prove this we may work locally on $\mathcal{X}$ and $\mathcal{S}$, and hence may assume that $\mathcal{X}$ is étale over $\mathbb{G}^r_{\mathcal{S}}$ for some integer r. Let $\frac{dx_1}{x_1}, \ldots, \frac{dx_r}{x_r}$ denote the standard basis for $\Omega^1_{\mathcal{X}_{\mathrm{et}}/\mathcal{S}}$, and let $x_1\frac{\partial}{\partial x_1}, \ldots, x_r\frac{\partial}{\partial x_r}$ the corresponding dual basis for $T_{\mathcal{X}_{\mathrm{et}}/\mathcal{S}}$. Then for any i and j we have $[x_i\frac{\partial}{\partial x_i}, x_j\frac{\partial}{\partial x_j}] = 0$. From this and the formula (2.2.16.2) it follows that ∇_s sends a local section $e \otimes (\frac{dx_{i_1}}{x_{i_1}} \wedge \cdots \wedge \frac{dx_{i_s}}{x_{i_s}})$ to

(2.2.17.1)
$$\nabla(e) \wedge \frac{dx_{i_1}}{x_{i_1}} \wedge \cdots \wedge \frac{dx_{i_s}}{x_{i_s}},$$

which agrees with the formula in the usual de Rham complex. □

2.2.18. — The preceding constructions are functorial in the following sense. Consider a 2-commutative diagram of algebraic stacks

(2.2.18.1)
$$\begin{array}{ccc} \mathcal{X}' & \xrightarrow{g} & \mathcal{X} \\ {\scriptstyle s}\downarrow & & \downarrow{\scriptstyle t} \\ \mathcal{S}' & \xrightarrow{f} & \mathcal{S} \end{array}$$

with s and t smooth, representable, and locally separated, and $\mathcal{X}$ and $\mathcal{X}'$ Deligne-Mumford stacks. Then there are natural functors

(2.2.18.2)
$$g^* : MC(\mathcal{X}_{\mathrm{et}}/\mathcal{S}) \longrightarrow MC(\mathcal{X}'_{\mathrm{et}}/\mathcal{S}'), \quad g^* : MIC(\mathcal{X}_{\mathrm{et}}/\mathcal{S}) \longrightarrow MIC(\mathcal{X}'_{\mathrm{et}}/\mathcal{S}')$$

defined as follows. Choose smooth covers $S' \to \mathcal{S}'$ and $S \to \mathcal{S}$ with S' and S schemes and a lifting $\tilde{f} : S' \to S$ of f. Let $S'_\bullet$ and $S_\bullet$ denote the 0-coskeletons of these maps, and write $X_\bullet$ (resp. $X'_\bullet$) for $\mathcal{X} \times_{\mathcal{S}} S_\bullet$ (resp. $\mathcal{X}' \times_{\mathcal{S}'} S'_\bullet$) so that we have a commutative diagram of simplicial algebraic spaces

(2.2.18.3)
$$\begin{array}{ccc} X'_\bullet & \xrightarrow{g_\bullet} & X_\bullet \\ \downarrow & & \downarrow \\ S'_\bullet & \longrightarrow & S_\bullet. \end{array}$$

As in [**44**, 1.1] there are pullback functors
(2.2.18.4)
$$g^*_\bullet : MC(X_\bullet/S_\bullet) \longrightarrow MC(X'_\bullet/S'_\bullet), \quad g^*_\bullet : MIC(X_\bullet/S_\bullet) \longrightarrow MIC(X'_\bullet/S'_\bullet).$$

The functors (2.2.18.2) are defined to be the functors obtained from $g^*_\bullet$ and the equivalences in 2.2.14.

That this definition of g^* does not depend on the choice of $\tilde{f} : S' \to S$ can be seen as follows. If $\tilde{f}^\heartsuit : S'^\heartsuit \to S^\heartsuit$ is a second choice of such data and if

(2.2.18.5)
$$\begin{array}{ccc} S'^\heartsuit & \xrightarrow{\tilde{f}^\heartsuit} & S^\heartsuit \\ {\scriptstyle \alpha'}\downarrow & & \downarrow{\scriptstyle \alpha} \\ S' & \xrightarrow{\tilde{f}} & S \end{array}$$

is a commutative diagram over f, then the diagrams
(2.2.18.6)

$$\begin{array}{ccc} MC(X_\bullet/S_\bullet) & \xrightarrow{\alpha^*} & MC(X^\heartsuit_\bullet/S^\heartsuit_\bullet) \\ {\scriptstyle g^*_\bullet}\downarrow & & \downarrow{\scriptstyle g^{\heartsuit *}_\bullet} \\ MC(X'_\bullet/S'_\bullet) & \xrightarrow{\alpha^*} & MC(X'^\heartsuit_\bullet/S'^\heartsuit_\bullet) \end{array} \qquad \begin{array}{ccc} MIC(X_\bullet/S_\bullet) & \xrightarrow{\alpha^*} & MIC(X^\heartsuit_\bullet/S^\heartsuit_\bullet) \\ {\scriptstyle g^*_\bullet}\downarrow & & \downarrow{\scriptstyle g^{\heartsuit *}_\bullet} \\ MIC(X'_\bullet/S'_\bullet) & \xrightarrow{\alpha^*} & MIC(X'^\heartsuit_\bullet/S'^\heartsuit_\bullet) \end{array}$$

commute, where $g^\heartsuit_\bullet : X'^\heartsuit_\bullet \rightarrow X^\heartsuit_\bullet$ denotes the morphism obtained from g by base change to $S'^\heartsuit_\bullet$ and $S^\heartsuit_\bullet$. This shows that $\tilde{f}^\heartsuit : S'^\heartsuit \rightarrow S^\heartsuit$ and $\tilde{f} : S' \rightarrow S$ define the same functor when there exists a diagram (2.2.18.5). The general case follows from this by considering the products $\tilde{f} \times \tilde{f}^\heartsuit : S' \times_{\mathcal{S}'} S'^\heartsuit \rightarrow S \times_{\mathcal{S}} S^\heartsuit$.

2.2.19. — The above results and definitions can be generalized to an arbitrary morphism of algebraic stacks $\mathcal{X} \rightarrow \mathcal{S}$ such that γ extends to $\mathcal{X}$ as follows.

Let $\omega^1_{\mathcal{X}/\mathcal{S}}$ denote the sheaf on $\mathcal{X}_{\text{lis-et}}$ which to any smooth $U \rightarrow \mathcal{X}$ with U a scheme associates $\Gamma(U, \Omega^1_{U_{\text{et}}/\mathcal{S}})$. The derivations $d : \mathcal{O}_{U_{\text{et}}} \rightarrow \Omega^1_{U_{\text{et}}/\mathcal{S}}$ induce a derivation $d : \mathcal{O}_{\mathcal{X}_{\text{lis-et}}} \rightarrow \omega^1_{\mathcal{X}/\mathcal{S}}$.

Remark 2.2.20. — The sheaf $\omega^1_{\mathcal{X}/\mathcal{S}}$ is not quasi-coherent as it is not cartesian. Let $L_{\mathcal{X}/\mathcal{S}}$ denote the cotangent complex defined in [**49**], and let $\tau_{\geq 0}L_{\mathcal{X}/\mathcal{S}}$ be the truncation. The complex $\tau_{\geq 0}L_{\mathcal{X}/\mathcal{S}}$ in the derived category of $\mathcal{O}_{\mathcal{X}_{\text{lis-et}}}$-modules can be explicitly described as the complex whose restriction to the étale site of any smooth $\mathcal{X}$-scheme $U \rightarrow \mathcal{X}$ is the complex $\Omega^1_{U/\mathcal{S}} \rightarrow \Omega^1_{U/\mathcal{X}}$ (concentrated in degrees 0 and 1). It follows that there is a natural morphism $\omega^1_{\mathcal{X}/\mathcal{S}} \rightarrow \tau_{\geq 0}L_{\mathcal{X}/\mathcal{S}}$ in the derived category of $\mathcal{O}_{\mathcal{X}_{\text{lis-et}}}$-modules. This map is not an isomorphism, however, since for example the complex $\tau_{\geq 0}L_{\mathcal{X}/\mathcal{S}}$ has cartesian cohomology sheaves.

Definition 2.2.21. — A *module with connection* $(\mathcal{E}, \nabla)$ on $\mathcal{X}_{\text{lis-et}}/\mathcal{S}$ is a quasi-coherent sheaf $\mathcal{E}$ on $\mathcal{X}_{\text{lis-et}}$ together with a map $\nabla : \mathcal{E} \rightarrow \mathcal{E} \otimes \omega^1_{\mathcal{X}/\mathcal{S}}$ such that for any local sections $e \in \mathcal{E}$ and $f \in \mathcal{O}_{\mathcal{X}_{\text{lis-et}}}$ we have

$$\nabla(fe) = f\nabla(e) + e \otimes df. \tag{2.2.21.1}$$

The pair $(\mathcal{E}, \nabla)$ is called integrable if for every smooth $\mathcal{X}$-scheme the module with connection on $U_{\text{et}}/\mathcal{S}$ obtained by restriction is integrable. We write $MC(\mathcal{X}_{\text{lis-et}}/\mathcal{S})$ (resp. $MIC(\mathcal{X}_{\text{lis-et}}/\mathcal{S})$) for the category of modules with connection (resp. integrable connection) on $\mathcal{X}_{\text{lis-et}}/\mathcal{S}$.

2.2.22. — If $\mathcal{X}$ is a Deligne-Mumford stack there are natural functors

$$MC(\mathcal{X}_{\text{lis-et}}/\mathcal{S}) \longrightarrow MC(\mathcal{X}_{\text{et}}/\mathcal{S}), \quad MIC(\mathcal{X}_{\text{lis-et}}/\mathcal{S}) \longrightarrow MIC(\mathcal{X}_{\text{et}}/\mathcal{S}) \tag{2.2.22.1}$$

obtained by restricting the sheaf $\mathcal{E}$ to $\mathcal{X}_{\text{et}}$ and nothing that $\omega^1_{\mathcal{X}/\mathcal{S}}|_{\mathcal{X}_{\text{et}}} \simeq \Omega^1_{\mathcal{X}_{\text{et}}/\mathcal{S}}$.

Proposition 2.2.23. — *The functors (2.2.22.1) are equivalences of categories.*

Proof. — Let $r : \mathcal{X}_{\text{lis-et}} \to \mathcal{X}_{\text{et}}$ be the natural morphism of topoi defined by the inclusion $\mathrm{Et}(\mathcal{X}) \subset \text{Lis-Et}(\mathcal{X})$. By [**68**, 6.12], the pullback r^* induces an equivalence of categories between the category of quasi-coherent sheaves on $\mathcal{X}_{\text{et}}$ and the category of quasi-coherent sheaves on $\mathcal{X}_{\text{lis-et}}$. Let $\mathcal{E}$ be a quasi-coherent sheaf in $\mathcal{X}_{\text{et}}$. Then giving a connection $\nabla : r^*\mathcal{E} \to r^*\mathcal{E} \otimes_{\mathcal{O}_{\mathcal{X}_{\text{lis-et}}}} \omega^1_{\mathcal{X}/\mathcal{S}}$ is equivalent to giving a map $\nabla^* : r^{-1}\mathcal{E} \to r^{-1}\mathcal{E} \otimes_{r^{-1}\mathcal{O}_{\mathcal{X}_{\text{et}}}} \omega^1_{\mathcal{X}/\mathcal{S}}$ such that for any local sections $e \in r^{-1}\mathcal{E}$ and $f \in r^{-1}\mathcal{O}_{\mathcal{X}_{\text{et}}}$ we have

$$\nabla^*(fe) = f\nabla(e) + e \otimes df. \tag{2.2.23.1}$$

Since r^{-1} is left adjoint to r_*, to give such a map ∇ is equivalent to giving a map

$$\bar{\nabla} : \mathcal{E} \longrightarrow r_*(r^{-1}\mathcal{E} \otimes_{r^{-1}\mathcal{O}_{\mathcal{X}_{\text{et}}}} \omega^1_{\mathcal{X}/\mathcal{S}}) \simeq \mathcal{E} \otimes_{\mathcal{O}_{\mathcal{X}_{\text{et}}}} \Omega^1_{\mathcal{X}_{\text{et}}/\mathcal{S}} \tag{2.2.23.2}$$

such that the Leibnitz rule (2.2.10.2) holds. This proves that the restriction functor $MC(\mathcal{X}_{\text{lis-et}}/\mathcal{S}) \to MC(\mathcal{X}_{\text{et}}/\mathcal{S})$ is an equivalence.

To complete the proof of the proposition it suffices to show that if $(r^*\mathcal{E}, \nabla) \in MC(\mathcal{X}_{\text{lis-et}}/\mathcal{S})$ then ∇ is integrable if and only if the restriction $(\mathcal{E}, \bar{\nabla}) \in MC(\mathcal{X}_{\text{et}}/\mathcal{S})$ is integrable. For this note that the preceding discussion shows that $(r^*\mathcal{E}, \nabla)$ is isomorphic to the module with integrable connection whose restriction to U_{et} for some smooth $\mathcal{X}$-scheme $f : U \to \mathcal{X}$ is equal to $g^*(\mathcal{E}, \bar{\nabla})$. Since the pullback of an integrable connection is again integrable, it follows that ∇ is integrable if and only if $\bar{\nabla}$ is integrable. □

2.3. Stratifications and differential operators

Fix a PD-stack $(\mathcal{S}, I, \gamma)$.

2.3.1. — Let $\mathcal{X} \to \mathcal{S}$ be a representable locally separated morphism such that γ extends to $\mathcal{X}$. Define $D_{\mathcal{X}}(1)$ (or $D_{\mathcal{X}/\mathcal{S}}(1)$ if we want the reference to $\mathcal{S}$ to be clear) to be the PD-envelope of $\mathcal{X}$ in $\mathcal{X} \times_{\mathcal{S}} \mathcal{X}$. We denote by $D^n_{\mathcal{X}}(1)$ the closed subspace defined by $\bar{I}^{[n+1]}$, where $\bar{I}$ is the PD-ideal defining $\mathcal{X}$ in $D_{\mathcal{X}}(1)$. As in the classical theory, the two projections

$$\mathrm{pr}_1, \mathrm{pr}_2 : \mathcal{X} \times_{\mathcal{S}} \mathcal{X} \longrightarrow \mathcal{X} \tag{2.3.1.1}$$

induce projections $\mathrm{pr}_i : D_{\mathcal{X}}(1) \to \mathcal{X}$ and $\mathrm{pr}_i : D^n_{\mathcal{X}}(1) \to \mathcal{X}$ $(i = 1, 2)$.

Lemma 2.3.2. — *The stacks $D_{\mathcal{X}}(1)$ and $D^n_{\mathcal{X}}(1)$ are affine over $\mathcal{X}$ via either projection* pr_i *$(i = 1, 2)$.*

Proof. — Since $D^n_{\mathcal{X}}(1)$ is a closed substack of $D_{\mathcal{X}}(1)$ it suffices to prove the result for $D_{\mathcal{X}}(1)$. Furthermore, since the formation of $D_{\mathcal{X}}(1)$ commutes with smooth base change $S \to \mathcal{S}$ by 1.2.10, it suffices to consider the case when S is a scheme. In this case $\mathcal{X}$ is an algebraic space since the structure map $\mathcal{X} \to \mathcal{S}$ is assumed representable. In this case the result is clear because $\mathcal{D}_{\mathcal{X}}(1)$ is a quasi-coherent sheaf on $\mathcal{X} \times_{\mathcal{S}} \mathcal{X}$ supported on the diagonal $\mathcal{X} \subset \mathcal{X} \times_{\mathcal{S}} \mathcal{X}$. □

Lemma 2.3.3. — *Let $\mathcal{X} \to \mathcal{S}$ be a smooth representable morphism of algebraic stacks. Then $D_{\mathcal{X}}(1)$ (resp. $D^n_{\mathcal{X}}(1)$ for $n \geq 0$) is flat (resp. finite and flat) over $\mathcal{X}$ via either projections.*

Proof. — This can also be verified after making a base change $S \to \mathcal{S}$, and hence it suffices to consider the case when $\mathcal{S}$ and $\mathcal{X}$ are algebraic spaces. In this case the result follows from [**7**, I.4.5.1]. □

Lemma 2.3.4. — *Let $S \to \mathcal{S}$ be a smooth morphism from a scheme and let $\pi : \mathcal{X}_S := \mathcal{X} \times_{\mathcal{S}} S \to \mathcal{X}$ be the natural projection. Then there are natural isomorphisms*

$$\mathcal{X}_S \times_{\mathcal{X},\mathrm{pr}_1} D^n_{\mathcal{X}}(1) \simeq D^n_{\mathcal{X}_S/S}(1) \simeq D^n_{\mathcal{X}}(1) \times_{\mathrm{pr}_2,\mathcal{X}} \mathcal{X}_S \tag{2.3.4.1}$$

and

$$\mathcal{X}_S \times_{\mathcal{X},\mathrm{pr}_1} D_{\mathcal{X}}(1) \simeq D_{\mathcal{X}_S/S}(1) \simeq D_{\mathcal{X}}(1) \times_{\mathrm{pr}_2,\mathcal{X}} \mathcal{X}_S. \tag{2.3.4.2}$$

Proof. — For $i = 1, 2$ we have a commutative diagram of cartesian squares

$$\begin{array}{ccccc} \mathcal{X}_S & \xrightarrow{\Delta_S} & \mathcal{X}_S \times_S \mathcal{X}_S & \longrightarrow & \mathcal{X}_S \\ \downarrow & & \downarrow & & \downarrow \\ \mathcal{X} & \xrightarrow{\Delta} & \mathcal{X} \times_{\mathcal{S}} \mathcal{X} & \xrightarrow{\mathrm{pr}_i} & \mathcal{X}. \end{array} \tag{2.3.4.3}$$

By the universal property of the PD-envelope this diagram induces maps

$$\mathcal{X}_S \times_{\mathcal{X},\mathrm{pr}_i} D^n_{\mathcal{X}/\mathcal{S}}(1) \longleftarrow D^n_{\mathcal{X}_S/S}(1), \quad \mathcal{X}_S \times_{\mathcal{X},\mathrm{pr}_i} D_{\mathcal{X}/\mathcal{S}}(1) \longleftarrow D_{\mathcal{X}_S/S}(1). \tag{2.3.4.4}$$

That these maps are isomorphisms follows from the fact that the formation of the PD-envelope commutes with smooth base change 1.2.3. □

2.3.5. — By the universal property of the PD-envelope, the divided power envelope of

$$\mathcal{X} \hookrightarrow \mathcal{X} \times_{\mathcal{S}} \mathcal{X} \times_{\mathcal{S}} \mathcal{X} \simeq (\mathcal{X} \times_{\mathcal{S}} \mathcal{X}) \times_{\mathcal{X}} (\mathcal{X} \times_{\mathcal{S}} \mathcal{X}) \tag{2.3.5.1}$$

is canonically isomorphic to $D_{\mathcal{X}}(1) \times_{\mathrm{pr}_2,\mathcal{X},\mathrm{pr}_1} D_{\mathcal{X}}(1)$. Define

$$\delta : D_{\mathcal{X}}(1) \times_{\mathrm{pr}_2,\mathcal{X},\mathrm{pr}_1} D_{\mathcal{X}}(1) \longrightarrow D_{\mathcal{X}}(1) \tag{2.3.5.2}$$

to be the map induced by $\mathrm{pr}_{13} : \mathcal{X} \times_{\mathcal{S}} \mathcal{X} \times_{\mathcal{S}} \mathcal{X} \to \mathcal{X} \times_{\mathcal{S}} \mathcal{X}$. Note that when $\mathcal{X}$ and $\mathcal{S}$ are algebraic spaces, then this definition of δ agrees with that in [**7**, II.1.1.4].

If $S \to \mathcal{S}$ is a flat morphism of algebraic stacks and $\mathcal{X}_S := \mathcal{X} \times_{\mathcal{S}} S$, then by construction the diagram

$$\begin{array}{ccc} (D_{\mathcal{X}}(1) \times_{\mathrm{pr}_2,\mathcal{X},\mathrm{pr}_1} D_{\mathcal{X}}(1)) \times_{\mathcal{S}} S & \xrightarrow{\delta \times \mathrm{id}} & D_{\mathcal{X}}(1) \times_{\mathcal{S}} S \\ \uparrow & & \uparrow \\ D_{\mathcal{X}_S/S}(1) \times_{\mathrm{pr}_2,\mathcal{X}_S,\mathrm{pr}_1} D_{\mathcal{X}_S/S}(1) & \xrightarrow{\delta_S} & D_{\mathcal{X}_S/S}(1), \end{array} \tag{2.3.5.3}$$

commutes, where the vertical arrows are isomorphisms by 1.2.3 and δ_S denotes the map obtained from taking $\mathcal{S} = S$ and $\mathcal{X} = \mathcal{X}_S$ in the above construction of δ.

Lemma 2.3.6. — *For all n and m, δ induces a map*

$$\delta^{m,n} : D^m_{\mathcal{X}}(1) \times_{\mathrm{pr}_2, \mathcal{X}, \mathrm{pr}_1} D^n_{\mathcal{X}}(1) \longrightarrow D^{n+m}_{\mathcal{X}}(1). \tag{2.3.6.1}$$

Proof. — This can be verified after base change $S \to \mathcal{S}$ and hence follows from the classical theory [**7**, I.1.1.5 (b)]. □

Definition 2.3.7. — Let $\mathcal{E}$ be quasi-coherent $\mathcal{O}_{\mathcal{X}_{\text{lis-et}}}$-module. A *PD-stratification on $\mathcal{E}$* is a compatible collection of isomorphisms on $D^n_{\mathcal{X}}(1)$

$$\epsilon_n : \mathrm{pr}_2^* \mathcal{E} \simeq \mathrm{pr}_1^* \mathcal{E} \tag{2.3.7.1}$$

for which $\epsilon_0 = \mathrm{id}$ and the diagram

(2.3.7.2)

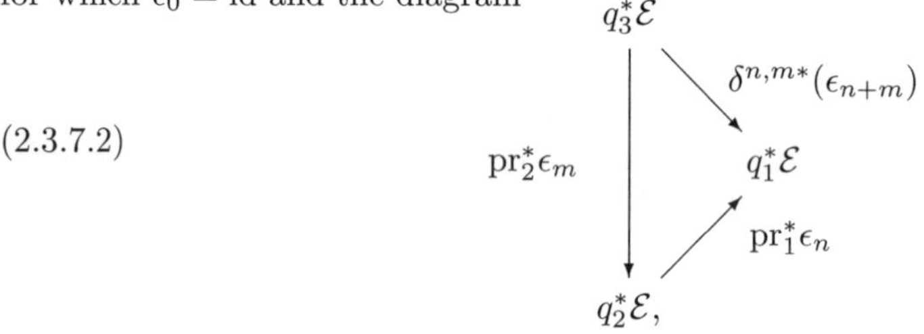

commutes for all m and n, where $q_i : D^n_{\mathcal{X}}(1) \times_{\mathrm{pr}_2, \mathcal{X}, \mathrm{pr}_1} D^m_{\mathcal{X}}(1) \to \mathcal{X}$ for $i = 1, 2, 3$ denotes the map obtained from the i-th projection $\mathcal{X} \times_{\mathcal{S}} \mathcal{X} \times_{\mathcal{S}} \mathcal{X} \to \mathcal{X}$.

An *HPD-stratification on $\mathcal{E}$* is an isomorphism of sheaves on $D_{\mathcal{X}}(1)$

$$\epsilon : \mathrm{pr}_2^* \mathcal{E} \simeq \mathrm{pr}_1^* \mathcal{E} \tag{2.3.7.3}$$

which reduces to the identity on $\mathcal{X}_{\text{lis-et}}$, and for which the diagram of sheaves on $D_{\mathcal{X}}(1) \times_{\mathrm{pr}_2, \mathcal{X}, \mathrm{pr}_1} D_{\mathcal{X}}(1)$

$$\begin{array}{ccc} q_3^*\mathcal{E} & & \\ & \searrow^{\delta^*(\epsilon)} & \\ \downarrow{\scriptstyle \mathrm{pr}_{23}^*\epsilon} & & q_1^*\mathcal{E}, \\ & \nearrow_{\mathrm{pr}_2^*(\epsilon)} & \\ q_2^*\mathcal{E} & & \end{array} \tag{2.3.7.4}$$

commutes, where as above $q_i : D_{\mathcal{X}}(1) \times_{\mathrm{pr}_2, \mathcal{X}, \mathrm{pr}_1} D_{\mathcal{X}}(1) \to \mathcal{X}$ denotes the map induced by the projection $\mathcal{X} \times_{\mathcal{S}} \mathcal{X} \times_{\mathcal{S}} \mathcal{X} \to \mathcal{X}$ onto the i-th factor.

Proposition 2.3.8. — *Let $S \to \mathcal{S}$ be a smooth cover, and denote by $\mathcal{X}_S$ the base change $\mathcal{X} \times_{\mathcal{S}} S$. Then the category of modules with PD-stratification on $\mathcal{X}_{\text{lis-et}}/\mathcal{S}$ is equivalent to the category of pairs $((\mathcal{E}, \epsilon_n), \sigma)$, where $(\mathcal{E}, \epsilon_n)$ is a module with PD-stratification on $\mathcal{X}_{S,\text{et}}/S$, and $\sigma : \mathrm{pr}_1^*(\mathcal{E}, \epsilon_n) \simeq \mathrm{pr}_2^*(\mathcal{E}, \epsilon_n)$ is an isomorphism of modules with PD-stratification on $\mathcal{X}_{S \times_{\mathcal{S}} S}/S \times_{\mathcal{S}} S$ such that $\mathrm{pr}_{13}^*(\sigma) = \mathrm{pr}_{23}^*(\sigma) \circ \mathrm{pr}_{12}^*(\sigma)$ on $\mathcal{X}_{S \times_{\mathcal{S}} S \times_{\mathcal{S}} S}$.*

Similarly, the category of modules with HPD-stratification on $\mathcal{X}_{\text{lis-et}}/\mathcal{S}$ is equivalent to the category of pairs $((\mathcal{E}, \epsilon), \sigma)$, where $(\mathcal{E}, \epsilon)$ is a module with HPD-stratification on $\mathcal{X}_{S,\text{et}}/S$ and $\sigma : \mathrm{pr}_1^*(\mathcal{E}, \epsilon) \simeq \mathrm{pr}_2^*(\mathcal{E}, \epsilon)$ is an isomorphism over $\mathcal{X}_{S \times_{\mathcal{S}} S}$ which satisfies the cocycle condition on $\mathcal{X}_{S \times_{\mathcal{S}} S \times_{\mathcal{S}} S}$.

Proof. — For any smooth $V \to \mathcal{S}$ there is by 1.2.3 a natural isomorphism

$$D^n_{\mathcal{X}}(1) \times_{\mathcal{S}} V \simeq D^n_{\mathcal{X}}(1) \times_{\mathcal{X}\times_{\mathcal{S}}\mathcal{X}} (\mathcal{X}_V \times_V \mathcal{X}_V) \simeq D^n_{\mathcal{X}_V/V}(1). \tag{2.3.8.1}$$

In particular the projection $\pi : D^n_{\mathcal{X}_S/S}(1) \to D^n_{\mathcal{X}}(1)$ is smooth and surjective,

$$D^n_{\mathcal{X}_S/S}(1) \times_{D^n_{\mathcal{X}}(1)} D^n_{\mathcal{X}_S/S}(1) \simeq D^n_{\mathcal{X}_{S\times_{\mathcal{S}} S}/S\times_{\mathcal{S}} S}(1), \tag{2.3.8.2}$$

and

$$D^n_{\mathcal{X}_S/S}(1) \times_{D^n_{\mathcal{X}}(1)} D^n_{\mathcal{X}_S/S}(1) \times_{D^n_{\mathcal{X}}(1)} D^n_{\mathcal{X}_S/S}(1) \simeq D^n_{\mathcal{X}_{S\times_{\mathcal{S}} S\times_{\mathcal{S}} S}/S\times_{\mathcal{S}} S\times_{\mathcal{S}} S}(1). \tag{2.3.8.3}$$

From this and descent theory for quasi-coherent sheaves it follows that if $\mathcal{E}$ is a quasi-coherent sheaf on $\mathcal{X}_{\text{et}}$ then to give an isomorphism $\epsilon_n : \text{pr}_1^*\mathcal{E} \simeq \text{pr}_2^*\mathcal{E}$ over $D^n_{\mathcal{X}}(1)$ is equivalent to giving a compatible collection of isomorphisms over each $D^n_{\mathcal{X}_{S_n}/S_n}(1)$. It follows that to give a PD-stratification on $\mathcal{E}$ is equivalent to giving a compatible collection of PD-stratifications on $\mathcal{E}|_{\mathcal{X}_{S_n}}$ over S_n. This implies the statement about PD-stratifications. The statement concerning HPD-stratifications follows by the same reasoning. □

2.3.9. — Let

$$\begin{array}{ccc} \mathcal{X}' & \xrightarrow{g} & \mathcal{X} \\ {\scriptstyle\alpha}\downarrow & & \downarrow{\scriptstyle\beta} \\ \mathcal{S}' & \xrightarrow{f} & \mathcal{S} \end{array} \tag{2.3.9.1}$$

be a 2-commutative diagram of algebraic stacks with α and β representable and locally separated, $\mathcal{X}'$ and $\mathcal{X}$ Deligne-Mumford stacks, and f obtained from a morphism of PD-stacks $(\mathcal{S}', I', \gamma') \to (\mathcal{S}, I, \gamma)$ such that γ' and γ extend to $\mathcal{X}'$ and $\mathcal{X}$ respectively. Then a PD-stratification $\{\epsilon_n\}$ on a quasi-coherent $\mathcal{O}_{\mathcal{X}_{\text{et}}}$-module $\mathcal{E}$ can be pulled back to a PD-stratification $\{g^*\epsilon_n\}$ on $g^*\mathcal{E}$. Indeed the universal property of the PD-envelope induces a map

$$\tilde{g} : D_{\mathcal{X}'/\mathcal{S}'}(1) \longrightarrow D_{\mathcal{X}/\mathcal{S}}(1) \tag{2.3.9.2}$$

which induces maps $\tilde{g}_n : D^n_{\mathcal{X}'/\mathcal{S}}(1) \to D^n_{\mathcal{X}/\mathcal{S}}(1)$ for all n. The stratification $\{g^*\epsilon_n\}$ is defined to be the one given by the isomorphisms

$$\text{pr}_2^* g^*\mathcal{E} \simeq \tilde{g}^*\text{pr}_2^*\mathcal{E} \xrightarrow{\tilde{g}^*(\epsilon_n)} \tilde{g}^*\text{pr}_1^*\mathcal{E} \simeq \text{pr}_1^* g^*\mathcal{E}. \tag{2.3.9.3}$$

Similarly HPD-stratifications can be pulled back via g.

2.3.10. — Using this definition of pullback, we can define a notion of module with PD-stratification and HPD-stratification for an arbitrary morphism of algebraic stacks $\mathcal{X} \to \mathcal{S}$. Let PD-Strat (resp. HPD-Strat) denote the fibered category over the lisse-étale site Lis-Et($\mathcal{X}$) which to any smooth $\mathcal{X}$-morphism $U \to \mathcal{X}$ with U a locally separated scheme associates the category of modules with PD-stratification (resp. HPD-stratification) on $U/\mathcal{S}$.

Definition 2.3.11. — The category of *modules with PD-stratification (resp. HPD-stratification)* on $\mathcal{X}/\mathcal{S}$ is the category of cartesian sections of the projection

$$\underline{\text{PD-Strat}} \longrightarrow \text{Lis-Et}(\mathcal{X}) \quad (\text{resp. } \underline{\text{HPD-Strat}} \longrightarrow \text{Lis-Et}(\mathcal{X})). \tag{2.3.11.1}$$

2.3.12. — More concretely, a module with PD-stratification (resp. with HPD-stratification) on $\mathcal{X}/\mathcal{S}$ consists of a quasi-coherent sheaf $\mathcal{E}$ on $\mathcal{X}_{\text{lis-et}}$ together with a PD-stratification (resp. HPD-stratification) on $\mathcal{E}_U$ relative to $U/\mathcal{S}$ for every smooth $\mathcal{X}$-scheme U such that for a morphism $\rho : V \to U$ in Lis-Et($\mathcal{X}$) the transition map $\rho^*\mathcal{E}_U \to \mathcal{E}_V$ is compatible with the PD-stratifications (resp. HPD-stratifications).

We can also generalize the relationship between stratifications and differential operators. Consider the case when $\mathcal{X} \to \mathcal{S}$ is representable and $\mathcal{X}$ is a Deligne-Mumford stack.

Definition 2.3.13. — Let $\mathcal{E}$ and $\mathcal{F}$ be quasi-coherent $\mathcal{O}_{\mathcal{X}_{\text{et}}}$-modules. A *PD-differential operator* $\mathcal{E} \to \mathcal{F}$ *of order* $\leq n$ is an $\mathcal{O}_{\mathcal{X}_{\text{et}}}$-linear map

$$\text{pr}^n_{1*}\text{pr}^{n*}_2\mathcal{E} \longrightarrow \mathcal{F}, \tag{2.3.13.1}$$

where pr^n_i denotes the i-th projection $D^n_{\mathcal{X}}(1) \to \mathcal{X}$ $(i = 1, 2)$. An *HPD-differential operator* $\mathcal{E} \to \mathcal{F}$ is an $\mathcal{O}_{\mathcal{X}_{\text{et}}}$-linear map

$$\text{pr}_{1*}\text{pr}^*_2\mathcal{E} \longrightarrow \mathcal{F}, \tag{2.3.13.2}$$

where $\text{pr}_i : D_{\mathcal{X}}(1) \to \mathcal{X}$ are the projections.

Remark 2.3.14. — As in the classical case, the stack $D^1_{\mathcal{X}}(1)$ is canonically isomorphic to the first infinitesimal neighborhood of the diagonal $\Delta : \mathcal{X} \to \mathcal{X} \times_{\mathcal{S}} \mathcal{X}$ since any square-zero ideal has a canonical divided power structure [**8**, 3.2 Example 4]. In particular, giving a PD-differential operator $\mathcal{E} \to \mathcal{F}$ of order ≤ 1 is equivalent to giving a map $\nabla : \mathcal{E} \to \mathcal{F}$ such that for any fixed local section $e_0 \in \mathcal{E}$ the induced map

$$\mathcal{O}_{\mathcal{X}_{\text{et}}} \longrightarrow \mathcal{F}, \quad g \longmapsto \nabla(g \cdot e_0) \tag{2.3.14.1}$$

is a derivation.

2.3.15. — If $\pi : U \to \mathcal{X}$ is a representable étale morphism of Deligne-Mumford stacks, and $\mathcal{E}$ and $\mathcal{F}$ are quasi-coherent sheaves on $\mathcal{X}_{\text{et}}$, then a PD-differential operator f : $\text{pr}^n_{1*}\text{pr}^{n*}_2\mathcal{E} \to \mathcal{F}$ induces a PD-differential operator $\pi^*(f) : \pi^*\mathcal{E} \to \pi^*\mathcal{F}$ on U_{et} of order $\leq n$. To see this note first that since $U \times_{\mathcal{S}} U \to U \times_{\mathcal{S}} \mathcal{X}$ is étale the divided power envelope $D^n_{U/\mathcal{S}}(1)$ is equal to the n-th order divided power envelope of the immersion $U \to U \times_{\mathcal{S}} \mathcal{X}$. This follows from the universal property of the divided power envelope and the observation that for any commutative diagram

$$\begin{array}{ccc} V & \xrightarrow{a} & T \\ \downarrow & & \downarrow h \\ U & \longrightarrow & U \times_{\mathcal{S}} \mathcal{X} \end{array} \tag{2.3.15.1}$$

where a is a PD-immersion, there exists a unique lifting of h to a morphism $\tilde{h}: T \to U \times_{\mathcal{S}} U$ such that the diagram

$$\begin{array}{ccc} V & \xrightarrow{a} & T \\ \downarrow & & \downarrow \tilde{h} \\ U & \xrightarrow{\Delta} & U \times_{\mathcal{S}} \mathcal{U} \end{array} \tag{2.3.15.2}$$

commutes. This is because the projection $U \times_{\mathcal{S}} U \to U \times_{\mathcal{S}} \mathcal{X}$ is étale and the map a is a nil-immersion. On the other hand, the n-th order divided power envelope of U in $U \times_{\mathcal{S}} \mathcal{X}$ is by 1.2.3 isomorphic to

$$U \times_{\mathcal{X}} D^n_{\mathcal{X}}(1) \simeq D^n_{\mathcal{X}}(1) \times_{\mathcal{X} \times_{\mathcal{S}} \mathcal{X}} (U \times_{\mathcal{S}} \mathcal{X}).. \tag{2.3.15.3}$$

This implies that $\pi^* \mathrm{pr}^n_{1*} \mathrm{pr}^{n*}_2 \mathcal{E}$ is isomorphic to $\mathrm{pr}^n_{U1*} \mathrm{pr}^{n*}_{U2} \pi^* \mathcal{E}$, where we write pr_{Ui} for the i-th projection $D^n_{U/\mathcal{S}}(1) \to U$ $(i = 1, 2)$. Applying π^* to a map $\mathrm{pr}^n_{1*} \mathrm{pr}^{n*}_2 \mathcal{E} \to \mathcal{F}$ we therefore obtain a map

$$\mathrm{pr}^n_{U1*} \mathrm{pr}^{n*}_{U2} \pi^* \mathcal{E} \longrightarrow \pi^* \mathcal{F}. \tag{2.3.15.4}$$

We can therefore define a sheaf

$$\mathit{Diff}^n_{\mathcal{X}/\mathcal{S}}(\mathcal{E}, \mathcal{F}) \tag{2.3.15.5}$$

by associating to every étale $U \to \mathcal{X}$ the group of differential operators $\mathcal{E}|_U \to \mathcal{F}|_U$ of order $\leq n$. If E is of finite presentation, the sheaf $\mathit{Diff}^n_{\mathcal{X}/\mathcal{S}}(\mathcal{E}, \mathcal{F})$ is a quasi-coherent sheaf of left $\mathcal{O}_{\mathcal{X}_{\mathrm{et}}}$-modules.

2.3.16. — If $S \to \mathcal{S}$ is a smooth morphism, and if $\pi : \mathcal{X}_S \to \mathcal{X}$ denotes the projection from the base change $\mathcal{X}_S := \mathcal{X} \times_{\mathcal{S}} S$, then if $\mathcal{E}$ is a quasi-coherent sheaf of finite presentation and $\mathcal{F}$ a quasi-coherent sheaf there is a natural isomorphism

$$\pi^* \mathit{Diff}^n_{\mathcal{X}/\mathcal{S}}(\mathcal{E}, \mathcal{F}) \longrightarrow \mathit{Diff}^n_{\mathcal{X}_S/S}(\pi^* \mathcal{E}, \pi^* \mathcal{F}). \tag{2.3.16.1}$$

Indeed by 2.3.4 we have

$$\pi^* \mathrm{pr}^n_{1*} \mathrm{pr}^{n*}_2 \mathcal{E} \simeq \widetilde{\mathrm{pr}}^n_{1*} \widetilde{\mathrm{pr}}^{n*}_2 \pi^* \mathcal{E}, \tag{2.3.16.2}$$

where $\widetilde{\mathrm{pr}}^n_i : D^n_{\mathcal{X}_S/S}(1) \to \mathcal{X}_S$ are the projections. Since $\mathcal{E}$ is of finite presentation the sheaf

$$\pi^* \mathit{Diff}^n_{\mathcal{X}/\mathcal{S}}(\mathcal{E}, \mathcal{F}) \simeq \pi^* \mathcal{H}om_{\mathcal{O}_{\mathcal{X}_{\mathrm{et}}}}(\mathrm{pr}^n_{1*} \mathrm{pr}^{n*}_2 \mathcal{E}, \mathcal{F}) \tag{2.3.16.3}$$

is isomorphic to

$$\begin{aligned} \mathcal{H}om_{\mathcal{O}_{\mathcal{X}_{S,\mathrm{et}}}}(\pi^* \mathrm{pr}^n_{1*} \mathrm{pr}^{n*}_2 \mathcal{E}, \pi^* \mathcal{F}) &\simeq \mathcal{H}om_{\mathcal{O}_{\mathcal{X}_{S,\mathrm{et}}}}(\widetilde{\mathrm{pr}}^n_{1*} \widetilde{\mathrm{pr}}^{n*}_2 \pi^* \mathcal{E}, \pi^* \mathcal{F}) \\ &= \mathit{Diff}^n_{\mathcal{X}_S/S}(\pi^* \mathcal{E}, \pi^* \mathcal{F}). \end{aligned}$$

2.3.17. — Composition of differential operators is defined using δ. If $f : \mathrm{pr}^n_{1*}\mathrm{pr}^{n*}_2\mathcal{E} \to \mathcal{F}$ and $g : \mathrm{pr}^m_{1*}\mathrm{pr}^{m*}_2\mathcal{F} \to \mathcal{G}$ are differential operators, then we define $g \circ f$ to be
(2.3.17.1)

$$\mathrm{pr}^{n+m}_{1*}\mathrm{pr}^{n+m*}_2\mathcal{E} \xrightarrow{\delta^{m,n}} \mathrm{pr}^m_{1*}(\pi_{1*}\pi_2^*\mathrm{pr}^{n*}_2\mathcal{E}) \simeq \mathrm{pr}^m_{1*}\mathrm{pr}^{m*}_2\mathrm{pr}^n_{1*}\mathrm{pr}^{n*}_2\mathcal{E} \xrightarrow{f} \mathrm{pr}^m_{1*}\mathrm{pr}^{m*}_2(\mathcal{F})$$

composed with $g : \mathrm{pr}^m_{1*}\mathrm{pr}^{m*}_2(\mathcal{F}) \to \mathcal{G}$, where $\pi_1 : D^m_{\mathcal{X}}(1) \times D^n_{\mathcal{X}}(1) \to D^m_{\mathcal{X}}(1)$ and $\pi_2 : D^m_{\mathcal{X}}(1) \times D^n_{\mathcal{X}}(1) \to D^n_{\mathcal{X}}(1)$ are the projections. In particular, for any quasi-coherent sheaf $\mathcal{E}$ there is a natural ring structure on

$$\bigcup_{n\geq 0} \mathit{Diff}^n_{\mathcal{X}/\mathcal{S}}(\mathcal{E},\mathcal{E}). \tag{2.3.17.2}$$

2.3.18. — Assume $\mathcal{X} \to \mathcal{S}$ is smooth. For $n = 1$, we have $\mathrm{pr}^1_{1*}\mathrm{pr}^{1*}_2\mathcal{O}_{\mathcal{X}_{\mathrm{et}}} \simeq \mathcal{O}_{\mathcal{X}_{\mathrm{et}}} \oplus \Omega^1_{\mathcal{X}_{\mathrm{et}}/\mathcal{S}}$, and hence there is a natural map $T_{\mathcal{X}_{\mathrm{et}}/\mathcal{S}} \to \mathit{Diff}^1_{\mathcal{X}/\mathcal{S}}(\mathcal{O}_{\mathcal{X}_{\mathrm{et}}}, \mathcal{O}_{\mathcal{X}_{\mathrm{et}}})$. Denote by $\mathcal{O}_{\mathcal{X}_{\mathrm{et}}}\langle T_{\mathcal{X}_{\mathrm{et}}/\mathcal{S}}\rangle$ the left $\mathcal{O}_{\mathcal{X}_{\mathrm{et}}}$-algebra which is the quotient of the associative algebra generated by $T_{\mathcal{X}_{\mathrm{et}}/\mathcal{S}}$ by the relations

$$\xi_1 \cdot \xi_2 + [\xi_2, \xi_1] = \xi_2 \cdot \xi_1 \tag{2.3.18.1}$$

for $\xi_1, \xi_2 \in T_{\mathcal{X}_{\mathrm{et}}/\mathcal{S}}$.

Proposition 2.3.19. — *The map $T_{\mathcal{X}_{\mathrm{et}}/\mathcal{S}} \to \mathit{Diff}^1_{\mathcal{X}/\mathcal{S}}(\mathcal{O}_{\mathcal{X}_{\mathrm{et}}}, \mathcal{O}_{\mathcal{X}_{\mathrm{et}}})$ induces a ring isomorphism*

$$\psi : \mathcal{O}_{\mathcal{X}_{\mathrm{et}}}\langle T_{\mathcal{X}_{\mathrm{et}}/\mathcal{S}}\rangle \longrightarrow \bigcup_{n\geq 0} \mathit{Diff}^n_{\mathcal{X}/\mathcal{S}}(\mathcal{O}_{\mathcal{X}_{\mathrm{et}}}, \mathcal{O}_{\mathcal{X}_{\mathrm{et}}}). \tag{2.3.19.1}$$

Proof. — Let $S \to \mathcal{S}$ be a smooth cover with S an algebraic space, and let $\pi : \mathcal{X}_S \to \mathcal{X}$ be the projection from $\mathcal{X}_S := \mathcal{X} \times_{\mathcal{S}} S$. Then by functoriality of the sheaf $T_{\mathcal{X}/\mathcal{S}}$ we have $\pi^* T_{\mathcal{X}_{\mathrm{et}}/\mathcal{S}} \simeq T_{\mathcal{X}_{S_{\mathrm{et}}}/S}$, and this isomorphism is compatible with the Lie-algebra structures. Also we have $\pi^*\mathit{Diff}^n_{\mathcal{X}/\mathcal{S}}(\mathcal{O}_{\mathcal{X}_{\mathrm{et}}}, \mathcal{O}_{\mathcal{X}_{\mathrm{et}}}) \simeq \mathit{Diff}^n_{\mathcal{X}_S/S}(\mathcal{O}_{\mathcal{X}_{S,\mathrm{et}}}, \mathcal{O}_{\mathcal{X}_{S,\mathrm{et}}})$. We conclude that it suffices to prove the proposition in the case when $\mathcal{S}$ and hence also $\mathcal{X}$ are algebraic spaces. Furthermore it suffices to prove the proposition after replacing $\mathcal{S}$ and $\mathcal{X}$ by étale covers which reduces the proof to the case of schemes which is [**7**, I.4.5.3]. □

Definition 2.3.20. — Let $\mathcal{X} \to \mathcal{S}$ be a representable locally separated morphism of algebraic stacks with $\mathcal{X}$ a Deligne-Mumford stack. A *$\mathcal{D}$-module on $\mathcal{X}_{\mathrm{et}}/\mathcal{S}$* is a quasi-coherent sheaf $\mathcal{E}$ of finite type on $\mathcal{X}_{\mathrm{et}}$ together with a compatible collection of maps

$$\rho_n : \mathit{Diff}^n_{\mathcal{X}/\mathcal{S}}(\mathcal{O}_{\mathcal{X}_{\mathrm{et}}}, \mathcal{O}_{\mathcal{X}_{\mathrm{et}}}) \longrightarrow \mathit{Diff}^n_{\mathcal{X}/\mathcal{S}}(\mathcal{E},\mathcal{E}) \tag{2.3.20.1}$$

which fit together to give a ring homomorphism

$$\rho : \bigcup_0^\infty \mathit{Diff}^n_{\mathcal{X}/\mathcal{S}}(\mathcal{O}_{\mathcal{X}_{\mathrm{et}}}, \mathcal{O}_{\mathcal{X}_{\mathrm{et}}}) \longrightarrow \bigcup_0^\infty \mathit{Diff}^n_{\mathcal{X}/\mathcal{S}}(\mathcal{E},\mathcal{E}). \tag{2.3.20.2}$$

Proposition 2.3.21. — *Let $\mathcal{X}/\mathcal{S}$ be as in 2.3.20 and assume in addition that $\mathcal{X} \to \mathcal{S}$ is smooth. Let $\mathcal{E}$ be a quasi-coherent sheaf on $\mathcal{X}_{\mathrm{et}}$. Then the structure of a $\mathcal{D}$-module on $\mathcal{E}$ is equivalent to the structure of a left $\mathcal{O}_{\mathcal{X}_{\mathrm{et}}}\langle T_{\mathcal{X}_{\mathrm{et}}/\mathcal{S}}\rangle$-module on $\mathcal{E}$ compatible with the left $\mathcal{O}_{\mathcal{X}_{\mathrm{et}}}$-module structure, and this is in turn equivalent to the data of an integrable connection on $\mathcal{E}$.*

Proof. — By 2.3.19 giving the maps ρ_n is equivalent to giving a map

$$\Psi : T_{\mathcal{X}_{\mathrm{et}}/\mathcal{S}} \longrightarrow \mathit{Diff}^1_{\mathcal{X}/\mathcal{S}}(\mathcal{E}, \mathcal{E}) \tag{2.3.21.1}$$

such that for any two local sections $\xi_1, \xi_2 \in T_{\mathcal{X}_{\mathrm{et}}/\mathcal{S}}$ the operators $\Psi(\xi_1)$ and $\Psi(\xi_2)$ satisfy the relation

$$\Psi(\xi_1) \circ \Psi(\xi_2) - \Psi(\xi_2) \circ \Psi(\xi_1) = \Psi([\xi_1, \xi_2]). \tag{2.3.21.2}$$

Let $\mathrm{pr}_2^* : \mathcal{E} \to \mathrm{pr}_{1*}\mathrm{pr}_2^*\mathcal{E}$ be the map obtained from the map $\mathrm{pr}_{1*}\mathrm{pr}_2^{-1}\mathcal{E} \to \mathrm{pr}_{1*}\mathrm{pr}_2^*\mathcal{E}$ and the observation that since the projections $D_{\mathcal{X}}(1) \to \mathcal{X}$ induce equivalences between the associated étale topoi there is a canonical isomorphism $\mathcal{E} \simeq \mathrm{pr}_{1*}\mathrm{pr}_2^{-1}\mathcal{E}$. For any $\xi \in T_{\mathcal{X}_{\mathrm{et}}/\mathcal{S}}$ denote by $\nabla_\xi : \mathcal{E} \to \mathcal{E}$ the composite

$$\mathcal{E} \xrightarrow{\mathrm{pr}_2^*} \mathrm{pr}^1_{1*}\mathrm{pr}_2^{1*}\mathcal{E} \xrightarrow{\rho_1(\xi)} \mathcal{E}. \tag{2.3.21.3}$$

Using the definition of composition of differential operators, it follows that the relation (2.3.21.2) is equivalent to the condition that

$$\nabla_{\xi_1} \circ \nabla_{\xi_2} - \nabla_{\xi_2} \circ \nabla_{\xi_1} = \nabla_{[\xi_1, \xi_2]}. \tag{2.3.21.4}$$

Note also that by the definition of $d : \mathcal{O}_{\mathcal{X}_{\mathrm{et}}} \to \Omega^1_{\mathcal{X}_{\mathrm{et}}/\mathcal{S}}$ the maps ∇_ξ satisfy the relation

$$\nabla_\xi(f \cdot e) = \xi(f) \cdot e + f\nabla_\xi(e). \tag{2.3.21.5}$$

In other words, giving the maps ρ_n is equivalent to giving an integrable connection ∇ on $\mathcal{E}$, and this is in turn equivalent to giving $\mathcal{E}$ the structure of a left $\mathcal{O}_{\mathcal{X}_{\mathrm{et}}}\langle T_{\mathcal{X}_{\mathrm{et}}/\mathcal{S}}\rangle$-module. □

Theorem 2.3.22. — *Let $\mathcal{X} \to \mathcal{S}$ be a smooth, representable, and locally separated morphism of algebraic stacks with $\mathcal{X}$ a Deligne-Mumford stack. Then the following categories are naturally equivalent.*

(i) *The category of modules with PD-stratification on $\mathcal{X}/\mathcal{S}$.*

(ii) *The category of $\mathcal{D}$-modules on $\mathcal{X}/\mathcal{S}$.*

(iii) *The category of modules with integrable connection on $\mathcal{X}_{\mathrm{et}}/\mathcal{S}$.*

Proof. — Note that all three categories can be described as categories of quasi-coherent sheaves on $\mathcal{X}$ with some additional structure. Thus it suffices to prove that if $\mathcal{E}$ is a quasi-coherent sheaf on $\mathcal{X}_{\mathrm{et}}$ then the set of PD-stratifications on $\mathcal{E}$ is naturally in bijection between the set of $\mathcal{D}$-module structures on $\mathcal{E}$ and this is in turn in natural bijection between the set of integrable connections on $\mathcal{E}$. The equivalence between (ii) and (iii) was established in 2.3.21.

For the equivalence between (i) and (ii), let $(\mathcal{E}, \{\epsilon_n\})$ be a module with PD-stratification, and define the maps ρ_n as follows. Given a morphism

$$f : \mathrm{pr}^n_{1*}\mathrm{pr}^{n*}_2 \mathcal{O}_{\mathcal{X}_{\mathrm{et}}} \longrightarrow \mathcal{O}_{\mathcal{X}_{\mathrm{et}}}, \tag{2.3.22.1}$$

we obtain by tensoring with $\mathcal{E}$ a map

$$\mathrm{pr}^n_{1*}\mathrm{pr}^{n*}_1 \mathcal{E} \simeq \mathcal{E} \otimes_{\mathcal{O}_{\mathcal{X}_{\mathrm{et}}}} \mathrm{pr}^n_{1*}\mathrm{pr}^{n*}_2 \mathcal{O}_{\mathcal{X}_{\mathrm{et}}} \longrightarrow \mathcal{E}, \tag{2.3.22.2}$$

and we define $\rho_n(f) : \mathrm{pr}^n_{1*}\mathrm{pr}^{n*}_2\mathcal{E} \to \mathcal{E}$ to be the map obtained by composing this map with the isomorphism

$$\mathrm{pr}^n_{1*}\mathrm{pr}^{n*}_1 \mathcal{E} \xrightarrow{\mathrm{pr}^n_{1*}(\epsilon_n)} \mathrm{pr}^n_{1*}\mathrm{pr}^{n*}_2 \mathcal{E}. \tag{2.3.22.3}$$

The same argument as in the classical case [**8**, 4.8] show that this gives a well-defined functor (i) $\to$ (ii).

To see that this functor is an equivalence note that by 2.2.15 and 2.3.8 it suffices to verify that it becomes an equivalence after base change $S \to \mathcal{S}$. This reduces the question to the case when $\mathcal{S}$ is a scheme and $\mathcal{X}$ is an algebraic space. Furthermore it suffices to show that this functor induces an equivalence after replacing $\mathcal{X}$ by an étale cover so we can also assume that $\mathcal{X}$ is a scheme. In this case the result is [**8**, 4.8]. □

Finally we describe the notion of HPD-stratification in terms of modules with integrable connection.

Lemma 2.3.23. — *Let $\mathcal{X} \to \mathcal{S}$ be a smooth morphism of algebraic stacks. Then the functor from the category of modules with HPD-stratification to the category of modules with PD-stratification which sends an HPD-stratified module $(\mathcal{E}, \epsilon)$ to the module $\mathcal{E}$ with the family of isomorphisms ϵ_n given by the reductions of ϵ is fully faithful.*

Proof. — By the definition of PD and HPD stratifications (2.3.11), it suffices to consider the case when $\mathcal{X}$ is an algebraic space. By 2.3.8 it suffices to verify the lemma after making a smooth base change $S \to \mathcal{S}$. It therefore suffices to consider the case when $\mathcal{S}$ is a scheme and $\mathcal{X}$ is an algebraic space. Since the assertion is also étale local on $\mathcal{X}$ we may in fact also assume that $\mathcal{X}$ is a scheme. In this case by [**7**, II.4.2.12] the data of a PD-stratification on a quasi-coherent sheaf $\mathcal{E}$ is equivalent to an integrable connection, and by [**7**, 4.3.11] the data of a HPD-stratification is equivalent to an integrable and quasi-nilpotent connection (see below for a review of the notion of a quasi-nilpotent connection). From this description the case of schemes, and hence also the general case, follows. □

2.3.24. — Recall [**8**, 4.10] that if $f : X \to S$ is a smooth morphism of schemes over $\mathbb{Z}/p^m$ for some prime p and integer $m \geq 1$, and $\nabla : \mathcal{E} \to \mathcal{E} \otimes \Omega^1_{X_{\mathrm{et}}/S}$ is a quasi-coherent sheaf with integrable connection, then $(\mathcal{E}, \nabla)$ is *quasi-nilpotent* if the following holds. Locally on X we can find an étale map $X \to \mathbb{A}^r_S$ for some r, which defines a basis $\{\frac{\partial}{\partial x_i}\}$ for $T_{X_{\mathrm{et}}/S}$. Then the condition is that for any open $U \subset X$ and section $s \in \Gamma(U, \mathcal{E})$ there exist an open covering $\{U_\alpha\}$ of U and integers $\{e_{i,\alpha}\}$ such that

$$(\nabla_{\frac{\partial}{\partial x_i}})^{e_{i,\alpha}}(s|_{U_\alpha}) = 0. \tag{2.3.24.1}$$

By [**8**, 4.12] this condition is independent of the choice of coordinate system.

Lemma 2.3.25. — *Let S be an algebraic space over $\mathbb{Z}/p^m$ for some $m \geq 1$, and consider a commutative diagram of algebraic spaces*

$$\begin{array}{ccc} X' & \xrightarrow{g} & X \\ {\scriptstyle\alpha}\downarrow & & \downarrow{\scriptstyle\beta} \\ S' & \xrightarrow{f} & S, \end{array} \tag{2.3.25.1}$$

with f smooth and surjective, β smooth, and the map $X' \to S' \times_S X$ induced by g smooth and surjective. Let $(\mathcal{E}, \nabla)$ be a module with integrable connection on X_{et}/S. Then $(\mathcal{E}, \nabla)$ is quasi-nilpotent if and only if $g^*(\mathcal{E}, \nabla)$ is quasi-nilpotent.

Proof. — Note first that for any smooth morphism $V \to Z$ of algebraic spaces the differential $d : \mathcal{O}_{V_{\mathrm{et}}} \to \Omega^1_{V_{\mathrm{et}}/Z}$ is a quasi-nilpotent integrable connection (here we use the assumption that $p^m \cdot \mathcal{O}_{V_{\mathrm{et}}} = 0$). In particular, if V is quasi-compact and if there exist an étale morphism $V \to \mathbb{A}^r_Z$ and $f \in \Gamma(V, \mathcal{O}_{V_{\mathrm{et}}})$ is a global section, then for each $1 \leq i \leq r$ there exists an integer e_i such that

$$\frac{\partial}{\partial x_i}^{e_i}(f) = 0. \tag{2.3.25.2}$$

Shrinking X we may assume that X is étale over $\mathbb{A}^r_S$ in which case X' is smooth over $\mathbb{A}^r_S \times_S S' \simeq \mathbb{A}^r_{S'}$. Shrinking X' we may further assume that the smooth morphism $X' \to \mathbb{A}^r_{S'}$ can be lifted to an étale morphism $X' \to \mathbb{A}^{r+r'}_{S'}$. Furthermore, we may assume that S, S', X, and hence also X' are all affine schemes. Let $\{x_1, \ldots, x_r\}$ be the coordinates on $\mathbb{A}^r_S$ and let $\{x_1, \ldots, x_r, y_1, \ldots, y_{r'}\}$ be the coordinates on $\mathbb{A}^{r+r'}_{S'}$.

The pullback of ∇ to X' is characterized by the condition that the diagram

$$\begin{array}{ccc} \Gamma(X, \mathcal{E}) & \xrightarrow{j} & \Gamma(X', g^*\mathcal{E}) \\ {\scriptstyle\nabla}\downarrow & & \downarrow{\scriptstyle g^*\nabla} \\ \Gamma(X, \mathcal{E} \otimes \Omega^1_{X_{\mathrm{et}}/S}) & \xrightarrow{j'} & \Gamma(X', g^*\mathcal{E} \otimes \Omega^1_{X'_{\mathrm{et}}/S'}) \end{array} \tag{2.3.25.3}$$

commutes, where j and j' are injections since g is smooth and surjective. If $x_1, \ldots, x_r$ denote the coordinates on $\mathbb{A}^r$ then it follows that the diagram

$$\begin{array}{ccc} \Gamma(X, \mathcal{E}) & \xrightarrow{j} & \Gamma(X', g^*\mathcal{E}) \\ {\scriptstyle\nabla_{\frac{\partial}{\partial x_i}}}\downarrow & & \downarrow{\scriptstyle g^*\nabla_{\frac{\partial}{\partial x_i}}} \\ \Gamma(X, \mathcal{E}) & \xrightarrow{j} & \Gamma(X', g^*\mathcal{E}) \end{array} \tag{2.3.25.4}$$

commutes. From this it follows that if $g^*(\mathcal{E}, \nabla)$ is quasi-nilpotent then so is $(\mathcal{E}, \nabla)$.

Conversely, every section $s' \in \Gamma(X', g^*\mathcal{E})$ is of the form $f' \cdot s$ with $f' \in \Gamma(X', \mathcal{O}_{X'})$ and $s \in \Gamma(X, \mathcal{E})$. Then

$$(\nabla_{\frac{\partial}{\partial y_i}})^e(s') = \frac{\partial^e}{\partial^e y_i}(f') \cdot s \tag{2.3.25.5}$$

and by the Leibnitz rule [**44**, 5.0.1] we have

$$(2.3.25.6) \qquad (\nabla_{\frac{\partial}{\partial x_i}})^e(s') = \sum_{j=0}^{e} \binom{e}{j} \Big(\frac{\partial}{\partial x_i}\Big)^j (f')(\nabla_{\frac{\partial}{\partial x_i}})^{e-j}(s).$$

Consequently, if $(\mathcal{E}, \nabla)$ is quasi-nilpotent the fact that $d : \mathcal{O}_{X'} \to \Omega^1_{X'_{\text{et}}/S'}$ is quasi-nilpotent implies that $g^*(\mathcal{E}, \nabla)$ is also quasi-nilpotent. □

Definition 2.3.26. — Let $f : \mathcal{X} \to \mathcal{S}$ be a smooth morphism of algebraic stacks. A module with integrable connection $(\mathcal{E}, \nabla) \in MIC(\mathcal{X}_{\text{lis-et}}/\mathcal{S})$ is *quasi-nilpotent* if for some diagram

$$(2.3.26.1) \qquad \begin{array}{ccc} X & \xrightarrow{g} & S \\ {\scriptstyle P}\downarrow & & \downarrow{\scriptstyle Q} \\ \mathcal{X} & \xrightarrow{f} & \mathcal{S} \end{array}$$

with P and Q smooth and surjective morphisms from algebraic spaces and $g \times P : X \to S \times_{\mathcal{S}} \mathcal{X}$ smooth, the pullback $P^*(\mathcal{E}, \nabla)|_{X_{\text{et}}} \in MIC(X_{\text{et}}/S)$ is quasi-nilpotent.

Remark 2.3.27. — By 2.3.25 if $(\mathcal{E}, \nabla) \in MIC(\mathcal{X}_{\text{lis-et}}/\mathcal{S})$ is quasi-nilpotent then for every diagram (2.3.26.1) the pullback $P^*(\mathcal{E}, \nabla)|_{X_{\text{et}}} \in MIC(X_{\text{et}}/S)$ is quasi-nilpotent. This lemma also implies that in the case when $\mathcal{X}$ and $\mathcal{S}$ are algebraic spaces then $(\mathcal{E}, \nabla) \in MIC(\mathcal{X}_{\text{lis-et}}/\mathcal{S})$ is quasi-nilpotent if and only if the restriction $(\mathcal{E}, \nabla)|_{\mathcal{X}_{\text{et}}} \in MIC(\mathcal{X}_{\text{et}}/\mathcal{S})$ is quasi-nilpotent in the usual sense.

Theorem 2.3.28. — *Let $\mathcal{X} \to \mathcal{S}$ be a smooth morphism of algebraic stacks with $\mathcal{X}$ Deligne-Mumford. Then the equivalence in 2.3.22 induces an equivalence of categories between the category of modules with HPD-stratification on $\mathcal{X}_{\text{et}}/\mathcal{S}$ (viewed as a subcategory of the category of modules with PD-stratification using 2.3.23), and the full subcategory of $MIC(\mathcal{X}_{\text{et}}/\mathcal{S})$ consisting of modules with integrable and quasi-nilpotent connection.*

Proof. — Let $S \to \mathcal{S}$ be a smooth cover with S an algebraic space. By 2.3.8 a module with PD-stratification $(\mathcal{E}, \{\epsilon_n\})$ on $\mathcal{X}_{\text{et}}/\mathcal{S}$ extends to an HPD-stratification if and only if the pullback to $\mathcal{X} \times_{\mathcal{S}} S/S$ admits an HPD-stratification. Similarly by 2.3.25 the module with integrable connection $(\mathcal{E}, \nabla) \in MIC(\mathcal{X}_{\text{et}}/\mathcal{S})$ associated to $(\mathcal{E}, \{\epsilon_n\})$ is quasi-nilpotent if and only if the pullback to $\mathcal{X} \times_{\mathcal{S}} S/S$ is quasi-nilpotent. From this it follows that it suffices to consider $\mathcal{X} \times_{\mathcal{S}} S \to S$, and hence we may assume that $\mathcal{S}$ and $\mathcal{X}$ are algebraic spaces. By working étale locally on $\mathcal{S}$ and $\mathcal{X}$ one further reduces to the case of schemes. In this case the result follows from [**7**, II.4.2.11]. □

Example 2.3.29. — Let k be a field and G/k a smooth group scheme of finite type. Let $\mathcal{S} = BG$ and $\mathcal{X} = \text{Spec}(k)$ with the morphism $\mathcal{X} \to \mathcal{S}$ defined by the trivial torsor.

We have a cartesian diagram

$$\begin{array}{ccc} \mathrm{Spec}(k) & \longleftarrow & G \\ \downarrow & & \downarrow \\ BG & \longleftarrow & \mathrm{Spec}(k). \end{array} \tag{2.3.29.1}$$

As mentioned in 2.2.8 one can deduce from this that $T_{\mathcal{X}/\mathcal{S}}$ is in this case isomorphic (as a Lie algebra) to $\mathrm{Lie}(G)$. In particular, a module with connection on $\mathcal{X}/\mathcal{S}$ is equivalent to the data of a k-vector space V and a k-linear map $\rho : \mathrm{Lie}(G) \to \mathrm{Aut}(V)$. Under this correspondence integrable connections correspond to maps ρ which are Lie algebra morphisms. The notion of quasi-nilpotence is for a general group G more difficult to describe. One description is simply to say that there is a canonical isomorphism $\mathrm{Lie}(G) \otimes_k \mathcal{O}_G \simeq T_{G/k}$ obtained by pullback from the diagram (2.3.29.1). A representation $\rho : \mathrm{Lie}(G) \to \mathrm{End}_k(V)$ therefore defines a module with integrable connection on G, and the condition that ρ is quasi-nilpotent is then equivalent to the condition that the resulting module with integrable connection on G/k is quasi-nilpotent (but see below for a better description in the case of $G = \mathbb{G}_m$).

The diagram (2.3.29.1) also shows that $D_{\mathcal{X}}(1)$ is in this case isomorphic to the divided power envelope of the identity section $e : \mathrm{Spec}(k) \hookrightarrow G$. Write G^{PD} for this divided power envelope. The multiplication $G \times G \to G$ induces a morphism $m : G^{PD} \times G^{PD} \to G^{PD}$. It follows from the construction of the morphism δ in (2.3.5.2) that in this special case the morphism δ is just this map induced by the group structure. If $p : G^{PD} \to \mathrm{Spec}(k)$ is the structural morphism, then it follows that a HPD-stratification on a k-vector space V is given by an isomorphism

$$\epsilon : p^*V \longrightarrow p^*V \tag{2.3.29.2}$$

over G^{PD} such that the isomorphism $m^*(\epsilon) : m^*V \to m^*V$ over $G^{PD} \times G^{PD}$ is equal to the isomorphism $\mathrm{pr}_1^*(\epsilon) \circ \mathrm{pr}_2^*(\epsilon)$. Similarly one can describe the notion of PD-stratification in terms of the n-th order divided power neighborhoods G_n^{PD} of the identity in G.

The ring of differential operators is given by

$$\mathit{Diff}_{\mathcal{X}/\mathcal{S}}(\mathcal{O}_{\mathcal{X}}, \mathcal{O}_{\mathcal{X}}) = \varinjlim \mathrm{Hom}_k(\mathcal{O}_{G_n^{PD}}, k) \tag{2.3.29.3}$$

with ring structure

$$\mathrm{Hom}_k(\mathcal{O}_{G_n^{PD}}, k) \otimes \mathrm{Hom}_k(\mathcal{O}_{G_m^{PD}}, k) \longrightarrow \mathrm{Hom}_k(\mathcal{O}_{G_{n+m}^{PD}}, k) \tag{2.3.29.4}$$

induced by the comultiplication

$$\mathcal{O}_{G_{n+m}^{PD}} \longrightarrow \mathcal{O}_{G_n^{PD}} \otimes \mathcal{O}_{G_m^{PD}}. \tag{2.3.29.5}$$

Example 2.3.30. — In the case when $G = \mathbb{G}_m = \mathrm{Spec}(k[x^{\pm}])$ we can describe things even more concretely. The differentials have as a basis $\mathrm{dlog}(x)$ which is translation invariant. A module with integrable connection on $\mathrm{Spec}(k)$ over $B\mathbb{G}_m$ in this case amounts to a vector space V with an endomorphism $N : V \to V$. The corresponding module with integrable connection on $\mathbb{G}_m$ obtained from the diagram (2.3.29.1) is $V \otimes_k k[x^{\pm}]$ with the unique connection ∇ with

$$\nabla(v \otimes 1) = (N(v) \otimes 1) \cdot \mathrm{dlog}(x). \tag{2.3.30.1}$$

From this one sees that the condition that ∇ is quasi-nilpotent is precisely equivalent to the condition that the endomorphism N is nilpotent. This example plays an important role in the construction of the monodromy operator occurring in Fontaine's theory.

Example 2.3.31. — For another example, we work again over a field k and consider $\mathcal{X} = \mathbb{A}^1$ and $\mathcal{S} = [\mathbb{A}^1/\mathbb{G}_m]$ with the projection map $\mathcal{X} \to \mathcal{S}$ the quotient map. In this case the diagonal map $\mathcal{X} \to \mathcal{X} \times_{\mathcal{S}} \mathcal{X}$ is the morphism

$$\mathrm{id} \times e : \mathbb{A}^1 \longrightarrow \mathbb{A}^1 \times \mathbb{G}_m. \tag{2.3.31.1}$$

If we write $\mathbb{G}_m = \mathrm{Spec}(k[u^{\pm}])$, then as mentioned in 2.2.9 if we let $\mathrm{dlog}(t)$ denote the element of $\Omega^1_{\mathcal{X}/\mathcal{S}}$ defined by the $u-1$ (an element in the ideal defining the diagonal) we have $\Omega^1_{\mathcal{X}/\mathcal{S}} \simeq k[x] \cdot \mathrm{dlog}(t)$ with $d(t) = t\mathrm{dlog}(t)$.

The ring of PD-differential operators $\mathit{Diff}_{\mathcal{X}/\mathcal{S}}(\mathcal{O}_{\mathcal{X}_{\mathrm{et}}}, \mathcal{O}_{\mathcal{X}_{\mathrm{et}}})$ is equal to the subring of the usual ring of differential operators $k[x, \partial]$, where $\partial x = 1 + x\partial$, generated by x and $x\partial$, and the coordinate ring of $D_{\mathcal{X}}(1)$ is the PD-polynomial algebra $k[x]\langle(u-1)\rangle$. The commutative diagram

$$\begin{array}{ccc} \mathbb{A}^1 & = & \mathbb{A}^1 \\ \downarrow & & \downarrow \\ [\mathbb{A}^1/\mathbb{G}_m] & \longrightarrow & \mathrm{Spec}(k) \end{array} \tag{2.3.31.2}$$

and functoriality of the divided power envelope furnishes a morphism $D_{\mathcal{X}/\mathcal{S}}(1) \to D_{\mathbb{A}^1/k}(1)$. This map of schemes is obtained from the map of coordinate rings

$$k[x]\langle(x_1 - x_2)\rangle \longrightarrow k[x]\langle(u-1)\rangle, \quad (x_1 - x_2)^{[n]} \longmapsto x^n(u-1)^{[n]}. \tag{2.3.31.3}$$

We conclude that the dual basis element of $k[x, x\partial]$ corresponding to $(u-1)^{[n]}$ is equal in $k[x, \partial]$ to $x^n\partial^n$ (since ∂^n is the dual of $(x_1 - x_2)^{[n]}$). From this and the formula

$$x^n\partial^n = \prod_{j=0}^{n-1}(x\partial - j) \tag{2.3.31.4}$$

it follows that an action of $k[x, x\partial]$ on a $k[x]$-module E induces a stratification if and only if for every $a \in E$ the sum

$$\sum_n \Big(\prod_{j=0}^{n-1}(x\partial - j)\Big)(a) \tag{2.3.31.5}$$

is a finite sum. In other words, if and only if for every $a \in E$ there exists an integer $n \geq 1$ such that

$$\left(\prod_{j=0}^{n-1}(x\partial - j)\right)(a) = 0. \tag{2.3.31.6}$$

This agrees with the notion of quasi-nilpotence in the logarithmic context [**40**, 6.2]. We discuss the relationship with logarithmic geometry in more detail in Chapter 9.

2.4. The $\mathcal{L}$-construction

Let $(\mathcal{S}, I, \gamma)$ be a PD-stack, and $i : \mathcal{X} \hookrightarrow \mathcal{Y}$ be a closed immersion of $\mathcal{S}$-stacks such that γ extends to $\mathcal{X}$ and $\mathcal{Y} \to \mathcal{S}$ is smooth.

Proposition 2.4.1. — *The functor i_* from abelian sheaves in $(\mathcal{X}_{\text{lis-et}}/\mathcal{S})_{\text{cris}}$ to abelian sheaves in $(\mathcal{Y}_{\text{lis-et}}/\mathcal{S})_{\text{cris}}$ is exact, and if E is a quasi-coherent crystal in $(\mathcal{X}_{\text{lis-et}}/\mathcal{S})_{\text{cris}}$ then i_*E is a quasi-coherent crystal of $\mathcal{O}_{\mathcal{Y}_{\text{lis-et}}/\mathcal{S}}$-modules.*

Proof. — Let $(U, T, \delta) \in \text{Cris}(\mathcal{Y}_{\text{lis-et}}/\mathcal{S})$ be an object, and let $\tilde{T} \in (\mathcal{Y}_{\text{lis-et}}/\mathcal{S})_{\text{cris}}$ be the associated sheaf. Then for any sheaf $F \in (\mathcal{X}_{\text{lis-et}}/\mathcal{S})_{\text{cris}}$ we have

$$i_*F(U, T, \delta) = \text{Hom}_{(\mathcal{Y}_{\text{lis-et}}/\mathcal{S})_{\text{cris}}}(\tilde{T}, i_*F) \simeq \text{Hom}_{(\mathcal{X}_{\text{lis-et}}/\mathcal{S})_{\text{cris}}}(i^*\tilde{T}, F). \tag{2.4.1.1}$$

On the other hand, for any $(V, Z, \epsilon) \in \text{Cris}(\mathcal{X}_{\text{lis-et}}/\mathcal{S})$ we have

$$i^*\tilde{T}(V, Z, \epsilon) = \text{Hom}_{g-PD}((V, Z, \epsilon), (U, T, \delta)). \tag{2.4.1.2}$$

Set $U_0 := \mathcal{X} \times_{\mathcal{Y}} U$, and let D denote the divided power envelope $D_{U_0,\gamma,\delta}(T)$ of U_0 in T with compatibilities with respect to γ and δ as in 1.2.12. Then $i^*\tilde{T}$ is represented by $U_0 \hookrightarrow D$. It follows that for any sheaf F we have

$$i_*F(U, T, \delta) = F(D). \tag{2.4.1.3}$$

Since the divided power envelope D is a thickening of U_0 and T is a thickening of U, the étale site of D (resp. T) is equivalent to the étale site of U_0 (resp. U). Since $U_0 \hookrightarrow U$ is a closed immersion it follows from this that the pushforward $D_{\text{et}} \to T_{\text{et}}$ is an exact functor. In particular, the functor i_* is exact.

To deduce that i_*E is a crystal, suppose $(U', T', \delta') \to (U, T, \delta)$ is a morphism in $\text{Cris}(\mathcal{Y}_{\text{lis-et}}/\mathcal{S})$, with U and U' affine schemes (this also implies that T and T' are affine schemes as well). Let D (resp. D') denote the divided power envelope of $U \times_{\mathcal{Y}} \mathcal{X}$ in T (resp. $U' \times_{\mathcal{Y}} \mathcal{X}$ in T') with compatibilities with γ and δ (resp. δ') as above, and let $t : T' \to T$ denote the morphism in question.

We need to show that the natural map

$$\mathcal{O}_{T'} \otimes_{t,\mathcal{O}_T} E_D \longrightarrow E_{D'} \tag{2.4.1.4}$$

is an isomorphism, where E_D (resp. $E_{D'}$) denotes the value of E on D (resp. D'). Since E is a crystal we have $E_{D'} \simeq \mathcal{O}_{D'} \otimes_{\mathcal{O}_D} E_D$. To prove the proposition it therefore suffices to show that the natural map

$$\mathcal{O}_{T'} \otimes_{t,\mathcal{O}_T} \mathcal{O}_D \longrightarrow \mathcal{O}_{D'} \tag{2.4.1.5}$$

is an isomorphism. Since the composite $U \to \mathcal{Y} \to \mathcal{S}$ is smooth and T is affine, there exists a retraction $\tilde{r} : T \to U$ of the inclusion $U \subset T$. Let $D_{\mathcal{X},\gamma}(\mathcal{Y})$ be the divided power envelope of $\mathcal{X}$ in $\mathcal{Y}$, and observe that by 1.2.3 the base change $D_{\mathcal{X},\gamma}(\mathcal{Y}) \times_{\mathcal{Y}} U$ is isomorphic to the divided power envelope $D_{U_0,\gamma}(U)$ of U_0 in U. We then have a commutative diagram

$$\begin{array}{ccc} U_0 & \longrightarrow & D_{\mathcal{X},\gamma}(\mathcal{Y}) \times_{\mathcal{Y}} U \\ \downarrow & & \downarrow \\ T & \xrightarrow{\tilde{r}} & U. \end{array} \tag{2.4.1.6}$$

From this and [**7**, I.2.8.2] it follows that the induced map

$$D_{U_0,\gamma,\delta}(T) \longrightarrow (D_{\mathcal{X},\gamma}(\mathcal{Y}) \times_{\mathcal{Y}} U) \times_U T \simeq D_{\mathcal{X},\gamma}(\mathcal{Y}) \times_{\mathcal{Y},r} T \tag{2.4.1.7}$$

is an isomorphism, where $r : T \to \mathcal{Y}$ is the composite of $\tilde{r}$ with the given map $U \to \mathcal{Y}$. Observe that this isomorphism with $D_{\mathcal{X},\gamma}(\mathcal{Y}) \times_{\mathcal{Y}} T$ depends only on the map r and not on the lifting $\tilde{r}$.

Applying the same discussion to (U', T', δ') we see that the choice of any retraction $r' : T' \to \mathcal{Y}$ induces an isomorphism

$$D_{U_0',\gamma,\delta'}(T') \simeq D_{\mathcal{X},\gamma}(\mathcal{Y}) \times_{\mathcal{Y}} T'. \tag{2.4.1.8}$$

In particular choosing an retraction $r : T \to \mathcal{Y}$ and setting r' equal to the composite $r \circ t$ the map (2.4.1.5) is identified with the map of rings induced by the projection

$$D_{\mathcal{X},\gamma}(\mathcal{Y}) \times_{\mathcal{Y},r'} T' \simeq (D_{\mathcal{X},\gamma}(\mathcal{Y}) \times_{\mathcal{Y},r} T) \times_T T' \longrightarrow D_{\mathcal{X},\gamma}(\mathcal{Y}) \times_{\mathcal{Y},r} T. \tag{2.4.1.9}$$

This implies the proposition. □

Remark 2.4.2. — We do not know if 2.4.1 holds for closed immersions $\mathcal{X} \hookrightarrow \mathcal{Y}$ with $\mathcal{Y} \to \mathcal{S}$ not necessarily smooth. For Deligne-Mumford stacks, however, this more general result does hold. That is, let $i : \mathcal{X} \hookrightarrow \mathcal{Y}$ be a closed immersion of $\mathcal{S}$-stacks which are Deligne-Mumford such that γ extends to $\mathcal{X}$ and $\mathcal{Y}$. Then $i_{\mathrm{cris}*}E$ is a crystal on $\mathrm{Cris}(\mathcal{Y}_{\mathrm{et}}/\mathcal{S})$.

The proof proceeds as in the proof of 2.4.1 with (U', T', δ') and (U, T, δ) in $\mathrm{Cris}(\mathcal{Y}_{\mathrm{et}}/\mathcal{S})$ except for the proof that (2.4.1.5) is an isomorphism. For this note that since U' and U are both étale over $\mathcal{Y}$, the map $U' \to U$ is étale. By the invariance of the étale site under infinitesimal thickenings [**15**, IV.18.1.2] there exists a unique étale lifting $\widetilde{T} \to T$ of U'. By 1.2.3 the base change $\widetilde{T} \times_T D_{U_0,\gamma,\delta}(T)$ is isomorphic to the divided power envelope $D_{U_0',\gamma,\delta}(\widetilde{T})$. Since the morphism $\widetilde{T} \to T$ is étale and $U' \hookrightarrow T'$ is a nil-immersion there exists a unique factorization of t through $\widetilde{T}$. This reduces the proof to the case of $(U', T', \delta') \to (U', \widetilde{T}, \delta)$ which follows from [**7**, I.2.8.2].

Corollary 2.4.3. — *With notation as in 2.4.1, assume in addition that $\mathcal{Y} \to \mathcal{S}$ is representable. Then the natural maps*
(2.4.3.1)
$$D_{\mathcal{X},\gamma}(\mathcal{Y} \times_{\mathcal{S}} \mathcal{Y}) \longrightarrow D_{\mathcal{X},\gamma}(\mathcal{Y}) \times_{\mathcal{Y},\mathrm{pr}_1} D_{\mathcal{Y},\gamma}(1), \ D_{\mathcal{X},\gamma}(\mathcal{Y} \times_{\mathcal{S}} \mathcal{Y}) \longrightarrow D_{\mathcal{X},\gamma}(\mathcal{Y}) \times_{\mathcal{Y},\mathrm{pr}_2} D_{\mathcal{Y},\gamma}(1)$$

are isomorphisms.

Proof. — Since the formation of PD-envelopes commutes with flat base change of the base (1.2.3), it suffices to prove the corollary after making a smooth base change $S \to \mathcal{S}$. It therefore suffices to consider the case when $\mathcal{X}$, $\mathcal{Y}$, and $\mathcal{S}$ are all algebraic spaces which follows from the classical theory [**8**, 6.3]. □

2.4.4. — Let $i : \mathcal{X} \hookrightarrow \mathcal{Y}$ be as at the beginning of 2.4, and assume in addition that $\mathcal{Y} \to \mathcal{S}$ is representable. Let $\mathcal{D}_{\mathcal{X},\gamma}(\mathcal{Y})$ denote the quasi-coherent sheaf algebras on $\mathcal{Y}_{\text{lis-et}}$ defining the affine stack $D_{\mathcal{X},\gamma}(\mathcal{Y})$ over $\mathcal{Y}$. The isomorphisms (2.4.3.1) induce an isomorphism

$$\epsilon : \mathrm{pr}_2^* \mathcal{D}_{\mathcal{X},\gamma}(\mathcal{Y}) \simeq \mathrm{pr}_1^* \mathcal{D}_{\mathcal{X},\gamma}(\mathcal{Y}) \tag{2.4.4.1}$$

of quasi-coherent sheaves on $D_{\mathcal{Y}}(1)$.

Lemma 2.4.5. — *The isomorphism ϵ defines an HPD-stratification on $\mathcal{D}_{\mathcal{X},\gamma}(\mathcal{Y})$.*

Proof. — We have to show that the diagram (2.3.7.4) commutes and that ϵ reduces to the identity on $\mathcal{X}$. Using the fact that the formation of PD-envelopes commutes with flat base change, we reduce as in the proof of 2.4.3 to the case when $\mathcal{X}$, $\mathcal{Y}$, and $\mathcal{S}$ are all algebraic spaces in which case the result follows from the classical theory [**7**, IV.1.3.5]. □

2.4.6. — As in the classical theory this HPD stratification ϵ on $\mathcal{D}_{\mathcal{X},\gamma}(\mathcal{Y})$ enables us to describe the category of crystals in $(\mathcal{X}_{\text{lis-et}}/\mathcal{Y})_{\text{cris}}$ in terms of $\mathcal{D}_{\mathcal{X},\gamma}(\mathcal{Y})$-modules on $\mathcal{Y}_{\text{lis-et}}$ with HPD-stratification.

For this let E denote a quasi-coherent crystal in $(\mathcal{X}_{\text{lis-et}}/\mathcal{S})_{\text{cris}}$, and let $E_{D_{\mathcal{X},\gamma}(\mathcal{Y})}$ denote the quasi-coherent sheaf on $D_{\mathcal{X},\gamma}(\mathcal{Y})_{\text{lis-et}}$ obtained by evaluating E. Since $D_{\mathcal{X},\gamma}(\mathcal{Y})$ is affine over $\mathcal{Y}$ we also view this as a quasi-coherent sheaf on $\mathcal{Y}_{\text{lis-et}}$ with a $\mathcal{D}_{\mathcal{X},\gamma}(\mathcal{Y})$-module structure. The sheaf $E_{D_{\mathcal{X},\gamma}}(\mathcal{Y})$ has a canonical HPD-stratification as an $\mathcal{O}_{\mathcal{Y}_{\text{lis-et}}}$-module compatible with the HPD-stratification ϵ on $\mathcal{D}_{\mathcal{X},\gamma}(\mathcal{Y})$. Indeed the isomorphisms (2.4.3.1) show that $\mathrm{pr}_i^* E_{D_{\mathcal{X},\gamma}(\mathcal{Y})}$ on $D_{\mathcal{Y}}(1)$ is canonically isomorphic to the quasi-coherent sheaf $E_{D_{\mathcal{X},\gamma}(\mathcal{Y} \times_{\mathcal{Y}} \mathcal{Y})}$ obtained by evaluating E on $D_{\mathcal{X},\gamma}(\mathcal{Y} \times_{\mathcal{S}} \mathcal{Y})$. These isomorphisms induce an isomorphism $\mathrm{pr}_2^* \mathcal{E}_{D_{\mathcal{X},\gamma}(\mathcal{Y})} \simeq \mathrm{pr}_1^* \mathcal{E}_{D_{\mathcal{X},\gamma}}(\mathcal{Y})$ compatible with the isomorphism ϵ. That this defines a stratification on $E_{D_{\mathcal{X},\gamma}(\mathcal{Y})}$ follows as in the proof of 2.4.5 from the classical theory.

Theorem 2.4.7. — *The preceding construction $E \mapsto E_{D_{\mathcal{X},\gamma}(\mathcal{Y})}$ induces an equivalence of categories between the category of quasi-coherent crystals in $(\mathcal{X}_{\text{lis-et}}/\mathcal{S})_{\text{cris}}$ and the category of quasi-coherent sheaves of $\mathcal{D}_{\mathcal{X},\gamma}(\mathcal{Y})$-modules on $\mathcal{Y}_{\text{lis-et}}$ with HPD-stratification compatible with the canonical HPD-stratification on $\mathcal{D}_{\mathcal{X},\gamma}(\mathcal{Y})$.*

Proof. — The construction of the functor $E \mapsto E_{D_{\mathcal{X},\gamma}(\mathcal{Y})}$ is functorial with respect to diagrams

$$\begin{array}{ccc} Y & \longrightarrow & \mathcal{Y} \\ \downarrow & & \downarrow \\ S & \longrightarrow & \mathcal{S} \end{array} \tag{2.4.7.1}$$

with Y/S smooth. In particular, it suffices to consider the case when $\mathcal{X}$ is an algebraic space since quasi-coherent crystals and modules with HPD-stratification can be constructed locally in the smooth topology on $\mathcal{Y}$ (2.3.10). Furthermore, by 2.3.8 and 2.1.11 it suffices to prove the theorem after making a smooth base change $S \to \mathcal{S}$ with S an algebraic space. This reduces the proof to the case when $\mathcal{X}$, $\mathcal{Y}$, and $\mathcal{S}$ are all algebraic spaces which follows from [**7**, IV.1.6.3]. □

2.4.8. — For a quasi-coherent sheaf $\mathcal{E}$ on $\mathcal{Y}_{\text{lis-et}}$, let $\mathcal{L}_{\mathcal{Y}}(\mathcal{E})$ denote the sheaf $\mathrm{pr}_{1*}\mathrm{pr}_2^*\mathcal{E}$, where $\mathrm{pr}_i : D_{\mathcal{Y}}(1) \to \mathcal{Y}$ denotes the projection. The sheaf $\mathcal{L}_{\mathcal{Y}}(\mathcal{E})$ has a canonical HPD-stratification defined as follows. For $i = 1, 2$ there is a commutative diagram (2.4.8.1)

$$\begin{array}{ccccccc} D_{\mathcal{Y},\gamma}(\mathcal{Y} \times_{\mathcal{S}} \mathcal{Y} \times_{\mathcal{S}} \mathcal{Y}) & \xrightarrow{\mathrm{pr}_{12} \times \mathrm{pr}_{i3}} & D_{\mathcal{Y}}(1) \times_{\mathrm{pr}_i, \mathcal{Y}, \mathrm{pr}_1} D_{\mathcal{Y}}(1) & \xrightarrow{\mathrm{pr}_2} & D_{\mathcal{Y}}(1) & \xrightarrow{\mathrm{pr}_2} & \mathcal{Y} \\ & & \downarrow{\scriptstyle \mathrm{pr}_1} & & \downarrow{\scriptstyle \mathrm{pr}_1} & & \\ & & D_{\mathcal{Y}}(1) & \xrightarrow{\mathrm{pr}_i} & \mathcal{Y}. & & \end{array}$$

Let $q_3 : D_{\mathcal{Y},\gamma}(\mathcal{Y} \times_{\mathcal{S}} \mathcal{Y} \times_{\mathcal{S}} \mathcal{Y}) \to \mathcal{Y}$ denote the map induced by the projection onto the third coordinate $\mathcal{Y} \times_{\mathcal{S}} \mathcal{Y} \times_{\mathcal{S}} \mathcal{Y} \to \mathcal{Y}$. Then the sheaf $\mathrm{pr}_i^*\mathcal{L}_{\mathcal{Y}}(\mathcal{E})$ on $D_{\mathcal{Y}}(1)$ is canonically isomorphic to the sheaf $\mathrm{pr}_{12*}q_3^*\mathcal{E}$, where $\mathrm{pr}_{12} : D_{\mathcal{Y},\gamma}(\mathcal{Y} \times_{\mathcal{S}} \mathcal{Y} \times_{\mathcal{S}} \mathcal{Y}) \to D_{\mathcal{Y}}(1)$ is the map induced by the projection onto the first two factors. This holds for $i = 1, 2$ so in particular there is a canonical isomorphism $\tilde{\epsilon} : \mathrm{pr}_2^*\mathcal{L}_{\mathcal{Y}}(\mathcal{E}) \simeq \mathrm{pr}_1^*\mathcal{L}_{\mathcal{Y}}(\mathcal{E})$. As above, to verify that this defines an HPD-stratification it suffices to verify that this is true after making a smooth base change $S \to \mathcal{S}$, and hence this holds by the classical theory [**8**, 6.9].

We write $L_{\mathcal{Y}}(\mathcal{E})$ for the crystal in $(\mathcal{Y}_{\text{lis-et}}/\mathcal{S})_{\text{cris}}$ corresponding via 2.4.7 to the module with HPD-stratification $\mathcal{L}_{\mathcal{Y}}(\mathcal{E})$, and write $i^*L_{\mathcal{Y}}(\mathcal{E})$ for the crystal in the topos $(\mathcal{X}_{\text{lis-et}}/\mathcal{S})_{\text{cris}}$ obtained by pulling back $L_{\mathcal{Y}}(\mathcal{E})$ to $\mathcal{X}/\mathcal{S}$.

2.4.9. — The cohomology of the crystal $i^*L_{\mathcal{Y}}(\mathcal{E})$ can be computed as follows. Let $\tilde{u}_{\mathcal{X}_{\text{lis-et}}/\mathcal{S}*}$ be the composite of the functor $u_{\mathcal{X}_{\text{lis-et}}/\mathcal{S}*}$ with the functor i_* from the category of sheaves on $\mathcal{X}_{\text{lis-et}}$ to sheaves on $\mathcal{Y}_{\text{lis-et}}$. Observe that since i_* is an exact functor, for any abelian sheaf $F \in (\mathcal{X}_{\text{lis-et}}/\mathcal{S})_{\text{cris}}$ and $s \geq 0$ we have $R^s\tilde{u}_{\mathcal{X}_{\text{lis-et}}/\mathcal{S}*}(F) \simeq i_*R^su_{\mathcal{X}_{\text{lis-et}}/\mathcal{S}*}(F)$.

Proposition 2.4.10. — *Let $\pi : D_{\mathcal{X},\gamma}(\mathcal{Y}) \to \mathcal{Y}$ be the projection, and let $\mathcal{E}$ be a quasi-coherent sheaf of $\mathcal{O}_{\mathcal{Y}_{\text{lis-et}}}$-modules. Then $R^s\tilde{u}_{\mathcal{X}_{\text{lis-et}}/\mathcal{S}*}(i^*L_{\mathcal{Y}}(\mathcal{E})) = 0$ for $s > 0$ and $\tilde{u}_{\mathcal{X}_{\text{lis-et}}/\mathcal{S}*}(i^*L_{\mathcal{Y}}(\mathcal{E})) \simeq \pi_*\pi^*\mathcal{E}$.*

The proof is in several steps 2.4.11–2.4.15.

2.4.11. — Let $\widetilde{\mathcal{Y}}$ denote the sheaf on $(\mathcal{X}_{\text{lis-et}}/\mathcal{S})_{\text{cris}}$ which associates to any $(U, T, \delta) \in \text{Cris}(\mathcal{X}_{\text{lis-et}}/\mathcal{S})$ the set of sections of the map of algebraic spaces (since $\mathcal{Y} \to \mathcal{S}$ is representable) $T \times_{\mathcal{S}} \mathcal{Y} \to T$, and consider the localized topos $(\mathcal{X}_{\text{lis-et}}/\mathcal{S})_{\text{cris}}|_{\widetilde{\mathcal{Y}}}$.

Recall [**5**, III.5.4] that $(\mathcal{X}_{\text{lis-et}}/\mathcal{S})_{\text{cris}}|_{\widetilde{\mathcal{Y}}}$ is equivalent to the topos associated to the site $\text{Cris}(\mathcal{X}_{\text{lis-et}}/\mathcal{S})|_{\widetilde{\mathcal{Y}}}$ whose objects are pairs $(j : U \hookrightarrow T, r)$, where $(j : U \hookrightarrow T) \in \text{Cris}(\mathcal{X}_{\text{lis-et}}/\mathcal{S})$ and $r : T \to \mathcal{Y}$ is a morphism over $\mathcal{S}$ such that $r \circ j : U \to \mathcal{Y}$ is the given map $U \to \mathcal{X} \to \mathcal{Y}$.

As in the classical case, there is a morphism of topoi

$$j_{\widetilde{\mathcal{Y}}} : (\mathcal{X}_{\text{lis-et}}/\mathcal{S})_{\text{cris}}|_{\widetilde{\mathcal{Y}}} \longrightarrow (\mathcal{X}_{\text{lis-et}}/\mathcal{S})_{\text{cris}} \tag{2.4.11.1}$$

for which $j_{\widetilde{\mathcal{Y}}}^*$ sends a sheaf $F \in (\mathcal{X}_{\text{lis-et}}/\mathcal{S})_{\text{cris}}$ to the sheaf $(T, r) \mapsto F(T)$. The pushforward $j_{\widetilde{\mathcal{Y}}*}$ sends a sheaf $G \in (\mathcal{X}_{\text{lis-et}}/\mathcal{S})_{\text{cris}}|_{\widetilde{\mathcal{Y}}}$ to the sheaf which associates to $(U, T, \delta) \in \text{Cris}(\mathcal{X}_{\text{lis-et}}/\mathcal{S})$ the value of G on the object of $\text{Cris}(\mathcal{X}_{\text{lis-et}}/\mathcal{S})|_{\widetilde{\mathcal{Y}}}$ given by

$$D_{U,\delta,\epsilon}(T \times_{\mathcal{S}} D_{\mathcal{X},\gamma}(\mathcal{Y})) \longrightarrow D_{\mathcal{X},\gamma}(\mathcal{Y}), \tag{2.4.11.2}$$

where ϵ denotes the divided power structure on the ideal of $\mathcal{X}$ in $D_{\mathcal{X},\gamma}(\mathcal{Y})$.

Lemma 2.4.12. — *Let $\mathcal{E}$ be a quasi-coherent sheaf on $\mathcal{Y}_{\text{lis-et}}$, and let $\mathcal{E}^\sharp$ be the sheaf in $(\mathcal{X}_{\text{lis-et}}/\mathcal{S})_{\text{cris}}|_{\widetilde{\mathcal{Y}}}$ which to any $(T, r : T \to \mathcal{Y})$ associates the global sections of the quasi-coherent sheaf $r^*\mathcal{E}$ on T_{et}. Then there is a natural isomorphism $i^*L_{\mathcal{Y}}(\mathcal{E}) \simeq j_{\widetilde{\mathcal{Y}}*}\mathcal{E}^\sharp$.*

Proof. — By adjunction, to give an arrow $i^*L_{\mathcal{Y}}(\mathcal{E}) \to j_{\widetilde{\mathcal{Y}}*}\mathcal{E}^\sharp$ is equivalent to giving an arrow $j_{\widetilde{\mathcal{Y}}}^* i^* L_{\mathcal{Y}}(\mathcal{E}) \to \mathcal{E}^\sharp$. The sheaf $j_{\widetilde{\mathcal{Y}}}^* i^* L_{\mathcal{Y}}(\mathcal{E})$ associates to any $(T, r : T \to \mathcal{Y})$ the global sections of $r^*\mathcal{L}_{\mathcal{Y}}(\mathcal{E})$. Thus what is needed is a map $r^*\mathcal{L}_{\mathcal{Y}}(\mathcal{E}) \to r^*\mathcal{E}$. In fact there is a canonical map $\mathcal{L}_{\mathcal{Y}}(\mathcal{E}) \to \mathcal{E}$ obtained from the diagonal map $\Delta : \mathcal{Y} \to D_{\mathcal{Y}}(1)$ which induces a map

$$\mathcal{L}_{\mathcal{Y}}(\mathcal{E}) = \text{pr}_{1*}\text{pr}_2^*\mathcal{E} \xrightarrow{\ \Delta^*\ } \text{pr}_{1*}\Delta_*\mathcal{E} \simeq \mathcal{E}. \tag{2.4.12.1}$$

This defines a map $i^*L_{\mathcal{Y}}(\mathcal{E}) \to j_{\widetilde{\mathcal{Y}}*}\mathcal{E}^\sharp$.

To prove that this map is an isomorphism, one reduces as in the classical [**8**, 6.10] to showing that the natural map

$$D_{U,\gamma}(T \times_{\mathcal{S}} \mathcal{Y}) \longrightarrow T \times_{r,\mathcal{Y},\text{pr}_1} D_{\mathcal{X},\gamma}(\mathcal{Y}) \tag{2.4.12.2}$$

is an isomorphism. Since the formation of divided power envelopes commutes with smooth base change $S \to \mathcal{S}$ (1.2.3), it suffices to verify that after making such a base change. In particular, we may assume that $\mathcal{S}$, and hence also $\mathcal{X}$ and $\mathcal{Y}$, is an algebraic space. In this case the result follows from the classical theory [**8**, 6.10]. □

Lemma 2.4.13. — *Let $\mathcal{E}$ be a quasi-coherent sheaf and $s > 0$ and integer. Then* $R^s j_{\widetilde{\mathcal{Y}}*}(\mathcal{E}^\sharp) = 0$.

Proof. — We show that for any object $(U, T, \delta) \in \mathrm{Cris}(\mathcal{X}_{\text{lis-et}}/\mathcal{S})$ and $s > 0$ the sheaf $R^s j_{\tilde{\mathcal{Y}}*}(\mathcal{E}^\sharp)|_{T_{\text{et}}}$ is zero.

By [**5**, V.5.1], to prove that $R^s j_{\tilde{\mathcal{Y}}*}(\mathcal{E}^\sharp)|_{T_{\text{et}}}$ is zero, it suffices to show that for any (U, T, δ) with T affine, the group

$$H^s(((\mathcal{X}_{\text{lis-et}}/\mathcal{S})_{\text{cris}}|_{\tilde{\mathcal{Y}}})|_{j^*_{\tilde{\mathcal{Y}}}\tilde{T}}, \mathcal{E}^\sharp) \tag{2.4.13.1}$$

is zero, where $\tilde{T}$ denotes the sheaf represented by T. The sheaf $j^*_{\tilde{\mathcal{Y}}}\tilde{T}$ is represented by the divided power envelope $D_{U,\gamma}(T \times_{\mathcal{S}} \mathcal{Y})$ with its natural projection pr_2 to $\mathcal{Y}$, and hence (2.4.13.1) is isomorphic to

$$H^s((\mathcal{X}_{\text{lis-et}}/\mathcal{S})_{\text{cris}}|_{D_{U,\gamma}(T \times_{\mathcal{S}} \mathcal{Y})}, \mathcal{E}^\sharp). \tag{2.4.13.2}$$

Let

$$\varphi : (\mathcal{X}_{\text{lis-et}}/\mathcal{S})_{\text{cris}}|_{D_{U,\gamma}(T \times_{\mathcal{S}} \mathcal{Y})} \longrightarrow D_{U,\gamma}(T \times_{\mathcal{S}} \mathcal{Y})_{\text{et}} \tag{2.4.13.3}$$

be the morphism of topoi for which φ_* sends a sheaf in $(\mathcal{X}_{\text{lis-et}}/\mathcal{S})_{\text{cris}}|_{D_{U,\gamma}(T \times_{\mathcal{S}} \mathcal{Y})}$ to its restriction to $D_{U,\gamma}(T \times_{\mathcal{S}} \mathcal{Y})_{\text{et}}$, and for which φ^{-1} sends a sheaf $G \in D_{U,\gamma}(T \times_{\mathcal{S}} \mathcal{Y})_{\text{et}}$ to the sheaf which to any pair (T, r) consisting of an object T of $\mathrm{Cris}(\mathcal{X}_{\text{lis-et}}/\mathcal{S})$ and a morphism $r : T \to D_{U,\gamma}(T \times_{\mathcal{S}} \mathcal{Y})$ associates $\Gamma(T_{\text{et}}, r^{-1}G)$. Note that $D_{U,\gamma}(T \times_{\mathcal{S}} \mathcal{Y})_{\text{et}}$ has finite projective limits and φ^{-1} commutes with them so that this really defines a morphism of topoi. Note also that φ is naturally a morphism of ringed topoi. The sheaf $\mathcal{E}^\sharp$ restricted to $(\mathcal{X}_{\text{lis-et}}/\mathcal{S})_{\text{cris}}|_{D_{U,\gamma}(T \times_{\mathcal{S}} \mathcal{Y})}$ is isomorphic to the sheaf $\varphi^*\mathrm{pr}_2^*\mathcal{E}$, and the group (2.4.13.2) is isomorphic to the s-th cohomology group of

$$R\Gamma \circ R\varphi_*(\varphi^*\mathrm{pr}_2^*\mathcal{E}). \tag{2.4.13.4}$$

Since the restriction functor φ_* is exact, we have

$$R\varphi_*(\varphi^*\mathrm{pr}_2^*\mathcal{E}) \simeq \varphi_*(\varphi^*\mathrm{pr}_2^*\mathcal{E}) \simeq \mathrm{pr}_2^*\mathcal{E}. \tag{2.4.13.5}$$

Thus (2.4.13.2) is isomorphic to

$$H^s(D_{U,\gamma}(T \times_{\mathcal{S}} \mathcal{Y}), \mathrm{pr}_2^*\mathcal{E}). \tag{2.4.13.6}$$

Since $U \hookrightarrow D_{U,\gamma}(T \times_{\mathcal{S}} \mathcal{Y})$ is a nil-immersion and U is an affine scheme, the space $D_{U,\gamma}(T \times_{\mathcal{S}} \mathcal{Y})$ is also an affine scheme. Since $\mathcal{E}$ is quasi-coherent this implies that the group (2.4.13.6) is zero for $s > 0$. □

2.4.14. — Let

$$\psi : \{\text{abelian sheaves in } (\mathcal{X}_{\text{lis-et}}/\mathcal{S})_{\text{cris}}|_{\tilde{\mathcal{Y}}}\} \longrightarrow \{\text{abelian sheaves in } D_{\mathcal{X},\gamma}(\mathcal{Y})_{\text{lis-et}}\} \tag{2.4.14.1}$$

be the functor which sends an abelian sheaf F in $(\mathcal{X}_{\text{lis-et}}/\mathcal{S})_{\text{cris}}|_{\tilde{\mathcal{Y}}}$ to its restriction to $D_{\mathcal{X},\gamma}(\mathcal{Y})_{\text{lis-et}}$, and let $\pi : D_{\mathcal{X},\gamma}(\mathcal{Y}) \to \mathcal{Y}$ denote the projection (note that because of the usual difficulty (1.4.17) with the lisse-étale topos ψ is not obtained from a morphism of topoi). Then there is a natural isomorphism of functors $\tilde{u}_{\mathcal{X}_{\text{lis-et}}/\mathcal{S}*} \circ j_{\tilde{\mathcal{Y}}*} \simeq \pi_* \circ \psi$.

Furthermore, since $j_{\tilde{\mathcal{Y}}*}$ is obtained from a morphism of topoi it takes injective abelian sheaves to injective abelian sheaves. From 2.4.12 and 2.4.13 it therefore follows that

$$\begin{aligned} R\tilde{u}_{\mathcal{X}_{\text{lis-et}}/\mathcal{S}*}(i^*L_{\mathcal{Y}}(\mathcal{E})) &\simeq R\tilde{u}_{\mathcal{X}_{\text{lis-et}}/\mathcal{S}*}(j_{\tilde{\mathcal{Y}}*}\mathcal{E}^\sharp) \simeq R\tilde{u}_{\mathcal{X}_{\text{lis-et}}/\mathcal{S}*} \circ Rj_{\tilde{\mathcal{Y}}*}(\mathcal{E}^\sharp) \\ &\simeq R(\tilde{u}_{\mathcal{X}_{\text{lis-et}}/\mathcal{S}*} \circ j_{\tilde{\mathcal{Y}}*})(\mathcal{E}^\sharp) \simeq R(\pi_* \circ \psi)(\mathcal{E}^\sharp). \end{aligned} \tag{2.4.14.2}$$

Since $\pi_*\psi(\mathcal{E}^\sharp) \simeq \pi_*\pi^*\mathcal{E}$, the following lemma completes the proof of 2.4.10. □

Lemma 2.4.15. — *For any $s > 0$, the group $R^s(\pi_* \circ \psi)(\mathcal{E}^\sharp)$ is zero.*

Proof. — Let $V \to \mathcal{Y}$ be a smooth morphism with V a scheme. The functor which sends a sheaf $F \in (\mathcal{X}_{\text{lis-et}}/\mathcal{S})_{\text{cris}}|_{\tilde{\mathcal{Y}}}$ to $(\pi_* \circ \psi)(F)(V)$ is isomorphic to the functor $F \mapsto F(\mathcal{X} \times_{\mathcal{Y}} V \hookrightarrow V \times_{\mathcal{Y}} D_{\mathcal{X},\gamma}(\mathcal{Y}))$, where $\mathcal{X} \times_{\mathcal{Y}} V \hookrightarrow V \times_{\mathcal{Y}} D_{\mathcal{X},\gamma}(\mathcal{Y})$ is viewed as an object of $\text{Cris}(\mathcal{X}_{\text{lis-et}}/\mathcal{S})|_{\tilde{\mathcal{Y}}}$ via the projection to $\mathcal{Y}$. It follows that $R^s(\pi_* \circ \psi)(\mathcal{E}^\sharp)(V)$ is isomorphic to the sheaf associated to the presheaf

$$V \longmapsto H^s((\mathcal{X}_{\text{lis-et}}/\mathcal{S})_{\text{cris}}|_{\tilde{V}}, \mathcal{E}^\sharp). \tag{2.4.15.1}$$

Let

$$\psi_{\text{et}} : (\mathcal{X}_{\text{lis-et}}/\mathcal{S})_{\text{cris}}|_{\tilde{V}} \longrightarrow V_{\text{et}} \tag{2.4.15.2}$$

be the morphism of topoi defined as in (2.4.13.3). Then the cohomology group in (2.4.15.1) is isomorphic to

$$H^s(V_{\text{et}}, R\psi_{\text{et}*}\mathcal{E}^\sharp). \tag{2.4.15.3}$$

Since $\psi_{\text{et}*}$ is exact this is just the cohomology of the sheaf $\psi_{\text{et}*}\mathcal{E}^\sharp \simeq \pi_*\pi^*\mathcal{E}|_{V_{\text{et}}}$ which in particular is quasi-coherent. It follows that if V is affine then these groups are zero which proves the lemma. □

2.4.16. — More generally, if $\mathcal{E}$ is a quasi-coherent sheaf in $D_{\mathcal{X},\gamma}(\mathcal{Y})_{\text{lis-et}}$, then the sheaf $j_{\tilde{\mathcal{Y}}*}\mathcal{E}^\sharp$ is a quasi-coherent crystal. Here $\mathcal{E}^\sharp$ is the sheaf on $(\mathcal{X}_{\text{lis-et}}/\mathcal{S})|_{\tilde{\mathcal{Y}}}$ which to any (T, r) associates $\tilde{r}^*\mathcal{E}(T)$, where $\tilde{r} : T \to D_{\mathcal{X},\gamma}(\mathcal{Y})$ denotes the map induced by r and the universal property of $D_{\mathcal{X},\gamma}(\mathcal{Y})$. The above argument then shows the following:

Corollary 2.4.17. — *There is a natural isomorphism $\pi_*\mathcal{E} \simeq R\tilde{u}_{\mathcal{X}/\mathcal{S}*}(i^*L_{\mathcal{Y}}(\mathcal{E}))$.*

2.4.18. — Assume in addition that $\mathcal{Y}$ is a Deligne-Mumford stack, and let $\mathcal{E}$ be a quasi-coherent sheaf in $D_{\mathcal{X},\gamma}(\mathcal{Y})_{\text{et}}$. Let $i^*_{\text{cris}}L_{\mathcal{Y}}(\mathcal{E})$ denote the quasi-coherent crystal in $(\mathcal{X}_{\text{et}}/\mathcal{S})_{\text{cris}}$ corresponding via 2.1.5 to $i^*L_{\mathcal{Y}}(\mathcal{E})$. Since $\mathcal{X} \hookrightarrow D_{\mathcal{X},\gamma}(\mathcal{Y})$ is a nil-immersion, reduction defines an equivalence between the étale site of $D_{\mathcal{X},\gamma}(\mathcal{Y})$ and the étale site of $\mathcal{X}$.

Corollary 2.4.19. — *Assume $\mathcal{Y}$ is a Deligne-Mumford stack. Then there is a natural isomorphism $\pi_*\mathcal{E} \simeq Ru_{\mathcal{X}_{\text{et}}/\mathcal{S}*}i^*_{\text{cris}}L_{\mathcal{Y}}(\mathcal{E})$.*

Proof. — Let $i_{\mathrm{et}} : \mathcal{X}_{\mathrm{et}} \to \mathcal{Y}_{\mathrm{et}}$ be the morphism of topoi induced by the closed immersion i, and let $\mathrm{res} : \mathcal{Y}_{\text{lis-et}} \to \mathcal{Y}_{\mathrm{et}}$ be the restriction functor.

Since $i_{\mathrm{et}}^{-1} \circ i_{\mathrm{et}*}$ is the identity functor and $i_{\mathrm{et}*}$ is exact, it suffices to exhibit an isomorphism $i_{\mathrm{et}*}\mathcal{E} \simeq i_{\mathrm{et}*} R u_{\mathcal{X}_{\mathrm{et}}/\mathcal{S}*} i^*_{\mathrm{cris}} L_{\mathcal{Y}}(\mathcal{E})$ which in turn is equivalent to an isomorphism

$$\text{(2.4.19.1)} \qquad \mathrm{res} \circ \pi_* \mathcal{E} \longrightarrow R(i_{\mathrm{et}*} \circ u_{\mathcal{X}_{\mathrm{et}}/\mathcal{S}*}) i^*_{\mathrm{cris}} L_{\mathcal{Y}}(\mathcal{E}).$$

For this note that $i_{\mathrm{et}*} \circ u_{\mathcal{X}_{\mathrm{et}}/\mathcal{S}*} \simeq \mathrm{res} \circ \tilde{u}_{\mathcal{X}_{\text{lis-et}}/\mathcal{S}*}$ and the functor res is exact. Hence

$$\text{(2.4.19.2)} \qquad R(i_{\mathrm{et}*} \circ u_{\mathcal{X}_{\mathrm{et}}/\mathcal{S}*}) i^*_{\mathrm{cris}} L_{\mathcal{Y}}(\mathcal{E}) \simeq \mathrm{res}(R\tilde{u}_{\mathcal{X}_{\text{lis-et}}/\mathcal{S}*} i^* L_{\mathcal{Y}}(\mathcal{E})).$$

The result therefore follows from 2.4.17. □

2.4.20. — As in the classical theory, these propositions enable us to construct resolutions for quasi-coherent crystals in $(\mathcal{X}_{\text{lis-et}}/\mathcal{S})_{\mathrm{cris}}$ which are acyclic for the functor $\tilde{u}_{\mathcal{X}_{\text{lis-et}}/\mathcal{S}*}$.

Recall that $\mathcal{Y} \to \mathcal{S}$ is assumed smooth and representable. Assume further that $\mathcal{Y} \to \mathcal{S}$ is locally separated and note that the quasi-coherent sheaf $\Omega^1_{\mathcal{Y}/\mathcal{S}}$ on $\mathcal{Y}_{\text{lis-et}}$ is a locally free sheaf of finite rank. Let $\Omega^i_{\mathcal{Y}/\mathcal{S}}$ be the i-th exterior power of $\Omega^1_{\mathcal{Y}/\mathcal{S}}$.

Let E be a quasi-coherent crystal in $(\mathcal{X}_{\text{lis-et}}/\mathcal{S})_{\mathrm{cris}}$, and let $\mathcal{E}$ denote the quasi-coherent sheaf with HPD-stratification on $\mathcal{Y}_{\text{lis-et}}$ associated to E by 2.4.7.

Lemma 2.4.21. — *Let Ω be a quasi-coherent sheaf of $\mathcal{O}_{\mathcal{Y}_{\text{lis-et}}}$-modules. Then there is a canonical isomorphism $\beta : \mathcal{E} \otimes_{\mathcal{O}_{\mathcal{Y}}} \mathcal{L}_{\mathcal{Y}}(\Omega) \to \mathcal{L}_{\mathcal{Y}}(\mathcal{E} \otimes_{\mathcal{O}_{\mathcal{Y}}} \Omega)$ of $\pi_* \mathcal{O}_{D_{\mathcal{X},\gamma}(\mathcal{Y})}$-modules with HPD-stratification.*

Proof. — By definition of the functor $\mathcal{L}_{\mathcal{Y}}(-)$, to define the arrow β it suffices to construct a map

$$\text{(2.4.21.1)} \qquad \mathcal{E} \otimes_{\mathcal{O}_{\mathcal{Y}}} \mathrm{pr}_{1*}\mathrm{pr}_2^*\Omega \longrightarrow \mathrm{pr}_{1*}\mathrm{pr}_2^*(\mathcal{E} \otimes \Omega).$$

For this note that

$$\text{(2.4.21.2)} \qquad \mathcal{E} \otimes_{\mathcal{O}_{\mathcal{Y}}} \mathrm{pr}_{1*}\mathrm{pr}_2^*\Omega \simeq (\mathcal{E} \otimes_{\mathcal{O}_{\mathcal{Y}}} \mathrm{pr}_{1*}(\mathcal{O}_{D_{\mathcal{X},\gamma}(\mathcal{Y})})) \otimes_{\mathrm{pr}_{1*}(\mathcal{O}_{D_{\mathcal{X},\gamma}(\mathcal{Y})})} \mathrm{pr}_{1*}\mathrm{pr}_2^*\Omega.$$

Thus the desired arrow follow from observing that the stratification on $\mathcal{E}$ gives an isomorphism

$$\text{(2.4.21.3)} \qquad \mathcal{E} \otimes_{\mathcal{O}_{\mathcal{Y}}} \mathrm{pr}_{1*}(\mathcal{O}_{D_{\mathcal{X},\gamma}(\mathcal{Y})}) \simeq \mathrm{pr}_{1*}\mathrm{pr}_2^*\mathcal{E}.$$

That the resulting morphism β is an isomorphism and is compatible with the HPD-stratifications can by 2.3.8 be verified after smooth base change $S \to \mathcal{S}$ in which case it follows from the classical theory [**8**, 6.15]. □

2.4.22. — For $i \geq 0$, the differential $d : \Omega^i_{\mathcal{Y}/\mathcal{S}} \to \Omega^{i+1}_{\mathcal{Y}/\mathcal{S}}$ induces a map $\mathcal{L}_{\mathcal{Y}}(\Omega^i_{\mathcal{Y}/\mathcal{S}}) \to \mathcal{L}_{\mathcal{Y}}(\Omega^{i+1}_{\mathcal{Y}/\mathcal{S}})$ of modules with HPD-stratification. A more direct definition of this map in the case when $\mathcal{Y}$ is Deligne-Mumford is given in 2.4.25 below. For general $\mathcal{Y}$ the map can be constructed as follows. By 2.3.8 to construct such a map it suffices to construct it after making a smooth base change $S \to \mathcal{S}$ with S an algebraic space. In this case the map is constructed in [**8**, 6.13]. From *loc. cit.* it also follows that the composite

$$\mathcal{L}_{\mathcal{Y}}(\Omega^i_{\mathcal{Y}/\mathcal{S}}) \longrightarrow \mathcal{L}_{\mathcal{Y}}(\Omega^{i+1}_{\mathcal{Y}/\mathcal{S}}) \longrightarrow \mathcal{L}_{\mathcal{Y}}(\Omega^{i+2}_{\mathcal{Y}/\mathcal{S}}) \tag{2.4.22.1}$$

is zero, so we obtain a complex $\mathcal{L}_{\mathcal{Y}}(\Omega^\bullet_{\mathcal{Y}/\mathcal{S}})$ in the category of $\mathcal{O}_{\mathcal{Y}}$-modules with HPD-stratification. Furthermore, by [**8**, 6.12] there is a canonical quasi-isomorphism $\mathcal{O}_{\mathcal{Y}_{\text{lis-et}}} \to \mathcal{L}_{\mathcal{Y}}(\Omega^\bullet_{\mathcal{Y}/\mathcal{S}})$ in the category of complexes of sheaves of $\mathcal{O}_{\mathcal{Y}_{\text{lis-et}}}$-modules with HPD-stratification.

For a quasi-coherent crystal E with associated module with HPD stratification $\mathcal{E}$, we can tensor the quasi-isomorphism $\mathcal{O}_{\mathcal{Y}_{\text{lis-et}}} \to \mathcal{L}_{\mathcal{Y}}(\Omega^\bullet_{\mathcal{Y}/\mathcal{S}})$ with $\mathcal{E}$ to obtain a morphism of complexes $\mathcal{E} \to \mathcal{E} \otimes \mathcal{L}_{\mathcal{Y}}(\Omega^\bullet_{\mathcal{Y}/\mathcal{S}})$ of modules with HPD stratification. Applying the equivalence (2.4.7) we obtain a morphism of complexes of crystals $E \to E \otimes i^*L_{\mathcal{Y}}(\Omega^\bullet_{\mathcal{Y}/\mathcal{S}})$.

Proposition 2.4.23. — *The map $E \to E \otimes i^*L_{\mathcal{Y}}(\Omega^\bullet_{\mathcal{Y}/\mathcal{S}})$ is a resolution of E in the category of $\mathcal{O}_{\mathcal{X}_{\text{lis-et}}/\mathcal{S}}$-modules in $(\mathcal{X}_{\text{lis-et}}/\mathcal{S})_{\text{cris}}$.*

Proof. — Let $(U, T, \delta) \in \text{Cris}(\mathcal{X}_{\text{lis-et}}/\mathcal{S})$ be an object, and let $S \to \mathcal{S}$ be a smooth surjection with S an algebraic space. Let $\mathcal{X}_S$ and $\mathcal{Y}_S$ the algebraic spaces obtained by base change, and let $(U_S, T_S, \delta) \in \text{Cris}(\mathcal{X}_{S,\text{lis-et}}/S)$ be the object obtained from (U, T, δ) by base change to S. Since the morphism $h : T_S \to T$ is faithfully flat, to verify that the morphism of complexes $E_T \to E_T \otimes i^*L_{\mathcal{Y}}(\Omega^\bullet_{\mathcal{Y}/\mathcal{S}})_T$ of sheaves of $\mathcal{O}_{T_{\text{et}}}$-modules is a quasi-isomorphism it suffices to verify that it becomes a quasi-isomorphism after pulling back to T_S. On the other hand, the pullback $h^*E_T \to h^*(E_T \otimes i^*L_{\mathcal{Y}}(\Omega^\bullet_{\mathcal{Y}/\mathcal{S}})_T)$ is canonically isomorphic to the morphism of complexes obtained by applying the above construction replacing $\mathcal{S}$ by S, $\mathcal{Y}$ by $\mathcal{Y}_S$, and $\mathcal{X}$ by $\mathcal{X}_S$. It follows that to prove the proposition it suffices to consider the case when $\mathcal{S}$, $\mathcal{X}$, and $\mathcal{Y}$ are all algebraic spaces.

Assuming this, we can further replace U by an étale cover and hence may assume that U, and hence also T, is an affine scheme. In this case there exists a map $r : T \to \mathcal{Y}$ such that the diagram

$$\begin{array}{ccc} U & \longrightarrow & T \\ \downarrow & & \downarrow r \\ \mathcal{X} & \longrightarrow & \mathcal{Y} \end{array} \tag{2.4.23.1}$$

commutes. From this it follows that the morphism $E_T \to E_T \otimes i^*L_{\mathcal{Y}}(\Omega^\bullet_{\mathcal{Y}/\mathcal{S}})_T$ is isomorphic to the morphism of complexes obtained by tensoring $r^*(\mathcal{O}_{\mathcal{Y}} \to \mathcal{L}_{\mathcal{Y}}(\Omega^\bullet_{\mathcal{Y}/\mathcal{S}}))$ with E_T. It follows from the construction of $\mathcal{L}_{\mathcal{Y}}(\Omega^\bullet_{\mathcal{Y}/\mathcal{S}})$ and 2.3.3 that $\mathcal{L}_{\mathcal{Y}}(\Omega^\bullet_{\mathcal{Y}/\mathcal{S}})$ is a complex of flat $\mathcal{O}_{\mathcal{Y}}$-modules, and hence to prove that $E_T \to E_T \otimes i^* L_{\mathcal{Y}}(\Omega^\bullet_{\mathcal{Y}/\mathcal{S}})_T$ is a quasi-isomorphism it suffices to show that $\mathcal{O}_{\mathcal{Y}} \to \mathcal{L}_{\mathcal{Y}}(\Omega^\bullet_{\mathcal{Y}/\mathcal{S}})$ is a quasi-isomorphism. This follows from [**8**, 6.12]. □

Remark 2.4.24. — The construction of the resolution in 2.4.23 is functorial with respect to 2-commutative diagrams

(2.4.24.1)
$$\begin{array}{ccc} \mathcal{X}' & \xrightarrow{i'} & \mathcal{Y}' \\ {\scriptstyle g}\downarrow & & \downarrow{\scriptstyle \tilde{g}} \\ \mathcal{X} & \xrightarrow{i} & \mathcal{Y} \end{array}$$

of algebraic stacks over $\mathcal{S}$, where i and i' are closed immersions and $\mathcal{Y}' \to \mathcal{S}$ and $\mathcal{Y} \to \mathcal{S}$ are representable, locally separated, and smooth. Precisely, there is a canonical morphism of complexes of crystals

(2.4.24.2)
$$g^*(E \to E \otimes i^* L_{\mathcal{Y}}(\Omega^\bullet_{\mathcal{Y}/\mathcal{S}})) \longrightarrow (g^*E \to g^*E \otimes i'^* L_{\mathcal{Y}'}(\Omega^\bullet_{\mathcal{Y}'/\mathcal{S}})).$$

Note here that since $\mathcal{Y} \to \mathcal{S}$ is representable, the map $\mathcal{X} \to \mathcal{S}$ is also representable and hence pullback of quasi-coherent crystals is well-behaved (2.1.3).

2.4.25. — If in the above the stack $\mathcal{Y}$ is a Deligne-Mumford stack, then the differentials in the complex $i^* L_{\mathcal{Y}}(\Omega^\bullet_{\mathcal{Y}/\mathcal{S}})$ can be described as in the classical case. An HPD-differential operator $\rho : \mathrm{pr}_{1*}\mathrm{pr}_2^*\mathcal{E} \to \mathcal{F}$ induces a map

(2.4.25.1)
$$\mathcal{L}_{\mathcal{Y}}(\rho) : \mathcal{L}_{\mathcal{Y}}(\mathcal{E}) \longrightarrow \mathcal{L}_{\mathcal{Y}}(\mathcal{F})$$

of $\mathcal{O}_{\mathcal{Y}_{\mathrm{et}}}$-modules with HPD-stratification. Since $\Delta : \mathcal{Y} \hookrightarrow D_{\mathcal{Y}}(1)$ is a nil-immersion, reduction defines an equivalence between the étale site of $D_{\mathcal{Y}}(1)$ and the étale site of $\mathcal{Y}$. In particular, the coordinate ring $\mathcal{D}_{\mathcal{Y}}(1)$ of $D_{\mathcal{Y}}(1)$ can be viewed as a sheaf on $\mathcal{Y}_{\mathrm{et}}$. The two-projections $D_{\mathcal{Y}}(1) \to \mathcal{Y}$ give $\mathcal{D}_{\mathcal{Y}}(1)$ the structure of a $\mathcal{O}_{\mathcal{Y}_{\mathrm{et}}}$-bi-algebra. The map $\mathcal{L}_{\mathcal{Y}}(\rho)$ is then the composite

(2.4.25.2)
$$\mathcal{D}_{\mathcal{Y}}(1) \otimes \mathcal{E} \xrightarrow{\delta \otimes \mathrm{id}} \mathcal{D}_{\mathcal{Y}}(1) \otimes \mathcal{D}_{\mathcal{Y}}(1) \otimes \mathcal{E} \xrightarrow{\mathrm{id} \otimes \rho} \mathcal{D}_{\mathcal{Y}}(1) \otimes \mathcal{F}.$$

In particular, if $(\mathcal{E}, \nabla)$ denotes the module with integrable connection on $\mathcal{Y}_{\mathrm{et}}$ corresponding to the crystal E by 2.3.28, and if $\mathcal{E} \otimes \Omega^\bullet_{\mathcal{Y}_{\mathrm{et}}/\mathcal{S}}$ denotes the de Rham-complex, then each map $\nabla_i : \mathcal{E} \otimes \Omega^i_{\mathcal{Y}_{\mathrm{et}}/\mathcal{S}} \to \mathcal{E} \otimes \Omega^{i+1}_{\mathcal{Y}_{\mathrm{et}}/\mathcal{S}}$ "is" a differential operator of order 1 (2.3.14). The differential $E \otimes i^* L_{\mathcal{Y}}(\Omega^i_{\mathcal{Y}/\mathcal{S}}) \to E \otimes i^* L_{\mathcal{Y}}(\Omega^{i+1}_{\mathcal{Y}/\mathcal{S}})$ is the map induced by this differential operator and the isomorphism in 2.4.21.

2.5. The cohomology of crystals

2.5.1. — Let $(\mathcal{S}, I, \gamma)$ be a PD-stack and let $\mathcal{X} \hookrightarrow \mathcal{Y}$ be a closed immersion of algebraic stacks over $\mathcal{S}$ such that γ extends to $\mathcal{X}$ and $\mathcal{Y}$. Assume further that the morphism $\mathcal{Y} \to \mathcal{S}$ is representable, locally separated, and smooth. Let $Y \to \mathcal{Y}$ be a smooth surjection with Y an algebraic space, and let $Y_\bullet$ denote the 0-coskeleton. Write $Y_\bullet^+$ for the associated strictly simplicial space. Let E be a quasi-coherent crystal in $(\mathcal{X}_{\text{lis-et}}/\mathcal{S})_{\text{cris}}$, and let $\mathcal{E}$ denote the quasi-coherent sheaf of $\mathcal{O}_{\mathcal{Y}_{\text{lis-et}}}$-modules obtained by pushing forward to $\mathcal{Y}$ the value $E_{D_{\mathcal{X},\gamma}(\mathcal{Y})}$ of E on the divided power envelope of $\mathcal{X}$ in $\mathcal{Y}$. By 2.4.6, the sheaf $\mathcal{E}$ has a canonical HPD-stratification. Let $\mathcal{E}_\bullet$ denote the sheaf of $\mathcal{O}_{Y^+_{\bullet,\text{et}}}$-modules obtained by pulling back $\mathcal{E}$ to the strictly simplicial space $Y_\bullet^+$, and note that each $\mathcal{E}_n$ has an HPD-stratification obtained by pullback (2.3.9) and that the transition morphisms for $\mathcal{E}_\bullet$ are compatible with these HPD-stratifications. By 2.3.21 these stratifications define an integrable connection ∇ on $\mathcal{E}_\bullet$ (*i.e.*, a compatible collection of connections $\nabla : \mathcal{E}_n \to \mathcal{E}_n \otimes_{\mathcal{O}_{Y_n}} \Omega^1_{Y_{n,\text{et}}/\mathcal{S}}$). Let $\mathcal{E}_\bullet \otimes \Omega^\bullet_{Y^+_{\bullet,\text{et}}/\mathcal{S}}$ denote the complex of sheaves on $Y^+_{\bullet,\text{et}}$ whose restriction to each Y_n is the de Rham complex of $\mathcal{E}_n$.

Theorem 2.5.2. — *There is a natural isomorphism*

$$R\tilde{u}_{\mathcal{X}_{\text{lis-et}}/\mathcal{S}*}(E)|_{Y^+_{\bullet,\text{et}}} \simeq \mathcal{E}_\bullet \otimes \Omega^\bullet_{Y^+_{\bullet,\text{et}}/\mathcal{S}}. \tag{2.5.2.1}$$

Proof. — Let $(\mathcal{X}^+_{\bullet,\text{lis-et}}/\mathcal{S})_{\text{cris}}$ denote the strictly simplicial topos

$$[n] \mapsto (\mathcal{X}_{n,\text{lis-et}}/\mathcal{S})_{\text{cris}} \tag{2.5.2.2}$$

which is defined since for any inclusion $[n] \hookrightarrow [m]$ the morphism $\mathcal{X}_m \to \mathcal{X}_n$ is smooth (1.4.19). Let res : $\mathcal{Y}_{\text{lis-et}} \to Y^+_{\bullet,\text{et}}$ be the restriction functor, r : $(\mathcal{X}^+_{\bullet,\text{lis-et}}/\mathcal{S})_{\text{cris}} \to (\mathcal{X}^+_{\bullet,\text{et}}/\mathcal{S})_{\text{cris}}$ the morphism of strictly simplicial topoi induced by the restriction maps $(\mathcal{X}_{n,\text{lis-et}}/\mathcal{S})_{\text{cris}} \to (\mathcal{X}_{n,\text{et}}/\mathcal{S})_{\text{cris}}$ defined in 1.5.1, let $i_\bullet : X_\bullet^+ \hookrightarrow Y_\bullet^+$ be the inclusion defined by i, and let $\pi : (X^+_{\bullet,\text{lis-et}}/\mathcal{S})_{\text{cris}} \to (\mathcal{X}_{\text{lis-et}}/\mathcal{S})_{\text{cris}}$ be the morphism of topoi defined as in 1.5.3. There is then a commutative diagram of functors

$$\begin{array}{ccccc} (X^+_{\bullet,\text{lis-et}}/\mathcal{S})_{\text{cris}} & \xrightarrow{r_*} & (X^+_{\bullet,\text{et}}/\mathcal{S})_{\text{cris}} & \xrightarrow{u_{X^+_{\bullet,\text{et}}/\mathcal{S}*}} & X^+_{\bullet,\text{et}} \\ \pi_* \downarrow & & & & \downarrow i_{\bullet*} \\ (\mathcal{X}_{\text{lis-et}}/\mathcal{S})_{\text{cris}} & \xrightarrow{\tilde{u}_{\mathcal{X}_{\text{lis-et}}/\mathcal{S}*}} & \mathcal{Y}_{\text{lis-et}} & \xrightarrow{\text{res}} & Y^+_{\bullet,\text{et}}. \end{array} \tag{2.5.2.3}$$

Since res is an exact functor, we have

$$\text{res}(R\tilde{u}_{\mathcal{X}_{\text{lis-et}}/\mathcal{S}*}(E)) \simeq R(\text{res} \circ \tilde{u}_{\mathcal{X}_{\text{lis-et}}/\mathcal{S}*})(E). \tag{2.5.2.4}$$

The adjunction map $E \to R\pi_*\pi^*E$ is an isomorphism by 1.5.4, and hence

$$R(\text{res} \circ \tilde{u}_{\mathcal{X}_{\text{lis-et}}/\mathcal{S}*})(E) \simeq R(\text{res} \circ \tilde{u}_{\mathcal{X}_{\text{lis-et}}/\mathcal{S}*}) \circ R\pi_*(\pi^*E). \tag{2.5.2.5}$$

Since π_* is obtained from a morphism of topoi it takes injective sheaves to injective sheaves. Therefore there is a canonical isomorphism

$$R(\mathrm{res}\circ \tilde{u}_{\mathcal{X}_{\text{lis-et}}/\mathcal{S}*})\circ R\pi_*(\pi^*E)\simeq R(\mathrm{res}\circ \tilde{u}_{\mathcal{X}_{\text{lis-et}}/\mathcal{S}*}\circ \pi_*)(\pi^*E). \tag{2.5.2.6}$$

From this, the commutativity of (2.5.2.3), and the fact that r_* and $i_{\bullet *}$ are exact and take injectives to injectives, we obtain an isomorphism

$$\mathrm{res}(R\tilde{u}_{\mathcal{X}_{\text{lis-et}}/\mathcal{S}*}(E))\simeq i_{\bullet *}Ru_{X^+_{\bullet,\text{et}}/\mathcal{S}*}(r_*\pi^*E). \tag{2.5.2.7}$$

Since the $\mathcal{L}_{\mathcal{Y}}$-construction is functorial (2.4.24), we can apply it to the complex $\mathcal{E}_\bullet\otimes\Omega^\bullet_{Y^+_{\bullet,\text{et}}/\mathcal{S}}$ on $Y^+_{\bullet,\text{et}}$. By 2.4.23, this gives a resolution $r_*\pi^*E\to i^*_{\text{cris}}L_{Y^+_\bullet}(\mathcal{E}_\bullet\otimes\Omega^\bullet_{Y^+_{\bullet,\text{et}}/\mathcal{S}})$ of $r_*\pi^*E$ by sheaves in $(X^+_{\bullet,\text{et}}/\mathcal{S})_{\text{cris}}$. Note that the terms in $i^*_{\text{cris}}L_{Y^+_\bullet}(\mathcal{E}_\bullet\otimes\Omega^\bullet_{Y^+_{\bullet,\text{et}}/\mathcal{S}})$ are not crystals in the sense of 2.1.8 since the pullback maps arising from morphisms $[n']\to[n]$ in Δ are not isomorphisms. However, it is still true that

$$i_{\bullet *}Ru_{X^+_{\bullet,\text{et}}/\mathcal{S}*}(i^*_{\text{cris}}L_{Y^+_\bullet}(\mathcal{E}_\bullet\otimes\Omega^\bullet_{Y^+_{\bullet,\text{et}}/\mathcal{S}}))\simeq \mathcal{E}_\bullet\otimes\Omega^\bullet_{Y^+_{\bullet,\text{et}}/\mathcal{S}} \tag{2.5.2.8}$$

since this can be verified for each $X_n/\mathcal{S}$ individually in which case it follows from 2.4.17 and the discussion in 2.4.25. Combining this with the isomorphism (2.5.2.7) we obtain an isomorphism
(2.5.2.9)

$$\mathrm{res}(R\tilde{u}_{\mathcal{X}_{\text{lis-et}}/\mathcal{S}*}(E))\simeq i_{\bullet *}Ru_{X^+_{\bullet,\text{et}}/\mathcal{S}*}(i^*_{\text{cris}}L_{Y^+_\bullet}(\mathcal{E}_\bullet\otimes\Omega^\bullet_{Y^+_{\bullet,\text{et}}/\mathcal{S}}))\simeq \mathcal{E}_\bullet\otimes\Omega^\bullet_{Y^+_{\bullet,\text{et}}/\mathcal{S}}.\qquad\square$$

Corollary 2.5.3. — *With notation as above, let $\Gamma_{\mathcal{X}_{\text{lis-et}}/\mathcal{S}}$ (resp. Γ) denote the global section functor on the category $(\mathcal{X}_{\text{lis-et}}/\mathcal{S})_{\text{cris}}$ (resp. $Y^+_{\bullet,\text{et}}$). Then there is a natural isomorphism*

$$R\Gamma_{\mathcal{X}_{\text{lis-et}}/\mathcal{S}}(E)\simeq R\Gamma(\mathcal{E}_\bullet\otimes\Omega^\bullet_{Y^+_{\bullet,\text{et}}/\mathcal{S}}). \tag{2.5.3.1}$$

Proof. — Since $E\to R\pi_*\pi^*E$ is an isomorphism (1.5.4), there is a natural isomorphism

$$R\Gamma_{\mathcal{X}_{\text{lis-et}}/\mathcal{S}}(E)\simeq R\Gamma_{X^+_{\bullet,\text{lis-et}}/\mathcal{S}}(\pi^*E). \tag{2.5.3.2}$$

Since the functor r_* is exact and takes injectives to injectives, there is also a natural isomorphism

$$R\Gamma_{X^+_{\bullet,\text{lis-et}}/\mathcal{S}}(\pi^*E)\simeq R\Gamma_{X^+_{\bullet,\text{et}}/\mathcal{S}}(r_*\pi^*E). \tag{2.5.3.3}$$

Since $i_{\bullet *}:X_{\bullet,\text{et}}\to Y_{\bullet,\text{et}}$ is an exact functor, we therefore have

$$R\Gamma_{\mathcal{X}_{\text{lis-et}}/\mathcal{S}}(E)\simeq R\Gamma(i_{\bullet *}Ru_{X^+_{\bullet,\text{et}}/\mathcal{S}*}(r_*\pi^*E)), \tag{2.5.3.4}$$

which by the isomorphism (2.5.2.8) is isomorphic to $R\Gamma(\mathcal{E}_\bullet\otimes\Omega^\bullet_{Y^+_{\bullet,\text{et}}/\mathcal{S}})$. $\square$

Corollary 2.5.4. — *Let $i : \mathcal{X} \hookrightarrow \mathcal{Y}$ be a closed immersion as in 2.5.1 with $\mathcal{Y}$ and $\mathcal{X}$ Deligne-Mumford stacks, let E be a quasi-coherent crystal in $(\mathcal{X}_{\mathrm{et}}/\mathcal{S})_{\mathrm{cris}}$, and let $(\mathcal{E}, \nabla)$ be the $\mathcal{O}_{\mathcal{Y}_{\mathrm{et}}}$-module with integrable connection corresponding to E via 2.4.7. Then there is a natural isomorphism*

$$Ru_{\mathcal{X}_{\mathrm{et}}/\mathcal{S}*}(E) \simeq i^{-1}(\mathcal{E} \otimes \Omega^{\bullet}_{\mathcal{Y}_{\mathrm{et}}/\mathcal{S}}), \tag{2.5.4.1}$$

where i^{-1} denotes the inverse image functor for the morphism of topoi $\mathcal{X}_{\mathrm{et}} \to \mathcal{Y}_{\mathrm{et}}$.

Proof. — With notation as in 2.4.18, there is a resolution $E \to i^*_{\mathrm{cris}} L_{\mathcal{Y}}(\mathcal{E} \otimes \Omega^{\bullet}_{\mathcal{Y}_{\mathrm{et}}/\mathcal{S}})$ by crystals, and by 2.4.19

$$Ru_{\mathcal{X}_{\mathrm{et}}/\mathcal{S}*}(i^*_{\mathrm{cris}} L_{\mathcal{Y}}(\mathcal{E} \otimes \Omega^{\bullet}_{\mathcal{Y}_{\mathrm{et}}/\mathcal{S}})) \simeq i^{-1}(\mathcal{E} \otimes \Omega^{\bullet}_{\mathcal{Y}_{\mathrm{et}}/\mathcal{S}}). \tag{2.5.4.2}$$ □

More generally we have the following:

Proposition 2.5.5. — *With notation as in 2.5.4, let D denote the divided power envelope of $\mathcal{X}$ in $\mathcal{Y}$ and let $\mathcal{D}$ be the coordinate ring of D viewed as a sheaf on $\mathcal{X}_{\mathrm{et}}$. Let $F^m(\mathcal{E} \otimes_{\mathcal{D}} \Omega^{\bullet}_{D/\mathcal{S}})$ be the subcomplex of $\mathcal{E} \otimes_{\mathcal{D}} \Omega^{\bullet}_{D/\mathcal{S}}$ which in degree q is $I^{[m-q]}\mathcal{E} \otimes_{\mathcal{D}} \Omega^q_{D/\mathcal{S}}$, where $I \subset \mathcal{D}$ is the ideal of $\mathcal{X}$ in $D := D_{\mathcal{X},\gamma}(\mathcal{Y})$. Then there is a natural isomorphism in the derived category of sheaves of abelian groups on $\mathcal{X}_{\mathrm{et}}$*

$$Ru_{\mathcal{X}_{\mathrm{et}}/\mathcal{S}*}(\mathcal{I}^{[m]}_{\mathcal{X}/\mathcal{S}} E) \longrightarrow F^m(\mathcal{E} \otimes_{\mathcal{D}} \Omega^{\bullet}_{D/\mathcal{S}}), \tag{2.5.5.1}$$

where $\mathcal{I}_{\mathcal{X}/\mathcal{S}}$ is the PD-ideal in $(\mathcal{X}_{\mathrm{et}}/\mathcal{S})_{\mathrm{cris}}$ which to any object $U \hookrightarrow T$ associates the ideal of U in T.

Proof. — This follows from the same argument as in the classical case [**8**, 7.2] once we generalize the filtered Poincaré lemma [**8**, 6.14] to the stack-theoretic setting. Restricting the natural exact sequence

$$0 \longrightarrow \mathcal{I}_{\mathcal{X}/\mathcal{S}} \longrightarrow \mathcal{O}_{\mathcal{X}/\mathcal{S}} \longrightarrow i_{\mathcal{X}/\mathcal{S}*}(\mathcal{O}_{\mathcal{X}}) \longrightarrow 0, \tag{2.5.5.2}$$

where $i : \mathcal{X}_{\mathrm{et}} \to (\mathcal{X}_{\mathrm{et}}/\mathcal{S})_{\mathrm{cris}}$ is the morphism of topoi induced by the morphism of sites $(U, T, \delta) \mapsto U$, to $(\mathcal{X}_{\mathrm{et}}/\mathcal{S})_{\mathrm{cris}}|_{\tilde{\mathcal{Y}}}$, we obtain an exact sequence

$$0 \longrightarrow j^*_{\tilde{Y}} \mathcal{I}_{\mathcal{X}/\mathcal{S}} \longrightarrow j^*_{\tilde{Y}} \mathcal{O}_{\mathcal{X}/\mathcal{S}} \longrightarrow j^*_{\tilde{Y}} i_{\mathcal{X}/\mathcal{S}*}(\mathcal{O}_{\mathcal{X}}) \longrightarrow 0. \tag{2.5.5.3}$$

The sheaf $j^*_{\tilde{Y}} i_{\mathcal{X}/\mathcal{S}*}(\mathcal{O}_{\mathcal{X}})$ is equal in the notation of 2.4.12 to $(i_* \mathcal{O}_{\mathcal{X}_{\mathrm{et}}})^\sharp$, and hence by 2.4.13 has no higher cohomology and $j_{\tilde{\mathcal{Y}}*} j^*_{\tilde{\mathcal{Y}}} i_{\mathcal{X}/\mathcal{S}*}(\mathcal{O}_{\mathcal{X}}) \simeq i_{\mathcal{X}/\mathcal{S}*}(\mathcal{O}_{\mathcal{X}})$. Define $\mathcal{K} := j_{\tilde{*}} j^*_{\tilde{Y}} \mathcal{I}_{\mathcal{X}/\mathcal{S}}$ so that there is an exact sequence

$$0 \longrightarrow \mathcal{K} \longrightarrow i^* L_{\mathcal{Y}}(\mathcal{O}_{\mathcal{Y}}) \longrightarrow i_{\mathcal{X}/\mathcal{S}*}(\mathcal{O}_{\mathcal{X}}) \longrightarrow 0. \tag{2.5.5.4}$$

From this and 2.4.12 it follows that $\mathcal{K}$ has a natural divided power structure.

Lemma 2.5.6. — *For any integer r, the map $i^* L_{\mathcal{Y}}(\Omega^q_{\mathcal{Y}/\mathcal{S}}) \to i^* L_{\mathcal{Y}}(\Omega^{q+1}_{\mathcal{Y}/\mathcal{S}})$ sends $\mathcal{K}^{[r]} i^* L_{\mathcal{Y}}(\Omega^q_{\mathcal{Y}/\mathcal{S}})$ to $\mathcal{K}^{[r-1]} i^* L_{\mathcal{Y}}(\Omega^{q+1}_{\mathcal{Y}/\mathcal{S}})$.*

Proof. — For any object T of $\mathrm{Cris}(\mathcal{X}_{\mathrm{et}}/\mathcal{S})$, to verify that $\mathcal{K}^{[r]}i^*L_{\mathcal{Y}}(\Omega^q_{\mathcal{Y}/\mathcal{S}})_T$ maps to the submodule $\mathcal{K}^{[r-1]}i^*L_{\mathcal{Y}}(\Omega^{q+1}_{\mathcal{Y}/\mathcal{S}})_T$ it suffices to do so after making a smooth base change $S \to \mathcal{S}$ (note this uses again that the formation of divided power envelopes commutes with flat base change (1.2.3)). This reduces the proof to the case when S is a scheme in which case it follows from the classical theory [**8**, 6.17]). □

Define $F^m(E \otimes i^*L_{\mathcal{Y}}(\Omega^\bullet_{\mathcal{Y}/\mathcal{S}}))$ to be the subcomplex of $E \otimes i^*L_{\mathcal{Y}}(\Omega^\bullet_{\mathcal{Y}/\mathcal{S}})$ which in degree q is $\mathcal{K}^{[m-q]}E \otimes i^*L_{\mathcal{Y}}(\Omega^q_{\mathcal{Y}/\mathcal{S}})$.

Lemma 2.5.7. — *The natural map $\mathcal{I}^{[m]}_{\mathcal{X}/\mathcal{S}}E \to F^m(E \otimes i^*L_{\mathcal{Y}}(\Omega^\bullet_{\mathcal{Y}/\mathcal{S}}))$ is a quasi-isomorphism.*

Proof. — As in the proof of 2.4.23 it suffices to consider the case when $\mathcal{X}$, $\mathcal{Y}$, and $\mathcal{S}$ are all algebraic spaces in which case it follows from the classical theory [**8**, 6.14.1]. □

By 2.4.21 there is a natural isomorphism between $F^m(E \otimes i^*L_{\mathcal{Y}}(\Omega^\bullet_{\mathcal{Y}/\mathcal{S}}))$ and the complex $F^m(i^*L_{\mathcal{Y}}(\mathcal{E} \otimes \Omega^\bullet_{\mathcal{Y}/\mathcal{S}}))$ which in degree q is $\mathcal{K}^{[m-q]}i^*L_{\mathcal{Y}}(\mathcal{E} \otimes \Omega^q_{\mathcal{Y}/\mathcal{S}})$. Hence there is a natural quasi-isomorphism

$$\mathcal{I}^{[m]}_{\mathcal{X}/\mathcal{S}}E \longrightarrow F^m(i^*L_{\mathcal{Y}}(\mathcal{E} \otimes \Omega^\bullet_{\mathcal{Y}/\mathcal{S}})). \tag{2.5.7.1}$$

Applying the functor $u_{\mathcal{X}_{\mathrm{et}}/\mathcal{S}*}$ and using 2.4.17 we obtain the isomorphism (2.5.5.1). □

2.5.8. — Let $\mathcal{X} \to \mathcal{S}$ be a smooth, locally separated, and representable morphism of algebraic stacks with $\mathcal{X}$ a locally noetherian Deligne-Mumford stack. The equivalence of categories (2.4.7 and 2.3.28) between modules with integrable connection on $\mathcal{X}_{\mathrm{et}}/\mathcal{S}$ and crystals defines an inclusion

$$M : MIC(\mathcal{X}_{\mathrm{et}}/\mathcal{S}) \subset \mathrm{Mod}((\mathcal{X}_{\mathrm{et}}/\mathcal{S})_{\mathrm{cris}}, \mathcal{O}_{\mathcal{X}_{\mathrm{et}}/\mathcal{S}}), \tag{2.5.8.1}$$

and the composite
(2.5.8.2)

$$MIC(\mathcal{X}_{\mathrm{et}}/\mathcal{S}) \xrightarrow{M} \mathrm{Mod}((\mathcal{X}_{\mathrm{et}}/\mathcal{S})_{\mathrm{cris}}, \mathcal{O}_{\mathcal{X}_{\mathrm{et}}/\mathcal{S}}) \xrightarrow{u_{\mathcal{X}_{\mathrm{et}}/\mathcal{S}*}} \{\text{abelian sheaves on } \mathcal{X}_{\mathrm{et}}\}$$

is the functor $(\mathcal{E}, \nabla) \mapsto \mathcal{E}^\nabla$.

Proposition 2.5.9. — *The category $MIC(\mathcal{X}_{\mathrm{et}}/\mathcal{S})$ has enough injectives, and every object of $MIC(\mathcal{X}_{\mathrm{et}}/\mathcal{S})$ can be embedded into an injective object of $MIC(\mathcal{X}_{\mathrm{et}}/\mathcal{S})$ whose image under M is acyclic for $u_{\mathcal{X}_{\mathrm{et}}/\mathcal{S}*}$.*

Proof. — Let $\mathcal{D}$ denote the sheaf on $\mathcal{X}_{\mathrm{et}}$ given by the divided power envelope of the diagonal $\mathcal{X} \hookrightarrow \mathcal{X} \times_{\mathcal{S}} \mathcal{X}$. The functor sending a quasi-coherent sheaf $\mathcal{F}$ to $\mathcal{D} \otimes \mathcal{F}$ with connection induced by the connection on $\mathcal{D}$ (2.4.5) is left adjoint to the functor which sends a quasi-coherent module with integrable connection $(\mathcal{G}, \nabla)$ to the underlying quasi-coherent sheaf $\mathcal{G}$. This can be seen for example as follows. Let $\widetilde{\mathcal{X}}$ denote the sheaf in $(\mathcal{X}_{\text{et}}/\mathcal{S})_{\text{cris}}$ which to any object (U, T, δ) associates the set of retractions $T \to \mathcal{X}$ over $\mathcal{S}$. Then as in 2.4.11 there is a natural morphism of topoi

$$j_{\mathcal{X}} : (\mathcal{X}_{\text{et}}/\mathcal{S})_{\text{cris}}|_{\tilde{\mathcal{X}}} \longrightarrow (\mathcal{X}_{\text{et}}/\mathcal{S})_{\text{cris}}. \tag{2.5.9.1}$$

The crystal corresponding to $\mathcal{D} \otimes \mathcal{F}$ is then in the notation of 2.4.12 equal to $j_{\mathcal{X}*}\mathcal{F}^{\sharp}$. If G is the crystal corresponding to $(\mathcal{G}, \nabla)$ we then have

$$\begin{aligned} \text{Hom}_{\text{MIC}}((\mathcal{G}, \nabla), \mathcal{D} \otimes \mathcal{F}) &\simeq \text{Hom}_{(\mathcal{X}/\mathcal{S})_{\text{cris}}}(G, j_{\mathcal{X}*}\mathcal{F}^{\sharp}) \\ &\simeq \text{Hom}_{(\mathcal{X}/\mathcal{S})_{\text{cris}}|_{\tilde{\mathcal{X}}}}(j_{\mathcal{X}}^* G, \mathcal{F}^{\sharp}) \\ &\simeq \text{Hom}_{\mathcal{O}_{\mathcal{X}}}(\mathcal{G}, \mathcal{F}). \end{aligned}$$

The functor $\mathcal{F} \mapsto \mathcal{D} \otimes \mathcal{F}$ is also exact since it has a left adjoint and hence is left exact (it is clearly right exact). Therefore for an injective quasi-coherent sheaf $\mathcal{F}$ the object $\mathcal{D} \otimes \mathcal{F} \in MIC(\mathcal{X}_{\text{et}}/\mathcal{S})$ is injective. Furthermore, for $(\mathcal{G}, \nabla) \in MIC(\mathcal{X}_{\text{et}}/\mathcal{S})$ the natural map $\mathcal{G} \to \mathcal{D} \otimes \mathcal{G}$ induced by the identity map $\mathcal{G} \to \mathcal{G}$ is an inclusion. This proves that $MIC(\mathcal{X}_{\text{et}}/\mathcal{S})$ has enough injectives.

That the modules with connection $\mathcal{D} \otimes \mathcal{F}$ are acyclic for $u_{\mathcal{X}_{\text{et}}/\mathcal{S}*}$ follows from 2.4.19. □

Warning 2.5.10. — Note that by our conventions $MIC(\mathcal{X}_{\text{et}}/\mathcal{S})$ is the category of quasi-coherent sheaves $\mathcal{F}$ with integrable connection $\nabla : \mathcal{F} \to \mathcal{F} \otimes \Omega^1_{\mathcal{X}/\mathcal{S}}$. If $\mathcal{X}$ is nonseparated an injective object in $MIC(\mathcal{X}_{\text{et}}/\mathcal{S})$ need not be injective in the category of all $\mathcal{O}_{\mathcal{X}_{\text{et}}}$-modules with connection.

2.5.11. — Let $g : \mathcal{S} \to S$ be a quasi-compact and quasi-separated morphism of finite type with S an algebraic space, and let $f : \mathcal{X} \to \mathcal{S}$ be a quasi-compact and quasi-separated morphism of algebraic stacks of finite presentation such that γ extends to $\mathcal{X}$. The composite $\mathcal{X} \to \mathcal{S} \to S$ induces a morphism of topoi $h : \mathcal{X}_{\text{lis-et}} \to S_{\text{et}}$ induced by the functor

$$\text{Et}(S) \longrightarrow \text{Lis-Et}(\mathcal{X}), \quad (V \to S) \longmapsto (V \times_S \mathcal{X} \to \mathcal{X}). \tag{2.5.11.1}$$

Let $h_{\mathcal{X}/S} : (\mathcal{X}_{\text{lis-et}}/\mathcal{S})_{\text{cris}} \to S_{\text{et}}$ denote the composite $h \circ u_{\mathcal{X}_{\text{lis-et}}/\mathcal{S}}$.

Proposition 2.5.12. — *Let $\mathcal{S}_0 \subset \mathcal{S}$ be the closed substack defined by I, and assume that $\mathcal{X} \to \mathcal{S}$ factors through a smooth morphism $\mathcal{X} \to \mathcal{S}_0$. Then for any quasi-coherent crystal E in $(\mathcal{X}_{\text{lis-et}}/\mathcal{S})_{\text{cris}}$ and integer i, the sheaf $R^i h_{\mathcal{X}/S*}(E)$ is a quasi-coherent sheaf on S_{et} whose formation is compatible with smooth base change $S' \to S$.*

Proof. — Let $X \to \mathcal{X}$ be a smooth surjection with X an algebraic space, let $X_\bullet$ denote the 0-coskeleton, $(X^+_{\bullet,\text{lis-et}}/\mathcal{S})_{\text{cris}}$ the resulting strictly simplicial topos, and $\pi : (\mathcal{X}^+_{\bullet,\text{lis-et}}/\mathcal{S})_{\text{cris}} \to (\mathcal{X}_{\text{lis-et}}/\mathcal{S})_{\text{cris}}$ the projection. By 1.5.4, the adjunction map $E \to R\pi_*\pi^*E$ is an isomorphism, and hence $R^i h_{\mathcal{X}/S*}(E)$ is isomorphic to the i-derived functor of $h_{\mathcal{X}/S*} \circ \pi_*$ applied to E. By [**68**, 2.7] there is consequently a spectral sequence

$$E_1^{st} = R^t h_{X_s/S*}(E|_{X_s}) \implies R^{s+t} h_{\mathcal{X}/S*}(E), \tag{2.5.12.1}$$

where $h_{X_s/S} : (X_s/\mathcal{S})_{\text{cris}} \to S_{\text{et}}$ denote the morphism of topoi induced by $h_{\mathcal{X}/\mathcal{S}}$ and the projection $X_s \to \mathcal{X}$. To prove that $R^i h_{\mathcal{X}/S*}(E)$ is quasi-coherent it therefore suffices to show that each $R^t h_{X_s/S*}(E|_{X_s})$ is quasi-coherent and compatible with smooth base change which reduces the proof to the case when $\mathcal{X}$ is an algebraic space.

Repeating the above argument with an étale cover of $\mathcal{X}$ it furthermore suffices to consider the case when $\mathcal{X}$ is a quasi-compact and quasi-separated scheme. Repeating the argument a third time with an affine cover of $\mathcal{X}$ we are finally reduced to the case when $\mathcal{X}$ is an affine scheme.

In this case there exists a smooth deformation $\mathcal{Y}$ of $\mathcal{X}$ to $\mathcal{S}$. Indeed by [**66**, 1.4] the obstruction to finding such a deformation lies in the second cohomology group of a certain quasi-coherent sheaf which is zero since $\mathcal{X}$ is affine. Choose a smooth lifting $h : \mathcal{Y} \to \mathcal{S}$ of $\mathcal{X}$ and let $(\mathcal{E}, \nabla)$ be the module integrable connection on $\mathcal{Y}/\mathcal{S}$ defined by E. Then

$$R^i h_{\mathcal{X}/S*}(E) \simeq R^i h_*(\mathcal{E} \otimes \Omega^\bullet_{\mathcal{Y}/\mathcal{S}}) \tag{2.5.12.2}$$

which is quasi-coherent and commutes with smooth base change by [**68**, 6.20]. □

Warning 2.5.13. — In general the sheaves $R^i h_{\mathcal{X}/S}(E)$ may be non-zero for infinitely many integers i. This is true even if one restricts to $\mathcal{S}$ a scheme and $\mathcal{X}$ a Deligne-Mumford stack.

For example, let p be a prime, set $\mathcal{S} = S = \text{Spec}(\mathbb{F}_p)$, let G be the group $\mathbb{Z}/(p)$, and set $\mathcal{X} := BG$. Let $\text{Spec}(k) \to \mathcal{X}$ be the smooth surjection corresponding to the trivial torsor. Then 2.5.3 and 2.5.4 show that $H^*((\mathcal{X}_{\text{et}}/\mathcal{S})_{\text{cris}}, \mathcal{O}_{\mathcal{X}_{\text{et}}/\mathcal{S}})$ is isomorphic to the cohomology of the normalized complex associated to the strictly simplicial module

$$[n] \longmapsto \text{Hom}_{G\text{-equivariant}}(G^{n+1}, \mathbb{F}_p), \tag{2.5.13.1}$$

where $\mathbb{F}_p$ is viewed as a trivial G-module. It follows that $H^*((\mathcal{X}_{\text{et}}/\mathcal{S})_{\text{cris}}, \mathcal{O}_{\mathcal{X}_{\text{et}}/\mathcal{S}})$ is isomorphic to the group cohomology $H^*(G, \mathbb{F}_p)$ which is isomorphic to $\mathbb{F}_p$ in every degree.

2.5.14. — Recall [**2**, 2.3.1] that a Deligne-Mumford stack $\mathcal{X}$ is *tame* if for every geometric point $x : \text{Spec}(k) \to \mathcal{X}$ the order of the group of automorphisms of x in $\mathcal{X}(k)$ is prime to the characteristic of k.

Theorem 2.5.15. — *With assumptions as in 2.5.12, assume in addition that $\mathcal{X}$ is a tame noetherian Deligne-Mumford stack. Then there exists an integer r such that $R^i h_{\mathcal{X}/S*}(E)$ is zero for all $i \geq r$ and any quasi-coherent crystal E on $(\mathcal{X}_{\text{et}}/\mathcal{S})_{\text{cris}}$.*

Proof. — By [**5**, V.5.1] the sheaf $R^i h_{\mathcal{X}/S*}(E)$ is the sheaf associated to the presheaf which to an étale morphism of schemes $V \to S$ associates

$$H^i((\mathcal{X}_{\text{lis-et}}/\mathcal{S})_{\text{cris}}|_{h^{-1}_{\mathcal{X}/S}\tilde{V}}, E), \tag{2.5.15.1}$$

where $h^{-1}_{\mathcal{X}/S}\tilde{V}$ denotes the pullback of the sheaf represented by V on S_{et}. This sheaf $h^{-1}_{\mathcal{X}/S}\tilde{V}$ is equal to the sheaf in $(\mathcal{X}_{\text{lis-et}}/\mathcal{S})_{\text{cris}}$ which to any $(U, T, \delta) \in \text{Cris}(\mathcal{X}_{\text{lis-et}}/\mathcal{S})$

associates the set of maps $T \to V$ over the map $U \to \mathcal{S} \to S$. From this and 1.4.23 it follows that (2.5.15.1) is isomorphic to

$$H^i\big(((\mathcal{X} \times_S V)_{\text{lis-et}}/\mathcal{S} \times_S V)_{\text{cris}}, E\big). \tag{2.5.15.2}$$

This reduces the proof to showing that if $\mathcal{S}$ and $\mathcal{X}$ are both quasi-compact then there exists an integer r such that

$$H^i((\mathcal{X}_{\text{lis-et}}/\mathcal{S})_{\text{cris}}, E) = 0 \tag{2.5.15.3}$$

for all $i \geq r$.

For this let $\pi : \mathcal{X} \to X$ be the coarse moduli space of $\mathcal{X}$ which exists by [**49**, 19.1], and let

$$\rho : (\mathcal{X}_{\text{lis-et}}/\mathcal{S})_{\text{cris}} \longrightarrow X_{\text{et}} \tag{2.5.15.4}$$

denote the composition of $u_{\mathcal{X}_{\text{lis-et}}/\mathcal{S}}$ with the natural morphism of topoi $\mathcal{X}_{\text{lis-et}} \to X_{\text{et}}$.

Lemma 2.5.16. — *Let Y be a noetherian separated algebraic space of dimension d, and assume p is a prime such that $p^e\mathcal{O}_{Y_{\text{et}}} = 0$ for some $e > 0$. Let G be a sheaf of abelian groups on Y killed by some power of p. Then $H^i(Y_{\text{et}}, G) = 0$ for $i > d + 1$.*

Proof. — Note first that the étale sites of Y and $Y \otimes_{\mathbb{Z}} (\mathbb{Z}/(p))$ are equivalent so it suffices to consider the case when $p \cdot \mathcal{O}_Y = 0$. Furthermore, by filtering G by the images of multiplication by p and considering the associated long exact sequences we see that it suffices to consider the case when $p \cdot G = 0$.

In this case when Y is a scheme the result follows from [**5**, X.5.1]. For the general case we proceed by noetherian induction. By [**46**, IV.3.1] there exists a dense open subspace $j : U \hookrightarrow Y$ which is a scheme and a proper morphism from a projective scheme $g : \tilde{Y} \to Y$ which is an isomorphism over U. Let $i : Z \hookrightarrow Y$ be the complement. Then it suffices to show that the result holds for X if it holds for Z. For this consider the excision sequence

$$0 \longrightarrow j_!j^*G \longrightarrow G \longrightarrow i_*i^*G \longrightarrow 0. \tag{2.5.16.1}$$

Consideration of the associated long exact sequence of cohomology groups shows that it suffices to prove the lemma for i_*i^*G and $j_!j^*G$. The result hold for i_*i^*G by induction so this reduces the proof to the case $G = j_!j^*G$.

Since g is proper there are natural isomorphisms

$$j_!j^*G \simeq Rj_!(j^*G) \simeq Rg_!R\tilde{j}_!(j^*G) \simeq Rg_*R\tilde{j}_!(j^*G) \simeq Rg_*(\tilde{j}_!(j^*G)), \tag{2.5.16.2}$$

where $\tilde{j} : U \hookrightarrow \tilde{Y}$ is the inclusion lifting j. In particular, there is a natural isomorphism

$$H^i(\tilde{Y}, \tilde{j}_!j^*G) \simeq H^i(Y, j_!j^*G). \tag{2.5.16.3}$$

Since $\dim(\tilde{Y}) = \dim(Y)$ the case when Y is a scheme applied to $\tilde{Y}$ implies that these groups are zero for $i > d + 1$. □

Lemma 2.5.17. — *There exists an integer r such that $R^i\rho_*(E) = 0$ for all $i \geq r$ and quasi-coherent crystals E in $(\mathcal{X}_{\text{et}}/\mathcal{S})_{\text{cris}}$.*

Proof. — It follows from [**5**, V.5.1] that $R^i\rho_*(E)$ is isomorphic to the sheaf associated to the presheaf on X_{et} which to an étale X-scheme V associates

$$H^*((\mathcal{X}_{V,\mathrm{et}}/\mathcal{S})_{\mathrm{cris}}, E), \tag{2.5.17.1}$$

where $\mathcal{X}_V := \mathcal{X} \times_X V$. We may therefore replace X by an affine étale cover and hence by [**2**, 2.2.3] can assume that $\mathcal{X} = [U/\Gamma]$ for some finite affine X-scheme U with action of a finite group Γ of order prime to p. Let $U \to [U/\Gamma]$ be the natural projection, and let $\pi : U_\bullet \to \mathcal{X}$ be the 0-coskeleton. For $[n] \in \Delta$ there is a canonical isomorphism

$$U_n \simeq \coprod_{\Gamma^{n+1}/\Gamma} U. \tag{2.5.17.2}$$

By 1.5.4, the adjunction map $E \to R\pi_*\pi^*E$ is an isomorphism. It follows that

$$H^*((\mathcal{X}_{\mathrm{et}}/\mathcal{S})_{\mathrm{cris}}, E) \simeq H^*((U_{\bullet,\mathrm{et}}/\mathcal{S})_{\mathrm{cris}}, \pi^*E). \tag{2.5.17.3}$$

By [**68**, 2.7], there is a spectral sequence

$$E_1^{st} = H^t((U_{s,\mathrm{et}}/\mathcal{S})_{\mathrm{cris}}, E|_{U_s}) \Longrightarrow H^{s+t}((U_{\bullet,\mathrm{et}}/\mathcal{S})_{\mathrm{cris}}, \pi^*E). \tag{2.5.17.4}$$

Via the isomorphism (2.5.17.2) the term E_1^{st} in this spectral sequence is identified with

$$\mathrm{Hom}_\Gamma(\Gamma^{s+1}, H^t((U_{\mathrm{et}}/\mathcal{S})_{\mathrm{cris}}, E|_{U_0})), \tag{2.5.17.5}$$

where the Γ-action on $H^t((U_{\mathrm{et}}/\mathcal{S})_{\mathrm{cris}}, E|_{U_0})$ is induced by the Γ-action on U. Moreover, the map $d_1 : E_1^{st} \to E_1^{(s+1)t}$ is simply the alternating sum of the maps induced by the projections $s_j : \Gamma^{s+1} \hookrightarrow \Gamma^s$ given by

$$(\gamma_0, \dots, \gamma_s) \longmapsto (\gamma_0, \dots, \hat{\gamma}_j, \gamma_{j+1}, \dots, \gamma_s). \tag{2.5.17.6}$$

From this description it follows that the E_2-terms are given by the group cohomology

$$E_2^{st} \simeq H^s(\Gamma, H^t((U_{\mathrm{et}}/\mathcal{S})_{\mathrm{cris}}, E|_U)). \tag{2.5.17.7}$$

Since Γ has order prime to p by assumption, these groups are zero for $s \neq 0$ and hence $E_2^{st} = 0$ for $s > 0$ and

$$E_2^{0t} \simeq H^t((U_{\mathrm{et}}/\mathcal{S})_{\mathrm{cris}}, E|_U)^\Gamma. \tag{2.5.17.8}$$

It follows that there is a natural isomorphism

$$H^t((U_{\bullet,\mathrm{et}}/\mathcal{S})_{\mathrm{cris}}, \pi^*E) \simeq H^t((U_{\mathrm{et}}/\mathcal{S})_{\mathrm{cris}}, E|_U)^\Gamma. \tag{2.5.17.9}$$

This reduces the proof to the case when $U = \mathcal{X}$ is an affine scheme. In this case there exists a smooth lifting $Y/\mathcal{S}$ of U and

$$H^t((U_{\mathrm{et}}/\mathcal{S})_{\mathrm{cris}}, E|_U) \simeq H^t(Y_{\mathrm{et}}, \mathcal{E} \otimes \Omega^\bullet_{Y_{\mathrm{et}}/\mathcal{S}}). \tag{2.5.17.10}$$

Let r be the relative dimension of $Y/\mathcal{S}$. Then $\mathcal{E}\otimes\Omega^i_{Y_{\rm et}/\mathcal{S}}=0$ for $i>r$ and since each term $\mathcal{E}\otimes\Omega^i_{Y_{\rm et}/\mathcal{S}}$ is quasi-coherent and Y is affine

(2.5.17.11)
$$H^i(Y_{\rm et},\mathcal{E}\otimes\Omega^i_{Y_{\rm et}/\mathcal{S}})=0$$

for $i>0$. From the spectral sequence of a filtered complex

(2.5.17.12)
$$E_1^{st}=H^t(Y_{\rm et},\mathcal{E}\otimes\Omega^s_{Y_{\rm et}/\mathcal{S}})\Longrightarrow H^{s+t}(Y_{\rm et},\mathcal{E}\otimes\Omega^\bullet_{Y_{\rm et}/\mathcal{S}})$$

it then follows that $H^t(Y_{\rm et},\mathcal{E}\otimes\Omega^\bullet_{Y_{\rm et}/\mathcal{S}})$ is zero for $t\geq r+1$. □

We can now complete the proof of 2.5.15. Since the global section functor is isomorphic to the composite $\Gamma_{X_{\rm et}}\circ\rho_*$ and ρ_* takes injectives to injectives since it has an exact left adjoint ρ^{-1}, there is a spectral sequence

(2.5.17.13)
$$E_2^{st}=H^s(X_{\rm et},R^t\rho_*(E))\Longrightarrow H^{s+t}((\mathcal{X}_{\rm et}/\mathcal{S})_{\rm cris},E).$$

Combining 2.5.16 and 2.5.17, theorem 2.5.15 follows. □

Remark 2.5.18. — The careful reader will note that even when $\mathcal{X}$ and $\mathcal{S}$ are algebraic spaces (2.5.15) does not follow from the arguments in [**8, 7**] as these proofs make essential use of the Zariski topology.

2.6. Base change theorems

2.6.1. — Let $u:(B',I',\gamma')\to(B,I,\gamma)$ be a morphism of PD-algebraic spaces, and let $\mathcal{S}/B$ be an algebraic stack. Set $\mathcal{S}':=\mathcal{S}\times_B B'$, and assume γ (resp. γ') extends to $\mathcal{S}$ (resp. $\mathcal{S}'$). Let $B'_0\subset B'$ and $B_0\subset B$ be closed subspaces defined by sub-PD-ideals such that the composite

(2.6.1.1)
$$B'_0\longrightarrow B'\longrightarrow B$$

factors through B_0. Set $\mathcal{S}_0$ denote $\mathcal{S}\times_B B_0$ and let $\mathcal{S}'_0$ denote $\mathcal{S}'\times_{B'}B'_0$.

Theorem 2.6.2. — *Let $f:\mathcal{X}\to\mathcal{S}_0$ be a smooth quasi-compact morphism of algebraic stacks with $\mathcal{X}$ a tame noetherian Deligne-Mumford stack, and let E be a quasi-coherent crystal in $(\mathcal{X}_{\rm et}/\mathcal{S})_{\rm cris}$. Assume either that the map $B'\to B$ is flat or that E is flat over B. Let $f':\mathcal{X}'\to\mathcal{S}'_0$ denote the base change $\mathcal{X}\times_{\mathcal{S}_0}\mathcal{S}'_0\to\mathcal{S}'_0$ and let $g:\mathcal{X}'\to\mathcal{X}$ be the projection. Then there is a natural isomorphism in the derived category of sheaves of $\mathcal{O}_{B'_{\rm et}}$-modules*

(2.6.2.1)
$$Lu^*Rh_{\mathcal{X}/B*}(E)\simeq Rh_{\mathcal{X}'/B'*}(g^*E).$$

Remark 2.6.3. — By 2.5.15 the complex $Rh_{\mathcal{X}/B*}(E)$ is bounded so Lu^* makes sense.

The proof is in several steps 2.6.4–2.6.6.

Lemma 2.6.4. — *For a quasi-coherent crystal F in $(\mathcal{X}'_{\rm et}/\mathcal{S}')_{\rm cris}$, there is a natural isomorphism*

(2.6.4.1)
$$R(u_*\circ h_{\mathcal{X}'/B'*})(F)\simeq Rh_{\mathcal{X}/B*}Rg_*(F).$$

Proof. — There is a natural isomorphism of functors $u_* \circ h_{\mathcal{X}'/B'*} \simeq h_{\mathcal{X}/B*} \circ g_*$ and hence a natural map

$$\theta : R(u_* \circ h_{\mathcal{X}'/B'*})(F) \simeq R(h_{\mathcal{X}/B*} \circ g_*)(F) \longrightarrow Rh_{\mathcal{X}/B*}Rg_*(F) \tag{2.6.4.2}$$

which we claim is an isomorphism.

To prove this it suffices to show that the map of cohomology sheaves

$$\theta : R^i(u_* \circ h_{\mathcal{X}'/B'*})(F) \simeq R(h_{\mathcal{X}/B*} \circ g_*)(F) \longrightarrow R^i h_{\mathcal{X}/B*}Rg_*(F) \tag{2.6.4.3}$$

is an isomorphism for every i. The sheaf $R^i(u_* \circ h_{\mathcal{X}'/B'*})(F)$ is by 1.6.2 the sheaf associated to the presheaf which to an étale morphism of algebraic spaces $V \to B$ associates

$$H^i((\mathcal{X}'_{V,\text{et}}/\mathcal{S}'_V)_{\text{cris}}, F). \tag{2.6.4.4}$$

Similarly, $R^i h_{\mathcal{X}/B*}Rg_*(F)$ is isomorphic to the sheaf associated to the presheaf

$$(V \to B) \longmapsto H^i((\mathcal{X}_{V,\text{et}}/\mathcal{S}_V)_{\text{cris}}, Rg_*F|_{(\mathcal{X}_{V,\text{et}}/\mathcal{S}_V)_{\text{cris}}}) \tag{2.6.4.5}$$

which by 1.4.25 is isomorphic to the sheaf associated to the presheaf

$$(V \to B) \longmapsto H^i((\mathcal{X}_{V,\text{et}}/\mathcal{S}_V)_{\text{cris}}, Rg^V_* F), \tag{2.6.4.6}$$

where $g^V : \mathcal{X}'_V \to \mathcal{X}_V$ denotes the morphism induced by base change.

To prove the lemma, it therefore suffices to show that the natural map

$$H^*((\mathcal{X}'_{\text{et}}/\mathcal{S}')_{\text{cris}}, F) \longrightarrow H^*((\mathcal{X}_{\text{et}}/\mathcal{S})_{\text{cris}}, Rg_*F) \tag{2.6.4.7}$$

is a isomorphism. Let $X \to \mathcal{X}$ be an étale surjection with X an algebraic space, and let $X_\bullet$ be the 0-coskeleton. Denote by $X'_\bullet$ the base change of $X_\bullet$ to $\mathcal{X}'$ so that there is a commutative square

$$\begin{array}{ccc} X'^+_\bullet & \xrightarrow{g_\bullet} & X^+_\bullet \\ \downarrow & & \downarrow \\ \mathcal{X}' & \xrightarrow{g} & \mathcal{X}. \end{array} \tag{2.6.4.8}$$

By 1.5.4 and 1.6.2, the natural maps

$$H^*((\mathcal{X}_{\text{et}}/\mathcal{S})_{\text{cris}}, Rg_*F) \longrightarrow H^*((X^+_{\bullet,\text{et}}/\mathcal{S})_{\text{cris}}, Rg_*(F)|_{X^+_{\bullet,\text{et}}}) \tag{2.6.4.9}$$

$$H^*((\mathcal{X}'_{\text{et}}/\mathcal{S}')_{\text{cris}}, F) \longrightarrow H^*((X'^+_{\bullet,\text{et}}/\mathcal{S}')_{\text{cris}}, F|_{X'^+_{\bullet,\text{et}}}) \tag{2.6.4.10}$$

are isomorphisms. Moreover, it follows from 1.4.25 that

$$Rg_*(F)|_{X^+_{\bullet,\text{et}}} \simeq Rg_{\bullet\text{cris}*}(F|_{X'^+_{\bullet,\text{et}}}), \tag{2.6.4.11}$$

where $g_{\text{cris}} : (X'^+_{\bullet,\text{et}}/\mathcal{S})_{\text{cris}} \to (X_{\bullet,\text{et}}/\mathcal{S})_{\text{cris}}$ denotes the morphism of topoi induced by $g_\bullet$. The functor $g_{\bullet\text{cris}*}$ takes injective abelian sheaves to injective abelian sheaves since it has an exact left adjoint, and consequently the map

$$H^*((X'^+_{\bullet,\text{et}}/\mathcal{S}')_{\text{cris}}, F|_{X'^+_{\bullet,\text{et}}}) \longrightarrow H^*((X^+_{\bullet,\text{et}}/\mathcal{S})_{\text{cris}}, Rg_*(F)|_{X^+_{\bullet,\text{et}}}) \tag{2.6.4.12}$$

is an isomorphism. □

2.6.5. — To define a morphism

$$Lu^*Rh_{\mathcal{X}/B*}(E) \longrightarrow Rh_{\mathcal{X}'/B'*}(g^*E) \tag{2.6.5.1}$$

it suffices by adjunction to define a morphism

$$Rh_{\mathcal{X}/B*}(E) \longrightarrow Ru_*Rh_{\mathcal{X}'/B'*}(g^*E). \tag{2.6.5.2}$$

By 2.6.4 there is a natural isomorphism

$$Ru_*Rh_{\mathcal{X}'/B'*}(g^*E) \simeq Rh_{\mathcal{X}/B*}Rg_*(g^*E), \tag{2.6.5.3}$$

and hence there is a natural map (2.6.5.2) obtained from the adjunction map $E \to Rg_*g^*E$. We define (2.6.5.1) to be the induced morphism.

2.6.6. — To prove that (2.6.5.1) is an isomorphism, we can without loss of generality assume that $B = \operatorname{Spec}(A)$ and $B' = \operatorname{Spec}(A')$ are affine schemes.

Consider first the special case when $\mathcal{X}$ is an affine scheme. In this case there exists a smooth lift $Y/\mathcal{S}$ of $\mathcal{X}$ and using 2.5.4 and our assumptions, the arrow (2.6.5.1) is identified with the morphism obtained from the isomorphism

$$A' \otimes_A (\mathcal{E} \otimes \Omega^\bullet_{\mathcal{X}/\mathcal{S}}) \longrightarrow \mathcal{E} \otimes_{\mathcal{O}_\mathcal{X}} \Omega^\bullet_{\mathcal{X}'/\mathcal{S}'}. \tag{2.6.6.1}$$

Thus the result holds in this case.

We deduce the general case from this special case as follows. Let $X \to \mathcal{X}$ be a smooth cover with X an algebraic space, let $X_\bullet$ be the 0-coskeleton, and let $X'_\bullet \to \mathcal{X}'$ be the base change to $\mathcal{X}'$. Then by 1.5.4 the projections induce isomorphisms

$$Rh_{\mathcal{X}/B*}(E) \simeq Rh_{X^+_\bullet/B*}(E|_{X^+_\bullet}), \quad Rh_{\mathcal{X}'/B'*}(g^*E) \simeq Rh_{X'^+_\bullet/B'*}(g^*_\bullet(E|_{X^+_\bullet})), \tag{2.6.6.2}$$

where $h_{X^+_\bullet/B}$ and $h_{X'^+_\bullet/B'}$ denote the augmentations

$$(X^+_{\bullet,\mathrm{et}}/\mathcal{S})_{\mathrm{cris}} \longrightarrow B_{\mathrm{et}} \quad \text{and} \quad (X'^+_{\bullet,\mathrm{et}}/\mathcal{S}')_{\mathrm{cris}} \longrightarrow B'_{\mathrm{et}} \tag{2.6.6.3}$$

defined by the morphisms $h_{X_n/B} : (X_{n,\mathrm{et}}/\mathcal{S})_{\mathrm{cris}} \to B_{\mathrm{et}}$ and $h_{X'_n/B'} : (X'_{n,\mathrm{et}}/\mathcal{S}')_{\mathrm{cris}} \to B'_{\mathrm{et}}$ and $g_\bullet : X'_\bullet \to X_\bullet$ denotes the projection. Let $B_\bullet$ (resp. $B'_\bullet$) denote the constant simplicial scheme defined by B (resp. B'), let $\pi : B_\bullet \to B$ (resp. $\pi' : B'_\bullet \to B'$) be the projection, and let $u_\bullet : B'_\bullet \to B_\bullet$ be the morphism induced by u. Let

$$h_{X^+_\bullet/B^+_\bullet} : (X^+_{\bullet,\mathrm{et}}/\mathcal{S})_{\mathrm{cris}} \longrightarrow B^+_{\bullet,\mathrm{et}}, \quad h_{X'^+_\bullet/B'^+_\bullet} : (X'^+_{\bullet,\mathrm{et}}/\mathcal{S}')_{\mathrm{cris}} \longrightarrow B'^+_{\bullet,\mathrm{et}} \tag{2.6.6.4}$$

be the morphisms of strictly simplicial topoi induced by the morphisms $h_{X_n/B}$ and $h_{X'_n/B'}$. The construction 2.6.5 then gives a morphism in the derived category of sheaves of $\mathcal{O}_{B'^+_{\bullet\mathrm{et}}}$-modules

$$Lu^*_\bullet Rh_{X^+_\bullet/B^+_\bullet*}(E|_{X^+_\bullet}) \longrightarrow Rh_{X'^+_\bullet/B'^+_\bullet*}(g^*_\bullet E|_{X^+_\bullet}) \tag{2.6.6.5}$$

such that (2.6.5.1) is obtained by applying $R\pi'_*$. It follows that it suffices to prove the theorem for each X_n.

This reduces the proof to the case when $\mathcal{X}$ is an algebraic space. Repeating the above argument with an étale cover of $\mathcal{X}$ by a scheme, we are further reduced to the case when $\mathcal{X}$ is a quasi-compact and quasi-separated scheme over B. Repeating the argument once again with a cover of $\mathcal{X}$ by affine opens reduces the problem to the case when $\mathcal{X}$ is quasi-affine scheme. In this case the intersection of two affines is again affine so repeating the argument one last time with a cover of $\mathcal{X}$ by affines we are reduced to the case when $\mathcal{X}$ is an affine scheme. This completes the proof of 2.6.2. □

Corollary 2.6.7. — *With notation as in 2.6.2, assume $\mathcal{X} \to B$ is proper, E is a coherent crystal flat over B, and that B is noetherian. Then for any integer i the sheaves $R^i h_{\mathcal{X}/B*}(E)$ are coherent sheaves on B_{et}.*

Proof. — Since B is noetherian and $B_0 \subset B$ is defined by a PD-ideal, the ideal J of B_0 in B is a nilpotent ideal. Let $B_n \subset B$ denote the subspace defined by J^{n+1}.

Consider first the sheaves $R^i h_{\mathcal{X}/B_0*}(E)$. By 2.5.4, this sheaf is isomorphic to $R^i h_*(\mathcal{E} \otimes \Omega^\bullet_{\mathcal{X}_{\mathrm{et}}/\mathcal{S}_0})$, where $\mathcal{E}$ denotes the module with integrable connection associated to E. The spectral sequence of a filtered complex gives a spectral sequence

$$E_1^{st} = R^t h_*(\mathcal{E} \otimes \Omega^s_{\mathcal{X}_{\mathrm{et}}/\mathcal{S}_0}) \Longrightarrow R^{s+t} h_*(\mathcal{E} \otimes \Omega^\bullet_{\mathcal{X}_{\mathrm{et}}/\mathcal{S}_0}), \tag{2.6.7.1}$$

and hence in this case the result follows from the fact that the sheaves $R^t h_*(\mathcal{E} \otimes \Omega^s_{\mathcal{X}_{\mathrm{et}}/\mathcal{S}_0})$ are coherent by [**68**, 7.13].

By induction for the general case it suffices to show that if $R^i h_{\mathcal{X}/B_j*}(E)$ is coherent for all $j < n$ then $R^i h_{\mathcal{X}/B_n}(E)$ is also coherent. For this note that the exact sequence

$$0 \longrightarrow J^n/J^{n+1} \longrightarrow \mathcal{O}_{B_n} \longrightarrow \mathcal{O}_{B_{n-1}} \longrightarrow 0 \tag{2.6.7.2}$$

induces a distinguished triangle
(2.6.7.3)

$$Rh_{\mathcal{X}/B_n*}(E) \otimes^{\mathbb{L}} J^n/J^{n+1} \longrightarrow Rh_{\mathcal{X}/B_n*}(E) \longrightarrow Rh_{\mathcal{X}/B_n*}(E) \otimes^{\mathbb{L}} \mathcal{O}_{B_{n-1}} \xrightarrow{+1} .$$

Since

$$Rh_{\mathcal{X}/B_n*}(E) \otimes^{\mathbb{L}} J^n/J^{n+1} \simeq Rh_{\mathcal{X}/B_{n-1}*}(E) \otimes^{\mathbb{L}}_{\mathcal{O}_{B_{n-1}}} J^n/J^{n+1} \tag{2.6.7.4}$$

and

$$Rh_{\mathcal{X}/B_n*}(E) \otimes^{\mathbb{L}} \mathcal{O}_{B_{n-1}} \simeq Rh_{\mathcal{X}/B_{n-1}*}(E) \tag{2.6.7.5}$$

by 2.6.2, consideration of the long exact sequence associated to (2.6.7.3) and induction implies that the cohomology sheaves of $Rh_{\mathcal{X}/B_n*}(E)$ are coherent. □

Corollary 2.6.8. — *With assumptions as in 2.6.7, the complex $Rh_{\mathcal{X}/B*}(E)$ on B_{et} is perfect.*

Proof. — This follows from 2.6.7 and 2.5.15. □

2.7. P-adic theory

2.7.1. — We extend the results of the previous sections to a p-adic theory just as in the classical case [**8**, Chapter 7].

Let (A, I, γ) be a PD-ring, and let $P \subset I$ be a sub-PD-ideal such that A is P-adically complete and separated. Assume that some prime p is contained in P. Denote by $B = \mathrm{Spec}(A)$, $\widehat{B} = \mathrm{Spf}(A)$, and $B_n = \mathrm{Spec}(A/P^{n+1})$. Let $\pi : \mathcal{S} \to B$ be an algebraic stack to which γ extends, and let $\mathcal{S}_n$ denote $\mathcal{S} \times_B B_n$.

Since we do not know of a theory of formal algebraic stacks, we will consider the $\mathcal{S}_n$ just as a system of algebraic stacks indexed by $n \in \mathbb{N}$ together with a closed immersion $\mathcal{S}_n \hookrightarrow \mathcal{S}_{n+1}$ for every n. We write $\widehat{\mathcal{S}}$ for this system.

Remark 2.7.2. — A more formal definition of $\widehat{\mathcal{S}}$ is the following. View $\mathbb{N}$ as a category in which $\mathrm{Hom}(n, m) = \varnothing$ if $m < n$ and $\mathrm{Hom}(n, m) = \{*\}$ if $m \geq n$. Then $\widehat{\mathcal{S}}$ can be viewed as the fibered category over $\mathbb{N}$ whose objects are triples (n, T, α) where $n \in \mathbb{N}$, T is a scheme and $\alpha \in \mathcal{S}_n(T)$. A morphism $(n, T, \alpha) \to (m, T', \beta)$ in $\widehat{\mathcal{S}}$ is a morphism $n \to m$ in $\mathbb{N}$ giving $j : \mathcal{S}_n \hookrightarrow \mathcal{S}_m$ (*i.e.*, $m \geq n$), a morphism $f : T \to T'$ of schemes and an isomorphism $f^* j^* \beta \simeq \alpha$ in $\mathcal{S}_n(T)$.

2.7.3. — For any morphism of algebraic stacks $\mathcal{X} \to \mathcal{S}_0$ such that γ extends to $\mathcal{X}$, define $\mathrm{Cris}(\mathcal{X}_{\text{lis-et}}/\widehat{\mathcal{S}})$ to be the site made up of two-commutative diagrams

$$\begin{array}{ccc} U & \xrightarrow{j} & T \\ \downarrow & & \downarrow h \\ \mathcal{X} & \longrightarrow & \mathcal{S}, \end{array} \tag{2.7.3.1}$$

where j is a PD-immersion of schemes, h is a PD-morphism, $U \to \mathcal{X}$ is smooth, and $P^n \mathcal{O}_T = 0$ for some n. The resulting topos $(\mathcal{X}_{\text{lis-et}}/\widehat{\mathcal{S}})_{\text{cris}}$ is naturally a ringed topos whose structure sheaf we denote by $\mathcal{O}_{\mathcal{X}_{\text{lis-et}}/\widehat{\mathcal{S}}}$. Let

$$u_{\mathcal{X}_{\text{lis-et}}/\widehat{\mathcal{S}}} : (\mathcal{X}/\widehat{\mathcal{S}})_{\text{cris}} \longrightarrow \mathcal{X}_{\text{lis-et}} \tag{2.7.3.2}$$

be the morphism of topoi with

$$u_{\mathcal{X}/\widehat{\mathcal{S}}*}(F)(U) = \Gamma((U/\widehat{\mathcal{S}})_{\text{cris}}, F) \tag{2.7.3.3}$$

as in [**8**, 7.27].

As usual, a sheaf F on $\mathrm{Cris}(\mathcal{X}/\widehat{\mathcal{S}})$ is equivalent to a sheaf F_T on T_{et} for each object T together with a map $u^{-1} F_T \to F_{T'}$ for each morphism $u : T' \to T$ satisfying the standard compatibility conditions.

2.7.4. — If $\mathcal{X}$ is a Deligne-Mumford stack let $\mathrm{Cris}(\mathcal{X}_{\text{et}}/\widehat{\mathcal{S}})$ denote the site consisting of the full subcategory of $\mathrm{Cris}(\mathcal{X}_{\text{lis-et}}/\widehat{\mathcal{S}})$ of objects with $U \to T$ étale with the topology induced by that on $\mathrm{Cris}(\mathcal{X}_{\text{lis-et}}/\widehat{\mathcal{S}})$. We write $(\mathcal{X}_{\text{et}}/\widehat{\mathcal{S}})_{\text{cris}}$ for the resulting topos. As in 1.5.1, there is a morphism of topoi

$$r_{\mathcal{X}} : (\mathcal{X}_{\text{lis-et}}/\widehat{\mathcal{S}})_{\text{cris}} \longrightarrow (\mathcal{X}_{\text{et}}/\widehat{\mathcal{S}})_{\text{cris}} \tag{2.7.4.1}$$

with $r_{\mathcal{X}*}$ exact. In particular, the cohomology of a sheaf in $(\mathcal{X}_{\text{lis-et}}/\widehat{\mathcal{S}})_{\text{cris}}$ can be computed in either topos.

Definition 2.7.5. — A sheaf F of $\mathcal{O}_{\mathcal{X}/\widehat{\mathcal{S}}}$-modules is a *crystal* if the map $u^*F_T \to F_{T'}$ is an isomorphism for each $u : T' \to T$. The sheaf F is *quasi-coherent* if each F_T is quasi-coherent.

Proposition 2.7.6. — *Let $Q : U \to \mathcal{X}$ be a smooth surjection with U a Deligne-Mumford stack, and let $U_\bullet$ be the 0-coskeleton of Q. Let $(U^+_{\bullet,\text{lis-et}}/\widehat{\mathcal{S}})_{\text{cris}}$ denote the strictly simplicial topos*

$$[n] \longmapsto (U^+_{n,\text{lis-et}}/\widehat{\mathcal{S}})_{\text{cris}}, \tag{2.7.6.1}$$

and let $\pi : (U^+_{\bullet,\text{lis-et}}/\widehat{\mathcal{S}})_{\text{cris}} \to (\mathcal{X}_{\text{lis-et}}/\widehat{\mathcal{S}})_{\text{cris}}$ be the projection induced by Q. Then as in 1.5.4 for any abelian sheaf $F \in (\mathcal{X}_{\text{lis-et}}/\widehat{\mathcal{S}})_{\text{cris}}$ the adjunction map

$$F \longrightarrow R\pi_*\pi^*F \tag{2.7.6.2}$$

is an isomorphism.

Proof. — As in the proof of 1.5.4, for any integer $i \geq 0$ the sheaf $R^i\pi_*(\pi^*F)$ is the sheaf associated to the presheaf which to an object $(V \hookrightarrow T, \delta) \in \text{Cris}(\mathcal{X}_{\text{lis-et}}/\widehat{\mathcal{S}})$ associates

$$H^i\big(((U_\bullet \times_{\mathcal{X}} V)_{\text{lis-et}}/T)_{\text{cris}}, \pi^*F\big). \tag{2.7.6.3}$$

The result therefore follows from 1.5.4. □

This proposition combined with the following two results in the case when $\mathcal{X}$ is a Deligne-Mumford stack enables one to compute cohomology of crystals in $(\mathcal{X}_{\text{lis-et}}/\mathcal{S})_{\text{cris}}$.

Theorem 2.7.7. — *Let $\mathcal{X} \to \mathcal{S}_0$ be a smooth representable morphism of algebraic stacks with $\mathcal{X}$ a Deligne-Mumford stack. Let $\mathcal{X} \hookrightarrow \mathcal{Y}$ be a closed immersion of Deligne-Mumford stacks with $\mathcal{Y} \to \mathcal{S}$ smooth, and let $\widehat{D}$ be the P-adic completion of the PD-envelope D of $\mathcal{X}$ in $\mathcal{Y}$. If E is a quasi-coherent crystal on $\text{Cris}(\mathcal{X}_{\text{et}}/\widehat{\mathcal{S}})$, then there exist a $\widehat{D}$-module $\mathcal{E}$ with integrable connection ∇ compatible with the natural connection on $\widehat{D}$ and a natural isomorphism*

$$Ru_{\mathcal{X}_{\text{et}}/\widehat{\mathcal{S}}*}E \simeq \mathcal{E} \otimes \Omega^\bullet_{\widehat{D}/\widehat{\mathcal{S}}}, \tag{2.7.7.1}$$

where $\Omega^\bullet_{\widehat{D}/\widehat{\mathcal{S}}}$ denotes the complex $\widehat{D} \otimes_{\mathcal{O}_{\mathcal{Y}}} \Omega^\bullet_{\mathcal{Y}/\mathcal{S}}$.

Proof. — This follows from the same argument used in [**8**, 7.23]. □

Remark 2.7.8. — Let $i : \mathcal{X} \hookrightarrow \mathcal{Z}$ be a second closed immersion of Deligne-Mumford stacks with $\mathcal{Z}$ smooth over $\mathcal{S}$, and let $g : \mathcal{Z} \to \mathcal{Y}$ be an $\mathcal{S}$-morphism such that $g \circ i = j$. Let $\widehat{D}_{\mathcal{Z}}$ denote the p-adic completion of the divided power envelope of $\mathcal{X}$ in $\mathcal{Z}$ and write also $g : \widehat{D}_{\mathcal{Z}} \to \widehat{D}$ for the morphism induced by g. If $\mathcal{E}_{\mathcal{Z}}$ denotes the module with integrable connection on $\widehat{D}_{\mathcal{Z}}$ obtained from 2.7.7, then it follows from the construction of $\mathcal{E}$ and $\mathcal{E}_{\mathcal{Z}}$ in [**8**, 7.23] that there is an induced map

$$g^{-1}(\mathcal{E} \otimes \Omega^\bullet_{\widehat{D}/\widehat{\mathcal{S}}}) \longrightarrow \mathcal{E}_{\mathcal{S}} \otimes \Omega^\bullet_{\widehat{D}_{\mathcal{Z}}/\widehat{\mathcal{S}}} \tag{2.7.8.1}$$

compatible with the isomorphisms with $Ru_{\mathcal{X}_{\mathrm{et}}/\widehat{\mathcal{S}}*}E$ (2.7.7.1) (in particular (2.7.8.1) is a quasi-isomorphism).

Lemma 2.7.9. — *Let $\mathcal{X} \to \mathcal{S}_0$ be a smooth morphism of algebraic stacks with $\mathcal{X}$ a Deligne-Mumford stack. Then after replacing $\mathcal{X}$ by an étale cover there exists an immersion $\mathcal{X} \hookrightarrow \mathcal{Y}$ into a smooth $\mathcal{S}$-stack $\mathcal{Y}$.*

Proof. — Let $U \to \mathcal{S}$ be a smooth surjection with U a scheme, and let $U_0 \to \mathcal{S}_0$ be the pullback to $\mathcal{S}_0$. By the existence of quasi-sections for smooth morphisms, there exists after replacing $\mathcal{X}$ by an étale cover a section $s : \mathcal{X} \to U_0$ over $\mathcal{S}_0$. After further étale localization on $\mathcal{X}$ we may also assume that $\mathcal{X}$ is an affine scheme. In this case there exist an integer r and a closed immersion $\mathcal{X}_0 \hookrightarrow \mathbb{A}^r_{U_0}$ over U_0. This defines an immersion $\mathcal{X} \hookrightarrow \mathbb{A}^r_U$ over $\mathcal{S}$. □

Corollary 2.7.10. — *Let $\mathcal{X} \to \mathcal{S}_0$ be a smooth morphism with $\mathcal{X}$ a Deligne-Mumford stack, and let E be a locally free finitely generated crystal on* $\mathrm{Cris}(\mathcal{X}_{\mathrm{et}}/\widehat{\mathcal{S}})$. *Then $Ru_{\mathcal{X}_{\mathrm{et}}/\widehat{\mathcal{S}}*}E$ is a bounded complex of A-modules on $\mathcal{X}_{\mathrm{et}}$.*

Proof. — The assertion is étale local on $\mathcal{X}$ so by 2.7.9 we may assume that there exists an embedding $\mathcal{X} \hookrightarrow \mathcal{Y}$ for some smooth Deligne-Mumford stack over $\mathcal{S}$. The result then follows from 2.7.7 since the right side of (2.7.7.1) is clearly a bounded complex. □

Theorem 2.7.11. — *With notation as in 2.7.7, let E be a locally free finitely generated crystal on* $\mathrm{Cris}(\mathcal{X}_{\mathrm{et}}/\widehat{\mathcal{S}})$, *and let E_n be the restriction to* $\mathrm{Cris}(\mathcal{X}_{\mathrm{et}}/\mathcal{S}_n)$. *Then for every n, the natural map*

$$A/P^{n+1} \otimes^{\mathbb{L}}_A Ru_{\mathcal{X}_{\mathrm{et}}/\widehat{\mathcal{S}}*}E \longrightarrow Ru_{\mathcal{X}_{\mathrm{et}}/\mathcal{S}_n*}E_n \tag{2.7.11.1}$$

is an isomorphism.

Proof. — This is the same as in [**8**, 7.24]. □

Corollary 2.7.12. — *Assume $P = (p)$ for some prime p. Then for any integer n there is a short exact sequence*

$$\begin{array}{c} 0 \longrightarrow H^n((\mathcal{X}_{\mathrm{et}}/\widehat{\mathcal{S}})_{\mathrm{cris}}, E) \otimes \mathbb{Z}/p^{n+1} \longrightarrow H^n((\mathcal{X}_{\mathrm{et}}/\mathcal{S}_n)_{\mathrm{cris}}, E_n) \\ \swarrow \\ \mathrm{Tor}^1_A(A/p^{n+1}, H^{n+1}((\mathcal{X}_{\mathrm{et}}/\widehat{\mathcal{S}})_{\mathrm{cris}}, E)) \longrightarrow 0. \end{array} \tag{2.7.12.1}$$

Proof. — Tensoring the short exact sequence

(2.7.12.2)
$$0 \longrightarrow A \xrightarrow{p^{n+1}} A \longrightarrow A_n \longrightarrow 0$$

with $K := Ru_{\mathcal{X}_{\mathrm{et}}/\widehat{\mathcal{S}}*}E$, we obtain a distinguished triangle

(2.7.12.3)
$$K \xrightarrow{p^{n+1}} K \longrightarrow K \otimes^{\mathbb{L}} A_n \longrightarrow K[1].$$

This triangle induces by the preceding lemma a long exact sequence

(2.7.12.4)
$$\cdots \longrightarrow H^n((\mathcal{X}_{\mathrm{et}}/\widehat{\mathcal{S}})_{\mathrm{cris}}, E) \xrightarrow{p^{n+1}} H^n((\mathcal{X}_{\mathrm{et}}/\widehat{\mathcal{S}})_{\mathrm{cris}}, E) \longrightarrow H^n((\mathcal{X}_{\mathrm{et}}/\mathcal{S}_n)_{\mathrm{cris}}, E_n) \longrightarrow H^{n+1}((\mathcal{X}_{\mathrm{et}}/\widehat{\mathcal{S}})_{\mathrm{cris}}, E) \longrightarrow \cdots.$$

From this the corollary follows. □

Remark 2.7.13. — In [**8**], the base ring A is assumed noetherian. For the results cited above, however, this assumption is unnecessary.

CHAPTER 3

THE CARTIER ISOMORPHISM AND APPLICATIONS

In this chapter we study the Cartier isomorphism in the stack theoretic context. Most of the results concern a smooth representable locally separated morphism $\mathcal{X} \to \mathcal{S}$ of algebraic stacks in characteristic $p > 0$, with $\mathcal{X}$ a Deligne-Mumford stack. This is rather restrictive assumptions. However, since for most applications the Cartier isomorphism will be used for local calculations this will enable us to deduce results for more general morphisms of algebraic stacks by working locally in the lisse-étale topology on the source. In particular, we generalize Ogus' generalization of Mazur's theorem [**58**, 7.3.1] and some of its consequences to arbitrary smooth morphisms of algebraic stacks.

3.1. Cartier descent

Let p be a fixed prime number, and let $\mathcal{S}$ be an algebraic stack over an $\mathbb{F}_p$-scheme T.

Definition 3.1.1. — The *Frobenius morphism* $F_{\mathcal{S}} : \mathcal{S} \to \mathcal{S}$ over the absolute Frobenius morphism $F_T : T \to T$ of T is the morphism of fibered categories which sends an object $x \in \mathcal{S}(X)$ over a T-scheme X to $F_X^*(x)$ and a morphism $\varphi : x \to x'$ in $\mathcal{S}(X)$ to $F_X^*(\varphi) : F_X^* x \to F_X^* x'$, where F_X denotes the absolute Frobenius morphism on the scheme X.

3.1.2. — If $x : \mathcal{X} \to \mathcal{S}$ is a morphism of algebraic stacks over $\mathbb{F}_p$, we can form the diagram

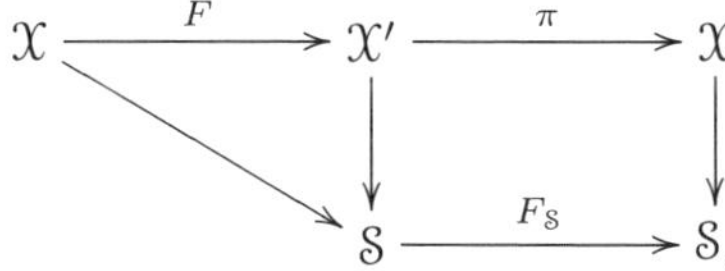

where $\pi \circ F$ is the absolute Frobenius morphism on $\mathcal{X}$, and the square is cartesian.

Warning 3.1.3. — Even if $\mathcal{X}$ is a scheme, the stack $\mathcal{X}'$ need not be a Deligne-Mumford stack.

For example, let $G/\mathbb{F}_p$ be a smooth group scheme, $\mathcal{X} = \mathrm{Spec}(\mathbb{F}_p)$, and let $\mathcal{X} \to BG$ be the map corresponding to the trivial torsor. Then there is a natural isomorphism $\mathcal{X}' \simeq BG^F$, where G^F denotes the kernel of Frobenius on G. To see this, note that $\mathcal{X}'$ is the stack which to any $\mathbb{F}_p$-scheme T associates the groupoid of pairs (P, ϵ), where P is a G-torsor on T and $\epsilon : F_T^*P \simeq G$ is an isomorphism of torsors. Given such a pair (P, ϵ), define P^a to be the G^F-torsor of trivializations $\rho : P \simeq G$ for which $F_T^*(\rho) = \epsilon$. The association $P \mapsto P^a$ defines a functor $\mathcal{X}' \to BG^F$ which is an equivalence.

Lemma 3.1.4. — *If $\mathcal{X} \to \mathcal{S}$ is smooth, then the morphism $F : \mathcal{X} \to \mathcal{X}'$ is flat and surjective.*

Proof. — Let $h : X \to \mathcal{X}$ be a smooth surjective morphism with X a scheme. There is then a commutative diagram

$$\begin{array}{ccccc} X & \xrightarrow{F} & X' & \longrightarrow & X \\ {\scriptstyle h}\downarrow & & \downarrow{\scriptstyle h'} & & \downarrow{\scriptstyle h} \\ \mathcal{X} & \xrightarrow{F} & \mathcal{X}' & \longrightarrow & \mathcal{X} \end{array} \tag{3.1.4.1}$$

over $F_{\mathcal{S}} : \mathcal{S} \to \mathcal{S}$, where h and h' are smooth and surjective. By [**15**, IV.2.2.11 (ii)], to verify that $F : \mathcal{X} \to \mathcal{X}'$ is flat and surjective it suffices to show that $F \circ h : X \to \mathcal{X}'$ is flat and surjective. Thus if $F : X \to X'$ is flat and surjective the result also holds for $\mathcal{X}$. This reduces the proof to the case when $\mathcal{X}$ is an algebraic space.

Let $S \to \mathcal{S}$ be a smooth surjection with S a scheme. To verify that $F : \mathcal{X} \to \mathcal{X}'$ is flat and surjective it suffices to show that the map

$$\mathcal{X} \times_{\mathcal{S}} S \longrightarrow \mathcal{X}' \times_{\mathcal{S}} S \simeq (\mathcal{X} \times_{\mathcal{S}} S) \times_{S, F_S} S \tag{3.1.4.2}$$

obtained by base change is flat and surjective. This reduces the proof to the case when both $\mathcal{S}$ and $\mathcal{X}$ are algebraic spaces.

Assuming this, note that by further replacing $\mathcal{S}$ and $\mathcal{X}$ by étale covers, we may assume that there exists an étale morphism $\mathcal{X} \to \mathbb{A}^r_{\mathcal{S}}$ for some integer r. There is then a cartesian diagram

$$\begin{array}{ccc} \mathcal{X} & \xrightarrow{F} & \mathcal{X}' \\ {\scriptstyle a}\downarrow & & \downarrow{\scriptstyle b} \\ \mathbb{A}^r_{\mathcal{S}} & \xrightarrow{F} & \mathbb{A}^r_{\mathcal{S}} \end{array} \tag{3.1.4.3}$$

with a and b étale. This further reduces the proof to the case when $\mathcal{X} = \mathbb{A}^r_{\mathcal{S}}$.

In this case the morphism is obtained by base change to $\mathcal{S}$ from the morphism $\mathbb{A}^r_{\mathbb{F}_p} \to \mathbb{A}^r_{\mathbb{F}_p}$ raising the coordinates to the p-th power. This implies the lemma. □

Example 3.1.5. — Consider the example in 3.1.3 with $\mathcal{X} = \mathrm{Spec}(\mathbb{F}_p)$, $\mathcal{S} = BG$, and $S \to \mathcal{S}$ also the map $\mathrm{Spec}(\mathbb{F}_p) \to BG$ defined by the trivial torsor. In this case the morphism (3.1.4.2) is equal to the absolute Frobenius morphism $F_G : G \to G$ of G. Since $G/\mathbb{F}_p$ is smooth the morphism F_G is finite and flat. Of course it is also clear directly that the morphism $\mathrm{Spec}(\mathbb{F}_p) \to BG^F$ is finite and flat since G^F is a finite flat group scheme over $\mathbb{F}_p$.

3.1.6. — For any smooth representable locally separated morphism of stacks $\mathcal{X} \to \mathcal{S}$ with $\mathcal{X}$ a Deligne-Mumford stack, the sheaf $T_{\mathcal{X}_{\mathrm{et}}/\mathcal{S}}$ on $\mathcal{X}_{\mathrm{et}}$ has a natural structure of a sheaf of restricted p-Lie algebras, characterized by the property that for any smooth morphism $S \to \mathcal{S}$ with S a scheme, the map $\mathrm{pr}_1^{-1} T_{\mathcal{X}_{\mathrm{et}}/\mathcal{S}} \to T_{\mathcal{X}_{S,\mathrm{et}}/S}$ is a map of sheaves of restricted p-Lie algebras, where $\mathrm{pr}_1 : \mathcal{X}_S := \mathcal{X} \times_{\mathcal{S}} S \to \mathcal{X}$ denotes the projection and $T_{\mathcal{X}_{S,\mathrm{et}}/S}$ is given a p-Lie algebra structure as in [**44**, 5.0]. This follows from the same reasoning used in the construction of the Lie-algebra structure in 2.2.6. We denote by

$$\partial \longmapsto \partial^{(p)} \tag{3.1.6.1}$$

the map $T_{\mathcal{X}_{\mathrm{et}}/\mathcal{S}} \to T_{\mathcal{X}_{\mathrm{et}}/\mathcal{S}}$ defined by the p-Lie algebra structure.

Definition 3.1.7. — The *p-curvature* of a module with integrable connection $(\mathcal{E}, \nabla)$ on $\mathcal{X}_{\mathrm{et}}/\mathcal{S}$ is the map

$$\psi : T_{\mathcal{X}_{\mathrm{et}}/\mathcal{S}} \longrightarrow \mathcal{E}nd_{\mathcal{O}_{\mathcal{X}_{\mathrm{et}}}}(\mathcal{E}) \tag{3.1.7.1}$$

defined by

$$\psi(\partial) = \nabla_\partial^p - \nabla_{\partial^{(p)}}. \tag{3.1.7.2}$$

Note that $\psi(\partial)$ is $\mathcal{O}_{\mathcal{X}_{\mathrm{et}}}$-linear, since this can be verified after base change $S \to \mathcal{S}$ in which case it follows from the classical theory [**44**, 5.0]. Similarly the morphism ψ is p-linear in the sense that for any local section $f \in \mathcal{O}_{\mathcal{X}_{\mathrm{et}}}$ and $\partial \in T_{\mathcal{X}_{\mathrm{et}}/\mathcal{S}}$ we have $\psi(f\partial) - f^p\psi(\partial)$ [**44**, 5.2].

The following is a stack-theoretic generalization of Cartier's fundamental theorem [**44**, 5.1]:

Theorem 3.1.8. — *Let B be an $\mathbb{F}_p$-scheme and let $x : \mathcal{X} \to \mathcal{S}$ be a smooth representable locally separated morphism of algebraic stacks over B, with $\mathcal{X}$ a Deligne-Mumford stack.*
(i) *For any quasi-coherent sheaf $\mathcal{E}'$ on $\mathcal{X}'$ there exists a unique connection ∇^{can} on $F^*\mathcal{E}'$ which is characterized by the following property: for any smooth morphism $S \to \mathcal{S}$ with S an algebraic space, the pullback $\mathrm{pr}_1^*\nabla^{\mathrm{can}}$ of ∇^{can} to a connection on $(F^*\mathcal{E}')|_{(\mathcal{X}\times_{\mathcal{S}} S)_{\mathrm{et}}}$ kills the image of the natural map*

$$F^{-1}\mathcal{E}'|_{(\mathcal{X}'\times_{\mathcal{S}} S)_{\mathrm{et}}} \longrightarrow (F^*\mathcal{E}')|_{(\mathcal{X}\times_{\mathcal{S}} S)_{\mathrm{et}}}. \tag{3.1.8.1}$$

Note that here the étale topos $(\mathcal{X}' \times_{\mathcal{S}} S)_{\mathrm{et}}$ is defined since $\mathcal{X}' \to \mathcal{S}$ is representable.
(ii) *The functor $\mathcal{E}' \mapsto (F^*\mathcal{E}', \nabla^{\mathrm{can}})$ induces an equivalence between the category of quasi-coherent sheaves on $\mathcal{X}'$ and the category of quasi-coherent modules with integrable connection and p-curvature 0 on $\mathcal{X}_{\mathrm{et}}/\mathcal{S}$.*

Warning 3.1.9. — If $(\mathcal{E}, \nabla)$ is a module with integrable connection and p-curvature 0, then we denote by $\mathcal{E}^{\nabla}_{\diamondsuit}$ the corresponding quasi-coherent sheaf on $\mathcal{X}'_{\text{lis-et}}$. As will be explained in section 3.3 below, there is a natural map $F_*\mathrm{Ker}(\nabla) \to \mathcal{E}^{\nabla}_{\diamondsuit}$ but this map is in general not an isomorphism.

Proof of 3.1.8. — Choose a smooth cover $S \to \mathcal{S}$ and let $S_\bullet$ be the associated simplicial algebraic space, $\mathcal{X}_\bullet = \mathcal{X} \times_{\mathcal{S}} S_\bullet$, $\mathcal{X}'_\bullet = \mathcal{X}' \times_{\mathcal{S}} S_\bullet$. The space $\mathcal{X}' \times_{\mathcal{S}} S_\bullet$ has the following description. For each i, we have

$$\mathcal{X}'_i \simeq (\mathcal{X} \times_{\mathcal{S}, F_{\mathcal{S}}} \mathcal{S}) \times_{\mathcal{S}} S_i \simeq (\mathcal{X} \times_{\mathcal{S}} S_i) \times_{S_i, F_{S_i}} S_i \tag{3.1.9.1}$$

and the various maps $\mathcal{X}'_i \to \mathcal{X}'_{i'}$ giving the simplicial structure are just those defined by functoriality.

Let $MIC(\mathcal{X}_\bullet/S_\bullet)$ be as in 2.2.13, and let $MIC^{\psi=0}(\mathcal{X}_\bullet/S_\bullet) \subset MIC(\mathcal{X}_\bullet/S_\bullet)$ be the full subcategory of modules with integrable connection $(\mathcal{E}_\bullet, \nabla_\bullet)$ such that each restriction $(\mathcal{E}_i, \nabla_i)$ has p-curvature 0. Since the pullback of a module with integrable connection and p-curvature 0 also has p-curvature 0 (this follows for example from [**44**, 5.1]), an object $(\mathcal{E}_\bullet, \nabla_\bullet) \in MIC(\mathcal{X}_\bullet/S_\bullet)$ lies in $MIC^{\psi=0}(\mathcal{X}_\bullet/S_\bullet)$ if and only if for some i_0 the restriction $(\mathcal{E}_{i_0}, \nabla_{i_0})$ to $\mathcal{X}_{i_0}/S_{i_0}$ has p-curvature 0.

By the definition of the p-curvature of an object $(\mathcal{E}, \nabla) \in MIC(\mathcal{X}_{\text{et}}/\mathcal{S})$, the equivalence (2.2.14) induces an equivalence between the category of objects $(\mathcal{E}, \nabla) \in MIC(\mathcal{X}_{\text{et}}/\mathcal{S})$ with p-curvature 0 and the category $MIC^{\psi=0}(\mathcal{X}_\bullet/S_\bullet)$. On the other hand, by the usual Cartier isomorphism [**44**, 5.1] the category $MIC^{\psi=0}(\mathcal{X}_\bullet/S_\bullet)$ is equivalent to the category of quasi-coherent sheaves on $\mathcal{X}'_{\bullet,\text{et}}$. Since $\mathcal{X}'_\bullet$ is the 0-coskeleton of the smooth surjection $\mathcal{X}'_0 \to \mathcal{X}'$, we also know by [**68**, 6.12] that the category of quasi-coherent sheaves on $\mathcal{X}'_{\bullet,\text{et}}$ is equivalent to the category of quasi-coherent sheaves on $\mathcal{X}'$. From this and 2.2.14 we deduce that ∇^{can} exists and that the resulting functor is an equivalence. □

Notational Remark 3.1.10. — If $(\mathcal{E}, \nabla)$ is any quasi-coherent sheaf with integrable connection on $\mathcal{X}_{\text{et}}/\mathcal{S}$, then we can form for each $i \geq 0$ a quasi-coherent sheaf $\mathcal{H}^i(\mathcal{E} \otimes \Omega^\bullet_{\mathcal{X}_{\text{et}}/\mathcal{S}})_\diamondsuit$ on $\mathcal{X}'$ as follows. Let $S \to \mathcal{S}$ be a smooth cover with S an algebraic space, and $S_\bullet$ and $\mathcal{X}_\bullet = \mathcal{X} \times_{\mathcal{S}} S_\bullet$ as above, then for each S_i let $\mathcal{H}^i(\mathcal{E} \otimes \Omega^\bullet_{\mathcal{X}_{\text{et}}/\mathcal{S}})_{\diamondsuit, \mathcal{X}'_{S_i,\text{et}}}$ be the quasi-coherent sheaf $\mathcal{H}^i(\mathcal{E} \otimes \Omega^\bullet_{\mathcal{X}_i/S_i})$ on $\mathcal{X}'_{S_i,\text{et}}$. For each map $\delta : S_{i'} \to S_i$ there is a natural isomorphism

$$\delta^* \mathcal{H}^i(\mathcal{E} \otimes \Omega^\bullet_{\mathcal{X}_{\text{et}}/\mathcal{S}})_{\diamondsuit, \mathcal{X}'_{S_i,\text{et}}} \simeq \mathcal{H}^i(\mathcal{E} \otimes \Omega^\bullet_{\mathcal{X}_{\text{et}}/\mathcal{S}})_{\diamondsuit, \mathcal{X}'_{S_{i'},\text{et}}}, \tag{3.1.10.1}$$

and hence we get a quasi-coherent sheaf $\mathcal{H}^i(\mathcal{E} \otimes \Omega^\bullet_{\mathcal{X}_{\text{et}}/\mathcal{S}})_\diamondsuit$ on $\mathcal{X}'_{\text{lis-et}}$. This sheaf is characterized by the property that for any smooth morphism $U \to \mathcal{S}$ with U an algebraic space, the restriction of $\mathcal{H}^i(\mathcal{E} \otimes \Omega^\bullet_{\mathcal{X}_{\text{et}}/\mathcal{S}})_\diamondsuit$ to $(\mathcal{X}' \times_{\mathcal{S}} U)_{\text{et}}$ is equal to the pushforward of $\mathcal{H}^i(\mathcal{E}|_{(\mathcal{X} \times_{\mathcal{S}} U)_{\text{et}}} \otimes \Omega^\bullet_{(\mathcal{X} \times_{\mathcal{S}} U)_{\text{et}}/U})$. In particular, the sheaf $\mathcal{H}^i(\mathcal{E} \otimes \Omega^\bullet_{\mathcal{X}_{\text{et}}/\mathcal{S}})_\diamondsuit$ is independent of the particular choice of smooth cover $S \to \mathcal{S}$ in the above. This sheaf will play an important role in section 3.3 when we discuss the Cartier isomorphism.

Remark 3.1.11. — It follows from the construction of the sheaf $\mathcal{H}^0(\mathcal{E}\otimes\Omega^\bullet)_\diamond$ in 3.1.10 that there is a natural morphism

$$\mathcal{H}^0(\mathcal{E}\otimes\Omega^\bullet)_\diamond \longrightarrow F_*\mathcal{E} \tag{3.1.11.1}$$

of quasi-coherent sheaves on $\mathcal{X}'$. In fact from the corresponding result for schemes one deduces that the functor

$$MIC(\mathcal{X}/\mathcal{S}) \longrightarrow \mathrm{Mod}_{\mathrm{qcoh}}(\mathcal{X}'),\quad (\mathcal{E},\nabla)\mapsto \mathcal{H}^0(\mathcal{E}\otimes\Omega^\bullet_{\mathcal{X}/\mathcal{S}})_\diamond \tag{3.1.11.2}$$

is right adjoint to the functor

$$\mathrm{Mod}_{\mathrm{qcoh}}(\mathcal{X}') \longrightarrow MIC(\mathcal{X}/\mathcal{S}),\quad \mathcal{E}'\longmapsto (F^*\mathcal{E}',\nabla^{\mathrm{can}}). \tag{3.1.11.3}$$

3.2. Frobenius acyclic stacks

In order to generalize to stacks the classical results relying on the Cartier isomorphism, it unfortunately seems necessary to make certain restrictions on the stacks considered.

Let B be a locally noetherian $\mathbb{F}_p$-scheme.

Definition 3.2.1. — An algebraic stack $\mathcal{S}$ locally of finite type over B is *Frobenius acyclic* if for every integer $i>0$ the sheaf $R^iF_{\mathcal{S}*}\mathcal{O}_{\mathcal{S}_{\text{lis-et}}}$ on $\mathcal{S}_{\text{lis-et}}$ is zero.

Example 3.2.2. — For an example of a stack which is not Frobenius acyclic, consider the stack $\mathcal{S} := B\mathbb{G}_a$ over $\mathbb{F}_p$. To compute $R^iF_{\mathcal{S}*}\mathcal{O}_{\mathcal{S}_{\text{lis-et}}}$, let $\mathrm{Spec}(\mathbb{F}_p)\to\mathcal{S}$ be the morphism corresponding to the trivial torsor. There is then an isomorphism of stacks

$$B\alpha_p \simeq \mathrm{Spec}(k)\times_{\mathcal{S},F_{\mathcal{S}}}\mathcal{S}, \tag{3.2.2.1}$$

where α_p denotes the kernel of Frobenius on $\mathbb{G}_a$. Therefore the restriction of

$$R^iF_{\mathcal{S}*}\mathcal{O}_{\mathcal{S}_{\text{lis-et}}} \tag{3.2.2.2}$$

to $\mathrm{Spec}(\mathbb{F}_p)$ is equal to the group cohomology $H^i(\alpha_p,\mathbb{F}_p)$ which need not vanish. For example, for $i=1$ we have a nonzero cohomology class provided by the extension E of the trivial representation $\mathbb{F}_p$ by itself corresponding to the 2-dimensional representation $\rho:\alpha_p\to GL_2$ sending $f\in\alpha_p$ to the matrix $\left(\begin{smallmatrix}1 & f\\ 0 & 1\end{smallmatrix}\right)$.

The main result of this section is the following result which often makes it easy to determine if a stack $\mathcal{S}$ is Frobenius acyclic.

Theorem 3.2.3. — *Let $\mathcal{S}/B$ be an algebraic stack locally of finite type. If for every geometric point $\bar{x}:\mathrm{Spec}(\bar{k})\to\mathcal{S}$ the stabilizer group scheme $\underline{\mathrm{Aut}}(\bar{x})$ over $\mathrm{Spec}(k(\bar{x}))$ is a diagonalizable group scheme, then $\mathcal{S}$ is Frobenius acyclic.*

The proof is in several steps 3.2.4–3.2.19.

3.2.4. — In [**1**] a theory of Artin stacks with linearly reductive stabilizers is developed. Since this paper is not yet available, let us develop the parts that we need here.

Let B be a locally noetherian scheme and $\mathcal{X}/B$ an Artin stack locally of finite type with finite diagonal. Let $\pi : \mathcal{X} \to X$ be the coarse moduli space of $\mathcal{X}$ (cf. [**2**, 2.2.1]). The space X is a separated algebraic space locally of finite type over B, and the morphism π is proper and quasi-finite. Moreover, for any algebraically closed field k the map $\mathcal{X}(k) \to X(k)$ induces a bijection between the set of isomorphism classes of objects in $\mathcal{X}(k)$ and $X(k)$.

Proposition 3.2.5. — *Let k be an algebraically closed field, and $\bar{x}$:* Spec$(k) \to \mathcal{X}$ *a geometric point such that the automorphism group scheme* $\underline{\text{Aut}}(\bar{x})$ *is a diagonalizable group scheme. Write also $\bar{x}$ for the geometric point of X obtained by composing with the projection $\mathcal{X} \to X$.*

(a) *After possibly replacing X by an fppf neighborhood of $\bar{x}$, there exist a finite X-scheme V and a finite diagonalizable group scheme G/B acting on V over X such that $\mathcal{X} \simeq [V/G]$.*

(b) *There exists an open neighborhood $U \subset X$ containing the image of $\bar{x}$ such that for any morphism of algebraic spaces $U' \to U$ the base change $\mathcal{X} \times_X U' \to U'$ identifies U' with the coarse moduli space of $\mathcal{X} \times_X U'$ (*i.e., *after replacing X by some neighborhood of $\bar{x}$ the formation of the coarse space commutes with arbitrary base change on X).*

Proof. — Note first that (b) follows from (a). Indeed if $X' \to X$ is an fppf neighborhood of $\bar{x}$ such that the pullback $X' \times_X \mathcal{X}$ is isomorphic to $[V/G]$ as in (a), then we claim that in (b) we can take U to be the (open) image of X'. For this let $U' \to U$ be any morphism of algebraic spaces. Since the formation of the coarse moduli space commutes with flat base change on the coarse space [**2**, 2.2.2], to verify that U' is the coarse space of $\mathcal{X} \times_X U'$ it suffices to do so after making the base change $U' \times_U X' \to U'$. This reduces the proof of (b) to showing that if X is an algebraic space and $\mathcal{X} = [V/G]$ is the stack quotient of a finite X-space V by the action of a finite diagonalizable group G, then for any morphism $g : X' \to X$ the map $\mathcal{X} \times_X X' \to X'$ is the coarse moduli space. For this let $\mathcal{A}$ denote the coherent sheaf of algebras on X corresponding to V. Then the coarse space of $\mathcal{X} \times_X X'$ is obtained by taking the subsheaf of G invariants of $g^*\mathcal{A}$. Therefore we need to show that the formation of G-invariants of the coherent sheaf $\mathcal{A}$ commutes with arbitrary pullbacks. This follows from the fact that G is diagonalizable which implies that as an $\mathcal{O}_X$-module we have a canonical decomposition $\mathcal{A} = \oplus_{\chi \in D} \mathcal{A}_\chi$, where the sum is taken over the characters of G. Moreover, this decomposition commutes with pullback.

By a standard limit argument, we can find a scheme B_0 of finite type over an excellent Dedekind ring and an algebraic stack $\mathcal{X}_0$ over B_0 with finite diagonal such that $\mathcal{X}$ is obtained by making a base change $B \to B_0$. Let $\mathcal{X}_0 \to X_0$ be the coarse space of $\mathcal{X}_0$, and let $\bar{x}_0 \to \mathcal{X}_0$ be the geometric point defined by $\bar{x}$. If we prove (a) for the pair $(\mathcal{X}_0, \bar{x}_0)$, then by (b) we get that after possibly shrinking on X_0 we have $\mathcal{X} = \mathcal{X}_0 \times_{X_0} X$, where the map $X \to X_0$ is the one obtained from the universal property of the coarse moduli space (the morphism $\mathcal{X} \to X$ is initial for morphisms to algebraic spaces). It follows that to prove (a) for $(\mathcal{X}, \bar{x})$ it suffices to prove (a) for $(\mathcal{X}_0, \bar{x}_0)$. We may therefore assume that B is of finite type over an excellent Dedekind ring (this is used in the proof of 3.2.6 below).

Since the formation of the coarse moduli space commutes with étale base change on the coarse space, we can, by base changing to the spectrum of a strictly henselian local ring of a point of X, assume that X is the spectrum of a strictly henselian local ring A (and B remains unchanged). Let k be the residue field, and let $\mathcal{X}_k$ denote the base change $\mathcal{X} \times_{\mathrm{Spec}(A)} \mathrm{Spec}(k)$. Choose a finite field extension $k \to k'$ such that $\mathcal{X}_k(k')$ is nonempty. Choose a lifting of $k \to k'$ to a finite flat morphism $A \to A'$ (which exists by [**15**, 0_{III}.10.3], and let $\mathcal{X}'$ denote the base change to $\mathrm{Spec}(A')$. Fix an element $x \in \mathcal{X}(k')$, and let G denote the group scheme of automorphisms of x. After making a further finite extension of k' we can assume that G is diagonalizable. This enables us to view G as a group scheme over B which we do in what follows. Replacing A by A' we may therefore assume that we have an object $x \in \mathcal{X}_k(k)$ whose automorphism group scheme G is diagonalizable. The choice of $x \in \mathcal{X}_k(k)$ defines a morphism $BG \to \mathcal{X}_k$. For any morphism $T \to \mathcal{X}_k$ corresponding to an object $t \in \mathcal{X}_k(T)$, the fiber product $T \times_{\mathcal{X}_k} BG$ is the quotient of $\underline{\mathrm{Isom}}(x|_T, t)$ by the action of G on the first factor (if $g \in G$ and $\sigma : x|_T \to t$ is an isomorphism then $g * \sigma$ is the isomorphism $\sigma \circ g$). Since G acts transitive on the set of isomorphisms $x|_T \to t$ it follows that $T \times_{\mathcal{X}_k} BG \to T$ is a proper monomorphism, and hence a closed immersion. This in turn implies that $BG \to \mathcal{X}_k$ is a closed immersion. Since $\pi : \mathcal{X} \to X$ is the coarse moduli space, for any algebraically closed field Ω the map $\mathcal{X}(\Omega) \to X(\Omega)$ identifies $X(\Omega)$ with the set of isomorphism classes in $\mathcal{X}(\Omega)$. It follows that $BG(\Omega) \to \mathcal{X}_k(\Omega)$ is also a bijection (since X is assumed strictly local). Since $\mathcal{X}_k$ is noetherian it follows that $BG \to \mathcal{X}_k$ is a closed immersion defined by a nilpotent ideal in $\mathcal{O}_{\mathcal{X}_k}$.

We then have a diagram

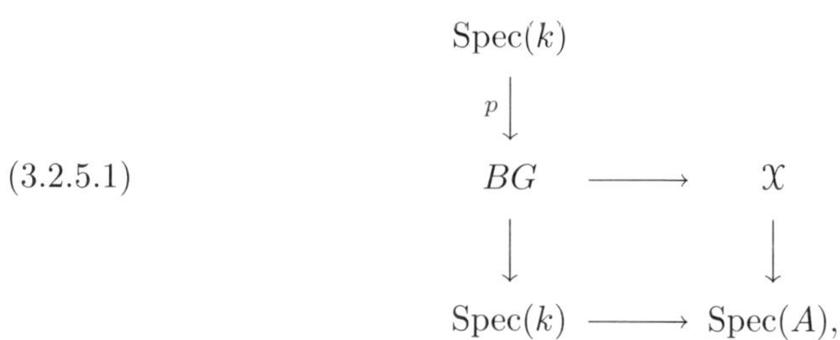

$$(3.2.5.1)\qquad \begin{array}{ccc} \mathrm{Spec}(k) & & \\ {\scriptstyle p}\downarrow & & \\ BG & \longrightarrow & \mathcal{X} \\ \downarrow & & \downarrow \\ \mathrm{Spec}(k) & \longrightarrow & \mathrm{Spec}(A), \end{array}$$

where p is the tautological G-torsor over BG.

To prove the proposition it suffices to show that we can find a G-torsor $P \to \mathcal{X}$ whose pullback to BG is the tautological torsor $\mathrm{Spec}(k) \to BG$. Let us explain why this suffices. If we have such a torsor then since $P \to \mathcal{X}$ and $\mathcal{X} \to \mathrm{Spec}(A)$ are proper and quasi-finite the morphism $P \to \mathrm{Spec}(A)$ is also proper and quasi-finite. Let $I \to P$ denote the inertia stack of P. By definition the stack I is the stack whose fiber over a scheme T is the groupoid whose objects are pairs (p, α), where $p \in P(T)$ and α is an automorphism of p in $P(T)$. A morphism $(p, \alpha) \to (p', \alpha')$ is an isomorphism $\iota : p \to p'$ in $P(T)$ such that $\iota \circ \alpha = \alpha' \circ \iota$. Geometrically the inertia stack can be described as the fiber product

$$\begin{array}{ccc} & & P \\ & & \big\downarrow{\scriptstyle\Delta} \\ P & \xrightarrow{\Delta} & P \times_{\mathrm{Spec}(A)} P, \end{array} \tag{3.2.5.2}$$

so in particular I since P has finite diagonal the stack I is representable and finite over P. By [**49**, 14.2.4] it follows that I is equal to the relative spectrum $\underline{\mathrm{Spec}}_P(\mathcal{A})$ for some coherent sheaf of $\mathcal{O}_P$-algebras on P. We claim that the map $\mathcal{O}_P \to \mathcal{O}_I$ is in fact an isomorphism. The map is injective because it admits a retraction given by the section $P \to I$ sending an object $p \in P(T)$ to the pair (p, id). To check that it is surjective it suffices by Nakayama's lemma to show that the map becomes surjective when pulled back to $\mathrm{Spec}(k)$. But the base change $P \times_{\mathrm{Spec}(A)} \mathrm{Spec}(k)$ is a thickening of the scheme $\mathrm{Spec}(k)$ and hence also a scheme. We conclude that the map $I \to P$ is an isomorphism and hence the objects of P admit no nontrivial automorphisms. Therefore P is an algebraic space proper and quasi-finite over $\mathrm{Spec}(A)$. By [**46**, II.6.16] this implies that P is in fact a finite $\mathrm{Spec}(A)$-scheme. Furthermore, since $P \to \mathcal{X}$ is a G-torsor the induced map $[P/G] \to \mathcal{X}$ is an isomorphism.

Lemma 3.2.6. — *To find a G-torsor $P \to \mathcal{X}$ whose pullback to $BG \hookrightarrow \mathcal{X}$ is the tautological torsor, it suffices to find such a torsor over the completion* $\mathrm{Spec}(\hat{A})$ *of* $\mathrm{Spec}(A)$ *along the closed point.*

Proof. — By "spreading out", we can find a finite type affine B-scheme $\mathrm{Spec}(A_0)$, an algebraic stack $\mathcal{X}_0/A_0$ with finite diagonal, and a morphism $\mathrm{Spec}(A) \to \mathrm{Spec}(A_0)$ such that $\mathcal{X} \simeq \mathcal{X}_0 \times_{\mathrm{Spec}(A_0)} \mathrm{Spec}(A)$.

Let F be the functor on the category of A_0-algebras which to any $A \to R$ associates the set of isomorphism classes of G-torsors on the stack

$$\mathcal{X}_R := \mathcal{X}_0 \times_{\mathrm{Spec}(A_0)} \mathrm{Spec}(R). \tag{3.2.6.1}$$

The functor F is clearly limit preserving so the lemma follows from [**3**, 1.12]. □

By 3.2.6 we may further assume that A is a complete local ring. Let $\mathcal{X}_n$ denote $\mathcal{X} \times_{\mathrm{Spec}(A)} \mathrm{Spec}(A/\mathfrak{m}_A^{n+1})$. By the Grothendieck existence theorem for stacks [**65**, 1.4] to give the G-torsor $P \to \mathcal{X}$ it suffices to find a compatible system $\{P_n \to \mathcal{X}_n\}$ of G-torsors over the reductions. Proposition 3.2.5 therefore follows from consideration of the sequence of closed immersions defined by nilpotent ideals

$$BG \hookrightarrow \mathcal{X}_0 \hookrightarrow \mathcal{X}_1 \hookrightarrow \cdots \tag{3.2.6.2}$$

and the following two lemmas, where we denote BG by $\mathcal{X}_{-1}$. □

Example 3.2.7. — In general the square in (3.2.5.1) is not cartesian. For example let $\mathcal{X}$ be the stack-theoretic quotient of $\mathbb{A}^1 = \mathrm{Spec}(k[z])$ by the multiplication action of μ_n for some $n \geq 2$. Then the coarse moduli space is isomorphic to $\mathbb{A}^1 = \mathrm{Spec}(k[t])$ with the map $k[t] \to k[z]$ given by $t \mapsto z^n$. The pullback of $\mathcal{X}$ to the point $\{t = 0\}$ of the coarse space is then isomorphic to the stack-theoretic quotient of $\mathrm{Spec}(k[z]/z^n)$ by the multiplicative action of μ_n. The μ_n-invariant ideal $(z) \subset k[z]/z^n$ defines a nilpotent ideal in the structure sheaf of $[\mathrm{Spec}(k[z]/z^n)/\mu_n]$ defining a closed immersion $B\mu_n \hookrightarrow [\mathrm{Spec}(k[z]/z^n)/\mu_n]$.

Lemma 3.2.8. — *For any integer $n \geq -1$ and quasi-coherent sheaf $\mathcal{F}$ on $\mathcal{X}_n$ we have $H^i(\mathcal{X}_n, \mathcal{F}) = 0$ for all $i > 0$.*

Proof. — Note that since the closed immersion $j : \mathcal{X}_{-1} \hookrightarrow \mathcal{X}_n$ is defined by a nilpotent ideal, any quasi-coherent sheaf $\mathcal{F}$ admits a finite filtration

$$0 = \mathcal{F}_n \subset \mathcal{F}_{n-1} \subset \cdots \subset \mathcal{F}_0 = \mathcal{V} \tag{3.2.8.1}$$

whose associated graded module is isomorphic to $j_*\mathcal{F}$ for some quasi-coherent sheaf $\mathcal{F}$ on BG. From this one deduces that it suffices to prove the lemma for the stack $\mathcal{X}_{-1} = BG$. But in this case the category of quasi-coherent sheaves on BG is equivalent to the category of linear k-representations of G, and if F is the G-representation corresponding to $\mathcal{F}$ then

$$H^*(BG, \mathcal{F}) \simeq H^*(G, F), \tag{3.2.8.2}$$

where the right side is group cohomology. Since G is diagonalizable we have $H^i(G, F) = 0$ for $i > 0$ by [**14**, I.5.3.3] and the lemma follows. □

Lemma 3.2.9. — *Let $j : \mathcal{Y}_0 \hookrightarrow \mathcal{Y}$ be a closed immersion of algebraic stacks over a base scheme B defined by a nilpotent ideal, let G be a diagonalizable group scheme over B, and let $P_0 \to \mathcal{Y}_0$ be a G-torsor. Assume that for any quasi-coherent sheaf $\mathcal{F}$ on $\mathcal{Y}_0$ we have $H^i(\mathcal{Y}_0, \mathcal{F}) = 0$ for all $i > 0$. Then P_0 lifts to a G-torsor $P \to \mathcal{Y}$.*

Proof. — Note first that if $I \subset \mathcal{O}_{\mathcal{Y}}$ is the ideal defining $\mathcal{Y}_0$ and if $\mathcal{Y}_n \subset \mathcal{Y}$ is the substack defined by I^n then $H^i(\mathcal{Y}_n, \mathcal{F}) = 0$ for all $i > 0$ and all quasi-coherent sheaves $\mathcal{F}$ on $\mathcal{Y}_n$. This follows from the same argument proving 3.2.8. Considering the successive immersions

$$\mathcal{Y}_0 \subset \mathcal{Y}_1 \subset \cdots \mathcal{Y} \tag{3.2.9.1}$$

we reduce to the case when $I^2 = 0$. By [**66**, 1.5], if $x_0 : \mathcal{Y}_0 \to BG$ is the morphism corresponding to P_0, then the obstruction to lifting P_0 to $\mathcal{Y}$ is a class in

$$\mathrm{Ext}^1(Lx^*L_{BG/B}, I). \tag{3.2.9.2}$$

We show that this group is zero. For this it suffices to show that the cotangent complex $L_{BG/B}$ is locally in the flat topology on BG isomorphic to a two-term complex of locally free sheaves of finite rank concentrated in degrees 0 and 1. For then $Lx^*L_{BG/B}$ is also locally in the flat topology on $\mathcal{Y}_0$ isomorphic to a two-term complex of locally free sheaves of finite rank concentrated in degrees 0 and 1, in which case $\mathcal{R}Hom(Lx^*L_{BG/B}, I)$ lies in $D^{[-1,0]}_{\mathrm{qcoh}}(\mathcal{Y}_0)$. From this and our assumption on vanishing of cohomology it then follows that (3.2.9.2) is zero.

To see that $L_{BG/B}$ has the desired form, it suffices to show that the pullback of $L_{BG/B}$ to B along the morphism $f : B \to BG$ defined by the trivial torsor is a two term complex of locally free sheaves of finite rank concentrated in degrees 0 and 1. From the distinguished triangle

$$Lf^*L_{BG/B} \longrightarrow L_{B/B} \longrightarrow L_{B/BG} \longrightarrow Lf^*L_{BG/B}[1] \tag{3.2.9.3}$$

arising from [**68**, 8.1 (ii)] applied to

$$B \xrightarrow{\ f\ } BG \longrightarrow B, \tag{3.2.9.4}$$

we see that $Lf^*L_{BG/B} \simeq L_{B/BG}[-1]$, and hence it suffices to show that $L_{B/BG}$ is isomorphic to a two term complex of locally free sheaves of finite rank concentrated in degrees -1 and 0. Consider the diagram

$$\begin{array}{ccccc} B & \xrightarrow{e} & G & \xrightarrow{a} & B \\ & & {\scriptstyle b}\downarrow & & \downarrow{\scriptstyle f} \\ & & B & \xrightarrow{f} & BG, \end{array} \tag{3.2.9.5}$$

where e is the identity section. Since f is a flat morphism we have $La^*L_{B/BG} \simeq L_{G/B}$ [**68**, 8.1 (ii)], and hence $L_{B/BG} \simeq Le^*L_{G/B}$. It therefore suffices to show that if G is a diagonalizable group scheme over B then $L_{G/B}$ is isomorphic to a two term complex of locally free sheaves of finite rank. Writing G as a product of group schemes of the form μ_n one reduces using the isomorphisms [**68**, 8.1.2] to the case when $G = \mu_n$. Consider the embedding $i : \mu_n \hookrightarrow \mathbb{G}_m$. This is a regular embedding, so if t denotes the standard coordinate on $\mathbb{G}_m$ we find that $L_{\mu_n/B}$ is represented by the complex

$$\mathcal{O}_{\mu_n} \cdot (t^n - 1)/(t^n - 1)^2 \longrightarrow \mathcal{O}_{\mu_n} \cdot \mathrm{dlog}(t), \quad (t^n - 1) \longmapsto n\mathrm{dlog}(t). \tag{3.2.9.6}$$

This completes the proof of 3.2.9. □

Remark 3.2.10. — In fact it is true that *étale* locally on the coarse space X the stack $\mathcal{X}$ is isomorphic to $[V/G]$ for a diagonalizable group scheme G and a finite X-scheme $V \to X$ with action of G. This is shown in [**1**].

Corollary 3.2.11. — *Let $\mathcal{X}/B$ be as in 3.2.5, and assume in addition that for any algebraically closed field k and morphism $\bar{x} : \mathrm{Spec}(k) \to \mathcal{X}$ the automorphism group scheme $\underline{\mathrm{Aut}}(\bar{x})$ is diagonalizable ($\underline{\mathrm{Aut}}(\bar{x})$ is automatically finite since $\mathcal{X}$ has finite diagonal). Let $\pi : \mathcal{X} \to X$ be the coarse moduli space. Then for any quasi-coherent sheaf $\mathcal{F}$ on $\mathcal{X}$ we have $R^i\pi_*\mathcal{F} = 0$ for $i > 0$.*

Proof. — We may work locally in the fppf topology on X and by 3.2.5 may therefore assume that $\mathcal{X} = [V/G]$ for some finite X-scheme V and diagonalizable group G. We can even assume that $X = \mathrm{Spec}(A)$ is affine and that $V = \mathrm{Spec}(R)$ for some finite A-algebra R with action of G. A quasi-coherent sheaf on $\mathcal{X}$ is then given by an R-module F with action of G covering the action on R, and there is an isomorphism of cohomology groups

$$H^*(\mathcal{X}, \mathcal{F}) \simeq H^*(G, F). \tag{3.2.11.1}$$

Indeed if $f : \mathcal{X} \simeq [V/G] \to BG$ denotes the morphism defined by the presentation of $\mathcal{X}$ as a quotient stack, then f is affine (since V is affine), so we have

$$H^*(\mathcal{X}, \mathcal{F}) \simeq H^*(BG, f_*\mathcal{F}), \tag{3.2.11.2}$$

for any quasi-coherent sheaf $\mathcal{F}$ on $\mathcal{X}$. Since X is affine the right side of (3.2.11.2) is canonically isomorphic to $H^*(G, F)$, where F denotes the G-representation corresponding to $\mathcal{F}$.

Again by [**14**, I.5.3.3] we have $H^i(G, F) = 0$ for $i > 0$ since G is diagonalizable, and the corollary follows. □

3.2.12. — Turning now to the proof of 3.2.3, we can without loss of generality assume that $\mathcal{S}$ is quasi-compact. Let $U \to \mathcal{S}$ be a smooth surjection with U an affine scheme. Let $\mathcal{U}'$ denote the base change $\mathcal{S} \times_{F_{\mathcal{S}}, \mathcal{S}} U$. In light of 3.2.11 it suffices to show the following assertions:

(a) $\mathcal{U}'$ has finite diagonal,

(b) the stabilizer group schemes of $\mathcal{U}'$ are diagonalizable,

(c) the morphism $p : \mathcal{U}' \to U$ is proper and quasi-finite (By definition p is quasi-finite if for any algebraically closed field k and $u : \mathrm{Spec}(k) \to U$ the set of isomorphism classes of k-valued points of the fiber product $\mathcal{U}'_u = \mathcal{U}' \times_{U,u} \mathrm{Spec}(k)$ is finite).

Indeed if these assertions hold, let $\eta : \mathcal{U}' \to \overline{\mathcal{U}}'$ be the coarse moduli space of $\mathcal{U}'$ and let $q : \overline{\mathcal{U}}' \to U$ be the projection so that we have a factorization of p

$$\mathcal{U}' \xrightarrow{\ \eta\ } \overline{\mathcal{U}}' \xrightarrow{\ q\ } U \tag{3.2.12.1}$$

with η proper and surjective. Since p is also proper this implies that q is proper [**67**, 2.7], and since p is quasi-finite and η is surjective the morphism q is also quasi-finite. Therefore q is finite. Using this and 3.2.11 we have for any quasi-coherent sheaf $\mathcal{F}$ on $\mathcal{U}'$ and $i > 0$

$$R^i p_*(\mathcal{V}) \simeq R^i q_*(\eta_*\mathcal{V}) = 0. \tag{3.2.12.2}$$

To see that $\mathcal{U}'$ has these properties, note first that the Frobenius morphism F_U on U induces a map $F_{U/\mathcal{S}} : U \to \mathcal{U}'$. The following 3.2.13–3.2.18 hold without any assumptions on $\mathcal{S}$ (other than the assumptions in 0.2.1).

Lemma 3.2.13. — *The morphism $F_{U/\mathcal{S}}$ is finite, flat, and surjective.*

Proof. — This follows from the proof of 3.1.4 in the case when $\mathcal{X}$ is a scheme.

We recall the argument. Let $S \to \mathcal{S}$ be a smooth morphism with S a scheme. Then it suffices to show that the base change of $F_{U/\mathcal{S}}$ to S is finite and flat. That is, the map

$$U \times_{\mathcal{S}} S \longrightarrow \mathcal{U}' \times_{\mathcal{S}} S \simeq (U \times_{\mathcal{S}} S) \times_{S, F_S} S \tag{3.2.13.1}$$

induced by $F_U \times \mathrm{id}$ is finite and flat. This reduces the proof to the case when S is a scheme. Proceeding as in the proof of 3.1.4 one further reduces to the case when U is affine space over $S = \mathrm{Spec}(\mathbb{F}_p)$ in which case the result is immediate. □

Corollary 3.2.14. — *The morphism $\mathcal{U}' \to U$ is surjective with geometrically connected fibers.*

Proof. — Since $U \to \mathcal{U}'$ is surjective this follows from the observation that the Frobenius $F_U : U \to U$ is surjective with geometrically connected fibers. □

Lemma 3.2.15. — *The product $U \times_{F_{U/\mathcal{S}},\mathcal{U}',F_{U/\mathcal{S}}} U$ is finite over $U \times_{F_U,U,F_U} U$.*

Proof. — The space $U \times_{F_U,U,F_U} U$ represents the functor which to any scheme T associates the set of pairs of morphisms $f_i : T \to U$ such that

$$F_U \circ f_1 = F_U \circ f_2. \tag{3.2.15.1}$$

Note that since $f_i \circ F_T = F_U \circ f_i$ $(i = 1, 2)$, the equality 3.2.15.1 is equivalent to

$$f_1 \circ F_T = f_2 \circ F_T. \tag{3.2.15.2}$$

The space $U \times_{F_{U/\mathcal{S}},\mathcal{U}',F_{U/\mathcal{S}}} U$ represents the functor which to any scheme T associates the set of triples (f_1, f_2, σ), where $f_1, f_2 : T \to U$ are morphisms defining a point of $U \times_{F_U,U,F_U} U$ and $\sigma : f_1^*u \to f_2^*u$ is an isomorphism in $\mathcal{S}(T)$ such that

$$F_T^*(\sigma) : F_T^* f_1^* u \longrightarrow F_T^* f_2^* u \tag{3.2.15.3}$$

is equal to the morphism obtained from the identification $F_T^* f_1^* = F_T^* f_2^*$. Here $u \in \mathcal{S}(U)$ denotes the object defining the morphism $U \to \mathcal{S}$.

Let $T \to U \times_{F_U,U,F_U} U$ be a morphism from a locally noetherian algebraic space T corresponding to morphisms $f_1, f_2 : T \to U$, and let $I \to T$ denote the T-space

$$I := \underline{\mathrm{Isom}}(f_1^*u, f_2^*u). \tag{3.2.15.4}$$

Note that $I \times_{T,F_T} T$ is canonically isomorphic to

$$\underline{\mathrm{Isom}}(F_T^* f_1^* u, F_T^* f_2^* u) \tag{3.2.15.5}$$

In particular, since $F_T^* f_1^* = F_T^* f_2^*$ there is a canonical section $\rho : T \to I \times_{T,F_T} T$. Furthermore, the relative Frobenius morphism $F_{I/T} : I \to I \times_{T,F_T} T$ is identified with the morphism sending an isomorphism $\sigma : g^* f_1^* u \to g^* f_2^* u$ over some T-scheme $g : W \to T$ to the isomorphism

$$g^* F_T^* f_1^* u \simeq F_W^* g^* f_1^* u \xrightarrow{F_W^* \sigma} F_W^* g^* f_2^* u \simeq g^* F_T^* f_2^* u. \tag{3.2.15.6}$$

From this it follows that the fiber product of the diagram

$$\begin{array}{ccc} & & T \\ & & \downarrow{\scriptstyle (f_1,f_2)} \\ U \times_{F_{U/\mathcal{S}},\mathcal{U}',F_{U/\mathcal{S}}} U & \longrightarrow & U \times_{F_U,U,F_U} U \end{array} \tag{3.2.15.7}$$

is isomorphic to the fiber product of the diagram

$$\begin{array}{ccc} & & T \\ (3.2.15.8) & & \downarrow \rho \\ I & \xrightarrow{F_{I/T}} & I \times_{T,F_T} T. \end{array}$$

Now since I is a locally noetherian algebraic space of finite type over T (since T was assumed locally noetherian), the morphism $F_{I/T}$ is finite, and hence the fiber product of the diagram (3.2.15.7) is finite over T. □

Corollary 3.2.16. — *The diagonal of $\mathcal{U}'$ over U is finite.*

Proof. — To verify that the diagonal $\Delta : \mathcal{U}' \to \mathcal{U}' \times_U \mathcal{U}'$ is finite it suffices to prove it after making the flat base change $U \times_{F_U,U,F_U} U \to \mathcal{U}' \times_U \mathcal{U}'$. Since the diagram

$$\begin{array}{ccc} U \times_{\mathcal{U}'} U & \longrightarrow & U \times_{F_U,U,F_U} U \\ \downarrow & & \downarrow \\ \mathcal{U}' & \xrightarrow{\Delta} & \mathcal{U}' \times_U \mathcal{U}' \end{array} \tag{3.2.16.1}$$

is cartesian, the corollary follows from 3.2.15. □

Corollary 3.2.17. — *The morphism $\mathcal{U}' \to U$ is proper, surjective, and quasi-finite. In particular by* [**49**, 7.13] *the sheaves $R^i\pi_*\mathcal{O}_{\mathcal{U}'}$ are quasi-coherent on U_{et} and coherent if U is locally noetherian.*

Proof. — Since the diagonal of $\mathcal{U}'$ over U is proper, the morphism $\mathcal{U}' \to U$ is separated.

Since the morphism $F_{U/\mathcal{S}} : U \to \mathcal{U}'$ is surjective, and $F_U : U \to U$ is quasi-finite, the morphism $\mathcal{U}' \to U$ is also quasi-finite. The statement that $\mathcal{U}' \to U$ is proper and surjective follows from the fact that $F_{U/\mathcal{S}} : U \to \mathcal{U}'$ is surjective, the fact that $F_U : U \to U$ is proper and surjective, and [**67**, 2.7]. □

Corollary 3.2.18. — *The morphism $F_{\mathcal{S}} : \mathcal{S} \to \mathcal{S}$ is proper. In particular by* [**49**, 7.13] *for any coherent sheaf $\mathcal{F}$ on $\mathcal{S}$ the sheaves $R^i\pi_*\mathcal{F}$ are quasi-coherent on $\mathcal{S}$ and coherent if $\mathcal{S}$ is locally noetherian.*

Proof. — This can be verified after making the smooth base change $U \to \mathcal{S}$ and so follows from 3.2.17. □

3.2.19. — To prove 3.2.3 it remains to see that for any field valued point $\operatorname{Spec}(\Omega) \to B$, with Ω algebraically closed, and $z \in \mathcal{U}'(\Omega)$ the stabilizer group scheme of z is diagonalizable.

Let $u \in \mathcal{S}(U)$ be the object defining the morphism $U \to \mathcal{S}$. The point $z \in \mathcal{U}'(\Omega)$ is then given by a morphism $x : \operatorname{Spec}(\Omega) \to U$, an object $v \in \mathcal{S}(\Omega)$ and an isomorphism $\iota : x^*u \simeq F^*v$ in $\mathcal{S}(\Omega)$, where $F : \operatorname{Spec}(\Omega) \to \operatorname{Spec}(\Omega)$ is the Frobenius morphism.

From this it follows that the automorphism group scheme of z is equal to the kernel of the morphism of diagonalizable group schemes

$$\underline{\mathrm{Aut}}(v) \longrightarrow \underline{\mathrm{Aut}}(x^*u), \quad \alpha \longmapsto \iota \circ F^*(\alpha) \circ \iota^{-1}. \tag{3.2.19.1}$$

By [**14**, VIII.3.4 (a)] this kernel is also diagonalizable. This completes the proof of 3.2.3.

3.3. The Cartier isomorphism

3.3.1. — Consider once again the diagram (3.1.2), where $\mathcal{X} \to \mathcal{S}$ is a smooth locally separated morphism of algebraic stacks with $\mathcal{X}$ a Deligne-Mumford stack.

By 3.2.18, the sheaf $F_{\mathcal{S}*}(\mathcal{O}_{\mathcal{S}_{\text{lis-et}}})$ is a quasi-coherent sheaf of $\mathcal{O}_{\mathcal{S}_{\text{lis-et}}}$-algebras on $\mathcal{S}_{\text{lis-et}}$ (and coherent in the locally noetherian case). We write $\overline{\mathcal{S}}$ for the relative spectrum $\mathrm{Spec}_{\mathcal{S}}(F_{\mathcal{S}*}(\mathcal{O}_{\mathcal{S}_{\text{lis-et}}}))$ [**49**, 14.2.4] and call the resulting factorization

$$\mathcal{S} \longrightarrow \overline{\mathcal{S}} \longrightarrow \mathcal{S} \tag{3.3.1.1}$$

the *Stein factorization* of $F_{\mathcal{S}} : \mathcal{S} \to \mathcal{S}$. Note that when $\mathcal{S}$ is an algebraic space the first map $\mathcal{S} \to \overline{\mathcal{S}}$ is an isomorphism.

Denote by $\overline{\mathcal{X}'}$ the base change $\overline{\mathcal{S}} \times_{\mathcal{S}} \mathcal{X}$. Since the formation of pushforwards commutes with flat base change (this follows for example from the proof of [**68**, 7.8] and the corresponding result for schemes), the stack $\overline{\mathcal{X}'}$ can also be described as the relative spectrum $\mathrm{Spec}_{\mathcal{X}}(\pi_*\mathcal{O}_{\mathcal{X}_{\text{lis-et}}})$ over $\mathcal{X}$. We write

$$P : \mathcal{X}' \longrightarrow \overline{\mathcal{X}}' \tag{3.3.1.2}$$

for the natural projection. Note that if $\mathcal{X}$ is a scheme then $\overline{\mathcal{X}}'$ is also a scheme.

Proposition 3.3.2. — *If $\mathcal{X}$ is a locally noetherian scheme, then the projection P is a coarse moduli space for $\mathcal{X}'$. In particular, it is universal for maps from $\mathcal{X}'$ to schemes.*

Proof. — Note first of all that by 3.2.16 the diagonal of $\mathcal{X}'$ is finite (taking $U \to \mathcal{S}$ to be $\mathcal{X} \to \mathcal{S}$) and hence the stack $\mathcal{X}'$ has a coarse moduli space $\gamma : \mathcal{X}' \to Y$ (see for example [**67**, 2.6] for a summary of the basic properties of coarse moduli spaces). The map γ is proper, surjective, quasi-finite, and the map $\mathcal{O}_Y \to \gamma_*\mathcal{O}_{\mathcal{X}'}$ is an isomorphism. The universal property of the coarse moduli space defines a unique factorization

$$\mathcal{X}' \xrightarrow{\ \gamma\ } Y \xrightarrow{\ \beta\ } \mathcal{X} \tag{3.3.2.1}$$

of the morphism $\pi : \mathcal{X}' \to \mathcal{X}$. Since π is proper and surjective by 3.2.17 and γ is proper and surjective, the morphism β is also proper and surjective [**67**, 2.7]. Since π is quasi-finite the morphism β is also quasi-finite hence finite. In particular β is an affine morphism. The fact that $\mathcal{O}_Y \to \gamma_*\mathcal{O}_{\mathcal{X}'}$ is an isomorphism then implies that β identifies Y with the relative spectrum $\mathrm{Spec}_{\mathcal{X}}(\pi_*\mathcal{O}_{\mathcal{X}'})$ which by definition is $\overline{\mathcal{X}}'$. □

Example 3.3.3. — To illustrate 3.3.2 consider the following example which will be discussed in greater detail and generality in section 9.3. Consider an integral and finitely generated monoid P, and let $P \to P^{\mathrm{gp}}$ be the universal map to a group. Let $k = \mathbb{F}_p$, set $\mathcal{X} = \mathrm{Spec}(k[P])$ (where $k[P]$ denotes the monoid algebra on P), and let G denote the diagonalizable group scheme $G := \mathrm{Spec}(k[P^{\mathrm{gp}}])$. Then G acts naturally on $\mathcal{X}$. Indeed $\mathcal{X}$ represents the functor on k-algebras

$$R \longmapsto \mathrm{Hom}_{\mathrm{monoids}}(P, R), \tag{3.3.3.1}$$

where R is viewed as a multiplicative monoid. An R-valued point $u \in G(R)$, which we view as a homomorphism $u : P^{\mathrm{gp}} \to R^*$, then acts on $\mathcal{X}(R)$ by sending a morphism of monoids $h : P \to R$ to the morphism $m \mapsto u(m) \cdot h(m)$. Let $\mathcal{S}$ denote the stack-theoretic quotient $[\mathcal{X}/G]$ and consider the quotient map $\mathcal{X} \to \mathcal{S}$. The stack $\pi : \mathcal{X}' \to \mathcal{X}$ is then the stack associating to any affine $\mathcal{X}$-scheme $\mathrm{Spec}(R) \to \mathcal{X}$ corresponding to a morphism of monoids $h : P \to R$ the groupoid whose objects are morphisms $\tilde{h} : P \to R$ such that the composite

$$P \xrightarrow{\times p} P \xrightarrow{\tilde{h}} R \tag{3.3.3.2}$$

is the morphism h and for which a morphism $\tilde{h} \to \tilde{h}'$ is an element $u \in G(R)$ such that the composite

$$P^{\mathrm{gp}} \xrightarrow{\times p} P^{\mathrm{gp}} \xrightarrow{u} R^* \tag{3.3.3.3}$$

is the identity and for which $\tilde{h}' = u * \tilde{h}$. In other words, let H denote the cokernel of the map $\times p : P^{\mathrm{gp}} \to P^{\mathrm{gp}}$ so that we have $D(H) \subset G$ (where $D(H)$ denotes the diagonalizable group corresponding to H). The group scheme $D(H)$ then acts on $\mathrm{Spec}(k[P])$ by restricting the G-action and $\pi : \mathcal{X}' \to \mathcal{X}$ is the morphism

$$[\mathrm{Spec}(k[P])/D(H)] \longrightarrow \mathrm{Spec}(k[P]) \tag{3.3.3.4}$$

induced by the map $\times p : P \to P$. The scheme $\overline{\mathcal{X}}'$ is equal to the spectrum of the ring of $D(H)$-invariants in $k[P]$. Let $\overline{P}' \subset P$ denote the submonoid of elements $m \in P$ such that the image of m in P^{gp} is in the image of $\times p : P^{\mathrm{gp}} \to P^{\mathrm{gp}}$. Then it follows that $\overline{\mathcal{X}}' \simeq \mathrm{Spec}(k[\overline{P}'])$.

Proposition 3.3.4. — *Let $\mathcal{X} \to \mathcal{S}$ be as in 3.3.1 and assume $\mathcal{X}$ is locally noetherian. Then the projection $\overline{\mathcal{X}}' \to \mathcal{X}$ is radicial and surjective.*

Proof. — The assertion is étale local on $\mathcal{X}$ so it suffices to consider the case when $\mathcal{X}$ is a scheme. This implies that $\overline{\mathcal{X}}'$ is also a scheme. The factorization

$$\mathcal{X} \longrightarrow \overline{\mathcal{X}}' \longrightarrow \mathcal{X} \tag{3.3.4.1}$$

of the Frobenius morphism of $\mathcal{X}$ (which is a radicial morphism) combined with the fact that $\mathcal{X} \to \overline{\mathcal{X}}'$ is surjective (since $\mathcal{X} \to \mathcal{X}'$ and $\mathcal{X}' \to \overline{\mathcal{X}}'$ are both surjective) implies that for any algebraically closed field Ω the map $\overline{\mathcal{X}}'(\Omega) \to \mathcal{X}(\Omega)$ is a bijection. The result then follows from [**15**, I.3.5.5]. □

Corollary 3.3.5. — *With assumptions as in 3.3.4 pullback along the projection $\overline{\mathcal{X}}' \to \mathcal{X}$ induces an equivalence between the category of algebraic spaces étale over $\mathcal{X}$ and the category of algebraic spaces étale over $\overline{\mathcal{X}}'$.*

Proof. — It suffices to prove the assertion after replacing $\mathcal{X}$ by an étale cover so we may assume that $\mathcal{X}$, and hence also $\overline{\mathcal{X}}'$, is a scheme. In this case, any algebraic space étale over $\mathcal{X}$ (resp. $\overline{\mathcal{X}}'$) is also a scheme by [**46**, II.6.16]. The result therefore follows from [**28**, IX.4.10]. □

Proposition 3.3.6. — *Let $\mathcal{X} \to \mathcal{S}$ be as in 3.3.1 with $\mathcal{S}$ locally noetherian, and let $\mathcal{E}$ be a flat quasi-coherent sheaf on $\overline{\mathcal{X}}'$. If $\mathcal{S}$ is Frobenius acyclic, then the adjunction map $\mathcal{E} \to RP_*P^*\mathcal{E}$ is an isomorphism.*

Proof. — If $U \to \mathcal{X}$ is an étale morphism, $\pi_U : \mathcal{U}' \to U$ denotes $\mathcal{S} \times_{F_{\mathcal{S}}, \mathcal{S}} U$ with the projection to U, and $P_U : \mathcal{U}' \to \overline{\mathcal{U}}'$ is the morphism to $\overline{\mathcal{U}}' := \mathrm{Spec}_U(\pi_{U*}\mathcal{O}_{\mathcal{U}'})$ then there is a commutative diagram

$$
(3.3.6.1)\qquad
\begin{array}{ccccc}
\mathcal{U}' & \xrightarrow{P_U} & \overline{\mathcal{U}}' & \longrightarrow & U \\
\downarrow & & \downarrow & & \downarrow \\
\mathcal{X}' & \xrightarrow{P} & \overline{\mathcal{X}}' & \longrightarrow & \mathcal{X},
\end{array}
$$

where all the diagrams are cartesian. It follows that to prove the proposition it suffices to prove the proposition for an étale cover of $\mathcal{X}$. In particular we may assume that $\mathcal{X}$ is an affine scheme. This implies that $\overline{\mathcal{X}}'$ is also an affine scheme. By [**16**, A.6.6], we can write $\mathcal{E}$ as a filtering direct limit of coherent flat sheaves of $\mathcal{O}_{\overline{\mathcal{X}}'}$-modules. Thus it suffices to consider the case when $\mathcal{E}$ is also coherent. Furthermore, since $\overline{\mathcal{X}}' \to \mathcal{X}$ is radicial by 3.3.4 and hence induces an isomorphism on underlying topological spaces $|\overline{\mathcal{X}}'| \to |\mathcal{X}|$, we can after shrinking on $\mathcal{X}$ some more assume that $\mathcal{E}$ is a free sheaf. This reduces the problem to showing that the direct image on $\mathcal{X}$ of the morphism $\mathcal{O}_{\overline{\mathcal{X}}'} \to RP_*\mathcal{O}_{\mathcal{X}'}$ is an isomorphism. But this map is equal to the restriction of the morphism $\mathcal{O}_{\mathcal{S}} \to RF_{\mathcal{S}*}(\mathcal{O}_{\mathcal{S}})$ to $\mathcal{X}_{\mathrm{et}}$. Thus the result follows from the definition of a Frobenius acyclic stack. □

Proposition 3.3.7. — *Let $\mathcal{X} \to \mathcal{S}$ be a smooth, locally separated, and representable morphism of algebraic stacks with $\mathcal{X}$ a locally noetherian Deligne-Mumford stack. Let $(\mathcal{E}, \nabla)$ be a quasi-coherent sheaf with integrable connection on $\mathcal{X}_{\mathrm{et}}/\mathcal{S}$. Then the map $\mathcal{H}^0(\mathcal{E} \otimes \Omega^\bullet_{\mathcal{X}_{\mathrm{et}}/\mathcal{S}})_\diamond \to F_*\mathcal{E}$ (3.1.11.1) induces an isomorphism*

$$
(3.3.7.1)\qquad P_*\mathcal{H}^0(\mathcal{E} \otimes \Omega^\bullet_{\mathcal{X}_{\mathrm{et}}/\mathcal{S}})_\diamond \longrightarrow (PF)_*\mathcal{E}^\nabla,
$$

where $\mathcal{E}^\nabla$ denotes the kernel (in the category of abelian sheaves on $\mathcal{X}_{\mathrm{et}}$) of the morphism $\nabla : \mathcal{E} \to \mathcal{E} \otimes \Omega^1_{\mathcal{X}/\mathcal{S}}$. In particular $(PF)_*\mathcal{E}^\nabla$ is a quasi-coherent sheaf.

Proof. — To prove that this map is an isomorphism, it suffices to show that for any étale morphism $\overline{U}' \to \overline{\mathcal{X}}'$ the map induces an isomorphism of between sections over $\overline{U}'$. By 3.3.5 any such étale morphism is obtained by base change from an étale morphism $U \to \mathcal{X}$. Since étale morphisms $U \to \mathcal{X}$ with U affine form a base for the étale topology, it therefore suffices to show that whenever $\mathcal{X}$ is an affine scheme the map $\mathcal{H}^0(\mathcal{E} \otimes \Omega^\bullet_{\mathcal{X}_{\mathrm{et}}/\mathcal{S}})_\diamond \to F_*\mathcal{E}$ induces an isomorphism on global sections

$$H^0(\mathcal{X}, \mathcal{E}^\nabla) = H^0(\mathcal{X}'_{\text{lis-et}}, \mathcal{H}^0(\mathcal{E} \otimes \Omega^\bullet_{\mathcal{X}_{\mathrm{et}}/\mathcal{S}})_\diamond). \tag{3.3.7.2}$$

Let $S \to \mathcal{S}$ be a smooth cover and let $S_1 = S \times_{\mathcal{S}} S$. Let $\mathcal{X}_S$ (resp. $\mathcal{X}_{S_1}$, $\mathcal{X}'_S$, $\mathcal{X}'_{S_1}$) denote $\mathcal{X} \times_{\mathcal{S}} S$ (resp. $\mathcal{X} \times_{\mathcal{S}} S_1$, $\mathcal{X}' \times_{\mathcal{S}} S$, $\mathcal{X}' \times_{\mathcal{S}} S_1$). Then there is a commutative diagram

(3.3.7.3)

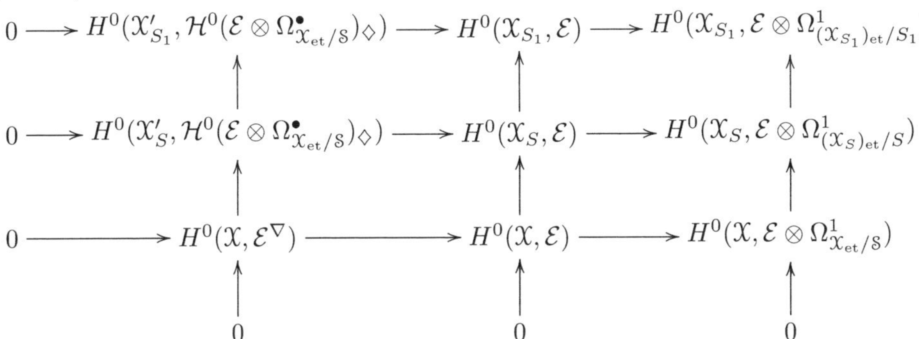

where the rows and the second two columns are exact. It follows that the first column is also exact which implies the proposition. □

3.3.8. — Let $\mathcal{X} \to \mathcal{S}$ be a smooth, locally separated, and representable morphism with $\mathcal{X}$ a locally noetherian Deligne-Mumford stack. Let

$$G : MIC(\mathcal{X}_{\mathrm{et}}/\mathcal{S}) \longrightarrow \mathrm{Mod}_{\mathrm{qcoh}}(\mathcal{X}') \tag{3.3.8.1}$$

be the functor sending $(\mathcal{E}, \nabla)$ to $\mathcal{H}^0(\mathcal{E} \otimes \Omega^\bullet_{\mathcal{X}/\mathcal{S}})_\diamond$. Note that G is a left exact functor.

Lemma 3.3.9. — *The functor G takes injective objects in $MIC(\mathcal{X}_{\mathrm{et}}/\mathcal{S})$ to injective objects in* $\mathrm{Mod}_{\mathrm{qcoh}}(\mathcal{X}')$.

Proof. — By 3.1.11 the functor G has a left adjoint which is exact since $F : \mathcal{X} \to \mathcal{X}'$ is flat. □

3.3.10. — Unfortunately we are interested in cohomology in the category of all $\mathcal{O}_{\mathcal{X}'}$-modules in the topos $\mathcal{X}'_{\text{lis-et}}$ and not cohomology in the smaller category $\mathrm{Mod}_{\mathrm{qcoh}}(\mathcal{X}')$. We therefore need to show that computing cohomology in either category yields the same answer.

Let $P^{\mathrm{qcoh}}_* : \mathrm{Mod}_{\mathrm{qcoh}}(\mathcal{X}') \to \mathrm{Mod}_{\mathrm{qcoh}}(\overline{\mathcal{X}}')$ be the pushforward of quasi-coherent sheaves, let $D^+(\mathcal{X}'_{\text{lis-et}})$ (resp. $D^+(\overline{\mathcal{X}}'_{\mathrm{et}})$) denote the derived category of bounded below complexes of $\mathcal{O}_{\mathcal{X}'}$-modules (resp. $\mathcal{O}_{\overline{\mathcal{X}}'}$-modules) in $\mathcal{X}'_{\text{lis-et}}$ (resp. $\overline{\mathcal{X}}'_{\text{et}}$), and let $\phi' : D^+(\text{Mod}_{\text{qcoh}}(\mathcal{X}')) \to D^+(\mathcal{X}')$ be the functor induced by the exact inclusion from $\text{Mod}_{\text{qcoh}}(\mathcal{X}')$ to all sheaves of $\mathcal{O}_{\mathcal{X}'}$-modules. We also have a functor $\bar{\phi}' : D^+(\text{Mod}_{\text{qcoh}}(\overline{\mathcal{X}}')) \to D^+(\overline{\mathcal{X}}'_{\text{et}})$. We then have derived functors

(3.3.10.1)
$$RP^{\text{qcoh}}_* : D^+(\text{Mod}_{\text{qcoh}}(\mathcal{X}')) \longrightarrow D^+(\text{Mod}_{\text{qcoh}}(\overline{\mathcal{X}}'_{\text{et}}))$$

and

(3.3.10.2)
$$RP_* : D^+(\mathcal{X}') \longrightarrow D^+(\overline{\mathcal{X}}'_{\text{et}}).$$

Since the two functors

(3.3.10.3)
$$\bar{\phi}' \circ P^{\text{qcoh}}_*, P_* \circ \phi' : \text{Mod}_{\text{qcoh}}(\mathcal{X}') \longrightarrow (\mathcal{O}_{\overline{\mathcal{X}}'}\text{-modules in } \overline{\mathcal{X}}'_{\text{et}})$$

are canonically isomorphic there is an induced morphism of derived functors

(3.3.10.4)
$$\bar{\phi}' \circ RP^{\text{qcoh}}_*(-) \longrightarrow RP_* \circ \phi'(-).$$

Proposition 3.3.11. — *The morphism (3.3.10.4) is an isomorphism of functors.*

Proof. — Since the relative Frobenius morphism $F : \mathcal{X} \to \mathcal{X}'$ is finite and faithfully flat and quasi-coherent sheaf on $\mathcal{X}'$ embeds into a quasi-coherent sheaf of the form $F_*\mathcal{G}$ for $\mathcal{G} \in \text{Mod}_{\text{qcoh}}(\mathcal{X})$ (for $\mathcal{F} \in \text{Mod}_{\text{qcoh}}(\mathcal{X}')$ consider $\mathcal{F} \to F_*F^*\mathcal{F}$). From this one deduces that it suffices to show that the morphism (3.3.10.4) is an isomorphism when evaluated on quasi-coherent sheaves $F_*\mathcal{G}$ for $\mathcal{G} \in \text{Mod}_{\text{qcoh}}(\mathcal{X})$.

Since F is affine and flat the functor $F^{\text{qcoh}}_* : \text{Mod}_{\text{qcoh}}(\mathcal{X}) \to \text{Mod}_{\text{qcoh}}(\mathcal{X}')$ is exact with an exact left adjoint. It follows that

(3.3.11.1)
$$RP^{\text{qcoh}}_*(F^{\text{qcoh}}_*\mathcal{G}) \simeq R(PF)^{\text{qcoh}}_*(\mathcal{G}).$$

Since PF is affine the functor $(PF)^{\text{qcoh}}_*$ is exact so we find that

(3.3.11.2)
$$RP^{\text{qcoh}}_*(F^{\text{qcoh}}_*\mathcal{G}) \simeq (PF)_*\mathcal{G}.$$

To prove the proposition it therefore suffices to show that for every $\mathcal{G} \in \text{Mod}_{\text{qcoh}}(\mathcal{X})$ and $i > 0$ we have

(3.3.11.3)
$$R^iP_*(\phi'(F^{\text{qcoh}}_*\mathcal{G})) = 0.$$

Let $U' \to \mathcal{X}'$ be a smooth surjection with U' a scheme and let $U'_\bullet \to \mathcal{X}'$ denote the 0-coskeleton. Let $\sigma : U_\bullet \to \mathcal{X}$ denote the pullback of $U'_\bullet$ by the morphism $F : \mathcal{X} \to \mathcal{X}'$. If $\pi : \mathcal{X}'_{\text{lis-et}|U'_\bullet} \to \mathcal{X}'_{\text{lis-et}}$ denotes the resulting simplicial localized topos there is then a commutative diagram of topoi

(3.3.11.4)
$$\begin{array}{ccc} \mathcal{X}'_{\text{lis-et}|U'_\bullet} & \xrightarrow{\xi} & U'_{\bullet,\text{et}} \\ {\scriptstyle\pi}\downarrow & & \downarrow{\scriptstyle P_{U'_\bullet}} \\ \mathcal{X}'_{\text{lis-et}} & \xrightarrow{P} & \overline{\mathcal{X}}'_{\text{et}}. \end{array}$$

Note that because of the usual problems with the lisse-étale topology [**68**] it does not make sense to speak about the simplicial topos $U'_{\bullet\text{lis-et}}$. Nonetheless, the localized topos $\mathcal{X}'_{\text{lis-et}|U'_\bullet}$ makes sense and is given by

$$[n] \longmapsto \mathcal{X}'_{\text{lis-et}|U'_n}, \tag{3.3.11.5}$$

where $\mathcal{X}'_{\text{lis-et}|U'_n}$ denotes the category of sheaves in $\mathcal{X}'_{\text{lis-et}}$ over the sheaf represented by U'_n.

We then compute

$$\begin{aligned} RP_*(\phi'(F_*^{\text{qcoh}}\mathcal{G})) &\simeq RP_*(R\pi_*\pi^*\phi'(F_*^{\text{qcoh}}\mathcal{G})) \\ &\simeq RP_{U'_\bullet *}R\xi_*\pi^*\phi'(F_*^{\text{qcoh}}\mathcal{G}) \\ &\simeq RP_{U'_\bullet *}\xi_*\pi^*\phi'(F_*^{\text{qcoh}}\mathcal{G}), \end{aligned} \tag{3.3.11.6}$$

where in the third isomorphism we use the fact that ξ_* is an exact functor. Let $\mathcal{G}_{U_\bullet}$ denote the pullback of $\mathcal{G}$ to $U_{\bullet\text{et}}$, and let $\widetilde{F} : U_{\bullet,\text{et}} \to U'_{\bullet,\text{et}}$ be the morphism induced by the projection. Then $\xi_*\pi^*\phi'(F_*^{\text{qcoh}}\mathcal{G})$ is simply the sheaf $\widetilde{F}_*\mathcal{G}_{U_\bullet}$.

Lemma 3.3.12. — *Let Z be a quasi-compact and quasi-separated Deligne-Mumford stack, and let $f : Y \to Z$ be an affine morphism of algebraic stacks. Then for any quasi-coherent sheaf $\mathcal{G}$ on Y we have $R^if_*\mathcal{G} = 0$ for all $i > 0$ (where f_* denotes the pushforward functor for the morphism of topoi $Y_{\text{et}} \to Z_{\text{et}}$).*

Proof. — It suffices to show that for any étale morphism $V \to Z$ with V an affine scheme the cohomology groups $H^i(Y \times_Z V, \mathcal{G})$ are zero for $i > 0$. This reduces the proof to showing that if Y is an affine scheme and $\mathcal{G}$ is a quasi-coherent sheaf on Y then $H^i(Y_{\text{et}}, \mathcal{G})$ is zero for $i > 0$. Let $\pi : Y_{\text{et}} \to Y_{\text{Zar}}$ be the projection morphism from the étale topos to the Zariski topos. Then since $\mathcal{G}$ is quasi-coherent the map $\pi^*\pi_*\mathcal{G} \to \mathcal{G}$ is an isomorphism and $R^i\pi_*\mathcal{G} = 0$ for $i > 0$. It follows that $H^i(Y_{\text{et}}, \mathcal{G}) \simeq H^i(Y_{\text{Zar}}, \mathcal{G}|_{Y_{\text{Zar}}})$ and the latter cohomology group vanishes for $i > 0$. □

The lemma implies that the natural map $\widetilde{F}_{U_\bullet *}\mathcal{G}_U \to R\widetilde{F}_{U_\bullet *}\mathcal{G}_U$ is an isomorphism since to verify this it suffices to show that for every natural number n the map

$$\widetilde{F}_{U_n *}\mathcal{G}_{U|U_n} \longrightarrow R\widetilde{F}_{U_n *}\mathcal{G}_{U|U_n}, \tag{3.3.12.1}$$

where F_{U_n} denotes the morphism of topoi $U_{n,\text{et}} \to U'_{n,\text{et}}$. Therefore

$$RP_{U'_\bullet *}\xi_*\pi^*\phi'(F_*^{\text{qcoh}}\mathcal{G}) \simeq RP_{U'_\bullet *}R\widetilde{F}_{U_\bullet *}\mathcal{G}_U \simeq R(P_{U'_\bullet}\widetilde{F}_{U_\bullet})_*(\mathcal{G}_U). \tag{3.3.12.2}$$

On the other hand $P_{U'_\bullet}\widetilde{F}_{U_\bullet}$ is also equal to the composite

$$U_{\bullet,\text{et}} \xrightarrow{\ \sigma\ } \mathcal{X}_{\text{et}} \xrightarrow{\ PF\ } \overline{\mathcal{X}}'_{|}Et. \tag{3.3.12.3}$$

It follows that

$$R(P_{U'_\bullet}\widetilde{F}_{U_\bullet})_*(\mathcal{G}_U) \simeq R(PF)_*R\sigma_*\sigma^*\mathcal{G} \simeq R(PF)_*\mathcal{G}, \tag{3.3.12.4}$$

where we use the fact that $\mathcal{G} \to R\sigma_*\sigma^*\mathcal{G}$ is an isomorphism by cohomological descent [**49**, 13.5.5] (see also [**68**, 6.14]). Combining (3.3.11.6), (3.3.12.2), and (3.3.12.4) we obtain an isomorphism

$$RP_*(\phi'(F_*^{\mathrm{qcoh}}\mathcal{G})) \simeq R(PF)_*\mathcal{G}. \tag{3.3.12.5}$$

Since $PF : \mathcal{X} \to \overline{\mathcal{X}}'$ is affine (in fact finite) another application of 3.3.12 shows that $R^i(PF)_*\mathcal{G} = 0$ for $i > 0$ and therefore we obtain the desired vanishing (3.3.11.3). This completes the proof of 3.3.11. □

3.3.13. — Let

$$\Psi : MIC(\mathcal{X}/\mathcal{S}) \longrightarrow (\mathcal{O}_{\overline{\mathcal{X}}'}\text{-modules in } \overline{\mathcal{X}}_{\mathrm{et}}) \tag{3.3.13.1}$$

be the functor sending $(\mathcal{E}, \nabla)$ to the quasi-coherent sheaf $(PF)_*\mathcal{E}^\nabla$. The functor Ψ factors as

$$MIC(\mathcal{X}/\mathcal{S}) \xrightarrow{G} \mathrm{Mod}_{\mathrm{qcoh}}(\mathcal{X}') \xrightarrow{P_*} (\mathcal{O}_{\overline{\mathcal{X}}'}\text{-modules in } \overline{\mathcal{X}}_{\mathrm{et}}). \tag{3.3.13.2}$$

From the spectral sequence of a composite of functors and using 3.3.11 we obtain for every $(\mathcal{E}, \nabla)$ a spectral sequence

$$E_2^{pq} = R^qP_*(R^pG(\mathcal{E}, \nabla)) \implies R^{p+q}\Psi(E, \nabla). \tag{3.3.13.3}$$

By 2.5.4 and 2.5.9, for any object $(\mathcal{E}, \nabla) \in MIC(\mathcal{X}/\mathcal{S})$ there is a natural isomorphism

$$R^i\Psi(\mathcal{E}, \nabla) \simeq \mathcal{H}^i(\mathcal{E} \otimes \Omega^\bullet_{\mathcal{X}_{\mathrm{et}}/\mathcal{S}}). \tag{3.3.13.4}$$

We can therefore rewrite (3.3.13.3) as

$$E_2^{pq} = R^qP_*(R^pG(\mathcal{E}, \nabla)) \implies (PF)_*\mathcal{H}^{p+q}(\mathcal{E} \otimes \Omega^\bullet_{\mathcal{X}/\mathcal{S}}). \tag{3.3.13.5}$$

3.3.14. — Let $\overline{\mathcal{E}}'$ be a quasi-coherent sheaf on $\overline{\mathcal{X}}'_{\mathrm{et}}$, $\mathcal{E}' := P^*\overline{\mathcal{E}}$, and let $(\mathcal{E}, \nabla)$ be the module with integrable connection on $\mathcal{X}_{\mathrm{et}}/\mathcal{S}$ associated by 3.1.8 to $\mathcal{E}'$. Let $S \to \mathcal{S}$ be a smooth surjection with S an algebraic space, and let $S_\bullet$ be the 0-coskeleton. Denote $\mathcal{X} \times_{\mathcal{S}} S_\bullet$ by $X_\bullet$ and $\mathcal{X}' \times_{\mathcal{S}} S_\bullet$ by $X'_\bullet$. Then there is a commutative diagram

$$\begin{array}{ccc} MIC(\mathcal{X}_{\mathrm{et}}/\mathcal{S}) & \xrightarrow{\alpha} & MIC(X_\bullet/S_\bullet) \\ {\scriptstyle G}\downarrow & & \downarrow{\scriptstyle G_\bullet} \\ \mathrm{Mod}_{\mathrm{qcoh}}(\mathcal{X}') & \xrightarrow{\beta} & \mathrm{Mod}_{\mathrm{qcoh}}(X'_\bullet), \end{array} \tag{3.3.14.1}$$

where α and β are the functors induced by pullback and $G_\bullet$ sends $(\mathcal{E}_\bullet, \nabla)$ to $\mathcal{E}_\bullet^\nabla$. By 2.2.14, the functor α is an equivalence of categories, and β is an equivalence by descent theory for quasi-coherent sheaves. The sheaf $\beta R^pG(\mathcal{E}, \nabla)$ is isomorphic to the sheaf $\mathcal{H}^p(\mathcal{E}_\bullet \otimes \Omega^\bullet_{X_\bullet/S_\bullet})$ which by the usual Cartier isomorphism is isomorphic to $\mathcal{E}'|_{X'_\bullet} \otimes \Omega^p_{X'_\bullet/S_\bullet}$. From this and descent theory we obtain an isomorphism $R^pG(\mathcal{E}, \nabla) \simeq P^*\overline{\mathcal{E}}' \otimes_{\mathcal{O}_{\mathcal{X}'}} \Omega^p_{\mathcal{X}'/\mathcal{S}}$. If $S' \to \mathcal{S}$ is a second smooth surjection, and

$f : S' \to S$ is a morphism over $\mathcal{S}$, then it follows from the construction that the isomorphism obtained using S' agrees with the isomorphism obtained using S. Since the category of smooth surjections $S \to \mathcal{S}$ is connected, it follows that the isomorphism $R^pG(\mathcal{E}, \nabla) \simeq P^*\mathcal{E}' \otimes_{\mathcal{O}_{\mathcal{X}'}} \Omega^p_{\mathcal{X}'/\mathcal{S}}$ is independent of the choice of S. In summary:

Corollary 3.3.15. — *For any $p \geq 0$ there is a natural isomorphism $R^pG(\mathcal{E}, \nabla) \simeq P^*\overline{\mathcal{E}}' \otimes_{\mathcal{O}_{\mathcal{X}'}} \Omega^p_{\mathcal{X}'/\mathcal{S}}$.*

3.3.16. — Combining 3.3.15, (3.3.13.5), and the projection formula

$$R^qP_*(\mathcal{E}' \otimes_{\mathcal{O}_{\mathcal{X}'}} \Omega^p_{\mathcal{X}'/\mathcal{S}}) \simeq R^qP_*(\mathcal{E}') \otimes_{\mathcal{O}_{\overline{\mathcal{X}}'_{\mathrm{et}}}} \Omega^p_{\overline{\mathcal{X}}'/\overline{\mathcal{S}}} \tag{3.3.16.1}$$

we obtain a spectral sequence

$$E_2^{pq} = R^qP_*(\mathcal{E}') \otimes_{\mathcal{O}_{\overline{\mathcal{X}}'_{\mathrm{et}}}} \Omega^p_{\overline{\mathcal{X}}'/\overline{\mathcal{S}}} \Longrightarrow (PF)_*\mathcal{H}^{p+q}(\mathcal{E} \otimes \Omega^\bullet_{\mathcal{X}/\mathcal{S}}), \tag{3.3.16.2}$$

which we call the *Cartier spectral sequence* (recall that $\mathcal{S} \to \overline{\mathcal{S}} \to \mathcal{S}$ is the Stein factorization of $F_{\mathcal{S}} : \mathcal{S} \to \mathcal{S}$ and that $\overline{\mathcal{X}}' \simeq \mathcal{X} \times_{\mathcal{S}} \overline{\mathcal{S}}$).

***Corollary 3.3.17* (The Cartier isomorphism)**. — *With notation as in 3.3.14, if $\mathcal{S}$ is Frobenius acyclic and $\overline{\mathcal{E}}'$ is a flat quasi-coherent sheaf on $\overline{\mathcal{X}}'$, then (3.3.16.2) degenerates and yields an isomorphism*

$$(PF)_*\mathcal{H}^p(\mathcal{E} \otimes \Omega^\bullet_{\mathcal{X}_{\mathrm{et}}/\mathcal{S}}) \simeq \overline{\mathcal{E}}' \otimes \Omega^p_{\overline{\mathcal{X}}'_{\mathrm{et}}/\overline{\mathcal{S}}}. \tag{3.3.17.1}$$

Proof. — This follows from 3.3.6 which shows that $E_2^{pq} = 0$ for $q > 0$. □

Remark 3.3.18. — In the case when $\mathcal{S}$ and $\mathcal{X}$ are schemes, we recover the usual Cartier isomorphism. Indeed in this case $\mathcal{X}' = \overline{\mathcal{X}}'$, $\mathcal{E}' = \overline{\mathcal{E}}'$, and in 3.3.14 we can take the constant hypercovers obtained by choosing $S = \mathcal{S}$. For a quasi-coherent sheaf $\mathcal{E}'$ on $\mathcal{X}'$ with associated module with integrable connection $(\mathcal{E}, \nabla)$, the isomorphism

$$\mathcal{E}' \otimes \Omega^p_{\mathcal{X}'/\mathcal{S}} \simeq F_*\mathcal{H}^p(\mathcal{E} \otimes \Omega^\bullet_{\mathcal{X}/\mathcal{S}}) \tag{3.3.18.1}$$

provided by 3.3.17 has the property that its pullback to the constant simplicial topos $\mathcal{X}'_\bullet$ is equal in each degree to the classical Cartier isomorphism. Therefore 3.3.17 coincides with the classical Cartier isomorphism in the schematic situation.

3.3.19. — One of the main applications of 3.3.17 that we will use later is the following. Let $\mathcal{S}'$ denote the stack $\mathcal{S} \times_{T,F_T} T$ so that $F_{\mathcal{S}}$ factors as

$$\mathcal{S} \xrightarrow{F_{\mathcal{S}/T}} \mathcal{S}' \longrightarrow \mathcal{S}. \tag{3.3.19.1}$$

Assumption 3.3.20. — *Assume that the induced map in $D_{\mathrm{qcoh}}(\mathcal{S}'_{\text{lis-et}})$*

$$\mathcal{O}_{\mathcal{S}'} \longrightarrow RF_{\mathcal{S}/T*}\mathcal{O}_{\mathcal{S}} \tag{3.3.20.1}$$

is an isomorphism.

Note that since the morphism $\mathcal{S}' \to \mathcal{S}$ is affine this assumption implies in particular that $\mathcal{S}$ is Frobenius acyclic.

Let $\mathcal{X} \to \mathcal{S}$ be a smooth locally separated morphism of algebraic stacks with $\mathcal{X}$ a Deligne-Mumford stack. Since the formation of cohomology commutes with smooth base change on the target, the assumption 3.3.20 implies that in the notation of 3.3.17 we have $\overline{\mathcal{S}} = \mathcal{S}'$ and that

$$\overline{\mathcal{X}}' = \mathcal{X} \times_{\mathcal{S}} \mathcal{S}' = \mathcal{X} \times_{T,F_T} T. \tag{3.3.20.2}$$

We will often write $\mathcal{X}^{(p)}$ for the stack $\mathcal{X} \times_{T,F_T} T$. Rewriting 3.3.17 in this situation we obtain the following:

Corollary 3.3.21. — *Let $\mathcal{E}^{(p)}$ be a flat quasi-coherent sheaf on $\mathcal{X}^{(p)}$, and let $(\mathcal{E}, \nabla)$ be the induced module with integrable connection on $\mathcal{X}/\mathcal{S}$. Then for every $q \geq 0$ there is a canonical isomorphism*

$$C^{-1} : \mathcal{E}^{(p)} \otimes \Omega^q_{\mathcal{X}^{(p)}/\mathcal{S}'} \longrightarrow F_{\mathcal{X}/T*}\mathcal{H}^q(\mathcal{E} \otimes \Omega^\bullet_{\mathcal{X}/\mathcal{S}}), \tag{3.3.21.1}$$

where $F_{\mathcal{X}/T} : \mathcal{X} \to \mathcal{X}^{(p)}$ is the relative Frobenius morphism.

3.4. Ogus' generalization of Mazur's theorem

3.4.1. — Let A be a p-adically complete and separated ring flat over $\mathbb{Z}_p$, and let $M = \mathrm{Spec}(A)$, $\widehat{M} = \mathrm{Spf}(A)$. Let $\mathcal{S}/M$ be a flat algebraic stack, and denote by $\mathcal{S}_n$, M_n, A_n etc., the reductions modulo p^{n+1}. Let $\mathcal{S}_0^{(1)} := \mathcal{S}_0 \times_{M_0, F_{M_0}} M_0$, and assume given a flat lifting $\mathcal{S}^{(1)}$ of $\mathcal{S}_0^{(1)}$ to M together with a quasi-compact and quasi-separated lifting

$$F_{\mathcal{S}/M} : \mathcal{S} \longrightarrow \mathcal{S}^{(1)} \tag{3.4.1.1}$$

of the relative Frobenius morphism $F_{\mathcal{S}_0/M_0} : \mathcal{S}_0 \to \mathcal{S}_0^{(1)}$. Also assume that $\mathcal{S}_0$ is Frobenius acyclic.

Example 3.4.2. — An important example that we will generalize in subsequent chapters is the following. Let $A = \mathbb{Z}_p$ and let $\mathcal{S}$ be the stack-theoretic quotient of $\mathrm{Spec}(\mathbb{Z}_p[x,y]/(xy-p))$ by the action of $\mathbb{G}_m$ in which a scheme-valued section $u \in \mathbb{G}_m$ acts by multiplication by u on x and multiplication by u^{-1} on y. In this case the stack $\mathcal{S}_0^{(1)}$ is isomorphic to the stack quotient

$$\mathcal{S}_0^{(1)} \simeq \big[\mathrm{Spec}(\mathbb{F}_p[x,y]/(xy))/\mathbb{G}_m\big], \tag{3.4.2.1}$$

where the action of $\mathbb{G}_m$ is as above. The relative Frobenius morphism $F_{\mathcal{S}_0/M_0} : \mathcal{S}_0 \to \mathcal{S}_0^{(1)}$ is given by the morphism of stacks over $\mathbb{F}_p$ given by the maps

$$\mathbb{F}_p[x,y]/(xy) \longrightarrow \mathbb{F}_p[x,y]/(xy), \quad x \longmapsto x^p, y \longmapsto y^p, \tag{3.4.2.2}$$

$$\mathbb{G}_m \longrightarrow \mathbb{G}_m, \quad u \longmapsto u^p. \tag{3.4.2.3}$$

We claim that in this situation if we take $\mathcal{S}^{(1)} = \mathcal{S}$ then there does not exist a morphism $\tilde{F} : \mathcal{S}_1 \to \mathcal{S}_1^{(1)}$ lifting $F_{\mathcal{S}_0/M_0}$. Suppose to the contrary that such a lifting

$$\tilde{F} : \big[\mathrm{Spec}(\mathbb{Z}/p^2[x,y]/(xy-p))/\mathbb{G}_m\big] \longrightarrow \big[\mathrm{Spec}(\mathbb{Z}/p^2[x,y]/(xy-p))/\mathbb{G}_m\big] \tag{3.4.2.4}$$

exists. Let $P \to [\mathrm{Spec}(\mathbb{Z}/p^2[x,y]/(xy-p))/\mathbb{G}_m]$ be the $\mathbb{G}_m$-torsor

$$\mathrm{Spec}(\mathbb{Z}/p^2[x,y]/(xy-p)) \longrightarrow \big[\mathrm{Spec}(\mathbb{Z}/p^2[x,y]/(xy-p))/\mathbb{G}_m\big]. \tag{3.4.2.5}$$

and let P' denote $\tilde{F}^*P$. Then P and P' are two $\mathbb{G}_m$-torsors over

$$\big[\mathrm{Spec}(\mathbb{Z}/p^2[x,y]/(xy-p))/\mathbb{G}_m\big] \tag{3.4.2.6}$$

whose reductions modulo p are isomorphic. Using the exponential sequence

$$0 \longrightarrow \mathcal{O}_{\mathcal{S}_0} \xrightarrow{a\mapsto 1+pa} \mathcal{O}^*_{\mathcal{S}_1} \longrightarrow \mathcal{O}^*_{\mathcal{S}_0} \longrightarrow 0 \tag{3.4.2.7}$$

one sees that such deformations are classified by the group

$$H^1(\mathcal{S}_0, \mathcal{O}_{\mathcal{S}_0}). \tag{3.4.2.8}$$

On the other hand, using the fact that $\mathbb{G}_m$ is a reductive group one sees that

$$H^1(\mathcal{S}_0, \mathcal{O}_{\mathcal{S}_0}) \simeq H^1\big(\mathrm{Spec}(\mathbb{F}_p[x,y]/(xy)), \mathcal{O}_{\mathbb{F}_p[x,y]/(xy)}\big)^{\mathbb{G}_m} = 0. \tag{3.4.2.9}$$

It follows that P and P' are isomorphic $\mathbb{G}_m$-torsors and therefore $\tilde{F}$ lifts to a morphism

$$\tilde{f} : \mathbb{Z}/(p^2)[x,y]/(xy-p) \longrightarrow \mathbb{Z}/(p^2)[x,y]/(xy-p) \tag{3.4.2.10}$$

whose reduction modulo p is the map (3.4.2.2). It follows that $\tilde{f}(x) = x^p + ph_1$ and $\tilde{f}(y) = y^p + ph_2$ for some $h_1, h_2 \in \mathbb{Z}/(p^2)[x,y]/(xy-p)$. On the other hand, since $\tilde{f}$ is a ring homomorphism we have

$$p = \tilde{f}(x)\tilde{f}(y) = (x^p + ph_1)(y^p + ph_2) = p^p + p(h_1y^p + h_2x^p) + p^2h_1h_2 \tag{3.4.2.11}$$

which is impossible.

Nonetheless, if we define $\mathcal{S}^{(1)}$ to be the stack quotient

$$\mathcal{S}^{(1)} := [\mathrm{Spec}(\mathbb{Z}_p[x,y]/(xy-p^p))/\mathbb{G}_m] \tag{3.4.2.12}$$

then we can lift the relative Frobenius map $\mathcal{S}_0 \to \mathcal{S}_0^{(1)}$ to a morphism $\mathcal{S} \to \mathcal{S}^{(1)}$ by the formulas

$$\mathbb{Z}_p[x,y]/(xy-p^p) \longrightarrow \mathbb{Z}_p[x,y]/(xy-p), \quad x \longmapsto x^p, y \longmapsto y^p, \tag{3.4.2.13}$$

$$\mathbb{G}_m \longrightarrow \mathbb{G}_m, \quad u \longmapsto u^p. \tag{3.4.2.14}$$

3.4.3. — Let $\mathcal{X} \to \mathcal{S}_0$ be a smooth locally separated morphism of algebraic stacks with $\mathcal{X}$ a Deligne-Mumford stack. Define $\overline{\mathcal{S}}$, $\overline{\mathcal{X}}'$, and $\mathcal{X}'$ as in 3.3.1. Denote by $F_{\mathcal{X}/M_0}$ the composite

(3.4.3.1)
$$\mathcal{X} \xrightarrow{F_{\mathcal{X}/\mathcal{S}_0}} \mathcal{X}' \longrightarrow \mathcal{X}^{(1)} := \mathcal{X} \times_{M_0, F_{M_0}} M_0.$$

For each $n \geq 0$ let $\overline{\mathcal{S}}_n$ denote the relative spectrum $\mathrm{Spec}_{\mathcal{S}_n^{(1)}}(F_{\mathcal{S}_n/M_n*}\mathcal{O}_{\mathcal{S}_n})$, where $F_{\mathcal{S}_n/M_n} : \mathcal{S}_n \to \mathcal{S}_n^{(1)}$ denotes the map obtained from $F_{\mathcal{S}/M}$ by reduction.

Lemma 3.4.4. — *Let n be a nonnegative integer, and let $j : \mathcal{S}_{n-1}^{(1)} \hookrightarrow \mathcal{S}_n^{(1)}$ be the inclusion. Then the natural map*

(3.4.4.1)
$$j^*(F_{\mathcal{S}_n/M_n*}\mathcal{O}_{\mathcal{S}_{n,\text{lis-et}}}) \to F_{\mathcal{S}_{n-1}/M_{n-1}*}\mathcal{O}_{\mathcal{S}_{n-1,\text{lis-et}}}$$

is an isomorphism

Proof. — Because $\mathcal{S}$ is flat over $\mathbb{Z}_p$ there is a natural exact sequence of quasi-coherent sheaves on $\mathcal{S}_{n,\text{lis-et}}$

(3.4.4.2)
$$0 \longrightarrow \mathcal{O}_{\mathcal{S}_0} \xrightarrow{\times p^n} \mathcal{O}_{\mathcal{S}_n} \longrightarrow j_*\mathcal{O}_{\mathcal{S}_{n-1}} \longrightarrow 0.$$

For any quasi-coherent sheaf $\mathcal{F}$ on $\mathcal{S}_{n-1}^{(1)}$ we have $j^*j_*\mathcal{F} = \mathcal{F}$. Hence it suffices to show that the natural map

(3.4.4.3)
$$\mathrm{Coker}(\times p^n : F_{\mathcal{S}_n/M_n*}\mathcal{O}_{\mathcal{S}_0} \to F_{\mathcal{S}_n/M_n*}\mathcal{O}_{\mathcal{S}_n}) \longrightarrow F_{\mathcal{S}_n/M_n*}j_*\mathcal{O}_{\mathcal{S}_{n-1}}$$

is an isomorphism. Thus the lemma follows from the fact that $R^1F_{\mathcal{S}_0/M_0*}\mathcal{O}_{\mathcal{S}_0} = 0$ by the definition of Frobenius acyclic stack (3.2.1). □

Corollary 3.4.5. — *For every $n \geq 0$ the sheaf $F_{\mathcal{S}_n/M_n*}\mathcal{O}_{\mathcal{S}_n}$ is a locally finite presented quasi-coherent sheaf on $\mathcal{S}^{(1)}_{n,\text{lis-et}}$.*

Proof. — By 3.4.4 it suffices to consider the case when $n = 0$ which follows from 3.2.13. □

3.4.6. — The lemma enables us to define a coherent crystal $\Theta_{\mathcal{S}^{(1)}/M}$ in $(\mathcal{X}^{(1)}_{\text{lis-et}}/\widehat{\mathcal{S}}^{(1)})_{\text{cris}}$ as follows (here $\widehat{\mathcal{S}}^{(1)}$ denotes the projective system $\{\mathcal{S}_n^{(1)}\}$ as in 2.7.1; see also 2.7.5 for the notion of crystal in this context). For any object

(3.4.6.1)
$$\begin{array}{ccc} U & \longrightarrow & T \\ \downarrow & & \downarrow r \\ \mathcal{X}^{(1)} & \longrightarrow & \mathcal{S}^{(1)} \end{array}$$

of $\mathrm{Cris}(\mathcal{X}^{(1)}_{\text{lis-et}}/\widehat{\mathcal{S}}^{(1)})$, the ideal of U in T is killed by p^n for some n by assumption, and hence the map r factors through a map $r_n : T \to \mathcal{S}_n^{(1)}$ for some n. We can then consider the quasi-coherent sheaf $r_n^* F_{\mathcal{S}_n/M_n *}\mathcal{O}_{\mathcal{S}_{n,\text{lis-et}}}$ on T_{et}. If $r_m : T \to \mathcal{S}_m^{(1)}$ is a second factorization of r then 3.4.4 yields an isomorphism

$$r_n^* F_{\mathcal{S}_n/M_n *}\mathcal{O}_{\mathcal{S}_{n,\text{lis-et}}} \simeq r_m^* F_{\mathcal{S}_m/M_m *}\mathcal{O}_{\mathcal{S}_{m,\text{lis-et}}} \tag{3.4.6.2}$$

and these isomorphisms satisfy the usual transitivity condition for three choices of factorization. It follows that we can define a crystal

$$\Theta_{\mathcal{S}^{(1)}/M} \tag{3.4.6.3}$$

on $(\mathcal{X}^{(1)}_{\text{lis-et}}/\widehat{\mathcal{S}}^{(1)})_{\text{cris}}$ by associating to $(U \hookrightarrow T)$ the global sections over T of

$$r_n^* F_{\mathcal{S}_n/M_n *}\mathcal{O}_{\mathcal{S}_{n,\text{lis-et}}} \tag{3.4.6.4}$$

for any choice of factorization r_n.

Remark 3.4.7. — Because we don't have a good notion of formal algebraic stack, we have adopted the above rather clumsy definition of $\Theta_{\mathcal{S}^{(1)}/M}$. In the case when $\mathcal{S} = S$ is a scheme we can give a better definition. Let $\widehat{S}^{(1)}$ denote the formal completion of $S^{(1)}$ along p. The compatible system of sheaves $F_{S_n/M_n *}\mathcal{O}_{S_n}$ over the reductions define a sheaf $\hat{E}$ of $\mathcal{O}_{\widehat{S}^{(1)}}$ modules on the formal scheme $\widehat{S}^{(1)}$. The crystal $\Theta_{\mathcal{S}^{(1)}/M}$ is then simply the pullback of the sheaf $\hat{E}$ under the natural morphism of ringed topoi

$$(\mathcal{X}^{(1)}_{\text{lis-et}}/S^{(1)})_{\text{cris}} \longrightarrow \widehat{S}^{(1)}_{\text{et}}. \tag{3.4.7.1}$$

Remark 3.4.8. — Let $A^\#$ be a second p-adically complete and separated flat $\mathbb{Z}_p$-algebra and $f : A \to A^\#$ a morphism of rings. Let $M^\#$ (resp. $\widehat{M}^\#$, etc.) denote $\mathrm{Spec}(A^\#)$ (resp. $\mathrm{Spf}(A^\#)$, etc.), and write also $f : M^\# \to M$ for the morphism induced by the morphism of rings f. Let $\mathcal{S}^\#/M^\#$ be a flat algebraic stack, and $\mathcal{S}^{\#(1)}/M^\#$ a flat lifting of $\mathcal{S}_0^{\#(1)} := \mathcal{S}_0^\# \times_{M_0^\#, F_{M_0^\#}} M_0^\#$ together with a lifting of the relative Frobenius morphism

$$F_{\mathcal{S}^\#/M^\#} : \mathcal{S}^\# \longrightarrow \mathcal{S}^{\#(1)}. \tag{3.4.8.1}$$

Suppose given a commutative diagram over f

$$\begin{array}{ccc} \mathcal{X}^\# & \xrightarrow{h} & \mathcal{X} \\ {\scriptstyle a^\#}\downarrow & & \downarrow{\scriptstyle a} \\ \mathcal{S}^\# & \xrightarrow{g} & \mathcal{S} \\ {\scriptstyle F_{\mathcal{S}^\#/M^\#}}\downarrow & & \downarrow{\scriptstyle F_{\mathcal{S}/M}} \\ \mathcal{S}^{\#(1)} & \xrightarrow{g^{(1)}} & \mathcal{S}^{(1)}, \end{array} \tag{3.4.8.2}$$

where a (resp. $a^\#$) is a smooth locally separated morphism of algebraic stacks which factors through $\mathcal{S}_0$ (resp. $\mathcal{S}_0^\#$), and $\mathcal{X}^\#$ and $\mathcal{X}$ are Deligne-Mumford stacks.

Then there is an induced morphism of crystals in $(\mathfrak{X}^{\#(1)}/\widehat{\mathcal{S}}^{\#(1)})_{\mathrm{cris}}$

$$h^*\Theta_{\mathcal{S}^{(1)}/M} \longrightarrow \Theta_{\mathcal{S}^{\#(1)}/M^{\#}} \tag{3.4.8.3}$$

defined as follows. For any object $(U^{\#} \hookrightarrow T^{\#}) \in \mathrm{Cris}(\mathfrak{X}^{\#(1)}/\widehat{\mathcal{S}}^{\#(1)})$ which admits a morphism

$$\begin{array}{ccc} U^{\#} & \xrightarrow{z} & U \\ \downarrow & & \downarrow \\ T^{\#} & \xrightarrow{z_T} & T \end{array} \tag{3.4.8.4}$$

to an object of $\mathrm{Cris}(\mathfrak{X}^{(1)}/\widehat{\mathcal{S}}^{(1)})$ over the diagram

$$\begin{array}{ccc} \mathfrak{X}^{\#(1)} & \xrightarrow{h} & \mathfrak{X}^{(1)} \\ \downarrow & & \downarrow \\ \mathcal{S}^{\#(1)} & \xrightarrow{g^{(1)}} & \mathcal{S}^{(1)}, \end{array} \tag{3.4.8.5}$$

the sheaf $h^*\Theta_{\mathcal{S}^{(1)}/M}|_{T^{\#}}$ on T_{et} is equal to $z_T^*\Theta_{\mathcal{S}^{(1)}/M}|_T$. Choose an integer n such that we have a commutative diagram

$$\begin{array}{ccc} T^{\#} & \xrightarrow{z_T} & T \\ {\scriptstyle r^{\#}}\downarrow & & \downarrow{\scriptstyle r} \\ \mathcal{S}_n^{\#(1)} & \xrightarrow{g^{(1)}} & \mathcal{S}_n^{(1)} \\ {\scriptstyle F_{\mathcal{S}^{\#}/M^{\#}}}\uparrow & & \uparrow{\scriptstyle F_{\mathcal{S}/M}} \\ \mathcal{S}_n^{\#} & \longrightarrow & \mathcal{S}_n. \end{array} \tag{3.4.8.6}$$

Then the map

$$h^*\Theta_{\mathcal{S}^{(1)}/M}|_{T^{\#}} \longrightarrow \Theta_{\mathcal{S}^{\#(1)}/M^{\#}}|_{T^{\#}} \tag{3.4.8.7}$$

is defined to be the map

$$z_T^* r^* F_{\mathcal{S}/M*}\mathcal{O}_{\mathcal{S}_n} = r^{\#*} g^{(1)*} F_{\mathcal{S}/M*}\mathcal{O}_{\mathcal{S}_n} \longrightarrow r^{\#*} F_{\mathcal{S}^{\#}/M^{\#}*}\mathcal{O}_{\mathcal{S}_n^{\#}}. \tag{3.4.8.8}$$

This defines the morphism (3.4.8.3) locally. We leave to the reader the task of verifying that the above constructed morphism is independent of the choices and therefore defined globally.

Remark 3.4.9. — From the commutative square

$$\begin{array}{ccc} \mathfrak{X} & \xrightarrow{F_{\mathfrak{X}/M_0}} & \mathfrak{X}^{(1)} \\ \downarrow & & \downarrow \\ \mathcal{S} & \xrightarrow{F_{\mathcal{S}/M}} & \mathcal{S}^{(1)} \end{array} \tag{3.4.9.1}$$

we obtain a morphism of topoi

$$F_{\mathcal{X}/M_0} : (\mathcal{X}_{\mathrm{et}}/\widehat{\mathcal{S}})_{\mathrm{cris}} \to (\mathcal{X}_{\mathrm{et}}^{(1)}/\widehat{\mathcal{S}}^{(1)})_{\mathrm{cris}} \tag{3.4.9.2}$$

such that the diagram

$$\begin{array}{ccc} (\mathcal{X}_{\mathrm{et}}/\widehat{\mathcal{S}})_{\mathrm{cris}} & \xrightarrow{F_{\mathcal{X}/M_0}} & (\mathcal{X}_{\mathrm{et}}^{(1)}/\widehat{\mathcal{S}}^{(1)})_{\mathrm{cris}} \\ {\scriptstyle u_{\mathcal{X}_{\mathrm{et}}/\widehat{\mathcal{S}}}}\downarrow & & \downarrow{\scriptstyle u_{\mathcal{X}_{\mathrm{et}}^{(1)}/\widehat{\mathcal{S}}^{(1)}}} \\ \mathcal{X}_{\mathrm{et}} & \xrightarrow{F_{\mathcal{X}/M_0}} & \mathcal{X}_{\mathrm{et}}^{(1)} \end{array} \tag{3.4.9.3}$$

commutes. We define a morphism of crystals in $(\mathcal{X}_{\mathrm{et}}/\widehat{\mathcal{S}})_{\mathrm{cris}}$

$$F_{\mathcal{X}/M_0}^{*}\Theta_{\mathcal{S}^{(1)}/M} \longrightarrow \mathcal{O}_{\mathcal{X}_{\mathrm{et}}/\widehat{\mathcal{S}}} \tag{3.4.9.4}$$

as follows. For an object $(U \hookrightarrow T) \in \mathrm{Cris}(\mathcal{X}_{\mathrm{et}}/\widehat{\mathcal{S}})$ which admits an $F_{\mathcal{S}/M}$-morphism

$$s : (U \hookrightarrow T) \longrightarrow (U' \hookrightarrow T') \tag{3.4.9.5}$$

to an object $(U' \hookrightarrow T') \in \mathrm{Cris}(\mathcal{X}_{\mathrm{et}}^{(1)}/\widehat{\mathcal{S}}^{(1)})$, the restriction of $F_{\mathcal{X}/M_0}^{*}\Theta_{\mathcal{S}^{(1)}/M}$ to T_{et} is equal to the pullback

$$s_T^{*}(F_{\mathcal{S}/M*}\mathcal{O}_{\mathcal{S}_n})|_{T'_{\mathrm{et}}} \tag{3.4.9.6}$$

where n is an integer such that $T \to \mathcal{S}$ and $T' \to \mathcal{S}^{(1)}$ factor through $\mathcal{S}_n$ and $\mathcal{S}_n^{(1)}$ respectively. From the commutative diagram

$$\begin{array}{ccc} T & \xrightarrow{s_T} & T' \\ \downarrow & & \downarrow \\ \mathcal{S} & \xrightarrow{F_{\mathcal{S}/M}} & \mathcal{S}^{(1)} \end{array} \tag{3.4.9.7}$$

one obtains a morphism $F_{\mathcal{S}/M*}\mathcal{O}_{\mathcal{S}_n}|_{T'_{\mathrm{et}}} \to s_{T*}\mathcal{O}_T$ which defines a morphism

$$s_T^{*}(F_{\mathcal{S}/M*}\mathcal{O}_{\mathcal{S}_n})|_{T'_{\mathrm{et}}} \longrightarrow s_T^{*}s_{T*}\mathcal{O}_T \xrightarrow{\text{adjunction}} \mathcal{O}_T. \tag{3.4.9.8}$$

This defines the morphism (3.4.9.4) locally. To define the map (3.4.9.4) globally it remains to show that the above locally constructed map is independent of the choices so that we can glue them to get a global map. This we leave to the reader.

By adjunction the map (3.4.9.4) induces a morphism $\Theta_{\mathcal{S}^{(1)}/M} \to F_{\mathcal{X}/M_0*}\mathcal{O}_{\mathcal{X}_{\text{lis-et}}/\widehat{\mathcal{S}}}$ which by the commutativity of (3.4.9.3) defines a morphism

(3.4.9.9)

$$Ru_{\mathcal{X}_{\mathrm{et}}^{(1)}/\widehat{\mathcal{S}}^{(1)}*}\Theta_{\mathcal{S}^{(1)}/M} \longrightarrow Ru_{\mathcal{X}_{\mathrm{et}}^{(1)}/\widehat{\mathcal{S}}^{(1)}*}F_{\mathcal{X}/M_0*}\mathcal{O}_{\mathcal{X}_{\mathrm{et}}/\widehat{\mathcal{S}}} \longrightarrow F_{\mathcal{X}_0/M_0*}Ru_{\mathcal{X}/\widehat{\mathcal{S}}*}\mathcal{O}_{\mathcal{X}_{\mathrm{et}}/\widehat{\mathcal{S}}}.$$

3.4.10. — Let $\mathcal{T}$ be a topos. As in [**8**, 8.20], for a complex $C^\bullet$ of sheaves of A-modules in $\mathcal{T}$ write $\eta C^\bullet$ for the subcomplex of $C^\bullet$ which in degree $i \geq 0$ is

$$\{f \in p^i C^i | df \in p^{i+1}C^{i+1}\}, \tag{3.4.10.1}$$

and in degrees $i < 0$ is equal to C^i. If $C^\bullet \to D^\bullet$ is a map of complexes with each C^i and D^i p-torsion free, then by [**8**, 8.19] the induced map $\eta C^\bullet \to \eta D^\bullet$ is a quasi-isomorphism. Since A is assumed flat over $\mathbb{Z}_p$, we can therefore define a functor

$$\mathbb{L}\eta : D^-_{\mathcal{T}}(A) \longrightarrow D^-_{\mathcal{T}}(A) \tag{3.4.10.2}$$

by sending a complex $C^\bullet$ to $\eta P^\bullet$ for any flat resolution $P^\bullet \to C^\bullet$ of $C^\bullet$. More generally $\mathbb{L}\eta C^\bullet$ is equal to $\eta P^\bullet$ where $P^\bullet$ is any complex of A-modules quasi-isomorphic to $C^\bullet$ and with each P^i p-torsion free. Note also that there is a canonical map

$$\mathbb{L}\eta \longrightarrow \mathrm{id} \tag{3.4.10.3}$$

induced by the inclusion $\eta P^\bullet \subset P^\bullet$.

Theorem 3.4.11. — *Let $(\mathcal{S}, F_{\mathcal{S}/M} : \mathcal{S} \to \mathcal{S}^{(1)})$ be as in 3.4.1, and let $\mathcal{X} \to \mathcal{S}_0$ be a smooth representable morphism of algebraic stacks. Then the composite morphism obtained from (3.4.9.9)*

$$Ru_{\mathcal{X}^{(1)}_{\mathrm{et}}/\widehat{\mathcal{S}}^{(1)}*}(\Theta_{\mathcal{S}^{(1)}/M}) \longrightarrow F_{\mathcal{X}/M_0*}Ru_{\mathcal{X}_{\mathrm{et}}/\widehat{\mathcal{S}}*}(\mathcal{O}_{\mathcal{X}_{\mathrm{et}}/\widehat{\mathcal{S}}}) \tag{3.4.11.1}$$

factors as an isomorphism

$$Ru_{\mathcal{X}^{(1)}_{\mathrm{et}}/\widehat{\mathcal{S}}^{(1)}*}(\Theta_{\mathcal{S}^{(1)}/M}) \longrightarrow F_{\mathcal{X}/M_0*}\mathbb{L}\eta Ru_{\mathcal{X}_{\mathrm{et}}/\widehat{\mathcal{S}}*}(\mathcal{O}_{\mathcal{X}_{\mathrm{et}}/\widehat{\mathcal{S}}}) \tag{3.4.11.2}$$

composed with the map

$$F_{\mathcal{X}/M_0*}\mathbb{L}\eta Ru_{\mathcal{X}_{\mathrm{et}}/\widehat{\mathcal{S}}*}(\mathcal{O}_{\mathcal{X}_{\mathrm{et}}/\widehat{\mathcal{S}}}) \to F_{\mathcal{X}/M_0*}Ru_{\mathcal{X}_{\mathrm{et}}/\widehat{\mathcal{S}}*}(\mathcal{O}_{\mathcal{X}_{\mathrm{et}}/\widehat{\mathcal{S}}}) \tag{3.4.11.3}$$

induced by (3.4.10.3).

The proof is in several steps 3.4.12–3.4.27.

3.4.12. — Consider first the case when there exists a commutative diagram of stacks

$$\begin{array}{ccccc} \mathcal{X} & \xrightarrow{i} & \mathcal{Y} & \xrightarrow{s} & \mathcal{S} \\ {\scriptstyle F_{\mathcal{X}/M_0}}\downarrow & & \downarrow{\scriptstyle F_{\mathcal{Y}/M}} & & \downarrow{\scriptstyle F_{\mathcal{S}/M}} \\ \mathcal{X}^{(1)} & \xrightarrow{i'} & \mathcal{Y}^{(1)} & \xrightarrow{s'} & \mathcal{S}^{(1)}, \end{array} \tag{3.4.12.1}$$

where s and s' are smooth, and i and i' are closed immersions, and $F_{\mathcal{Y}/M} : \mathcal{Y} \to \mathcal{Y}^{(1)}$ is a lifting of the relative Frobenius morphism of $\mathcal{Y}_0/M_0$, and $\mathcal{Y}$ and $\mathcal{Y}^{(1)}$ are Deligne-Mumford stacks.

Let $D^{(1)}$ denote the p-adic completion of the divided power envelope of $\mathcal{X}^{(1)}$ in $\mathcal{Y}^{(1)}$, and let $\mathcal{D}^{(1)}$ be its coordinate ring viewed as a sheaf on $\mathcal{Y}^{(1)}_{\text{et}}$. The crystal $\Theta_{\mathcal{S}^{(1)}/M}$ in $(\mathcal{X}^{(1)}_{\text{et}}/\mathcal{S}^{(1)})_{\text{cris}}$ then corresponds to a $\mathcal{D}^{(1)}$-module $\mathcal{E}$ with integrable connection

$$\nabla : \mathcal{E} \longrightarrow \mathcal{E} \otimes_{\mathcal{O}_{\mathcal{Y}^{(1)}_{\text{et}}}} \Omega^1_{\mathcal{Y}^{(1)}/\mathcal{S}^{(1)}} \tag{3.4.12.2}$$

compatible with the canonical connection on $\mathcal{D}^{(1)}$. This module with integrable connection $(\mathcal{E}, \nabla)$ can be described as follows.

For each n, let $\overline{\mathcal{Y}}^{(1)}_n$ denote the base change of $\mathcal{Y}^{(1)}_n$ to $\overline{\mathcal{S}}^{(1)}_n := \operatorname{Spec}_{\mathcal{S}^{(1)}_n}(F_{\mathcal{S}_n/M_n*}\mathcal{O}_{\mathcal{S}_n})$, and let $\pi : \overline{\mathcal{Y}}^{(1)}_n \to \mathcal{Y}^{(1)}_n$ denote the projection. Then the module with integrable connection $(\mathcal{E}_n, \nabla)$ obtained from $(\mathcal{E}, \nabla)$ by reduction modulo p^{n+1} is equal to the pullback to $D^{(1)}$ of $\pi_*\mathcal{O}_{\overline{\mathcal{Y}}^{(1)}_n}$ with connection

$$\pi_*(d) : \pi_*\mathcal{O}_{\overline{\mathcal{Y}}^{(1)}_n} \longrightarrow \pi_*\Omega^1_{\overline{\mathcal{Y}}^{(1)}_n/\overline{\mathcal{S}}^{(1)}_n} \simeq (\pi_*\mathcal{O}_{\overline{\mathcal{Y}}^{(1)}_n}) \otimes_{\mathcal{O}_{\mathcal{Y}^{(1)}_n}} \Omega^1_{\mathcal{Y}^{(1)}_n/\mathcal{S}^{(1)}_n}. \tag{3.4.12.3}$$

The right hand side of (3.4.11.2) can be described as follows. Let D be the p-adic completion of the divided power envelope of $\mathcal{X}$ in $\mathcal{Y}$, and let $\mathcal{D}$ be the coordinate ring of D viewed as a sheaf on $\mathcal{Y}_{\text{et}}$. Let $\Omega^\bullet_{\mathcal{D}_{\text{et}}/\mathcal{S}}$ denote the complex $\mathcal{D} \otimes_{\mathcal{O}_{\mathcal{Y}_{\text{et}}}} \Omega^\bullet_{\mathcal{Y}_{\text{et}}/M}$ with differential induced by the canonical connection on $\mathcal{D}$. Let $N^\bullet \subset \Omega^\bullet_{\mathcal{D}_{\text{et}}/\mathcal{S}}$ be the subcomplex with N^i equal to the sheaf of sections of the form $p^i\omega \in \Omega^i_{\mathcal{D}_{\text{et}}/\mathcal{S}}$ with $dp^i\omega \in p^{i+1}\Omega^{i+1}_{\mathcal{D}_{\text{et}}/\mathcal{S}}$.

Lemma 3.4.13. — *The sheaf of rings $\mathcal{D}$ is p-torsion free.*

Proof. — Since the formation of divided power envelopes is compatible with flat base change on $\mathcal{S}$, we can by base changing to a smooth cover $S \to \mathcal{S}$ by a scheme assume that $\mathcal{S}$ is a scheme. Furthermore, for a cartesian diagram

$$\begin{array}{ccc} \mathcal{X}_1 & \longrightarrow & \mathcal{Y}_1 \\ \downarrow & & \downarrow \rho \\ \mathcal{X} & \longrightarrow & \mathcal{Y} \end{array} \tag{3.4.13.1}$$

with ρ étale, the induced morphism of divided power envelopes $D_1 \to D$ is flat (in fact the construction of the divided power envelope D in the proof of 1.2.3 is an étale local construction on $\mathcal{Y}$). It follows that the assertion of the lemma is étale local on $\mathcal{Y}$, and hence we may also assume that $\mathcal{X}$ and $\mathcal{Y}$ are schemes. In this case the result follows from [**8**, 3.32]. □

Corollary 3.4.14. — *The complex $N^\bullet$ represents $\mathbb{L}\eta Ru_{\mathcal{X}_{\text{et}}/\widehat{\mathcal{S}}*}(\mathcal{O}_{\mathcal{X}_{\text{et}}/\widehat{\mathcal{S}}})$.*

Proof. — This follows from 3.4.13 and [**8**, 8.19]. □

Lemma 3.4.15. — *The natural map $\mathcal{E} \otimes \Omega^\bullet_{\mathcal{Y}^{(1)}/\mathcal{S}^{(1)}} \to F_{\mathcal{Y}/M*}\Omega^\bullet_{\mathcal{D}_{\text{et}}/\mathcal{S}}$ factors through $F_{\mathcal{Y}/M*}N^\bullet$.*

Proof. — Since $\mathcal{D}$ is p-torsion free, it suffices to show that if $\omega \in \mathcal{E} \otimes \Omega^i_{\mathcal{Y}^{(1)}/\mathcal{S}^{(1)}}$ is a local section, then the image of ω (resp. $d\omega$) in $F_{\mathcal{Y}_{i-1}/M_{i-1}*}\Omega^i_{\mathcal{D}_{\mathrm{et}}\otimes\mathbb{Z}/p^i/\mathcal{S}_{i-1}}$ (resp. $F_{\mathcal{Y}_i/M_i*}\Omega^{i+1}_{\mathcal{D}_{\mathrm{et}}\otimes\mathbb{Z}/(p^{i+1})/\mathcal{S}_i}$) is zero. Furthermore, this can be verified after making a smooth base change $S \to \mathcal{S}$, and hence it suffices to consider the case when $\mathcal{S}$ is a scheme. In this case the result follows from [**8**, 8.21]. □

Proposition 3.4.16. — *The induced map* $\mathcal{E} \otimes \Omega^\bullet_{\mathcal{Y}^{(1)}/\mathcal{S}^{(1)}} \to F_{\mathcal{Y}/M*}N^\bullet$ *is a quasi-isomorphism.*

The proof of this proposition occupies 3.4.17–3.4.24.

3.4.17. — Let $W \to \mathcal{Y}^{(1)}$ be an étale morphism with W an affine scheme, and let $V \to \mathcal{Y}$ denote $W \times_{\mathcal{Y}^{(1)}, F_{\mathcal{Y}/M}} \mathcal{Y}$ so that there is a cartesian diagram

$$\begin{array}{ccc} V & \longrightarrow & W \\ \downarrow & & \downarrow \\ \mathcal{Y} & \xrightarrow{F_{\mathcal{Y}/M}} & \mathcal{Y}^{(1)}. \end{array} \tag{3.4.17.1}$$

Since $V_0 \to \mathcal{Y}_0$ is étale, the diagram

$$\begin{array}{ccc} V_0 & \xrightarrow{F_{V_0/M_0}} & V_0^{(1)} \\ \downarrow & & \downarrow \\ \mathcal{Y}_0 & \xrightarrow{F_{\mathcal{Y}_0/M_0}} & \mathcal{Y}_0^{(1)} \end{array} \tag{3.4.17.2}$$

is cartesian. It follows that both W_0 and $V_0^{(1)}$ are étale $\mathcal{Y}_0^{(1)}$-schemes whose pullbacks along the relative Frobenius morphism $F_{\mathcal{Y}_0/M_0} : \mathcal{Y}_0 \to \mathcal{Y}_0^{(1)}$ come with isomorphism to V_0. Since the map $F_{\mathcal{Y}_0/M_0}$ is radicial and surjective the pullback functor

$$F^*_{\mathcal{S}_0/M_0} : (\text{étale } \mathcal{Y}_0^{(1)}\text{-spaces }) \longrightarrow (\text{étale } \mathcal{Y}_0\text{-spaces}) \tag{3.4.17.3}$$

is fully faithful (this follows from the same argument used in [**28**, IX.4.10] in the setting of schemes). Consequently, there is a unique isomorphism $W_0 \simeq V_0^{(1)}$ whose pullback to $\mathcal{Y}_0$ induces the identity morphism $V_0 \to V_0$. Write $V^{(1)}$ for W viewed as a lifting of $V_0^{(1)}$ using this isomorphism, and let $F_{V/M} : V \to V^{(1)}$ be the projection so that there is a commutative diagram

$$\begin{array}{ccc} V & \xrightarrow{F_{V/M}} & V^{(1)} \\ \downarrow & & \downarrow \\ \mathcal{Y} & \xrightarrow{F_{\mathcal{Y}/M}} & \mathcal{Y}^{(1)}. \end{array} \tag{3.4.17.4}$$

Let U denote $\mathcal{X} \times_{\mathcal{Y}} V$. Using a similar argument to the above, one sees that $U^{(1)}$ is canonically isomorphic to $\mathcal{X}^{(1)} \times_{\mathcal{Y}^{(1)}} V^{(1)}$. We then obtain a commutative diagram

$$\begin{array}{ccccc} U & \longrightarrow & V & \longrightarrow & \mathcal{S} \\ {\scriptstyle F_{U/M_0}}\downarrow & & \downarrow{\scriptstyle F_{V/M}} & & \downarrow{\scriptstyle F_{\mathcal{S}/M}} \\ U^{(1)} & \longrightarrow & V^{(1)} & \longrightarrow & \mathcal{S}^{(1)} \end{array} \tag{3.4.17.5}$$

mapping to (3.4.12.1), and with U, V, and $V^{(1)}$ affine schemes. To prove the proposition it suffices to prove the proposition for (3.4.17.5), and hence we may assume that both $\mathcal{Y}$, $\mathcal{X}$, and $\mathcal{Y}^{(1)}$ are affine.

After replacing S by an étale covering we may also assume that there exists a smooth lifting $\widetilde{\mathcal{Y}}$ of $\mathcal{X}$ to $\mathcal{S}$. We reduce to the case when $\mathcal{Y} = \widetilde{\mathcal{Y}}$ as follows. Choose a lift

$$F_{\widetilde{\mathcal{Y}}/M} : \widetilde{\mathcal{Y}} \longrightarrow \widetilde{\mathcal{Y}}^{(1)} \tag{3.4.17.6}$$

of the relative Frobenius of $\mathcal{X}$, and let $\mathcal{Z}$ denote $\mathcal{Y} \times_{\mathcal{S}} \widetilde{\mathcal{Y}}$ with lifting of the relative Frobenius

$$F_{\mathcal{Z}/M} : \mathcal{Z} \longrightarrow \mathcal{Z}^{(1)} \simeq \mathcal{Y}^{(1)} \times_{\mathcal{S}^{(1)}} \widetilde{\mathcal{Y}}^{(1)} \tag{3.4.17.7}$$

given by $F_{\mathcal{Y}/M} \times F_{\widetilde{\mathcal{Y}}/M}$. Let $N_{\mathcal{Y}}$ (resp. $N_{\widetilde{\mathcal{Y}}}$, $N_{\mathcal{Z}}$) denote the complex obtained from the above construction using $\mathcal{Y}$ (resp. $\widetilde{\mathcal{Y}}$, $\mathcal{Z}$). Let $\mathcal{E}_{\mathcal{Y}}$ (resp. $\mathcal{E}_{\widetilde{\mathcal{Y}}}$, $\mathcal{E}_{\mathcal{Z}}$) denote the module with integrable connection defined by $\Theta_{\mathcal{S}^{(1)}/M}$ on $\mathcal{Y}_{\mathrm{et}}$ (resp. $\widetilde{\mathcal{Y}}_{\mathrm{et}}$, $\mathcal{Z}_{\mathrm{et}}$). Then there are commutative diagrams

$$\begin{array}{ccc} \mathrm{pr}_{1*}\mathcal{E}_{\mathcal{Z}} \otimes \Omega^\bullet_{\mathcal{Z}^{(1)}/\mathcal{S}} & \longrightarrow & \mathrm{pr}_{1*}F_{\mathcal{Z}/M*}N^\bullet_{\mathcal{Z}} \\ \uparrow & & \uparrow \\ \mathcal{E}_{\mathcal{Y}} \otimes \Omega^\bullet_{\mathcal{Y}^{(1)}/\mathcal{S}^{(1)}} & \longrightarrow & N^\bullet_{\mathcal{Y}}, \end{array} \tag{3.4.17.8}$$

and

$$\begin{array}{ccc} \mathrm{pr}_{2*}\mathcal{E}_{\mathcal{Z}} \otimes \Omega^\bullet_{\mathcal{Z}^{(1)}/\mathcal{S}} & \longrightarrow & \mathrm{pr}_{2*}F_{\mathcal{Z}/M*}N^\bullet_{\mathcal{Z}} \\ \uparrow & & \uparrow \\ \mathcal{E}_{\widetilde{\mathcal{Y}}} \otimes \Omega^\bullet_{\widetilde{\mathcal{Y}}^{(1)}/\mathcal{S}^{(1)}} & \longrightarrow & N^\bullet_{\widetilde{\mathcal{Y}}}. \end{array} \tag{3.4.17.9}$$

The same argument used in the proof of 2.3.2, shows that the p-adic completion of the divided power envelope of $\mathcal{X}$ in $\mathcal{Z}$ is affine. From this and 2.7.8 it follows that the vertical arrows in (3.4.17.8) and (3.4.17.9) are quasi-isomorphisms, and that in order to prove that $\mathcal{E}_{\mathcal{Z}} \otimes \Omega^\bullet_{\mathcal{Z}^{(1)}/\mathcal{S}} \rightarrow F_{\mathcal{Z}/M*}N^\bullet_{\mathcal{Z}}$ is a quasi-isomorphism it suffices to show that the bottom horizontal arrow in either of the above two diagrams is a quasi-isomorphism. It therefore suffices to consider the case when $\mathcal{Y} = \widetilde{\mathcal{Y}}$.

3.4.18. — Assuming that the reduction modulo p of $\mathcal{Y}$ is $\mathcal{X}$, we reduce the proof to the classical version as follows. Since the assertion is étale local on $\mathcal{Y}$ we may also assume that $\mathcal{Y}$ is an affine scheme.

Let $S \to \mathcal{S}$ be a smooth surjection with S an affine scheme, and let $S_\bullet$ denote the 0-coskeleton. Choose a morphism $F_{S/M} : S \to \overline{S} := \overline{\mathcal{S}} \times_{\mathcal{S}^{(1)}} S^{(1)}$ such that the diagram

(3.4.18.1)
$$\begin{array}{ccc} S & \xrightarrow{F_{S/M}} & \overline{S} \\ \downarrow & & \downarrow \\ \mathcal{S} & \xrightarrow{F_{\mathcal{S}/M}} & \overline{\mathcal{S}} \end{array}$$

commutes. Let $X'_\bullet$ (resp. $X_\bullet$) denote the simplicial algebraic space obtained from $\mathcal{X}' \to \mathcal{S}_0$ (resp. $\mathcal{X} \to \mathcal{S}_0$) by base change to $S_{0,\bullet}$, and let $F_{X_\bullet/S_{0,\bullet}} : X_\bullet \to X'_\bullet$ be the relative Frobenius morphism. Let $\overline{\mathcal{X}}'$ denote the relative spectrum $\mathrm{Spec}_{\mathcal{X}^{(1)}}(P_*\mathcal{O}_{\mathcal{X}'})$ as in 3.3.1, and let $\Lambda' : X'_{\bullet,\mathrm{et}} \to \overline{\mathcal{X}}'_{\mathrm{et}}$ be the natural morphism of topoi. Write $\overline{F}_{\mathcal{X}/\mathcal{S}_0} : \mathcal{X} \to \overline{\mathcal{X}}'$ for the composite $P \circ F_{\mathcal{X}/\mathcal{S}_0}$.

For an integer $i \geq 0$, let $Z^i_{\mathcal{X}/\mathcal{S}_0}$ (resp. $Z^i_{X_\bullet/S_\bullet}$) denote the kernel of the map $d : \Omega^i_{\mathcal{X}_{\mathrm{et}}/\mathcal{S}_0} \to \Omega^{i+1}_{\mathcal{X}_{\mathrm{et}}/\mathcal{S}_0}$ (resp. $d : \Omega^i_{X_{\bullet,\mathrm{et}}/S_{0,\bullet}} \to \Omega^{i+1}_{X_{\bullet,\mathrm{et}}/S_{0,\bullet}}$), and let $B^i_{\mathcal{X}/\mathcal{S}_0}$ (resp. $B^i_{X_\bullet/S_{0,\bullet}}$) be the image of $\Omega^{i-1}_{\mathcal{X}_{\mathrm{et}}/\mathcal{S}_0}$ (resp. $\Omega^{i-1}_{X_{\bullet,\mathrm{et}}/S_{0,\bullet}}$).

Lemma 3.4.19. — *The natural maps*

(3.4.19.1)
$$\overline{F}_{\mathcal{X}/\mathcal{S}_0*}Z^i_{\mathcal{X}/\mathcal{S}_0} \longrightarrow R\Lambda'_*Z^i_{X_\bullet/S_{0,\bullet}}, \quad \overline{F}_{\mathcal{X}/\mathcal{S}_0*}B^i_{\mathcal{X}/\mathcal{S}_0} \longrightarrow R\Lambda'_*B^i_{X_\bullet/S_{0,\bullet}}$$

are isomorphisms.

Proof. — The proof is by induction on i. If $i = 0$, the statement for the B^i is trivial. For the Z^0's, note that by the Cartier isomorphism there is a canonical isomorphism $Z^0_{\mathcal{X}/\mathcal{S}_0} \simeq \mathcal{O}_{\overline{\mathcal{X}}'}$ and $Z^0_{X_\bullet/S_{0,\bullet}} = \mathcal{O}_{X'_\bullet}$. With these identifications the map in question becomes the canonical map

(3.4.19.2)
$$\mathcal{O}_{\overline{\mathcal{X}}'_{\mathrm{et}}} \longrightarrow R\Lambda'_*\mathcal{O}_{X'_\bullet} \simeq RP_*\mathcal{O}_{\mathcal{X}'}$$

which is an isomorphism by the definition of a Frobenius acyclic stack.

Next we assume the result holds for i and prove it for $i+1$. To obtain the result for the B^{i+1}, observe that there are exact sequences

(3.4.19.3)
$$0 \longrightarrow Z^i_{\mathcal{X}/\mathcal{S}_0} \longrightarrow \Omega^i_{\mathcal{X}/\mathcal{S}_0} \longrightarrow B^{i+1}_{\mathcal{X}/\mathcal{S}_0} \longrightarrow 0$$

(3.4.19.4)
$$0 \longrightarrow Z^i_{X_\bullet/S_{0,\bullet}} \longrightarrow \Omega^i_{X_\bullet/S_{0,\bullet}} \longrightarrow B^{i+1}_{X_\bullet/S_{0,\bullet}} \longrightarrow 0,$$

and hence the result for the B^{i+1} follows from the result for the Z^i. To get the result for the Z^{i+1}, note that the Cartier isomorphism (3.3.17) yields exact sequences

(3.4.19.5)
$$0 \longrightarrow B^{i+1}_{\mathcal{X}/\mathcal{S}_0} \longrightarrow Z^{i+1}_{\mathcal{X}/\mathcal{S}_0} \longrightarrow \mathcal{O}_{\overline{\mathcal{X}}'} \otimes_{\mathcal{O}_{\mathcal{X}}} \Omega^{i+1}_{\mathcal{X}/\mathcal{S}_0} \longrightarrow 0$$

(3.4.19.6)
$$0 \longrightarrow B^{i+1}_{X_\bullet/S_{0,\bullet}} \longrightarrow Z^{i+1}_{X_\bullet/S_{0,\bullet}} \longrightarrow \Omega^{i+1}_{X'_\bullet/S_{0,\bullet}} \longrightarrow 0.$$

Consideration of the distinguished triangles arising from these exact sequences together with the definition of a Frobenius acyclic morphism then yields the result for Z^{i+1}. □

3.4.20. — Let $\widehat{\mathcal{Y}}$ denote the formal algebraic space [**46**, V.2.1] obtained by completing $\mathcal{Y}$ along $(p) \subset \mathcal{O}_{\mathcal{Y}}$, and let $\widehat{Y}_\bullet$ denote the p-adic completion of the simplicial algebraic space $Y_\bullet := \mathcal{Y} \times_{\mathcal{S}} S_\bullet$. Note that the underlying topos of $\widehat{\mathcal{Y}}$ (resp. $\widehat{Y}_\bullet$) is canonically identified with the topos $\mathcal{X}_{\mathrm{et}}$ (resp. $X_{\bullet,\mathrm{et}}$).

Lemma 3.4.21. — *Let $\mathcal{F}$ be a coherent sheaf on $\widehat{\mathcal{Y}}$, and let $\mathcal{F}_\bullet$ be the pullback to $\widehat{Y}_\bullet$. Then the natural map*

$$\overline{F}_{\mathcal{X}/\mathcal{S}_0*}\mathcal{F} \longrightarrow R\Lambda'_*(F_{X_\bullet/S_{0,\bullet}*}\mathcal{F}_\bullet) \tag{3.4.21.1}$$

is an isomorphism.

Proof. — Consideration of the commutative diagram of topoi

$$\begin{array}{ccc} X_{\bullet,\mathrm{et}} & \xrightarrow{F_{X_\bullet/\mathcal{S}_\bullet}} & X'_{\bullet,\mathrm{et}} \\ \Lambda\downarrow & & \downarrow\Lambda' \\ \mathcal{X}_{\mathrm{et}} & \xrightarrow{\overline{F}_{\mathcal{X}/\mathcal{S}_0}} & \overline{\mathcal{X}}'_{\mathrm{et}}, \end{array} \tag{3.4.21.2}$$

implies that it suffices to show that

$$F_{X_\bullet/S_\bullet*}\mathcal{F}_\bullet \simeq RF_{X_\bullet/S_\bullet*}\mathcal{F}_\bullet, \quad \overline{F}_{\mathcal{X}/\mathcal{S}_0*}\mathcal{F} \simeq R\overline{F}_{\mathcal{X}/\mathcal{S}_0*}\mathcal{F}, \tag{3.4.21.3}$$

and that the natural map

$$\mathcal{F} \longrightarrow R\Lambda_*\mathcal{F}_\bullet \tag{3.4.21.4}$$

is a quasi-isomorphism.

To prove that the maps in (3.4.21.3) are quasi-isomorphisms, define topoi $\mathbb{N} \times X_{\bullet,\mathrm{et}}$ and $\mathbb{N} \times X'_{\bullet,\mathrm{et}}$ as in [**8**, proof of 7.20]. Recall that the topos $\mathbb{N} \times X_{\bullet,\mathrm{et}}$ (resp. $\mathbb{N} \times X'_{\bullet,\mathrm{et}}$) is the topos associated to a site made of pairs (n, U), where $n \in \mathbb{N}$ and $U \in \mathrm{Et}(X_\bullet)$ (resp. $U \in \mathrm{Et}(X'_\bullet)$). A sheaf on $\mathbb{N} \times X_{\bullet,\mathrm{et}}$ is simply a projective system of sheaves on $X_{\bullet,\mathrm{et}}$, and similarly sheaves on $\mathbb{N} \times X'_{\bullet,\mathrm{et}}$ are projective systems of sheaves on $X'_{\bullet,\mathrm{et}}$. There is a natural commutative diagram of topoi

$$\begin{array}{ccc} \mathbb{N} \times X_{\bullet,\mathrm{et}} & \xrightarrow{F_{\mathbb{N}\times X_\bullet/S_\bullet}} & \mathbb{N} \times X'_{\bullet,\mathrm{et}} \\ j_\bullet\downarrow & & \downarrow j'_\bullet \\ X_{\mathrm{et}} & \xrightarrow{F_{X/S}} & X'_{\mathrm{et}}, \end{array} \tag{3.4.21.5}$$

where $j_{\bullet*}$ and $j'_{\bullet*}$ send a projective system of sheaves $\mathcal{F}_\bullet$ to $\varprojlim \mathcal{F}_\bullet$. Let $\{\mathcal{F}_{n,\bullet}\}$ be the sheaf on $\mathbb{N} \times X_{\bullet,\mathrm{et}}$ whose value on $(n, ([r], U))$ is the value of the pullback of $\mathcal{F}$ to the unique lifting of U to an étale $\mathcal{Y}^r \times_{\mathcal{S}} \mathcal{S}_n$-scheme. Then $\mathcal{F}_\bullet \simeq j^\bullet_*\{\mathcal{F}_{n,\bullet}\}$, and by the argument of [**8**, 7.20] $R^i j_{\bullet*}\{\mathcal{F}_{n,\bullet}\} = 0$ for $i > 0$. Similarly, $R^i j'_{\bullet*}(F_{\mathbb{N}\times X_\bullet/S_\bullet*}\{\mathcal{F}_{n,\bullet}\}) = 0$

for $i > 0$. Now to verify that $R^iF_{\mathbb{N}\times X_\bullet/S_\bullet *}\{\mathcal{F}_{n,\bullet}\} = 0$ for $i > 0$, it suffices to show that each of the $\mathcal{F}_{n,\bullet}$ are acyclic for $F_{X_\bullet/S_\bullet *}$ which is clear since $X_{n,\mathrm{et}} \to X'_{n,\mathrm{et}}$ is an equivalence of topoi. Thus we find that the map

$$F_{X_\bullet/S_\bullet *}\mathcal{F}_\bullet \longrightarrow RF_{X_\bullet/S_\bullet *}\mathcal{F}_\bullet \tag{3.4.21.6}$$

is a quasi-isomorphism, and the same argument gives that $\overline{F}_{\mathfrak{X}/\mathcal{S}_0 *}\mathcal{F} \simeq R\overline{F}_{\mathfrak{X}/\mathcal{S}_0 *}\mathcal{F}$.

To show that the map (3.4.21.4) is a quasi-isomorphism, we proceed as follows. Consider the commutative diagram

$$\begin{array}{ccc} \mathbb{N}\times X_{\bullet,\mathrm{et}} & \xrightarrow{j_\bullet} & X_{\bullet,\mathrm{et}} \\ {\scriptstyle \mathbb{N}\times\Lambda}\downarrow & & \downarrow{\scriptstyle\Lambda} \\ \mathbb{N}\times\mathfrak{X}_{\mathrm{et}} & \xrightarrow{j} & \mathfrak{X}_{\mathrm{et}}, \end{array} \tag{3.4.21.7}$$

Define $\{\mathcal{F}_n\}$ to be the sheaf on $\mathbb{N}\times\mathfrak{X}_{\mathrm{et}}$ whose value on (n, U) is equal to the value on the unique lifting of U to Y_n of the pullback of $\mathcal{F}$ to Y_n. Then $\mathcal{F} = j_*\{\mathcal{F}_n\}$ and $\mathcal{F}_\bullet = j_{\bullet *}\{\mathcal{F}_{n,\bullet}\}$. Moreover, as above $\{\mathcal{F}_n\}$ is acyclic for j_*. Thus to show that (3.4.21.4) is an isomorphism, it is enough to show that

$$\{\mathcal{F}_n\} \simeq R(\mathbb{N}\times\Lambda)_*\{\mathcal{F}_{n,\bullet}\}. \tag{3.4.21.8}$$

For this it suffices to show that for each n the natural map $\mathcal{F}_n \to R\Lambda_*\mathcal{F}_{n,\bullet}$ is an isomorphism, which follows from [**49**, 13.5.5]. □

3.4.22. — Define a formal algebraic space $\widehat{\overline{\mathcal{Y}}}'$ as follows. The underlying topos is $\overline{\mathfrak{X}}'_{\mathrm{et}}$, and the structure sheaf is the sheaf which to any étale scheme $U \to \overline{\mathfrak{X}}'$ associates

$$\varprojlim_n \Gamma(U_n, \mathcal{O}_{U_n}), \tag{3.4.22.1}$$

where U_n denotes the unique lifting of U to an étale scheme over

$$\overline{\mathcal{Y}}'_n := \mathrm{Spec}_{\mathcal{Y}^{(1)}_n}(F_{\mathcal{Y}_n/M_n *}\mathcal{O}_{\mathcal{Y}_n}). \tag{3.4.22.2}$$

It follows from 3.4.4 that this sheaf of rings defines a p-adic formal space $\widehat{\overline{\mathcal{Y}}}'$ and that there is a canonical essentially affine morphism of p-adic formal spaces $P : \widehat{\overline{\mathcal{Y}}}' \to \widehat{\mathcal{Y}}^{(1)}$. There is also a natural projection $\widehat{Y}'_\bullet \to \widehat{\overline{\mathcal{Y}}}'$.

Lemma 3.4.23. — *Let $\mathcal{G}$ be a locally free coherent sheaf on $\widehat{\overline{\mathcal{Y}}}'$, and let $\mathcal{G}_\bullet$ be the pullback of $\mathcal{G}$ to $\widehat{Y}'_\bullet$. Then the natural map $\mathcal{G} \to R\Lambda'_*\mathcal{G}_\bullet$ is an isomorphism.*

Proof. — Consider the commutative diagram of topoi

$$\begin{array}{ccc} \mathbb{N}\times X'_{\bullet,\mathrm{et}} & \xrightarrow{j'_\bullet} & X'_{\bullet,\mathrm{et}} \\ {\scriptstyle \mathbb{N}\times\Lambda'}\downarrow & & \downarrow{\scriptstyle\Lambda'} \\ \mathbb{N}\times\overline{\mathfrak{X}}'_{\mathrm{et}} & \xrightarrow{\bar{j}} & \overline{\mathfrak{X}}'_{\mathrm{et}}, \end{array} \tag{3.4.23.1}$$

and let $\{\mathcal{G}_n\}$ be the sheaf on $\mathbb{N} \times \overline{\mathfrak{X}}'_{\text{et}}$ which to any (n, U) associates the value of the pullback of $\mathcal{G}$ to $\overline{\mathcal{Y}}'_n$ on the unique lifting of U to an étale $\overline{\mathcal{Y}}'_n$-scheme, $\{\mathcal{G}_n^\bullet\}$ the sheaf on $\mathbb{N} \times X'_{\bullet,\text{et}}$ whose value on $(n, ([r], U))$ is the value of the pullback of $\mathcal{G}$ to the unique lifting of U to an étale $\mathcal{Y}'^r \times_{\mathcal{S}} \mathcal{S}_n$-scheme. Then as above, to prove (ii) it suffices to show that the natural map $\{\mathcal{G}_n\} \to R(\mathbb{N} \times \Lambda')_*\{\mathcal{G}_{n,\bullet}\}$ is an isomorphism, and for this it suffices to show that each of the maps $\mathcal{G}_n \to R\Lambda'_*\mathcal{G}_{n,\bullet}$ is an isomorphism. Now the case $n = 0$ follows from the definition of a Frobenius-acyclic morphism (3.2.1), and the case of general n follows from this by induction using the exact sequences

(3.4.23.2) $$0 \longrightarrow \mathcal{G}_1 \xrightarrow{\times p^n} \mathcal{G}_n \longrightarrow \mathcal{G}_{n-1} \longrightarrow 0. \quad \square$$

3.4.24. — Let $\Omega^\bullet_{\widehat{\overline{\mathcal{Y}}}/\overline{\mathcal{S}}}$ denote the complex whose i-th term is $\varprojlim_n \Omega^i_{\overline{\mathcal{Y}}_n/\overline{\mathcal{S}}_n}$ and whose differential is induced by differentiation. Then by construction of $\Theta_{\mathcal{S}^{(1)}/M}$ there is a canonical isomorphism

(3.4.24.1) $$\mathcal{E} \otimes \Omega^\bullet_{\mathcal{Y}^{(1)}/\mathcal{S}^{(1)}} \simeq P_*\Omega^\bullet_{\widehat{\overline{\mathcal{Y}}}/\overline{\mathcal{S}}}.$$

Let $M^\bullet \subset F_{X_\bullet/S_\bullet *}\Omega^\bullet_{\widehat{Y}_\bullet/\widehat{S}_\bullet}$ be the maximal sub-complex with M^k contained in $p^k F_{X_\bullet/S_\bullet *}\Omega^\bullet_{\widehat{Y}_\bullet/\widehat{S}_\bullet}$ for every k. By [**9**, 1.5], the natural map $\Omega^\bullet_{\widehat{Y}'_\bullet/\widehat{S}_\bullet} \to F_{X_\bullet/S_\bullet *}\Omega^\bullet_{\widehat{Y}_\bullet/\widehat{S}_\bullet}$ induces a quasi-isomorphism $\Omega^\bullet_{\widehat{Y}'_\bullet/\widehat{S}_\bullet} \simeq M^\bullet$. We then have a commutative diagram

(3.4.24.2) $$\begin{array}{ccc} \Omega^\bullet_{\widehat{\overline{\mathcal{Y}}}/\overline{\mathcal{S}}} & \longrightarrow & N^\bullet \\ {\scriptstyle a}\downarrow & & \downarrow{\scriptstyle b} \\ R\Lambda'_*\Omega^\bullet_{\widehat{Y}'_\bullet/\widehat{S}_\bullet} & \longrightarrow & R\Lambda'_*M^\bullet, \end{array}$$

where the map a is a quasi-isomorphism by 3.4.23. Thus to complete the proof of 3.4.16 it suffices to prove that the natural map $b : N^\bullet \to R\Lambda'_*M^\bullet$ is a quasi-isomorphism. This follows from 3.4.19, the definition of a Frobenius acyclic stack, and the exact sequences

(3.4.24.3) $$0 \longrightarrow N^i \longrightarrow p^i\Omega^i_{\widehat{\overline{\mathcal{Y}}}/\widehat{\mathcal{S}}} \xrightarrow{\pi} B^{i+1}_{X/\mathcal{S}_0} \longrightarrow 0$$

(3.4.24.4) $$0 \longrightarrow M^i \longrightarrow p^i\Omega^i_{\widehat{Y}^\bullet/\widehat{S}^\bullet} \xrightarrow{\pi^\bullet} B^{i+1}_{X^\bullet/S_0^\bullet} \longrightarrow 0,$$

where π and $\pi^\bullet$ are defined by sending $p^i\omega$ to the image under d of the reduction of ω.

This completes the proof of 3.4.16 and hence also 3.4.11 in the lifted situation. $\square$

3.4.25. — It remains to deduce 3.4.11 in general from the result in the lifted situation. This is done as in the classical case [**8**, proof of 8.20] using cohomological descent.

Since $\mathfrak{X}$ is a Deligne-Mumford stack, there exist an étale cover $U \to \mathfrak{X}$ and embedding $U \hookrightarrow \mathfrak{U}$ of U into a smooth $\mathcal{S}$-scheme $\mathfrak{U}$. Furthermore, we may assume that U is an affine scheme, in which case there also exists a lifting $F_{\mathfrak{U}/M} : \mathfrak{U} \to \mathfrak{U}^{(1)}$ over $F_{\mathcal{S}/M}$. Let $U_\bullet$ be the 0-coskeleton of $U \to \mathfrak{X}$, and let $\mathfrak{U}_\bullet$ be the 0-coskeleton of $\mathfrak{U} \to \mathcal{S}$. Then there is a canonical closed immersion $U_\bullet \hookrightarrow \mathfrak{U}_\bullet$ of simplicial algebraic spaces.

Let $\pi : U_\bullet \to \mathfrak{X}$ and $\pi^{(1)} : U_\bullet^{(1)}$ be the projections.

Lemma 3.4.26. — *The natural maps*

$$Ru_{\mathcal{X}^{(1)}_{\mathrm{et}}/\widehat{\mathcal{S}}*}(\Theta_{\mathcal{S}^{(1)}/M}) \longrightarrow R\pi^{(1)}_* Ru_{U^{(1)}_{\bullet\mathrm{et}}/\widehat{\mathcal{S}}^{(1)}*}(\Theta_{\mathcal{S}^{(1)}/M}) \tag{3.4.26.1}$$

and

$$\mathbb{L}\eta Ru_{\mathcal{X}_{\mathrm{et}}/\widehat{\mathcal{S}}*}(\mathcal{O}_{\mathcal{X}_{\mathrm{et}}/\widehat{\mathcal{S}}}) \longrightarrow R\pi_*\mathbb{L}\eta Ru_{U_{\bullet\mathrm{et}}/\widehat{\mathcal{S}}*}(\mathcal{O}_{U_{\bullet\mathrm{et}}/\widehat{\mathcal{S}}}) \tag{3.4.26.2}$$

are isomorphisms.

Proof. — That (3.4.26.1) and the natural map

$$Ru_{\mathcal{X}_{\mathrm{et}}/\widehat{\mathcal{S}}*}(\mathcal{O}_{\mathcal{X}_{\mathrm{et}}/\widehat{\mathcal{S}}}) \longrightarrow R\pi_* Ru_{U_{\bullet\mathrm{et}}/\widehat{\mathcal{S}}*}(\mathcal{O}_{U_{\bullet\mathrm{et}}/\widehat{\mathcal{S}}}) \tag{3.4.26.3}$$

are isomorphisms follows from 1.4.24. In addition, by [**8**, 8.19] for any p-torsion free complex $A^\bullet$ the complex $\mathbb{L}\eta A^\bullet$ is represented by the subcomplex of $A^\bullet$ which in degree i is equal to $\{a \in p^iA^i | da \in p^{i+1}A^{i+1}\}$. From this it follows that (3.4.26.2) is also an isomorphism. □

3.4.27. — To construct the desired isomorphism in 3.4.11, it therefore suffices to construct an isomorphism

$$Ru_{U^{(1)}_{\bullet\mathrm{et}}/\widehat{\mathcal{S}}^{(1)}*}(\Theta_{\mathcal{S}^{(1)}/M}) \longrightarrow F_{U_\bullet/M*}\mathbb{L}\eta Ru_{U_{\bullet\mathrm{et}}/\widehat{\mathcal{S}}}(\mathcal{O}_{U_{\bullet\mathrm{et}}/\widehat{\mathcal{S}}}). \tag{3.4.27.1}$$

Let $\mathcal{D}_\bullet$ (resp. $\mathcal{D}^{(1)}_\bullet$) denote the coordinate ring of the divided power envelope of $U_\bullet$ in $\mathcal{U}_\bullet$ (resp. $U^{(1)}_\bullet$ in $\mathcal{U}^{(1)}_\bullet$), and let $\mathcal{E}$ denote the $\mathcal{D}^{(1)}_\bullet$-module with integrable connection on $U^{(1)}_{\bullet\mathrm{et}}$ corresponding to $\Theta_{\mathcal{S}^{(1)}/M}$. Let $N^\bullet \subset \Omega^\bullet_{\mathcal{D}_\bullet/\widehat{\mathcal{S}}}$ denote the subcomplex defined in 3.4.12. By 3.4.14 the complex $N^\bullet$ represents $\mathbb{L}\eta Ru_{U_{\bullet\mathrm{et}}/\widehat{\mathcal{S}}}(\mathcal{O}_{U_{\bullet\mathrm{et}}/\widehat{\mathcal{S}}})$, and by 3.4.16 the natural map

$$\mathcal{E} \otimes \Omega^\bullet_{\mathcal{U}^{(1)}_\bullet/\mathcal{S}^{(1)}} \longrightarrow F_{\mathcal{U}_\bullet/M*}\Omega^\bullet_{\mathcal{U}_\bullet/\mathcal{S}} \tag{3.4.27.2}$$

induces a quasi-isomorphism $\mathcal{E} \otimes \Omega^\bullet_{\mathcal{U}^{(1)}_\bullet/\mathcal{S}^{(1)}} \to F_{\mathcal{U}_\bullet/M*}N^\bullet$. Since

$$\mathcal{E} \otimes \Omega^\bullet_{\mathcal{U}^{(1)}_\bullet/\mathcal{S}^{(1)}} \simeq Ru_{U^{(1)}_{\bullet\mathrm{et}}/\widehat{\mathcal{S}}^{(1)}*}(\Theta_{\mathcal{S}^{(1)}/M}) \tag{3.4.27.3}$$

this defines the arrow in 3.4.11 for general $\mathcal{X}$.

Finally if $U' \to \mathcal{X}$ is a second choice of covering with lifting $U' \hookrightarrow \mathcal{U}'$, then consideration of the product $U \times_{\mathcal{X}} U' \hookrightarrow \mathcal{U} \times_{\mathcal{S}} \mathcal{U}'$ shows that the isomorphism in 3.4.11 defined by $U \hookrightarrow \mathcal{U}$ and $U' \hookrightarrow \mathcal{U}'$ are the same. This completes the proof of 3.4.11. □

3.4.28. — The complex $Ru_{\mathcal{X}^{(1)}/\mathcal{S}^{(1)}*}(\Theta_{\mathcal{S}^{(1)}/M})$ can be described in terms of the stack $\mathcal{X}'$.

Assume that for every integer n the natural map

$$j_n^* F_{\mathcal{S}/M*}\mathcal{O}_{\mathcal{S}_{\text{lis-et}}} \longrightarrow F_{\mathcal{S}_n/M_n*}\mathcal{O}_{\mathcal{S}_{n,\text{lis-et}}} \tag{3.4.28.1}$$

is an isomorphism.

Remark 3.4.29. — The assumption that (3.4.28.1) is an isomorphism is not essential (see 3.4.36 below), but it holds in every example we consider. In particular, if the morphism $F_{\mathcal{S}/M} : \mathcal{S} \to \mathcal{S}^{(1)}$ is proper then it holds (note that we know by 3.2.17 that the morphism $F_{\mathcal{S}_0/M_0} : \mathcal{S}_0 \to \mathcal{S}_0^{(1)}$ is proper). To see this, consider the short exact sequence (using the flatness of $\mathcal{S}/M$)

$$(3.4.29.1) \qquad 0 \longrightarrow \mathcal{O}_{\mathcal{S}} \xrightarrow{\times p^n} \mathcal{O}_{\mathcal{S}} \longrightarrow j_{n*}\mathcal{O}_{\mathcal{S}_n} \longrightarrow 0.$$

Taking cohomology we obtain an exact sequence

$$(3.4.29.2) \qquad 0 \to F_{\mathcal{S}/M*}\mathcal{O}_{\mathcal{S}} \xrightarrow{\times p^n} F_{\mathcal{S}/M*}\mathcal{O}_{\mathcal{S}} \longrightarrow j_{n*}F_{\mathcal{S}_n/M_n*}\mathcal{O}_{\mathcal{S}_n} \longrightarrow R^1F_{\mathcal{S}/M*}\mathcal{O}_{\mathcal{S}}.$$

Since $F_{\mathcal{S}/M}$ is proper and $R^1F_{\mathcal{S}_n/M_n*}\mathcal{O}_{\mathcal{S}_n/M_n} = 0$ for all n since $\mathcal{S}_0$ is Frobenius acyclic, it follows from the stack-version of the theorem on formal functions [**68**, 11.1] that $R^1F_{\mathcal{S}/M*}\mathcal{O}_{\mathcal{S}} = 0$ over some open substack of $\mathcal{S}^{(1)}$ containing $\mathcal{S}_0^{(1)}$. In particular the boundary map

$$(3.4.29.3) \qquad j_{n*}F_{\mathcal{S}_n/M_n*}\mathcal{O}_{\mathcal{S}_n} \longrightarrow R^1F_{\mathcal{S}/M*}\mathcal{O}_{\mathcal{S}}$$

in (3.4.29.2) is zero which gives an isomorphism

$$(3.4.29.4) \qquad j_{n*}j_n^*F_{\mathcal{S}/M*}\mathcal{O}_{\mathcal{S}} = \operatorname{Coker}\big(F_{\mathcal{S}/M*}\mathcal{O}_{\mathcal{S}} \xrightarrow{\times p^n} F_{\mathcal{S}/M*}\mathcal{O}_{\mathcal{S}}\big) \simeq j_{n*}F_{\mathcal{S}_n/M_n*}\mathcal{O}_{\mathcal{S}_n}.$$

Applying j_n^* it follows that (3.4.28.1) is an isomorphism.

3.4.30. — Define $\overline{\mathcal{S}}$ to be the relative spectrum $\operatorname{Spec}_{\mathcal{S}^{(1)}}(F_{\mathcal{S}/M*}\mathcal{O}_{\mathcal{S}_{\text{lis-et}}})$. Then there is a natural commutative diagram

$$(3.4.30.1) \qquad \begin{array}{ccc} \overline{\mathcal{X}}' & \xrightarrow{\beta} & \mathcal{X}^{(1)} \\ \downarrow & & \downarrow \\ \overline{\mathcal{S}} & \longrightarrow & \mathcal{S}^{(1)}, \end{array}$$

and hence also a morphism of topoi

$$(3.4.30.2) \qquad \beta : (\overline{\mathcal{X}}'_{\text{et}}/\widehat{\overline{\mathcal{S}}})_{\text{cris}} \longrightarrow (\mathcal{X}^{(1)}/\mathcal{S}^{(1)})_{\text{cris}}.$$

Lemma 3.4.31. — *For any object $(U, T, \delta) \in \operatorname{Cris}(\mathcal{X}_{\text{et}}^{(1)}/\mathcal{S}^{(1)})$, the divided power structure δ extends to the ideal of $\overline{U} := U \times_{\mathcal{S}^{(1)}} \overline{\mathcal{S}}$ in $\overline{T} := T \times_{\mathcal{S}^{(1)}} \overline{\mathcal{S}}$.*

Proof. — The assertion is étale local on U so we may assume there exist a smooth lifting $\mathcal{U} \to \mathcal{S}^{(1)}$ of U to $\mathcal{S}^{(1)}$ and a retraction $r : T \to \mathcal{U}$. By [**7**, I.2.8.2], the divided power envelope of $\overline{U}$ in $\overline{T}$ is equal to the base change to $\overline{T}$ of the divided power envelope of $\overline{U}$ in $\overline{\mathcal{U}} := \mathcal{U} \times_{\mathcal{S}^{(1)}} \overline{\mathcal{S}}$. It follows that is suffices to consider the case when $T = \mathcal{U}$ which is immediate since $\overline{\mathcal{S}}$ is flat over $\mathbb{Z}_p$. □

Lemma 3.4.32. — *There is a natural isomorphism $\Theta_{\mathcal{S}^{(1)}/M} \simeq R\beta_*\mathcal{O}_{\overline{\mathcal{X}}'_{\text{et}}/\widehat{\overline{\mathcal{S}}}}$.*

Proof. — For any object $T \in \mathrm{Cris}(\mathcal{X}^{(1)}_{\mathrm{et}}/\mathcal{S}^{(1)})$, the preceding lemma shows that the inverse image sheaf $\beta^{-1}(\tilde{T})$ is represented by $\overline{U} \hookrightarrow \overline{T}$. Since $\overline{T} \to T$ is affine this implies that $R^i\beta_*\mathcal{O}_{\overline{\mathcal{X}}'_{\mathrm{et}}/\widehat{\overline{\mathcal{S}}}} = 0$ for $i > 0$ and that $R^0\beta_*\mathcal{O}_{\overline{\mathcal{X}}'_{\mathrm{et}}/\overline{\mathcal{S}}}$ is equal to the sheaf which to any T associates the pullback of $F_{\mathcal{S}/M*}(\mathcal{O}_{\mathcal{S}_{\text{lis-et}}})$ which by definition is $\Theta_{\mathcal{S}^{(1)}/M}$. □

3.4.33. — Let

$$\alpha : (\mathcal{X}'_{\text{lis-et}}/\widehat{\mathcal{S}})_{\mathrm{cris}} \longrightarrow (\overline{\mathcal{X}}'_{\mathrm{et}}/\widehat{\overline{\mathcal{S}}})_{\mathrm{cris}} \tag{3.4.33.1}$$

be the natural morphism of topoi defined as in 1.4.15.

Lemma 3.4.34. — *The adjunction map $\mathcal{O}_{\overline{\mathcal{X}}'_{\mathrm{et}}/\widehat{\overline{\mathcal{S}}}} \to R\alpha_*\mathcal{O}_{\mathcal{X}'_{\text{lis-et}}/\widehat{\mathcal{S}}}$ induces an isomorphism*

$$Ru_{\overline{\mathcal{X}}'_{\mathrm{et}}/\widehat{\overline{\mathcal{S}}}*}\mathcal{O}_{\overline{\mathcal{X}}'_{\mathrm{et}}/\widehat{\overline{\mathcal{S}}}} \longrightarrow R\overline{P}_*Ru_{\mathcal{X}'_{\text{lis-et}}/\widehat{\mathcal{S}}*}\mathcal{O}_{\mathcal{X}'_{\text{lis-et}}/\widehat{\mathcal{S}}}|_{\overline{\mathcal{X}}'_{\mathrm{et}}}. \tag{3.4.34.1}$$

Proof. — The assertion is étale local on $\overline{\mathcal{X}}'$, and hence we may assume that there exists a lifting $\overline{\mathcal{Y}}'$ of $\overline{\mathcal{X}}'$ to a p-adically complete formal scheme $\overline{\mathcal{Y}}'$ formally smooth over $\overline{\mathcal{S}}$. Let $S \to \mathcal{S}$ be a smooth cover by a locally separated scheme, set

$$X' := \mathcal{X}' \times_{\mathcal{S}} S, \quad Y' := \overline{\mathcal{Y}}' \times_{\overline{\mathcal{S}}} S, \tag{3.4.34.2}$$

and let $X'_\bullet$ (resp. $Y'_\bullet$) be the 0-coskeleton of the morphism $X' \to \mathcal{X}'$ (resp. the simplicial formal algebraic space sending $[n] \in \Delta$ to $\varprojlim_n (Y'_n \times_{\mathcal{S}_n} Y'_n \cdots \times_{\mathcal{S}_n} Y'_n))$.

For any $[n] \in \Delta$, let $\widetilde{S}_\bullet[n]$ denote the sheaf on $\mathrm{Cris}(\mathcal{X}'_{\text{lis-et}}/\widehat{\mathcal{S}})$ which to any object (U, T, δ) associates the set of liftings $T \to S_\bullet[n]$ (where we write $S_\bullet[n]$ instead of the more customary S_n so as not to get confused with reduction modulo p^{n+1}) of the morphism $T \to \mathcal{S}$, and let $(\mathcal{X}'_{\text{lis-et}}/\widehat{\mathcal{S}})_{\mathrm{cris}}|_{\widetilde{S}_\bullet[n]}$ denote the associated topos. The formation of this localized topos is functorial so we obtain a simplicial topos $(\mathcal{X}'_{\text{lis-et}}/\widehat{\mathcal{S}})_{\mathrm{cris}}|_{\widetilde{S}_\bullet}$ by

$$[n] \longmapsto (\mathcal{X}'_{\text{lis-et}}/\widehat{\mathcal{S}})_{\mathrm{cris}}|_{\widetilde{S}_\bullet[n]}. \tag{3.4.34.3}$$

For any $[n] \in \Delta$, there is a natural inclusion $\mathrm{Cris}(X'_{n,\mathrm{et}}/\widehat{S}_\bullet[n]) \subset (\mathcal{X}'_{\text{lis-et}}/\widehat{\mathcal{S}})_{\mathrm{cris}}|_{\widetilde{S}_\bullet[n]}$ which induces a morphism of topoi

$$(\mathcal{X}'_{\text{lis-et}}/\widehat{\mathcal{S}})_{\mathrm{cris}}|_{\widetilde{S}_\bullet[n]} \longrightarrow (X'_{n,\mathrm{et}}/\widehat{S}_\bullet[n])_{\mathrm{cris}} \tag{3.4.34.4}$$

and also a morphism of simplicial topoi

$$b : (\mathcal{X}'_{\text{lis-et}}/\widehat{\mathcal{S}})_{\mathrm{cris}}|_{\widetilde{S}_\bullet} \longrightarrow (X'_{\bullet,\mathrm{et}}/\widehat{S}_\bullet)_{\mathrm{cris}}. \tag{3.4.34.5}$$

There is a commutative diagram of topoi

$$\begin{array}{ccccc} (\mathcal{X}'_{\text{lis-et}}/\widehat{\mathcal{S}})_{\mathrm{cris}}|_{\widetilde{S}_\bullet} & \xrightarrow{b} & (X'_{\bullet,\mathrm{et}}/\widehat{S}_\bullet)_{\mathrm{cris}} & \xrightarrow{u_{X'_{\bullet,\mathrm{et}}/\widehat{S}_\bullet}} & X'_{\bullet,\mathrm{et}} \\ {\scriptstyle a}\downarrow & & & & \downarrow{\scriptstyle c} \\ (\mathcal{X}'_{\text{lis-et}}/\widehat{\mathcal{S}})_{\mathrm{cris}} & \xrightarrow{\alpha} & (\overline{\mathcal{X}}'_{\mathrm{et}}/\widehat{\overline{\mathcal{S}}})_{\mathrm{cris}} & \xrightarrow{u_{\overline{\mathcal{X}}'_{\mathrm{et}}/\widehat{\overline{\mathcal{S}}}}} & \overline{\mathcal{X}}'_{\mathrm{et}}. \end{array} \tag{3.4.34.6}$$

Since $\widetilde{S}$ covers the initial object of the topos $(\mathcal{X}'_{\text{lis-et}}/\widehat{\mathcal{S}})_{\text{cris}}$, the adjunction map $\mathcal{O}_{\mathcal{X}'_{\text{lis-et}}/\widehat{\mathcal{S}}} \to Ra_*\mathcal{O}_{(\mathcal{X}'_{\text{lis-et}}/\widehat{\mathcal{S}})_{\text{cris}}|_{\widetilde{S}_\bullet}}$ is an isomorphism (1.4.24). Thus to prove the lemma it suffices to show that the natural map

$$Ru_{\overline{\mathcal{X}}'_{\text{et}}/\widehat{\mathcal{S}}*}\mathcal{O}_{\overline{\mathcal{X}}'_{\text{et}}/\widehat{\mathcal{S}}} \longrightarrow Rc_*Ru_{X'_{\bullet,\text{et}}/\widehat{S}_\bullet *}Rb_*\mathcal{O}_{(\mathcal{X}'_{\text{lis-et}}/\widehat{\mathcal{S}})_{\text{cris}}|_{\widetilde{S}_\bullet}} \tag{3.4.34.7}$$

is an isomorphism. Since b_* is exact and has an exact left adjoint, the functor b_* takes injectives to injectives and

$$Rc_*Ru_{X'_{\bullet,\text{et}}/\widehat{S}_\bullet *}\mathcal{O}_{X'_{\bullet,\text{et}}/S_\bullet} \simeq Rc_*Ru_{X'_{\bullet,\text{et}}/S_\bullet *}Rb_*\mathcal{O}_{(\mathcal{X}'_{\text{lis-et}}/\widehat{\mathcal{S}})_{\text{cris}}|_{\widetilde{S}_\bullet}}. \tag{3.4.34.8}$$

By 2.5.2, the map (3.4.34.1) is therefore identified with the natural map

$$\Omega^\bullet_{\overline{\mathcal{Y}}'/\overline{\mathcal{S}}} \longrightarrow Rc_*\Omega^\bullet_{Y'_\bullet/S_\bullet}. \tag{3.4.34.9}$$

To prove that this map is an isomorphism, it suffices to show that for any integer i the natural map

$$\Omega^i_{\overline{\mathcal{Y}}'/\overline{\mathcal{S}}} \longrightarrow Rc_*\Omega^i_{Y'_\bullet/S_\bullet} \tag{3.4.34.10}$$

is an isomorphism. Since $\Omega^i_{Y'_\bullet/S_\bullet}$ is equal to the pullback of $\Omega^i_{\overline{\mathcal{Y}}'/\overline{\mathcal{S}}}$ and this second sheaf is locally free, to prove that (3.4.34.10) is an isomorphism it suffices to show that the natural map

$$\mathcal{O}_{\overline{\mathcal{Y}}'} \longrightarrow Rc_*\mathcal{O}_{Y'_\bullet} \tag{3.4.34.11}$$

is an isomorphism. This follows from 3.4.23. □

We can now restate 3.4.11 as follows.

Theorem 3.4.35. — *Fix data $F_{\mathcal{S}/M} : \mathcal{S} \to \mathcal{S}^{(1)}$ as in 3.4.1, and let $\mathcal{X} \to \mathcal{S}_0$ be a smooth representable morphism of algebraic stacks with $\mathcal{X}$ a Deligne-Mumford stack. Assume further that the assumption in 3.4.28 holds. Let $\overline{P} : \mathcal{X}'_{\text{lis-et}} \to \overline{\mathcal{X}}_{\text{et}}$ be the natural morphism of topoi, and let $\overline{F}_{\mathcal{X}/\mathcal{S}_0} : \mathcal{X} \to \overline{\mathcal{X}}'$ denote the natural map. Then there are natural isomorphisms*

(3.4.35.1)
$$\overline{F}_{\mathcal{X}/\mathcal{S}_0 *}\mathbb{L}\eta Ru_{\mathcal{X}_{\text{et}}/\widehat{\mathcal{S}}}(\mathcal{O}_{\mathcal{X}_{\text{et}}/\widehat{\mathcal{S}}}) \simeq R\overline{P}_*Ru_{\mathcal{X}'_{\text{lis-et}}/\widehat{\mathcal{S}}*}\mathcal{O}_{\mathcal{X}'_{\text{lis-et}}/\widehat{\mathcal{S}}} \simeq Ru_{\overline{\mathcal{X}}'_{\text{et}}/\widehat{\mathcal{S}}*}(\mathcal{O}_{\overline{\mathcal{X}}'/\widehat{\mathcal{S}}}).$$

Proof. — This follows from 3.4.11, 3.4.32, and 3.4.34. □

Remark 3.4.36. — The assumption that the map (3.4.28.1) is an isomorphism can be avoided as follows. Define a site $\text{Cris}(\overline{\mathcal{X}}'/\widehat{\overline{\mathcal{S}}})$ as follows. For any integer $n \geq 0$ there is a natural inclusion

$$\text{Cris}(\overline{\mathcal{X}}'/\overline{\mathcal{S}}_n) \subset \text{Cris}(\overline{\mathcal{X}}'/\overline{\mathcal{S}}_{n+1}), \tag{3.4.36.1}$$

and we define

$$\text{Cris}(\overline{\mathcal{X}}'/\widehat{\overline{\mathcal{S}}}) = \varinjlim_n \text{Cris}(\overline{\mathcal{X}}'/\overline{\mathcal{S}}_n), \tag{3.4.36.2}$$

where the direct limit on the right is taken with respect to the inclusions (3.4.36.1). A collection of morphisms $\{(U_i, T_i, \delta_i) \to (U, T, \delta)\}$ is defined to be a covering if the map $\coprod T_i \to T$ is étale and surjective. With this definition of $(\overline{\mathcal{X}}'/\widehat{\mathcal{S}})_{\mathrm{cris}}$ the isomorphism $\Theta_{\mathcal{S}^{(1)}/M} \simeq R\beta_* \mathcal{O}_{\overline{\mathcal{X}}'/\widehat{\mathcal{S}}}$ still holds with no assumptions on $\mathcal{S}$.

However, since we do not need these more general results in what follows we make the simplifying assumption that the map (3.4.28.1) is an isomorphism.

3.4.37. — As in [**58**, 7.3.1], theorem 3.4.11 can be generalized as follows.

Recall that if (K, P) is a filtered complex of objects in some abelian category $\mathcal{A}$, then the *décalage* of P, denoted Dec P, is the filtration on K given by

$$(\mathrm{Dec}\ P)^q K^i = d^{-1}(P^{i+q+1}K^{i+1}) \cap P^{i+q}K^i, \tag{3.4.37.1}$$

where d denotes the differential $K^i \to K^{i+1}$. As discussed in [**58**, p. 133], this operation passes to the filtered derived category.

Let $\mathcal{S}_0$ be as in 3.4.28, $\mathcal{X} \to \mathcal{S}_0$ a smooth representable morphism of algebraic stacks with $\mathcal{X}$ a Deligne-Mumford stack, and let E be a locally free crystal on $\overline{\mathcal{X}}'/\widehat{\mathcal{S}}$. Denote by $\overline{F}_{\mathcal{X}/\mathcal{S}_0} : \mathcal{X} \to \overline{\mathcal{X}}'$ the natural morphism.

Theorem 3.4.38. — *There is a natural isomorphism in the filtered derived category*

$$\overline{F}_{\mathcal{X}/\mathcal{S}_0 *}(Ru_{\mathcal{X}_{\mathrm{et}}/\widehat{\mathcal{S}}*}(\overline{F}^*_{\mathcal{X}/\widehat{\mathcal{S}}}E), \mathrm{Dec}\ P) \simeq (Ru_{\overline{\mathcal{X}}'/\widehat{\mathcal{S}}*}(E), P), \tag{3.4.38.1}$$

*where P denotes the filtration (*i.e.*, structure of an object in the filtered derived category) on $Ru_{\mathcal{X}_{\mathrm{et}}/\widehat{\mathcal{S}}*}(\overline{F}^*_{\mathcal{X}/\widehat{\mathcal{S}}}E)$ (resp. $Ru_{\overline{\mathcal{X}}'_{\mathrm{et}}/\widehat{\mathcal{S}}*}(E)$) obtained by taking cohomology of $\overline{F}^*_{\mathcal{X}/\widehat{\mathcal{S}}}E$ (resp. E) viewed as an object in the filtered derived category using the filtration given by the images of multiplication by p^i.*

Proof. — The proof is essentially the same as the proof of 3.4.11. The structure sheaf should be replaced by the crystal E and the following modifications should be made to the proof:

▷ In the proof of 3.4.15, the reference to [**8**, 8.21] must be replaced by [**58**, 7.3.6].

▷ The reference in 3.4.24 to [**9**, 1.5] should be replaced by [**58**, 7.3.6]. □

Remark 3.4.39. — If in 3.4.38 we take $E = \mathcal{O}_{\overline{\mathcal{X}}'/\widehat{\mathcal{S}}}$, then

$$\overline{F}_{\mathcal{X}/\mathcal{S}_0 *}(\mathrm{Dec}\ P)^0 Ru_{\mathcal{X}_{\mathrm{et}}/\widehat{\mathcal{S}}*}(\overline{F}^*_{\mathcal{X}/\widehat{\mathcal{S}}}E) = F_{\mathcal{X}/M_0 *}\mathbb{L}\eta Ru_{\mathcal{X}_{\mathrm{et}}/\widehat{\mathcal{S}}*}(\mathcal{O}_{\mathcal{X}_{\mathrm{et}}/\widehat{\mathcal{S}}}) \tag{3.4.39.1}$$

and

$$P^0 Ru_{\overline{\mathcal{X}}'/\widehat{\mathcal{S}}*}(E) = Ru_{\mathcal{X}^{(1)}_{\mathrm{et}}/\widehat{\mathcal{S}}^{(1)}*}(\Theta_{\mathcal{S}^{(1)}/M}). \tag{3.4.39.2}$$

Therefore we recover 3.4.11 from 3.4.38.

Corollary 3.4.40. — *With notation as in 3.4.38, let d be the relative dimension of $\mathcal{X}$ over $\mathcal{S}_0$. Then there exists a map*

$$V : \overline{F}_{\mathcal{X}/\mathcal{S}_0 *} Ru_{\mathcal{X}_{\mathrm{et}}/\widehat{\mathcal{S}}*}(\overline{F}^*_{\mathcal{X}/\widehat{\mathcal{S}}} E) \longrightarrow Ru_{\overline{\mathcal{X}}'/\widehat{\widehat{\mathcal{S}}}*} E \tag{3.4.40.1}$$

such that if

$$\psi : Ru_{\overline{\mathcal{X}}_{\mathrm{et}}/\widehat{\widehat{\mathcal{S}}}*} E \longrightarrow \overline{F}_{\mathcal{X}/\mathcal{S}_0 *} Ru_{\mathcal{X}_{\mathrm{et}}/\widehat{\mathcal{S}}*}(\overline{F}^*_{\mathcal{X}/\widehat{\mathcal{S}}} E) \tag{3.4.40.2}$$

denotes the natural map, then the composites $\psi \circ V$ and $V \circ \psi$ are multiplication by p^d.

Proof. — This follows from the same argument used in [**9**, 1.6]. □

Corollary 3.4.40 can be generalized as follows. Let $\mathcal{X} \to \mathcal{S}_0$ be as in 3.4.28.

Definition 3.4.41. — An *F-span of width b* on $\mathcal{X}/\mathcal{S}$ is a triple (E, E', Φ), where E is a crystal $(\mathcal{X}_{\mathrm{et}}/\widehat{\mathcal{S}})_{\mathrm{cris}}$, E' is a crystal in $(\overline{\mathcal{X}}'_{\mathrm{et}}/\widehat{\widehat{\mathcal{S}}})_{\mathrm{cris}}$, and $\Phi : \overline{F}^*_{\mathcal{X}/\mathcal{S}} E' \to E$ is a morphism of crystals such that there exists a morphism $V : E \to \overline{F}^*_{\mathcal{X}/\mathcal{S}} E'$ such that the composites $\Phi \circ V$ and $V \circ \Phi$ are multiplication by p^b.

Corollary 3.4.42. — *If (E, E', Φ) is an F-span of width b and if d is the relative dimension of $\mathcal{X}$ over $\mathcal{S}_0$, then there exists a map*

$$V : \overline{F}_{\mathcal{X}/\mathcal{S}_0 *} Ru_{\mathcal{X}_{\mathrm{et}}/\widehat{\mathcal{S}}*}(E) \longrightarrow Ru_{\overline{\mathcal{X}}'/\widehat{\widehat{\mathcal{S}}}*} E' \tag{3.4.42.1}$$

such that if

$$\psi : Ru_{\overline{\mathcal{X}}'/\widehat{\widehat{\mathcal{S}}}*} E' \longrightarrow \overline{F}_{\mathcal{X}/\mathcal{S}_0 *} Ru_{\mathcal{X}_{\mathrm{et}}/\widehat{\mathcal{S}}*}(E) \tag{3.4.42.2}$$

denotes the map induced by Φ, then the composites $\psi \circ V$ and $V \circ \psi$ are multiplication by p^{d+b}.

Proof. — This follows from the same reasoning used in [**58**, 7.3.7]. □

3.4.43. — Finally we record a consequence of 3.4.38 which will be used in Chapter 7. Let M be the spectrum of a complete discrete valuation ring R of mixed characteristic $(0, p)$. For a filtered complex (K, F) of R-modules and n an integer, let $(K, \mathrm{Dec}_n\ F)$ denote the filtered complex obtained by applying the functor Dec iteratively n times to (K, F).

Let $\mathcal{S}/M$ be a flat algebraic stack, and assume given for each integer $i \in [0, n]$ a flat lifting $\mathcal{S}^{(i)}$ of $\mathcal{S}_0 \times_{M_0, F^i_{M_0}} M_0$ with $\mathcal{S}^{(0)} = \mathcal{S}$, and maps $\tilde{F}_{\mathcal{S}^{(i)}/M} : \mathcal{S}^{(i-1)} \to \mathcal{S}^{(i)}$ lifting the relative Frobenius map. Assume that the reduction moduli p of each $\mathcal{S}^{(i)}$ is Frobenius acyclic, and that the maps $\tilde{F}_{\mathcal{S}^{(i)}/M}$ satisfy the assumption in 3.4.28 for each i. For each $0 \leq j \leq i \leq n$, let $\Lambda^{(i)}_j : \mathcal{S}^{(j)} \to \mathcal{S}^{(i)}$ denote the map $\tilde{F}_{\mathcal{S}^{(i)}/M} \circ \tilde{F}_{\mathcal{S}^{(i-1)}/M} \circ \cdots \circ \tilde{F}_{\mathcal{S}^{(j+1)}/M}$. Let $\mathcal{X} \to \mathcal{S}_0$ be a smooth, locally separated, and representable morphism of algebraic stacks with $\mathcal{X}$ a Deligne-Mumford stack, and set $\overline{\mathcal{X}}^{(n)} := \mathcal{X} \times_{\mathcal{S}_0} \mathrm{Spec}(F^n_{\mathcal{S}_0 *}\mathcal{O}_{\mathcal{S}_0})$ with induced map $\overline{\Lambda}_{\mathcal{X}} : \mathcal{X} \to \overline{\mathcal{X}}^{(n)}$. Assume also that the formation of $\Lambda^{(i)}_{j*}(\mathcal{O}_{\mathcal{S}^{(j)}})$ is compatible with arbitrary base change $M' \to M$. Let $\overline{\mathcal{S}}^i_j := \mathrm{Spec}(\Lambda^{(i)}_{j*}\mathcal{O}_{\mathcal{S}^{(j)}})$. Then there is a commutative diagram

(3.4.43.1)

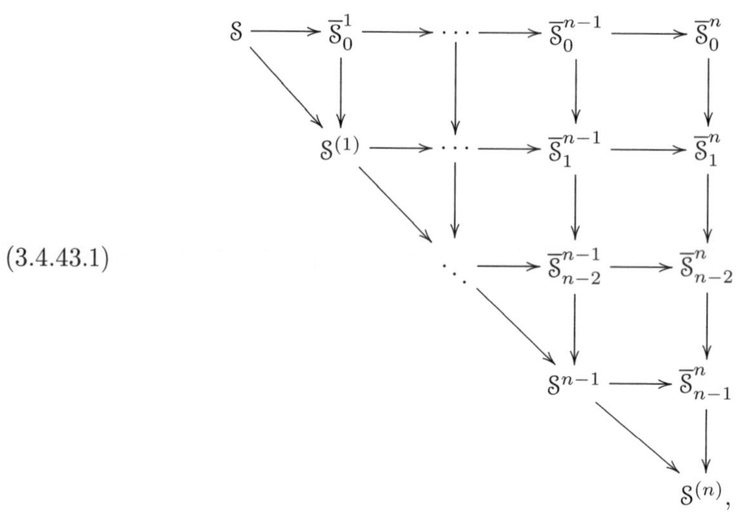

where the maps

$$\tilde{F}_{(\overline{\mathcal{S}}^1_0)^{(i)}/M} : \overline{\mathcal{S}}^{i+1}_i \longrightarrow \overline{\mathcal{S}}^{i+2}_{i+1}, \tag{3.4.43.2}$$

are induced by the maps $\Lambda^{(i)}_j$. For $0 \leq j \leq n$ let $\overline{\mathcal{X}}^{(j)}$ denote the fiber product of the diagram

$$\begin{array}{ccc} & & \mathcal{X} \\ & & \downarrow \\ \overline{\mathcal{S}}^j_0 \times_M M_0 & \xrightarrow{\tau} & \mathcal{S} \times_M M_0, \end{array} \tag{3.4.43.3}$$

where τ is the composite

$$\overline{\mathcal{S}}^j_0 \times_M M_0 \longrightarrow \mathcal{S}^{(j)} \times_M M_0 \simeq \mathcal{S} \times_{M_0, F^j_{M_0}} M_0 \xrightarrow{\text{projection}} \mathcal{S} \times_M M_0. \tag{3.4.43.4}$$

Equivalently, if $\mathcal{X}^{(j)}$ denotes $\mathcal{X} \times_{M_0, F^j_{M_0}} M_0$ then $\overline{\mathcal{X}}^{(j)}$ is equal to the fiber product

$$\overline{\mathcal{X}}^{(j)} = \overline{\mathcal{S}}^j_0 \times_{\mathcal{S}^{(j)}} \mathcal{X}^{(j)}. \tag{3.4.43.5}$$

Observe also that for $0 \leq j \leq i \leq n$ the stack $\overline{\mathcal{S}}_{j+1}^{i+1}$ is a flat lifting of $(\overline{\mathcal{S}}_j^i \times_M M_0)^{(1)}$, and that for $i = 0, \ldots, n-1$ there is a commutative diagram

(3.4.43.6)

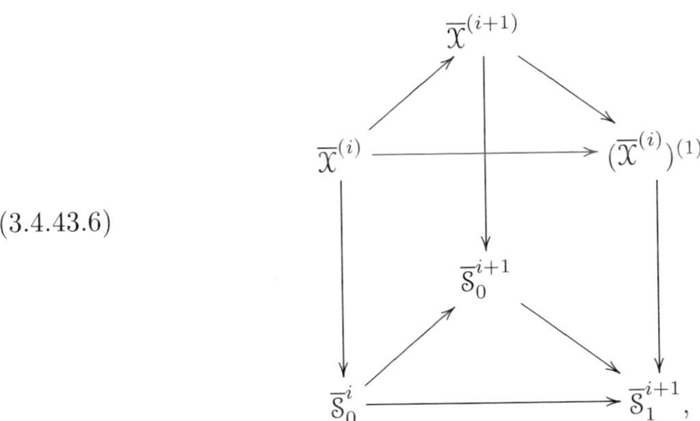

where we make the convention that $\overline{\mathcal{S}}_0^0 = \mathcal{S}$.

For $0 \leq i < n$, we can then apply 3.4.38 to the morphism of stacks

$$\overline{\mathcal{X}}^{(i)} \longrightarrow \overline{\mathcal{S}}_0^i \tag{3.4.43.7}$$

with the lifting of the relative Frobenius (3.4.43.2).

Let E be a locally free crystal $(\overline{\mathcal{X}}_{\mathrm{et}}^{(n)}/\widehat{\overline{\mathcal{S}}}_0^n)_{\mathrm{cris}}$, and for $i = 0, \ldots, n-1$ let $E|_{\overline{\mathcal{X}}^{(i)}}$ denote the crystal in $(\overline{\mathcal{X}}_{\mathrm{et}}^{(i)}/\widehat{\overline{\mathcal{S}}}_0^i)_{\mathrm{cris}}$ obtained from E by pullback using the commutative diagram

$$\begin{array}{ccc} \overline{\mathcal{X}}^{(i)} & \longrightarrow & \overline{\mathcal{X}}^{(n)} \\ \downarrow & & \downarrow \\ \overline{\mathcal{S}}_0^i & \longrightarrow & \overline{\mathcal{S}}_0^n. \end{array} \tag{3.4.43.8}$$

If

$$F^i : \overline{\mathcal{X}}^{(i)} \longrightarrow \overline{\mathcal{X}}^{(i+1)} \tag{3.4.43.9}$$

denotes the morphism induced by Frobenius, we get from 3.4.38 an isomorphism in the filtered derived category

$$\rho_i : \big(F_*^i Ru_{\overline{\mathcal{X}}^{(i)}/\widehat{\overline{\mathcal{S}}}_0^i *} E|_{\overline{\mathcal{X}}^{(i)}}, \mathrm{Dec}\ P\big) \simeq \big(Ru_{\overline{\mathcal{X}}^{(i+1)}/\widehat{\overline{\mathcal{S}}}_0^{i+1} *} E|_{\overline{\mathcal{X}}^{(i+1)}}, P\big). \tag{3.4.43.10}$$

For any $1 \leq r \leq n-i$ we then get an isomorphism

$$\rho_i^r : \big(F_*^{i,r} Ru_{\overline{\mathcal{X}}^{(i)}/\widehat{\overline{\mathcal{S}}}_0^i *} E|_{\overline{\mathcal{X}}^{(i)}}, \mathrm{Dec}\ _r P\big) \simeq \big(Ru_{\overline{\mathcal{X}}^{(i+r)}/\widehat{\overline{\mathcal{S}}}_0^{i+r} *} E|_{\overline{\mathcal{X}}^{(i+r)}}, P\big) \tag{3.4.43.11}$$

from the composite

$$(3.4.43.12)\qquad \begin{array}{c} \big(F_*^{i,r} Ru_{\overline{\mathfrak{X}}^{(i)}/\widehat{\mathcal{S}}_0^i *} E|_{\overline{\mathfrak{X}}^{(i)}}, \mathrm{Dec}\ _r P\big) \\ \downarrow \rho_i \\ \big(F_*^{i+1,r-1} Ru_{\overline{\mathfrak{X}}^{(i+1)}/\widehat{\mathcal{S}}_0^{i+1} *} E|_{\overline{\mathfrak{X}}^{(i+1)}}, \mathrm{Dec}\ _{r-1} P\big) \\ \downarrow \rho_{i+1} \\ \vdots \\ \downarrow \rho_{i+r-1} \\ \big(Ru_{\overline{\mathfrak{X}}^{(i+r)}/\widehat{\mathcal{S}}_0^{i+r} *} E|_{\overline{\mathfrak{X}}^{(i+r)}}, P\big), \end{array}$$

where

$$(3.4.43.13)\qquad F^{i,r} = F^{i+r-1}\circ F^{i+r-2}\circ\cdots\circ F^i : \overline{\mathfrak{X}}^{(i)} \to \overline{\mathfrak{X}}^{(i+r)}.$$

By the construction, these isomorphisms ρ_i^r are compatible in the sense that given two integers $r, l \geq 1$ such that $r + l \leq n - i$ the composite morphism

$$(3.4.43.14)\qquad \begin{array}{c} \big(F_*^{i,r+l} Ru_{\overline{\mathfrak{X}}^{(i)}/\widehat{\mathcal{S}}_0^i *} E|_{\overline{\mathfrak{X}}^{(i)}}, \mathrm{Dec}\ _{r+l} P\big) \\ \downarrow \rho_i^r \\ \big(F_*^{i,l} Ru_{\overline{\mathfrak{X}}^{(i+r)}/\widehat{\mathcal{S}}_0^{i+r} *} E|_{\overline{\mathfrak{X}}^{(i+r)}}, \mathrm{Dec}\ _l P\big) \\ \downarrow \rho_{i+r}^l \\ \big(Ru_{\overline{\mathfrak{X}}^{(i+r+l)}/\widehat{\mathcal{S}}_0^{i+r+l} *} E|_{\overline{\mathfrak{X}}^{(i+r+l)}}, P\big) \end{array}$$

is equal to ρ_i^{r+l}.

Taking $i = 0$ and $r = n$ we obtain the following corollary:

Corollary 3.4.44. — *Let $\mathcal{S}/M$ be as in 3.4.28, and assume given for each integer $i \in [0, n]$ a flat lifting $\mathcal{S}^{(i)}$ of $\mathcal{S}_0 \times_{M_0, F^i_{M_0}} M_0$ with $\mathcal{S}^{(0)} = \mathcal{S}$, and maps $\tilde{F}_{\mathcal{S}^{(i)}/M} : \mathcal{S}^{(i-1)} \to \mathcal{S}^{(i)}$ lifting the relative Frobenius map. For each $0 \leq j \leq i \leq n$, let $\Lambda_j^{(i)} : \mathcal{S}^{(j)} \to \mathcal{S}^{(i)}$ denote the map $\tilde{F}_{\mathcal{S}^{(i)}/M}\circ \tilde{F}_{\mathcal{S}^{(i-1)}/M}\circ\cdots\circ\tilde{F}_{\mathcal{S}^{(j+1)}/M}$. Let $\mathfrak{X} \to \mathcal{S}_0$ be a smooth, locally separated, and representable morphism of algebraic stacks with $\mathfrak{X}$ a Deligne-Mumford stack, and set $\overline{\mathfrak{X}}^{(n)} := \mathfrak{X}\times_{\mathcal{S}_0} \mathrm{Spec}(F^n_{\mathcal{S}_0 *}\mathcal{O}_{\mathcal{S}_0})$ with induced map $\overline{\Lambda}_{\mathfrak{X}} : \mathfrak{X} \to \overline{\mathfrak{X}}^{(n)}$. Assume also that the formation of $\Lambda_{j*}^{(i)}(\mathcal{O}_{\mathcal{S}^{(j)}})$ is compatible with arbitrary base change $M' \to M$. Then if E is a locally free crystal on $\overline{\mathfrak{X}}^{(n)}/\mathrm{Spec}(\Lambda_{0*}^{(n)}\mathcal{O}_{\mathcal{S}})$ there is a natural isomorphism in the filtered derived category*

$$(3.4.44.1)\qquad \overline{\Lambda}_{\mathfrak{X}*}(Ru_{\mathfrak{X}_{\mathrm{et}}/\widehat{\mathcal{S}}*}\overline{\Lambda}_{\mathfrak{X}}^* E, \mathrm{Dec}_n\ P) \simeq \big(Ru_{\mathfrak{X}^{(n)}/\mathrm{Spec}(\widehat{\Lambda_{0*}^{(n)}\mathcal{O}_{\mathcal{S}}})*} E, P\big).$$

Corollary 3.4.45. — *Let the notation and hypotheses be as in 3.4.44, and let d be the relative dimension of $\mathfrak{X}$ over $\mathcal{S}_0$. Then there exists a map*

$$V : \overline{\Lambda}_{\mathfrak{X}*} Ru_{\mathfrak{X}_{\mathrm{et}}/\widehat{\mathcal{S}}*} \overline{\Lambda}^*_{\mathfrak{X}} E \longrightarrow Ru_{\overline{\mathfrak{X}}^{(n)}/\mathrm{Spec}(\widehat{\Lambda^{(n)}_{0*}\mathcal{O}_{\mathcal{S}}})*} E, \tag{3.4.45.1}$$

such that if

$$\psi : Ru_{\overline{\mathfrak{X}}^{(n)}/\mathrm{Spec}(\widehat{\Lambda^{(n)}_{0*}\mathcal{O}_{\mathcal{S}}})*} E \longrightarrow \overline{\Lambda}_{\mathfrak{X}*} Ru_{\mathfrak{X}_{\mathrm{et}}/\widehat{\mathcal{S}}*} \overline{\Lambda}^*_{\mathfrak{X}} E \tag{3.4.45.2}$$

is the natural map then $\psi \circ V$ and $V \circ \psi$ are both equal to multiplication by p^{nd}.

Proof. — As in 3.4.40 this follows from the same argument used in [**9**, 1.6]. □

CHAPTER 4

DE RHAM-WITT THEORY

Throughout this chapter, unless otherwise noted we work with the étale crystalline topos, and hence often omit from the notation the reference to the étale topology.

4.1. The algebra $\mathcal{A}^{\bullet}_{n,X/T}$

Let p denote a fixed prime number. The construction that follows is a generalization of those sketched in [**36**, III (1.5)] (suggested by Katz), and made in the log context in [**31**].

4.1.1. — Let S be an algebraic space over $\mathbb{F}_p$, and let $S \hookrightarrow T$ be a closed immersion defined by a divided power ideal into a p-adically complete formal algebraic space $T/\mathbb{Z}_p$. We assume that multiplication by p is injective on $\mathcal{O}_T$ and denote by $(\mathcal{I}, \gamma) \subset \mathcal{O}_T$ the divided power ideal defining S in T. Assume further that there exists a lifting $\sigma : T \to T$ of Frobenius to T, and fix one such lifting σ. For $n \geq 0$ denote by T_n the reduction of T modulo p^{n+1}.

4.1.2. — Let $X \to S$ be a smooth morphism of algebraic spaces. For each $n \geq 1$ and $q \geq 0$, define

$$\mathcal{A}^q_{n,X/T} := R^q u_{X/T_{n-1}*} \mathcal{O}_{X/T_{n-1}}. \tag{4.1.2.1}$$

We define operators

$$d : \mathcal{A}^q_{n,X/T} \longrightarrow \mathcal{A}^{q+1}_{n,X/T}, \tag{4.1.2.2}$$

$$F : \mathcal{A}^q_{n+1,X/T} \longrightarrow \mathcal{A}^q_{n,X/T}, \tag{4.1.2.3}$$

$$V : \mathcal{A}^q_{n,X/T} \longrightarrow \mathcal{A}^q_{n+1,X/T} \tag{4.1.2.4}$$

as follows.

The map F is simply the natural restriction map. The maps d and V are defined locally as follows. First assume that we can embed X in a formally smooth T-space Y/T, and let D be the divided power envelope of X in Y. Observe that D is flat over

$\mathbb{Z}_p$ by [8, 3.32]. Write Y_n and D_n for the reductions modulo p^{n+1}. Then there is a canonical isomorphism

$$Ru_{X/T_{n-1}*}\mathcal{O}_{X/T_{n-1}} \simeq \Omega^\bullet_{D_{n-1}/T_{n-1}}, \tag{4.1.2.5}$$

and d is defined to be the connecting homomorphism coming from the exact sequence (4.1.2.6)

$$0 \longrightarrow \Omega^\bullet_{D_{n-1}/T_{n-1}} \xrightarrow{p^n} \Omega^\bullet_{D_{2n-1}/T_{2n-1}} \longrightarrow \Omega^\bullet_{D_{n-1}/T_{n-1}} \longrightarrow 0,$$

where we write $\Omega^\bullet_{D_n/T_n}$ for $\mathcal{O}_{D_n} \otimes_{\mathcal{O}_{Y_n}} \Omega^\bullet_{Y_n/T_n}$. The map V is defined to be the map induced by $\times p : \Omega^\bullet_{D_{n-1}/T_{n-1}} \to \Omega^\bullet_{D_n/T_n}$.

If $X \subset Y'$ is a second embedding of X into a smooth T-space, and if $f : Y' \to Y$ is a morphism compatible with the inclusions of X, then it follows from the construction that the maps d and V obtained from Y and Y' are equal. That d and V are independent of the lifting Y then follows from observing that for any two embeddings $X \subset Y_1$ and $X \subset Y_2$ we can form $X \subset Y_1 \widehat{\times}_T Y_2$ which maps to both Y_1 and Y_2. Consequently the maps d and V are independent of the choices and defined globally.

Lemma 4.1.3. — *The operators d, F, V satisfy the following equations*

$$d^2 = 0, \quad FV = VF = p, \quad dF = pFd, \quad Vd = pdV, \quad FdV = d. \tag{4.1.3.1}$$

Furthermore, if $\omega \in \mathcal{A}^\bullet_{n-1,X/T}$ and $\eta \in \mathcal{A}^\bullet_{n,X/T}$ then $V(\omega)\cdot\eta = V(\omega\cdot F(\eta))$.

Proof. — All but the formula $d^2 = 0$ follow immediately from the definitions. To see this formula, we may work locally and may assume that we have a smooth lifting Y/T of X, and furthermore that we have units $y_i \in \mathcal{O}_Y^*$ such that the forms $\{\mathrm{dlog}(y_i)\}$ form a basis for $\Omega^1_{Y/T}$. Now suppose $\omega = \sum_{\underline{i}} f_{\underline{i}}\mathrm{dlog}(y_{\underline{i}})$ is a closed r-form, where $\underline{i}$ denotes a multi-index $(i_1,\ldots,i_r)$, $\mathrm{dlog}(y_{\underline{i}}) = \mathrm{dlog}(y_{i_1})\wedge\cdots\wedge\mathrm{dlog}(y_{i_r})$, and $f_{\underline{i}} \in \mathcal{O}_{Y_{n-1}}$. We show that $d^2([\omega]) = 0$. Let $\tilde{f}_{\underline{i}} \in \mathcal{O}_{Y_{3n-1}}$ denote liftings of the $f_{\underline{i}}$. From the commutative diagram
(4.1.3.2)

$$\begin{array}{ccccccccc}
0 & \longrightarrow & \Omega^\bullet_{Y_{n-1}/T_{n-1}} & \xrightarrow{\times p^n} & \Omega^\bullet_{Y_{2n-1}/T_{2n-1}} & \longrightarrow & \Omega^\bullet_{Y_{n-1}/T_{n-1}} & \longrightarrow & 0 \\
 & & \downarrow{=} & & \downarrow{\times p^n} & & \downarrow{\times p^n} & & \\
0 & \longrightarrow & \Omega^\bullet_{Y_{n-1}/T_{n-1}} & \xrightarrow{\times p^{2n}} & \Omega^\bullet_{Y_{3n-1}/T_{3n-1}} & \longrightarrow & \Omega^\bullet_{Y_{2n-1}/T_{2n-1}} & \longrightarrow & 0,
\end{array}$$

we see that the class $d^2([\omega])$ is equal to the image of the class of $\sum_{\underline{i}} d\tilde{f}_{\underline{i}} \wedge \mathrm{dlog}(y_{\underline{i}})$ under the boundary map

$$\mathcal{H}^{r+1}(\Omega^\bullet_{Y_{2n-1}/T_{2n-1}}) \longrightarrow \mathcal{H}^{r+2}(\Omega^\bullet_{Y_{n-1}/T_{n-1}}). \tag{4.1.3.3}$$

Since $\sum_{\underline{i}} d\tilde{f}_{\underline{i}} \wedge \mathrm{dlog}(y_{\underline{i}}) \in \Omega^{r+1}_{Y_{2n-1}/T_{2n-1}}$ lifts to a closed 1-form in $\Omega^{r+1}_{Y_{3n-1}/T_{3n-1}}$, it follows that the image under (4.1.3.3) is zero. ☐

Notational Remark 4.1.4. — We will also sometimes consider the situation of a closed immersion $S \hookrightarrow T$ of algebraic spaces defined by a divided power ideal with T flat over $\mathbb{Z}_p$. If $\widehat{T}$ denotes the p-adic completion of T, and if we are given a lifting of Frobenius $\sigma : \widehat{T} \to \widehat{T}$, then the data $(S \hookrightarrow \widehat{T}, \sigma)$ satisfies the conditions of 4.1.1. If $X \to S$ is a smooth morphism of algebraic spaces we will often write $\mathcal{A}^{\bullet}_{n,X/T}$ instead of $\mathcal{A}^{\bullet}_{n,X/\widehat{T}}$ if no confusion seems likely to arise.

4.1.5. — Denote the by $X^{(p^n)}$ the pullback of X via the n-th power of Frobenius $F^n_S : S \to S$. By [**8**, theorem 8.20] there is a natural quasi-isomorphism

$$Ru_{X^{(p)}/T*}\mathcal{O}_{X^{(p)}/T} \longrightarrow \mathbb{L}\eta Ru_{X/T*}\mathcal{O}_{X/T} \tag{4.1.5.1}$$

where the notation on the right hand side is as in 3.4.11 (note that though [**8**, Chapter 8] is written only for schemes, the same argument gives the result for algebraic spaces). In terms of a smooth lifting Y/T of X, $\mathbb{L}\eta Ru_{X/T*}\mathcal{O}_{X/T}$ is the subcomplex of $\Omega^{\bullet}_{Y/T}$ whose i-th component is

$$\tilde{E}^i_{\infty} := \{\omega \in p^i\Omega^i_{Y/T} |\ d\omega \in p^{i+1}\Omega^{i+1}_{Y/T}\}. \tag{4.1.5.2}$$

Since each $\Omega^i_{Y/T}$ is locally free, we obtain:

Corollary 4.1.6. — *If Y/T is a smooth lifting of X, then*

$$\mathcal{O}_{T_{n-1}} \otimes^{\mathbb{L}}_{\mathcal{O}_T} \mathbb{L}\eta Ru_{X/T*}\mathcal{O}_{X/T} \simeq \tilde{E}^{\bullet}_{\infty} \otimes_{\mathcal{O}_T} \mathcal{O}_T/(p^n). \tag{4.1.6.1}$$

Lemma 4.1.7. — *For $m \geq 0$, define a subcomplex $\tilde{E}^{\cdot}_m \subset \Omega^{\cdot}_{Y_{m-1}/T_{m-1}}$ by*

$$\tilde{E}^q_m := \big\{\omega \in p^q\Omega^q_{Y_{m-1}/T_{m-1}} |\ d\omega \in p^{q+1}\Omega^{q+1}_{Y_{m-1}/T_{m-1}}\big\}. \tag{4.1.7.1}$$

Then if $m > n + q$, the module

$$E^q_n := \tilde{E}^q_m / p^n \tilde{E}^q_m \tag{4.1.7.2}$$

is independent of the choice of m, and by varying q we obtain a complex $E^{\bullet}_n$ with differential induced by the differential on $\tilde{E}^{\bullet}_m$. Moreover, the isomorphism (4.1.6.1) composed with the projection

$$\tilde{E}^{\bullet}_{\infty} \otimes_{\mathcal{O}_T} \mathcal{O}_T/(p^n) \longrightarrow E^{\bullet}_n \tag{4.1.7.3}$$

is an isomorphism

$$E^{\bullet}_n \simeq \mathcal{O}_{T_{n-1}} \otimes^{\mathbb{L}}_{\mathcal{O}_T} \mathbb{L}\eta Ru_{X/T*}\mathcal{O}_{X/T}. \tag{4.1.7.4}$$

Proof. — We first show that E^q_n is independent of the choice of m. If $m' > m$, then there is a natural map

$$\tilde{E}^q_{m'} \longrightarrow \tilde{E}^q_m. \tag{4.1.7.5}$$

If $\omega \in \tilde{E}^q_m$, then by definition of $\tilde{E}^q_m$ we can write $\omega = p^q\omega'$ for some $\omega' \in \Omega^q_{Y_{m-1}/T_{m-1}}$ with $d\omega'$ congruent to 0 modulo p. It follows that if $\tilde{\omega}' \in \Omega^q_{Y_{m'-1}/T_{m'-1}}$ is a lifting of ω' then $p^q\tilde{\omega}' \in \tilde{E}^q_{m'}$ is a lifting of ω. Hence the map (4.1.7.5) is surjective.

On the other hand, if $\omega \in \Omega^q_{Y_{m'-1}/T_{m'-1}}$ is a section such that $p^q\omega$ defines an element of $\tilde{E}^q_{m'}$ which maps to zero in $\tilde{E}^q_m/p^n\tilde{E}^q_m$, then there exist elements $b, c \in \Omega^q_{X_{m'-1}/T_{m'-1}}$, with dc congruent to zero modulo p, such that

$$p^q\omega = p^m b + p^n(p^q c). \tag{4.1.7.6}$$

Since $m > n+q$, $p^q\omega = p^{n+q}(p^{m-n-q}b + c)$ and $d(p^{m-n-q}b + c)$ is congruent to zero modulo p. Hence $p^q\omega \in p^n\tilde{E}^q_{m'}$. This proves that E^q_n is independent of the choice of m.

To prove the last statement in the proposition, note first that since $\Omega^i_{Y/T}$ is locally free, there is a natural isomorphism

$$\mathcal{O}_{T_{n-1}} \otimes^{\mathbb{L}}_{\mathcal{O}_T} \mathbb{L}\eta Ru_{X/T*}\mathcal{O}_{X/T} \simeq \tilde{E}^\bullet_\infty \otimes_{\mathcal{O}_T} \mathcal{O}_T/(p^n). \tag{4.1.7.7}$$

There is a natural map

$$\tilde{E}^\bullet_\infty \otimes_{\mathcal{O}_T} \mathcal{O}_T/(p^n) \longrightarrow E^\bullet_n \tag{4.1.7.8}$$

which we claim is a quasi-isomorphism. Suppose $\omega \in E^q_n$ maps to zero in E^{q+1}_n and let $p^q\eta \in \Omega^q_{Y/T}$ be a representative for ω. Then $d(p^q\eta) \in p^n(p^{q+1}\Omega^{q+1}_{Y/T})$, and hence (4.1.7.8) induces a surjection on cohomology.

Conversely, suppose

$$p^q\eta \in \operatorname{Ker}\big(\tilde{E}^q_\infty \otimes_{\mathcal{O}_T} \mathcal{O}_T/(p^n) \longrightarrow \tilde{E}^{q+1}_\infty \otimes_{\mathcal{O}_T} \mathcal{O}_T/(p^n)\big) \tag{4.1.7.9}$$

represents a class in the kernel of

$$\mathcal{H}^q(\tilde{E}^\bullet_\infty \otimes_{\mathcal{O}_T} \mathcal{O}_T/(p^n)) \longrightarrow \mathcal{H}^q(E^\bullet_n). \tag{4.1.7.10}$$

Then we can write

$$p^q\eta = p^n(p^q\eta') + p^m\lambda + d(p^{q-1}\epsilon) \tag{4.1.7.11}$$

where $d\eta' \in p^{q+1}\Omega^{q+1}_{Y/T}$, $d\epsilon \equiv 0 \pmod p$, and $m > q+n$. From this it follows that (4.1.7.8) is injective. □

Corollary 4.1.8. — *The composition of (4.1.7.4) and the isomorphism*

$$\mathcal{O}_{T_{n-1}} \otimes^{\mathbb{L}}_{\mathcal{O}_T} Ru_{X^{(p)}/T*}\mathcal{O}_{X^{(p)}/T} \longrightarrow \mathcal{O}_{T_{n-1}} \otimes^{\mathbb{L}}_{\mathcal{O}_T} \mathbb{L}\eta Ru_{X/T*}\mathcal{O}_{X/T} \tag{4.1.8.1}$$

obtained from (4.1.5.1) is an isomorphism

$$E^\bullet_n \simeq Ru_{X^{(p)}/T_{n-1}*}\mathcal{O}_{X^{(p)}/T_{n-1}}. \tag{4.1.8.2}$$

4.1.9. — The lemma allows us to define a map

$$\pi_n : \mathcal{A}^q_{n+1,X/T} \longrightarrow \mathcal{A}^q_{n,X^{(p)}/T} \tag{4.1.9.1}$$

as follows. Locally we can choose a smooth lifting Y/T. For a closed form $\omega \in \Omega^q_{Y_n/T_n}$, $p^q\omega$ defines a class in

$$\mathrm{Ker}\big(\tilde{E}^q_{n+q+1} \to E^q_n \xrightarrow{d} E^{q+1}_n\big), \tag{4.1.9.2}$$

and hence an element in $\mathcal{H}^q(E^{\bullet}_n) \simeq R^q u_{X^{(p)}/T_{n-1}*}\mathcal{O}_{X^{(p)}/T_{n-1}}$. If ω is equal to $d\omega'$ for some $\omega' \in \Omega^{q-1}_{Y_n/T_n}$, then $p^q\omega'$ defines an element of $\tilde{E}^{q-1}_{n+q+1}$ mapping to $p^q\omega$, and hence for any closed form $\omega \in \Omega^q_{Y_n/T_n}$ the class

$$[p^q\omega] \in \mathcal{H}^q(E^{\bullet}_n) \simeq R^q u_{X^{(p)}/T_{n-1}*}\mathcal{O}_{X^{(p)}/T_{n-1}} \tag{4.1.9.3}$$

depends only on the image of ω in $\mathcal{H}^q(\Omega^{\bullet}_{Y_n/T_n})$, where the isomorphism $\mathcal{H}^q(E^{\bullet}_n) \simeq R^q u_{X^{(p)}/T_{n-1}*}\mathcal{O}_{X^{(p)}/T_{n-1}}$ is obtained from (4.1.8.2). The map π_n is the induced map

$$\mathcal{A}^q_{n+1,X/T} = \mathcal{H}^q(\Omega^{\bullet}_{Y_n/T_n}) \longrightarrow R^q u_{X^{(p)}/T_{n-1}*}\mathcal{O}_{X^{(p)}/T_{n-1}} = \mathcal{A}^q_{n,X^{(p)}/T}. \tag{4.1.9.4}$$

Considerations as in 4.1.2 show that this map is independent of the choice of Y and hence π_n is defined globally.

4.1.10. — As in [**34**, 0.2.2] define a chain of submodules

$$0 \subset B_1\Omega^i_{X/S} \subset \cdots \subset B_n\Omega^i_{X/S} \subset \cdots \subset Z_{n+1}\Omega^i_{X/S} \subset \cdots \subset Z_1\Omega^i_{X/S} \subset \Omega^i_{X/S} \tag{4.1.10.1}$$

by the formulas

$$B_0\Omega^i_{X/S} = 0, \quad Z_0\Omega^i_{X/S} = \Omega^i_{X/S}, \tag{4.1.10.2}$$

$$B_1\Omega^i_{X/S} = \mathrm{Im}(d : \Omega^{i-1}_{X/S} \to \Omega^i_{X/S}), \quad Z_1\Omega^i_{X/S} = \mathrm{Ker}(d : \Omega^i_{X/S} \to \Omega^{i+1}_{X/S}), \tag{4.1.10.3}$$

$$\begin{aligned} B_n\Omega^i_{X^{(p)}/S} &\xrightarrow{C^{-1}_{X/S}} B_{n+1}\Omega^i_{X/S}/B_1\Omega^i_{X/S}, \\ Z_n\Omega^i_{X^{(p)}/S} &\xrightarrow{C^{-1}_{X/S}} Z_{n+1}\Omega^i_{X/S}/B_1\Omega^i_{X/S}, \end{aligned} \tag{4.1.10.4}$$

where $C^{-1}_{X/S}$ denotes the map induced by the inverse Cartier isomorphism.

Remark 4.1.11. — It follows from the definitions that for every n there are natural maps

$$C^n : B_{n+1}\Omega^q_{X/S} \longrightarrow B_1\Omega^q_{X^{(p^n)}/S}, \quad dC^n : Z_n\Omega^q_{X/S} \longrightarrow B_1\Omega^q_{X^{(p^n)}/S}, \tag{4.1.11.1}$$

whose kernels are equal to $B_n\Omega^q_{X/S}$ and $Z_{n+1}\Omega^q_{X/S}$ respectively.

Theorem 4.1.12 ([**31**, 4.4]). — *The map* $\pi_n : \mathcal{A}^q_{n+1,X/T} \to \mathcal{A}^q_{n,X^{(p)}/T}$ *is surjective with kernel the image of* $(V^n, dV^n) : \mathcal{A}^q_{1,X/T} \oplus \mathcal{A}^{q-1}_{1,X/T} \to \mathcal{A}^q_{n+1,X/T}$. *If* $T_0 = S$ *(so the ideal of* S *in* T *is* $p\mathcal{O}_T$*) then the composite map*

$$s_n : \Omega^q_{X^{(p)}/S} \oplus \Omega^{q-1}_{X^{(p)}/S} \xrightarrow{C^{-1}} \mathcal{A}^q_{1,X/T} \oplus \mathcal{A}^{q-1}_{1,X/T} \xrightarrow{(V^n, dV^n)} \mathcal{A}^q_{n+1,X/T} \tag{4.1.12.1}$$

induces an isomorphism

$$\big(\Omega^q_{X^{(p)}/S} \oplus \Omega^{q-1}_{X^{(p)}/S}\big)/R^q_{n,X^{(p)}/S} \simeq \mathrm{Ker}(\pi_n), \tag{4.1.12.2}$$

where $R^q_{n,X^{(p)}/S}$ *is defined by the exact sequence*
(4.1.12.3)

$$0 \longrightarrow R^q_{n,X^{(p)}/S} \longrightarrow B_{n+1}\Omega^q_{X^{(p)}/S} \oplus Z_n\Omega^{q-1}_{X^{(p)}/S} \xrightarrow{(C^n, dC^n)} B_1\Omega^q_{X^{(p^{n+1})}/S} \longrightarrow 0.$$

The proof is in steps 4.1.13–4.1.25.

4.1.13. — All the assertions are étale local on X, and hence we may assume that we have a smooth lift Y/T of X. Furthermore, by replacing S by T_0 and X by Y_0 we may assume that $S = T_0$. Define $\tilde{E}^q_m$, $\tilde{E}^q_\infty$, and $E^\bullet_n$ as in 4.1.5 and 4.1.7, so that the map π_n can be described as in 4.1.9.

4.1.14. — If $\omega \in \tilde{E}^q_{n+q+2}$ defines (via (4.1.8.2)) a class $[\omega] \in \mathcal{A}^q_{n,X^{(p)}/T}$, then we can write

$$\omega = p^q\eta, \quad p^q d\eta \in p^n\tilde{E}^{q+1}_{n+q+2} \tag{4.1.14.1}$$

for some $\eta \in \Omega^q_{Y_{n+q+1}/T_{n+q+1}}$. Hence $p^q d\eta = p^n(p^{q+1}\lambda)$ for some $\lambda \in \Omega^{q+1}_{Y_{n+q+1}/T_{n+q+1}}$, and so $d\eta$ is zero modulo p^{n+1}. Therefore if $[\eta] \in \mathcal{A}^q_{n,X/T}$ denotes the class defined by the image of η in $\Omega^q_{Y_{n-1}/T_{n-1}}$ then $\pi_n([\eta]) = [\omega]$, and so π_n is surjective.

4.1.15. — If $\pi_n(\omega) = 0$, then there exist $b \in \Omega^{q-1}_{Y_{n+q}/T_{n+q}}$, and $c \in \Omega^q_{Y_{n+q}/T_{n+q}}$, such that $db \in p\Omega^q_{Y_{n+q}/T_{n+q}}$, $dc \in p\Omega^{q+1}_{Y_{n+q}/T_{n+q}}$, and

$$p^q\omega = d(p^{q-1}b) + p^n(p^qc). \tag{4.1.15.1}$$

It follows that if $\bar{b}$ and $\bar{c}$ denote the reductions of b and c modulo p, then

$$[\omega] = dV^n(\bar{b}) + V^n(\bar{c}). \tag{4.1.15.2}$$

Hence the kernel of π_n is contained in $\mathrm{Im}(s_n)$.

4.1.16. — To see that $\mathrm{Im}(s_n) \subset \mathrm{Ker}(\pi_n)$, we can without loss of generality assume that X is étale over $\mathbb{G}^r_{m,S}$ for some integer r. Let $\mathrm{dlog}(x_1), \ldots, \mathrm{dlog}(x_r)$ be the standard basis for $\Omega^1_{X/S}$ induced by the choice of a map to $\mathbb{G}^r_{m,S}$. Then it follows from the definitions, that for any $f \in \mathcal{O}_{X^{(p)}}$ with image $h \in \mathcal{O}_X$ and integers $i_1, \ldots, i_q$ we have

$$V^nC^{-1}([f\mathrm{dlog}(x_{i_1}) \wedge \cdots \wedge \mathrm{dlog}(x_{i_q})]) = [p^n\tilde{h}\mathrm{dlog}(\tilde{x}_{i_1}) \wedge \cdots \wedge \mathrm{dlog}(\tilde{x}_{i_q})] \tag{4.1.16.1}$$

in $\mathcal{H}^q(\Omega^{\bullet}_{Y_n/T_n}) = \mathcal{A}^q_{n+1,X/T}$, where $\tilde{h} \in \mathcal{O}_{Y_{n-1}}$ and $\tilde{x}_i \in \mathcal{O}^*_{Y_{n-1}}$ are liftings of h and the x_i. Note that the right side of (4.1.16.1) is independent of the choices of the liftings. Then

$$(4.1.16.2)\quad \pi_n([p^n\tilde{h}\mathrm{dlog}(\tilde{x}_{i_1}) \wedge \cdots \wedge \mathrm{dlog}(\tilde{x}_{i_q})]) = [p^n\tilde{f}\mathrm{dlog}(\tilde{x}_{i_1}) \wedge \cdots \wedge \mathrm{dlog}(\tilde{x}_{i_q})] = 0,$$

where $\tilde{f} \in \mathcal{O}_{Y_{n-1}} \otimes_{\mathcal{O}_T,\sigma} \mathcal{O}_T$ is a lifting of f.

To see that the image of $dV^nC^{-1}(-)$ is also in the kernel of π_n, let $\tilde{h} \in \mathcal{O}_Y$ be a lifting of h. Since h is in the image of the relative Frobenius we have $d\tilde{h} \equiv 0 \pmod{p}$. If $\omega \in \Omega^1_{Y/T}$ is a section such that $d\tilde{h} = p\omega$, then

$$(4.1.16.3)\quad dV^nC^{-1}([f\mathrm{dlog}(x_{i_1}) \wedge \cdots \wedge \mathrm{dlog}(x_{i_q})]) = [\omega \wedge \mathrm{dlog}(\tilde{x}_{i_1}) \wedge \cdots \wedge \mathrm{dlog}(\tilde{x}_{i_q})].$$

Indeed as in (4.1.16.1) we have

$$(4.1.16.4)\quad V^nC^{-1}([f\mathrm{dlog}(x_{i_1}) \wedge \cdots \wedge \mathrm{dlog}(x_{i_q})]) = [p^n\tilde{h}\mathrm{dlog}(\tilde{x}_{i_1}) \wedge \cdots \wedge \mathrm{dlog}(\tilde{x}_{i_q})]$$

so $dV^nC^{-1}([f\mathrm{dlog}(x_{i_1}) \wedge \cdots \wedge \mathrm{dlog}(x_{i_q})])$ is equal to the image in $\mathcal{A}^{q+1}_{n+1,X/T}$ of the class

$$(4.1.16.5)\qquad [p^n\tilde{h}\mathrm{dlog}(\tilde{x}_{i_1}) \wedge \cdots \wedge \mathrm{dlog}(\tilde{x}_{i_q})] \in \mathcal{H}^q(\Omega^{\bullet}_{Y_n/T_n}) = \mathcal{A}^q_{n+1,X/T}$$

under the boundary map

$$(4.1.16.6)\qquad \mathcal{H}^q(\Omega^{\bullet}_{Y_n/T_n}) \longrightarrow \mathcal{H}^{q+1}(\Omega^{\bullet}_{Y_n/T_n})$$

arising from the exact sequence of complexes

$$(4.1.16.7)\qquad 0 \longrightarrow \Omega^{\bullet}_{Y_n/T_n} \xrightarrow{\times p^{n+1}} \Omega^{\bullet}_{Y_{2n+1}/T_{2n+1}} \longrightarrow \Omega^{\bullet}_{Y_n/T_n} \longrightarrow 0.$$

Therefore $dV^nC^{-1}([f\mathrm{dlog}(x_{i_1}) \wedge \cdots \wedge \mathrm{dlog}(x_{i_q})])$ is equal to the class of any closed form $\tau \in \Omega^{q+1}_{Y_n/T_n}$ for which $p^{n+1}\tau$ is equal to $d\alpha$ for some lifting $\alpha \in \Omega^q_{Y_{2n+1}/T_{2n+1}}$ of $p^n\tilde{h}\mathrm{dlog}(\tilde{x}_{i_1}) \wedge \cdots \wedge \mathrm{dlog}(\tilde{x}_{i_q})$. Taking $\alpha = p^n\tilde{h}\mathrm{dlog}(\tilde{x}_{i_1}) \wedge \cdots \wedge \mathrm{dlog}(\tilde{x}_{i_q})$ and $\tau = \omega \wedge \mathrm{dlog}(\tilde{x}_{i_1}) \wedge \cdots \wedge \mathrm{dlog}(\tilde{x}_{i_q})$ we obtain (4.1.16.3).

Since

$$(4.1.16.8)\quad p^{q+1}\omega \wedge \mathrm{dlog}(x_{i_1}) \wedge \cdots \wedge \mathrm{dlog}(x_{i_q}) = d(p^q\tilde{h} \wedge \mathrm{dlog}(x_{i_1}) \wedge \cdots \wedge \mathrm{dlog}(x_{i_q})),$$

it follows that $\pi_n(dV^n(C^{-1}([f\mathrm{dlog}(x_{i_1}) \wedge \cdots \wedge \mathrm{dlog}(x_{i_q})]))) = 0$. Thus $\mathrm{Im}(s_n) \subset \mathrm{Ker}(\pi_n)$.

4.1.17. — In order to prove that $R^q_{n,X^{(p)}/S}$ is equal to $\mathrm{Ker}(s_n)$, we proceed by induction on n. The case $n = 0$ follows from the Cartier isomorphism. Thus we assume the result holds for $n-1$ and prove it for n. The key ingredient is the following lemma of Illusie.

Lemma 4.1.18 ([**34**, 0.2.2.8]). — *For any integer r, identify $\Omega^q_{X^{(p^r)}/S}$ with $\mathcal{O}_S \otimes_{F^r_S, \mathcal{O}_S} \Omega^q_{X/S}$.*

(i) *The sheaf $B_{n+1}\Omega^q_{X^{(p)}/S}$ is generated locally by elements of the form*

$$\lambda \otimes g_1^{p^r-1} \cdots g_q^{p^r-1} dg_1 \wedge \cdots \wedge dg_q, \tag{4.1.18.1}$$

where $\lambda \in \mathcal{O}_S$, $g_i \in \mathcal{O}_X$, and $0 \leq r \leq n$.

(ii) *The sheaf $Z_n\Omega^q_{X^{(p)}/S}$ is generated locally by elements of $B_n\Omega^q_{X^{(p)}/S}$ and elements of the form*

$$\lambda \otimes f^{p^n} g_1^{p^n-1} \cdots g_q^{p^n-1} dg_1 \wedge \cdots \wedge dg_q, \tag{4.1.18.2}$$

where $\lambda \in \mathcal{O}_S$ and $g_i, f \in \mathcal{O}_X$.

Lemma 4.1.19

(i) *For any local section (4.1.18.1) of $B_{n+1}\Omega^q_{X^{(p)}/S}$ we have*

$$C^n(\lambda \otimes g_1^{p^r-1} \cdots g_q^{p^r-1} dg_1 \wedge \cdots \wedge dg_q) \in \Omega^q_{X^{(p^{n+1})}/S} \tag{4.1.19.1}$$

is zero unless $r = n$ in which case it is equal to $\lambda \otimes dg_1 \wedge \cdots \wedge dg_q$.

(ii) *For any local section (4.1.18.2) of $Z_n\Omega^q_{X^{(p)}/S}$ we have*

(4.1.19.2)
$$dC^n(\lambda \otimes f^{p^n} g_1^{p^n-1} \cdots g_q^{p^n-1} dg_1 \wedge \cdots \wedge dg_q) = \lambda \otimes df \wedge dg_1 \wedge \cdots \wedge dg_q \in \Omega^{q+1}_{X^{(p^{n+1})}/S}.$$

Proof. — For (i), note that by definition of the Cartier operator C, we have with our identifications $\mathcal{O}_S \otimes_{F^r_S, \mathcal{O}_S} \Omega^q_{X/S} \simeq \Omega^q_{X^{(p^r)}/S}$

$$C(\lambda \otimes g_1^{p^r-1} \cdots g_q^{p^r-1} dg_1 \wedge \cdots \wedge dg_q) = \begin{cases} \lambda \otimes g_1^{p^{r-1}-1} \cdots g_q^{p^{r-1}-1} dg_1 \wedge \cdots \wedge dg_q & r \geq 1 \\ 0 & r = 0. \end{cases}$$

From this and induction (i) follows.

For (ii), note that

(4.1.19.3)
$$C(\lambda \otimes f^{p^n} g_1^{p^n-1} \cdots g_q^{p^n-1} dg_1 \wedge \cdots \wedge dg_q) = \lambda \otimes f^{p^{n-1}} g_1^{p^{n-1}-1} \cdots g_q^{p^{n-1}-1} dg_1 \wedge \cdots \wedge dg_q.$$

By induction this gives

$$C^n(\lambda \otimes f^{p^n} g_1^{p^n-1} \cdots g_q^{p^n-1} dg_1 \wedge \cdots \wedge dg_q) = \lambda \otimes f dg_1 \wedge \cdots \wedge dg_q \tag{4.1.19.4}$$

which implies (ii). □

Corollary 4.1.20. — *For any local section $\omega \in B_n\Omega^q_{X^{(p)}/S}$, we have $V^nC^{-1}(\omega) = 0$.*

Proof. — By the lemma it suffices to consider $\omega = g_1^{p^r-1}\cdots g_q^{p^r-1}dg_1\wedge\cdots\wedge dg_q$ with $r < n$. Then

$$\begin{aligned} V^nC^{-1}(\omega) &= V^n[g_1^{p^{r+1}-1}\cdots g_q^{p^{r+1}-1}dg_1\wedge\cdots\wedge dg_q] \\ &= [p^n g_1^{p^{r+1}-1}\cdots g_q^{p^{r+1}-1}dg_1\wedge\cdots\wedge dg_q] \\ &= [p^{n-r-1}d(g_1^{p^{r+1}}g_2^{p^{r+1}-1}\cdots g_q^{p^{r+1}-1}dg_2\wedge\cdots\wedge dg_q)] \\ &= 0. \quad \square \end{aligned}$$

Corollary 4.1.21. — *For any local section $\omega \in Z_{n+1}\Omega^{q-1}_{X^{(p)}/S}$, the class $dV^n(C^{-1}(\omega))$ is zero.*

Proof. — By the lemma and 4.1.20 it suffices to consider the case when

$$\omega = f^{p^{n+1}}g_1^{p^{n+1}-1}\cdots g_{q-1}^{p^{n+1}-1}dg_1\wedge\cdots\wedge dg_{q-1}. \tag{4.1.21.1}$$

In this case

$$\begin{aligned} dV^n(C^{-1}(\omega)) &= dV^n\big([f^{p^{n+2}}g_1^{p^{n+2}-1}\cdots g_{q-1}^{p^{n+2}-1}dg_1\wedge\cdots\wedge dg_{q-1}]\big) \\ &= d\big[p^n f^{p^{n+2}}g_1^{p^{n+2}-1}\cdots g_{q-1}^{p^{n+2}-1}dg_1\wedge\cdots\wedge dg_{q-1}\big] \\ &= \big[p^{n+1}f^{p^{n+2}-1}g_1^{p^{n+2}-1}\cdots g_{q-1}^{p^{n+2}-1}df\wedge dg_1\wedge\cdots\wedge dg_{q-1}\big] \\ &= 0. \quad \square \end{aligned}$$

Corollary 4.1.22 — *The sheaf $R^q_{n,X^{(p)}/S}$ is generated locally by sections of the form $(0,b)$ with $b \in Z_{n+1}\Omega^{q-1}_{X^{(p)}/S}$, $(a,0)$ with $a \in B_n\Omega^q_{X^{(p)}/S}$, and elements of the form*

$$(\lambda\otimes g_1^{p^n-1}\cdots g_q^{p^n-1}dg_1\wedge\cdots\wedge dg_q, -\lambda\otimes g_1^{p^n}g_2^{p^n-1}\cdots g_q^{p^n-1}dg_2\wedge\cdots\wedge dg_q). \tag{4.1.22.1}$$

Proof. — Let

(4.1.22.2)

$$c = (c_1,c_2) \in \mathrm{Ker}\big(B_{n+1}\Omega^q_{X^{(p)}/S}\oplus Z_n\Omega^{q-1}_{X^{(p)}/S} \xrightarrow{(C^n,dC^n)} B_1\Omega^q_{X^{(p^{n+1})}/S}\big) = R^q_{n,X^{(p)}/S}$$

be a local section. Writing $C^n(c_1) = -dC^n(c_2)$ as a sum of terms of the form

$$\lambda\otimes dg_1\wedge\cdots\wedge dg_q, \tag{4.1.22.3}$$

we see that by subtracting elements of the form (4.1.22.1) from c we can assume that $C^n(c_1) = 0$ and $dC^n(c_2) = 0$. In this case by 4.1.11 we have $c_1 \in B_n\Omega^q_{X^{(p)}/S}$ and $c_2 \in Z_{n+1}\Omega^{q-1}_{X^{(p)}/S}$. $\square$

4.1.23. — We continue the proof of 4.1.12. That $R^q_{n,X^{(p)}/S} \subset \mathrm{Ker}(s_n)$ now follows from the above three corollaries and the computation

$$\begin{aligned}
&s_n(g_1^{p^n-1}\cdots g_q^{p^n-1}dg_1\cdots dg_q, -g_1^{p^n}g_2^{p^n-1}\cdots g_q^{p^n-1}dg_2\cdots dg_q)\\
&= V^n[g_1^{p^{n+1}-1}\cdots g_q^{p^{n+1}-1}dg_1\cdots dg_q] - dV^n[g_1^{p^{n+1}}g_2^{p^{n+1}-1}\cdots g_q^{p^{n+1}-1}dg_2\cdots dg_q]\\
&= [p^n g_1^{p^{n+1}-1}\cdots g_q^{p^{n+1}-1}dg_1\cdots dg_q] - [p^n g_1^{p^{n+1}-1}g_2^{p^{n+1}-1}\cdots g_q^{p^{n+1}-1}dg_1dg_2\cdots dg_q]\\
&= 0.
\end{aligned}$$

4.1.24. — To prove that $R^q_{n,X^{(p)}/S}$ is the whole kernel, suppose $(a,b) \in \mathrm{Ker}(s_n)$. From the commutative diagram

$$\begin{array}{ccc}
\mathcal{A}^q_{1,X/T} \oplus \mathcal{A}^{q-1}_{1,X/T} & \xrightarrow{(V^n, dV^n)} & \mathcal{A}^q_{n+1,X/T} \\
\downarrow{\scriptstyle \mathrm{pr}_2} & & \downarrow{\scriptstyle F} \\
\mathcal{A}^{q-1}_{1,X/T} & \xrightarrow{dV^{n-1}} & \mathcal{A}^q_{n,X/T}
\end{array} \tag{4.1.24.1}$$

we deduce that

$$dV^{n-1}(C^{-1}(b)) = F(s_n(a,b)) = 0 \tag{4.1.24.2}$$

which by induction on n implies that $(0,b) \in R^q_{n-1,X^{(p)}/S}$. Therefore

$$b \in \mathrm{Ker}\big(dC^{n-1} : Z_{n-1}\Omega^q_{X^{(p)}/S} \to B_1\Omega^q_{X^{(p^n)}/S}\big), \tag{4.1.24.3}$$

which implies that $b \in Z_n\Omega^q_{X^{(p)}/S}$.

Since the map $C^n : B_{n+1}\Omega^q_{X^{(p)}/S} \to B_1\Omega^1_{X^{(p^{n+1})}/S}$ is surjective, we can, after subtracting an element of $R^q_{n,X^{(p)}/S}$ from (a,b), assume that $b = 0$.

The commutative diagram

(4.1.24.4)

$$\begin{array}{ccccccccc}
0 & \longrightarrow & \Omega^\bullet_{Y_{n-1}/T_{n-1}} & \xrightarrow{\times p} & \Omega^\bullet_{Y_n/T_n} & \longrightarrow & \Omega^\bullet_{X/S} & \longrightarrow & 0 \\
 & & \downarrow{\scriptstyle \mathrm{id}} & & \downarrow{\scriptstyle \times p^{n-1}} & & \downarrow{\scriptstyle \times p^{n-1}} & & \\
0 & \longrightarrow & \Omega^\bullet_{Y_{n-1}/T_{n-1}} & \xrightarrow{\times p^n} & \Omega^\bullet_{Y_{2n-1}/T_{2n-1}} & \longrightarrow & \Omega^\bullet_{Y_{n-1}/T_{n-1}} & \longrightarrow & 0
\end{array}$$

shows that the cohomology sequence of the upper row can be written as

$$\mathcal{H}^{q-1}(\Omega^\bullet_{X/S}) \xrightarrow{dV^{n-1}} \mathcal{H}^q(\Omega^\bullet_{Y_{n-1}/T_{n-1}}) \xrightarrow{V} \mathcal{H}^q(\Omega^\bullet_{Y_n/T_n}). \tag{4.1.24.5}$$

Hence if $V^n(C^{-1}(a)) = 0$, there exists $c \in \Omega^{q-1}_{X^{(p)}/S}$ such that

$$V^{n-1}(C^{-1}(a)) = dV^{n-1}(C^{-1}(c)). \tag{4.1.24.6}$$

Therefore

$$0 = FV^{n-1}(C^{-1}(a)) = FdV^{n-1}(C^{-1}(c)) = dV^{n-2}(C^{-1}(c)), \tag{4.1.24.7}$$

and by induction

$$c \in \operatorname{Ker}\big(Z_{n-2}\Omega^{q-1}_{X^{(p)}/S} \xrightarrow{dC^{n-2}} B_1\Omega^q_{X^{(p^{n-1})}/S}\big) = Z_{n-1}\Omega^{q-1}_{X^{(p)}/S}. \tag{4.1.24.8}$$

By the following lemma, after subtracting an element of $B_n\Omega^q_{X^{(p)}/S}$ from a we may assume that $V^{n-1}(C^{-1}(a)) = 0$. Induction therefore completes the proof of 4.1.12. □

Lemma 4.1.25. — *The image of $Z_{n-1}\Omega^{q-1}_{X^{(p)}/S}$ in $\mathcal{A}^q_{n,X/T}$ under $dV^{n-1}(C^{-1}(-))$ is equal to the image of $B_n\Omega^q_{X^{(p)}/S}$ under $V^{n-1}C^{-1}(-)$.*

Proof. — Since $dV^{n-1}(C^{-1}(B_{n-1}\Omega^q_{X^{(p)}/S})) = 0$, it suffices by 4.1.21 to consider the image of an element ω of the form

$$\omega = f^{p^{n-1}} g_1^{p^{n-1}-1} \cdots g_{q-1}^{p^{n-1}-1} dg_1 \wedge \cdots \wedge dg_{q-1} \tag{4.1.25.1}$$

We have

$$\begin{aligned} dV^{n-1}(C^{-1}(\omega)) &= dV^{n-1}[f^{p^n} g_1^{p^n-1} \cdots g_{q-1}^{p^n-1} dg_1 \wedge \cdots \wedge dg_{q-1}] \\ &= d[p^{n-1} f^{p^n} g_1^{p^n-1} \cdots g_{q-1}^{p^n-1} dg_1 \wedge \cdots \wedge dg_{q-1}]. \end{aligned}$$

Now observe that

$$d[p^{n-1} f^{p^n} g_1^{p^n-1} \cdots g_{q-1}^{p^n-1} dg_1 \wedge \cdots \wedge dg_{q-1}] \tag{4.1.25.2}$$

is equal to

$$[p^{n-1} f^{p^n-1} g_1^{p^n-1} \cdots g_{q-1}^{p^n-1} df \wedge dg_1 \wedge \cdots \wedge dg_{q-1}] \tag{4.1.25.3}$$

Indeed it suffices to prove this equality étale locally on X, so we may assume we have a smooth lifting Y/T of X and liftings $\tilde{f}, \tilde{g}_i \in \mathcal{O}_Y$ of f and the g_i. By definition $d[p^{n-1} f^{p^n} g_1^{p^n-1} \cdots g_{q-1}^{p^n-1} dg_1 \wedge \cdots \wedge dg_{q-1}] \in \mathcal{H}^{q+1}(\Omega^\bullet_{Y_{n-1}/T_{n-1}})$ is equal to the class of a closed form $\omega \in \Omega^{q+1}_{Y_{n-1}/T_{n-1}}$ such that $p^n\omega \in \Omega^{q+1}_{Y_{2n-1}/T_{2n-1}}$ is equal to

$$d(p^{n-1} \tilde{f}^{p^n} \tilde{g}_1^{p^n-1} \cdots \tilde{g}_{q-1}^{p^n-1} d\tilde{g}_1 \wedge \cdots \wedge d\tilde{g}_{q-1}) \tag{4.1.25.4}$$

which equals

$$p^n(p^{n-1} \tilde{f}^{p^n-1} \tilde{g}_1^{p^n-1} \cdots \tilde{g}_{q-1}^{p^n-1} d\tilde{f} \wedge d\tilde{g}_1 \wedge \cdots \wedge d\tilde{g}_{q-1}). \tag{4.1.25.5}$$

From this description it follows that we can take $\omega = p^{n-1} f^{p^n-1} g_1^{p^n-1} \cdots g_{q-1}^{p^n-1} df \wedge dg_1 \wedge \cdots \wedge dg_{q-1}$ giving the equality of (4.1.25.2) and (4.1.25.3).

Therefore

$$dV^{n-1}(C^{-1}(\omega)) = [p^{n-1}f^{p^n-1}g_1^{p^n-1}\cdots g_{q-1}^{p^n-1}df\wedge dg_1\wedge\cdots\wedge dg_{q-1}]$$
$$= V^{n-1}C^{-1}(f^{p^{n-1}-1}g_1^{p^{n-1}-1}\cdots g_{q-1}^{p^{n-1}-1}df\wedge dg_1\wedge\cdots\wedge dg_{q-1}). \quad \square$$

4.1.26. — Next we gather together some results about the behavior of $\mathcal{A}^q_{n,X/T}$ under base change $T'\to T$. Fix a smooth lifting Y/T of X, and set $Y_n := Y\times_T T_n$. Define $Z^q_{Y_n}$ to be the kernel of $d:\Omega^q_{Y_n/T_n}\to\Omega^{q+1}_{Y_n/T_n}$, and let $B^{q+1}_{Y_n}$ be the image. We denote by Y' and Y'_n the spaces obtained by base change to T'.

Proposition 4.1.27 **([34, 0.2.2.8] in the case $n=0$).** — *The natural maps*

$$Z^q_{Y_n}\otimes_{\mathcal{O}_{T_n}}\mathcal{O}_{T'_n}\longrightarrow Z^q_{Y'_n} \tag{4.1.27.1}$$

$$B^q_{Y_n}\otimes_{\mathcal{O}_{T_n}}\mathcal{O}_{T'_n}\longrightarrow B^q_{Y'_n} \tag{4.1.27.2}$$

are isomorphisms.

Proof. — Replacing X by Y_0 we may assume that $S=T_0$. We first prove the result for $Z^q_{Y_n}$ by induction on n. The case of $n=0$ follows from [**34**, 0.2.8.8]. Thus assume true for $n-1$. Let $\times p: Z^q_{Y_{n-1}}\to Z^q_{Y_n}$ be the natural inclusion, and let Q_n be the cokernel. To prove that the formation of $Z^q_{Y_n}$ is compatible with base change, it suffices to show that the formation of Q_n commutes with base change. Let $\iota: Q_n\to Z^q_X$ be the map sending $\omega\in Z^q_{Y_n}$ to its reduction. The map ι is injective and its image contains B^q_X. Define $\overline{Q}_n$ to be the quotient sheaf $\iota(Q_n)/B^q_X$. Then there is a natural isomorphism

$$\overline{Q}_n\simeq\operatorname{Ker}\big(dV^{n-1}C^{-1}:\Omega^q_{X^{(p)}/T_0}\to\mathcal{A}^{q+1}_{n,X/T}\big)=Z_n\Omega^q_{X^{(p)}/T_0}. \tag{4.1.27.3}$$

Indeed, as in (4.1.24.5) this kernel is by construction of d and V equal to the kernel of the coboundary map

$$\Omega^q_{X^{(p)}/T_0}\xrightarrow{C^{-1}}\mathcal{H}^q(\Omega^\bullet_{X/S})\longrightarrow\mathcal{H}^{q+1}(\Omega^\bullet_{Y_{n-1}/T_{n-1}}) \tag{4.1.27.4}$$

arising from the exact sequence

$$0\longrightarrow\Omega^\bullet_{Y_{n-1}/T_{n-1}}\xrightarrow{\times p}\Omega^\bullet_{Y_n/T_n}\longrightarrow\Omega^\bullet_{X/S}\longrightarrow 0. \tag{4.1.27.5}$$

Thus we have an exact sequence

$$0\longrightarrow B^q_X\longrightarrow Q_n\longrightarrow Z_n\Omega^q_{X^{(p)}/S}\longrightarrow 0, \tag{4.1.27.6}$$

and hence by [**34**, 0.2.8.8] the formation of Q_n commutes with base change. This proves the result for $Z^q_{Y_n}$.

To get the result for $B^q_{Y_n}$, note that there is a commutative diagram
(4.1.27.7)

$$\begin{array}{ccccccccc} & & Z^q_{Y_n}\otimes_{\mathcal{O}_{T_n}}\mathcal{O}_{T'_n} & \longrightarrow & \Omega^q_{Y_n/T_n}\otimes_{\mathcal{O}_{T_n}}\mathcal{O}_{T'_n} & \longrightarrow & B^{q+1}_{Y_n}\otimes_{\mathcal{O}_{T_n}}\mathcal{O}_{T'_n} & \longrightarrow & 0 \\ & & \downarrow{\scriptstyle\alpha} & & \downarrow{\scriptstyle\beta} & & \downarrow{\scriptstyle\gamma} & & \\ 0 & \longrightarrow & Z^q_{Y'_n} & \longrightarrow & \Omega^q_{Y'_n/T'_n} & \longrightarrow & B^{q+1}_{Y'_n} & \longrightarrow & 0 \end{array}$$

where α and β are isomorphisms. □

Corollary 4.1.28. — *The formation of $\mathcal{A}^{\bullet}_{n,X/T}$ is compatible with base change $T' \to T$.*

4.1.29. — We conclude this section by making some observations about the sheaves $B_n\Omega^i_{X/S}$ and $Z_n\Omega^i_{X/S}$ which will be used in what follows.

Assume that S is reduced and let $f : S' \to S$ be a surjective finite morphism (for example $S' = S$ and f a power of the Frobenius map). Denote by X'/S' the base change $X \times_S S'$ and let $g : X' \to X$ be the projection.

Lemma 4.1.30. — *The maps $\mathcal{O}_S \to f_*\mathcal{O}_{S'}$ and $\Omega^i_{X/S} \to g_*\Omega^i_{X'/S'}$ are injective.*

Proof. — Since $g_*\Omega^i_{X'/S'} = \Omega^i_{X/S} \otimes_{\mathcal{O}_S} f_*\mathcal{O}_{S'}$ and $\Omega^i_{X/S}$ is flat over $\mathcal{O}_S$ it suffices to prove the injectivity of the map $\mathcal{O}_S \to f_*\mathcal{O}_{S'}$.

For the injectivity of $\mathcal{O}_S \to f_*\mathcal{O}_{S'}$ we may work étale locally on S and hence may assume that S is a scheme. Let $s \in \mathcal{O}_S$ be a nonzero local section which is in the kernel. Since S is reduced, there exists a point $p \in S$ such that the image of s in $k(p)$ is non-zero. Replacing S by $\mathrm{Spec}(k(p))$ and S' by $S' \times_S \mathrm{Spec}(k(p))$ we see that it suffices to consider the case when S is a field which is immediate. □

Proposition 4.1.31. — *As subsheaves of $f_*\Omega^i_{X'/S'}$ we have*

$$B_n\Omega^i_{X/S} = f_*(B_n\Omega^i_{X'/S'}) \cap \Omega^i_{X/S}, \quad Z_n\Omega^i_{X/S} = f_*(Z_n\Omega^i_{X'/S'}) \cap \Omega^i_{X/S}. \tag{4.1.31.1}$$

Proof. — By induction on n.

The case $n = 0$ is trivial from the definitions.

For the case $n = 1$ note that we have a commutative square

$$\begin{array}{ccc} \Omega^i_{X/S} & \xrightarrow{d} & \Omega^{i+1}_{X/S} \\ \downarrow & & \downarrow \\ f_*\Omega^i_{X'/S'} & \xrightarrow{d} & f_*\Omega^{i+1}_{X'/S'}, \end{array} \tag{4.1.31.2}$$

where the vertical arrows are injections. This implies the result for $Z_1\Omega^i_{X/S}$.

For $B_1\Omega^i_{X/S}$, consider the commutative diagram
(4.1.31.3)

$$\begin{array}{ccccccccc} 0 & \longrightarrow & B_1\Omega^i_{X/S} & \longrightarrow & Z_1\Omega^i_{X/S} & \xrightarrow{C^{-1}} & \Omega^i_{X^{(p)}/S} & \longrightarrow & 0 \\ & & \downarrow{\scriptstyle j_1} & & \downarrow{\scriptstyle j_2} & & \downarrow{\scriptstyle j_3} & & \\ 0 & \longrightarrow & f_*B_1\Omega^i_{X'/S'} & \longrightarrow & f_*Z_1\Omega^i_{X'/S'} & \xrightarrow{C^{-1}} & f_*\Omega^i_{X^{(p)}_{S'}/S'} & \longrightarrow & 0. \end{array}$$

From this diagram and the result for Z_1 we see that

$$B_1\Omega^i_{X/S} = f_*(B_1\Omega^i_{X'/S'}) \cap Z_1\Omega^i_{X/S} = f_*(B_1\Omega^i_{X'/S'}) \cap \Omega^i_{X/S}. \tag{4.1.31.4}$$

For the inductive step we assume the result for $n-1$ and prove it for n. Consider the commutative diagram
(4.1.31.5)

$$\begin{array}{ccccccccc} 0 & \longrightarrow & B_1\Omega^i_{X/S} & \longrightarrow & Z_n\Omega^i_{X/S} & \xrightarrow{C^{-1}} & Z_{n-1}\Omega^i_{X^{(p)}/S} & \longrightarrow & 0 \\ & & \downarrow & & \downarrow & & \downarrow & & \\ 0 & \longrightarrow & f_*B_1\Omega^i_{X'/S'} & \longrightarrow & f_*Z_n\Omega^i_{X'/S'} & \xrightarrow{C^{-1}} & f_*Z_{n-1}\Omega^i_{X^{(p)}_{S'}/S'} & \longrightarrow & 0. \end{array}$$

Given $\lambda \in f_*(Z_n\Omega^i_{X'/S'}) \cap \Omega^i_{X/S}$, the image $\bar{\lambda}$ in $f_*Z_{n-1}\Omega^i_{X^{(p)}_{S'}/S'}$ is in (using the induction hypothesis)

$$f_*\big(Z_{n-1}\Omega^i_{X^{(p)}_{S'}/S'}\big) \cap \Omega^i_{X^{(p)}/S} = Z_{n-1}\Omega^i_{X^{(p)}/S}. \tag{4.1.31.6}$$

Thus we can lift $\bar{\lambda}$ to an element $\lambda' \in Z_n\Omega^i_{X/S}$. Replacing λ by $\lambda - \lambda'$ we may therefore assume that $\lambda \in f_*B_1\Omega^i_{X'/S'} \cap \Omega^i_{X/S}$ which is $B_1\Omega^i_{X/S}$ by the $n=1$ case. This proves that $Z_n\Omega^i_{X/S} = f_*(Z_n\Omega^i_{X'/S'}) \cap \Omega^i_{X/S}$.

A similar argument using the commutative diagram
(4.1.31.7)

$$\begin{array}{ccccccccc} 0 & \longrightarrow & B_1\Omega^i_{X/S} & \longrightarrow & B_n\Omega^i_{X/S} & \xrightarrow{C^{-1}} & B_{n-1}\Omega^i_{X^{(p)}/S} & \longrightarrow & 0 \\ & & \downarrow & & \downarrow & & \downarrow & & \\ 0 & \longrightarrow & f_*B_1\Omega^i_{X'/S'} & \longrightarrow & f_*B_n\Omega^i_{X'/S'} & \xrightarrow{C^{-1}} & f_*B_{n-1}\Omega^i_{X^{(p)}_{S'}/S'} & \longrightarrow & 0 \end{array}$$

implies that $B_n\Omega^i_{X/S} = f_*(B_n\Omega^i_{X'/S'}) \cap \Omega^i_{X/S}$. □

Corollary 4.1.32. — *Let $R^q_{n,X/S}$ and $R^q_{n,X'/S'}$ be as in 4.1.12. Then in $f_*\Omega^q_{X'/S'} \oplus f_*\Omega^{q-1}_{X'/S}$ we have*

$$R^q_{n,X/S} = (f_*R^q_{n,X'/S'}) \cap (\Omega^q_{X/S} \oplus \Omega^{q-1}_{X'/S'}). \tag{4.1.32.1}$$

Proof. — The commutative diagram

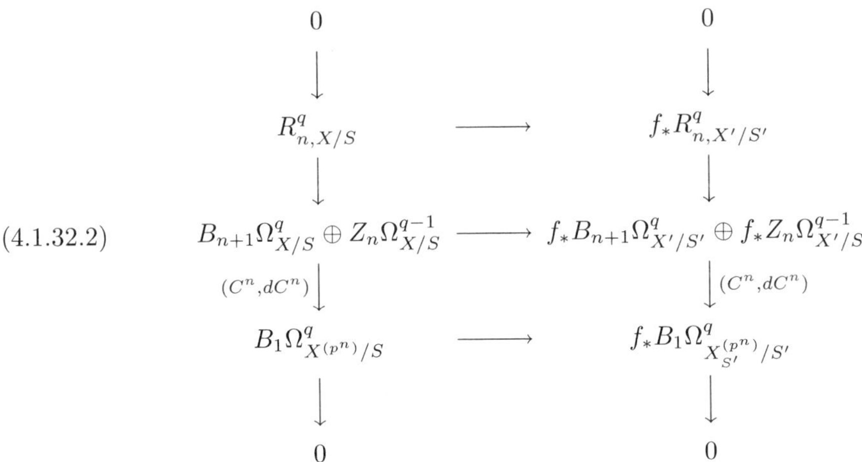

implies that

$$(4.1.32.3)\qquad R^q_{n,X/S} = (f_*R^q_{n,X'/S'}) \cap (B_n\Omega^q_{X/S} \oplus Z_n\Omega^{q-1}_{X/S}).$$

Since

$$(4.1.32.4)\qquad B_n\Omega^q_{X/S} \oplus Z_n\Omega^{q-1}_{X/S} = (f_*B_n\Omega^q_{X'/S'} \oplus f_*Z_n\Omega^{q-1}_{X'/S'}) \cap (\Omega^q_{X/S} \oplus \Omega^{q-1}_{X/S})$$

by 4.1.31 the result follows. □

4.2. Comparison with the Langer-Zink de Rham-Witt complex I

4.2.1. Let S be an algebraic space over $\mathbb{F}_p$, and let $X \to S$ be a smooth morphism of algebraic spaces. Assume given a closed immersion $S \hookrightarrow T$ into a p-adically complete formal flat $\mathbb{Z}_p$-space defined by a divided power ideal in $\mathcal{O}_T$ and a lifting $F_T : T \to T$ of Frobenius. Denote by T_n the reduction of T modulo p^{n+1}.

In [**47**], Langer and Zink introduced the *de Rham-Witt complex* of X/S generalizing the de Rham-Witt complex of Illusie [**34**] (in fact in [**47**] the de Rham-Witt complex is defined for more general $X \to S$ but we will only consider the above situation). In what follows we will denote the Langer-Zink de Rham-Witt complex of X/S by $W^{\mathrm{LZ}}_\bullet\Omega^\bullet_{X/S}$. This is a projective system of differential graded algebras

$$(4.2.1.1)\qquad \cdots \xrightarrow{\pi_{n+1}} W^{\mathrm{LZ}}_{n+1}\Omega^\bullet_{X/S} \xrightarrow{\pi_n} W^{\mathrm{LZ}}_n\Omega^\bullet_{X/S} \xrightarrow{\pi_{n-1}} \cdots$$

which comes equipped with operators

$$(4.2.1.2)\qquad F : W^{\mathrm{LZ}}_n\Omega^\bullet_{X/S} \longrightarrow W^{\mathrm{LZ}}_{n-1}\Omega^\bullet_{X/S},\quad V : W^{\mathrm{LZ}}_n\Omega^\bullet_{X/S} \longrightarrow W^{\mathrm{LZ}}_{n+1}\Omega^\bullet_{X/S}$$

satisfying $FV = p$ and $FdV = d$ and $V(\omega \cdot F(\eta)) = (V(\omega)) \cdot \eta$ for $\eta \in W^{\mathrm{LZ}}_{n+1}\Omega^\bullet_{X/S}$ and $\omega \in W^{\mathrm{LZ}}_n\Omega^\bullet_{X/S}$ (this implies in particular that $VF = p$).

4.2.2. — By the construction of $W_n^{\mathrm{LZ}}\Omega^\bullet_{X/S}$ reviewed below in 4.2.4–4.2.6 (see also [**47**, Introduction]) for a description in terms of a universal property), there is a natural isomorphism $W_n(\mathcal{O}_X) \simeq W_n^{\mathrm{LZ}}\Omega^0_{X/S}$. There is also a map

$$\rho_n : W_n(\mathcal{O}_X) \longrightarrow \mathcal{A}^0_{n,X/T} \tag{4.2.2.1}$$

defined as follows. First we define the map locally when we fix an embedding $X \hookrightarrow Y$ with Y/T_{n-1} smooth. Let D denote the divided power envelope of X in Y. The map ρ_n is defined by

$$\rho_n(a_0, \dots, a_{n-1}) := \sum_{i=0}^{n-1} p^i \tilde{a}_i^{p^{n-i}} \in \mathcal{H}^0(\Omega^\bullet_{D/T_{n-1}}) \simeq \mathcal{A}^0_{n,X/T}, \tag{4.2.2.2}$$

where $\tilde{a}_i$ is any lifting of a_i to $\mathcal{O}_D$. This map is well-defined because the kernel of $\mathcal{O}_D \to \mathcal{O}_X$ has divided powers and hence for any $h \in \mathrm{Ker}(\mathcal{O}_D \to \mathcal{O}_X)$ we have

$$(\tilde{a}_i + h)^{p^{n-i}} = \sum_{j=0}^{p^{n-i}} \binom{p^{n-i}}{j} j! h^{[j]} \tilde{a}_i^{p^{n-i}-j} \equiv \tilde{a}_i^{p^{n-i}} \pmod{p^{n-i}}. \tag{4.2.2.3}$$

That ρ_n is a ring homomorphism follows from the definition of the ring structure on $W_n(\mathcal{O}_X)$ [**71**, Chapter II §6].

If Y'/T_{n-1} is a second lifting of X to T_{n-1} we can form $Y'' := Y \times_S Y'$ to obtain a commutative diagram

$$\begin{array}{ccccc} D & \longleftarrow & D'' & \longrightarrow & D' \\ \downarrow & & \downarrow & & \downarrow \\ Y & \longleftarrow & Y'' & \longrightarrow & Y', \end{array} \tag{4.2.2.4}$$

where D' (resp. D'') denotes the divided power envelope of X in Y' (resp. Y''). If $\rho_{n,Y}$ (resp. $\rho_{n,Y'}$, $\rho_{n,Y''}$) denote the map (4.2.2.2) then it follows from the definition that the diagram

$$\begin{array}{ccc} W_n(\mathcal{O}_X) & \xrightarrow{\rho_{n,Y}} & \mathcal{H}^0(\Omega^\bullet_{D/T_{n-1}}) \\ {\scriptstyle \mathrm{id}}\downarrow & & \downarrow{\scriptstyle \simeq} \\ W_n(\mathcal{O}_X) & \xrightarrow{\rho_{n,Y''}} & \mathcal{H}^0(\Omega^\bullet_{D''/T_{n-1}}) \\ {\scriptstyle \mathrm{id}}\uparrow & & \uparrow{\scriptstyle \simeq} \\ W_n(\mathcal{O}_X) & \xrightarrow{\rho_{n,Y'}} & \mathcal{H}^0(\Omega^\bullet_{D'/T_{n-1}}) \end{array} \tag{4.2.2.5}$$

commutes. It follows that ρ_n is independent of the choice of Y/T_{n-1} and hence is defined globally. Observe also that if follows from the construction of ρ_n

that the diagrams

$$\begin{array}{ccc} W_n(\mathcal{O}_X) & \xrightarrow{\rho_n} & \mathcal{A}^0_{n,X/T} \\ {\scriptstyle F}\downarrow & & \downarrow{\scriptstyle F} \\ W_{n-1}(\mathcal{O}_X) & \xrightarrow{\rho_{n-1}} & \mathcal{A}^0_{n-1,X/T} \end{array} \tag{4.2.2.6}$$

and

$$\begin{array}{ccc} W_n(\mathcal{O}_X) & \xrightarrow{\rho_n} & \mathcal{A}^0_{n,X/T} \\ {\scriptstyle V}\downarrow & & \downarrow{\scriptstyle V} \\ W_{n+1}(\mathcal{O}_X) & \xrightarrow{\rho_{n+1}} & \mathcal{A}^0_{n+1,X/T} \end{array} \tag{4.2.2.7}$$

commute.

The main result of this section is the following:

Theorem 4.2.3. — *There is a unique morphism $\rho_n : W_n^{\mathrm{LZ}}\Omega^\bullet_{X/S} \to \mathcal{A}^\bullet_{n,X/T}$ of differential graded algebras which in degree 0 is equal to the above defined map ρ_n. This map is compatible with the operators F and V. If S is reduced then the map is injective.*

The proof will be in several steps 4.2.4–4.2.14.

First we recall the definition of $W_n^{\mathrm{LZ}}\Omega^\bullet_{X/S}$ in [**47**].

4.2.4. — Let A be a ring and B an A-algebra with a divided power ideal $\mathfrak{b} \subset B$. If M is a B-module, a *pd-derivation* $\partial : B \to M$ *over* A is an A-linear B-derivation such that

$$\partial(\gamma_n(b)) = \gamma_{n-1}(b)\partial(b) \tag{4.2.4.1}$$

for every $n \geq 1$ and $b \in \mathfrak{b}$. Here $\gamma_n : \mathfrak{b} \to \mathfrak{b}$ denotes the divided power structure.

There is a universal pd-derivation $d : B \to \breve{\Omega}^1_{B/A}$ [**47**, 1.1]. The B-module $\breve{\Omega}^1_{B/A}$ is the quotient of the module $\Omega^1_{B/A}$ by the sub-B-module generated by elements of the form

$$d(\gamma_n(b)) - \gamma_{n-1}(b)db. \tag{4.2.4.2}$$

Defining $\breve{\Omega}^i_{B/A} := \bigwedge^i \breve{\Omega}^1_{B/A}$ we obtain a differential graded algebra $\breve{\Omega}^\bullet_{B/A}$ with differential determined by the formula

$$d(\omega\eta) = (d\omega)\eta + (-1)^i\omega d\eta, \tag{4.2.4.3}$$

for $\omega \in \breve{\Omega}^i_{B/A}$ and $\eta \in \breve{\Omega}^i_{B/A}$. The differential graded algebra $\breve{\Omega}^\bullet_{B/A}$ has the following universal property: If $\mathcal{A}^\bullet$ is a differential graded A-algebra with a ring homomorphism $\rho : B \to \mathcal{A}^0$ such that the composite with the differential $B \to \mathcal{A}^0 \to \mathcal{A}^1$ is a pd-derivation, then there exists a unique map of differential graded algebras $\breve{\Omega}^\bullet_{B/A} \to \mathcal{A}^\bullet$ extending ρ.

4.2.5. — Let $R \to S$ be a map of $\mathbb{F}_p$-algebras and let $W_n(R)$ and $W_n(S)$ be the Witt-vectors of R and S respectively. Recall that there are operators F and V on these rings of Witt-vectors and that the ideal $V(W_{n-1}(S)) \subset W_n(S)$ has a natural divided power structure [**34**, 0.1.4]. Denote by $\breve{\Omega}^\bullet_{W_n(S)/W_n(R)}$ the de Rham-complex obtained from the construction of the preceding paragraph. The de Rham-Witt complex $W_n^{\mathrm{LZ}}\Omega^\bullet_{S/R}$ is a quotient of $\breve{\Omega}^\bullet_{W_n(S)/W_n(R)}$ which can be described inductively as follows.

For $n = 1$ we take simply $W_1^{\mathrm{LZ}}\Omega^\bullet_{S/R} := \Omega^\bullet_{S/R} = \breve{\Omega}^\bullet_{W_1(S)/W_1(R)}$.

The complex $W_{n+1}^{\mathrm{LZ}}\Omega^\bullet_{S/R}$ is constructed from $W_n^{\mathrm{LZ}}\Omega^\bullet_{S/R}$ as follows. First define the ideal $I \subset \breve{\Omega}^\bullet_{W_{n+1}(S)/W_{n+1}(R)}$ as follows. Consider all relations

$$\sum_{l=1}^{M} \xi^{(l)} d\eta_1^{(l)} \cdots d\eta_i^{(l)} = 0 \tag{4.2.5.1}$$

in $W_n^{\mathrm{LZ}}\Omega^i_{S/R}$, where i and M are integers ≥ 1 and $\xi^{(l)}, \eta_k^{(l)} \in W_n(S)$. Then I is the smallest ideal of $\breve{\Omega}^\bullet_{W_{n+1}(S)/W_{n+1}(R)}$ with $d(I) \subset I$ containing the elements

$$\sum_k V(\xi^{(l)}) dV(\eta_1^{(l)}) \cdots dV(\eta_i^{(l)}). \tag{4.2.5.2}$$

Let $\bar{\Omega}^\bullet_{n+1}$ denote the quotient of $\breve{\Omega}^\bullet_{W_n(S)/W_n(R)}$ by I. By the definition of I there is a map of abelian groups

$$V : W_n\Omega^\bullet_{S/R} \longrightarrow \bar{\Omega}^\bullet_{S/R}, \quad \xi d\eta_1 \cdots d\eta_i \longmapsto V(\xi)dV(\eta_1) \cdots dV(\eta_i). \tag{4.2.5.3}$$

The algebra $W_{n+1}^{\mathrm{LZ}}\Omega^\bullet_{S/R}$ is then obtained as the quotient of $\bar{\Omega}^\bullet_{n+1}$ by the smallest ideal $J \subset \bar{\Omega}^\bullet_{n+1}$ with $d(J) \subset J$ containing all elements

$$V(\omega \cdot F(\eta)) - V(\omega) \cdot \eta, \tag{4.2.5.4}$$

where $\omega \in W_n^{\mathrm{LZ}}\Omega^\bullet_{S/R}$ and $\eta \in \bar{\Omega}^\bullet_{n+1}$.

4.2.6. — Let $f : X \to S$ be a morphism of algebraic spaces over $\mathbb{F}_p$. We define the complex $W_n^{\mathrm{LZ}}\Omega^\bullet_{X/S}$ as follows. First consider the case when $S = \mathrm{Spec}(R)$ is affine. Then we define $W_n^{\mathrm{LZ}}\Omega^\bullet_{X/S}$ by associating to any affine étale X-scheme $\mathrm{Spec}(T) \to X$ the complex $W_n^{\mathrm{LZ}}\Omega^\bullet_{T/R}$. By [**47**, 1.7] this is a sheaf. If $\mathrm{Spec}(R) \to \mathrm{Spec}(R')$ is an étale morphism then by [**47**, 1.11] there is a natural isomorphism $W_n^{\mathrm{LZ}}\Omega^\bullet_{T/R} \simeq W_n^{\mathrm{LZ}}\Omega^\bullet_{T/R'}$. It follows that for general S we can construct $W_n^{\mathrm{LZ}}\Omega^\bullet_{X/S}$ working étale locally on S and hence we obtain the complex $W_n^{\mathrm{LZ}}\Omega^\bullet_{X/S}$ in general.

In what follows we will also need to consider the completed de Rham-Witt complex $W^{\mathrm{LZ}}\Omega^\bullet_{X/S}$ which by definition is $\varprojlim_n W_n^{\mathrm{LZ}}\Omega^\bullet_{X/S}$.

For a sheaf $\mathcal{M}$ of $W_n(\mathcal{O}_X)$-modules on X_{et}, we define a $W_n(S)$*-linear pd-derivation* $\partial : W_n(\mathcal{O}_X) \to \mathcal{M}$ to be a $W_n(S)$-linear derivation such that for every étale $U \to X$ the derivation

$$\Gamma(U, W_n(U)) \longrightarrow \mathcal{M}(U) \tag{4.2.6.1}$$

is a pd-derivation. There is a universal pd-derivation $d : W_n(\mathcal{O}_X) \to \breve{\Omega}^\bullet_{W_n(X)/W_n(S)}$ obtained by sheafifying the universal pd-derivation constructed in 4.2.4. Denoting by $\breve{\Omega}^\bullet_{W_n(X)/W_n(S)}$ the associated de Rham complex, we obtain a unique surjection $\breve{\Omega}^\bullet_{W_n(X)/W_n(S)} \to W_n^{\mathrm{LZ}}\Omega^\bullet_{X/S}$ of differential graded $W_n(\mathcal{O}_X)$-algebras.

Lemma 4.2.7. — *With notation as in 4.2.6 and $X \to S$ smooth, the composite*

$$\partial : W_n(\mathcal{O}_X) \xrightarrow{\rho_n} \mathcal{A}^0_{n,X/T} \xrightarrow{d} \mathcal{A}^1_{n,X/T} \tag{4.2.7.1}$$

is a $W_n(\mathcal{O}_S)$-linear pd-derivation. Therefore the map ρ_n extends to a unique map of differential graded algebras $\lambda_n : \breve{\Omega}^\bullet_{W_n(X)/W_n(S)} \to \mathcal{A}^\bullet_{n,X/T}$.

Proof. — The assertion is étale local on X so we may assume that we have a smooth lifting Y/T so that we can identify $\mathcal{A}^\bullet_{n,X/T}$ with $\mathcal{H}^\bullet(\Omega^\bullet_{Y_{n-1}/T_{n-1}})$. By [**47**, bottom of p. 240] to prove that ∂ is a pd-derivation it suffices to show that

$$p^{p-2}\partial(V(\xi^p)) = p^{p-2}\rho_n(V(\xi^{p-1}))\partial(V(\xi)) \tag{4.2.7.2}$$

for any $\xi \in W_{n-1}(\mathcal{O}_X)$. By the commutativity of (4.2.2.7) this is equivalent to the equality

$$p^{p-2}d(V(\rho_{n-1}(\xi)^p)) = p^{p-2}V(\rho_{n-1}(\xi)^{p-1})d(V(\rho_{n-1}(\xi))). \tag{4.2.7.3}$$

For this consider first the case when $p = 2$. In this case (4.2.7.3) is equivalent to the equality

$$d(V(\rho_{n-1}(\xi)^2)) = V(\rho_{n-1}(\xi))d(V(\rho_{n-1}(\xi))). \tag{4.2.7.4}$$

Let $f \in \mathcal{O}_{Y_{n-2}}$ be $\rho_{n-1}(\xi)$, and let $\tilde{f} \in \mathcal{O}_Y$ be a lifting of f. Let $\omega \in \Omega^1_{Y_{n-1}/T_{n-1}}$ be a form with $p^n\omega \in \Omega^1_{Y_{2n-1}/T_{2n-1}}$ equal to $d(p\tilde{f})$. Then $V(\rho_{n-1}(\xi))d(V\rho_{n-1}(\xi)) \in \mathcal{H}^1(\Omega^\bullet_{Y_{n-1}/T_{n-1}})$ is represented by the form

$$p\tilde{f}\omega \in \Omega^1_{Y_{n-1}/T_{n-1}}. \tag{4.2.7.5}$$

On the other hand, $dV(\rho_{n-1}(\xi)^2)$ is represented by any closed form $\alpha \in \Omega^1_{Y_{n-1}/T_{n-1}}$ for which $p^n\alpha \in \Omega^1_{Y_{2n-1}/T_{2n-1}}$ is equal to $d(p\tilde{f}^2) = p^2\tilde{f}d\tilde{f}$ (since $p = 2$). The equality (4.2.7.4) then follows from noting that

$$p^n(p\tilde{f}\omega) = p\tilde{f}d(p\tilde{f}) = p^2\tilde{f}d\tilde{f} \tag{4.2.7.6}$$

in $\Omega^1_{Y_{2n-1}/T_{2n-1}}$ so we can take $\alpha = p\tilde{f}\omega$.

For the case $p \geq 3$, note that by 4.1.3 we have $pdV = Vd$ and so

$$\begin{aligned} p^{p-2}V(\rho_{n-1}(\xi)^{p-1})d(V(\rho_{n-1}(\xi))) &= p^{p-3}(V(\rho_{n-1}(\xi)^{p-1}))V(d(\rho_{n-1}(\xi))) \\ &= p^{p-2}V(\rho_{n-1}(\xi)^{p-1}d(\rho_{n-1}(\xi))) \\ &= p^{p-3}V(d(\rho_{n-1}(\xi)^p)) \\ &= p^{p-2}d(V(\rho_{n-1}(\xi)^p)). \end{aligned} \tag{4.2.7.7}$$

To see that ∂ is $W_n(\mathcal{O}_S)$-linear let $(a_0, \ldots, a_{n-1}) \in W_n(\mathcal{O}_S)$ be a local section, and choose a smooth lifting Y/S of X. Then $d(\rho_n(a_0, \ldots, a_{n-1}))$ is equal to the class of

$$\sum_{i=0}^{n-1} \tilde{a}_i^{p^{n-i}-1} d\tilde{a}_i, \tag{4.2.7.8}$$

where $\tilde{a}_i$ is a lifting of a_i to $\mathcal{O}_Y$. Since $\mathcal{O}_T \to \mathcal{O}_S$ is surjective we can choose $\tilde{a}_i \in \mathcal{O}_T$ and hence $d\tilde{a}_i = 0$, and the class is zero. □

Lemma 4.2.8. — *The map* $\lambda_n : \breve{\Omega}^\bullet_{W_n(X)/W_n(S)} \to \mathcal{A}^\bullet_{n,X/T}$ *factors through* $W_n^{\mathrm{LZ}}\Omega^\bullet_{X/S}$.

Proof. — The proof is by induction on n. The case $n = 1$ is clear. So we prove the result for n assuming it holds for $n - 1$. First we show that the map factors through $\bar{\Omega}^\bullet_n$ (defined as in 4.2.5). For this it suffices to show that if i and M are integers ≥ 1 and $\xi^{(l)}$ and η_k are elements of $W_{n-1}(\mathcal{O}_X)$ such that

$$\sum_{l=1}^{M} \xi^{(l)} d\eta_1^{(l)} \cdots d\eta_i^{(l)} = 0 \tag{4.2.8.1}$$

in $W_{n-1}^{\mathrm{LZ}}\Omega^\bullet_{X/S}$, then

$$\begin{aligned}\lambda_n\Big(\sum_k V(\xi^{(l)})dV(\eta_1^{(l)}) \cdots dV(\eta_i^{(l)})\Big) \\ = \sum_k V(\rho_{n-1}(\xi^{(l)}))dV(\rho_{n-1}(\eta_1^{(l)})) \cdots dV(\rho_{n-1}(\eta_i^{(l)}))\end{aligned} \tag{4.2.8.2}$$

is zero in $\mathcal{A}^\bullet_{n,X/T}$. By 4.1.3 for any ξ and η we have $V(\xi) \cdot dV(\eta) = V(\xi \cdot FdV(\eta)) = V(\xi \cdot d\eta)$ and so

$$V(\rho_{n-1}(\xi^{(l)}))dV(\rho_{n-1}(\eta_1^{(l)})) \cdots dV(\rho_{n-1}(\eta_i^{(l)})) = V\lambda_{n-1}(\xi^{(l)} d\eta_1^{(l)} \cdots d\eta_i^{(l)}). \tag{4.2.8.3}$$

Therefore (4.2.8.2) is equal to

$$V\Big(\lambda_{n-1}\Big(\sum_{l=1}^{M} \xi^{(l)} d\eta_1^{(l)} \cdots d\eta_i^{(l)}\Big)\Big). \tag{4.2.8.4}$$

By induction this is zero and hence λ_n factors through $\bar{\Omega}^\bullet_n$. To prove that λ_n factors through $W_n^{\mathrm{LZ}}\Omega^\bullet_{X/S}$ it suffices by the definition of the ideal $J \subset \bar{\Omega}^\bullet_n$ in 4.2.5 that for any elements $\omega \in \mathcal{A}^\bullet_{n-1,X/T}$ and $\eta \in \mathcal{A}^\bullet_{n,X/T}$ we have $V(\omega \cdot F(\eta)) = V(\omega) \cdot \eta$. This is shown in 4.1.3. □

4.2.9. — The commutativity of the diagrams (4.2.2.6) and (4.2.2.7) and the universal property of $\breve{\Omega}^\bullet_{W_n(X)/W_n(S)}$ implies that the diagrams

$$\begin{array}{ccc} W_n^{\mathrm{LZ}}\Omega^\bullet_{X/S} & \xrightarrow{\rho_n} & \mathcal{A}^\bullet_{n,X/T} \\ {\scriptstyle F}\downarrow & & \downarrow{\scriptstyle F} \\ W_{n-1}^{\mathrm{LZ}}\Omega^\bullet_{X/S} & \xrightarrow{\rho_{n-1}} & \mathcal{A}^\bullet_{n-1,X/T} \end{array} \tag{4.2.9.1}$$

and

$$\text{(4.2.9.2)} \qquad \begin{array}{ccc} W_n^{\mathrm{LZ}}\Omega^\bullet_{X/S} & \xrightarrow{\rho_n} & \mathcal{A}^\bullet_{n,X/T} \\ {\scriptstyle V}\downarrow & & \downarrow{\scriptstyle V} \\ W_{n+1}^{\mathrm{LZ}}\Omega^\bullet_{X/S} & \xrightarrow{\rho_{n+1}} & \mathcal{A}^\bullet_{n+1,X/T} \end{array}$$

commute. This therefore completes the proof of 4.2.3 except for the injectivity of ρ_n when S is reduced.

To prove that ρ_n is injective when S is reduced we first make some reductions.

Lemma 4.2.10. — *If S is reduced, then $W(S)$ is flat over $\mathbb{Z}_p$.*

Proof. — For any local section $(a_0, a_1, \dots) \in W(\mathcal{O}_S)$ we have

$$\text{(4.2.10.1)} \qquad p \cdot (a_0, a_1, \dots) = (0, a_0^p, a_1^p, \dots),$$

which is nonzero if $(a_0, a_1, \dots)$ is nonzero and S is reduced. □

4.2.11. — This lemma enables us to reduce to the case when $T = W(S)$. For as explained in [**34**, 0.1.3] there exists a canonical map $W(S) \to T$ compatible with the liftings of Frobenius and the inclusions of S. It follows that the map $W_n^{\mathrm{LZ}}\Omega^\bullet_{X/S} \to \mathcal{A}^\bullet_{n,X/W(S)}$ factors through the map $W_n^{\mathrm{LZ}}\Omega^\bullet_{X/S} \to \mathcal{A}^\bullet_{n,X/T}$. For the rest of the proof we will therefore assume that $T = W(S)$.

We can further reduce to the case when $X = \mathbb{A}^r_S$ for some integer r as follows. For this reduction note first that to prove that ρ_n is injective we may replace S and X by étale covers. Therefore we may assume that S is affine, say $S = \mathrm{Spec}(R)$. Furthermore, we may assume that X is étale over $\mathbb{A}^r_R$ for some integer r. To show that it suffices to consider $X = \mathbb{A}^r_S$, it therefore suffices to show that if $X' \to X$ is an étale morphism of smooth affine R-schemes, and if the result holds for X then it also holds for X'.

Choose a smooth lifting Y of X to $W(R)$ and a lift of Frobenius $F_Y : Y \to Y$ compatible with the canonical lift of Frobenius on $W(R)$. Let $Y' \to Y$ be the unique lifting of X' to an étale Y-scheme and let $F_{Y'} : Y' \to Y'$ be the unique lifting of Frobenius on Y' to a morphism over F_Y. Let $m \geq n$ be an integer so that $p^m W_n(R) = 0$ (for example take $m = n$). By [**47**, Remark 1.8] we have an isomorphism of differential graded algebras

$$\text{(4.2.11.1)} \qquad W_{m+n}(X') \otimes_{W_{m+n}(X), F^m} W_n^{\mathrm{LZ}}\Omega^\bullet_{X/R} \simeq W_n^{\mathrm{LZ}}\Omega^\bullet_{X'/R}.$$

Here $W_n^{\mathrm{LZ}}\Omega^\bullet_{X/R}$ is viewed as a $W_{m+n}(X)$-linear algebra via the map $F^m : W_{m+n}(X) \to W_n(X)$ and the observation that $dF^m(\xi) = p^m F^m d(\xi) = 0$ by [**47**, 1.19]. On the other hand, the lifts F_Y and $F_{Y'}$ of Frobenius define by [**34**, bottom of p. 508] a canonical commutative diagram

$$\begin{array}{ccc} \mathcal{O}_{Y'} & \longrightarrow & W_n(X') \\ \uparrow & & \uparrow \\ \mathcal{O}_Y & \xrightarrow{\delta} & W_n(X). \end{array} \tag{4.2.11.2}$$

Since both $Y' \times_Y W_n(X)$ and $W_n(X')$ are étale liftings of X' to $W_n(X)$ [**47**, A.14], we conclude that the map

$$\mathcal{O}_{Y'} \otimes_{\mathcal{O}_Y} W_n(X) \longrightarrow W_n(X') \tag{4.2.11.3}$$

induced by (4.2.11.2) is an isomorphism. Combining this with (4.2.11.1) we obtain

$$\mathcal{O}_{Y'_{m+n}} \otimes_{\mathcal{O}_{Y_{m+n}}, F_Y^m} W_n^{\mathrm{LZ}}\Omega^\bullet_{X/S} \simeq W_n^{\mathrm{LZ}}\Omega^\bullet_{X'/S}. \tag{4.2.11.4}$$

On the other hand, the diagram

$$\begin{array}{ccc} Y' & \xrightarrow{F_{Y'}^{m+n}} & Y' \\ \downarrow & & \downarrow \\ Y & \xrightarrow{F_Y^{m+n}} & Y \end{array} \tag{4.2.11.5}$$

is cartesian since $Y' \to Y$ is étale. Therefore

$$\mathcal{A}^\bullet_{n,X'/T} \simeq \mathcal{O}_{Y'_{m+n}} \otimes_{\mathcal{O}_{Y_{m+n}}, F_Y^{m+n}} \mathcal{A}^\bullet_{n,X/T}. \tag{4.2.11.6}$$

To complete the reduction to $X = \mathbb{A}^r_R$, it now suffices to prove that the isomorphisms (4.2.11.5) and (4.2.11.4) are compatible. By the universal property of $\breve{\Omega}^\bullet_{W_n(X')/W_n(S)}$ it suffices to show that the diagram

$$\begin{array}{ccc} \mathcal{O}_{Y_{m+n}} & \xrightarrow{\delta} & W_{n+m}(X) \\ {\scriptstyle F_Y^{m+n}}\downarrow & & \downarrow{\scriptstyle F^m} \\ \mathcal{A}^0_{n,X/W(S)} & \xleftarrow{\rho_n} & W_n(X) \end{array} \tag{4.2.11.7}$$

commutes. If $h \in \mathcal{O}_{Y_{m+n}}$ then $\delta(h)$ is a vector $(a_0, \ldots, a_{n+m-1})$ where

$$F_Y^n(h) \equiv \sum_{i=0}^{n-1} p^i \tilde{a}_i^{p^{n-i}} \pmod{p^n} \tag{4.2.11.8}$$

for any liftings $\tilde{a}_i \in \mathcal{O}_Y$ of the a_i. Thus we find that

$$\rho_n \circ F^m \circ \delta(h) = \sum_{i=0}^{n-1} p^i (\tilde{a}_i^{p^m})^{p^{n-i}} = \sum_{i=0}^{n-1} p^i \tilde{a}_i^{p^{n+m-i}} = F_Y^{m+n}(h). \tag{4.2.11.9}$$

This completes the reduction to $X = \mathbb{A}^r_R$.

4.2.12. — For the case $X = \mathbb{A}^r_R$, let $\mathbb{D}_R$ be the *Cartier-Raynaud ring* defined in [**47**, 2.7]. The ring $\mathbb{D}_R$ is constructed as follows. First let $\mathbb{D}^0_R$ be the ring whose elements are finite formal sums

$$\sum_{s\geq 0} V^s\xi_s + \sum_{s>0} \eta_s F^s + \sum_{s\geq 0} dV^s\xi'_s + \sum_{s>0} \eta'_s F^s d, \tag{4.2.12.1}$$

where $\xi_s, \xi'_s, \eta_s, \eta'_s \in W(R)$. The addition in $\mathbb{D}^0_R$ is componentwise and the ring structure is determined by the following rules

$$\begin{aligned} FV = p = V^0 p,\quad V\xi F = V(\xi),\quad F\xi = F(\xi)F,\quad \xi\, \mathbb{V} = VF(\xi),\\ d\xi = \xi d,\quad d^2 = 0,\quad FdV = d,\quad Vd = dVp,\quad dF = pFd. \end{aligned} \tag{4.2.12.2}$$

For any integer $n \geq 0$ consider the right ideal $\mathbb{D}^0_R(n) = V^n\mathbb{D}^0_R + dV^n\mathbb{D}^0_R$. By [**47**, 2.20] the right ideal $\mathbb{D}^0_R(n)$ consists of elements (4.2.12.1) for which $\xi_s, \xi'_s \in V^{n-s}(W(R))$ for $n > s$ and $\eta_s, \eta'_s \in V^n(W(R))$ for $s > 0$. Denote the quotient $\mathbb{D}^0_R/\mathbb{D}^0_R(n)$ by $\mathbb{D}^{(n)}_R$, and let $\mathbb{D}_R = \varprojlim_n \mathbb{D}^{(n)}_R$. The module $\mathbb{D}_R$ inherits a ring structure for $\mathbb{D}^0_R$ by [**47**, 2.21]. Observe also that as a set $\mathbb{D}^{(n)}_R$ is isomorphic to the set of formal sums

$$\sum_{s=0}^{n-1} V^s\xi_s + \sum_{s>0} \eta_s F^s + \sum_{s=0}^{n-1} dV^s\xi'_s + \sum_{s>0} \eta'_s F^s d, \tag{4.2.12.3}$$

with $\xi_s, \xi'_s \in W_{n-s}(R)$ and $\eta_s, \eta'_s \in W_n(R)$. The abelian group structure on $\mathbb{D}^{(n)}_R$ is given by componentwise addition.

The operators F and V on $W^{\mathrm{LZ}}\Omega^\bullet_{\mathbb{A}^r_R/R}$ gives $W^{\mathrm{LZ}}\Omega^\bullet_{\mathbb{A}^r_R/R}$ the structure of a left $\mathbb{D}_R$-module. By [**47**, 2.25] the natural map $W^{\mathrm{LZ}}\Omega^\bullet_{\mathbb{A}^r_{\mathbb{F}_p}/\mathbb{F}_p} \to W^{\mathrm{LZ}}_n\Omega^\bullet_{\mathbb{A}^r_S/S}$ induces an isomorphism

$$W^{\mathrm{LZ}}_n\Omega^\bullet_{\mathbb{A}^r_S/S} \simeq \mathbb{D}^{(n)}_R \otimes_{\mathbb{D}_{\mathbb{F}_p}} W^{\mathrm{LZ}}\Omega^\bullet_{\mathbb{A}^r_{\mathbb{F}_p}/\mathbb{F}_p}. \tag{4.2.12.4}$$

On the other hand by 4.1.28, we have

$$\mathcal{A}^\bullet_{n,\mathbb{A}^r_S/T} \simeq W_n(R) \otimes_{W_n(\mathbb{F}_p)} \mathcal{A}^\bullet_{n,\mathbb{A}^r_{\mathbb{F}_p}/W(\mathbb{F}_p)}. \tag{4.2.12.5}$$

Define a map $\Theta : \mathbb{D}^{(n)}_R \to W_n(R) \otimes_{W_n(\mathbb{F}_p)} \mathbb{D}^{(n)}_{\mathbb{F}_p}$ by sending an expression (4.2.12.3) to

$$\sum_{s=0}^{n-1} \varphi^n(\tilde{\xi}_s) \otimes V^s + \sum_{s>0} \varphi^n(\eta_s) \otimes F^s + \sum_{s=0}^{n-1} \varphi^n(\tilde{\xi}'_s) \otimes dV^s + \sum_{s>0} \varphi^n(\eta'_s) \otimes F^s d, \tag{4.2.12.6}$$

where $\tilde{\xi}_s$ and $\tilde{\xi}'_s$ are liftings of ξ_s and ξ'_s to $W_n(R)$ and $\varphi : W_n(R) \to W_n(R)$ denotes the canonical lift of Frobenius. That this map is well-defined can be seen as follows.

If $\tilde{\xi}_{s,1}$ and $\tilde{\xi}_{s,2}$ are two liftings of ξ_s to $W_n(R)$, then their difference $\tilde{\xi}_{s,1} - \tilde{\xi}_{s,2}$ is of the form $V^{n-s}(h)$ for some $h \in W_s(R)$. Write $h = (a_0, \ldots, a_{s-1})$ and observe that

$$\varphi^n(V^{n-s}(a_0, \ldots, a_{s-1})) = \sum_{i=n-s}^{n-1} p^i \big[a_{i-(n-s)}^{p^{n-i}}\big], \tag{4.2.12.7}$$

where for $b \in R$ we write $[b] \in W_n(R)$ for its Teichmuller lifting. It follows that $\varphi^n(V^{n-s}(h))$ is a multiple of p^{n-s}. Since $V^s \in \mathbb{D}^{(n)}_{\mathbb{F}_p}$ is killed by p^{n-s} this implies that $\varphi^n(\tilde{\xi}_s) \otimes V^s$ is independent of the choice of the lifting $\tilde{\xi}_s$. A similar argument shows that $\varphi^n(\tilde{\xi}'_s) \otimes dV^s$ is independent of the choice of lifting $\tilde{\xi}'_s$, and hence the map Θ is well-defined.

Lemma 4.2.13. — *The diagram*

(4.2.13.1)

$$\begin{array}{ccc}
W_n^{\mathrm{LZ}}\Omega^\bullet_{\mathbb{A}^r_S/S} & \xleftarrow{(4.2.12.4)} & \mathbb{D}^{(n)}_R \otimes_{\mathbb{D}_{\mathbb{F}_p}} W^{\mathrm{LZ}}\Omega^\bullet_{\mathbb{A}^r_{\mathbb{F}_p}/\mathbb{F}_p} \\
\downarrow{\scriptstyle \rho_n} & & \downarrow{\scriptstyle \Theta\otimes 1} \\
\mathcal{A}^\bullet_{n,\mathbb{A}^r_S/T} & & (W_n(R)\otimes_{W_n(\mathbb{F}_p)} \mathbb{D}^{(n)}_{\mathbb{F}_p}) \otimes_{\mathbb{D}_{\mathbb{F}_p}} W^{\mathrm{LZ}}\Omega^\bullet_{\mathbb{A}^r_{\mathbb{F}_p}/\mathbb{F}_p} \\
\uparrow{\scriptstyle (4.2.12.5)} & & \downarrow{\scriptstyle \simeq} \\
W_n(R)\otimes_{W_n(\mathbb{F}_p)} \mathcal{A}^\bullet_{\mathbb{A}^r_{\mathbb{F}_p}/W(\mathbb{F}_p)} & \xleftarrow{1\otimes\rho_n} & W_n(R)\otimes_{W_n(\mathbb{F}_p)} W_n^{\mathrm{LZ}}\Omega^\bullet_{\mathbb{A}^r_{\mathbb{F}_p}/\mathbb{F}_p}
\end{array}$$

commutes.

Proof. — The composite $\rho_n \circ$ (4.2.12.4) sends an element

$$\Big(\sum_{s\geq 0} V^s\xi_s + \sum_{s>0} \eta_s F^s + \sum_{s\geq 0} dV^s\xi'_s + \sum_{s>0}\eta'_s F^s d\Big)\otimes\omega \in \mathbb{D}^{(n)}_R \otimes_{\mathbb{D}_{\mathbb{F}_p}} W^{\mathrm{LZ}}\Omega^\bullet_{\mathbb{A}^r_{\mathbb{F}_p}/\mathbb{F}_p} \tag{4.2.13.2}$$

to the element

(4.2.13.3)

$$\sum_{s=0}^{n-1} V^s(\varphi^n(\xi_s)\cdot\rho_n(\omega)) + \sum_{s>0}\varphi^n(\eta_s)F^s(\rho_n(\omega)) + \sum_{s=0}^{n-1} dV^s(\varphi^n(\xi'_s)\cdot\rho_n(\omega)) + \sum_{s>0}\varphi^n(\eta'_s)F^s d(\rho_n(\omega)),$$

where we abusively write ω for the image of ω in $W_n^{\mathrm{LS}}\Omega^\bullet_{\mathbb{A}^r_R/R}$. Thus to prove the lemma it suffices to show the equalities

(4.2.13.4)

$$V^s(\varphi^n(\xi_s)\cdot\rho_n(\omega)) = \varphi^n(\tilde{\xi}_s)V^s(\rho_n(\omega)), \quad dV^s(\varphi^n(\xi'_s)\cdot\rho_n(\omega)) = \varphi^n(\tilde{\xi}'_s)dV^s(\rho_n(\omega))$$

in $\mathcal{A}^\bullet_{n,\mathbb{A}^r_R/W(R)}$, which follow from the definition of V and d. □

4.2.14. — By [**47**, 2.24] there exists a collection $\{e_i\}_{i\in I}$ of elements $e_i \in W^{\mathrm{LZ}}\Omega^\bullet_{\mathbb{A}^r_{\mathbb{F}_p}/\mathbb{F}_p}$ such that every element of $\mathbb{D}_R^{(n)} \otimes_{\mathbb{D}_{\mathbb{F}_p}} W^{\mathrm{LZ}}\Omega^\bullet_{\mathbb{A}^r_{\mathbb{F}_p}/\mathbb{F}_p}$ can be written uniquely as

$$\xi + \sum_{i\in I} \theta_i e_i, \tag{4.2.14.1}$$

where $\xi \in W_n(R)$ and $\theta_i \in \mathbb{D}_R^{(n)}$ and almost all θ_i are zero. The map $\Theta \otimes 1$ sends such an element to

$$\varphi^n(\xi) \otimes 1 + \sum_{i\in I} \Theta(\theta_i) \otimes e_i. \tag{4.2.14.2}$$

Since R is reduced, the Frobenius map $\varphi : W_n(R) \to W_n(R)$ is injective. From the definition of Θ if follows that Θ is also injective, and hence the map $\Theta \otimes 1$ is injective as well.

To complete the proof of 4.2.3, it is therefore sufficient to show that the map $\rho_n : W_n^{\mathrm{LZ}}\Omega^\bullet_{\mathbb{A}^r_{\mathbb{F}_p}/\mathbb{F}_p} \to \mathcal{A}^\bullet_{n,\mathbb{A}^r_{\mathbb{F}_p}/W(\mathbb{F}_p)}$ is an isomorphism. But in this case the map ρ_n is the map obtained from the higher Cartier isomorphism defined in [**36**, III.1.5] and in particular ρ_n is an isomorphism. This completes the proof of 4.2.3. □

In what follows we will also need the following analogue of 4.1.12 for the Langer-Zink de Rham-Witt complex.

Theorem 4.2.15. — *Let $S/\mathbb{F}_p$ be an algebraic space and let $X \to S$ be a smooth morphism of algebraic spaces. Assume that étale locally on S there exist a flat lifting $T/\mathbb{Z}_p$ of S and a lifting $F_T : T \to T$ of Frobenius. Then for any $q \geq 0$ and $n \geq 1$, the sequence*

$$\begin{aligned} 0 \longrightarrow R^q_{n,X/S} &\longrightarrow W_1^{\mathrm{LZ}}\Omega^q_{X/S} \oplus W_1^{\mathrm{LZ}}\Omega^{q-1}_{X/S} \\ &\xrightarrow{(V^n, dV^n)} W_{n+1}^{\mathrm{LZ}}\Omega^q_{X/S} \xrightarrow{\pi_n} W_n^{\mathrm{LZ}}\Omega^q_{X/S} \longrightarrow 0 \end{aligned} \tag{4.2.15.1}$$

is exact, where as in 4.1.12 the sheaf $R^q_{n,X/S}$ is defined by the exact sequence

$$0 \longrightarrow R^q_{n,X/S} \longrightarrow B_{n+1}\Omega^q_{X/S} \oplus Z_n\Omega^{q-1}_{X/S} \xrightarrow{(C^n, dC^n)} B_1\Omega^q_{X^{(p^n)}/S} \longrightarrow 0, \tag{4.2.15.2}$$

and the inclusion $R^q_{n,X/S} \subset W_1^{\mathrm{LZ}}\Omega^q_{X/S} \oplus W_1^{\mathrm{LZ}}\Omega^{q-1}_{X/S}$ is the composite of the inclusion $R^q_{n,X/S} \subset \Omega^q_{X/S} \oplus \Omega^{q-1}_{X/S}$ and the isomorphism

$$\Omega^q_{X/S} \oplus \Omega^{q-1}_{X/S} \simeq W_1^{\mathrm{LZ}}\Omega^q_{X/S} \oplus W_1^{\mathrm{LZ}}\Omega^{q-1}_{X/S} \tag{4.2.15.3}$$

provided by the construction in 4.2.5.

The proof of 4.2.15 occupies the remainder of the section.

Remark 4.2.16. — Note that the operator

$$B_{n+1}\Omega^q_{X/S} \oplus Z_n\Omega^{q-1}_{X/S} \xrightarrow{(C^n, dC^n)} B_1\Omega^q_{X^{(p^n)}/S} \tag{4.2.16.1}$$

is $\mathcal{O}_S$-linear, so $R^q_{n,X/S}$ is an $\mathcal{O}_S$-submodule of $W_1^{\mathrm{LZ}}\Omega^q_{X/S} \oplus W_1^{\mathrm{LZ}}\Omega^{q-1}_{X/S}$.

Remark 4.2.17. — In the case when S is perfect 4.2.15 can be deduced from the classical theory of de Rham-Witt complexes [**34**] and the comparison between the Langer-Zink de Rham-Witt complex and the classical de Rham-Witt complex. To see this, note that by [**34**, I.3.10 (b)], the maps dV^n and V^n induce isomorphisms

$$dV^n : Z_n\Omega^{q-1}_{X/S}/Z_{n+1}\Omega^{q-1}_{X/S} \longrightarrow V^n\Omega^q_{X/S} \cap dV^n\Omega^{q-1}_{X/S}, \tag{4.2.17.1}$$

and

$$V^n : B_{n+1}\Omega^q_{X/S}/B_n\Omega^q_{X/S} \longrightarrow V^n\Omega^q_{X/S} \cap dV^n\Omega^{q-1}_{X/S}. \tag{4.2.17.2}$$

The resulting isomorphism

$$\rho : Z_n\Omega^{q-1}_{X/S}/Z_{n+1}\Omega^{q-1}_{X/S} \longrightarrow B_{n+1}\Omega^q_{X/S}/B_n\Omega^q_{X/S} \tag{4.2.17.3}$$

is by *loc. cit.* equal to the isomorphism [**34**, 0.2.2.6.2]. Unwinding the definitions one finds that this isomorphism ρ can also be described as the isomorphism induced by the isomorphisms

$$C^n : B_{n+1}\Omega^q_{X/S}/B_n\Omega^q_{X/S} \longrightarrow B_1\Omega^q_{X^{(p^n)}/S} \tag{4.2.17.4}$$

and

$$dC^n : Z_n\Omega^{q-1}_{X/S}/Z_{n+1}\Omega^{q-1}_{X/S} \longrightarrow B_1\Omega^q_{X^{(p^n)}/S}. \tag{4.2.17.5}$$

Therefore $R^q_{n,X/S}$ is equal to the kernel of the map

$$(V^n, dV^n) : W_1^{\mathrm{LZ}}\Omega^q_{X/S} \oplus W_1^{\mathrm{LZ}}\Omega^{q-1}_{X/S} \xrightarrow{(V^n, dV^n)} W_{n+1}^{\mathrm{LZ}}\Omega^q_{X/S}. \tag{4.2.17.6}$$

The exactness of the remaining part of (4.2.15.1) follows from [**34**, I.3.2].

4.2.18. — The assertion of the theorem is étale local on both S and X. Hence we may assume that we have a lifting (T, F_T) defined globally.

Lemma 4.2.19. — *The kernel of π_n is equal to the image of (V^n, dV^n).*

Proof. — The assertion is local on S and X and hence we may assume that S is affine, say $S = \mathrm{Spec}(R)$. By [**47**, 2.25] we have for any integer n an isomorphism

$$\mathbb{D}_R^{(n)} \otimes_{\mathbb{D}_R} W^{\mathrm{LZ}}\Omega^\bullet_{X/R} \simeq W_n^{\mathrm{LZ}}\Omega^\bullet_{X/R}. \tag{4.2.19.1}$$

Since the kernel of $\mathbb{D}_R^{(n+1)} \to \mathbb{D}_R^{(n)}$ is equal to the image of $V^n\mathbb{D}_R + dV^n\mathbb{D}_R$ it follows that the kernel of $W_{n+1}^{\mathrm{LZ}}\Omega^\bullet_{X/S} \to W_n^{\mathrm{LZ}}\Omega^\bullet_{X/S}$ is equal to the image of $(V^n\mathbb{D}_R + dV^n\mathbb{D}_R) \otimes_{\mathbb{D}_R} W^{\mathrm{LZ}}\Omega^\bullet_{X/S}$. □

4.2.20. — To complete the proof of 4.2.15, it remains only to see that the kernel of (V^n, dV^n) is $R^q_{n,X/S}$.

We already know this in the case when S is perfect by the classical theory (4.2.17). We reduce the proof of 4.2.15 for general S to the case when S is perfect (in fact $S = \mathrm{Spec}(\mathbb{F}_p)$ and $X = \mathbb{A}^r_{\mathbb{F}_p}$).

Let $X' \to X$ be an étale morphism over S. By [**47**, Remark 1.8] for any $m \geq n$ the map

$$W_{n+m}(X') \otimes_{W_{n+m}(X), F^m} W^{\mathrm{LZ}}_n \Omega^\bullet_{X/S} \longrightarrow W^{\mathrm{LZ}}_n \Omega^\bullet_{X'/S} \tag{4.2.20.1}$$

is an isomorphism.

Consider the sequence

(4.2.20.2)

$$0 \longrightarrow R^q_{n,X/S} \longrightarrow W^{\mathrm{LZ}}_1 \Omega^q_{X/S} \oplus W^{\mathrm{LZ}}_1 \Omega^{q-1}_{X/S} \overset{(V^n, dV^n)}{\longrightarrow} W^{\mathrm{LZ}}_{n+1} \Omega^q_{X/S} \overset{\pi_n}{\longrightarrow} W^{\mathrm{LZ}}_n \Omega^q_{X/S} \longrightarrow 0.$$

We will view this as a sequence of $W_{2n+2}(X)$-modules, where $W_{2n+2}(X)$ acts on $W^{\mathrm{LZ}}_1 \Omega^q_{X/S} \oplus W^{\mathrm{LZ}}_1 \Omega^{q-1}_{X/S}$ through the projection $F^{2n+1} : W_{2n+2}(X) \to W_1(X)$, and on $W^{\mathrm{LZ}}_{n+1} \Omega^q_{X/S}$ through the map $F^{n+1} : W_{2n+2}(X) \to W_{n+1}(X)$. Note that with this definition of the action, the sequence (4.2.20.2) is a sequence of $W_{2n+2}(X)$-modules. Indeed for $f \in W_{2n+2}(X)$ and $\omega \in W^{\mathrm{LZ}}_1 \Omega^q_{X/S}$ (resp. $\omega \in W^{\mathrm{LZ}}_1 \Omega^{q-1}_{X/S}$) we have by (4.2.12.2)

$$V^n(F^{2n+1}(f) \cdot \omega) = F^{n+1}(f) V^n \omega, \tag{4.2.20.3}$$

and

$$\begin{aligned} dV^n(F^{2n+1}(f) \cdot \omega) &= d(F^{n+1}(f) \cdot V^n \omega) \\ &= (dF^{n+1}(f)) \cdot V^n \omega + F^{n+1}(f) \cdot dV^n \omega \\ &= F^{n+1}(f) \cdot dV^n \omega \quad (\text{since } dF^{n+1}(f) = p^{n+1} F^{n+1} df = 0). \end{aligned}$$

If M is an $\mathcal{O}_X$-module, then there is a canonical isomorphism

$$W_{2n+2}(X') \otimes_{W_{2n+2}(X), F^{2n+1}} M \simeq \mathcal{O}_{X'} \otimes_{\mathcal{O}_X, F^{2n+1}} M. \tag{4.2.20.4}$$

Indeed if $\pi : W_{2n+2}(X) \to \mathcal{O}_X$ denotes the canonical projection, then the two maps

$$F^{2n+1}_X \circ \pi, F^{2n+1} : W_{2n+2}(X) \longrightarrow \mathcal{O}_X \tag{4.2.20.5}$$

are equal, so (4.2.20.4) can also be written as

$$(W_{2n+2}(X') \otimes_{W_{2n+2}(X), \pi} \mathcal{O}_X) \otimes_{\mathcal{O}_X, F^{2n+1}_X} M. \tag{4.2.20.6}$$

Since the canonical projection $W_{2n+2}(X') \to \mathcal{O}_{X'}$ induces an isomorphism (by [**34**, 0.1.5.8])

$$(W_{2n+2}(X') \otimes_{W_{2n+2}(X), \pi} \mathcal{O}_X) \longrightarrow \mathcal{O}_{X'}, \tag{4.2.20.7}$$

this gives the isomorphism (4.2.20.4).

Applying the functor

$$W_{2n+2}(X') \otimes_{W_{2n+2}(X)} (-) \tag{4.2.20.8}$$

to (4.2.20.2), and using the isomorphism (4.2.20.1) we conclude that if (4.2.20.2) is exact then so is the sequence

$$\begin{array}{l} 0 \longrightarrow \mathcal{O}_{X'} \otimes_{\mathcal{O}_X, F_X^{2n+1}} R^q_{n,X/S} \longrightarrow W_1^{\mathrm{LZ}}\Omega^q_{X'/S} \oplus W_1^{\mathrm{LZ}}\Omega^{q-1}_{X'/S} \xrightarrow{(V^n, dV^n)} \\ W^{\mathrm{LZ}}_{n+1}\Omega^q_{X'/S} \xrightarrow{\pi_n} W_n^{\mathrm{LZ}}\Omega^q_{X'/S} \longrightarrow 0. \end{array} \tag{4.2.20.9}$$

The following lemma then implies that if 4.2.15 holds for X/S then it also holds for X'/S.

Lemma 4.2.21. — *The induced map*

$$\mathcal{O}_{X'} \otimes_{\mathcal{O}_X, F_X^{2n+1}} R^q_{n,X/S} \longrightarrow R^q_{n,X'/S} \tag{4.2.21.1}$$

is an isomorphism.

Proof. — Recall that the Cartier operator

$$C_{X/S} : B_{n+1}\Omega^q_{X/S} \longrightarrow B_n\Omega^q_{X^{(p)}/S} \tag{4.2.21.2}$$

is Frobenius semilinear in the sense that for a section $f \in \mathcal{O}_X$ we have

$$C_{X/S}(F(f)\omega) = fC_{X/S}(\omega). \tag{4.2.21.3}$$

The sequence

$$0 \longrightarrow R^q_{n,X/S} \longrightarrow B_{n+1}\Omega^q_{X/S} \oplus Z_n\Omega^{q-1}_{X/S} \xrightarrow{(C^n, dC^n)} B_1\Omega^q_{X^{(p^n)}/S} \longrightarrow 0 \tag{4.2.21.4}$$

becomes a sequence of $\mathcal{O}_X$-modules, where $\mathcal{O}_X$ acts on $R^q_{n,X/S}$ and $B_{n+1}\Omega^q_{X/S} \oplus Z_n\Omega^{q-1}_{X/S}$ through the Frobenius map $F^{2n+1} : \mathcal{O}_X \to \mathcal{O}_X$ and on $B_1\Omega^1_{X^{(p^n)}/S}$ through the map $F^{n+1} : \mathcal{O}_X \to \mathcal{O}_X$. As in [**34**, 0.2.2.7] the maps

$$\mathcal{O}_{X'} \otimes_{\mathcal{O}_X, F^{2n+1}} (B_{n+1}\Omega^q_{X/S} \oplus Z_n\Omega^{q-1}_{X/S}) \longrightarrow B_{n+1}\Omega^q_{X'/S} \oplus Z_n\Omega^{q-1}_{X'/S}, \tag{4.2.21.5}$$

and

$$\mathcal{O}_{X'} \otimes_{\mathcal{O}_X, F^{n+1}} B_1\Omega^q_{X^{(p^n)}/S} \longrightarrow B_1\Omega^q_{X'^{(p^n)}/S} \tag{4.2.21.6}$$

are isomorphisms. Applying $\mathcal{O}_{X'} \otimes_{\mathcal{O}_X} (-)$ to the sequence (4.2.21.4) it follows that the map

$$\mathcal{O}_{X'} \otimes_{\mathcal{O}_X, F^{2n+1}} R^q_{n,X/S} \longrightarrow R^q_{n,X'/S} \tag{4.2.21.7}$$

is an isomorphism. □

4.2.22. — Since étale locally on X and S, the space X is étale over $\mathbb{A}^r_S$ for some integer r, it suffices to prove 4.2.15 for $X = \mathbb{A}^r_S$. So we assume $X = \mathbb{A}^r_S$ for the remainder of this section. Let X_0 denote $\mathbb{A}^r_{\mathbb{F}_p}$. By the case when S is perfect we know that 4.2.15 holds for $X_0/\mathbb{F}_p$.

Lemma 4.2.23. — *The map*

$$\mathscr{O}_S \otimes_{\mathbb{F}_p} R^q_{n,X_0/\mathbb{F}_p} \to R^q_{n,X/S} \tag{4.2.23.1}$$

induced by the $\mathscr{O}_S$-structure on $R^q_{n,X/S}$ (4.2.16) is an isomorphism.

Proof. — Consider the commutative diagram with exact columns

$$\begin{array}{ccc}
0 & & 0 \\
\downarrow & & \downarrow \\
\mathscr{O}_S \otimes_{\mathbb{F}_p} R^q_{n,X_0/\mathbb{F}_p} & \xrightarrow{a} & R^q_{n,X/S} \\
\downarrow & & \downarrow \\
\mathscr{O}_S \otimes_{\mathbb{F}_p} (B_{n+1}\Omega^q_{X_0/\mathbb{F}_p} \oplus Z_n\Omega^{q-1}_{X_0/\mathbb{F}_p}) & \xrightarrow{b} & B_{n+1}\Omega^q_{X/S} \oplus Z_n\Omega^{q-1}_{X/S} \\
{\scriptstyle (C^n, dC^n)}\downarrow & & \downarrow{\scriptstyle (C^n, dC^n)} \\
\mathscr{O}_S \otimes_{\mathbb{F}_p} B_1\Omega^q_{X_0/\mathbb{F}_p} & \xrightarrow{c} & B_1\Omega^q_{X^{(p^n)}/S} \\
\downarrow & & \downarrow \\
0 & & 0,
\end{array} \tag{4.2.23.2}$$

where the horizontal arrows are induced by the $\mathscr{O}_S$-module structure of the right column. By [**34**, 0.2.2.8] the maps labelled b and c are isomorphisms, and therefore the map a is also an isomorphism. □

To prove that $R^q_{n,X/S}$ is equal to the kernel of (V^n, dV^n), we need to again study the Cartier-Raynaud ring $\mathbb{D}_R$ of an $\mathbb{F}_p$-algebra R (4.2.12). Recall that as a set $\mathbb{D}^{(n+1)}_R$ is isomorphic to the set of finite formal sums

$$\sum_{s=0}^{n} V^s\xi_s + \sum_{s>0} \eta_s F^s + \sum_{s=0}^{n} dV^s\xi'_s + \sum_{s>0} \eta'_s F^s d, \tag{4.2.23.3}$$

with $\xi_s, \xi'_s \in W_{n+1-s}(R)$ and $\eta_s, \eta'_s \in W_{n+1}(R)$. In particular an element $x \in \mathbb{D}^{(1)}_R$ can be written uniquely as

$$x = \xi_0 + \sum_{s>0} \eta_s F^s + d\xi'_0 + \sum_{s>0} \eta'_s F^s d. \tag{4.2.23.4}$$

To ease the notation, for any element $f \in R$ let ${}^{V^s}\!f$ denote $V^s(f) \in W_s(R)$.

Lemma 4.2.24. — *In terms of the description (4.2.23.3) of* $\mathbb{D}_R^{(n+1)}$ *we have*

$$V^n(x) = V^n\xi_0 + \sum_{s=1}^{n} V^{n-s\,V^s}\eta_s + \sum_{s>n} {}^{V^n}\eta_s F^{s-n} + d^{V^n}\eta'_n + \sum_{s>n} {}^{V^n}\eta'_s F^{s-n}d \tag{4.2.24.1}$$

and

$$dV^n(x) = dV^n\xi_0 + \sum_{s=1}^{n} dV^{n-s\,V^s}\eta_s. \tag{4.2.24.2}$$

Proof. — Computing

$$V^n(x) = V^n\xi_0 + \sum_{s>0} V^n\eta_s F^s + V^n d\xi'_0 + \sum_{s>0} V^n\eta'_s F^s d. \tag{4.2.24.3}$$

For $1 \le s \le n$

$$\begin{aligned} V^n\eta'_s F^s d &= V^{n-s\,V^s}\eta'_s d \quad (V\xi F = {}^V\xi) \\ &= V^{n-s} d^{V^s}\eta'_s \quad (d\xi = \xi d) \\ &= dV^{n-s} p^{n-s\,V^s}\eta'_s \quad (Vd = dVp). \end{aligned}$$

For $s < n$ this last expression is 0 because in the description (4.2.23.3) of $\mathbb{D}_R^{(n+1)}$ the coefficient of dV^{n-s} is in $W_{s+1}(R)$, and $p^{n-s\,V^s}\eta'_s = 0$ in this ring. If $s = n$ we get the term $d^{V^n}\eta'_n$.

Also the relation $Vd = dVp$ shows that

$$V^n d\xi'_0 = dV^n p^n \xi'_0 \tag{4.2.24.4}$$

which is zero since the coefficient of dV^n in (4.2.23.3) is in $W_1(R)$.

It follows that

$$\begin{aligned} V^n(x) &= V^n\xi_0 + \sum_{s>0} V^n\eta_s F^s + d^{V^n}\eta'_n + \sum_{s>n} V^n\eta'_s F^s d \\ &= V^n\xi_0 + \sum_{s=1}^{n} V^{n-s\,V^s}\eta_s + \sum_{s>n} {}^{V^n}\eta_s F^{s-n} + d^{V^n}\eta'_n + \sum_{s>n} {}^{V^n}\eta'_s F^{s-n}d, \end{aligned}$$

where we use the relation $V\xi F = {}^V\xi$. This proves (4.2.24.1).

Formula (4.2.24.2) follows by applying d to (4.2.24.1), and noting that $dF = Fdp$ so that

$$\begin{aligned} \sum_{s>n} d^{V^n}\eta_s F^{s-n} &= \sum_{s>n} {}^{V^n}\eta_s dF^{s-n} \quad (\text{since } d\xi = \xi d) \\ &= \sum_{s>n} p^{s-n\,V^n}\eta_s F^{s-n} d \\ &= 0, \end{aligned}$$

and

$$\begin{aligned}\sum_{s>n} d^{V^n}\eta'_s F^{s-n}d &= \sum_{s>n} {}^{V^n}\eta'_s dF^{s-n}d \quad (\text{since } d\xi = \xi d) \\ &= \sum_{s>n} {}^{V^n}\eta'_s p^{s-n}F^{s-n}dd \\ &= 0 \quad (\text{since } d^2 = 0). \quad \square\end{aligned}$$

Corollary 4.2.25. — *In terms of the description (4.2.23.4) of elements of* $\mathbb{D}_R^{(1)}$*, we have*

$$\mathrm{Ker}\big(V^n : \mathbb{D}_R^{(1)} \to \mathbb{D}_R^{(n+1)}\big) = \Big\{ d\xi'_0 + \sum_{s=1}^{n-1} \eta'_s F^s d \Big\}, \tag{4.2.25.1}$$

and

$$\mathrm{Ker}\big(dV^n : \mathbb{D}_R^{(1)} \to \mathbb{D}_R^{(n+1)}\big) = \Big\{ \sum_{s>n} \eta_s F^s + d\xi'_0 + \sum_{s>0} \eta'_s F^s d \Big\}. \tag{4.2.25.2}$$

Moreover, if $\mathbb{L} \subset \mathbb{D}_R^{(1)} \oplus \mathbb{D}_R^{(1)}$ *denotes the left* R*-submodule of elements*

$$(\eta F^n d, -\eta F^n) \tag{4.2.25.3}$$

then the map

$$\begin{array}{c} \mathbb{L} \oplus \mathrm{Ker}\big(V^n : \mathbb{D}_R^{(1)} \to \mathbb{D}_R^{(n+1)}\big) \oplus \mathrm{Ker}\big(dV^n : \mathbb{D}_R^{(1)} \to \mathbb{D}_R^{(n+1)}\big) \\ \downarrow \\ \mathrm{Ker}\big((V^n, dV^n) : \mathbb{D}_R^{(1)} \oplus \mathbb{D}_R^{(1)} \to \mathbb{D}_R^{(n+1)}\big) \end{array} \tag{4.2.25.4}$$

is an isomorphism.

Proof. — The equalities (4.2.25.1 and (4.2.25.2) follow immediately from 4.2.24.

To see that (4.2.25.4) is an isomorphism, note that by (4.2.24.1) and (4.2.24.2) any element of $\mathbb{D}_R^{(n+1)}$ in the intersection of the images of V^n and dV^n is of the form $d^{V^n}\eta$ for some $\eta \in R$. Thus any element $(\alpha, \beta) \in \mathbb{D}_R^{(1)} \oplus \mathbb{D}_R^{(1)}$ in the kernel of (V^n, dV^n) is after subtracting an element of $\mathbb{L}$ equal to a sum $(\alpha, 0) + (0, \beta)$ of elements in the kernel. This proves the surjectivity of (4.2.25.4). The injectivity follows from the explicit descriptions (4.2.25.1) and (4.2.25.2) in terms of the basis (4.2.23.3). $\square$

4.2.26. — Let $\mathcal{K}_R$ denote the kernel of the map

$$(V^n, dV^n) : \mathbb{D}_R^{(1)} \oplus \mathbb{D}_R^{(1)} \longrightarrow \mathbb{D}_R^{(n+1)} \tag{4.2.26.1}$$

so that there is an exact sequence of right $\mathbb{D}_{\mathbb{F}_p}$-modules

$$0 \longrightarrow \mathcal{K}_R \longrightarrow \mathbb{D}_R^{(1)} \oplus \mathbb{D}_R^{(1)} \xrightarrow{(V^n, dV^n)} \mathbb{D}_R^{(n+1)} \longrightarrow \mathbb{D}_R^{(n)} \longrightarrow 0. \tag{4.2.26.2}$$

By the above computations $\mathcal{K}_R$ is a left R-submodule of $\mathbb{D}_R^{(1)} \oplus \mathbb{D}_R^{(1)}$.

Corollary 4.2.27. — *The map*

(4.2.27.1) $$R \otimes_{\mathbb{F}_p} \mathcal{K}_{\mathbb{F}_p} \longrightarrow \mathcal{K}_R$$

induced by the R-module structure on $\mathcal{K}_R$ is an isomorphism.

Proof. — Immediate from the descriptions (4.2.25.1) and (4.2.25.2) of the kernels. □

4.2.28. — By [**47**, 2.23], for any integer n the map of left $\mathbb{D}_R^{(n)}$-modules

(4.2.28.1) $$\mathbb{D}_R^{(n)} \otimes_{\mathbb{D}_{\mathbb{F}_p}} W(\mathbb{F}_p) \longrightarrow W_n(R)$$

induced by the map $W(\mathbb{F}_p) \to W_n(R)$ is an isomorphism. In particular applying the functor

(4.2.28.2) $$(-) \otimes_{\mathbb{D}_{\mathbb{F}_p}} W(\mathbb{F}_p)$$

to the exact sequence (4.2.26.2) gives a sequence

(4.2.28.3) $$\mathcal{K}_R \otimes_{\mathbb{D}_{\mathbb{F}_p}} W(\mathbb{F}_p) \to R \oplus R \xrightarrow{(V^n,0)} W_{n+1}(R) \longrightarrow W_n(R) \to 0.$$

Lemma 4.2.29. — *The sequence (4.2.28.3) is exact.*

Proof. — The only nontrivial point is that $\mathcal{K}_R \otimes_{\mathbb{D}_{\mathbb{F}_p}} W(\mathbb{F}_p)$ surjects onto the kernel

(4.2.29.1) $$(0, R) = \mathrm{Ker}((V^n, 0) : R \oplus R \longrightarrow W_{n+1}(R)).$$

This follows from noting that the map

(4.2.29.2) $$\mathcal{K}_R \otimes_{\mathbb{D}_{\mathbb{F}_p}} W(\mathbb{F}_p) \longrightarrow R$$

sends $\eta F^{n+1} \otimes 1 \in \mathrm{Ker}(dV^n : \mathbb{D}_R^{(1)} \to \mathbb{D}_R^{(n+1)}) \otimes_{\mathbb{D}_{\mathbb{F}_p}} W(\mathbb{F}_p)$ to η. □

4.2.30. — Let I be a set, and let M denote the free left $\mathbb{D}_{\mathbb{F}_p}$-module generated by I

(4.2.30.1) $$M := \oplus_{i \in I} \mathbb{D}_{\mathbb{F}_p} \cdot e_i.$$

Set

(4.2.30.2) $$M^{(n)} := \oplus_{i \in I} \mathbb{D}_{\mathbb{F}_p}^{(n)} \cdot e_i,$$

and let $\widehat{M}$ denote the projective limit

(4.2.30.3) $$\widehat{M} = \varprojlim_n M^{(n)}.$$

An element of $\widehat{M}$ is given by a formal sum $\sum_{i \in I} a_i \cdot e_i$, such that for every $n \geq 1$ almost all a_i are contained in $V^n \mathbb{D}_{\mathbb{F}_p} + dV^n \mathbb{D}_{\mathbb{F}_p}$.

Define a map

(4.2.30.4) $$\rho : \mathbb{D}_R^{(n)} \otimes_{\mathbb{D}_{\mathbb{F}_p}} \widehat{M} \longrightarrow \oplus_{i \in I} \mathbb{D}_R^{(n)} = \mathbb{D}_R^{(n)} \otimes_{\mathbb{D}_{\mathbb{F}_p}} M$$

by

$$\delta \otimes \Big(\sum_{i \in I} a_i e_i\Big) \longmapsto \sum_i \delta \cdot a_i. \tag{4.2.30.5}$$

Note that this is well-defined, because by [**47**, 2.21] for any $\delta \in \mathbb{D}_R^{(n)}$ there exists a positive integer c such that δ is annihilated by $V^c\mathbb{D}_R + dV^c\mathbb{D}_R$. There is also a map

$$\lambda : \mathbb{D}_R^{(n)} \otimes_{\mathbb{D}_{\mathbb{F}_p}} M \longrightarrow \mathbb{D}_R^{(n)} \otimes_{\mathbb{D}_{\mathbb{F}_p}} \widehat{M} \tag{4.2.30.6}$$

induced by the map $M \to \widehat{M}$.

Lemma 4.2.31

(a) *For every integer $n \geq 1$ the map*

$$\widehat{M}/(V^n\widehat{M} + dV^n\widehat{M}) \longrightarrow M^{(n)} \tag{4.2.31.1}$$

induced by the projection $\widehat{M} \to M^{(n)}$ is an isomorphism.

(b) *The maps ρ and λ are inverse isomorphisms.*

Proof. — For (a), define

$$K_r := \mathrm{Ker}(V^n + dV^n : M^{(r)} \oplus M^{(r)} \to M^{(r+n)}), \tag{4.2.31.2}$$

so we have an exact sequence of projective systems
(4.2.31.3)

$$0 \longrightarrow K_\cdot \longrightarrow M^{(\cdot)} \oplus M^{(\cdot)} \xrightarrow{V^n + dV^n} M^{(\cdot+n)} \longrightarrow M^{(n)} \longrightarrow 0,$$

where $M^{(n)}$ is viewed as a constant projective system.

For every $r \geq 1$, the map $K_{r+1} \to K_r$ is surjective. To see this let $(\alpha, \beta) \in K_r$ be an element, and choose any lifting $(\tilde{\alpha}, \tilde{\beta}) \in M^{(r+1)} \oplus M^{(r+1)}$. Then $V^n\tilde{\alpha} + dV^n\tilde{\beta}$ is in the kernel of $M^{(r+n+1)} \to M^{(r+n)}$ and hence there exist $\gamma, \delta \in M^{(1)}$ such that

$$V^n\tilde{\alpha} + dV^n\tilde{\beta} = V^{n+r}\gamma + dV^{n+r}\delta. \tag{4.2.31.4}$$

The element $(\tilde{\alpha} - V^r\gamma, \tilde{\beta} - V^r\delta)$ is then an element in K_{r+1} mapping to (α, β) in K_r.

It follows that (4.2.31.3) is an exact sequence of projective systems which satisfy the Mittag-Leffler condition, and therefore applying $\varprojlim$ to (4.2.31.3) we obtain an exact sequence
(4.2.31.5)

$$0 \longrightarrow \varprojlim_r K_r \longrightarrow \widehat{M} \oplus \widehat{M} \xrightarrow{V^n + dV^n} \widehat{M} \longrightarrow M^{(n)} \longrightarrow 0.$$

This implies (a).

For (b), note that since $\rho \circ \lambda$ is an isomorphism it suffices to show that λ is surjective. Consider a tensor $\delta \otimes \hat{m} \in \mathbb{D}_R^{(n)} \otimes_{\mathbb{D}_k} \widehat{M}$, and let c be an integer such that

$$\delta \cdot (V^c\mathbb{D}_R + dV^c\mathbb{D}_R) = 0. \tag{4.2.31.6}$$

By (a), we can write $\hat{m} = m + V^c\alpha + dV^c\beta$, where $m \in M$ and $\alpha, \beta \in \widehat{M}$. Then

$$\begin{aligned}\delta \otimes \hat{m} &= \lambda(\delta \otimes m) + \delta \otimes V^c\alpha + \delta \otimes dV^c\beta \\ &= \lambda(\delta \otimes m) + (\delta V^c) \otimes \alpha + (\delta dV^c) \otimes \beta \\ &= \lambda(\delta \otimes m) \quad (\text{by } (4.2.31.6)). \quad \square\end{aligned}$$

4.2.32. — Now returning to the situation of 4.2.22, assume further that S is affine and write $S = \mathrm{Spec}(R)$. By [**47**, 2.25], for any integer n the map of left $\mathbb{D}_R^{(n)}$-modules

$$\mathbb{D}_R^{(n)} \otimes_{\mathbb{D}_{\mathbb{F}_p}} W^{\mathrm{LZ}}\Omega^\bullet_{X_0/\mathbb{F}_p} \longrightarrow W_n^{\mathrm{LZ}}\Omega^\bullet_{X/R} \tag{4.2.32.1}$$

induced by the map $W^{\mathrm{LZ}}\Omega^\bullet_{X_0/\mathbb{F}_p} \to W_n^{\mathrm{LZ}}\Omega^\bullet_{X/R}$ is an isomorphism. Applying

$$(-) \otimes_{\mathbb{D}_{\mathbb{F}_p}} W^{\mathrm{LZ}}\Omega^\bullet_{X_0/\mathbb{F}_p} \tag{4.2.32.2}$$

to the sequence (4.2.26.2) we obtain a sequence
(4.2.32.3)

$$\mathcal{K}_R \otimes_{\mathbb{D}_{\mathbb{F}_p}} W^{\mathrm{LZ}}\Omega^\bullet_{X_0/\mathbb{F}_p} \longrightarrow W_1^{\mathrm{LZ}}\Omega^\bullet_{X/S} \oplus W_1^{\mathrm{LZ}}\Omega^\bullet_{X/S} \xrightarrow{V^n, dV^n} W_{n+1}\Omega^\bullet_{X/S} \longrightarrow W_n\Omega^\bullet_{X/S} \longrightarrow 0.$$

Corollary 4.2.33. — *The sequence (4.2.32.3) is exact.*

Proof. — By [**47**, equation 2.54] the $\mathbb{D}_{\mathbb{F}_p}$-module $W^{\mathrm{LZ}}\Omega^\bullet_{X_0/\mathbb{F}_p}$ is isomorphic to a direct sum

$$W^{\mathrm{LZ}}\Omega^\bullet_{X_0/\mathbb{F}_p} \simeq W(\mathbb{F}_p) \oplus \widehat{M}, \tag{4.2.33.1}$$

where $\widehat{M}$ is as in 4.2.30 for some set I. Using 4.2.31 (b), it follows that (4.2.32.3) decomposes as a direct sum of sequences of the form (4.2.26.2), and one copy of the sequence (4.2.28.3). The result therefore follows from 4.2.29. $\square$

4.2.34. — From this and 4.2.15 in the case when S is perfect it follows that if

$$\Gamma := \mathrm{Ker}\big((V^n, dV^n) : W_1^{\mathrm{LZ}}\Omega^\bullet_{X/S} \oplus W_1^{\mathrm{LZ}}\Omega^\bullet_{X/S} \to W_{n+1}^{\mathrm{LZ}}\Omega^\bullet_{X/S}\big) \tag{4.2.34.1}$$

then there is a commutative diagram

$$\begin{array}{ccc} R \otimes_{\mathbb{F}_p} \mathcal{K}_{\mathbb{F}_p} \otimes_{\mathbb{D}_{\mathbb{F}_p}} W^{\mathrm{LZ}}\Omega^\bullet_{X_0/\mathbb{F}_p} & \xrightarrow{\simeq} & \mathcal{K}_R \otimes_{\mathbb{D}_{\mathbb{F}_p}} W^{\mathrm{LZ}}\Omega^\bullet_{X_0/\mathbb{F}_p} \\ \downarrow u & & \downarrow \epsilon \\ \oplus_q R \otimes_{\mathbb{F}_p} R^q_{n,X_0/\mathbb{F}_p} & & \Gamma \\ \downarrow & & \downarrow \\ R \otimes_{\mathbb{F}_p} (\Omega^\bullet_{X_0/\mathbb{F}_p} \oplus \Omega^\bullet_{X_0/\mathbb{F}_p}) & \xrightarrow{\simeq} & \Omega^\bullet_{X/R} \oplus \Omega^\bullet_{X/R}, \end{array} \tag{4.2.34.2}$$

where ϵ and u are surjections. It follows that the image of $\oplus_q R \otimes_{\mathbb{F}_p} R^q_{n,X_0/\mathbb{F}_p}$ in $\Omega^\bullet_{X/R} \oplus \Omega^\bullet_{X/R}$ is equal to Γ. On the other hand this image is also equal to $\oplus_q R^q_{n,X/S}$ by 4.2.23 so this completes the proof of 4.2.15. $\square$

4.3. The algebra $\mathcal{A}_{n,\mathcal{X}/\mathcal{S}}$ over an algebraic stack

In this section we generalize the results of 4.1 to a theory over algebraic stacks.

4.3.1. — Let $T \to \mathrm{Spec}(\mathbb{Z}_p)$ be a flat morphism of algebraic spaces, and let $\widehat{T}$ denote the p-adic completion of T. Let $\overline{T}$ be an algebraic space over $\mathbb{F}_p$, and let $\overline{T} \hookrightarrow \widehat{T}$ be a closed immersion defined by a divided power ideal $(\mathcal{I}, \gamma) \subset \mathcal{O}_T$. Assume further that there exists a lifting $\sigma : T \to T$ of Frobenius to T, and fix one such lifting σ. For $n \geq 0$ denote by T_n the reduction of T modulo p^{n+1}.

Let $\mathcal{S}/T$ be a flat algebraic stack together with a lifting $F_{\mathcal{S}} : \mathcal{S} \to \mathcal{S}$ of Frobenius to $\mathcal{S}$ compatible with the lifting $\sigma : T \to T$, and let $\mathcal{S}_{\overline{T}}$ be the base change to $\overline{T}$.

For any integer $n > 0$, let $\mathcal{S}^{(n)}$ denote the base change $\mathcal{S} \times_{T,\sigma^n} T$. For any integer $n > 0$ we then have a factorization of $F^n_{\mathcal{S}} : \mathcal{S} \to \mathcal{S}$

$$\mathcal{S} \xrightarrow{F^n_{\mathcal{S}/T}} \mathcal{S}^{(n)} \xrightarrow{\pi_n} \mathcal{S}, \tag{4.3.1.1}$$

where $\pi_n : \mathcal{S}^{(n)} = \mathcal{S} \times_{T,\sigma^n} T \to \mathcal{S}$ is the projection. We therefore obtain a morphism

$$\mathcal{O}_{\mathcal{S}^{(n)}} \longrightarrow RF^n_{\mathcal{S}/T*}\mathcal{O}_{\mathcal{S}}. \tag{4.3.1.2}$$

More generally for any morphism of algebraic stacks $W \to \mathcal{S}$ we obtain by base change a diagram

$$\widetilde{W}^{(n)} \xrightarrow{P_n} W^{(n)} \longrightarrow W, \tag{4.3.1.3}$$

where $\widetilde{W}^{(n)}$ denotes $\mathcal{S} \times_{F^n_{\mathcal{S}},\mathcal{S}} W$ and $W^{(n)}$ denotes $W \times_{T,\sigma^n} T$.

Assumption 4.3.2. — *Assume that for every $n > 0$ and morphism of algebraic stacks $W \to \mathcal{S}$ the map*

$$\mathcal{O}_{W^{(n)}} \longrightarrow RP_{n*}\mathcal{O}_{\widetilde{W}^{(n)}} \tag{4.3.2.1}$$

is an isomorphism.

Remark 4.3.3. — Since the morphism $\mathcal{S}^{(n)} \to \mathcal{S}$ is affine, this assumption implies in particular that for any $i, n > 0$ the sheaf $R^iF^n_{\mathcal{S}*}\mathcal{O}_{\mathcal{S}}$ on $\mathcal{S}_{\text{lis-et}}$ is zero, and the same holds for $\mathcal{S}_0$ and $\mathcal{S}_{\overline{T}}$. In particular the stacks $\mathcal{S}_0$ and $\mathcal{S}_{\overline{T}}$ are Frobenius acyclic (3.2.1).

4.3.4. — Let $f : \mathcal{X} \to \mathcal{S}_{\overline{T}}$ be a smooth representable morphism of algebraic stacks with $\mathcal{X}$ a Deligne-Mumford stack. For every $n \geq 1$ and $q \geq 0$, define

$$\mathcal{A}^q_{n,\mathcal{X}/\mathcal{S}} := R^q u_{\mathcal{X}_{\text{et}}/\mathcal{S}_{n-1}*}\mathcal{O}_{\mathcal{X}_{\text{et}}/\mathcal{S}_{n-1}}. \tag{4.3.4.1}$$

For $n \leq 0$, we define $\mathcal{A}^q_{n,\mathcal{X}_{\text{et}}/\mathcal{S}}$ to be zero. The sheaves $\mathcal{A}^q_{n,\mathcal{X}/\mathcal{S}}$ are sheaves of $\mathcal{O}_{T_{n,\text{et}}}|_{\mathcal{X}_{\text{et}}}$-modules on $\mathcal{X}_{\text{et}}$.

For any integer $n \geq 0$ set

$$\mathcal{X}^{(p^n)} := \mathcal{S}^{(n)} \times_{\mathcal{S}} \mathcal{X}. \tag{4.3.4.2}$$

The Frobenius morphism on $\mathcal{X}$ induces a canonical map $\overline{F}^n_{\mathcal{X}/\mathcal{S}} : \mathcal{X} \to \mathcal{X}^{(p^n)}$.

4.3.5. — If $\mathcal{X} \hookrightarrow \mathcal{Y}$ is a closed immersion over $\mathcal{S}$ with $\mathcal{Y} \to \mathcal{S}$ smooth, then by 2.5.4 there are canonical isomorphisms

$$\mathcal{A}^q_{n,\mathcal{X}/\mathcal{S}} \simeq \mathcal{H}^q(\Omega^\bullet_{D_{n-1}/\mathcal{S}_{n-1}}), \tag{4.3.5.1}$$

where D denotes the divided power envelope of $\mathcal{X}$ in $\mathcal{Y}$.

We can also generalize 4.1.7. Assume $\mathcal{Y}/\mathcal{S}$ is a smooth lifting of $\mathcal{X}$, and note that 3.4.35 yields in the present situation (since we are under the assumptions of 4.3.1) a canonical isomorphism

$$Ru_{\mathcal{X}^{(p)}/\mathcal{S}^{(p)}*}\mathcal{O}_{\mathcal{X}^{(p)}/\mathcal{S}^{(p)}} \longrightarrow \mathbb{L}\eta Ru_{\mathcal{X}/\mathcal{S}*}\mathcal{O}_{\mathcal{X}/\mathcal{S}}. \tag{4.3.5.2}$$

For $m \geq 0$, define a subcomplex $\tilde{E}^{\cdot}_\infty \subset \Omega^{\cdot}_{\mathcal{Y}/\mathcal{S}}$ by

$$\tilde{E}^q_\infty := \{\omega \in p^q\Omega^q_{\mathcal{Y}/\mathcal{S}} |\ d\omega \in p^{q+1}\Omega^{q+1}_{\mathcal{Y}/\mathcal{S}}\}. \tag{4.3.5.3}$$

Then by definition of $\mathbb{L}\eta Ru_{\mathcal{X}/\mathcal{S}*}\mathcal{O}_{\mathcal{X}/\mathcal{S}}$ the isomorphism (4.3.5.2) gives an isomorphism

$$E^\bullet_\infty \simeq Ru_{\mathcal{X}^{(p)}/\mathcal{S}^{(p)}*}\mathcal{O}_{\mathcal{X}^{(p)}/\mathcal{S}^{(p)}}. \tag{4.3.5.4}$$

We then have the following generalization of 4.3.6:

Lemma 4.3.6. — *For $m \geq 0$, define a subcomplex $\tilde{E}^\bullet_m \subset \Omega^{\cdot}_{\mathcal{Y}_{m-1}/\mathcal{S}_{m-1}}$ by*

$$\tilde{E}^q_m := \{\omega \in p^q\Omega^q_{\mathcal{Y}_{m-1}/\mathcal{S}_{m-1}} |\ d\omega \in p^{q+1}\Omega^{q+1}_{\mathcal{Y}_{m-1}/\mathcal{S}_{m-1}}\}. \tag{4.3.6.1}$$

Then if $m > n+q$, the module

$$E^q_n := \tilde{E}^q_m / p^n \tilde{E}^q_m \tag{4.3.6.2}$$

is independent of the choice of m, and by varying q we obtain a complex $E^\bullet_n$ with differential induced by the differential on $\tilde{E}^{\cdot}_m$. Moreover, the isomorphism (4.3.5.4) composed with the projection

$$\tilde{E}^\bullet_\infty \otimes_{\mathcal{O}_T} \mathcal{O}_T/(p^n) \longrightarrow E^\bullet_n \tag{4.3.6.3}$$

is an isomorphism

$$E^\bullet_n \simeq \mathcal{O}_{T_{n-1}} \otimes^{\mathbb{L}}_{\mathcal{O}_T} \mathbb{L}\eta Ru_{X/T*}\mathcal{O}_{X/T}. \tag{4.3.6.4}$$

Proof. — This follows from the same argument proving 4.1.7. □

4.3.7. — Just as in 4.1.2 and 4.1.9, this local description enables us to define operators (4.3.7.1)

$$d : \mathcal{A}^q_{n,\mathcal{X}/\mathcal{S}} \longrightarrow \mathcal{A}^{q+1}_{n,\mathcal{X}/\mathcal{S}},\quad F : \mathcal{A}^q_{n+1,\mathcal{X}/\mathcal{S}} \longrightarrow \mathcal{A}^q_{n,\mathcal{X}/\mathcal{S}},\quad V : \mathcal{A}^q_{n,\mathcal{X}/\mathcal{S}} \longrightarrow \mathcal{A}^q_{n+1,\mathcal{X}/\mathcal{S}}$$

and using 3.4.35

$$\pi_n : \mathcal{A}^q_{n+1,\mathcal{X}/\mathcal{S}} \longrightarrow \mathcal{A}^q_{n,\mathcal{X}^{(p)}/\mathcal{S}}. \tag{4.3.7.2}$$

By the same reasoning used in 4.1.3, all the formulas (4.1.3.1) hold, except possibly $d^2 = 0$ which will become apparent in 4.3.19 below.

4.3.8. — For any $q \geq 0$ let

$$C^{-1} : \Omega^q_{\mathcal{X}^{(p)}_{\mathrm{et}}/\mathcal{S}_{\overline{T}}} \longrightarrow \mathcal{H}^q(\Omega^\bullet_{\mathcal{X}_{\mathrm{et}}/\mathcal{S}_{\overline{T}}}) \tag{4.3.8.1}$$

be the Cartier isomorphism (3.3.21). Define $B_n\Omega^q_{\mathcal{X}_{\mathrm{et}}/\mathcal{S}_{\overline{T}}}$ and $Z_n\Omega^q_{\mathcal{X}_{\mathrm{et}}/\mathcal{S}_{\overline{T}}}$ inductively as follows:

$$B_0\Omega^i_{\mathcal{X}_{\mathrm{et}}/\mathcal{S}_{\overline{T}}} = 0, \quad Z_0\Omega^i_{\mathcal{X}_{\mathrm{et}}/\mathcal{S}_{\overline{T}}} = \Omega^i_{\mathcal{X}_{\mathrm{et}}/\mathcal{S}_{\overline{T}}}, \tag{4.3.8.2}$$

$$B_1\Omega^i_{\mathcal{X}_{\mathrm{et}}/\mathcal{S}_{\overline{T}}} = \mathrm{Im}\big(d : \Omega^{i-1}_{\mathcal{X}_{\mathrm{et}}/\mathcal{S}_{\overline{T}}} \to \Omega^i_{\mathcal{X}_{\mathrm{et}}/\mathcal{S}_{\overline{T}}}\big), \tag{4.3.8.3}$$

$$Z_1\Omega^i_{\mathcal{X}_{\mathrm{et}}/\mathcal{S}_{\overline{T}}} = \mathrm{Ker}\big(d : \Omega^i_{\mathcal{X}_{\mathrm{et}}/\mathcal{S}_{\overline{T}}} \to \Omega^{i+1}_{\mathcal{X}_{\mathrm{et}}/\mathcal{S}_{\overline{T}}}\big), \tag{4.3.8.4}$$

$$\begin{aligned} B_n\Omega^i_{\mathcal{X}^{(p)}_{\mathrm{et}}/\mathcal{S}_{\overline{T}}} &\xrightarrow{C^{-1}_{\mathcal{X}_{\mathrm{et}}/\mathcal{S}_{\overline{T}}}} B_{n+1}\Omega^i_{\mathcal{X}_{\mathrm{et}}/\mathcal{S}_{\overline{T}}}/B_1\Omega^i_{\mathcal{X}_{\mathrm{et}}/\mathcal{S}_{\overline{T}}} \\ Z_n\Omega^i_{\mathcal{X}^{(p)}_{\mathrm{et}}/\mathcal{S}_{\overline{T}}} &\xrightarrow{C^{-1}_{\mathcal{X}_{\mathrm{et}}/\mathcal{S}_{\overline{T}}}} Z_{n+1}\Omega^i_{\mathcal{X}_{\mathrm{et}}/\mathcal{S}_{\overline{T}}}/B_1\Omega^i_{\mathcal{X}_{\mathrm{et}}/\mathcal{S}_{\overline{T}}}\,. \end{aligned} \tag{4.3.8.5}$$

We also define $R^q_{n,\mathcal{X}^{(p)}_{\mathrm{et}}}$ by the sequence (where by convention C^0 is the identity map)
(4.3.8.6)

$$0 \longrightarrow R^q_{n,\mathcal{X}^{(p)}_{\mathrm{et}}} \longrightarrow B_{n+1}\Omega^q_{\mathcal{X}^{(p)}_{\mathrm{et}}/\mathcal{S}_{\overline{T}}} \oplus Z_n\Omega^{q-1}_{\mathcal{X}^{(p)}_{\mathrm{et}}/\mathcal{S}_{\overline{T}}} \xrightarrow{(C^n,dC^n)} B_1\Omega^q_{\mathcal{X}^{(p^{n+1})}_{\mathrm{et}}/\mathcal{S}^{(n+1)}_{\overline{T}}} \longrightarrow 0,$$

and if $\mathcal{I} = p\mathcal{O}_T$ let s_n be the map

$$s_n : \Omega^q_{\mathcal{X}^{(p)}_{\mathrm{et}}/\mathcal{S}_{\overline{T}}} \oplus \Omega^{q-1}_{\mathcal{X}^{(p)}_{\mathrm{et}}/\mathcal{S}_{\overline{T}}} \xrightarrow{C^{-1}} \mathcal{A}^q_{1,\mathcal{X}/\mathcal{S}} \oplus \mathcal{A}^{q-1}_{1,\mathcal{X}/\mathcal{S}} \xrightarrow{(V^n,dV^n)} \mathcal{A}^q_{n+1,\mathcal{X}/\mathcal{S}}. \tag{4.3.8.7}$$

4.3.9. — Let $S \to \mathcal{S}$ be a smooth surjection with S an algebraic space, and let $\widehat{S}$ denote the p-adic completion of S. Assume given a lifting of Frobenius $F_S : \widehat{S} \to \widehat{S}$ such that for every integer $n > 0$ the diagram

$$\begin{array}{ccc} S_n & \xrightarrow{F_S} & S_n \\ \downarrow & & \downarrow \\ \mathcal{S}_n & \xrightarrow{F_{\mathcal{S}}} & \mathcal{S}_n \end{array} \tag{4.3.9.1}$$

commutes. Denote by $S_\bullet$ the 0-coskeleton of $S \to \mathcal{S}$ and let $\widehat{S}_\bullet$ denote the simplicial formal scheme obtained by taking the p-adic completion of each S_n. The lifting of Frobenius $F_S : \widehat{S} \to \widehat{S}$ defines a morphism of simplicial formal schemes

$$F_{S_\bullet} : \widehat{S}_\bullet \longrightarrow \widehat{S}_\bullet \tag{4.3.9.2}$$

Denote by $\overline{S}_\bullet$ the simplicial algebraic space $S_\bullet \times_T \overline{T}$, and let $X_\bullet$ denote $\mathcal{X} \times_{\mathcal{S}} S_\bullet$.

Remark 4.3.10. — The pair (S, F_S) can be constructed as follows. Choose any smooth surjection $c : S \to \mathcal{S}$ with S an affine scheme. The lifting of Frobenius $F_S : \widehat{S} \to \widehat{S}$ is then determined by the liftings of Frobenius $F_{S_n} : S_n \to S_n$ for each n. These liftings can be constructed inductively. If $F_{S_{n-1}} : S_{n-1} \to S_{n-1}$ has been constructed, consider the diagram

$$\begin{array}{ccccc} S_{n-1} & \xrightarrow{F_{S_{n-1}}} & S_{n-1} & \xhookrightarrow{c} & S_n \\ {\scriptstyle c}\downarrow & & & \nearrow & \downarrow \\ S_n & \longrightarrow & \mathcal{S} & \xrightarrow{F_{\mathcal{S}}} & \mathcal{S}. \end{array} \tag{4.3.10.1}$$

By [**66**, 1.5], the obstruction to filling in the diagram with the dotted arrow is a class in

$$H^1(S_0, F_{S_0}^* c^* \Omega^1_{S_n/\mathcal{S}_n}) \tag{4.3.10.2}$$

which is zero since S_0 is affine. Thus the pair (S, F_S) exists.

4.3.11. — For any integer $r \geq 0$, there is a canonical map $X_\bullet^{(p^r)} := X_\bullet \times_{\overline{S}_\bullet, F^r_{\overline{S}_\bullet}} \overline{S}_\bullet \to \mathcal{X} \times_{\mathcal{S}_{\overline{T}}, F^r_{\mathcal{S}_{\overline{T}}}} \mathcal{S}_{\overline{T}}$ which identifies $X_\bullet \times_{\overline{S}_\bullet, F^r_{\overline{S}_\bullet}} \overline{S}_\bullet$ with the 0-coskeleton of the smooth surjection $(\mathcal{X} \times_{\mathcal{S}_{\overline{T}}} \overline{S}) \times_{\overline{S}, F^r_{\overline{S}}} \overline{S} \to \mathcal{X} \times_{\mathcal{S}_{\overline{T}}, F^r_{\mathcal{S}_{\overline{T}}}} \mathcal{S}_{\overline{T}}$. In particular there is a canonical morphism

$$G : X_\bullet^{(p^r)} \longrightarrow \mathcal{X}^{(p^r)}. \tag{4.3.11.1}$$

Note that $X_\bullet^{(p^r)}$ is *not* equal to the 0-coskeleton of some covering $U \to \mathcal{X}^{(p^r)}$. The simplicial space $X_\bullet^{(p^r)}$ is equal to the fiber product of the diagram

$$\begin{array}{ccc} \mathcal{X} \times_{\mathcal{S}, F^n_{\mathcal{S}}} \mathcal{S} & & \\ {\scriptstyle \mathrm{pr}_2}\downarrow & & \\ \mathcal{S} & \longleftarrow & S_\bullet, \end{array} \tag{4.3.11.2}$$

so $X_\bullet^{(p^r)}$ is a hypercover of the stack $\mathcal{X} \times_{\mathcal{S}, F^n_{\mathcal{S}}} \mathcal{S}$, and G is obtained by composing the projection $X_\bullet^{(p^r)} \to \mathcal{X} \times_{\mathcal{S}, F_{\mathcal{S}}} \mathcal{S}$ with the projection

$$\mathcal{X} \times_{\mathcal{S}, F^n_{\mathcal{S}}} \mathcal{S} \longrightarrow \mathcal{X}^{(p^r)} = \mathcal{X} \times_{T, \sigma^n} T. \tag{4.3.11.3}$$

In what follows we view $X_\bullet^{(p^r)}$ as a simplicial algebraic space over $\overline{S}_\bullet$ via the projection $\mathrm{pr}_2 : X_\bullet^{(p^r)} \to \overline{S}_\bullet$. Define $B_n\Omega^q_{X_\bullet^{(p^r)}/\overline{S}_\bullet}$ (resp. $Z_n\Omega^q_{X_\bullet^{(p^r)}/\overline{S}_\bullet}$) to be the sheaf on $X_{\bullet,\mathrm{et}}^{(p^r)}$ whose restriction to each $X_i^{(p^r)}$ is equal to $B_n\Omega^q_{X_i^{(p^r)}/\overline{S}_i}$ (resp. $Z_n\Omega^q_{X_i^{(p^r)}/\overline{S}_i}$) defined in 4.1.10. The map G induces by adjunction canonical maps

$$B_n\Omega^q_{\mathcal{X}^{(p^r)}/\mathcal{S}^{(r)}_{\overline{T}}} \longrightarrow RG_*(B_n\Omega^q_{X_\bullet^{(p^r)}/\overline{S}_\bullet}) \tag{4.3.11.4}$$

$$Z_n\Omega^q_{\mathcal{X}^{(p^r)}/\mathcal{S}^{(r)}_{\overline{T}}} \longrightarrow RG_*(Z_n\Omega^q_{X_\bullet^{(p^r)}/\overline{S}_\bullet}) \tag{4.3.11.5}$$

Proposition 4.3.12. — *The maps (4.3.11.4) and (4.3.11.5) are isomorphisms.*

Proof. — Note first that the assertion is étale local on $\mathcal{X}$, and hence we may assume that $\mathcal{X}$ is a scheme. Since the relative Frobenius morphism $F_{\mathcal{X}/B} : \mathcal{X} \to \mathcal{X}^{(1)}$ is radicial, proper and surjective, this implies by [**5**, VIII.1.1] that $F_{\mathcal{X}/B}$ induces an isomorphism of topoi $\mathcal{X}_{\text{et}} \to \mathcal{X}^{(1)}_{\text{et}}$. Proceeding inductively it follows from this that the relative Frobenius morphism induces an isomorphism of topoi $\mathcal{X}_{\text{et}} \simeq \mathcal{X}^{(r)}_{\text{et}}$ for all $r \geq 0$. In the following calculations we can therefore without loss of generality view all sheaves as sheaves on $\mathcal{X}_{\text{et}}$ (this eases the notation).

Lemma 4.3.13. — *The Cartier isomorphism (3.3.21) induces for every $r \geq 0$ an isomorphism*

(4.3.13.1)
$$C^{-1} : \Omega^q_{\mathcal{X}^{(p^{r+1})}/\mathcal{S}^{(r+1)}_{\overline{T}}} \simeq Z_1\Omega^q_{\mathcal{X}^{(p^r)}/\mathcal{S}^{(r)}_{\overline{T}}}/B_1\Omega^q_{\mathcal{X}^{(p^r)}/\mathcal{S}^{(r)}_{\overline{T}}}.$$

Proof. — Note first that by our assumptions on $\mathcal{S}$, the relative Frobenius morphism $F_{\mathcal{S}^{(r)}/B} : \mathcal{S}^{(r)} \to \mathcal{S}^{(r)} \times_{B,\sigma} B \simeq \mathcal{S}^{(r+1)}$ is obtained from $F_{\mathcal{S}/B} : \mathcal{S} \to \mathcal{S} \times_{B,\sigma} B$ by the base change $\sigma^r : B \to B$. In particular, the natural map $\mathcal{O}_{\mathcal{S}^{(r)}} \to RF_{\mathcal{S}^{(r)}/B*}(\mathcal{O}_{\mathcal{S}^{(r)}})$ is an isomorphism, and this remains true after arbitrary base change $W \to B$, so 3.3.21 applies. Moreover, if in *loc. cit.* we take $\mathcal{X} \to \mathcal{S}$ to be the present $\mathcal{X}^{(p^r)} \to \mathcal{S}^{(r)}$ then the stack denoted $\overline{\mathcal{X}}'$ in *loc. cit.* is equal to $\mathcal{X}^{(p^{r+1})}$. The lemma therefore follows from *loc. cit.*. □

The proof of 4.3.12 now proceeds by induction on n.

For the case $n = 0$ note that both sides of (4.3.11.4) are 0, and (4.3.11.5) is identified with the map

(4.3.13.2)
$$\Omega^q_{\mathcal{X}^{(p^r)}_{\text{et}}/\mathcal{S}^{(r)}_{\overline{T}}} \longrightarrow RG_*(G^*\Omega^q_{\mathcal{X}^{(p^r)}_{\text{et}}/\mathcal{S}^{(r)}_{\overline{T}}}) \simeq RG_*(\mathcal{O}_{X^{(p^r)}_\bullet}) \otimes \Omega^q_{\mathcal{X}^{(p^r)}_{\text{et}}/\mathcal{S}^{(r)}_{\overline{T}}},$$

where the second isomorphism follows from the projection formula and the fact that $\Omega^q_{\mathcal{X}^{(p^r)}_{\text{et}}/\mathcal{S}^{(r)}_{\overline{T}}}$ is locally free since $\mathcal{X} \to \mathcal{S}_{\overline{T}}$ is smooth. Let

(4.3.13.3)
$$P_r : \widetilde{\mathcal{X}}^{(p^r)} := \mathcal{X} \times_{\mathcal{S},F^r_{\mathcal{S}}} \mathcal{S} \to \mathcal{X} \times_{T,\sigma^r} T = \mathcal{X}^{(p^r)}$$

denote the projection so that G factors as

(4.3.13.4)
$$X^{(p^r)}_\bullet \xrightarrow{\ \widetilde{G}\ } \widetilde{\mathcal{X}}^{(p^r)} \xrightarrow{\ P_r\ } \mathcal{X}^{(p^r)}.$$

Since $\widetilde{G}$ is a hypercover, we find that

(4.3.13.5)
$$RG_*\mathcal{O}_{X^{(p^r)}_\bullet} \simeq RP_{r*}R\widetilde{G}_*\mathcal{O}_{X^{(p^r)}_\bullet} \simeq RP_{r*}\mathcal{O}_{\widetilde{\mathcal{X}}^{(p^r)}}.$$

Since $\mathcal{O}_{\mathcal{X}^{(p^r)}} \to RP_{r*}(\mathcal{O}_{\widetilde{\mathcal{X}}^{(p^r)}})$ is an isomorphism by our assumptions on $\mathcal{S}$ (4.3.2), we conclude that (4.3.13.2) is an isomorphism proving the case $n = 0$.

The case $n = 1$ is proven by induction on q by showing that if the result holds for the $B_1\Omega^q$'s and $Z_1\Omega^{q-1}$'s, then the result also holds for the $B_1\Omega^{q+1}$'s and $Z_1\Omega^q$'s.

Once this inductive step is shown the proof of the case $n = 1$ follows by noting that the statement for the $B_1\Omega^0$'s is trivial.

So assume the result holds for the $B_1\Omega^q$'s and $Z_1\Omega^{q-1}$'s. To get the result for the $Z_1\Omega^q$'s, observe that the short exact sequences

(4.3.13.6)
$$0 \longrightarrow B_1\Omega^q_{\mathcal{X}^{(p^r)}/\mathcal{S}^{(r)}_{\overline{T}}} \longrightarrow Z_1\Omega^q_{\mathcal{X}^{(p^r)}/\mathcal{S}^{(r)}_{\overline{T}}} \xrightarrow{C} \Omega^q_{\mathcal{X}^{(p^{r+1})}/\mathcal{S}^{(r+1)}_{\overline{T}}} \longrightarrow 0$$

(4.3.13.7)
$$0 \longrightarrow B_1\Omega^q_{X_\bullet^{(p^r)}/\overline{S}_\bullet} \longrightarrow Z_1\Omega^q_{X_\bullet^{(p^r)}/\overline{S}_\bullet} \xrightarrow{C} \Omega^q_{X_\bullet^{(p^{r+1})}/\overline{S}_\bullet} \longrightarrow 0.$$

give rise to a morphism of distinguished triangles
(4.3.13.8)
$$\begin{array}{ccccccc} B_1\Omega^q_{\mathcal{X}^{(p^r)}/\mathcal{S}^{(r)}_{\overline{T}}} & \longrightarrow & Z_1\Omega^q_{\mathcal{X}^{(p^r)}/\mathcal{S}^{(r)}_{\overline{T}}} & \xrightarrow{C} & \Omega^q_{\mathcal{X}^{(p^{r+1})}/\mathcal{S}^{(r+1)}_{\overline{T}}} & \xrightarrow{+1} \\ \alpha\downarrow & & \downarrow\beta & & \downarrow\gamma & \\ RG_*B_1\Omega^q_{X_\bullet^{(p^r)}/\overline{S}_\bullet} & \longrightarrow & RG_*Z_1\Omega^q_{X_\bullet^{(p^r)}/\overline{S}_\bullet} & \xrightarrow{C} & RG_*\Omega^q_{X_\bullet^{(p^{r+1})}/\overline{S}_\bullet} & \xrightarrow{+1} & . \end{array}$$

The map α is an isomorphism by the induction hypothesis, and the map γ is an isomorphism by the same argument used to prove the case $n = 0$. It follows that the map β is also an isomorphism.

Similarly, if the result holds for the $Z_1\Omega^q$'s and $B_1\Omega^q$'s then the result holds for the $B_1\Omega^{q+1}$'s by considering the morphism of distinguished triangles obtained from the short exact sequences

(4.3.13.9)
$$0 \longrightarrow Z_1\Omega^q_{\mathcal{X}^{(p^r)}/\mathcal{S}^{(r)}_{\overline{T}}} \longrightarrow \Omega^q_{\mathcal{X}^{(p^r)}/\mathcal{S}^{(r)}_{\overline{T}}} \longrightarrow B_1\Omega^{q+1}_{\mathcal{X}^{(p^r)}/\mathcal{S}^{(r)}_{\overline{T}}} \longrightarrow 0$$

(4.3.13.10)
$$0 \longrightarrow Z_1\Omega^q_{X_\bullet^{(p^r)}/\overline{S}_\bullet} \longrightarrow \Omega^q_{X_\bullet^{(p^r)}/\overline{S}_\bullet} \longrightarrow B_1\Omega^{q+1}_{X_\bullet^{(p^r)}/\overline{S}_\bullet} \longrightarrow 0$$

This completes the proof for $n = 1$.

The result for general n follows from the case $n = 1$ by induction and consideration of the morphisms of distinguished triangles obtained from the short exact sequences

(4.3.13.11)
$$0 \longrightarrow B_1\Omega^q_{\mathcal{X}^{(p^r)}/\mathcal{S}^{(r)}_{\overline{T}}} \longrightarrow B_{n+1}\Omega^q_{\mathcal{X}^{(p^r)}/\mathcal{S}^{(r)}_{\overline{T}}} \xrightarrow{C} B_n\Omega^q_{\mathcal{X}^{(p^{r+1})}/\mathcal{S}^{(r+1)}_{\overline{T}}} \longrightarrow 0,$$

(4.3.13.12)
$$0 \longrightarrow B_1\Omega^q_{\mathcal{X}^{(p^r)}/\mathcal{S}^{(r)}_{\overline{T}}} \longrightarrow Z_{n+1}\Omega^q_{\mathcal{X}^{(p^r)}/\mathcal{S}^{(r)}_{\overline{T}}} \xrightarrow{C} Z_n\Omega^q_{\mathcal{X}^{(p^{r+1})}/\mathcal{S}^{(r+1)}_{\overline{T}}} \longrightarrow 0,$$

(4.3.13.13)
$$0 \longrightarrow B_1\Omega^q_{X_\bullet^{(p^r)}/\overline{S}_\bullet} \longrightarrow B_{n+1}\Omega^q_{X_\bullet^{(p^r)}/\overline{S}_\bullet} \xrightarrow{C} B_n\Omega^q_{X_\bullet^{(p^{r+1})}/\overline{S}_\bullet} \longrightarrow 0,$$

(4.3.13.14)
$$0 \longrightarrow B_1\Omega^q_{X_\bullet^{(p^r)}/\overline{S}_\bullet} \longrightarrow Z_{n+1}\Omega^q_{X_\bullet^{(p^r)}/\overline{S}_\bullet} \xrightarrow{C} Z_n\Omega^q_{X_\bullet^{(p^{r+1})}/\overline{S}_\bullet} \longrightarrow 0. \quad \square$$

4.3.14. — By functoriality of crystalline cohomology there is also a natural map for every $r \geq 0$

$$\mathcal{A}^q_{n,\mathcal{X}^{(p^r)}/\mathcal{S}^{(r)}} \longrightarrow RG_*(\mathcal{A}^q_{n,X^{(p^r)}_\bullet/\widehat{S}_\bullet}). \tag{4.3.14.1}$$

Theorem 4.3.15. — *The morphism (4.3.14.1) is an isomorphism.*

4.3.16. — It suffices to prove 4.3.15 in the local situation when $\mathcal{X}$ is a scheme and there exists a smooth lifting $\mathcal{Y}/\mathcal{S}$ of $\mathcal{X}$ with a lift of Frobenius $F_\mathcal{Y} : \mathcal{Y} \to \mathcal{Y}$ compatible with $F_\mathcal{S}$. Replacing $\mathcal{X}$ by $\mathcal{Y}\otimes_\mathbb{Z}\mathbb{Z}/(p)$ and $\overline{T}$ by $T\otimes_\mathbb{Z}\mathbb{Z}/(p)$ we may furthermore assume that the ideal of $\overline{T}$ in T is equal to $p\mathcal{O}_T$ so $\overline{T} = T_0$.

In this case it is convenient to prove 4.3.15 in conjunction with several other statements which we summarize in the following proposition. Let $\mathcal{Y}^{(p^r)}$ denote the fiber product $\mathcal{Y} \times_{T,F^r_T} T$, and let $Y^{(p^r)}_\bullet$ denote the base change to $S_\bullet$. Define $Z^q_{\mathcal{Y}^{(p^r)}_{n-1}}$ (resp. $Z^q_{Y^{(p^r)}_{n-1,\bullet}}$) to be the kernel of

$$d : \Omega^q_{\mathcal{Y}^{(p^r)}_{n-1}/\mathcal{S}^{(r)}_{n-1}} \longrightarrow \Omega^q_{\mathcal{Y}^{(p^r)}_{n-1}/\mathcal{S}^{(r)}_{n-1}} \quad (\text{resp. } d : \Omega^q_{Y^{(p^r)}_{n-1,\bullet}/S_{n-1,\bullet}} \longrightarrow \Omega^q_{Y^{(p^r)}_{n-1,\bullet}/S_{n-1,\bullet}}), \tag{4.3.16.1}$$

and let $B\Omega^{q+1}_{\mathcal{Y}^{(p^r)}_{n-1}}$ (resp. $B^q_{Y^{(p^r)}_{n-1,\bullet}}$) be the image. Note that by definition there is an isomorphism

$$\mathcal{A}^q_{n,Y^{(p^r)}_\bullet/\widehat{S}_\bullet} \simeq Z^q_{Y^{(p^r)}_{n-1,\bullet}}/B^q_{Y^{(p^r)}_{n-1,\bullet}}. \tag{4.3.16.2}$$

Proposition 4.3.17. — *For every $r \geq 0$, the natural maps*

$$\mathcal{A}^q_{n,\mathcal{X}^{(p^r)}} \longrightarrow RG_*(\mathcal{A}^q_{n,X^{(p^r)}_\bullet/\widehat{S}_\bullet}), \tag{4.3.17.1}$$

$$Z^q_{\mathcal{Y}^{(p^r)}_{n-1}} \longrightarrow RG_*(Z^q_{Y^{(p^r)}_{n-1,\bullet}}), \tag{4.3.17.2}$$

$$B^q_{\mathcal{Y}^{(p^r)}_{n-1}} \longrightarrow RG_*(B^q_{Y^{(p^r)}_{n-1,\bullet}}) \tag{4.3.17.3}$$

are isomorphisms. Moreover, the sequence

$$0 \longrightarrow R^q_{n-1,\mathcal{X}^{(p)}} \longrightarrow \Omega^q_{\mathcal{X}^{(p)}/\mathcal{S}^{(1)}_0} \oplus \Omega^{q-1}_{\mathcal{X}^{(p)}/\mathcal{S}^{(1)}_0} \xrightarrow{s_{n-1}} \mathcal{A}^q_{n,\mathcal{X}/\mathcal{S}_0} \xrightarrow{\pi_n} \mathcal{A}^q_{n-1,\mathcal{X}^{(p)}/\mathcal{S}^{(1)}_0} \longrightarrow 0 \tag{4.3.17.4}$$

is exact, where $R^q_{n-1,\mathcal{X}^{(p)}}$ is defined in (4.3.8.6) and s_{n-1} is defined in (4.3.8.7).

Proof. — The proof is by induction on n.

The equalities (4.3.17.1)–(4.3.17.3) in the case of $n = 1$ follow from 4.3.12. The exactness of (4.3.17.4) in the case of $n = 1$ follows from the Cartier isomorphism (3.3.21).

For the induction step we assume that the result is true for n and prove it for $n+1$. Define

$$\Lambda^q_{n,\mathcal{X}^{(p^r)}} := \mathrm{Ker}(B^q_{\mathcal{Y}^{(p^r)}_n} \to B^q_{\mathcal{Y}^{(p^r)}_{n-1}}) \tag{4.3.17.5}$$

$$\Lambda^q_{n,X^{(p^r)}_\bullet} := \mathrm{Ker}(B^q_{Y^{(p^r)}_{n,\bullet}} \to B^q_{Y^{(p^r)}_{n-1,\bullet}}). \tag{4.3.17.6}$$

To prove that (4.3.17.3) is a quasi-isomorphism for $B^q_{\mathcal{Y}_n^{(p^r)}}$, it suffices by induction to show that the map

$$\Lambda^q_{n,\mathcal{X}^{(p^r)}} \longrightarrow RG_*\Lambda^q_{n,X_\bullet^{(p^r)}}. \tag{4.3.17.7}$$

is a quasi-isomorphism for all r.

Consider the exact sequence

$$0 \longrightarrow \Omega^q_{\mathcal{X}^{(p^r)}/\mathcal{S}_0^{(r)}} \xrightarrow{\times p^n} \Omega^q_{\mathcal{Y}^{(p^r)}/\mathcal{S}_n^{(r)}} \longrightarrow \Omega^q_{\mathcal{Y}^{(p^r)}/\mathcal{S}_n^{(r)}} \longrightarrow 0. \tag{4.3.17.8}$$

If $\alpha \in \Lambda^q_{n,\mathcal{X}^{(p)}}$ is a local section, we can write $\alpha = p^n\omega$ for a unique section $\omega \in \Omega^q_{\mathcal{X}^{(p^r)}/\mathcal{S}^{(r)}}$. Moreover, since α is a boundary we have

$$0 = d\alpha = p^n d\omega. \tag{4.3.17.9}$$

Therefore ω is a closed form. Note, however, that ω itself need not be a boundary, but there is a short exact sequence

(4.3.17.10)

$$0 \longrightarrow B^q_{\mathcal{X}^{(p^r)}} \xrightarrow{\times p^n} \Lambda^q_{n,\mathcal{X}^{(p^r)}} \longrightarrow \{\omega \in Z^q_{\mathcal{X}^{(p^r)}} | p^n\omega \in B^q_{\mathcal{Y}_n^{(p^r)}}\}/B^q_{\mathcal{X}^{(p^r)}} \longrightarrow 0.$$

Similarly there is a short exact sequence

(4.3.17.11)

$$0 \longrightarrow B^q_{X_\bullet^{(p^r)}} \xrightarrow{\times p^n} \Lambda^q_{n,X_\bullet^{(p^r)}} \longrightarrow \{\omega \in Z^q_{X_\bullet^{(p^r)}} | p^n\omega \in B^q_{Y_{n,\bullet}^{(p^r)}}\}/B^q_{\mathcal{X}_\bullet^{(p^r)}} \longrightarrow 0.$$

The group

$$\{\omega \in Z^q_{\mathcal{X}^{(p^r)}} | p^n\omega \in B^q_{\mathcal{Y}_n^{(p^r)}}\}/B^q_{\mathcal{X}^{(p^r)}} \tag{4.3.17.12}$$

can also be described as follows. By definition of V^n the map

$$V^n : \mathcal{A}^q_{1,\mathcal{X}^{(p^r)}} \simeq \mathcal{H}^q(\Omega^\bullet_{\mathcal{X}^{(p^r)}/\mathcal{S}_0^{(r)}}) \to \mathcal{H}^q(\Omega^\bullet_{\mathcal{Y}^{(p^r)}/\mathcal{S}_n^{(r)}}) \simeq \mathcal{A}^q_{n+1,\mathcal{X}^{(p^r)}/\mathcal{S}^{(r)}} \tag{4.3.17.13}$$

sends the class of a closed form $\omega \in Z^q_{\mathcal{X}^{(p^r)}}$ to the class of $p^n\omega \in Z^q_{\mathcal{Y}_n^{(p^r)}/\mathcal{S}_n^{(r)}}$. Therefore the kernel of V^n is isomorphic to (4.3.17.12). Via the Cartier isomorphism

$$C^{-1} : \Omega^q_{\mathcal{X}^{(p^{r+1})}/\mathcal{S}_0^{(r+1)}} \to \mathcal{H}^q(\Omega^\bullet_{\mathcal{X}^{(p^r)}/\mathcal{S}_0^{(r)}}) \tag{4.3.17.14}$$

the sheaf (4.3.17.12) is identified with the kernel of the map

$$s_n|_{\Omega^q_{\mathcal{X}^{(p^{r+1})}/\mathcal{S}_0^{(r+1)}}} : \Omega^q_{\mathcal{X}^{(p^{r+1})}/\mathcal{S}_0^{(r+1)}} \longrightarrow \mathcal{A}^q_{n+1,\mathcal{X}^{(p^r)}/\mathcal{S}^{(r)}}. \tag{4.3.17.15}$$

Similarly, the sheaf

$$\{\omega \in Z^q_{X_\bullet^{(p^r)}} | p^n\omega \in B^q_{Y_{n,\bullet}^{(p^r)}}\}/B^q_{\mathcal{X}_\bullet^{(p^r)}} \tag{4.3.17.16}$$

is canonically isomorphic to the kernel of the map

$$s_n|_{\Omega^q_{X_\bullet^{(p^{r+1})}/S_{0,\bullet}}} : \Omega^q_{X_\bullet^{(p^{r+1})}/S_{0,\bullet}^{(r+1)}} \longrightarrow \mathcal{A}^q_{n+1,X_\bullet^{(p^r)}/\widehat{S}_\bullet^{(r)}}. \tag{4.3.17.17}$$

Also, by 4.1.12 the kernel of $s_n|_{\Omega^q_{X_\bullet^{(p^{r+1})}/S_{0,\bullet}}}$ is $B_n\Omega^q_{X_\bullet^{(p^{r+1})}/S_{0,\bullet}}$.

Lemma 4.3.18. — *The kernel* $\mathrm{Ker}(s_n)|_{\Omega^q_{\mathcal{X}^{(p^{r+1})}/\mathcal{S}_0^{(r+1)}}}$ *is equal to* $B_n\Omega^q_{\mathcal{X}^{(p^{r+1})}/\mathcal{S}_0^{(r+1)}}$.

Proof. — If $\omega \in \mathrm{Ker}(s_n)$ is an element in the kernel, then $V^{n-1}C^{-1}(\omega)$ defines a class in $\mathcal{A}^q_{n,X_\bullet^{(p^r)}/\widehat{S}_\bullet}$ which is locally in the topos $X_{\bullet,\mathrm{et}}$ in the image of $B_n\Omega^q_{X_\bullet^{(p^{r+1})}/S_{0,\bullet}}$ (recall that this simply means that for every $\delta \in \mathbb{N}$ the restriction of the class to $\mathcal{A}^q_{n,X_\delta/\widehat{S}_\delta}$ is locally on X_δ in the image of $B_n\Omega^q_{X_\delta^{(p^{r+1})}/S_{0,\delta}}$). By 4.1.12 the set of sections of $B_n\Omega^q_{X_\bullet^{(p^{r+1})}/S_{0,\bullet}}$ whose image in $\mathcal{A}^q_{n,X_\bullet^{(p^{r+1})}/\widehat{S}_\bullet}$ is equal to the class of $V^{n-1}C^{-1}(\omega)$ is a torsor under $B_{n-1}\Omega^q_{X_\bullet^{(p^{r+1})}/S_{0,\bullet}}$. By 4.3.12 we have

$$R^1G_*(B_{n-1}\Omega^q_{X_\bullet^{(p^{r+1})}/S_{0,\bullet}}) = 0 \tag{4.3.18.1}$$

and

$$B_n\Omega^q_{\mathcal{X}^{(p^{r+1})}/\mathcal{S}_0^{(r)}} \longrightarrow R^0G_*(B_n\Omega^q_{X_\bullet^{(p^{r+1})}/S_{0,\bullet}}) \tag{4.3.18.2}$$

is an isomorphism. It follows that étale locally on $\mathcal{X}$ there exists a section $\omega' \in B_n\Omega^q_{\mathcal{X}^{(p^{r+1})}/\mathcal{S}_0^{(r+1)}}$ such that

$$V^{n-1}C^{-1}(\omega') = V^{n-1}C^{-1}(\omega). \tag{4.3.18.3}$$

The difference $\omega' - \omega$ is then in the kernel of $s_{n-1}|_{\Omega^q_{\mathcal{X}^{(p^{r+1})}/\mathcal{S}_0^{(r+1)}}}$ which by induction is $B_{n-1}\Omega^q_{\mathcal{X}^{(p^{r+1})}/\mathcal{S}_0^{(r+1)}}$. It follows that $\mathrm{Ker}(s_n)|_{\Omega^q_{\mathcal{X}^{(p^{r+1})}/\mathcal{S}_0^{(r+1)}}} \subset B_n\Omega^q_{\mathcal{X}^{(p^{r+1})}/\mathcal{S}_0^{(r+1)}}$.

A similar argument shows that if $\omega \in B_n\Omega^q_{\mathcal{X}^{(p^{r+1})}/\mathcal{S}_0^{(r+1)}}$, then there exists an element $a \in \Omega^{q-1}_{\mathcal{X}^{(p^{r+1})}/\mathcal{S}_0^{(r+1)}}$ such that

$$V^{n-1}(C^{-1}(\omega)) = dV^{n-1}(C^{-1}(a)). \tag{4.3.18.4}$$

But then

$$V^n(C^{-1}(\omega)) = VdV^{n-1}(C^{-1}(a)) = dpV^n(C^{-1}(a)) = 0, \tag{4.3.18.5}$$

and so $\mathrm{Ker}(s_n)|_{\Omega^q_{\mathcal{X}^{(p^{r+1})}/\mathcal{S}_0^{(r+1)}}} = B_n\Omega^q_{\mathcal{X}^{(p^{r+1})}/\mathcal{S}_0^{(r+1)}}$. □

The exact sequences

$$(4.3.18.6)\quad 0 \longrightarrow B^q_{\mathcal{X}^{(p^r)}} \xrightarrow{\times p^n} \Lambda^q_{n,\mathcal{X}^{(p^r)}} \longrightarrow B_n\Omega^q_{\mathcal{X}^{(p^r+1)}/\mathcal{S}_0^{(r+1)}} \longrightarrow 0$$

and
(4.3.18.7)

$$0 \longrightarrow B^q\Omega_{X_\bullet^{(p^r)}/S_{0,\bullet}} \xrightarrow{\times p^n} \Lambda^q_{n,X_\bullet^{(p^r)}} \longrightarrow B_n\Omega^q_{X_\bullet^{(p^r+1)}/S_{0,\bullet}} \longrightarrow 0$$

induces a morphism of distinguished triangles
(4.3.18.8)

$$\begin{array}{ccccccc} B^q_{\mathcal{X}^{(p^r)}} & \xrightarrow{\times p^n} & \Lambda^q_{n,\mathcal{X}^{(p^r)}} & \longrightarrow & B_n\Omega^q_{\mathcal{X}^{(p^r+1)}/\mathcal{S}_0^{(r+1)}} & \xrightarrow{+1} & \\ \alpha\downarrow & & \downarrow\beta & & \downarrow\gamma & & \\ RG_*B^q\Omega_{X_\bullet^{(p^r)}/S_{0,\bullet}} & \xrightarrow{\times p^n} & RG_*\Lambda^q_{n,X_\bullet^{(p^r)}} & \longrightarrow & RG_*B_n\Omega^q_{X_\bullet^{(p^r+1)}/S_{0,\bullet}} & \xrightarrow{+1} & . \end{array}$$

Since the maps α and γ are isomorphisms by 4.3.12 it follows that the map β is also an isomorphism.

From this, induction, and a similar argument using the exact sequences

$$(4.3.18.9)\qquad 0 \longrightarrow \Lambda^q_{n,\mathcal{X}^{(p^r)}} \longrightarrow B^q_{\mathcal{Y}_n^{(p^r)}} \longrightarrow B^q_{\mathcal{Y}_{n-1}^{(p^r)}} \longrightarrow 0$$

$$(4.3.18.10)\qquad 0 \longrightarrow \Lambda^q_{n,X_\bullet^{(p^r)}} \longrightarrow B^q_{Y_{n,\bullet}^{(p^r)}} \longrightarrow B^q_{Y_{n-1,\bullet}^{(p^r)}} \longrightarrow 0$$

it follows that the map

$$(4.3.18.11)\qquad B^q_{\mathcal{Y}_n^{(p^r)}} \longrightarrow RG_*(B^q_{Y_{n,\bullet}^{(p^r)}})$$

in (4.3.17.3) is an isomorphism.

There is also a natural exact sequence
(4.3.18.12)

$$0 \longrightarrow Z^q_{\mathcal{X}^{(p^r)}} \xrightarrow{p^n} Z^q_{\mathcal{Y}_n^{(p^r)}} \longrightarrow Z^q_{\mathcal{Y}_{n-1}^{(p^r)}} \xrightarrow{t} \Lambda^{q+1}_{n,\mathcal{X}^{(p^r)}}/B^{q+1}_{\mathcal{X}^{(p^r)}} \longrightarrow 0,$$

where t is the map which sends a form $\omega \in Z^q_{Y_{n-1}^{(p^r)}}$ to the class of a form $\eta \in Z^{q+1}_{\mathcal{X}^{(p^r)}}$ such that $p^n\eta = d\tilde{\omega}$, where $\tilde{\omega} \in \Omega^q_{\mathcal{Y}_n^{(p^r)}/\mathcal{S}_n^{(r)}}$ is a lifting of ω.

Now we showed above that there is a natural isomorphism

$$(4.3.18.13)\qquad \Lambda^{q+1}_{n,\mathcal{X}^{(p^r)}}/B^{q+1}_{\mathcal{X}^{(p^r)}} \simeq B_n\Omega^{q+1}_{\mathcal{X}^{(p^r+1)}/\mathcal{S}_0^{(r+1)}}.$$

The same argument shows that there is an exact sequence
(4.3.18.14)

$$0 \longrightarrow Z^q_{X_\bullet^{(p^r)}} \xrightarrow{p^n} Z^q_{Y_{n,\bullet}^{(p^r)}} \longrightarrow Z^q_{Y_{n-1,\bullet}^{(p^r)}} \xrightarrow{t} B_n\Omega^{q+1}_{X_\bullet^{(p^r+1)}/S_{0,\bullet}} \longrightarrow 0.$$

By comparing the sequences (4.3.18.12) and (4.3.18.14), we deduce that equation (4.3.17.2) holds for $n+1$.

Combining (4.3.17.2) and (4.3.17.3) we deduce (4.3.17.1) for $n+1$. Finally (4.3.17.4) follows from applying RG_* to the exact sequence
(4.3.18.15)

$$0 \longrightarrow R^q_{n,X_\bullet^{(p)}} \longrightarrow \Omega^q_{X_\bullet^{(p)}/S_{0,\bullet}} \oplus \Omega^{q-1}_{X_\bullet^{(p)}/S_{0,\bullet}} \xrightarrow{s_n} \mathcal{A}^q_{n+1,X_\bullet/\widehat{S}_\bullet} \xrightarrow{\pi_n} \mathcal{A}^q_{n,X_\bullet^{(p)}/\widehat{S}_\bullet} \longrightarrow 0.$$

obtained from 4.1.12. □

Corollary 4.3.19. — *With assumptions as in 4.3.12, the map $d^2 : \mathcal{A}^q_{n,\mathcal{X}/\mathcal{S}} \to \mathcal{A}^{q+2}_{n,\mathcal{X}/\mathcal{S}}$ is zero.*

Proof. — The map in question is obtained by applying G_* to the map

(4.3.19.1) $$d^2 : \mathcal{A}^q_{n,X_\bullet/\widehat{S}_\bullet} \longrightarrow \mathcal{A}^{q+2}_{n,X_\bullet/\widehat{S}_\bullet}$$

which is zero by 4.1.3. □

Corollary 4.3.20. — *Let $G_0 : X_0 \to \mathcal{X}$ be the projection. Then the natural map*

(4.3.20.1) $$\mathcal{A}^q_{n,\mathcal{X}/\mathcal{S}} \longrightarrow G_{0*}\mathcal{A}^q_{n,X_0/\widehat{S}_0}$$

is injective.

Proof. — Let $G_i : X_i \to \mathcal{X}$ denote the projection. Recall (see for example [**13**, 5.2.2]) that RG_* is the derived functor of the functor G_* sending an abelian sheaf $F_\bullet$ in the topos $X_{\bullet,\text{et}}$ to

(4.3.20.2) $$\text{Ker}(d : G_{0*}F_0 \to G_{1*}F_1),$$

where d denotes the map obtained by taking the difference of the two maps $G_{0*}F_0 \to G_{1*}F_1$ defined by the simplicial structure on $F_\bullet$ and the two inclusions $[0] \hookrightarrow [1]$. It follows that there is a natural inclusion

(4.3.20.3) $$R^0G_*(F_\bullet) \hookrightarrow G_{0*}F_0.$$

The corollary then follows by noting that the map (4.3.20.1) factors as

(4.3.20.4) $$\mathcal{A}^q_{n,\mathcal{X}/\mathcal{S}} \xrightarrow{\simeq} R^0G_*\mathcal{A}^q_{n,X_\bullet/\widehat{S}_\bullet} \hookrightarrow G_{0*}\mathcal{A}^q_{n,X_0/\widehat{S}_0},$$

where the first map is an isomorphism by 4.3.15 and the second map is an inclusion by the above discussion. □

4.4. De Rham-Witt theory for algebraic stacks

Definition 4.4.1. — An algebraic stack $\mathcal{Y}$ over $\mathbb{F}_p$ is *perfect* if the natural map $\mathcal{O}_{\mathcal{Y}} \to RF_{\mathcal{Y}*}\mathcal{O}_{\mathcal{Y}}$ is an isomorphism, where $F_{\mathcal{Y}} : \mathcal{Y} \to \mathcal{Y}$ is the Frobenius morphism.

Remark 4.4.2. — A perfect stack $\mathcal{Y}$ is Frobenius acyclic in the sense of 3.2.1. Also, for any integer $n > 0$ the map

$$\mathcal{O}_{\mathcal{Y}} \longrightarrow RF^n_{\mathcal{Y}*}\mathcal{O}_{\mathcal{Y}} \tag{4.4.2.1}$$

is an isomorphism. This follows by induction on n (the case $n = 1$ being by assumption): If the result holds for $n-1$ then one gets that the map

$$\mathcal{O}_{\mathcal{Y}} \longrightarrow RF^{n-1}_{\mathcal{Y}*}\mathcal{O}_{\mathcal{Y}} = RF^{n-1}_{\mathcal{Y}*}RF_{\mathcal{Y}*}\mathcal{O}_{\mathcal{Y}} = RF^n_{\mathcal{Y}*}\mathcal{O}_{\mathcal{Y}} \tag{4.4.2.2}$$

is also an isomorphism.

Note also that in the context of 4.3.1, if $\overline{T} \hookrightarrow T$ is the inclusion $\mathrm{Spec}(k) \hookrightarrow \mathrm{Spec}(W)$, where k is a perfect ring and W is its ring of Witt vectors with its canonical lifting of Frobenius, and $\mathcal{S}/W$ is a flat algebraic stack satisfying the conditions in 4.3.1, then the reduction $\mathcal{S}_0 := \mathcal{S} \times_{\mathrm{Spec}(W)} \mathrm{Spec}(k)$ is a perfect stack.

Example 4.4.3. — If $\mathcal{Y}$ is a scheme, then $\mathcal{Y}$ is perfect if and only if the Frobenius morphism $F_{\mathcal{Y}} : \mathcal{Y} \to \mathcal{Y}$ is an isomorphism since the map on underlying topological spaces of $F_{\mathcal{Y}}$ is an isomorphism. If $\mathcal{Y}$ is a perfect stack, however, the Frobenius morphism is not necessarily an isomorphism.

For example, let $G/\mathbb{F}_p$ be a finite type smooth group scheme and $X/\mathbb{F}_p$ a scheme of finite type on which G acts. Let $\mathcal{Y} := [X/G]$ be the stack-theoretic quotient, and let $t : X \to \mathcal{Y}$ be the projection. In this case the Frobenius morphism on $\mathcal{Y}$ is induced by the Frobenius morphisms F_X and F_G on X and G respectively.

Let $\mathcal{P}$ denote the fiber product of the diagram

$$\begin{array}{ccc} & & X \\ & & \downarrow{\scriptstyle t} \\ \mathcal{Y} & \xrightarrow{F_{\mathcal{Y}}} & \mathcal{Y}. \end{array} \tag{4.4.3.1}$$

The stack $\mathcal{P}$ associates to any $\mathbb{F}_p$-scheme T the groupoid of triples

$$(W \longrightarrow T, h : W \longrightarrow X, \iota : G_T \longrightarrow F_T^*W), \tag{4.4.3.2}$$

where $W \to T$ is a G-torsor, h is a G-equivariant map, and ι is an isomorphism of G-torsors (*i.e.*, a trivialization of F_T^*W). The first projection $t' : \mathcal{P} \to \mathcal{Y}$ sends such a triple to (W, h), and the second projection to X sends (W, h, ι) to the composite

$$T \xrightarrow{e} G_T \xrightarrow{\iota} F_T^*W \xrightarrow{\pi_{W/T}} W \xrightarrow{h} X, \tag{4.4.3.3}$$

where $\pi_{W/T}$ is the projection $F_T^*W = T \times_{F_T,T} W \to W$ and e is the identity section. It follows that the fiber product $\mathcal{P} \times_{t',\mathcal{Y},t} X$ associates to any scheme T the set of isomorphism classes of quadruples (W, h, ι, σ), where (W, h, ι) is as above and σ is a trivialization of W. Equivalently $\mathcal{P} \times_{t',\mathcal{Y},t} X$ classifies pairs (f, u), where $f \in X(T)$ is the image of e under the map

$$G_T \xrightarrow{\sigma} W \xrightarrow{h} X_T, \tag{4.4.3.4}$$

and $u \in G(T)$ is the image of e under the isomorphisms

$$G_T \xrightarrow{\iota} F_T^* W \xrightarrow{F_T^*(\sigma)} F_T^* G_T \simeq G_T, \tag{4.4.3.5}$$

where the last isomorphism is the canonical isomorphism following from the assumption that G is defined over $\mathbb{F}_p$. The action of a scheme-theoretic point $v \in G$ on $\mathcal{P} \times_{t',\mathcal{Y},t} X$ induced by the action on the second factor is given by sending such a pair (f, u) to the pair $(vf, F_G(v)u)$. Therefore $\mathcal{P}$ is isomorphic to the stack-theoretic quotient of $X \times G$ by the action of G given by the usual action on the first factor and the action defined by F_G on the second factor.

Let $\mu \subset G$ be the kernel of Frobenius on G. Since the homomorphism $F_G : G \to G$ is surjective (since $G/\mathbb{F}_p$ is smooth), it follows that the inclusion

$$X \hookrightarrow X \times G, \quad f \longmapsto (f, e) \tag{4.4.3.6}$$

induces an isomorphism

$$\mathcal{P} \simeq [X/\mu], \tag{4.4.3.7}$$

where μ acts on X through the embedding $\mu \subset G$ and the action of G on X. With these identifications, the second projection

$$q : [X/\mu] \longrightarrow X \tag{4.4.3.8}$$

is the map induced by the Frobenius map on X. The assertion that $\mathcal{Y}$ is perfect is then equivalent to the assertion that the natural map

$$\mathcal{O}_X \longrightarrow Rq_*\mathcal{O}_{[X/\mu]} \tag{4.4.3.9}$$

is an isomorphism. To verify this, one in turn can work locally on X. If X is affine, then $R^i q_* \mathcal{O}_{[X/\mu]}$ is equal to the group cohomology

$$H^i(\mu, \Gamma(X, \mathcal{O}_X)). \tag{4.4.3.10}$$

Thus $\mathcal{Y}$ is perfect if and only if these cohomology groups vanish for $i > 0$ and if the Frobenius morphism $\Gamma(X, \mathcal{O}_X) \to \Gamma(X, \mathcal{O}_X)$ induces a bijection onto the μ-invariants $\Gamma(X, \mathcal{O}_X)^\mu$.

For a simple example of this, take $G = \mathbb{G}_m$, and $X = \mathbb{A}^1$ with the standard action. In this case $\mu = \mu_p$ which is linearly reductive so the groups 4.4.3.10 are zero for $i > 0$. Also the subring of μ_p-invariants in $\Gamma(X, \mathcal{O}_X) = \mathbb{F}_p[x]$ is the subalgebra generated by x^p which is equal to the image of the injective Frobenius morphism

$$\mathbb{F}_p[x] \longrightarrow \mathbb{F}_p[x], \quad x \longmapsto x^p. \tag{4.4.3.11}$$

Therefore the stack-theoretic quotient $[\mathbb{A}^1/\mathbb{G}_m]$ is perfect.

Example 4.4.4. — More generally if P is a fine saturated p-torsion free monoid with associated affine toric variety $X = \mathrm{Spec}(\mathbb{F}_p[P])$ and torus $T = \mathrm{Spec}(\mathbb{F}_p[P^{\mathrm{gp}}])$, then the stack quotient $[X/T]$ is a perfect stack. Indeed in this case the kernel of Frobenius on T is equal to the diagonalizable group D corresponding to the cokernel of the map $\times p : P^{\mathrm{gp}} \to P^{\mathrm{gp}}$. This implies the vanishing of the groups 4.4.3.10 for $i > 0$. Also the image of Frobenius on the algebra $\mathbb{F}_p[P]$ is equal to the subalgebra generated by elements $m \in P$ for which the image of m in P^{gp} is in the image of $\times p : P^{\mathrm{gp}} \to P^{\mathrm{gp}}$. Since P is saturated such an element m is necessarily in the image of $\times p : P \to P$. Therefore $[X/T]$ is perfect.

4.4.5. — Let A_0 be a perfect ring, and $A := W(A_0)$ the ring of Witt vectors of A_0. Let $\mathcal{S}/A$ be a flat algebraic stack, and let $\mathcal{S}_0$ denote its reduction to A_0. Assume there exists a lifting of Frobenius $F_{\mathcal{S}} : \mathcal{S} \to \mathcal{S}$ compatible with the canonical lifting of Frobenius σ to A, and fix one such lifting $F_{\mathcal{S}}$. Assume that the conditions in 4.3.1 are satisfied with $\overline{T} \hookrightarrow T$ equal to $\mathrm{Spec}(A_0) \hookrightarrow \mathrm{Spec}(A)$ (this implies in particular that $\mathcal{S}_0$ is perfect as noted in 4.4.2). As usual, for a stack $\mathcal{Y} \to \mathcal{S}$ we denote by $\mathcal{Y}_n$ the reduction of $\mathcal{Y}$ modulo p^{n+1}. Since $\mathcal{S}$ is flat over A the ideal (p) has a canonical divided power structure [**8**, 3.3]. In what follows we view $\mathcal{S}$ as a PD-stack using this divided power ideal.

Lemma 4.4.6. — *Let $\mathcal{X} \to \mathcal{S}_0$ be a smooth (not necessarily representable) morphism of algebraic stacks. Then for any integer $r \geq 0$ the natural map*

$$\mathcal{O}_{\mathcal{X}_{\text{lis-et}}} \longrightarrow R\mathrm{pr}_{1*}(\mathcal{O}_{(\mathcal{X} \times_{\mathcal{S}_0, F^r_{\mathcal{S}_0}} \mathcal{S}_0)_{\text{lis-et}}}) \tag{4.4.6.1}$$

is an isomorphism.

Proof. — By induction on r it suffices to consider the case when $r = 1$. For any $i \geq 0$ the sheaf $R^i\mathrm{pr}_{1*}(\mathcal{O}_{(\mathcal{X} \times_{\mathcal{S}_0, F^r_{\mathcal{S}_0}} \mathcal{S}_0)_{\text{lis-et}}})$ is by [**68**, 6.20] equal to the restriction of $R^i F_{\mathcal{S}_0 *} \mathcal{O}_{\mathcal{S}_{0,\text{lis-et}}}$ to $\mathcal{X}_{\text{lis-et}}$. The result therefore follows from the definition of a perfect stack. □

4.4.7. — Let $\mathcal{X} \to \mathcal{S}_0$ be a smooth representable morphism of algebraic stacks, with $\mathcal{X}$ a Deligne-Mumford stack. We apply the construction of 4.3.4 with $\overline{T} \hookrightarrow T$ the inclusion $\mathrm{Spec}(A_0) \hookrightarrow \mathrm{Spec}(A)$, $\sigma : T \to T$ the canonical lifting of Frobenius, and $F_{\mathcal{S}}$ the lifting of Frobenius fixed in 4.4.5. Since $\sigma : T \to T$ is an isomorphism (because A_0 is perfect), the projections

$$\mathcal{X}^{(p^r)} := \mathcal{X} \times_{T, \sigma^r} T \longrightarrow \mathcal{X} \tag{4.4.7.1}$$

and

$$\mathcal{S}^{(r)} := \mathcal{S} \times_{T, \sigma^r} T \longrightarrow \mathcal{S} \tag{4.4.7.2}$$

are isomorphisms. In particular, we can view the maps π_n in (4.3.7.2) as σ^{-1}-linear maps denoted by the same letter

$$\pi_n : \mathcal{A}^{\bullet}_{n+1, \mathcal{X}/\mathcal{S}} \longrightarrow \mathcal{A}^{\bullet}_{n, \mathcal{X}/\mathcal{S}}. \tag{4.4.7.3}$$

We refer to these maps π_n as the *canonical projections.*

Definition 4.4.8. — Let $\mathcal{X} \to \mathcal{S}_0$ be a smooth representable morphism of algebraic stacks with $\mathcal{X}$ Deligne-Mumford. The *de Rham-Witt complex of level* n, denoted $W_n\Omega^\bullet_{\mathcal{X}_{\mathrm{et}}/\mathcal{S}}$, is the differential graded $W_n(A_0)$-algebras $\sigma^{-n*}\mathcal{A}^\bullet_{n,\mathcal{X}/\mathcal{S}}$. The *de Rham-Witt pro-complex*, denoted $W_\bullet\Omega^\bullet_{\mathcal{X}/\mathcal{S}}$, is the projective systems of differential graded A-algebras

(4.4.8.1)
$$\cdots \longrightarrow W_{n+1}\Omega^\bullet_{\mathcal{X}_{\mathrm{et}}/\mathcal{S}} \xrightarrow{\pi_n} W_n\Omega^\bullet_{\mathcal{X}_{\mathrm{et}}/\mathcal{S}} \xrightarrow{\pi_{n-1}} W_{n-1}\Omega^\bullet_{\mathcal{X}_{\mathrm{et}}/\mathcal{S}} \longrightarrow \cdots.$$

The *de Rham-Witt complex*, denote $W\Omega^\bullet_{\mathcal{X}/\mathcal{S}}$ is the complex of differential graded A-algebras $\varprojlim_n W_n\Omega^\bullet_{\mathcal{X}/\mathcal{S}}$.

Remark 4.4.9. — The groups $\mathcal{A}^q_{n,\mathcal{X}/\mathcal{S}}$ as well as the operators d, V, and F do not depend on the choice of $F_\mathcal{S}$. The definition of the map π_n, however, depends on the choice of the lifting of Frobenius $F_\mathcal{S}$. In what follows this lifting of Frobenius will always be fixed and so we suppress it from the notation.

4.4.10. — The operators F and V on $\mathcal{A}^\bullet_{n,\mathcal{X}/\mathcal{S}}$ induce operators

$$F : W_{n+1}\Omega^\bullet_{\mathcal{X}_{\mathrm{et}}/\mathcal{S}} \longrightarrow W_n\Omega^\bullet_{\mathcal{X}_{\mathrm{et}}/\mathcal{S}}, \quad V : W_n\Omega^\bullet_{\mathcal{X}_{\mathrm{et}}/\mathcal{S}} \longrightarrow W_{n+1}\Omega^\bullet_{\mathcal{X}_{\mathrm{et}}/\mathcal{S}} \tag{4.4.10.1}$$

satisfying relations as in 4.1.3.

Proposition 4.4.11 (**[34**, I.3.4 and I.3.17] in the case of schemes)

If $m \geq n$, *then* $\times p^n : W_m\Omega^q_{\mathcal{X}/\mathcal{S}} \to W_m\Omega^q_{\mathcal{X}/\mathcal{S}}$ *factors through the canonical projection* $W_m\Omega^q_{\mathcal{X}/\mathcal{S}} \to W_{m-n}\Omega^q_{\mathcal{X}/\mathcal{S}}$. *The induced map*

$$\text{“}p^n\text{”} : W_{m-n}\Omega^q_{\mathcal{X}/\mathcal{S}} \longrightarrow W_m\Omega^q_{\mathcal{X}/\mathcal{S}} \tag{4.4.11.1}$$

is injective and the natural map

$$W_m\Omega^\bullet_{\mathcal{X}/\mathcal{S}}/\text{“}p^n\text{”}W_{m-n}\Omega^\bullet_{\mathcal{X}/\mathcal{S}} \longrightarrow W_n\Omega^\bullet_{\mathcal{X}/\mathcal{S}} \tag{4.4.11.2}$$

induced by the canonical projection is a quasi-isomorphism.

Proof. — We may work étale locally on $\mathcal{X}$ (recall that $\mathcal{X}$ is assumed to be Deligne-Mumford) and so may assume that $\mathcal{X}$ is an affine scheme. In this case we can choose a p-adically complete formal scheme $\mathcal{Y}/T$ and a compatible collection of maps $\mathcal{Y}_n \to \mathcal{S}_n$ (where $\mathcal{Y}_n$ denotes the reduction modulo p^{n+1} of $\mathcal{Y}$) such that each $\mathcal{Y}_n \to \mathcal{S}_n$ is a smooth lifting of $\mathcal{X} \to \mathcal{S}_0$.

First we consider the case $n = 1$. If $\omega \in Z^q_{\mathcal{Y}_{m-1}}$ defines a class $[\omega]$ in the kernel of multiplication by p, then there exists an element $\Lambda \in \Omega^{q-1}_{\mathcal{Y}_{m-1}/\mathcal{S}_{m-1}}$ such that $p\omega = d\Lambda$. Then $p^q\omega = d(p^{q-1}\Lambda)$, and hence $p^{q-1}\Lambda$ defines an element in $\tilde{E}^{q-1}_{m+q}$ (defined as in 4.3.6). The induced class in E^{q-1}_m then maps to the class $\pi([\omega])$ under the differential, and hence $\pi([\omega]) = 0$. Therefore $\mathrm{Ker}(\times p) \subset \mathrm{Ker}(\pi)$.

On the other hand, if $\omega \in Z^q_{\mathcal{Y}_{m-1}}$ represents a class in $W_m\Omega^q_{\mathcal{X}/\mathcal{S}}$ which is killed by π, then there exist elements ω' and η such that

$$p^q\omega = p^{m-1}(p^q\omega') + p^{q-1}d\eta \tag{4.4.11.3}$$

in $\Omega^q_{\mathcal{Y}_{m+q-1}/\mathcal{S}_{m+q-1}}$. Therefore $p\omega = d\eta$ in $\Omega^q_{\mathcal{Y}_{m-1}/\mathcal{S}_{m-1}}$ and hence $\mathrm{Ker}(\pi) \subset \mathrm{Ker}(\times p)$.

This shows that $\times p$ factors through an inclusion $W_{m-1}\Omega^q_{\mathcal{X}/\mathcal{S}} \hookrightarrow W_m\Omega^q_{\mathcal{X}/\mathcal{S}}$, and hence by induction we find that $\times p^n$ factors through an inclusion $W_{m-n}\Omega^q_{\mathcal{X}/\mathcal{S}} \hookrightarrow W_m\Omega^q_{\mathcal{X}/\mathcal{S}}$.

To prove that (4.4.11.2) is a quasi-isomorphism, we begin with the case $n = m-1$. Suppose $a \in W_{m-1}\Omega^q_{\mathcal{X}/\mathcal{S}}$ is a class annihilated by d defining an element in $\mathcal{H}^q(W_{m-1}\Omega^\bullet_{\mathcal{X}/\mathcal{S}})$, and let $\tilde{a}$ be a lifting of the class to $W_m\Omega^q_{\mathcal{X}/\mathcal{S}}$ (this is possible since the canonical projection is surjective by 4.3.17). Then $d\tilde{a}$ is in the kernel of the canonical projection $W_m\Omega^{q+1}_{\mathcal{X}/\mathcal{S}} \to W_{m-1}\Omega^{q+1}_{\mathcal{X}/\mathcal{S}}$, and therefore by 4.3.17 there exist $b \in \Omega^{q+1}_{\mathcal{X}/\mathcal{S}_0}$ and $c \in \Omega^q_{\mathcal{X}/\mathcal{S}_0}$ such that

$$d\tilde{a} = V^{m-1}(C^{-1}(b)) + dV^{m-1}(C^{-1}(c)), \tag{4.4.11.4}$$

and hence after changing the lifting $\tilde{a}$ by $V^{m-1}(C^{-1}(c))$ we find a lifting $\tilde{a}$ such that

$$d\tilde{a} \in p^{m-1}W_1\Omega^{q+1}_{\mathcal{X}/\mathcal{S}_0}. \tag{4.4.11.5}$$

Therefore the map

$$\mathcal{H}^q(W_m\Omega^\bullet_{\mathcal{X}/\mathcal{S}}/\text{“}p^{m-1}\text{”}W_1\Omega^\bullet_{\mathcal{X}/\mathcal{S}}) \longrightarrow \mathcal{H}^q(W_{m-1}\Omega^\bullet_{\mathcal{X}/\mathcal{S}}) \tag{4.4.11.6}$$

is surjective.

On the other hand, suppose $a \in W_m\Omega^q_{\mathcal{X}/\mathcal{S}}$ defines the zero class in $\mathcal{H}^q(W_{m-1}\Omega^\bullet_{\mathcal{X}/\mathcal{S}})$. Then there exists an element $b \in W_m\Omega^{q-1}_{\mathcal{X}/\mathcal{S}}$ such that $a = db + a'$ where a' is in the kernel of the canonical projection. But then

$$a' = V^{m-1}(C^{-1}(e)) + dV^{m-1}(C^{-1}(c)) \tag{4.4.11.7}$$

for some e and c. Therefore, a defines the zero class in

$$\mathcal{H}^q(W_m\Omega^\bullet_{\mathcal{X}/\mathcal{S}}/\text{“}p^{m-1}\text{”}W_1\Omega^\bullet_{\mathcal{X}/\mathcal{S}}). \tag{4.4.11.8}$$

This proves that (4.4.11.2) is a quasi-isomorphism in the case when $n = m-1$.

Now to prove that (4.4.11.2) is a quasi-isomorphism in general proceed by induction on m. Thus assume true that (4.4.11.2) is a quasi-isomorphism for $m < m_0$. Then to prove the result for m_0, we proceed by descending induction on n. The case of $n = m_0 - 1$ was done above. Thus assume true for $n+1$ and that $n < m_0 - 1$. In that case, there is a quasi-isomorphism

$$W_m\Omega^\bullet_{\mathcal{X}/\mathcal{S}}/\text{“}p^{n+1}\text{”}W_{m-n-1}\Omega^\bullet_{\mathcal{X}/\mathcal{S}} \simeq W_{n+1}\Omega^\bullet_{\mathcal{X}/\mathcal{S}} \tag{4.4.11.9}$$

which identifies the map (4.4.11.2) with the map

$$W_{n+1}\Omega^\bullet_{\mathcal{X}/\mathcal{S}}/\text{“}p\text{”}W_n\Omega^\bullet_{\mathcal{X}/\mathcal{S}} \to W_n\Omega^\bullet_{\mathcal{X}/\mathcal{S}} \tag{4.4.11.10}$$

which is a quasi-isomorphism since $n+1 < m_0$. □

Corollary 4.4.12. — *The sheaf* $\varprojlim_n W_n\Omega^q_{\mathcal{X}/\mathcal{S}}$ *is p-torsion free.*

Proof. — For every integer r, the map $\pi^r : \varprojlim_n W_n\Omega^q_{\mathcal{X}/\mathcal{S}} \to \varprojlim_n W_n\Omega^q_{\mathcal{X}/\mathcal{S}}$ obtained from the morphism of projective systems induced by the canonical projections $W_n\Omega^q_{\mathcal{X}/\mathcal{S}} \to W_{n-r}\Omega^q_{\mathcal{X}/\mathcal{S}}$ is an isomorphism. On the other hand, 4.4.11 implies that a p-torsion element in $\varprojlim_n W_n\Omega^q_{\mathcal{X}/\mathcal{S}}$ is killed by π^r for some r. □

Proposition 4.4.13. — *For every $n \geq 1$ and integer i, the canonical projection induces an exact sequence*

(4.4.13.1)
$$0 \longrightarrow VW_{n-1}\Omega^{i-1}_{\mathcal{X}/\mathcal{S}}/pW_n\Omega^{i-1}_{\mathcal{X}/\mathcal{S}} \xrightarrow{d} W_n\Omega^i_{\mathcal{X}/\mathcal{S}}/VW_{n-1}\Omega^i_{\mathcal{X}/\mathcal{S}} \longrightarrow \Omega^i_{\mathcal{X}/\mathcal{S}_0} \longrightarrow 0.$$

Proof. — All but the injectivity of d follows from the surjectivity of the canonical projection, and the exact sequence (4.3.17.4).

For the injectivity of d, we may as in the proof of 4.4.11 work étale locally on $\mathcal{X}$ and hence may assume there exist a p-adically complete formal scheme $\mathcal{Y}/T$ and a compatible collection of maps $\mathcal{Y}_n \to \mathcal{S}_n$ such that each $\mathcal{Y}_n \to \mathcal{S}_n$ is a smooth lifting of $\mathcal{X} \to \mathcal{S}_0$. Define $Z^i_{\mathcal{Y}_n}$ and $B^i_{\mathcal{Y}_n}$ as in 4.3.16. Let $\omega \in Z^{i-1}_{\mathcal{Y}_{n-2}}$ be a closed form representing a class $[\omega] \in W_{n-1}\Omega^{i-1}_{\mathcal{X}/\mathcal{S}}$ such that $dV[\omega] \in VW_{n-1}\Omega^i_{\mathcal{X}/\mathcal{S}}$. After shrinking $\mathcal{X}$ some more, we may assume that the form ω lifts to a class $\tilde{\omega} \in \Omega^{i-1}_{\mathcal{Y}_{2n-1}/\mathcal{S}_{2n-1}}$. That $dV[\omega] \in VW_{n-1}\Omega^i_{\mathcal{X}/\mathcal{S}}$ then means that there exist forms $\lambda \in Z^i_{\mathcal{Y}_{n-2}}$ and $\gamma \in Z^{i-1}_{\mathcal{Y}_{n-1}}$ such that

$$dp\tilde{\omega} = p^{n+1}\lambda + p^n d\gamma. \tag{4.4.13.2}$$

Replacing $\tilde{\omega}$ by $\tilde{\omega} - p^{n-1}\gamma$ we may assume that $\gamma = 0$. From this it follows that $d\tilde{\omega} \equiv 0 \pmod{p^n}$, and hence ω lifts to a closed form in $Z^{i-1}_{\mathcal{Y}_{n-1}}$. It follows that $[\omega] = F[\tilde{\omega}]$ for some class $[\tilde{\omega}] \in W_n\Omega^{i-1}_{\mathcal{X}/\mathcal{S}}$. Since $VF = p$ this implies the injectivity of d. □

Corollary 4.4.14. — *For every $n \geq 1$ and all integers i, the canonical projection induces a quasi-isomorphism*

(4.4.14.1)
$$\begin{array}{ccccccccc} 0 \longrightarrow & W_n\Omega^0_{\mathcal{X}/\mathcal{S}}/p & \xrightarrow{d} \cdots \longrightarrow & W_n\Omega^{i-1}_{\mathcal{X}/\mathcal{S}}/p & \longrightarrow & W_n\Omega^i_{\mathcal{X}/\mathcal{S}}/VW_{n-1}\Omega^i_{\mathcal{X}/\mathcal{S}} & \longrightarrow 0 \\ & \downarrow & & \downarrow & & \downarrow & \\ 0 \longrightarrow & \mathcal{O}_{\mathcal{X}} & \longrightarrow \cdots \longrightarrow & \Omega^{i-1}_{\mathcal{X}/\mathcal{S}} & \longrightarrow & \Omega^i_{\mathcal{X}/\mathcal{S}} & \longrightarrow 0. \end{array}$$

Proof. — This follows from 4.4.13 and the same argument used in [**34**, I.3.20]. □

4.4.15. — By construction, the de Rham-Witt complex is functorial with respect to $\mathcal{S}_0$-morphisms $\mathcal{X}' \to \mathcal{X}$ between Deligne-Mumford stacks with representable, locally separated, and smooth structure morphisms to $\mathcal{S}_0$. This enables us to define the de Rham-Witt complex for more general algebraic stacks (recall that by 0.2.2 any morphism from a locally separated T-scheme to $\mathcal{S}$ is locally separated).

Definition 4.4.16. — Let $\mathcal{S}/A$ be as in 4.4.5, and let $\mathcal{X} \to \mathcal{S}_0$ be a smooth morphism of algebraic stacks (not necessarily representable and $\mathcal{X}$ not necessarily a Deligne-Mumford stack). Define the *de Rham-Witt pro-complex of* $\mathcal{X}/\mathcal{S}$, denoted $W_\bullet\Omega^\bullet_{\mathcal{X}_{\text{lis-et}}/\mathcal{S}}$, to be the pro-complex of sheaves in $\mathcal{X}_{\text{lis-et}}$ whose restriction to the étale site of any smooth affine $\mathcal{X}$-scheme U is the de Rham-Witt pro-complex $W_\bullet\Omega^\bullet_{U/\mathcal{S}}$. The *de Rham-Witt complex of* $\mathcal{X}/\mathcal{S}$ is the complex of sheaves $\varprojlim W_\bullet\Omega^\bullet_{\mathcal{X}_{\text{lis-et}}/\mathcal{S}}$ on $\mathcal{X}_{\text{lis-et}}$, and the *de Rham-Witt complex of level* n *of* $\mathcal{X}/\mathcal{S}$ is the complex $W_n\Omega^q_{\mathcal{X}_{\text{lis-et}}/\mathcal{S}}$.

If $\mathcal{X}$ is a Deligne-Mumford stack we write $W_\bullet\Omega^\bullet_{\mathcal{X}/\mathcal{S}}$ (resp. $W\Omega^\bullet_{\mathcal{X}/\mathcal{S}}$, $W_n\Omega^\bullet_{\mathcal{X}/\mathcal{S}}$) for the restriction of $W_\bullet\Omega^\bullet_{\mathcal{X}_{\text{lis-et}}/\mathcal{S}}$ (resp. $W\Omega^\bullet_{\mathcal{X}_{\text{lis-et}}/\mathcal{S}}$, $W_n\Omega^\bullet_{\mathcal{X}_{\text{lis-et}}/\mathcal{S}}$) to $\mathcal{X}_{\text{et}}$.

Next we generalize the comparison with crystalline cohomology [**34**, II.1.4].

Theorem 4.4.17. — *Let $\mathcal{X} \to \mathcal{S}$ be a smooth morphism of algebraic stacks with $\mathcal{X}$ a Deligne-Mumford stack. Then there is a canonical isomorphism*

$$I_n : Ru_{\mathcal{X}_{\text{et}}/\mathcal{S}_{n-1}*}\mathcal{O}_{\mathcal{X}_{\text{et}}/\mathcal{S}_{n-1}} \simeq W_n\Omega^\bullet_{\mathcal{X}/\mathcal{S}}. \tag{4.4.17.1}$$

Proof. — We consider first the local situation when $\mathcal{X}$ is an algebraic space and there exists a closed immersion $\mathcal{X} \hookrightarrow \mathcal{Y}$ of $\mathcal{X}$ into a smooth $\mathcal{S}$-space $\mathcal{Y}$. We further assume that there exists a lifting $F_{\mathcal{Y}} : \mathcal{Y} \to \mathcal{Y}$ of Frobenius compatible with the lifting $F_{\mathcal{S}}$.

Choose a smooth surjection $S \to \mathcal{S}$ and a lifting of Frobenius $F_S : \widehat{S} \to \widehat{S}$ to the p-adic completion $\widehat{S}$ of S such that for every integer n the diagram

$$\begin{array}{ccc} S_n & \xrightarrow{F_{S_n}} & S_n \\ \downarrow & & \downarrow \\ \mathcal{S}_n & \xrightarrow{F_{\mathcal{S}_n}} & \mathcal{S}_n \end{array} \tag{4.4.17.2}$$

commutes (as explained in 4.3.10 such a pair (S, F_S) exists). Denote by $S_\bullet$ the 0-coskeleton of $S \to \mathcal{S}$, and let $X_\bullet$ (resp. $Y_\bullet$) denote $\mathcal{X} \times_{\mathcal{S}} S_\bullet$ (resp. $\mathcal{Y} \times_{\mathcal{S}} S_\bullet$). Also let $\widehat{S}_\bullet$ denote the p-adic completion of the simplicial space $S_\bullet$. Let D denote the divided power envelope of $\mathcal{X}$ in $\mathcal{Y}$, $\mathcal{D}$ the sheaf on $\mathcal{X}_{\text{et}}$ obtained from the coordinate ring on D, and let $\mathcal{D}_\bullet$ be the sheaf on $X_{\bullet,\text{et}}$ whose restriction to $X_{i,\text{et}}$ is the coordinate ring of the divided power envelope of X_i in Y_i. To define the map I_n it then suffices to define a morphism

$$\mathcal{D} \otimes_{\mathcal{O}_{\mathcal{Y}_{n-1}}} \Omega^\bullet_{\mathcal{Y}_{n-1}/\mathcal{S}_{n-1}} \longrightarrow W_n\Omega^\bullet_{\mathcal{X}/\mathcal{S}}. \tag{4.4.17.3}$$

By descent theory for quasi-coherent sheaves and 4.3.15 there are canonical isomorphisms

$$\pi_*(\mathcal{D}_\bullet \otimes_{\mathcal{O}_{Y_{n-1,\bullet}}} \Omega^\bullet_{Y_{n-1}/S_{n-1,\bullet}}) \simeq \mathcal{D} \otimes_{\mathcal{O}_{\mathcal{Y}_{n-1}}} \Omega^\bullet_{\mathcal{Y}_{n-1}/\mathcal{S}_{n-1}}, \tag{4.4.17.4}$$

$$\sigma^{-n*}\pi_*(\mathcal{A}^\bullet_{n,X_\bullet/\widehat{S}_\bullet}) \simeq W_n\Omega^\bullet_{\mathcal{X}/\mathcal{S}}, \tag{4.4.17.5}$$

where $\pi : X_{\bullet,\mathrm{et}} \to \mathcal{X}_{\mathrm{et}}$ is the projection. To define the arrow (4.4.17.3) it therefore suffices to define a morphism of differential graded algebras

$$\phi : \mathcal{D}_\bullet \otimes_{\mathcal{O}_{Y_{n-1,\bullet}}} \Omega^\bullet_{Y_{n-1}/S_{n-1,\bullet}} \longrightarrow \mathcal{A}^\bullet_{n,X_\bullet/\widehat{S}_\bullet}. \tag{4.4.17.6}$$

Furthermore, since $\mathcal{D}_\bullet \otimes_{\mathcal{O}_{Y_{n-1,\bullet}}} \Omega^\bullet_{Y_{n-1}/S_{n-1,\bullet}}$ is generated as a differential graded algebra in degree 1, to define the map of algebras ϕ it suffices to define a ring homomorphism

$$\tau : \mathcal{D}_\bullet \longrightarrow \mathcal{A}^0_{n,X_\bullet/\widehat{S}_\bullet}, \tag{4.4.17.7}$$

and a τ-linear map

$$\epsilon : \mathcal{D}_\bullet \otimes_{\mathcal{O}_{Y_{n-1,\bullet}}} \Omega^1_{Y_{n-1}/S_{n-1,\bullet}} \longrightarrow \mathcal{A}^1_{n,X_\bullet/\widehat{S}_\bullet} \tag{4.4.17.8}$$

such that the diagram

$$\begin{array}{ccc} \mathcal{D}_\bullet & \xrightarrow{d} & \mathcal{D}_\bullet \otimes_{\mathcal{O}_{Y_{n-1,\bullet}}} \Omega^1_{Y_{n-1}/S_{n-1,\bullet}} \\ \tau\downarrow & & \downarrow\epsilon \\ \mathcal{A}^0_{n,X_\bullet/\widehat{S}_\bullet} & \xrightarrow{d} & \mathcal{A}^1_{n,X_\bullet/\widehat{S}_\bullet} \end{array} \tag{4.4.17.9}$$

commutes.

The map τ is constructed as follows. The sheaf $W_n(\mathcal{O}_{X^\bullet})$ has a canonical divided power structure [**34**, 0.1.4]. Now the lifting of Frobenius F_Y defines by [**34**, 0.1.3.20] a map $\mathcal{O}_{Y_\bullet} \to W_n(\mathcal{O}_{Y_{0,\bullet}})$ which when composed with the canonical map $W_n(\mathcal{O}_{Y_{0,\bullet}}) \to W_n(\mathcal{O}_{X_\bullet})$ gives a map $\mathcal{O}_{Y_\bullet} \to W_n(\mathcal{O}_{X_\bullet})$. Therefore by the universal property of $\mathcal{D}_\bullet$ there is a canonical map

$$\mathcal{D}_\bullet \longrightarrow W_n(\mathcal{O}_{X_\bullet}). \tag{4.4.17.10}$$

The map τ is defined to be the composite of this map with the map

$$W_n(\mathcal{O}_{X_\bullet}) \longrightarrow \mathcal{H}^0(\mathcal{D}_\bullet \otimes_{\mathcal{O}_{Y_\bullet}} \Omega^\bullet_{Y_{n-1,\bullet}/S_{n-1,\bullet}}), \tag{4.4.17.11}$$

which sends

$$(a_0, \dots, a_{n-1}) \longmapsto \sum_{i=0}^{n-1} p^i \tilde{a}_i^{p^{n-i}}, \tag{4.4.17.12}$$

for some liftings $\tilde{a}_i \in \mathcal{D}_\bullet$ of the a_i. As in 4.2.2, it follows from the binomial theorem that this map is well-defined, and from the definition of Witt vectors that it is a ring homomorphism. To define ϵ, it is enough to exhibit a τ-derivation

$$\partial_\epsilon : \mathcal{D}_\bullet \longrightarrow \mathcal{H}^1(\mathcal{D}_\bullet \otimes_{\mathcal{O}_{Y_\bullet}} \Omega^\bullet_{Y_{n-1,\bullet}/S_{n-1,\bullet}}). \tag{4.4.17.13}$$

We define ∂_ϵ to be the map induced by the composite of the map (4.4.17.10) and the map

$$W_n(\mathcal{O}_{X_\bullet}) \longrightarrow \mathcal{A}^1_{n,X_\bullet/\widehat{S}_\bullet}, \quad (a_0, \dots, a_{n-1}) \longmapsto \left[\sum_{i=0}^{n-1} \tilde{a}_i^{p^{n-i}-1} d\tilde{a}_i\right], \tag{4.4.17.14}$$

where as before $\tilde{a}_i \in \mathcal{D}_\bullet$ denotes a lifting of a_i. The verification that ∂_ϵ is well-defined and is a τ-derivation is left to the reader.

This defines the map I_n in the local setting. To obtain the morphism I_n in general one proceeds using cohomological descent as in [**34**, proof of II.1.4].

It remains only to show that the map I_n is an isomorphism. For this note that it follows from the construction that the diagram

$$\begin{array}{ccc} Ru_{\mathcal{X}/\mathcal{S}_{n-1}*}\mathcal{O}_{\mathcal{X}/\mathcal{S}_{n-1}} & \xrightarrow{I_n} & W_n\Omega^\bullet_{\mathcal{X}/\mathcal{S}} \\ {\scriptstyle r}\downarrow & & \downarrow{\scriptstyle \pi_{n-1}} \\ Ru_{\mathcal{X}/\mathcal{S}_{n-2}*}\mathcal{O}_{\mathcal{X}/\mathcal{S}_{n-2}} & \xrightarrow{I_{n-1}} & W_{n-1}\Omega^\bullet_{\mathcal{X}/\mathcal{S}} \end{array} \tag{4.4.17.15}$$

commutes, where r denotes the reduction map. Combining this with 4.4.11 it follows that to prove that I_n is an isomorphism it suffices to consider the case $n = 1$ in which case I_1 is the Cartier isomorphism (3.3.21). □

Let us also note the following corollary which will be used in what follows:

Corollary 4.4.18. — *There is a canonical map $\rho : W_n(\mathcal{O}_\mathcal{X}) \to W_n\Omega^0_{\mathcal{X}/\mathcal{S}}$. For every integer $i \geq 0$ the $W_n(\mathcal{O}_\mathcal{X})$-module $W_n\Omega^i_{\mathcal{X}/\mathcal{S}}$ is of finite type.*

Proof. — The map ρ is the map obtained from (4.4.17.11). That $W_n\Omega^i_{\mathcal{X}/\mathcal{S}}$ is of finite type is shown by induction on n.

For $n = 1$ the module $W_1\Omega^i_{\mathcal{X}/\mathcal{S}}$ is isomorphic as a $\mathcal{O}_\mathcal{X} \simeq W_1(\mathcal{O}_\mathcal{X})$-module to $\Omega^i_{\mathcal{X}/\mathcal{S}}$. Thus in the case $n = 1$ the result is immediate.

To prove the result for $n + 1$ assuming the result for n, note that the kernel of the canonical projection $\pi_n : W_{n+1}\Omega^i_{\mathcal{X}/\mathcal{S}} \to W_n\Omega^i_{\mathcal{X}/\mathcal{S}}$ is by 4.3.17 a finite type $\mathcal{O}_\mathcal{X}$-module. □

4.4.19. — Let $\varphi : W\Omega^\bullet_{\mathcal{X}/\mathcal{S}} \to W\Omega^\bullet_{\mathcal{X}/\mathcal{S}}$ be the endomorphism which in degree i is equal to p^iF. Since $dF = pFd$ the map φ is a morphism of complexes. It follows from the proof of 4.4.17 that the map φ induces via I_n the Frobenius endomorphism of $Ru_{\mathcal{X}_{\text{et}}/\mathcal{S}_{n-1}}\mathcal{O}_{\mathcal{X}_{\text{et}}/\mathcal{S}_{n-1}}$.

4.5. The slope spectral sequence and finiteness results

In this section we use the arguments of [**34**, II.2] to study the finiteness properties of the de Rham-Witt complex.

4.5.1. — Let k be a perfect field of characteristic $p > 0$, W the ring of Witt vectors of k, and let $\mathcal{S}/W$ be an algebraic stack with a lift of Frobenius $F_\mathcal{S} : \mathcal{S} \to \mathcal{S}$ compatible with the canonical lift σ of Frobenius to W. Assume $\mathcal{S}$ satisfies the assumptions in 4.3.1

(by 4.4.2 this implies that $\mathcal{S}_0$ is perfect). Let $\mathcal{X} \to \mathcal{S}_0$ be a smooth locally separated morphism of algebraic stacks with $\mathcal{X}$ a tame Deligne-Mumford stack (see 2.5.14 for the definition of a tame stack). Assume further that $\mathcal{X}$ is proper over k. Denote by Γ the global section functor on $\mathcal{X}_{\mathrm{et}}$.

Proposition 4.5.2 **(Generalization of [34, II.2.1]).** — *Let $i, j \in \mathbb{Z}$ be integers.*

(i) *For every n the W-module $H^j(\mathcal{X}_{\mathrm{et}}, W_n\Omega^i_{\mathcal{X}/\mathcal{S}})$ is of finite length and the canonical maps*

$$R\Gamma(W\Omega^i_{\mathcal{X}/\mathcal{S}}) \longrightarrow R\varprojlim_n R\Gamma(W_n\Omega^i_{\mathcal{X}/\mathcal{S}}), \tag{4.5.2.1}$$

$$H^j(\mathcal{X}_{\mathrm{et}}, W\Omega^i_{\mathcal{X}/\mathcal{S}}) \longrightarrow \varprojlim_n H^j(\mathcal{X}_{\mathrm{et}}, W_n\Omega^i_{\mathcal{X}/\mathcal{S}}) \tag{4.5.2.2}$$

are isomorphisms.

(ii) *Let $r \geq 0$ be an integer. For every n the W-modules*

$$\begin{gathered} H^j(\mathcal{X}_{\mathrm{et}}, W_{n+r}\Omega^i_{\mathcal{X}/\mathcal{S}}/V^rW_n\Omega^i_{\mathcal{X}/\mathcal{S}}), \quad H^j(\mathcal{X}_{\mathrm{et}}, W_n\Omega^i_{\mathcal{X}/\mathcal{S}}/p^rW_n\Omega^i_{\mathcal{X}/\mathcal{S}}), \\ \text{and} \quad H^j(\mathcal{X}_{\mathrm{et}}, W_n\Omega^i_{\mathcal{X}/\mathcal{S}}/F^rW_{n+r}\Omega^i_{\mathcal{X}/\mathcal{S}}) \end{gathered} \tag{4.5.2.3}$$

are of finite length and the canonical maps

$$H^j(\mathcal{X}_{\mathrm{et}}, W\Omega^i_{\mathcal{X}/\mathcal{S}}/V^rW\Omega^i_{\mathcal{X}/\mathcal{S}}) \longrightarrow \varprojlim_n H^j(\mathcal{X}_{\mathrm{et}}, W_{n+r}\Omega^i_{\mathcal{X}/\mathcal{S}}/V^rW_n\Omega^i_{\mathcal{X}/\mathcal{S}}), \tag{4.5.2.4}$$

$$H^j(\mathcal{X}_{\mathrm{et}}, W\Omega^i_{\mathcal{X}/\mathcal{S}}/p^rW\Omega^i_{\mathcal{X}/\mathcal{S}}) \longrightarrow \varprojlim_n H^j(\mathcal{X}_{\mathrm{et}}, W_n\Omega^i_{\mathcal{X}/\mathcal{S}}/p^rW_n\Omega^i_{\mathcal{X}/\mathcal{S}}), \tag{4.5.2.5}$$

$$H^j(\mathcal{X}_{\mathrm{et}}, W\Omega^i_{\mathcal{X}/\mathcal{S}}/F^rW\Omega^i_{\mathcal{X}/\mathcal{S}}) \longrightarrow \varprojlim_n H^j(\mathcal{X}_{\mathrm{et}}, W_n\Omega^i_{\mathcal{X}/\mathcal{S}}/F^rW_{n+r}\Omega^i_{\mathcal{X}/\mathcal{S}}) \tag{4.5.2.6}$$

are isomorphisms.

Proof. — For (i), observe that by 4.3.17 the canonical projection

$$\pi_n : W_{n+1}\Omega^i_{\mathcal{X}/\mathcal{S}} \longrightarrow W_n\Omega^i_{\mathcal{X}/\mathcal{S}} \tag{4.5.2.7}$$

is surjective for every n. It follows that

$$W\Omega^i_{\mathcal{X}/\mathcal{S}} \simeq R\varprojlim_n W_n\Omega^i_{\mathcal{X}/\mathcal{S}}, \tag{4.5.2.8}$$

and hence

$$R\Gamma(W\Omega^i_{\mathcal{X}/\mathcal{S}}) \simeq R\Gamma R\varprojlim_n W_n\Omega^i_{\mathcal{X}/\mathcal{S}} \simeq R\varprojlim_n R\Gamma W_n\Omega^i_{\mathcal{X}/\mathcal{S}}. \tag{4.5.2.9}$$

This shows that (4.5.2.1) is an isomorphism.

The statement that $H^j(\mathcal{X}_{\mathrm{et}}, W_n\Omega^i_{\mathcal{X}/\mathcal{S}})$ is of finite length follows from 4.4.18 and the following lemma:

Lemma 4.5.3. — *Let $\mathcal{F}$ be a sheaf of $W_n(\mathcal{O}_{\mathcal{X}})$-modules on $\mathcal{X}_{\mathrm{et}}$ of finite type. Then $H^j(\mathcal{X}_{\mathrm{et}}, \mathcal{F})$ is of finite length over W_n.*

Proof. — Let $W_n(\mathcal{X})$ be the stack defined in 0.2.7. Then the closed immersion $\mathcal{X} \subset W_n(\mathcal{X})$ is defined by a nilpotent ideal and hence $W_n(\mathcal{X})$ is proper and tame over W_n since to verify both these conditions it suffices to verify that $\mathcal{X}$ is proper and tame over $\mathrm{Spec}(W_n)$ which is clear since $\mathcal{X}$ is proper and tame over $\mathrm{Spec}(k)$. From this and [**68**, 7.13] the result follows. □

This implies in particular that the projective systems $H^j(\mathcal{X}_{\mathrm{et}}, W_n\Omega^i_{\mathcal{X}/\mathcal{S}})$ satisfy the Mittag-Leffler condition so $R^q \varprojlim_n H^j(\mathcal{X}_{\mathrm{et}}, W_n\Omega^i_{\mathcal{X}/\mathcal{S}}) = 0$ for $q > 0$. This combined with the isomorphism (4.5.2.1) implies that (4.5.2.2) is an isomorphism.

The statements in (ii) follows from a similar argument. It follows from 4.4.18 that $W_{n+r}\Omega^i_{\mathcal{X}/\mathcal{S}}/V^rW_n\Omega^i_{\mathcal{X}/\mathcal{S}}$ (resp. $W_n\Omega^i_{\mathcal{X}/\mathcal{S}}/p^rW_n\Omega^i_{\mathcal{X}/\mathcal{S}}$, $W_n\Omega^i_{\mathcal{X}/\mathcal{S}}/F^rW_{n+r}\Omega^i_{\mathcal{X}/\mathcal{S}}$) is of finite type over $W_{n+r}(\mathcal{O}_\mathcal{X})$ (resp. $W_n(\mathcal{O}_\mathcal{X})$). From this and 4.5.3 it follows that the groups

(4.5.3.1)
$$H^j(\mathcal{X}_{\mathrm{et}}, W_{n+r}\Omega^i_{\mathcal{X}/\mathcal{S}}/V^rW_n\Omega^i_{\mathcal{X}/\mathcal{S}}),\quad H^j(\mathcal{X}_{\mathrm{et}}, W_n\Omega^i_{\mathcal{X}/\mathcal{S}}/p^rW_n\Omega^i_{\mathcal{X}/\mathcal{S}}),\\ \text{and}\quad H^j(\mathcal{X}_{\mathrm{et}}, W_n\Omega^i_{\mathcal{X}/\mathcal{S}}/F^rW_{n+r}\Omega^i_{\mathcal{X}/\mathcal{S}})$$

are of finite type. From this and the same argument used in (i) the isomorphisms (4.5.2.4), (4.5.2.5), and (4.5.2.6) follow. □

Definition 4.5.4. — The *standard topology* on $H^j(\mathcal{X}_{\mathrm{et}}, W\Omega^i_{\mathcal{X}/\mathcal{S}})$ is the topology defined by the *canonical filtration*

(4.5.4.1)
$$\mathrm{Fil}^n H^j(\mathcal{X}_{\mathrm{et}}, W\Omega^i_{\mathcal{X}/\mathcal{S}}) := \mathrm{Ker}\big(H^j(\mathcal{X}_{\mathrm{et}}, W\Omega^i_{\mathcal{X}/\mathcal{S}}) \to H^j(\mathcal{X}_{\mathrm{et}}, W_n\Omega^i_{\mathcal{X}/\mathcal{S}})\big).$$

By 4.5.2 (i), the standard topology is separated and complete. It is useful to also consider other topologies. If M is an abelian group and $\tau : M \to M$ is an endomorphism, define the *τ-adic topology* on M to be the topology defined by the filtration $\tau^n(M) \subset M$ $(n \geq 0)$.

Corollary 4.5.5. — *For every $r \geq 0$, the endomorphisms V^r, p^r, and F^r of $H^j(\mathcal{X}_{\mathrm{et}}, W\Omega^i_{\mathcal{X}/\mathcal{S}})$ have closed image.*

Proof. — This follows from 4.5.2 (ii), which shows that the images of V^r, p^r, and F^r are equal to the inverse limit of the kernels of the morphisms of projective systems

(4.5.5.1)
$$H^j(\mathcal{X}_{\mathrm{et}}, W_{n+r}\Omega^i_{\mathcal{X}/\mathcal{S}}) \longrightarrow H^j(\mathcal{X}_{\mathrm{et}}, W_{n+r}\Omega^i_{\mathcal{X}/\mathcal{S}}/V^rW_n\Omega^i_{\mathcal{X}/\mathcal{S}}),$$

(4.5.5.2)
$$H^j(\mathcal{X}_{\mathrm{et}}, W_n\Omega^i_{\mathcal{X}/\mathcal{S}}) \longrightarrow H^j(\mathcal{X}_{\mathrm{et}}, W_n\Omega^i_{\mathcal{X}/\mathcal{S}}/p^rW_n\Omega^i_{\mathcal{X}/\mathcal{S}}),$$

(4.5.5.3)
$$H^j(\mathcal{X}_{\mathrm{et}}, W_n\Omega^i_{\mathcal{X}/\mathcal{S}}) \longrightarrow H^j(\mathcal{X}_{\mathrm{et}}, W_n\Omega^i_{\mathcal{X}/\mathcal{S}}/F^rW_{n+r}\Omega^i_{\mathcal{X}/\mathcal{S}}).$$ □

Corollary 4.5.6. — *The group $H^j(\mathcal{X}_{\mathrm{et}}, W\Omega^i_{\mathcal{X}/\mathcal{S}})$ is separated and complete for the p-adic topology (resp. V-adic topology).*

Proof. — This follows from the argument used in the proof of [**34**, II.2.5]. □

4.5.7. — We can also consider the naive truncation $W\Omega^{\leq i}_{\mathcal{X}/\mathcal{S}}$ which is the quotient of $W\Omega^{\bullet}_{\mathcal{X}/\mathcal{S}}$ which in degrees $j \leq i$ is equal to $W\Omega^{j}_{\mathcal{X}/\mathcal{S}}$ and is 0 for $j > i$. The formula $Vd = pdV$ implies that if $V_i : W\Omega^{\leq i}_{\mathcal{X}/\mathcal{S}} \to W\Omega^{\leq i}_{\mathcal{X}/\mathcal{S}}$ is the operator which is equal to $p^{i-j}V$ in degree $j \leq i$ then V_i is a morphism of complexes.

***Proposition 4.5.8* (Generalization of [34, II.2.10]).** — *Let $i, j \in \mathbb{Z}$ be integers.*

(ii) *The W-module $H^j(\mathcal{X}_{\mathrm{et}}, W\Omega^{\leq i}_{\mathcal{X}/\mathcal{S}}/V_iW\Omega^{\leq i}_{\mathcal{X}/\mathcal{S}})$ is of finite length.*

(ii) *The group $H^j(\mathcal{X}_{\mathrm{et}}, W\Omega^{\leq i}_{\mathcal{X}/\mathcal{S}})$ is separated and complete for the p-adic topology (resp. V_i-adic topology).*

Proof. — This follows from the same argument used in [**34**, proof of II.2.10] combined with 4.4.14. □

4.5.9. — Let $W_\sigma[\![V]\!]$ denote the ring of non-commutative formal power series in one variable V with relation $aV = Va^\sigma$ for $a \in W$. We view $W_\sigma[\![V]\!]$ as a topological ring with the V-adic topology. The endomorphism V_i gives by 4.5.8 (ii) the module $H^j(\mathcal{X}_{\mathrm{et}}, W\Omega^{\leq i}_{\mathcal{X}/\mathcal{S}})$ the structure of a $W_\sigma[\![V]\!]$-module complete with respect to the V-adic topology.

Corollary 4.5.10. — *The module $H^j(\mathcal{X}_{\mathrm{et}}, W\Omega^{\leq i}_{\mathcal{X}/\mathcal{S}})$ is of finite type over $W_\sigma[\![V]\!]$ and the quotient $H^j(\mathcal{X}_{\mathrm{et}}, W\Omega^{\leq i}_{\mathcal{X}/\mathcal{S}})/V_iH^j(\mathcal{X}_{\mathrm{et}}, W\Omega^{\leq i}_{\mathcal{X}/\mathcal{S}})$ is of finite length over W.*

Proof. — Since $H^j(\mathcal{X}_{\mathrm{et}}, W\Omega^{\leq i}_{\mathcal{X}/\mathcal{S}})$ is complete with respect to the V-adic topology, it suffices to prove the second statement. For this note that there is a natural inclusion

$$H^j(\mathcal{X}_{\mathrm{et}}, W\Omega^{\leq i}_{\mathcal{X}/\mathcal{S}})/V_iH^j(\mathcal{X}_{\mathrm{et}}, W\Omega^{\leq i}_{\mathcal{X}/\mathcal{S}}) \subset H^j(\mathcal{X}_{\mathrm{et}}, W\Omega^{\leq i}_{\mathcal{X}/\mathcal{S}}/V_i\Omega^{\leq i}_{\mathcal{X}/\mathcal{S}}) \tag{4.5.10.1}$$

so in particular by 4.5.8 (ii), the module $H^j(\mathcal{X}_{\mathrm{et}}, W\Omega^{\leq i}_{\mathcal{X}/\mathcal{S}})/V_iH^j(\mathcal{X}_{\mathrm{et}}, W\Omega^{\leq i}_{\mathcal{X}/\mathcal{S}})$ is of finite length over W. □

Corollary 4.5.11. — *For all $i, j \in \mathbb{Z}$, the module $H^j(\mathcal{X}_{\mathrm{et}}, W\Omega^{\leq i}_{\mathcal{X}/\mathcal{S}})$ is isomorphic to a direct sum of a finitely generated free W-module and a torsion module killed by some power of p.*

Proof. — Let $\varphi : W\Omega^{\bullet}_{\mathcal{X}/\mathcal{S}} \to W\Omega^{\bullet}_{\mathcal{X}/\mathcal{S}}$ denote the endomorphism described in 4.4.19. This endomorphism induces by restriction an endomorphism, which we denote by the same letter φ, of $W\Omega^{\leq i}_{\mathcal{X}/\mathcal{S}}$. Since $FV = VF = p$ we have $\varphi V_i = V_i\varphi = p^{i+1}$. This implies that the V_i-torsion in $H^j(\mathcal{X}_{\mathrm{et}}, W\Omega^{\leq i}_{\mathcal{X}/\mathcal{S}})$ is contained in the p-torsion submodule. From this and [**10**, III.2.4] it follows that the quotient of $H^j(\mathcal{X}_{\mathrm{et}}, W\Omega^{\leq i}_{\mathcal{X}/\mathcal{S}})$ by its p-torsion subgroup is a finitely generated free W-module. On the other hand, the module $H^j(\mathcal{X}_{\mathrm{et}}, W\Omega^{\leq i}_{\mathcal{X}/\mathcal{S}})$ is a $W_\sigma[\![V]\!]$-module of finite type, and hence in particular its p-torsion subgroup (which is a sub-$W_\sigma[\![V]\!]$-module) is killed by some power of p. □

Theorem 4.5.12. — *For every $i, j \in \mathbb{Z}$, the torsion submodule $T^{ij} \subset H^j(\mathcal{X}_{\text{et}}, W\Omega^i_{\mathcal{X}/\mathcal{S}})$ is killed by a power of p, and the quotient $H^j(\mathcal{X}_{\text{et}}, W\Omega^i_{\mathcal{X}/\mathcal{S}})/T^{ij}$ is a free W-module of finite type.*

Proof. — For all integers $i > 0$ there is a short exact sequence

$$0 \longrightarrow W\Omega^i_{\mathcal{X}/\mathcal{S}}[-i] \longrightarrow W\Omega^{\leq i}_{\mathcal{X}/\mathcal{S}} \longrightarrow W\Omega^{\leq i-1}_{\mathcal{X}/\mathcal{S}} \longrightarrow 0. \tag{4.5.12.1}$$

The theorem then follows from considerations of the associated long exact sequence and 4.5.11. □

Definition 4.5.13. — The *slope spectral sequence* is the spectral sequence

$$E_1^{ij} = H^j(\mathcal{X}_{\text{et}}, W\Omega^i_{\mathcal{X}/\mathcal{S}}) \implies H^{i+j}((\mathcal{X}_{\text{et}}/\widehat{\mathcal{S}})_{\text{cris}}, \mathcal{O}_{\mathcal{X}_{\text{et}}/\mathcal{S}}) \tag{4.5.13.1}$$

obtained from the spectral sequence of a filtered complex and the isomorphism I_n (4.4.17).

4.5.14. — Let $W\Omega^{\geq i}_{\mathcal{X}/\mathcal{S}} \subset W\Omega^{\bullet}_{\mathcal{X}/\mathcal{S}}$ be the subcomplex which is zero in degrees less than i and in degree $j \geq i$ is equal to $W\Omega^j_{\mathcal{X}/\mathcal{S}}$. Then the subcomplex $W\Omega^{\geq i}_{\mathcal{X}/\mathcal{S}}$ is stable under the Frobenius endomorphism φ defined in 4.4.19, and hence φ induces an endomorphism of the slope spectral sequence inducing the Frobenius endomorphism on crystalline cohomology on the abutment.

Theorem 4.5.15. — *The spectral sequence*

$$E_1^{ij} = H^j(\mathcal{X}_{\text{et}}, W\Omega^i_{\mathcal{X}/\mathcal{S}}) \otimes \mathbb{Q} \implies H^{i+j}((\mathcal{X}_{\text{et}}/\widehat{\mathcal{S}})_{\text{cris}}, \mathcal{O}_{\mathcal{X}_{\text{et}}/\widehat{\mathcal{S}}}) \otimes \mathbb{Q} \tag{4.5.15.1}$$

induced by the slope spectral sequence degenerates at E_1.

Proof. — This follows from the same argument used to prove [**34**, II.3.2]. □

4.5.16. — Following a suggestion of the referee, the above finiteness results (and more) can also be deduced from Ekedahls' approach to coherence of the cohomology of the de Rham-Witt complex [**17, 18**] (see also the excellent survey [**35**] and in the logarithmic context [**51**]). Following the notation in [**35**], let R denote the Cartier-Raynaud ring of k (denoted $\mathbb{D}^0_k$ in 4.2.12). We view R as a graded ring with F and V of degree 0 and d of degree 1. Let R_n denote the quotient $R/V^nR + dV^nR$, which is a graded$(W_n[d], R)$-bimodule, where $W_n[d]$ denotes the graded W_n-algebra with d in degree 1 and $d^2 = 0$. The R_n form a projective system of $(W[d], R)$-bimodules, which we denote by $R_\bullet$. As in [**35**] we will only consider graded R-modules, and refer to these simply as R-modules. Let $D(R)$ denote the derived category of R-modules, and recall [**35**, 2.4.6] that there is a triangulated subcategory $D^b_c(R) \subset D(R)$ consisting of bounded complexes of R-modules whose cohomology modules are *coherent* in the sense [**35**, 2.2.2].

If T is a topos we can also talk about sheaves of R-modules in T and get a derived category $D(T, R)$. The global section functor derives to give a functor

$$R\Gamma : D(T, R) \longrightarrow D(R). \tag{4.5.16.1}$$

In [**35**, 2.4.7] a very useful criterion for verifying that an object $M \in D(R)$ lies in $D^b_c(R)$ is given. We review this criterion. Define an $R_\bullet$*-module* to be a projective system of graded $W_n[d]$-modules

$$M_\bullet = (M_1 \longleftarrow \cdots \longleftarrow M_n \longleftarrow M_{n+1} \longleftarrow \cdots) \tag{4.5.16.2}$$

together with respectively σ and σ^{-1}-linear maps

$$F : M_{n+1} \longrightarrow M_n, \quad V : M_n \longrightarrow M_{n+1} \tag{4.5.16.3}$$

satisfying

$$FV = VF = p, \quad FdV = d. \tag{4.5.16.4}$$

The category of $R_\bullet$-modules is an abelian category, and we write $D(R_\bullet)$ for its derived category.

If M is an R-module, the projective system $M_n = R_n \otimes_R M$ has a natural structure of an $R_\bullet$-module. This functor can be derived to a functor

$$R_\bullet \otimes^{\mathbb{L}}_R - : D(R) \longrightarrow D(R_\bullet). \tag{4.5.16.5}$$

If $M_\bullet$ is an $R_\bullet$-module, then the inverse limit $\varprojlim M_\bullet$ is naturally an R-module. This functor $\varprojlim$ can be derived and gives a functor

$$R\varprojlim : D(R_\bullet) \longrightarrow D(R). \tag{4.5.16.6}$$

There is a natural morphism of functors

$$\mathrm{id} \longrightarrow R\varprojlim(R_\bullet \otimes^{\mathbb{L}}_R -). \tag{4.5.16.7}$$

An object $M \in D(R)$ is called *complete* if the natural map

$$M \longrightarrow R\varprojlim(R_\bullet \otimes^{\mathbb{L}}_R M) \tag{4.5.16.8}$$

is an isomorphism.

The notion of an $R_\bullet$-module extends immediately to a notion of an $R_\bullet$-module in a topos T, and one obtains a derived category $D(T, R_\bullet)$ and functors

$$R\varprojlim : D(T, R_\bullet) \longrightarrow D(T, R) \tag{4.5.16.9}$$

and

$$R_\bullet \otimes^{\mathbb{L}}_R - : D(T, R) \longrightarrow D(T, R_\bullet). \tag{4.5.16.10}$$

The forgetful functor from $R_\bullet$-modules to $W_n[d]$-modules sending $M_\bullet$ to M_n can also be derived to give a functor

$$e_n : D(R_\bullet) \longrightarrow D(W_n[d]). \tag{4.5.16.11}$$

The composite functor

$$D(R) \xrightarrow{R_\bullet \otimes^{\mathbb{L}}_R -} D(R_\bullet) \xrightarrow{e_n} D(W_n[d]) \tag{4.5.16.12}$$

is denoted

$$R_n \otimes^{\mathbb{L}}_R - : D(R) \longrightarrow D(W_n[d]). \tag{4.5.16.13}$$

The underlying W-modules of a complex of R-modules M naturally form a bicomplex [**35**, 2.1]. Let sM denote the associated total complex. The functor $M \mapsto sM$ extends to the derived category

$$s : D(R) \longrightarrow D(W). \tag{4.5.16.14}$$

Similarly, there is a total complex functor

$$s : D(W_n[d]) \longrightarrow D(W_n). \tag{4.5.16.15}$$

As explained in [**35**, equation 2.3.6] the diagram

$$\begin{array}{ccc} D^+(R) & \xrightarrow{R_n \otimes^{\mathbb{L}}_R -} & D^+(W_n[d]) \\ {\scriptstyle s}\downarrow & & \downarrow{\scriptstyle s} \\ D^+(W) & \xrightarrow{W_n \otimes^{\mathbb{L}}_W -} & D^+(W_n) \end{array} \tag{4.5.16.16}$$

commutes, where $W_n \otimes^{\mathbb{L}}_W -$ denotes the usual derived functor of $W_n \otimes_W -$ (which extends to the unbounded derived category since W_n has finite tor dimension as a W-module).

We can now state the key criterion for verifying that an object $M \in D(R)$ is in $D^b_c(R)$:

Proposition 4.5.17 ([**35**, 2.4.7]). — *Let $M \in D^b(R)$. The following are equivalent:*

(i) $M \in D^b_c(R)$;

(ii) *M is complete and $R_n \otimes^{\mathbb{L}}_R M \in D^b_c(W_n[d])$ for every $n \geq 1$, where $D^b_c(W_n[d])$ denotes the subcategory of $D^b(W_n[d])$ consisting of objects N such that $H^i(N)$ is finitely generated over W_n for all i;*

(iii) *M is complete and $R_1 \otimes^{\mathbb{L}}_R M \in D^b_c(k[d])$.*

4.5.18. — We now return to the situation in 4.5.1.

The operators F, V, and d give the procomplex $W_\bullet\Omega^\bullet_{\mathcal{X}/\mathcal{S}}$ the structure of an object in $D(\mathcal{X}_{\text{et}}, R_\bullet)$, and hence $R\Gamma(W_\bullet\Omega^\bullet_{\mathcal{X}/\mathcal{S}})$ has the structure of an object of $D(R_\bullet)$. Similarly the complex $W\Omega^\bullet_{\mathcal{X}/\mathcal{S}}$ has the structure of an object of $D(\mathcal{X}_{\text{et}}, R)$, and hence $R\Gamma(W\Omega^\bullet_{\mathcal{X}/\mathcal{S}})$ is an object of $D(R)$.

Theorem 4.5.19. — *The object $R\Gamma(W\Omega^\bullet_{\mathcal{X}/S}) \in D(R)$ lies in $D^b_c(R)$.*

Proof. — By 4.5.2 (i), the natural map

$$R\Gamma(W\Omega^\bullet_{\mathcal{X}/\mathcal{S}}) \longrightarrow \varprojlim R\Gamma(W_\bullet\Omega^\bullet_{\mathcal{X}/\mathcal{S}}) \tag{4.5.19.1}$$

is an isomorphism, since this can be verified after applying the total complex functor $s : D(R) \to D(W)$. Since each $R\Gamma(W_n\Omega^\bullet_{\mathcal{X}/\mathcal{S}})$ is in $D^b_c(W_n[d])$ by 4.5.2 (i) and the diagram (4.5.16.16) commutes, it follows (see for example [**8**, Appendix B, Prop. B.5]) that the natural map

(4.5.19.2) $$R_n \otimes^{\mathbb{L}}_R R\Gamma(W\Omega^\bullet_{\mathcal{X}/\mathcal{S}}) \longrightarrow R\Gamma(W_n\Omega^\bullet_{\mathcal{X}/\mathcal{S}})$$

is an isomorphism. This implies that $R\Gamma(W\Omega^\bullet_{\mathcal{X}/\mathcal{S}})$ is complete, and also shows that $R_1 \otimes^{\mathbb{L}}_R R\Gamma(W\Omega^\bullet_{\mathcal{X}/\mathcal{S}})$ is isomorphic to the Hodge cohomology $H^*(\mathcal{X}_{\text{et}}, \Omega^\bullet_{\mathcal{X}/\mathcal{S}_0})$. Since $\mathcal{X}$ is proper this verifies 4.5.17 (iii). □

Remark 4.5.20. — Theorem 4.5.12 follows from 4.5.19 by [**35**, 2.5 (b1)], as does 4.5.15 by [**35**, 2.5.4].

4.6. Comparison with the Langer-Zink de Rham-Witt complex II

We continue with the notation of 4.4.5. Let $\mathcal{X} \to \mathcal{S}_0$ be a smooth locally separated morphism of algebraic stacks with $\mathcal{X}$ a Deligne-Mumford stack.

4.6.1. — The de Rham-Witt complex $W_n\Omega^\bullet_{\mathcal{X}_{\text{et}}/\mathcal{S}}$ can also be described in terms of the Langer-Zink de Rham-Witt complex as follows. Let $S \to \mathcal{S}$ be a smooth cover with S an algebraic space, and let $S_\bullet$ be the 0-coskeleton. Denote by $X_\bullet$ the simplicial algebraic space $\mathcal{X} \times_{\mathcal{S}} S_\bullet$, and let $\pi : X_{\bullet,\text{et}} \to \mathcal{X}_{\text{et}}$ be the projection. Denote by $W_n^{\text{LZ}}\Omega^\bullet_{X_\bullet/S_\bullet}$ the differential graded algebra in $X_{\bullet,\text{et}}$ whose restriction to $X_{i,\text{et}}$ is the Langer-Zink de Rham Witt complex $W_n^{\text{LZ}}\Omega^\bullet_{X_i/S_i}$. Let $R^q_{n,X_\bullet/S_\bullet}$ be as in 4.2.15.

Proposition 4.6.2. — *For any integer $r \geq 0$, the natural map*

(4.6.2.1) $$R^q_{n,\mathcal{X}/\mathcal{S}} \simeq R^q_{n,\mathcal{X}^{(p^r)}/\mathcal{S}^{(r)}} \longrightarrow R\pi_* R^q_{n,X_\bullet^{(p^r)}/S_\bullet}$$

is an isomorphism.

Proof. — Consideration of the exact sequences

(4.6.2.2) $$0 \longrightarrow R^q_{n,X_\bullet^{(p^r)}/S_\bullet} \longrightarrow B_{n+1}\Omega^q_{X_\bullet^{(p^r)}/S_\bullet} \oplus Z_n\Omega^{q-1}_{X_\bullet^{(p^r)}/S_\bullet} \longrightarrow B_1\Omega^q_{X_\bullet^{(p^{n+r})}/S_\bullet} \longrightarrow 0$$

(4.6.2.3) $$0 \longrightarrow R^q_{n,\mathcal{X}_{\text{et}}} \longrightarrow B_{n+1}\Omega^q_{\mathcal{X}_{\text{et}}/\mathcal{S}_0} \oplus Z_n\Omega^{q-1}_{\mathcal{X}_{\text{et}}/\mathcal{S}_0} \longrightarrow B_1\Omega^q_{\mathcal{X}^{(p^n)}_{\text{et}}/\mathcal{S}^{(n)}_0} \longrightarrow 0$$

shows that it suffices to prove that the natural maps

(4.6.2.4) $$B_{n+1}\Omega^q_{\mathcal{X}_{\text{et}}/\mathcal{S}_0} \oplus Z_n\Omega^{q-1}_{\mathcal{X}_{\text{et}}/\mathcal{S}_0} \longrightarrow R\pi_*(B_{n+1}\Omega^q_{X_\bullet^{(p^r)}/S_\bullet} \oplus Z_n\Omega^{q-1}_{X_\bullet^{(p^r)}/S_\bullet})$$

and

(4.6.2.5) $$B_1\Omega^q_{\mathcal{X}^{(p^n)}_{\text{et}}/\mathcal{S}^{(n)}_0} \longrightarrow R\pi_* B_1\Omega^q_{X_\bullet^{(p^{n+r})}/S_\bullet}$$

are isomorphisms. This follows from 4.3.12. □

4.6.3. — Define $K^q_{n,X_\bullet^{(p^r)}}$ and $K^q_{n,\mathfrak{X}^{(p^r)}}$ by

$$K^q_{n,X_\bullet^{(p^r)}} := (\Omega^q_{X_\bullet^{(p^r)}/S_{0,\bullet}} \oplus \Omega^{q-1}_{X_\bullet^{(p^r)}/S_{0,\bullet}})/R^q_{n,X_\bullet^{(p^r)}/S_\bullet} \tag{4.6.3.1}$$

$$K^q_{n,\mathfrak{X}^{(p^r)}} := (\Omega^q_{\mathfrak{X}^{(p^r)}/\mathcal{S}_0^{(r)}} \oplus \Omega^{q-1}_{\mathfrak{X}^{(p^r)}/\mathcal{S}_0^{(r)}})/R^q_{n,\mathfrak{X}^{(p^r)}/\mathcal{S}^{(r)}}. \tag{4.6.3.2}$$

Corollary 4.6.4. — *For any $r, n \in \mathbb{Z}$, the natural map $K^q_{n,\mathfrak{X}^{(p^r)}} \to R\pi_* K^q_{n,X_\bullet^{(p^r)}}$ is an isomorphism.*

Proof. — This follows from 4.6.2 and descent theory. □

Corollary 4.6.5. — *For any integers n, s, and $i > 0$ the groups $R^i\pi_* W_n^{\mathrm{LZ}}\Omega^s_{X_\bullet/S_\bullet}$ are zero.*

Proof. — This follows from the same argument used in the proof of the preceding corollary using the exact sequence

$$0 \longrightarrow K^q_{n,X_\bullet} \longrightarrow W^{\mathrm{LZ}}_{n+1}\Omega^q_{X_\bullet/S_\bullet} \longrightarrow W^{\mathrm{LZ}}_n\Omega^q_{X_\bullet/S_\bullet} \longrightarrow 0 \tag{4.6.5.1}$$

provided by 4.2.15. □

4.6.6. — By 4.2.3 (taking $T = S$), there is a canonical map of differential graded algebras

$$W_n^{\mathrm{LZ}}\Omega^\bullet_{X_\bullet/S_\bullet} \longrightarrow \mathcal{A}^\bullet_{n,X_\bullet/S_\bullet} \tag{4.6.6.1}$$

which induces a map

$$\pi_* W_n^{\mathrm{LZ}}\Omega^\bullet_{X_\bullet/S_\bullet} \longrightarrow \pi_*\mathcal{A}^\bullet_{n,X_\bullet/S_\bullet} \xrightarrow{4.3.15} \mathcal{A}^\bullet_{n,\mathfrak{X}/\mathcal{S}}. \tag{4.6.6.2}$$

By the construction this map is σ^n-linear, and hence induces a W-linear map

$$\psi : \pi_* W_n^{\mathrm{LZ}}\Omega^\bullet_{X_\bullet/S_\bullet} \longrightarrow W_n\Omega^\bullet_{\mathfrak{X}/\mathcal{S}} \tag{4.6.6.3}$$

of differential graded algebras compatible with the operators F and V as well as the canonical projections.

Theorem 4.6.7. — *The morphism ψ is an isomorphism.*

Proof. — The proof is by induction on n. For $n = 1$, the map is identified via the Cartier isomorphism with the map $\Omega^\bullet_{\mathfrak{X}/\mathcal{S}_0} \to \pi_*\Omega^\bullet_{X_\bullet/S_{0,\bullet}}$ which is an isomorphism by descent theory for quasi-coherent sheaves.

Next we prove the result for $n+1$ assuming the result for n. By 4.6.4 the exact sequence obtained from 4.2.15

$$0 \longrightarrow K^q_{n,X_\bullet/S_\bullet} \longrightarrow W^{\mathrm{LZ}}_{n+1}\Omega^q_{X_\bullet/S_\bullet} \longrightarrow W^{\mathrm{LZ}}_n\Omega^q_{X_\bullet/S_\bullet} \longrightarrow 0 \tag{4.6.7.1}$$

induces an exact sequence

$$0 \longrightarrow \pi_* K^q_{n,X_\bullet/S_\bullet} \longrightarrow \pi_* W^{\mathrm{LZ}}_{n+1}\Omega^q_{X_\bullet/S_\bullet} \longrightarrow \pi_* W^{\mathrm{LZ}}_n\Omega^q_{X_\bullet/S_\bullet} \longrightarrow 0. \tag{4.6.7.2}$$

Using 4.3.17, we therefore obtain a morphism of exact sequences

(4.6.7.3)

$$\begin{array}{ccccccccc} 0 & \longrightarrow & \pi_*K^q_{n,X_\bullet} & \longrightarrow & \pi_*W^{\mathrm{LZ}}_{n+1}\Omega^q_{X_\bullet/S_\bullet} & \longrightarrow & \pi_*W^{\mathrm{LZ}}_n\Omega^q_{X_\bullet/S_\bullet} & \longrightarrow & 0 \\ & & \downarrow{\scriptstyle a} & & \downarrow{\scriptstyle b} & & \downarrow{\scriptstyle c} & & \\ 0 & \longrightarrow & K^q_{n,\mathcal{X}^{(p)}} & \longrightarrow & W_{n+1}\Omega^q_{\mathcal{X}/\mathcal{S}} & \longrightarrow & W_n\Omega^q_{\mathcal{X}/\mathcal{S}} & \longrightarrow & 0. \end{array}$$

Since the map c is an isomorphism by induction and the map a is an isomorphism by 4.6.4 (and the fact that $\mathcal{X}^{(p)} \simeq \mathcal{X}$ since $\mathcal{S}_0$ is perfect), it follows that the map c is also an isomorphism. □

4.6.8. — This description of $W_n\Omega^q_{\mathcal{X}/\mathcal{S}}$ in terms of the Langer-Zink de Rham-Witt complex has the following important consequence which will be used later in 6.4.

Let T/W be a flat scheme with a lifting of Frobenius $F_T : T \to T$ compatible with the canonical lift σ of Frobenius to W, and let $j : \mathrm{Spec}(W) \hookrightarrow T$ be a closed immersion defined by a divided power ideal. Let $\mathcal{S}_T$ be a flat algebraic stack over T such that the reduction $\mathcal{S} := \mathcal{S}_T \times_{T,j} \mathrm{Spec}(W)$ satisfies the assumptions of 4.4.5.

Let $\mathcal{X} \to \mathcal{S}_0$ be a smooth morphism of algebraic stacks with $\mathcal{X}$ a Deligne-Mumford stack, and let $S_T \to \mathcal{S}_T$ be a smooth surjection with S an algebraic space. Define $S_{T,\bullet}$ to be the 0-coskeleton of $S_T \to \mathcal{S}_T$, let $S_\bullet$ denote $\mathcal{S} \times_{j,\mathcal{S}_T} S_{T,\bullet}$, and let $X_\bullet$ denote $\mathcal{X} \times_{\mathcal{S}} S_\bullet$.

By 4.2.3 for every $n \geq 1$ there is a canonical σ^n-linear map

$$\rho_n : W^{\mathrm{LZ}}_n\Omega^\bullet_{X_\bullet/S_\bullet} \longrightarrow \mathcal{A}^\bullet_{n,X_\bullet/S_{T,\bullet}} \tag{4.6.8.1}$$

which induces a map

$$\pi_*W^{\mathrm{LZ}}_n\Omega^\bullet_{X_\bullet/S_\bullet} \longrightarrow \pi_*\mathcal{A}^\bullet_{n,X_\bullet/S_{T,\bullet}}. \tag{4.6.8.2}$$

Define

$$\iota : \mathcal{A}^\bullet_{n,\mathcal{X}/\mathcal{S}} \longrightarrow \mathcal{A}^\bullet_{n,\mathcal{X}/\mathcal{S}_T} \tag{4.6.8.3}$$

to be the unique morphism making the following diagram commute

$$\begin{array}{ccc} \pi_*W^{\mathrm{LZ}}_n\Omega^\bullet_{X_\bullet/S_\bullet} & \xrightarrow{(4.6.8.2)} & \pi_*\mathcal{A}^\bullet_{n,X_\bullet/S_{T,\bullet}} \\ {\scriptstyle \psi^{-1}}\uparrow & & \uparrow{\scriptstyle 4.3.15} \\ \mathcal{A}^\bullet_{n,\mathcal{X}/\mathcal{S}} & \xrightarrow{\iota} & \mathcal{A}^\bullet_{n,\mathcal{X}/\mathcal{S}_T}. \end{array} \tag{4.6.8.4}$$

Here ψ is as in 4.6.7. The map ι is compatible with the operators F, V, and the canonical projections giving a *section* of the map $\mathcal{A}^\bullet_{n,\mathcal{X}/\mathcal{S}_T} \to \mathcal{A}^\bullet_{n,\mathcal{X}/\mathcal{S}}$ induced by j.

Theorem 4.6.9. — *The map ι induces an isomorphism*

$$\mathcal{A}^\bullet_{n,\mathcal{X}/\mathcal{S}} \otimes_W \mathcal{O}_T \simeq \mathcal{A}^\bullet_{n,\mathcal{X}/\mathcal{S}_T}. \tag{4.6.9.1}$$

Proof. — The assertion is étale local on $\mathcal{X}$, and hence we may assume that there exists a smooth lifting $\mathcal{Y}/T_0$ of $\mathcal{X}$.

Let $J \subset \mathcal{O}_T$ be the ideal defining $\mathrm{Spec}(W)$ in T, and let $J_0 \subset \mathcal{O}_{T_0}$ be the reduction. Since J_0 has divided powers, for any element $m \in J_0$ we have $m^p = p \cdot m^{[p]}$, and hence the Frobenius map

$$F_{T_0} : T_0 \longrightarrow T_0 \tag{4.6.9.2}$$

factors through $\mathrm{Spec}(k) \subset T_0$. This implies in particular that there is a canonical isomorphism

$$\Omega^\bullet_{\mathcal{Y}^{(p)}/\mathcal{S}^{(1)}_{T_0}} \simeq \Omega^\bullet_{\mathcal{X}^{(p)}/\mathcal{S}^{(1)}} \otimes_k \mathcal{O}_{T_0}. \tag{4.6.9.3}$$

It follows from the construction that this is the map induced by ι for $n = 1$. Furthermore, this shows that ι also induces canonical isomorphisms

$$B_n\Omega^q_{\mathcal{Y}^{(p)}/\mathcal{S}^{(1)}_{T_0}} \simeq B_n\Omega^q_{\mathcal{X}^{(p)}/\mathcal{S}^{(1)}_0} \otimes_k \mathcal{O}_{T_0}, \quad Z_n\Omega^q_{\mathcal{Y}^{(p)}/\mathcal{S}^{(1)}_{T_0}} \simeq Z_n\Omega^q_{\mathcal{X}^{(p)}_{\mathrm{et}}/\mathcal{S}_0} \otimes_k \mathcal{O}_{T_0}. \tag{4.6.9.4}$$

From the definition (4.3.8.6) this in turn implies that ι induces a canonical isomorphism

$$R^q_{n,\mathcal{Y}^{(p)}} \simeq R^q_{n,\mathcal{X}^{(p)}} \otimes_k \mathcal{O}_{T_0}. \tag{4.6.9.5}$$

By induction on n it therefore suffices to show that the theorem holds for $n+1$ if it holds for n. This follows from the fact that $\mathcal{O}_T$ is flat over W, by consideration of the morphism of exact sequences

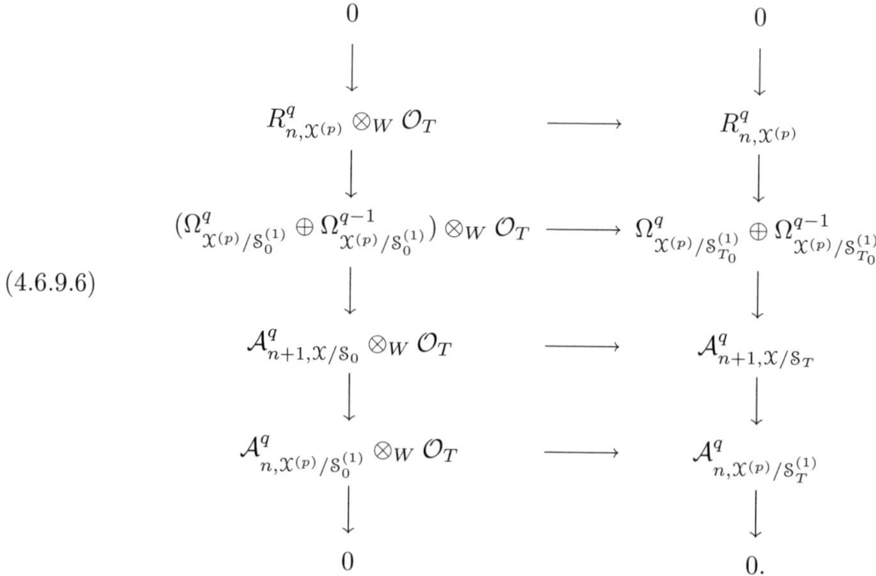

$$\begin{array}{ccc} 0 & & 0 \\ \downarrow & & \downarrow \\ R^q_{n,\mathcal{X}^{(p)}} \otimes_W \mathcal{O}_T & \longrightarrow & R^q_{n,\mathcal{X}^{(p)}} \\ \downarrow & & \downarrow \\ (\Omega^q_{\mathcal{X}^{(p)}/\mathcal{S}^{(1)}_0} \oplus \Omega^{q-1}_{\mathcal{X}^{(p)}/\mathcal{S}^{(1)}_0}) \otimes_W \mathcal{O}_T & \longrightarrow & \Omega^q_{\mathcal{X}^{(p)}/\mathcal{S}^{(1)}_{T_0}} \oplus \Omega^{q-1}_{\mathcal{X}^{(p)}/\mathcal{S}^{(1)}_{T_0}} \\ \downarrow & & \downarrow \\ \mathcal{A}^q_{n+1,\mathcal{X}/\mathcal{S}_0} \otimes_W \mathcal{O}_T & \longrightarrow & \mathcal{A}^q_{n+1,\mathcal{X}/\mathcal{S}_T} \\ \downarrow & & \downarrow \\ \mathcal{A}^q_{n,\mathcal{X}^{(p)}/\mathcal{S}^{(1)}_0} \otimes_W \mathcal{O}_T & \longrightarrow & \mathcal{A}^q_{n,\mathcal{X}^{(p)}/\mathcal{S}^{(1)}_T} \\ \downarrow & & \downarrow \\ 0 & & 0. \end{array} \tag{4.6.9.6}$$

obtained from 4.1.12. □

Remark 4.6.10. — In the logarithmic context the map (4.6.8.3) has been constructed by Hyodo and Kato directly using an explicit description of the logarithmic de Rham-Witt complex [**31**, 4.8]. We prove the equivalence of the two constructions in 9.4. In our general setup, however, we have not been able to give a direct definition.

CHAPTER 5

THE ABSTRACT HYODO-KATO ISOMORPHISM

5.1. Projective systems

Definition 5.1.1. — An abelian group M has *bounded p-torsion* if there exists an integer r such that the p-torsion subgroup of M is annihilated by p^r.

5.1.2. — Let p be a prime and R a p-torsion free and p-adically complete and separated ring. As usual we write R_n for $R/p^{n+1}R$, and $\mathrm{Mod}(R)$ for the category of R-modules.

Let $\mathrm{ps}(R)$ denote the category of projective systems $L_. = \{L_n\}$ of R-modules such that $p^{n+1}L_n = 0$ (and hence L_n can be viewed as a R_n-module). The category $\mathrm{ps}(R)$ is abelian. If $f : L_. \to L'_.$ is a morphism in $\mathrm{ps}(R)$ given by a compatible collection of maps $f_n : L_n \to L'_n$, then $\mathrm{Ker}(f)$ (resp. $\mathrm{Coker}(f)$) is equal to the projective system $\{\mathrm{Ker}(f_n)\}$ (resp. $\{\mathrm{Coker}(f_n)\}$).

Remark 5.1.3. — Note that R is not necessarily noetherian.

5.1.4. — Denote by $\mathrm{ps}(R)_{\mathbb{Q}}$ (resp. $\mathrm{Mod}(R)_{\mathbb{Q}}$) the category whose objects are the same as those of $\mathrm{ps}(R)$ (resp. $\mathrm{Mod}(R)$) and whose morphisms are given by

$$\begin{aligned} &\mathrm{Hom}_{\mathrm{ps}(R)_{\mathbb{Q}}}(M,N) := \mathrm{Hom}_{\mathrm{ps}(R)}(M,N)\otimes_{\mathbb{Z}}\mathbb{Q} \\ &(\text{resp. } \mathrm{Hom}_{\mathrm{Mod}(R)_{\mathbb{Q}}}(M,N) := \mathrm{Hom}_{\mathrm{Mod}(R)}(M,N)\otimes_{\mathbb{Z}}\mathbb{Q}). \end{aligned} \tag{5.1.4.1}$$

The category $\mathrm{ps}(R)_{\mathbb{Q}}$ (resp. $\mathrm{Mod}(R)_{\mathbb{Q}}$) can also be viewed as the quotient category $\mathrm{ps}(R)/\mathcal{T}$ (resp. $\mathrm{Mod}(R)/\mathcal{T}$), where $\mathcal{T}$ denotes the full subcategory of objects annihilated by some non-zero integer $n \in \mathbb{Z}$. This follows from the definition of the quotient of a category by a Serre subcategory [**26**, Chapitre III, §1]. In particular, by *loc. cit.*, Chapitre III, Proposition 1, the categories $\mathrm{ps}(R)_{\mathbb{Q}}$ and $\mathrm{Mod}(R)_{\mathbb{Q}}$ are abelian and the canonical functors

$$\mathrm{ps}(R) \longrightarrow \mathrm{ps}(R)_{\mathbb{Q}}, \quad \mathrm{Mod}(R) \longrightarrow \mathrm{Mod}(R)_{\mathbb{Q}} \tag{5.1.4.2}$$

are exact.

Remark 5.1.5. — In the language of quotient categories, for any $u \in \mathrm{Hom}_{\mathrm{ps}(R)}(M, N)$ and $n \geq 0$ the element

$$u \otimes p^{-n} \in \mathrm{Hom}_{\mathrm{ps}(R)_{\mathbb{Q}}}(M, N) \tag{5.1.5.1}$$

can be described as follows. Let $M[p^n] \subset M$ and $N[p^n] \subset N$ be the kernels of multiplication by p^n on M and N respectively. Let $M' \subset M$ be the image of multiplication by p^n on M, and let N' denote $N/N[p^n]$. The composite morphism

$$M \xrightarrow{\ u\ } N \xrightarrow{\text{projection}} N' \tag{5.1.5.2}$$

then factors through a morphism $\bar{u} : M' \simeq M/M[p^n] \to N'$. In the notation of [**26**, Chapitre III, §1]) the map $u \otimes p^{-n}$ in the quotient category is represented by the diagram

$$\begin{array}{ccc} M' & \longrightarrow & M \\ {\scriptstyle \bar{u}}\downarrow & & \\ N'. & & \end{array} \tag{5.1.5.3}$$

Definition 5.1.6. — A morphism $f : M \to N$ in $\mathrm{Mod}(R)$ is an *isomorphism mod* $\mathcal{T}$ if the induced map in $\mathrm{Mod}(R)_{\mathbb{Q}}$ is an isomorphism. An object $M \in \mathrm{Mod}(R)$ is *free of finite type mod* $\mathcal{T}$ if there exists a map $f : M \to M'$ which is an isomorphism mod $\mathcal{T}$ with M' a free R-module of finite rank.

Remark 5.1.7. — Observe that an R-module M which is free of finite type mod $\mathcal{T}$ has bounded p-torsion.

Lemma 5.1.8. — *If R is a Cohen ring in the sense of* [**15**, 0.19.8.4], *then an object $M \in \mathrm{Mod}(R)$ is free of finite type mod $\mathcal{T}$ if and only if M has bounded p-torsion and the quotient $\overline{M}$ of M by its p-torsion is of finite type (recall that R is a Cohen ring if R is noetherian, local, p-adically complete, flat over $\mathbb{Z}_p$, and R/pR is a field).*

Proof. — For the only if direction, note that if M is free of finite type mod $\mathcal{T}$ and $f : M \to M'$ is an isomorphism mod $\mathcal{T}$ with M' free of finite type, then f descends to a map $\bar{f} : \overline{M} \to M'$ which becomes an isomorphism after inverting p. Therefore $\bar{f}$ is an injection and since R is noetherian it follows that $\overline{M}$ is of finite type over R.

Conversely, if the quotient $\overline{M}$ is of finite type over R then $\overline{M}$ is a free R-module of finite rank since the maximal ideal of R is generated by p (which implies that $\overline{M}$ is flat over the local ring R), and if M also has bounded p-torsion then the projection $M \to \overline{M}$ is an isomorphism mod $\mathcal{T}$. □

Remark 5.1.9. — Tensoring with $\mathbb{Q}$ we obtain functor $\mathrm{Mod}(R) \to \mathrm{Mod}(R \otimes \mathbb{Q})$ which induces a functor

$$\mathrm{Mod}(R)_{\mathbb{Q}} \longrightarrow \mathrm{Mod}(R \otimes \mathbb{Q}). \tag{5.1.9.1}$$

If R is a Cohen ring then $R \otimes \mathbb{Q}$ is a field, and it follows from 5.1.8 that in this case the functor (5.1.9.1) induces an equivalence of categories between the full subcategory of $\mathrm{Mod}(R)_{\mathbb{Q}}$ consisting of objects which are free of finite type mod $\mathcal{T}$ and the category of finite dimensional $R \otimes \mathbb{Q}$-vector spaces.

Lemma 5.1.10. — *A morphism $f : M \to N$ in* $\mathrm{Mod}(R)$ *is an isomorphism mod* $\mathcal{T}$ *if and only if there exists an integer r such that p^r annihilates* $\mathrm{Ker}(f)$ *and* $\mathrm{Coker}(f)$.

Proof. — This follows from [**26**, Chapitre III Lemma 4]. □

Remark 5.1.11. — The property of a module M being free of finite type mod $\mathcal{T}$ depends only on the image of M in $\mathrm{Mod}(R)_{\mathbb{Q}}$. In what follows we will therefore also sometimes speak of an object $M \in \mathrm{Mod}(R)_{\mathbb{Q}}$ being free of finite type mod $\mathcal{T}$.

Lemma 5.1.12. — *An R-module M is free of finite type mod $\mathcal{T}$ if and only if there exists a morphism $g : M' \to M$ with M' free of finite type such that* $\mathrm{Ker}(g)$ *and* $\mathrm{Coker}(g)$ *are annihilated by some power of p.*

Proof. — Let M be a module which is free of finite type mod $\mathcal{T}$, and let $f : M \to M'$ be a morphism to a free of finite type module M'. If p^{r_1} annihilates $\mathrm{Ker}(f)$ and p^{r_2} annihilates $\mathrm{Coker}(f)$, then there exists a map $g : M' \to M$ such that the composite $f \circ g$ is equal to multiplication by $p^{r_1+r_2}$. Indeed for any $m' \in M'$ there exists an element $m \in M$ such that $f(m) = p^{r_2}m'$. This element m is not unique, but if $\tilde{m}$ is a second lifting then $p^{r_1}(m - \tilde{m}) = 0$ and hence there is a well-defined map $g : M' \to M$ sending m' to $p^{r_1}m$. This prove the "only if" direction.

Conversely, if M is an R-module and there exists a morphism $g : M' \to M$ as in the lemma, then by the same argument if p^{r_1} annihilates $\mathrm{Ker}(g)$ and p^{r_2} annihilates $\mathrm{Coker}(g)$ then there exists a map $f : M \to M'$ such that $g \circ f$ is multiplication by $p^{r_1+r_2}$. If $m \in M$ is in $\mathrm{Ker}(f)$, then $p^{r_1}m$ is equal to $g(m')$ for some $m' \in M'$ with $p^{r_1+r_2}m' = fg(m') = p^{r_1}f(m) = 0$. It follows that $p^{2r_1+r_2}$ annihilates $\mathrm{Ker}(f)$. Also, this shows that the cokernel of f is a quotient of $M'/p^{r_1+r_2}M'$ and hence is also annihilated by some power of p. □

Lemma 5.1.13. — *Let $0 \to M' \to M \to M'' \to 0$ be an exact sequence of R-modules. If M' and M'' are free of finite type mod $\mathcal{T}$ then M is free of finite type mod $\mathcal{T}$.*

Proof. — Replacing the sequence first by the pushout via a map $M' \to \tilde{M}'$ with $\tilde{M}'$ free, and then by the pullback via a map $\tilde{M}'' \to M''$ with $\tilde{M}''$ free, we may assume that M' and M'' are free modules of finite rank. In this case the statement that M is free of finite type mod $\mathcal{T}$ is immediate since the exact sequence is split. □

Definition 5.1.14. — A projective system $L. \in \mathrm{ps}(R)$ is *free of finite type mod* $\mathcal{T}$ if $L := \varprojlim L_n$ is free of finite type mod $\mathcal{T}$ and the canonical map

$$c_L : \{L/p^n L\} \longrightarrow \{L_n\} \tag{5.1.14.1}$$

induces an isomorphism in $\mathrm{ps}(R)_{\mathbb{Q}}$. We write $\mathrm{ps}^{\mathrm{fft}}(R) \subset \mathrm{ps}(R)$ for the full subcategory of objects which are free of finite type mod $\mathcal{T}$.

5.1.15. — Note that the condition that a system $L. \in \mathrm{ps}(R)$ is free of finite type mod $\mathcal{T}$ depends only on the image of $L.$ in $\mathrm{ps}(R)_{\mathbb{Q}}$. We therefore extend the notion to $\mathrm{ps}(R)_{\mathbb{Q}}$, and write $\mathrm{ps}^{\mathrm{fft}}(R)_{\mathbb{Q}} \subset \mathrm{ps}(R)_{\mathbb{Q}}$ for the full subcategory of objects that are free of finite type mod $\mathcal{T}$.

Lemma 5.1.16. — *Let $L_{\cdot} \in \mathrm{ps}(R)$ be a projective system free of finite type mod $\mathcal{T}$. Then $R^1 \varprojlim\{L_n\}$ is annihilated by some power of p.*

Proof. — Let $I_n \subset L_n$ be the image of L, and let Q_n denote L_n/I_n. Since c_L induces an isomorphism in $\mathrm{ps}(R)_{\mathbb{Q}}$, there exists an integer r such that p^r annihilates all Q_n. On the other hand, since the system $\{I_n\}$ satisfies the Mittag-Leffler condition, we have $R^1 \varprojlim I_n = 0$. Consideration of the long exact sequence of derived functors arising from

$$0 \longrightarrow \{I_n\} \longrightarrow \{L_n\} \longrightarrow \{Q_n\} \longrightarrow 0 \tag{5.1.16.1}$$

then shows that

$$R^1 \varprojlim L_n \simeq R^1 \varprojlim Q_n, \tag{5.1.16.2}$$

and hence $R^1 \varprojlim L_n$ is also annihilated by p^r. □

Corollary 5.1.17. — *Let*

$$0 \longrightarrow A_{\cdot} \longrightarrow B_{\cdot} \longrightarrow C_{\cdot} \longrightarrow 0 \tag{5.1.17.1}$$

be an exact sequence in $\mathrm{ps}(R)$. *If $A_{\cdot}$ and $C_{\cdot}$ are free of finite type mod $\mathcal{T}$ then so is $B_{\cdot}$.*

Proof. — Set $A = \varprojlim A_n$, $B = \varprojlim B_n$, and $C = \varprojlim C_n$. Let r_1 (resp. r_2) be an integer such that the cokernel of $A \to A_n$ (resp. $C \to C_n$) is annihilated by p^{r_1} (resp. p^{r_2}) for all n, and let k_1 (resp. k_2) be an integer such that p^{k_1} (resp. p^{k_2}) annihilates $\mathrm{Ker}(A/p^nA \to A_n)$ (resp. $\mathrm{Ker}(C/p^nC \to C_n)$).

Consider the commutative diagram

$$\begin{array}{ccccccccc} 0 & \longrightarrow & A & \longrightarrow & B & \longrightarrow & C & \longrightarrow & R^1 \varprojlim A_n \\ & & \downarrow{\scriptstyle\alpha} & & \downarrow{\scriptstyle\beta} & & \downarrow{\scriptstyle\gamma} & & \\ 0 & \longrightarrow & A_n & \longrightarrow & B_n & \longrightarrow & C_n & \longrightarrow & 0 \end{array} \tag{5.1.17.2}$$

and let l be an integer so that p^l annihilates $R^1 \varprojlim A_n$. If $\bar{b} \in B_n$ is a section with image $\bar{c} \in C_n$, then $p^{r_2}\bar{c}$ lifts to an element in C, and hence $p^{l+r_2}\bar{c}$ lifts to an element $\tilde{b}$ in B. The difference $\beta(\tilde{b}) - p^{l+r_2}\bar{b}$ lies in A_n. Hence $p^{r_1}(\beta(\tilde{b}) - p^{l+r_2}\bar{b})$ lifts to an element of A. It follows that $p^{r_1+r_2+l}\bar{b}$ lifts to an element of B, and hence $\mathrm{Coker}(B \to B_n)$ is annihilated by $p^{r_1+r_2+l}$.

Conversely, suppose $b \in B$ maps to zero in B_n. Then the image $c \in C$ of b maps to zero in C_n so $p^{k_2}c = p^n c'$ for some c'. Since $p^l c'$ lifts to B, it follows that there exists an element $b' \in B$ such that $p^{k_2+l}b - p^n b'$ is in A, and maps to zero in A_n. This implies that there exists an element $a' \in A$ such that $p^{k_1}(p^{k_2+l}b - p^n b') = p^n a'$. Consequently

$$p^{k_1+k_2+l}b = p^n(a' + p^{k_1}b') \tag{5.1.17.3}$$

so $\mathrm{Ker}(B/p^nB \to B_n)$ is annihilated by $p^{k_1+k_2+l}$.

This shows that c_B induces an isomorphism in $\mathrm{ps}(R)_{\mathbb{Q}}$. Furthermore, the top row of the above commutative diagram and 5.1.13 shows that B is free of finite type mod $\mathcal{T}$. □

5.1.18. — Let

$$\varprojlim : \mathrm{ps}(R) \longrightarrow \mathrm{Mod}(R) \tag{5.1.18.1}$$

be the functor sending $L.$ to $\varprojlim L_n$. The functor $\varprojlim$ has a left adjoint $r : \mathrm{Mod}(R) \to \mathrm{ps}(R)$ sending M to $\{M/p^n M\}$. These functors induce adjoint functors which we denote by the same letters

$$\varprojlim : \mathrm{ps}(R)_{\mathbb{Q}} \longrightarrow \mathrm{Mod}(R)_{\mathbb{Q}}, \quad r : \mathrm{Mod}(R)_{\mathbb{Q}} \longrightarrow \mathrm{ps}(R)_{\mathbb{Q}}. \tag{5.1.18.2}$$

Proposition 5.1.19. — *The functor $\varprojlim$ restricted to $\mathrm{ps}^{\mathrm{fft}}(R)_{\mathbb{Q}}$ is fully faithful, and induces an equivalence of categories between the category $\mathrm{ps}^{\mathrm{fft}}(R)_{\mathbb{Q}}$ and the category of objects in $\mathrm{Mod}(R)_{\mathbb{Q}}$ which are free of finite type mod $\mathcal{T}$.*

Proof. — It follows from the definition 5.1.14 that for any two objects $L., L'_. \in \mathrm{ps}^{\mathrm{fft}}(R)$ there are isomorphisms
(5.1.19.1)

$$\mathrm{Hom}_{\mathrm{ps}(R)_{\mathbb{Q}}}(L., L'_.) \simeq \mathrm{Hom}_{\mathrm{ps}(R)}(\{L/p^n L\}, \{L'/p^n L'\}) \otimes \mathbb{Q} \simeq \mathrm{Hom}_{\mathrm{Mod}(R)}(L, L') \otimes \mathbb{Q},$$

where $L := \varprojlim L_n$ and $L' := \varprojlim L'_n$. This proves the full faithfulness. That the essential image image consists of modules free of finite type mod $\mathcal{T}$ follows from the definition of free of finite type mod $\mathcal{T}$. Finally if M is an R-module free of finite type mod $\mathcal{T}$, then we claim that the map $M \to \varprojlim M/p^n M$ is an isomorphism mod $\mathcal{T}$. To see this choose (using 5.1.12) a map $s : M' \to M$ whose kernel and cokernel are in $\mathcal{T}$ with M' a free R-module of finite type. Then for some $k \geq 0$ there exists a map $t : M \to M'$ such that the composites $s \circ t$ and $t \circ s$ are both equal to multiplication by p^k. Passing first to the reductions and then to the projective limits we obtain maps

$$\hat{s} : \varprojlim M'/p^n M' \longrightarrow \varprojlim M/p^n M \tag{5.1.19.2}$$

and

$$\hat{t} : \varprojlim M/p^n M \longrightarrow \varprojlim M'/p^n M' \tag{5.1.19.3}$$

such that the composites $\hat{s} \circ \hat{t}$ and $\hat{t} \circ \hat{s}$ are both equal to multiplication by p^k. This implies that $\hat{s}$ is an isomorphism mod $\mathcal{T}$, and from the commutative diagram

$$\begin{array}{ccc} M' & \xrightarrow{s} & M \\ {\scriptstyle a}\downarrow & & \downarrow{\scriptstyle b} \\ \varprojlim M'/p^n M' & \xrightarrow{\hat{s}} & \varprojlim M/p^n M \end{array} \tag{5.1.19.4}$$

we deduce that the map labelled b is also an isomorphism mod $\mathcal{T}$, since s is an isomorphism mod $\mathcal{T}$ by assumption, a is an isomorphism since M' is free of finite type and R is p-adically complete, and $\hat{s}$ is an isomorphism mod $\mathcal{T}$ by the above. □

Example 5.1.20. — An important example of projective systems free of finite type mod $\mathcal{T}$ arises as follows. Let k be a perfect field of characteristic $p > 0$ and let W denote the Witt vectors of k. If $C^\bullet \in D(W)$ is a perfect complex of W-modules, then the system

$$L^*_\cdot := \{H^*(C^\bullet \otimes^{\mathbb{L}}_W (W/p^n))\} \tag{5.1.20.1}$$

is a projective system free of finite type mod $\mathcal{T}$.

This can be seen as follows. Let $\tilde{L}^*$ denote the graded group $H^*(C^\bullet)$. For any integer n, there is an exact sequence

$$0 \longrightarrow W \xrightarrow{p^n} W \longrightarrow W/p^n \longrightarrow 0 \tag{5.1.20.2}$$

which gives a distinguished triangle

$$C^\bullet \xrightarrow{p^n} C^\bullet \longrightarrow C^\bullet \otimes^{\mathbb{L}} W/p^n \longrightarrow C^\bullet[1]. \tag{5.1.20.3}$$

Looking at the associated long exact sequence of cohomology we obtain short exact sequences for all m

$$0 \longrightarrow \tilde{L}^m \otimes W/p^n \longrightarrow L^m_n \longrightarrow \mathrm{Tor}^1_W(W/p^n, \tilde{L}^{m+1}) \longrightarrow 0. \tag{5.1.20.4}$$

Since $C^\bullet$ is perfect, the group $\tilde{L}^{m+1}$ is finitely generated and in particular has bounded p-torsion. It follows that there exists an integer r such that p^r annihilates $\mathrm{Tor}^1_W(W/p^n, \tilde{L}^{m+1})$ for all n. Consequently the map

$$s : \{\tilde{L}^m/p^n\tilde{L}^m\} \longrightarrow L^m_\cdot \tag{5.1.20.5}$$

induces an isomorphism in $\mathrm{ps}(W)_{\mathbb{Q}}$. This implies that there exists a map $t : L^m_\cdot \to \{\tilde{L}^m/p^n\tilde{L}^m\}$ such that $s \circ t$ and $t \circ s$ are both equal to multiplication by p^k for some $k \geq 1$. Therefore the map

$$\varprojlim \tilde{L}^m/p^n\tilde{L}^m \longrightarrow \varprojlim L^m_n \tag{5.1.20.6}$$

is an isomorphism mod $\mathcal{T}$. Since $\tilde{L}^m$ is of finite type over W (and hence free of finite type mod $\mathcal{T}$), we also have

$$\tilde{L}^m \simeq \varprojlim \tilde{L}^m/p^n\tilde{L}^m, \tag{5.1.20.7}$$

and therefore $\tilde{L}^m$ is isomorphic to $\varprojlim L_n$ mod $\mathcal{T}$. In particular, $\varprojlim L_n$ is free of finite type mod $\mathcal{T}$.

Remark 5.1.21. — The argument given in 5.1.20 is fairly standard, as similar reasonings are often used in the development of the theory of $\mathbb{Q}_\ell$-coefficients in étale cohomology. In this theory, however, one usually works over a noetherian ring. The main challenge of this chapter is to study what happens over a non-noetherian ring (in particular the ring $W\langle t\rangle$ considered in the next section) as these arise naturally in the crystalline theory.

Example 5.1.22. — Let k be a perfect field of characteristic p, and let W be its ring of Witt vectors. Then by [**34**, II.2.13] for a smooth proper k-scheme X the W-modules $H^j(X, W\Omega^i_X)$ are free of finite type mod $\mathcal{T}$, though often the torsion subgroups of these modules are infinite. This example (and the stack-theoretic generalizations considered in section 4.5) plays a key role in what follows.

5.2. Ogus' twisted inverse limit construction

5.2.1. — Let $M. = \{M_n\}$ be an inverse system of abelian groups and let $m \in \mathbb{N}$ be a natural number. Define $\varprojlim^m M.$ to be the subgroup of $\prod_n M_n$ consisting of elements (y_n) with $\pi_n(y_{n+1}) = p^m y_n$ for all n, where $\pi_n : M_{n+1} \to M_n$ denotes the projection map.

5.2.2. — The case of interest in this work is the following. Let R be a p-adically complete and separated ring, and let ν_n be a sequence of natural numbers such that $\{\nu_n - nm\}$ is eventually increasing and $\lim_n(\nu_n - nm) = \infty$.

Define a functor

$$\Omega_\nu : \mathrm{Mod}(R) \longrightarrow \mathrm{Mod}(R) \tag{5.2.2.1}$$

as follows. Choose $n_0 \in \mathbb{N}$ such that $\nu_{n+1} - (n+1)m \geq \nu_n - nm$ for all $n \geq n_0$. This implies that $\nu_{n+1} \geq \nu_n$ for all $n \geq n_0$, and therefore we get a projective system $\{M/p^{\nu_n}M\}_{n\geq n_0}$ with transition maps the natural projections

$$M/p^{\nu_{n+1}}M \longrightarrow M/p^{\nu_n}M. \tag{5.2.2.2}$$

Define Ω_ν by sending $M \in \mathrm{Mod}(R)$ to

$$\varprojlim{}^m \{M/p^{\nu_n}M\}_{n\geq n_0} \tag{5.2.2.3}$$

Note that this is independent of the choice of n_0.

Denote by $\xi : M \to \Omega_\nu(M)$ the canonical map induced by the maps $\times p^{nm} : M/p^{\nu_n}M \to M/p^{\nu_n}M$.

Lemma 5.2.3. — *Let M be a p-adically complete and separated R-module. If M has bounded p-torsion, then so does $\Omega_\nu(M)$.*

Proof. — Let r be an integer such that the torsion subgroup of M is annihilated by p^r.

Let τ be a natural number such that $\nu_{n+1} - \nu_n > m$ for all $n > \tau$, and let $(y_n) \in \Omega_\nu(M)$ be an element annihilated by p^s for some s. We claim that if $s > \sup\{\tau, r+m\}$ then (y_n) is also annihilated by p^{s-1}.

For each n choose a lifting $\tilde{y}_n \in M$ of y_n. Then $p^s\tilde{y}_n = p^{\nu_n}\tilde{y}'_n$ for some $\tilde{y}'_n \in M$. Since the torsion subgroup of M is annihilated by p^r, this implies that if $\nu_n \geq s$, then $p^r\tilde{y}_n = p^{\nu_n - s + r}\tilde{y}'_n$. On the other hand, we have

$$p^{\nu_{n+1}-s+r}\tilde{y}'_{n+1} = p^r\tilde{y}_{n+1} \equiv p^{r+m}\tilde{y}_n = p^{\nu_n - s+r+m}\tilde{y}'_n \pmod{p^{\nu_n}}. \tag{5.2.3.1}$$

It follows that if $s > r+m$ and $\nu_{n+1} - \nu_n > m$ (and still $\nu_n \geq s$), then $\tilde{y}'_n = p\lambda_n + t$, where $\lambda_n \in M$ and $t \in M$ is p-torsion. From this it follows that if $s > r+m$, $\nu_{n+1} - \nu_n > m$, and $\nu_n \geq s$ then

$$p^{(s-1)-r}p^r\tilde{y}_n = p^{(s-1)-r}p^{\nu_n - s+r}p\lambda_n = p^{\nu_n}\lambda_n \equiv 0 \pmod{p^{\nu_n}}. \tag{5.2.3.2}$$

Therefore, y_n is annihilated by p^{s-1} whenever $\nu_n \geq s$. Since y_n is trivially annihilated by p^{s-1} when $\nu_n < s$ this implies the lemma. □

Proposition 5.2.4 (**[59**, Lemma 18]). — *Let M be a p-adically complete and separated R-module with bounded p-torsion. Then $\xi : M \to \Omega_\nu(M)$ is an isomorphism mod $\mathcal{T}$.*

Proof. — As above let r be an integer such that the torsion subgroup of M is annihilated by p^r.

If $x \in M$ is in the kernel of ξ, then $p^{nm}x \equiv 0 \pmod{p^{\nu_n}M}$ for all n. Let x'_n denote an element such that $p^{nm}x = p^{\nu_n}x'_n$. Let n_0 denote an integer such that $\nu_n \geq nm$ for all $n \geq n_0$. If $n \geq n_0$ then the element $x - p^{\nu_n - nm}x'_n$ is annihilated by p^{nm} and hence also killed by p^r. It follows that $p^r x = p^{\nu_n - nm+r}x'_n$. In particular, $p^r x \in \cap_n p^{\nu_n - nm+r}M$ and $\nu_n - nm + r$ tends to infinity by assumption. Since M is p-adically separated and complete it follows that $p^r x = 0$, so the kernel of ξ is annihilated by p^r.

To see that the cokernel of ξ has bounded p-torsion, let $(y_n) \in \Omega_\nu(M)$ be an element, and choose for each n a lifting $\tilde{y}_n \in M$ of y_n. Let n_0 be an integer such that $\nu_n \geq nm + m$ for all $n \geq n_0$.

Lemma 5.2.5. — *For all $n \geq n_0$, $\tilde{y}_n \in p^{(n-n_0)m}M$.*

Proof. — The proof is by induction on n, the case $n = n_0$ being trivial. So we prove the result for $n+1$ assuming it holds for n. Since $\pi(y_{n+1}) = p^m y_n$, we can write $\tilde{y}_{n+1} = p^m\tilde{y}_n + p^{\nu_n}\lambda_n$ for some $\lambda_n \in M$. By induction, there exists $z \in M$ such that $\tilde{y}_n = p^{(n-n_0)m}z$, and hence

$$\tilde{y}_{n+1} = p^{(n+1-n_0)m}z + p^{\nu_n}\lambda_n. \tag{5.2.5.1}$$

Since $n \geq n_0$ we have $\nu_n \geq nm + m$, and so

$$\tilde{y}_{n+1} = p^{((n+1)-n_0)m}(z + p^{\nu_n + mn_0 - nm - m}\lambda). \tag{5.2.5.2}$$

□

For each $n \geq n_0$ choose an element $\tilde{x}_n \in M$ such that $\tilde{y}_n = p^{(n-n_0)m}\tilde{x}_n$. Then

$$p^{(n+1-n_0)m}\tilde{x}_{n+1} \equiv p^{(n+1-n_0)m}\tilde{x}_n \pmod{p^{\nu_n}}. \tag{5.2.5.3}$$

Equivalently there exists $\lambda \in M$ such that

$$p^{(n+1-n_0)m}\tilde{x}_{n+1} = p^{(n+1-n_0)m}\tilde{x}_n + p^{\nu_n}\lambda. \tag{5.2.5.4}$$

Since $\nu_n \geq nm + m$ by assumption, this implies that

$$\tilde{x}_{n+1} = \tilde{x}_n + p^{\nu_n - (n+1-n_0)m}\lambda + t, \tag{5.2.5.5}$$

where $t \in M$ is a torsion element. Therefore

$$p^r\tilde{x}_{n+1} = p^r\tilde{x}_n + p^{\nu_n+r-(n+1-n_0)m}\lambda. \tag{5.2.5.6}$$

In particular, since $\varinjlim \nu_n - nm = \infty$ the elements $\{p^r x_n\}$ define a Cauchy sequence in M. Let $\tilde{x} \in M$ be the limit.

We claim that $\xi(\tilde{x}) - p^{n_0m+r}(y_n) \in \Omega_\nu(M)$ is a torsion element. For this note that by construction for $n \geq n_0$ we have

$$\tilde{x} \equiv p^r\tilde{x}_n \pmod{p^{\nu_n+r-(n+1-n_0)m}}, \tag{5.2.5.7}$$

and hence

$$p^{nm}\tilde{x} \equiv p^{nm+r}\tilde{x}_n \equiv p^{n_0m+r}\tilde{y}_n \pmod{p^{\nu_n}} \tag{5.2.5.8}$$

It follows that if $(z_n) \in \Omega_\nu(M)$ denotes $\xi(\tilde{x}) - p^{n_0m+r}(y_n)$, then $z_n = 0$ for $n \geq n_0$. In particular, $p^{\nu_{n_0}}$ annihilates (z_n). Consequently, $\mathrm{Coker}(\xi)$ is annihilated by $p^{n_0m+\nu_{n_0}+r}$. □

Remark 5.2.6. — If δ_n is a second sequence of natural numbers such that $\{\delta_n - nm\}$ is eventually increasing and such that $\lim_n(\delta_n - nm) = \infty$, then the sequence $\{\delta_n + \nu_n\}$ is also such a sequence. Furthermore, if $\delta_n \geq \nu_n$ for all n, then for any p-adically complete and separated R-module M with bounded p-torsion the natural maps

$$M/p^{\delta_n}M \longrightarrow M/p^{\nu_n}M \tag{5.2.6.1}$$

induce an isomorphism mod $\mathcal{T}$

$$\Omega_\delta(M) \longrightarrow \Omega_\nu(M) \tag{5.2.6.2}$$

by 5.2.4.

5.3. The main results on F-crystals over $W\langle t\rangle$

5.3.1. — Let k be a perfect field, W the ring of Witt vectors of k, and let σ denote the canonical lift of Frobenius to W. Let $W\langle t\rangle$ denote the p-adic completion of the divided power envelope of the closed immersion corresponding to the surjection $W[t] \to k$ sending t to 0. Denote by $F : W\langle t\rangle \to W\langle t\rangle$ the lifting of Frobenius induced by σ on W and $t \mapsto t^p$. Denote by $\mathcal{I} \subset W\langle t\rangle$ the divided power ideal generated by t, so that $W\langle t\rangle/\mathcal{I} \simeq W$, and by $\mathcal{I}^{[r]} \subset \mathcal{I}$ the ideal generated by elements $h^{[r']}$ for $h \in \mathcal{I}$ and $r' \geq r$.

The ring $W\langle t\rangle$ can be described explicitly as the subring of $K[\![t]\!]$, where K denotes the field of fractions of W, consisting of power series $\sum_{i\geq 0} a_i t^i/i!$, where $a_i \in W$ and the sequence $\{a_i\}$ tends to 0 with respect to the p-adic norm.

Definition 5.3.2. — Let R be a p-adically complete and separated ring with a lifting $F : R \to R$ of Frobenius. An *F-structure* on an object $M \in \mathrm{Mod}(R)_{\mathbb{Q}}$ is a morphism $\varphi : F^*M \to M$. Denote by $F\text{-}\mathrm{Mod}(R)_{\mathbb{Q}}$ the category of pairs (M, φ), where $M \in \mathrm{Mod}(R)_{\mathbb{Q}}$ and φ is an F-structure on M. Denote by $F\text{-}\mathrm{Mod}^{fft}(R)_{\mathbb{Q}} \subset F\text{-}\mathrm{Mod}(R)_{\mathbb{Q}}$ the full subcategory of pairs (M, φ), where M is free of finite type mod $\mathcal{T}$ and $\varphi : F^*M \to M$ is an isomorphism.

Remark 5.3.3. — The category $F\text{-}\mathrm{Mod}(R)_{\mathbb{Q}}$ is abelian. The kernel (resp. cokernel) of a morphism $f : (M, \varphi_M) \to (N, \varphi_N)$ is given by the kernel (resp. cokernel) of the underlying morphism $M \to N$ with the induced F-structure. If in addition $F : R \to R$ is flat and R is noetherian, then $F\text{-}\mathrm{Mod}^{fft}(R)_{\mathbb{Q}} \subset F\text{-}\mathrm{Mod}(R)_{\mathbb{Q}}$ is an abelian subcategory. In general, however, there is no reason to expect the subcategory $F\text{-}\mathrm{Mod}^{fft}(R)_{\mathbb{Q}}$ to be abelian (note that $W\langle t\rangle$ is not noetherian and the lifting of Frobenius $W\langle t\rangle \to W\langle t\rangle$ is not flat as the Frobenius morphism $k\langle t\rangle \to k\langle t\rangle$ is not flat). However, we will show that $F\text{-}\mathrm{Mod}^{fft}(W\langle t\rangle)_{\mathbb{Q}}$ is abelian in 5.3.16 below.

The following is the main result of this section:

Theorem 5.3.4. — *The functor*

$$P : F\text{-}\mathrm{Mod}^{fft}(W)_{\mathbb{Q}} \longrightarrow F\text{-}\mathrm{Mod}(W\langle t\rangle)_{\mathbb{Q}}, \quad (N, \varphi) \longmapsto (N \otimes_W W\langle t\rangle, \varphi) \tag{5.3.4.1}$$

is fully faithful with essential image $F\text{-}\mathrm{Mod}^{fft}(W\langle t\rangle)_{\mathbb{Q}}$. In particular, by 5.1.13 the essential image of P is closed under extensions.

The proof is in several steps 5.3.5–5.3.15.

5.3.5. — For $M \in \mathrm{Mod}(W\langle t\rangle)$ let $\overline{M} \in \mathrm{Mod}(W)$ denote $M/\mathcal{I}M$. The composite

$$\mathrm{Mod}(W\langle t\rangle) \xrightarrow{M \mapsto \overline{M}} \mathrm{Mod}(W) \longrightarrow \mathrm{Mod}(W)_{\mathbb{Q}} \tag{5.3.5.1}$$

factors uniquely through a functor

$$\mathrm{Mod}(W\langle t\rangle)_{\mathbb{Q}} \longrightarrow \mathrm{Mod}(W)_{\mathbb{Q}} \tag{5.3.5.2}$$

which we again denote by $M \mapsto \overline{M}$.

For $M \in \mathrm{Mod}(W\langle t\rangle)$ there is a canonical isomorphism $\overline{F^*M} \simeq \sigma^*\overline{M}$. It follows that $M \mapsto \overline{M}$ induces a functor

$$Q : F\text{-}\mathrm{Mod}(W\langle t\rangle)_{\mathbb{Q}} \longrightarrow F\text{-}\mathrm{Mod}(W)_{\mathbb{Q}}, \quad (M, \varphi_M) \longmapsto (\overline{M}, \varphi_{\overline{M}}). \tag{5.3.5.3}$$

The composite functor

$$F\text{-}\mathrm{Mod}(W)_{\mathbb{Q}} \xrightarrow{P} F\text{-}\mathrm{Mod}(W\langle t\rangle)_{\mathbb{Q}} \xrightarrow{Q} F\text{-}\mathrm{Mod}(W)_{\mathbb{Q}} \tag{5.3.5.4}$$

is canonically isomorphic to the identity functor. In particular P is faithful.

5.3.6. — To see that P is fully faithful, let $(M,\varphi_M),(N,\varphi_N)$ be two objects of $F\text{-Mod}^{fft}(W)_{\mathbb{Q}}$, and let $f : P(M,\varphi_M)\to P(N,\varphi_N)$ be a map in $F\text{-Mod}(W\langle t\rangle)_{\mathbb{Q}}$. Denote by $\bar f : (M,\varphi_M)\to(N,\varphi_N)$ the map obtained by applying Q. We need to show that $f=P(\bar f)$.

For this note that after replacing M and N by the quotients by their p-torsion subgroups we may assume that M and N are finitely generated free W-modules. In addition, after replacing φ_M by $p^d\varphi_M$ and φ_N by $p^d\varphi_N$ for some $d\geq 1$, we may assume that φ_M and φ_N are represented by maps, which we denote by the same letters, in $\text{Mod}(W)$. Finally note that after replacing f by p^sf for some s we may also assume that f is represented by a map in $\text{Mod}(W)$. To prove that $f=P(\bar f)$, it suffices to show that the two maps

$$f,P(\bar f) : (M\otimes_W W\langle t\rangle)\otimes\mathbb{Q}\longrightarrow(N\otimes_W W\langle t\rangle)\otimes\mathbb{Q} \tag{5.3.6.1}$$

are equal, and for this it suffices to show that for every r they are equal modulo the ideal $\mathcal{I}^{[r]}\cdot(W\langle t\rangle\otimes\mathbb{Q})$. Let $\psi : M\to M$ be a map such that $\varphi_M\circ\psi=p^d$ and $\psi\circ\varphi_M=p^d$ for some d. Then for $m\in M$ we have
(5.3.6.2)

$$f(m)=\frac{1}{p^{dr}}\varphi_N^r(f(\psi^r(m)))\equiv\frac{1}{p^{dr}}\varphi_N^r(P(\bar f)(\psi^r(m)))\equiv P(\bar f)(m)\pmod{\varphi_N^r(\mathcal{I})}.$$

Since $\varphi_N^r(\mathcal{I})\subset\mathcal{I}^{[r]}$ this proves that $f\equiv P(\bar f)\pmod{\mathcal{I}^{[r]}}$ and hence also that $f=P(\bar f)$.

5.3.7. — Note that the essential image of P is clearly contained in $F\text{-Mod}^{fft}(W\langle t\rangle)_{\mathbb{Q}}$. To prove that P is essentially surjective onto $F\text{-Mod}^{fft}(W\langle t\rangle)_{\mathbb{Q}}$, we need to show that any object $(M,\varphi_M)\in F\text{-Mod}^{fft}(W\langle t\rangle)_{\mathbb{Q}}$ is in the essential image. Since M is free of finite type mod $\mathcal{T}$, we can without loss of generality assume that M is actually a free $W\langle t\rangle$-module of finite type and that φ_M is given by a map $\varphi_M : F^*M\to M$ in $\text{Mod}(W\langle t\rangle)$. Denote by $(\overline{M},\varphi_{\overline{M}})\in F\text{-Mod}^{fft}(W)_{\mathbb{Q}}$ the image of (M,φ_M) under Q. We construct a section $s:\overline{M}\to M$ in $\text{Mod}(W)_{\mathbb{Q}}$ compatible with φ_M and $\varphi_{\overline{M}}$ such that the induced map

$$P(\overline{M},\varphi_{\overline{M}})\longrightarrow(M,\varphi_M) \tag{5.3.7.1}$$

is an isomorphism mod $\mathcal{T}$.

5.3.8. — Let H denote $\text{Hom}_{W\langle t\rangle}(\overline{M}\otimes W\langle t\rangle,M)$. The $W\langle t\rangle$-module H is free of finite rank. There is a natural isomorphism

$$\varphi_H : (F^*H)\otimes\mathbb{Q}\simeq\text{Hom}_{W\langle t\rangle}(\sigma^*\overline{M}\otimes W\langle t\rangle,F^*M)\otimes\mathbb{Q}\to H\otimes\mathbb{Q} \tag{5.3.8.1}$$

defined as follows. Let $\psi : M\to F^*M$ be a map such that $\varphi_M\circ\psi=p^d$ and $\psi\circ\varphi_M=p^d$ for some d, and let $\bar\psi:\overline{M}\to\sigma^*\overline{M}$ denote the map obtained by reduction. Then φ_H is defined by sending $f:\sigma^*\overline{M}\otimes W\langle t\rangle\to F^*M$ to $1/p^d$ times the composite

$$\overline{M}\otimes W\langle t\rangle\xrightarrow{\ \bar\psi\ }\sigma^*\overline{M}\otimes W\langle t\rangle\xrightarrow{\ f\ }F^*M\xrightarrow{\ \varphi_M\ }M \tag{5.3.8.2}$$

If $\psi' : M \to F^*M$ is a second map such that $\varphi_M \circ \psi' = p^{d'}$ and $\psi' \circ \varphi_M = p^{d'}$ for some d', then $p^{d'}\psi = p^d\psi'$, and hence φ_H is independent of the choice of ψ.

To give a map $s : \overline{M} \otimes W\langle t\rangle \to M$ in $\mathrm{Mod}(W\langle t\rangle)_{\mathbb{Q}}$ compatible with $\varphi_{\overline{M}}$ and φ_M is then equivalent to giving an element $h \in H \otimes \mathbb{Q}$ invariant under the endomorphism

$$\Lambda_H : H \otimes \mathbb{Q} \xrightarrow{\text{can}} F^*H \otimes \mathbb{Q} \xrightarrow{\varphi_H} H \otimes \mathbb{Q}, \tag{5.3.8.3}$$

where can denotes the canonical map $H \to F^*H = H \otimes_{W\langle t\rangle, F} W\langle t\rangle$ sending $h \in H$ to $h \otimes 1$.

5.3.9. — Fix a map $\psi : M \to F^*M$ such that $\varphi_M \circ \psi = p^m$ and $\psi \circ \varphi_M = p^m$ for some m, and let $\bar{\psi} : \overline{M} \to \sigma^*\overline{M}$ be the reduction. Define a sequence ν_n as in 5.2.2 by $\nu_n := \mathrm{ord}_p(p^n!) = (p^n - 1)/(p-1)$, and for $M \in \mathrm{Mod}(W\langle t\rangle)$ define $\Omega_\nu(M)$ as in 5.2.2.

For H as above, the elements of $\Omega_\nu(H)$ can be described as a collection of maps

$$h_n : (\overline{M} \otimes W\langle t\rangle) \otimes_{\mathbb{Z}} \mathbb{Z}/(p^{\nu_n}) \longrightarrow M \otimes_{\mathbb{Z}} \mathbb{Z}/(p^{\nu_n}) \tag{5.3.9.1}$$

such that $\pi_n(h_{n+1}) = p^m h_n$, where π_n denotes the reduction map.

Lemma 5.3.10. — *For any integer $n \geq 1$, there is a canonical isomorphism $F^{n*}M \otimes \mathbb{Z}/(p^n!) \simeq (\sigma^{n*}\overline{M}) \otimes_W (W\langle t\rangle/(p^n!))$.*

Proof. — For every integer n, we have $t^{p^n} = p^n! t^{[p^n]}$ and hence the map

$$F^n : \mathrm{Spec}(W\langle t\rangle/(p^n!)) \longrightarrow \mathrm{Spec}(W\langle t\rangle/(p^n!)) \tag{5.3.10.1}$$

factors through a map $\rho : \mathrm{Spec}(W\langle t\rangle/(p^n!)) \to \mathrm{Spec}(W/(p^n!))$ over

$$\sigma^n : \mathrm{Spec}(W/(p^n!)) \longrightarrow \mathrm{Spec}(W/(p^n!)). \tag{5.3.10.2}$$

It follows that

$$F^{n*}M \otimes \mathbb{Z}/(p^n!) \simeq \rho^*\overline{M} \otimes (\mathbb{Z}/p^n!) \simeq (\sigma^{n*}\overline{M}) \otimes_W (W\langle t\rangle/(p^n!)). \tag{5.3.10.3}$$ □

5.3.11. — Define h_n to be the composite

$$\begin{aligned}(\overline{M} \otimes W\langle t\rangle) \otimes_{\mathbb{Z}} \mathbb{Z}/(p^{\nu_n}) &\xrightarrow{\bar{\psi}^n} (\sigma^{n*}\overline{M}) \otimes_W (W\langle t\rangle/(p^n!)) \\ &\simeq F^{n*}M \otimes \mathbb{Z}/(p^n!) \xrightarrow{\varphi^n_M} M \otimes \mathbb{Z}/(p^n!).\end{aligned} \tag{5.3.11.1}$$

Here we are abusing notation and writing $\bar{\psi}^n$ for the composite
(5.3.11.2)

$$\overline{M} \otimes W\langle t\rangle \xrightarrow{\bar{\psi}} \sigma^*M \otimes W\langle t\rangle \xrightarrow{\sigma^*(\bar{\psi})} \cdots \xrightarrow{\sigma^{n-1*}(\bar{\psi})} \sigma^{n*}\overline{M} \otimes W\langle t\rangle,$$

and similarly for φ^n_M. The equalities $\varphi_M \circ \psi = p^m$ and $\psi \circ \varphi_M = p^m$ imply that h_{n+1} reduces to $p^m h_n$ modulo $p^n!$. We thus obtain an element $h \in \Omega_\nu(H)$. Since the map $\xi : H \to \Omega_\nu(H)$ is an isomorphism in $\mathrm{Mod}(W\langle t\rangle)_{\mathbb{Q}}$ by 5.2.4, we obtain a morphism

$$f \in \mathrm{Hom}_{W\langle t\rangle}(\overline{M} \otimes W\langle t\rangle, M) \otimes \mathbb{Q} \tag{5.3.11.3}$$

with $\xi(f) = h$. To complete the proof of 5.3.4, we show that f is an isomorphism compatible with φ_M and $\varphi_{\overline{M}}$ and that f reduces modulo $\mathcal{J}$ to the identity.

5.3.12. — To see that the diagram

$$\begin{array}{ccc} (\overline{M}\otimes W\langle t\rangle)\otimes\mathbb{Q} & \xrightarrow{f} & M\otimes\mathbb{Q} \\ \tilde{\varphi}_{\overline{M}}\downarrow & & \downarrow\tilde{\varphi}_M \\ (\overline{M}\otimes W\langle t\rangle)\otimes\mathbb{Q} & \xrightarrow{f} & M\otimes\mathbb{Q} \end{array} \tag{5.3.12.1}$$

commutes, where $\tilde{\varphi}_{\overline{M}}$ and $\tilde{\varphi}_M$ denote the semi-linear maps obtained from $\varphi_{\overline{M}}$ and φ_M, let w be an integer such that $p^w f$ is obtained from a map g in $\mathrm{Mod}(W\langle t\rangle)$. To show that the diagram (5.3.12.1) commutes, it suffices to show that

$$g\circ\tilde{\varphi}_{\overline{M}}=\tilde{\varphi}_M\circ g. \tag{5.3.12.2}$$

Since $\varinjlim_n(\nu_n-nm)=\infty$, it suffices to show that

$$p^{nm}g\circ\tilde{\varphi}_{\overline{M}}\equiv\tilde{\varphi}_M\circ p^{nm}g \pmod{p^n!}. \tag{5.3.12.3}$$

Since $\xi(f)=h\in\Omega_\nu(H)$, the reduction of $p^{nm}g$ modulo $p^n!$ is equal to the map $p^w h_n$. Therefore, to prove that (5.3.12.1) commutes it suffices to show that the diagram

$$\begin{array}{ccc} (\overline{M}\otimes W\langle t\rangle)\otimes\mathbb{Z}/p^n! & \xrightarrow{h_n} & M\otimes\mathbb{Z}/p^n! \\ \tilde{\varphi}_{\overline{M}}\downarrow & & \downarrow\tilde{\varphi}_M \\ (\overline{M}\otimes W\langle t\rangle)\otimes\mathbb{Z}/p^n! & \xrightarrow{h_n} & M\otimes\mathbb{Z}/p^n! \end{array} \tag{5.3.12.4}$$

commutes. Consider the diagram
(5.3.12.5)

$$\begin{array}{ccccc} \sigma^*(\overline{M}\otimes W\langle t\rangle)\otimes\mathbb{Z}/p^n! & \xrightarrow{\mathrm{id}} & \sigma^*(\overline{M}\otimes W\langle t\rangle)\otimes\mathbb{Z}/p^n! & \xrightarrow{\varphi_{\overline{M}}} & (\overline{M}\otimes W\langle t\rangle)\otimes\mathbb{Z}/p^n! \\ \sigma^*(\bar{\psi}^{n-1})\downarrow & & \sigma^*(\bar{\psi}^n)\downarrow & & \bar{\psi}^n\downarrow \\ \sigma^{n*}(\overline{M}\otimes W\langle t\rangle)\otimes\mathbb{Z}/p^n! & \xrightarrow{\sigma^{n*}(\bar{\psi})} & \sigma^{n+1*}(\overline{M}\otimes W\langle t\rangle)\otimes\mathbb{Z}/p^n! & & \sigma^{n*}(\overline{M}\otimes W\langle t\rangle)\otimes\mathbb{Z}/p^n! \\ \simeq\downarrow & & \downarrow\simeq & & \downarrow\simeq \\ F^{n*}M\otimes\mathbb{Z}/p^n! & \xrightarrow{\psi} & F^{n+1*}M\otimes\mathbb{Z}/p^n! & & F^{n*}(M)\otimes\mathbb{Z}/p^n! \\ p^mF^*(\varphi_M^{n-1})\downarrow & & F^*(\varphi^n)\downarrow & & \varphi_M^n\downarrow \\ F^*M\otimes\mathbb{Z}/p^n! & \xrightarrow{\mathrm{id}} & F^*M\otimes\mathbb{Z}/p^n! & \xrightarrow{\varphi} & M\otimes\mathbb{Z}/p^n!. \end{array}$$

Since $\varphi_M\circ\psi=p^m$ and $\psi\circ\varphi_M=p^m$, the left three small squares commute and the large outside square commutes. It follows that the right rectangle also commutes. From this and the definition of h_n it follows that (5.3.12.4) commutes.

5.3.13. — Similarly, to prove that the reduction of f modulo $\mathcal{J}$ is equal to the identity, it suffices to show that the reduction of g is multiplication by p^w (where g and w are as in 5.3.12). For this it suffices to show that g reduces to multiplication by p^w modulo $p^n!/p^{nm}$ for all n sufficiently large, and for this in turn it suffices to show that h_n reduces to multiplication by p^{nm} modulo $\mathcal{J}$. This follows from the definition of h_n and the relation $\varphi_{\overline{M}} \circ \bar{\psi} = p^m$.

5.3.14. — Finally to see that f is an isomorphism, we define an inverse as follows. Define

$$H' := \mathrm{Hom}_{W\langle t\rangle}(M, \overline{M} \otimes W\langle t\rangle) \qquad (5.3.14.1)$$

Let $h' \in \Omega_\nu(H')$ be the element corresponding to the maps $h'_n : M \otimes \mathbb{Z}/p^n! \to (\overline{M} \otimes W\langle t\rangle) \otimes \mathbb{Z}/p^n!$ defined to be the composite

$$\begin{aligned} M\otimes\mathbb{Z}/(p^n!) &\xrightarrow{\psi_M^n} F^{n*}M \otimes \mathbb{Z}/(p^n!) \\ &\simeq (\sigma^{n*}\overline{M}) \otimes_W (W\langle t\rangle/(p^n!)) \xrightarrow{\varphi_{\overline{M}}^n} (\overline{M} \otimes W\langle t\rangle) \otimes_{\mathbb{Z}} \mathbb{Z}/(p^{\nu_n}). \end{aligned} \qquad (5.3.14.2)$$

Since H' is a free $W\langle t\rangle$-module of finite rank, the map $\xi : H' \otimes \mathbb{Q} \to \Omega_\nu(H') \otimes \mathbb{Q}$ is an isomorphism (5.2.4), and we let f' denote $\xi^{-1}(h') \in H' \otimes \mathbb{Q}$. We claim that f' is an inverse to f.

5.3.15. — To see that $f \circ f' = \mathrm{id}$ and $f' \circ f = \mathrm{id}$, note first that $\varinjlim(\nu_n - 2nm) = \infty$ and the sequence $\nu_n - 2nm$ is also eventually increasing. For a module $N \in \mathrm{Mod}(W\langle t\rangle)$ define $\widetilde{\Omega}_\nu(N)$ to be $\varprojlim^{2m} N/p^{\nu_n}N$. It follows from the definitions that if

$$c : H \times H' \longrightarrow \mathrm{End}(\overline{M} \otimes W\langle t\rangle), \quad d : H' \times H \longrightarrow \mathrm{End}(M) \qquad (5.3.15.1)$$

are the maps defined by composition, then the diagrams
(5.3.15.2)

$$\begin{array}{ccc} H \times H' & \xrightarrow{c} & \mathrm{End}(\overline{M} \otimes W\langle t\rangle) \\ \downarrow{\scriptstyle \xi\times\xi} & & \downarrow{\scriptstyle \xi} \\ \Omega_\nu(H) \times \Omega_\nu(H') & \xrightarrow{\Omega_\nu(c)} & \widetilde{\Omega}_\nu(\mathrm{End}(\overline{M} \otimes W\langle t\rangle)) \end{array} \qquad \begin{array}{ccc} H' \times H & \xrightarrow{d} & \mathrm{End}(M) \\ \downarrow{\scriptstyle \xi\times\xi} & & \downarrow{\scriptstyle \xi} \\ \Omega_\nu(H') \times \Omega_\nu(H) & \xrightarrow{\Omega_\nu(d)} & \widetilde{\Omega}_\nu(\mathrm{End}(M)) \end{array}$$

commute, where $\Omega_\nu(c)$ (resp. $\Omega_\nu(d)$) is the map sending $((k_n, l_n))$ to $(l_n \circ k_n)$. It follows that to prove that f' is an inverse to f it suffices to show that for all n we have $h_n \circ h'_n = p^{2nm}$ and $h'_n \circ h_n = p^{2nm}$ which follows from the relations $\varphi_M \circ \psi = p^{nm}$ and $\psi \circ \varphi_M = p^{nm}$. This completes the proof of 5.3.4. □

Corollary 5.3.16. — *The subcategory F-$\mathrm{Mod}^{fft}(W\langle t\rangle)_{\mathbb{Q}} \subset F$-$\mathrm{Mod}(W\langle t\rangle)_{\mathbb{Q}}$ is an abelian subcategory.*

Proof. — That F-$\mathrm{Mod}^{fft}(W\langle t\rangle)$ is an abelian category follows from the observation in 5.3.3 that F-$\mathrm{Mod}^{fft}(W)$ is abelian.

To see the exactness of the inclusion functor

$$F\text{-Mod}^{fft}(W\langle t\rangle)_{\mathbb{Q}} \subset F\text{-Mod}(W\langle t\rangle)_{\mathbb{Q}}, \tag{5.3.16.1}$$

it suffices to show that the forgetful functor $F\text{-Mod}^{fft}(W\langle t\rangle)_{\mathbb{Q}} \to \text{Mod}(W\langle t\rangle)_{\mathbb{Q}}$ sending (M, φ_M) to M is exact (since the forgetful functor $F\text{-Mod}(W\langle t\rangle)_{\mathbb{Q}} \to \text{Mod}(W\langle t\rangle)_{\mathbb{Q}}$ is exact and faithful).

To see the exactness of the forgetful functor to $\text{Mod}(W\langle t\rangle)_{\mathbb{Q}}$, note that there is a commutative diagram

$$\begin{array}{ccc} F\text{-Mod}(W\langle t\rangle)_{\mathbb{Q}} & \xrightarrow{\text{forget}} & \text{Mod}(W\langle t\rangle)_{\mathbb{Q}} \\ {\scriptstyle 5.3.4}\big\uparrow & & \big\uparrow{\scriptstyle \otimes W\langle t\rangle} \\ F\text{-Mod}^{fft}(W)_{\mathbb{Q}} & \xrightarrow{\text{forget}} & \text{Mod}(W)_{\mathbb{Q}}. \end{array} \tag{5.3.16.2}$$

Since the forgetful functor

$$F\text{-Mod}^{fft}(W)_{\mathbb{Q}} \longrightarrow \text{Mod}(W)_{\mathbb{Q}} \tag{5.3.16.3}$$

is exact, and the functor

$$\otimes W\langle t\rangle : \text{Mod}(W)_{\mathbb{Q}} \longrightarrow \text{Mod}(W\langle t\rangle)_{\mathbb{Q}} \tag{5.3.16.4}$$

is exact since $W\langle t\rangle$ is flat over W, it follows that the forgetful functor

$$F\text{-Mod}^{fft}(W\langle t\rangle)_{\mathbb{Q}} \longrightarrow \text{Mod}(W\langle t\rangle)_{\mathbb{Q}} \tag{5.3.16.5}$$

is also exact. □

Remark 5.3.17. — Theorem 5.3.4 is equivalent to the following two assertions:

(i) For any object $(M, \varphi_M) \in F\text{-Mod}^{fft}(W\langle t\rangle)_{\mathbb{Q}}$ with reduction $(\overline{M}, \varphi_{\overline{M}}) \in F\text{-Mod}^{fft}(W)_{\mathbb{Q}}$ (notation as in 5.3.5), and any morphism

$$f : (N, \varphi_N) \longrightarrow (\overline{M}, \varphi_{\overline{M}}) \tag{5.3.17.1}$$

in $F\text{-Mod}^{fft}(W)_{\mathbb{Q}}$, there exists a unique morphism

$$\tilde{f} : N \longrightarrow M \tag{5.3.17.2}$$

in $\text{Mod}(W)_{\mathbb{Q}}$ (where M is viewed as an object of $\text{Mod}(W)$ by forgetting the $W\langle t\rangle$-structure) such that the diagram

(5.3.17.3)

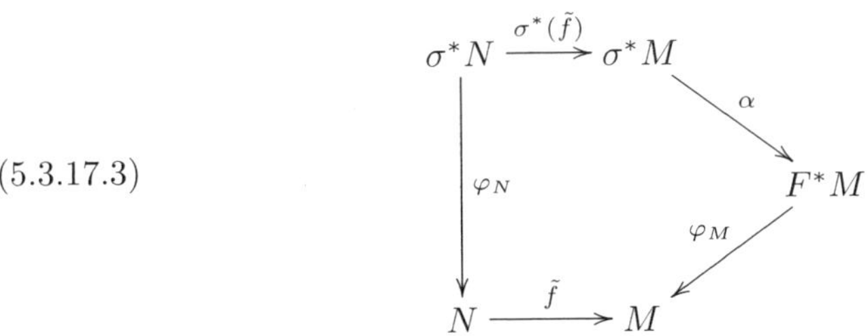

commutes, where

$$\alpha : M \otimes_{W,\sigma} W \longrightarrow M \otimes_{W\langle t\rangle,F} W\langle t\rangle \tag{5.3.17.4}$$

is the canonical map.

(ii) For f the identity morphism $(\overline{M}, \varphi_{\overline{M}}) \to (\overline{M}, \varphi_{\overline{M}})$ the morphism

$$\overline{M} \otimes_W W\langle t\rangle \longrightarrow M \tag{5.3.17.5}$$

in $\mathrm{Mod}(W\langle t\rangle)_{\mathbb{Q}}$ obtained from the section $\widetilde{\mathrm{id}_{\overline{M}}} : \overline{M} \to M$ by extending scalars is an isomorphism.

The equivalence of these two conditions with 5.3.4 can be seen as follows. If $(N, \varphi_N) \in F\text{-}\mathrm{Mod}^{fft}(W)_{\mathbb{Q}}$, then $\overline{(N \otimes W\langle t\rangle)} = N$ so clearly there exists a section as in (ii). Conversely (ii) implies that any object of $F\text{-}\mathrm{Mod}^{fft}(W\langle t\rangle)_{\mathbb{Q}}$ is in the essential image of $F\text{-}\mathrm{Mod}^{fft}(W)_{\mathbb{Q}}$.

To obtain the full faithfulness from the above conditions, note that if (N, φ_N) and $(N', \varphi_{N'})$ are objects of $F\text{-}\mathrm{Mod}^{fft}(W)_{\mathbb{Q}}$, then a morphism

$$g : (N \otimes_W W\langle t\rangle, \varphi_N) \longrightarrow (M \otimes_W W\langle t\rangle, \varphi_M) \tag{5.3.17.6}$$

in $F\text{-}\mathrm{Mod}^{fft}(W\langle t\rangle)_{\mathbb{Q}}$, with reduction $\bar{g}$ in $F\text{-}\mathrm{Mod}^{fft}(W)_{\mathbb{Q}}$, is determined by the induced morphism

$$g' : N \longrightarrow M \otimes_W W\langle t\rangle \tag{5.3.17.7}$$

in $\mathrm{Mod}(W)_{\mathbb{Q}}$, and by (i) this morphism is equal to the composition of $\bar{g}$ with the section

$$M \longrightarrow M \otimes_M W\langle t\rangle, \quad m \longmapsto m \otimes 1. \tag{5.3.17.8}$$

It follows that g is equal to the map

$$\bar{g} \otimes 1 : (N \otimes_W W\langle t\rangle, \varphi_N) \longrightarrow (M \otimes_W W\langle t\rangle, \varphi_M). \tag{5.3.17.9}$$

Conversely, the full faithfulness of P in 5.3.4 clearly implies (i).

5.3.18. — Define $\mathrm{Isoc}(W\langle t\rangle)_{\mathbb{Q}}$ to be the category of pairs (M, ∇), where $M \in \mathrm{Mod}(W\langle t\rangle)_{\mathbb{Q}}$ is free of finite type mod $\mathcal{T}$, and $\nabla : M \to M$ is a map in $\mathrm{Mod}(W)_{\mathbb{Q}}$ (where M is viewed as an object of $\mathrm{Mod}(W)_{\mathbb{Q}}$ by forgetting the $W\langle t\rangle$-structure) such that for any integer $i \geq 0$ the formula in $\mathrm{Hom}_{\mathrm{Mod}(W)_{\mathbb{Q}}}(M, M)$

$$\nabla \circ t^{[i]} = it^{[i]} + t^{[i]}\nabla \tag{5.3.18.1}$$

holds. We call such a map $\nabla : M \to M$ a *connection* on M. Note that for such a pair (M, ∇), the map $\nabla^{(s)} := p^s\nabla$ no longer satisfies (5.3.18.1), but rather the equation

$$\nabla^{(s)} \circ t^{[i]} = p^s it^{[i]} + t^{[i]}\nabla^{(s)}. \tag{5.3.18.2}$$

It follows that any object of $\mathrm{Isoc}(W\langle t\rangle)_{\mathbb{Q}}$ can be represented by a free $W\langle t\rangle$-module M of finite type with a map $\nabla^{(s)} : M \to M$ in $\mathrm{Mod}(W)$ satisfying (5.3.18.2) (note that since M is free and $W\langle t\rangle$ is p-torsion free, the equality (5.3.18.2) holds in $\mathrm{Mod}(W)_{\mathbb{Q}}$ if and only if it holds in $\mathrm{Mod}(W)$).

Lemma 5.3.19. — *For any* $(M,\nabla)\in \mathrm{Isoc}(W\langle t\rangle)_{\mathbb{Q}}$*, there exists a unique connection* $F^*\nabla : F^*M\to F^*M$ *such that the diagram in* $\mathrm{Mod}(W)_{\mathbb{Q}}$

$$\begin{array}{ccc} \sigma^*M & \xrightarrow{\sigma^*\nabla} & \sigma^*M \\ {\scriptstyle\tau}\downarrow & & \downarrow{\scriptstyle p\tau} \\ F^*M & \xrightarrow{F^*\nabla} & F^*M \end{array} \tag{5.3.19.1}$$

commutes, where $\tau:\sigma^*M\to F^*M$ *denotes the map*

$$W\otimes_{\sigma,W} M\longrightarrow W\langle t\rangle\otimes_{F,W\langle t\rangle} M,\quad a\otimes m\longmapsto a\otimes m. \tag{5.3.19.2}$$

Proof. — We can without loss of generality assume that M is a free $W\langle t\rangle$-module of finite type. Then for s sufficiently big, there exists a map

$$\widetilde{\nabla}^{(s)}: M\longrightarrow M \tag{5.3.19.3}$$

in $\mathrm{Mod}(W\langle t\rangle)$ inducing $\nabla^{(s)}$ in $\mathrm{Mod}(W\langle t\rangle)_{\mathbb{Q}}$, and satisfying the equations (5.3.18.2) in $\mathrm{Mod}(W)$. Let $e_1,\dots,e_r\in M$ be a basis, and write

$$\widetilde{\nabla}^{(s)}(e_i)=\sum_{j=1}^{r} a_{ij}e_j,\quad a_{ij}\in W\langle t\rangle. \tag{5.3.19.4}$$

Then define

$$F^*\widetilde{\nabla}^{(s)}: F^*M\simeq \oplus_{i=1}^{r} W\langle t\rangle\cdot F^*(e_i)\longrightarrow \oplus_{i=1}^{r} W\langle t\rangle\cdot F^*(e_i)\simeq F^*M \tag{5.3.19.5}$$

to be the W-linear map sending $t^{[j]}\cdot F^*(e_i)$ to

$$p^s j t^{[j]}\cdot F^*(e_i)+t^{[j]}\cdot\sum_{i=1}^{r} pF^*(a_{ij})\cdot F^*(e_j). \tag{5.3.19.6}$$

One verifies immediately that (5.3.18.2) holds, and we define $F^*\nabla$ to be the map $F^*\widetilde{\nabla}^{(s)}\otimes p^{-s}$ in $\mathrm{Mod}(W)_{\mathbb{Q}}$. The diagram (5.3.19.1) commutes since the diagram in $\mathrm{Mod}(W)$

$$\begin{array}{ccc} \sigma^*M & \xrightarrow{\sigma^*\widetilde{\nabla}^{(s)}} & \sigma^*M \\ {\scriptstyle\tau}\downarrow & & \downarrow{\scriptstyle p\tau} \\ F^*M & \xrightarrow{F^*\widetilde{\nabla}^{(s)}} & F^*M \end{array} \tag{5.3.19.7}$$

commutes by construction.

The uniqueness of $F^*\nabla$ follows from the observation that if $\beta: W\langle t\rangle\otimes_W \sigma^*M\to F^*M$ is the map obtained from τ by extension of scalars to $W\langle t\rangle$, and if

$D : W\langle t\rangle \to W\langle t\rangle$ is the W-linear map sending $t^{[j]}$ to $jt^{[j]}$, then the equality (5.3.18.1) forces the diagram

$$\begin{array}{ccc} W\langle t\rangle \otimes_W \sigma^* M & \xrightarrow{D\otimes 1+1\otimes p\sigma^*\nabla} & W\langle t\rangle \otimes_W \sigma^* M \\ \beta\downarrow & & \downarrow\beta \\ F^*M & \xrightarrow{F^*\nabla} & F^*M \end{array} \tag{5.3.19.8}$$

to commute, and the map β is an epimorphism in $\mathrm{Mod}(W\langle t\rangle)_{\mathbb{Q}}$. □

5.3.20. — Define the pullback functor

$$F^* : \mathrm{Isoc}(W\langle t\rangle)_{\mathbb{Q}} \longrightarrow \mathrm{Isoc}(W\langle t\rangle)_{\mathbb{Q}} \tag{5.3.20.1}$$

by sending (M,∇) to $(F^*M, F^*\nabla)$.

Define F-$\mathrm{Isoc}(W\langle t\rangle)$ to be the category of triples (M,φ_M,∇), where $(M,\nabla)\in \mathrm{Isoc}(W\langle t\rangle)_{\mathbb{Q}}$ and $\varphi_M : (F^*M, F^*\nabla) \to (M,\nabla)$ is an isomorphism in $\mathrm{Isoc}(W\langle t\rangle)_{\mathbb{Q}}$.

Also define F-$\mathrm{Isoc}(W)$ to be the category of triples (N,φ_N,∇), where $(N,\varphi_N)\in F\text{-}\mathrm{Mod}^{fft}(W)_{\mathbb{Q}}$ and $\nabla : N\to N$ is a morphism in $\mathrm{Mod}(W)_{\mathbb{Q}}$ such that the diagram

$$\begin{array}{ccc} \sigma^*N & \xrightarrow{\varphi_N} & N \\ p\sigma^*(\nabla)\downarrow & & \downarrow\nabla \\ \sigma^*N & \xrightarrow{\varphi_N} & N \end{array} \tag{5.3.20.2}$$

commutes.

Remark 5.3.21. — Following [**24**, 4.2.1], define $\mathrm{Mod}(\varphi,\mathcal{N})$ to be the category of triples $(M,\varphi_M,\mathcal{N})$, where M is a finite dimensional K-vector space (where K is the field of fractions of W), $\varphi_M : M\to M$ is semi-linear isomorphism, and $\mathcal{N} : M\to M$ is a linear map such that the diagram

$$\begin{array}{ccc} M & \xrightarrow{\varphi_M} & M \\ p\mathcal{N}\downarrow & & \downarrow\mathcal{N} \\ M & \xrightarrow{\varphi_M} & M \end{array} \tag{5.3.21.1}$$

commutes in the category of K-vector spaces. From 5.1.9 it follows that the functor

$$F\text{-}\mathrm{Isoc}(W) \longrightarrow \mathrm{Mod}(\varphi,\mathcal{N}),\quad (N,\varphi_N,\nabla)\longmapsto (N\otimes\mathbb{Q},\varphi_N\otimes\mathbb{Q},\nabla\otimes\mathbb{Q}) \tag{5.3.21.2}$$

is an equivalence of categories.

5.3.22. — Reduction defines a functor

$$F\text{-}\mathrm{Isoc}(W\langle t\rangle) \longrightarrow F\text{-}\mathrm{Isoc}(W). \tag{5.3.22.1}$$

There is also a functor

$$F\text{-}\mathrm{Isoc}(W) \longrightarrow F\text{-}\mathrm{Isoc}(W\langle t\rangle) \tag{5.3.22.2}$$

which sends (N, φ_N, ∇) to $N \otimes W\langle t\rangle$ with Frobenius induced by φ_N and connection defined by

$$\nabla(t^{[i]} \otimes n) := it^{[i]} \otimes n + t^{[i]} \otimes \nabla(n). \quad (5.3.22.3)$$

Remark 5.3.23. — If M is a $W\langle t\rangle$-module and

$$\nabla : M \longrightarrow M \cdot \mathrm{dlog}(t) \quad (5.3.23.1)$$

is a logarithmic connection with corresponding endomorphism $\nabla_t : M \to M$, then the map

$$M/\mathcal{I} \longrightarrow M/\mathcal{I} \quad (5.3.23.2)$$

induced by ∇_t is the *residue* of the connection ∇.

Theorem 5.3.24. — *The reduction functor (5.3.22.1) is an equivalence of categories with quasi-inverse given by (5.3.22.2).*

Proof. — By 5.3.4 it suffices to show that if $(N, \varphi_N) \in F\text{-Isoc}(W)$ is an object with N a free W-module of finite type and

$$\nabla, \nabla' : N \otimes W\langle t\rangle \longrightarrow N \otimes W\langle t\rangle \quad (5.3.24.1)$$

are two maps in $\mathrm{Mod}(W)_{\mathbb{Q}}$ giving $(N \otimes W\langle t\rangle, \varphi_N \otimes W\langle t\rangle)$ the structure of an object in $F\text{-Isoc}(W\langle t\rangle)$ such that $\nabla \equiv \nabla' \pmod{\mathcal{I}}$, then $\nabla = \nabla'$. After replacing φ_N by $p^k\varphi_N$ for some k, we may also assume that φ_N is given by a map $\sigma^*N \to N$ in $\mathrm{Mod}(W)$.

Choose s sufficiently big so that the maps $\nabla^{(s)}$ and $\nabla'^{(s)}$ can be represented by maps $\widetilde{\nabla}^{(s)}$ and $\widetilde{\nabla}'^{(s)}$ respectively in $\mathrm{Mod}(W)$ satisfying (5.3.18.2), compatible with φ_N, and with the same reductions modulo $\mathcal{I}$. Let $\psi : N \to F^*N$ be a map such that $\varphi_N \circ \psi = p^d$ and $\psi \circ \varphi_N = p^d$ for some integer d. Then for any $r \geq 0$ and $n \in N$ we have

$$p^{dr}\widetilde{\nabla}^{(s)}(n) = p^r\varphi_N^r(\widetilde{\nabla}^{(s)}(\psi^r(n))), \quad (5.3.24.2)$$

and similarly for $\widetilde{\nabla}'^{(s)}$. It follows that the two maps

$$\widetilde{\nabla}^{(s)}, \widetilde{\nabla}'^{(s)} : (N \otimes W\langle t\rangle \otimes \mathbb{Q}) \longrightarrow (N \otimes W\langle t\rangle \otimes \mathbb{Q}) \quad (5.3.24.3)$$

agree modulo $\varphi_N^r(\mathcal{I})$. Since $\varphi_N^r(\mathcal{I}) \subset \mathcal{I}^{[r]}$ and $N \otimes W\langle t\rangle$ is p-torsion free, it follows that $\widetilde{\nabla}^{(s)} = \widetilde{\nabla}'^{(s)}$ and hence $\nabla = \nabla'$. □

Remark 5.3.25. — The category $F\text{-Isoc}(W\langle t\rangle)$ is closely related to log F-isocrystals on k/W (the reader not familiar with the logarithmic language can omit this remark). Let M_k be the log structure on $\mathrm{Spec}(k)$ given by the map $\mathbb{N} \to k$ sending 1 to 0, and consider the log crystalline topos $((\mathrm{Spec}(k), M_k)/W)_{\mathrm{crys}}$ defined in [**40**, 5.2]. Let $M_{W\langle t\rangle}$ denote the log structure on $\mathrm{Spec}(W\langle t\rangle)$ induced by the map $\mathbb{N} \to W\langle t\rangle$ sending 1 to t, and let

$$i : (\mathrm{Spec}(k), M_k) \hookrightarrow (\mathrm{Spec}(W\langle t\rangle), M_{W\langle t\rangle}) \quad (5.3.25.1)$$

be the natural exact closed immersion. Let $\mathscr{C}_{k/W}$ denote the category of pairs (E, φ_E), where E is a locally free finite rank crystal in $((\mathrm{Spec}(k), M_k)/W)_{\mathrm{crys}}$ and $\varphi_E : F^*E \to E$ is a morphism of crystals for which there exist a map $\psi : E \to F^*E$ and an integer r such that the two composites $\varphi_E \circ \psi$ and $\psi \circ \varphi_E$ are both equal to multiplication by p^r (φ_E is a *p-isogeny*). The category $\mathscr{C}_{k/W}$ is a W-linear category so we can form the quotient category $\mathscr{C}_{k/W} \otimes \mathbb{Q}$ which we refer to as the *category of log F-isocrystals on k/W*. By [**40**, 6.2] the category $\mathscr{C}_{k/W}$ is equivalent to a full subcategory of the category of triples $(\mathcal{E}, \varphi_{\mathcal{E}}, \nabla)$ where:

(i) $\mathcal{E}$ is a locally free $W\langle t\rangle$-module of finite rank;

(ii) $\varphi_{\mathcal{E}} : F^*\mathcal{E} \to \mathcal{E}$ is a $W\langle t\rangle$-linear map which is an isomorphism in $\mathrm{Mod}(W\langle t\rangle)_{\mathbb{Q}}$;

(iii) ∇ is a W-linear map $\mathcal{E} \to \mathcal{E}$ (we fix the basis $\mathrm{dlog}(t)$ for the module of differentials) such that for all $i \geq 1$ and $e \in \mathcal{E}$ we have

$$\nabla(t^{[i]} \cdot e) = it^{[i]} \cdot e + t^{[i]} \cdot \nabla(e) \tag{5.3.25.2}$$

and the diagram in $\mathrm{Mod}(W\langle t\rangle)$

$$\begin{array}{ccc} F^*\mathcal{E} & \xrightarrow{\varphi_{\mathcal{E}}} & \mathcal{E} \\ {\scriptstyle F^*\nabla}\downarrow & & \downarrow{\scriptstyle \nabla} \\ F^*\mathcal{E} & \xrightarrow{\varphi_{\mathcal{E}}} & \mathcal{E} \end{array} \tag{5.3.25.3}$$

commutes.

In particular, there is a natural functor

$$A : \mathscr{C}_{k/W} \otimes \mathbb{Q} \longrightarrow F\text{-}\mathrm{Isoc}(W\langle t\rangle), \quad (E, \varphi_E) \longmapsto (\mathcal{E}, \varphi_{\mathcal{E}}, \nabla), \tag{5.3.25.4}$$

which by the above description of $\mathscr{C}_{k/W}$ is fully faithful.

Theorem 5.3.24 implies that the functor A induces an equivalence between $F\text{-}\mathrm{Isoc}(W\langle t\rangle)$ and the category obtained from $\mathscr{C}_{k/W} \otimes \mathbb{Q}$ by formally inverting the Tate object in $\mathscr{C}_{k/W}$ (the F-crystal $(\mathcal{O}_{k/W}, \times p : F^*\mathcal{O}_{k/W} = \mathcal{O}_{k/W} \to \mathcal{O}_{k/W})$). To verify this, it suffices by 5.3.24 to show that for any object $(N, \varphi_N, \nabla_N) \in F\text{-}\mathrm{Isoc}(W)$ with image (M, φ_M, ∇_M) in $F\text{-}\mathrm{Isoc}(W\langle t\rangle)$, the object (M, φ_M, ∇_M) is in the essential image of A, after possibly replacing φ_N by $p^k\varphi_N$ for some integer k. As noted for example in [**24**, 4.2.2] and recalled in 6.5.9 below, the relation $\nabla_N\varphi_N = p\varphi_N\nabla_N$ implies that ∇_N is nilpotent. This in turn implies that there exist a free W-module L and a nilpotent operator ∇_L on L inducing the pair (N, ∇_N) after tensoring with $\mathbb{Q}$. After replacing φ_N by $p^k\varphi_N$ for suitable k, we may also assume that φ_N extends to a semi-linear p-isogeny $\varphi_L : L \to L$. Now set $\mathcal{E} = W\langle t\rangle \otimes_W L$, $\varphi_{\mathcal{E}} : F^*\mathcal{E} \to \mathcal{E}$ the map induced by φ_L, and $\nabla : \mathcal{E} \to \mathcal{E}$ the map defined by

$$\nabla(t^{[i]} \otimes \ell) = it^{[i]} \otimes \ell + t^{[i]} \otimes \nabla_L(\ell). \tag{5.3.25.5}$$

Then for any section $\alpha \in \mathcal{E}$ and integer $\nu \geq 1$ there exist natural numbers $\{m_i\}_{i=1}^k$ and $\{n_i\}_{i=1}^k$ such that

$$\left(\prod_{j=1}^{k} (\nabla - m_j)^{n_j}\right)(\alpha) \equiv 0 \pmod{p^\nu}. \tag{5.3.25.6}$$

Indeed to verify this it suffices by additivity to consider $\alpha = t^{[i]} \otimes \ell$ where we have

$$(5.3.25.7) \qquad (\nabla - i)^n(\alpha) = t^{[i]} \otimes \nabla_L^n(\ell)$$

which is zero for n sufficiently big since ∇_L is nilpotent. From this and [**40**, 6.2] it follows that $(\mathcal{E}, \varphi_{\mathcal{E}}, \nabla_{\mathcal{E}})$ defines an object of $\mathscr{C}_{k/W}$ mapping to (M, φ_M, ∇_M).

It is convenient to restate 5.3.4 using projective systems and the equivalence 5.1.19.

Lemma 5.3.26. — *Let $M. \in \mathrm{ps}(W\langle t\rangle)$ be a projective system with $M = \varprojlim M_n$, and let $\overline{M}. \in \mathrm{ps}(W)$ be the projective system $\{M_n/\mathfrak{I}M_n\}$. If $M.$ is free of finite type mod $\mathfrak{T}$, then $\overline{M}.$ is free of finite type mod $\mathfrak{T}$ and the natural map*

$$(5.3.26.1) \qquad \{M/(p^n, \mathfrak{I})M\} \longrightarrow \{\overline{M}_n\}$$

induces an isomorphism in $\mathrm{ps}(W)_{\mathbb{Q}}$.

Proof. — Let $f : M \to M'$ be a map with M' a free $W\langle t\rangle$-module of finite rank and $\mathrm{Ker}(f)$ and $\mathrm{Coker}(f)$ annihilated by some power of p. Since the map

$$(5.3.26.2) \qquad h : \{M_n\} \longrightarrow \{M'/p^n M'\}$$

becomes an isomorphism in $\mathrm{ps}(W\langle t\rangle)_{\mathbb{Q}}$, there exists a map $g : \{M'/p^n M'\} \to \{M_n\}$ in $\mathrm{ps}(W\langle t\rangle)$ such that $h \circ g = p^d$ and $g \circ h = p^d$ for some integer d. In particular, the maps

$$(5.3.26.3) \qquad \{M/(p^n, \mathfrak{I})M\} \longrightarrow \{M'/(p^n, \mathfrak{I})M'\}, \quad \{\overline{M}_n\} \longrightarrow \{\overline{M}'/p^n\overline{M}'\}$$

induce isomorphisms in $\mathrm{ps}(W)_{\mathbb{Q}}$. It follows that it suffices to consider the case when $M. = \{M'/p^n M'\}$ with M' a free module in which case the result is immediate. □

The lifting of Frobenius $F : W\langle t\rangle \to W\langle t\rangle$ induces a pullback functor $\mathrm{ps}(W\langle t\rangle) \to \mathrm{ps}(W\langle t\rangle)$ which we denote by F^*. Similarly there is also a functor $\sigma^* : \mathrm{ps}(W) \to \mathrm{ps}(W)$.

Definition 5.3.27. — An *F-structure* on an object $M. \in \mathrm{ps}(W\langle t\rangle)_{\mathbb{Q}}$ is a morphism $\varphi : F^*M. \to M.$ in $\mathrm{ps}(W\langle t\rangle)_{\mathbb{Q}}$. Denote by $F\text{-}\mathrm{ps}(W\langle t\rangle)_{\mathbb{Q}}$ the category of pairs $(M., \varphi)$, where $M \in \mathrm{ps}(W\langle t\rangle)_{\mathbb{Q}}$ and φ is an F-structure on $M.$. Denote by $F\text{-}\mathrm{ps}^{fft}(W\langle t\rangle)_{\mathbb{Q}} \subset F\text{-}\mathrm{ps}(W\langle t\rangle)_{\mathbb{Q}}$ the full subcategory of pairs $(M., \varphi)$, where M is free of finite type mod $\mathfrak{T}$ and $\varphi : F^*M. \to M.$ is an isomorphism.

Similarly define $F\text{-}\mathrm{ps}(W)_{\mathbb{Q}}$ to be the category of pairs $(N., \varphi)$, where $N. \in \mathrm{ps}(W)_{\mathbb{Q}}$ and $\varphi : \sigma^*N. \to N.$ is a morphism in $\mathrm{ps}(W)_{\mathbb{Q}}$. Denote by $F\text{-}\mathrm{ps}^{fft}(W)_{\mathbb{Q}} \subset F\text{-}\mathrm{ps}(W)_{\mathbb{Q}}$ the full subcategory of pairs $(N., \varphi)$, where $N.$ is free of finite type mod $\mathfrak{T}$, and $\varphi : \sigma^*N. \to N.$ is an isomorphism.

5.3.28. — By 5.3.26, if $(M_., \varphi_.) \in F\text{-ps}^{fft}(W\langle t\rangle)_{\mathbb{Q}}$, then the reduction $\overline{M}_. := \{M_n/\mathcal{I}M_n\}$ with the map $\varphi_{\overline{M}_.} : \sigma^*\overline{M}_. \to \overline{M}_.$ is an object of $F\text{-ps}^{fft}(W)_{\mathbb{Q}}$. Furthermore, there is a natural commutative diagram

$$
\begin{array}{ccc}
F\text{-ps}^{fft}(W\langle t\rangle)_{\mathbb{Q}} & \xrightarrow{\varprojlim} & F\text{-Mod}^{fft}(W\langle t\rangle)_{\mathbb{Q}} \\
{\scriptstyle (M_.,\varphi_{M_.})\mapsto(\overline{M}_.,\varphi_{\overline{M}_.})}\downarrow & & \downarrow{\scriptstyle (M,\varphi_M)\mapsto(\overline{M},\varphi_{\overline{M}})} \\
F\text{-ps}^{fft}(W)_{\mathbb{Q}} & \xrightarrow{\varprojlim} & F\text{-Mod}^{fft}(W)_{\mathbb{Q}}.
\end{array}
\tag{5.3.28.1}
$$

From 5.3.4 it follows that the reduction functor

$$F\text{-ps}^{fft}(W\langle t\rangle)_{\mathbb{Q}} \longrightarrow F\text{-ps}^{fft}(W)_{\mathbb{Q}} \tag{5.3.28.2}$$

is an equivalence of categories.

Corollary 5.3.29. — *Let $(M_., \varphi_M)$ and $(N_., \varphi_N)$ be objects of $F\text{-ps}^{fft}(W\langle t\rangle)_{\mathbb{Q}}$, and let $f : (M_., \varphi_M) \to (N_., \varphi_N)$ be a morphism in $F\text{-ps}^{fft}(W\langle t\rangle)_{\mathbb{Q}}$. Then the kernel and cokernels of the underlying morphism of projective systems $M_. \to N_.$ are free of finite type mod $\mathcal{T}$, and φ_M and φ_N give them the structure of objects $F\text{-ps}^{fft}(W\langle t\rangle)_{\mathbb{Q}}$.*

Proof. — Set $M = \varprojlim M_.$ and $N = \varprojlim N_.$ and write also φ_M and φ_N for the F-structures on M and N obtained by passage to the limit. By the definition of "free of finite type mod $\mathcal{T}$", it suffices to consider the case when M and N are free modules of finite rank and $M_. = \{M/p^nM\}$ and $N_. = \{N/p^nN\}$. Furthermore, by 5.3.4 we may assume that $(M, \varphi_M) = (\tilde{M}, \varphi_{\tilde{M}}) \otimes W\langle t\rangle$ and $(N, \varphi_N) = (\tilde{N}, \varphi_{\tilde{N}}) \otimes W\langle t\rangle$ for some $(\tilde{M}, \varphi_{\tilde{M}}), (\tilde{N}, \varphi_{\tilde{N}}) \in F\text{-Mod}^{fft}(W)$ and that f is induced by a map $\tilde{f} : \tilde{M} \to \tilde{N}$. Let $\tilde{I} \subset \tilde{N}$ denote the image of $\tilde{f}$, $\tilde{K}$ the kernel, and $\tilde{Q}$ the cokernel. Since $\tilde{I}$ is p-torsion free, for every $n \geq 0$ there are exact sequences

$$0 \longrightarrow \tilde{K}/p^n \longrightarrow \tilde{M}/p^n \longrightarrow \tilde{I}/p^n \longrightarrow 0, \tag{5.3.29.1}$$

and

$$0 \longrightarrow \mathrm{Tor}^1_{\mathbb{Z}}(\mathbb{Z}/p^n, \tilde{Q}) \longrightarrow \tilde{I}/p^n \longrightarrow \tilde{N}/p^n \longrightarrow \tilde{Q}/p^n \longrightarrow 0. \tag{5.3.29.2}$$

Since $\tilde{Q}$ is a finitely generated W-module and consequently has bounded p-torsion, it follows that $\{\tilde{I}/p^n\} \simeq \mathrm{Ker}(\{\tilde{N}/p^n\} \to \{\tilde{Q}/p^n\})$ in $\mathrm{ps}(W)_{\mathbb{Q}}$. Hence

$$\mathrm{Ker}(f) \simeq \{\tilde{K}/p^n\} \otimes W\langle t\rangle, \quad \mathrm{Coker}(f) \simeq \{\tilde{Q}/p^n\} \otimes W\langle t\rangle \tag{5.3.29.3}$$

in $\mathrm{ps}(W\langle t\rangle)_{\mathbb{Q}}$. □

5.3.30. — We now explain a technical result which will be used in what follows.

Let $M^*_.$ be an $\mathbb{N}$-graded object in $\mathrm{ps}(W\langle t\rangle)$, and let $\varphi_{M_.} : M^*_. \to M^*_.$ be a semi-linear (with respect to $F : W\langle t\rangle \to W\langle t\rangle$) endomorphism of $M^*_.$ preserving the grading. Assume

$$E_1^{pq} = A_.^{pq} \implies M_.^{p+q} \tag{5.3.30.1}$$

is a first quadrant spectral sequence in $\mathrm{ps}(W\langle t\rangle)$, and $\varphi^{\cdot\cdot} : E_1^{\cdot\cdot} \to E_1^{\cdot\cdot}$ is a semi-linear endomorphism of the spectral sequence compatible with $\varphi_{M.}$ on the abutment. Assume further that (A^{pq}, φ^{pq}) defines an object of $F\text{-}\mathrm{ps}^{fft}(W\langle t\rangle)$ for all $p, q \in \mathbb{Z}$.

Proposition 5.3.31. — *With the above notation and assumptions, the projective system $M.$ as well as each term E_r^{pq} in the spectral sequence is free of finite type mod $\mathfrak{T}$. Moreover, the map φ_M gives the image of $M.$ in $\mathrm{ps}(W\langle t\rangle)_{\mathbb{Q}}$ the structure of an object of $F\text{-}\mathrm{ps}^{fft}(W\langle t\rangle)_{\mathbb{Q}}$ and the spectral sequence (5.3.30.1) with its endomorphism $\varphi^{\cdot\cdot}$ gives rise to a spectral sequence in $F\text{-}\mathrm{ps}^{fft}(W\langle t\rangle)_{\mathbb{Q}}$*

$$E_1^{pq} = (A^{pq}_{\cdot}, \varphi^{pq}) \implies (M^{p+q}_{\cdot}, \varphi_M). \tag{5.3.31.1}$$

Proof. — It follows from 5.3.29 that each term E_r^{pq} is free of finite type mod $\mathfrak{T}$ and that $\varphi^{\cdot\cdot}$ gives E_r^{pq} the structure of an object of $F\text{-}\mathrm{ps}^{fft}(W\langle t\rangle)_{\mathbb{Q}}$. From this it follows that $M.$ admits a φ_M stable exhaustive and finite filtration Fil such that each $\mathrm{gr}^i_{\mathrm{Fil}}(M.)$ with its endomorphism induced by φ_M is an object of $F\text{-}\mathrm{ps}^{fft}(W\langle t\rangle)_{\mathbb{Q}}$. From this and 5.1.17 it follows that $M.$ is also free of finite type mod $\mathfrak{T}$ and that the map φ_M induces an isomorphism in $\mathrm{ps}(W\langle t\rangle)_{\mathbb{Q}}$. □

5.3.32. — With notation and assumptions as in 5.3.30, let $(N^*_{\cdot}, \varphi_N)$ be a graded object of $F\text{-}\mathrm{ps}^{fft}(W)_{\mathbb{Q}}$ and let $\rho : (\overline{M}., \varphi_{\overline{M}}) \to (N., \varphi_N)$ be a morphism in $F\text{-}\mathrm{ps}(W)_{\mathbb{Q}}$. Assume further given a first quadrant spectral sequence

$$E_1^{pq} = (B^{pq}_{\cdot}, \varphi_B^{pq}) \implies (N., \varphi_N) \tag{5.3.32.1}$$

in $F\text{-}\mathrm{ps}(W)_{\mathbb{Q}}$ and a morphism of spectral sequences of projective systems of W-modules

$$(E_1^{pq} = A^{pq}_{\cdot} \implies M^{p+q}_{\cdot}) \xrightarrow{\ \gamma^{\cdot\cdot}\ } (E_1^{pq} = B^{pq}_{\cdot} \implies N.) \tag{5.3.32.2}$$

compatible with the Frobenius endomorphisms and ρ.

***Corollary 5.3.33* (The abstract Hyodo-Kato isomorphism).** — *Assume that for each $p, q \in \mathbb{Z}$, the map $\overline{A}^{pq}_{\cdot} \to B^{pq}_{\cdot}$ induced by $\gamma^{\cdot\cdot}$ is an isomorphism. Then the map $\rho : M. \to N.$ admits a unique section s in $\mathrm{ps}(W)_{\mathbb{Q}}$ compatible with φ_M and φ_N, and the induced morphism*

$$(N. \otimes W\langle t\rangle, \varphi_N \otimes F) \longrightarrow (M., \varphi_M) \tag{5.3.33.1}$$

is an isomorphism in $F\text{-}\mathrm{ps}^{fft}(W\langle t\rangle)_{\mathbb{Q}}$.

Proof. — Let

$$Q : F\text{-}\mathrm{ps}^{fft}(W\langle t\rangle)_{\mathbb{Q}} \longrightarrow F\text{-}\mathrm{ps}^{fft}(W)_{\mathbb{Q}} \tag{5.3.33.2}$$

be the reduction functor, which is exact (in fact an equivalence of categories).

Let $E^{pq}_{r,M}$ (resp. $E^{pq}_{r,N}$) denote the spectral sequence for $M.$ (resp. $N.$). Then by 5.3.31, for every p, q, and r the projection

$$E^{pq}_{r,M} \longrightarrow E^{pq}_{r,N} \tag{5.3.33.3}$$

identifies $E^{pq}_{r,N}$ with $Q(E^{pq}_{r,M})$. Let Fil_M (resp. Fil_N) denote the filtration on $M.$ (resp. $N.$) defined by $E^{pq}_{r,M}$ (resp. $E^{pq}_{r,N}$). By assumption $\rho : M. \to N.$ is compatible with the filtrations, and the map

$$\bar{\rho} : Q(M.) \longrightarrow N. \tag{5.3.33.4}$$

defined by ρ induces an isomorphism on the associated graded objects. It follows that $\bar{\rho}$ is also an isomorphism. The corollary therefore follows from the remarks in 5.3.17. □

CHAPTER 6

THE (φ, N, G)-STRUCTURE ON DE RHAM COHOMOLOGY

6.1. The stack $\mathcal{S}_H(\alpha)$

6.1.1. — Fix an integer $r \geq 1$ and a collection $\alpha = (\alpha_1, \dots, \alpha_r)$ of positive natural numbers. Define

$$U(\alpha) := \operatorname{Spec}(\mathbb{Z}[t][X_1, \dots, X_r, V^{\pm}]/(X_1^{\alpha_1} \cdots X_r^{\alpha_r} V = t)), \tag{6.1.1.1}$$

where $\mathbb{Z}[t]$ is the polynomial ring in one variable. In what follows the collection α will usually be fixed, and we write just U for $U(\alpha)$ if no confusion seems likely to arise.

The scheme U represents the sheaf on the category of $\mathbb{Z}[t]$-schemes

(6.1.1.2)
$$T \longmapsto \{(x_1, \dots, x_r, v) | x_i \in \Gamma(T, \mathcal{O}_T), v \in \Gamma(T, \mathcal{O}_T^*), \text{ such that } x_1^{\alpha_1} \cdots x_r^{\alpha_r} v = t\}.$$

Let $H \subset \mathrm{S}_r$ be a subgroup of the symmetric group on r letters contained in the subgroup of elements $\sigma \in \mathrm{S}_r$ for which $\alpha_{\sigma(i)} = \alpha_i$ for all i, and let G be the semi-direct product $\mathbb{G}_m^r \rtimes H$ with product structure given by

$$(u_1, \dots, u_r, h) \cdot (u_1', \dots, u_r', h') = ((u_{h'(i)} u_i')_i, h \circ h'). \tag{6.1.1.3}$$

An element $(u, h) \in G$ acts on U by

$$(x, v) \longmapsto \Big(u_{h^{-1}(1)} x_{h^{-1}(1)}, \dots, u_{h^{-1}(r)} x_{h^{-1}(r)}, \Big(\prod_i u_i^{-\alpha_i}\Big) v\Big). \tag{6.1.1.4}$$

Let $[U/G]$ denote the stack-theoretic quotient, and let $\tilde{R} = U \times_{[U/G]} U$, so that we have a groupoid in algebraic spaces [**49**, 2.4.3]

$$s, b : \tilde{R} \longrightarrow U, \quad m : \tilde{R} \times_U \tilde{R} \longrightarrow \tilde{R}. \tag{6.1.1.5}$$

In what follows, we shall denote this and other groupoids simply by $\tilde{R} \rightrightarrows U$. The scheme $\tilde{R}$ represents the functor which to a $\mathbb{Z}[t]$-scheme T associates the set of triples

$$\{(x, v), (x', v'), (u, h)\}, \tag{6.1.1.6}$$

where (x, v) and (x', v') are objects of $U(T)$ and $(u, h) \in G(T)$ such that

$$x'_{h(i)} = u_i x_i, \quad v' = \Big(\prod_i u_i^{-\alpha_i}\Big) v. \tag{6.1.1.7}$$

6.1.2. — Define an equivalence relation Γ on $\tilde{R}$ by

$$((x, v), (x', v'), (u, h)) \sim ((y, w), (y', w'), (u', h')) \tag{6.1.2.1}$$

if

6.1.2 (i) $(x, v) = (y, w)$, and $(x', v') = (y', w')$;
6.1.2 (ii) $h(i) = h'(i)$ for all i with $x_i \notin \mathcal{O}_T^*$;
6.1.2 (iii) $u_i = u'_i$ for all i with $x_i \notin \mathcal{O}_T^*$.

It is immediate that Γ is an equivalence relation, and it is shown in [**63**, 2.3] that Γ is an étale equivalence relation. Therefore the quotient $R := [\tilde{R}/\Gamma]$ exists as an algebraic space, and is smooth over U via either projection. Furthermore, as explained in *loc. cit.* the groupoid structure on $\tilde{R}$ descends to a groupoid structure $R \rightrightarrows U$. We write

$$\mathcal{S}_H(\alpha) \tag{6.1.2.2}$$

for the resulting algebraic stack (or just $\mathcal{S}$ if the reference to H and α is clear). Note that the groupoid $R \rightrightarrows U$ is defined over $\mathbb{Z}[t]$, and therefore there is a natural morphism $f : \mathcal{S}_H(\alpha) \to \mathrm{Spec}(\mathbb{Z}[t])$.

Lemma 6.1.3. — *The structure morphism $f : \mathcal{S}_H(\alpha) \to \mathrm{Spec}(\mathbb{Z}[t])$ is flat.*

Proof. — Since the projection $U \to \mathcal{S}_H(\alpha)$ is smooth and surjective, it suffices to show that U is flat over $\mathbb{Z}[t]$ which is immediate. □

Lemma 6.1.4. — *The projection $\pi : [U/G] \to \mathcal{S}_H(\alpha)$ is representable by Deligne-Mumford stacks 0.2.1 and étale.*

Proof. — By [**49**, 8.1], it suffices to show that π is formally étale in the sense of [**62**, 4.5]. Consider a commutative diagram

$$\begin{array}{ccc} \mathrm{Spec}(A_0) & \xrightarrow{i_0} & [U/G] \\ {\scriptstyle j}\downarrow & & \downarrow{\scriptstyle \pi} \\ \mathrm{Spec}(A) & \xrightarrow{i} & \mathcal{S}_H(\alpha), \end{array} \tag{6.1.4.1}$$

where $j^* : A \to A_0$ is a surjective map of rings with nilpotent kernel. We can without loss of generality assume that i_0 and i factor through morphisms $\tilde{i}_0$ and $\tilde{i}$ to U. Then any lifting τ of i to $[U/G]$ also lifts to U, and hence π is formally smooth in the sense of [**49**, 4.15 (ii)]. To show that π is formally étale, it therefore suffices to show that given a point $\sigma \in R(A)$ and a lifting $\tilde{\sigma}_0 \in \tilde{R}(A_0)$ of the reduction of σ to A_0, there exists a unique lifting $\tilde{\sigma}$ of σ to $\tilde{R}(A)$ inducing $\tilde{\sigma}_0$. This follows from the fact that $\tilde{R} \to R$ is étale and hence also formally étale. □

Definition 6.1.5. — If $x = (x_1, \ldots, x_r)$ are elements of $\Gamma(T, \mathcal{O}_T)$ for some scheme T, define *the essential set of* x, denoted $E(x)$, to be

$$E(x) := \{i | x_i \notin \Gamma(T, \mathcal{O}_T^*)\}. \tag{6.1.5.1}$$

Note that if $g : T' \to T$ is a morphism of schemes and if $g^*x := (g^*x_1, \ldots, g^*x_r)$, then there is a natural inclusion $E(g^*x) \hookrightarrow E(x)$.

Lemma 6.1.6. — *Let $(x, v), (x', v') \in U(T)$ be T-valued points, for some scheme T. Then the sheaf*

$$\underline{\mathrm{Isom}}_{\mathcal{S}_H(\alpha)}((x, v), (x', v')) \tag{6.1.6.1}$$

is naturally isomorphic to the sheaf associated to the presheaf $F((x, v), (x', v'))$ over T which to any $g : T' \to T$ associates the set of pairs $((u_i)_{i \in E(g^*x)}, h)$ where

(i) *$h : E(g^*x) \to E(g^*x')$ is a bijection which is the restriction of an element of H.*

(ii) *$(u_i)_{i \in E(g^*x)}$ is a set of element of $\Gamma(T', \mathcal{O}_{T'}^*)$ such that $x'_{h(i)} = u_i x_i$ for all $i \in E(g^*x)$.*

(iii) *$(\prod_{i \notin E(g^*x')} x_i'^{\alpha_i}) v' = (\prod_{i \in E(g^*x)} u_i^{-\alpha_i})(\prod_{i \notin E(g^*x)} x_i^{\alpha_i}) v$.*

Proof. — There is a natural map of presheaves

$$\underline{\mathrm{Isom}}_{[U/G]}((x, v), (x', v')) \longrightarrow F((x, v), (x', v')) \tag{6.1.6.2}$$

sending $(u, h) \in G(\alpha)$ to $((u_i)_{i \in E(x)}, h|_{E(x)})$. Moreover, it follows from the definition of the equivalence relation Γ that two isomorphisms map to the same element if and only if they are equivalent. Thus we obtain an injection of presheaves

$$\underline{\mathrm{Isom}}_{[U/G]}((x, v), (x', v'))/\Gamma \hookrightarrow F((x, v), (x', v')). \tag{6.1.6.3}$$

To see that (6.1.6.2) is surjective, suppose $((u_i)_{i \in E(x)}, h) \in F((x, v), (x', v'))$ is a section. We construct an element $(\tilde{u}_i, \tilde{h})$ inducing $((u_i)_{i \in E(x)}, h)$ by defining $\tilde{h}$ to be any element of H which agrees with h on $E(x)$, and letting $\tilde{u}_i = u_i$ if $i \in E(x)$ and $\tilde{u}_i = x'_{\tilde{h}(i)} x_i^{-1}$ otherwise. □

6.1.7. — Note that the composition law of morphisms exists already at the level of the presheaves $F((x, v), (x', v'))$. More precisely, suppose given

$$(x, v), (x', v'), (x'', v'') \in U(T) \tag{6.1.7.1}$$

for some scheme T. Then there is a natural map

$$F((x, v), (x', v')) \times F((x', v'), (x'', v'')) \longrightarrow F((x, v), (x'', v'')) \tag{6.1.7.2}$$

obtained by

$$(u, h) \times (u', h') \longmapsto ((u'_{h(i)} u_i, h' \circ h)). \tag{6.1.7.3}$$

This map induces the composition law of morphisms in $\mathcal{S}_H(\alpha)$.

Corollary 6.1.8. — *The stack $\mathcal{S}_H(\alpha)$ is naturally equivalent to the stack associated to the prestack $\mathcal{S}_H^{\mathrm{ps}}(\alpha)$ whose objects over a scheme T are elements of $U(T)$ and whose* $\underline{\mathrm{Isom}}$*-functors are given by the sheaves associated to the presheaves $F((x,v),(x',v'))$ defined above.*

6.1.9. — The stack $\mathcal{S}_H(\alpha)$ over $\mathbb{A}^1 = \mathrm{Spec}(\mathbb{Z}[t])$ descends in a natural way to a stack $\overline{\mathcal{S}}_H(\alpha)$ over $[\mathbb{A}^1/\mathbb{G}_m]$, where $\mathbb{G}_m$ acts on $\mathbb{A}^1$ by multiplication. For this note that there is a natural action of $\mathbb{G}_m$ on U compatible with the action on $\mathbb{Z}[t]$ for which a section $\lambda \in \mathbb{G}_m(T)$ over some scheme T acts on $T \times_{\mathbb{Z}} U$ by $X_i \mapsto X_i$ and $V \mapsto \lambda V$. For an element $(x,v) \in U(T)$ we write $(x, \lambda v) \in U(T)$ for the T-valued point obtained by applying λ. For two points $(x',v'), (x,v) \in U(T)$ and a section $\lambda \in \mathbb{G}_m(T)$ there is a natural map

$$F((x,v),(x',v')) \longrightarrow F((x,\lambda v),(x',\lambda v')) \tag{6.1.9.1}$$

obtained by sending $((u_i)_{i\in E(x)}, h)$ to $((u_i)_{i\in E(x)}, h)$. For a third section $(x'',v'') \in U(T)$ the composition law (6.1.7.2) extends to a map
(6.1.9.2)

$$(F((x,v),(x',v')) \times \mathbb{G}_m) \times (F((x',v'),(x'',v'')) \times \mathbb{G}_m) \longrightarrow F((x,v),(x'',v'')) \times \mathbb{G}_m$$

obtained by sending

$$((u,h),\lambda) \times ((u',h'),\lambda') \longmapsto ((u'_{h(i)}u_i, h' \circ h), \lambda \cdot \lambda'). \tag{6.1.9.3}$$

Define $\overline{\mathcal{S}}_H^{\mathrm{ps}}(\alpha)$ to be the prestack over $\mathbb{Z}$ which to any scheme T associates the groupoid whose objects are $U(T)$ and for which a morphism $(x,v) \to (x',v')$ is an element of

$$(F((x,v),(x',v')) \times \mathbb{G}_m)(T). \tag{6.1.9.4}$$

Composition of morphisms is defined using the map (6.1.9.2). We define $\overline{\mathcal{S}}_H(\alpha)$ to be the stack associated to the prestack $\overline{\mathcal{S}}_H^{\mathrm{ps}}(\alpha)$. The map $U \to \mathbb{A}^1$ induces a map from $\mathcal{S}_H^{\mathrm{ps}}(\alpha)$ to the prestack $[\mathbb{A}^1/\mathbb{G}_m]^{\mathrm{ps}}$ whose objects are point $z \in \mathbb{A}^1(T)$ and for which a morphism $z \to z'$ is an element $\lambda \in \mathbb{G}_m(T)$ such that $z' = \lambda z$. Passing to the associated stacks we obtain a map $\overline{\mathcal{S}}_H(\alpha) \to [\mathbb{A}^1/\mathbb{G}_m]$. There is also a natural map $\mathcal{S}_H^{\mathrm{ps}}(\alpha) \to \mathcal{S}_H^{\mathrm{ps}}(\alpha)$ induced by the identity map $U \to U$ and the maps

$$\mathrm{id} \times e : F((x,v),(x',v')) \longrightarrow F((x,v),(x',v')) \times \mathbb{G}_m. \tag{6.1.9.5}$$

Lemma 6.1.10. — *The stack $\overline{\mathcal{S}}_H(\alpha)$ is algebraic, and the diagram*

$$\begin{array}{ccc} \mathcal{S}_H(\alpha) & \longrightarrow & \overline{\mathcal{S}}_H(\alpha) \\ \downarrow & & \downarrow \\ \mathbb{A}^1 & \longrightarrow & [\mathbb{A}^1/\mathbb{G}_m] \end{array} \tag{6.1.10.1}$$

is cartesian.

Proof. — That the diagram (6.1.10.1) is cartesian follows from the definitions.

To see that $\overline{\mathcal{S}}_H(\alpha)$ is algebraic, note first that for any two sections $(x,v),(x',v')\in U(T)$, the sheaf

$$\underline{\mathrm{Isom}}_{\overline{\mathcal{S}}_H(\alpha)}((x,v),(x',v')) \tag{6.1.10.2}$$

is a $\mathbb{G}_m$-torsor over $\underline{\mathrm{Isom}}_{\mathcal{S}_H(\alpha)}((x,v),(x',v'))$. Since $\underline{\mathrm{Isom}}_{\mathcal{S}_H(\alpha)}((x,v),(x',v'))$ is representable by an algebraic space it follows that $\underline{\mathrm{Isom}}_{\mathcal{S}_H(\alpha)}((x,v),(x',v'))$ is also representable by an algebraic space. From this it follows that for any two objects $o_1,o_2\in\overline{\mathcal{S}}_H(\alpha)(T)$ the sheaf $\underline{\mathrm{Isom}}_{\overline{\mathcal{S}}_H(\alpha)}(o_1,o_2)$ is representable by an algebraic space. For by construction of the stack associated to a prestack the objects o_1 and o_2 are étale locally obtained from sections $(x,v),(x',v')\in U(T)$. It follows that étale locally on T the sheaf $\underline{\mathrm{Isom}}_{\overline{\mathcal{S}}_H(\alpha)}(o_1,o_2)$ is representable by an algebraic space, and since representability by an algebraic space over T is an étale local condition a sheaf it follows that $\underline{\mathrm{Isom}}_{\overline{\mathcal{S}}_H(\alpha)}(o_1,o_2)$ is representable globally. To conclude the proof note that since (6.1.10.1) is cartesian the composite map $U\to\mathcal{S}_H(\alpha)\to\overline{\mathcal{S}}_H(\alpha)$ is smooth and surjective. □

6.1.11. — For a scheme T, the category $[\mathbb{A}^1/\mathbb{G}_m](T)$ is equivalent to the category of pairs $(\mathcal{L},\rho)$, where $\mathcal{L}$ is a line bundle on T and $\rho:\mathcal{L}\to\mathcal{O}_T$ is a map of line bundles. To see this note that $\mathbb{A}^1$ can be viewed as representing the functor which to any scheme T associates the set of triples $(\mathcal{L},e,\rho)$, where $\mathcal{L}$ is a line bundle on T, $e:\mathcal{O}_T\to\mathcal{L}$ is an isomorphism of line bundles, and $\rho:\mathcal{L}\to\mathcal{O}_T$ is a morphism of line bundles. The action of $\mathbb{G}_m$ on $\mathbb{A}^1$ translates into the action on this functor for which a section $u\in\mathbb{G}_m(T)$ sends $(\mathcal{L},e,\rho)$ to $(\mathcal{L},u\cdot e,\rho)$. From this it follows that $[\mathbb{A}^1/\mathbb{G}_m]^{\mathrm{ps}}$ is the prestack which to any T associates the groupoid of pairs $(\mathcal{L},\rho)$, where $\mathcal{L}$ is a trivial line bundle on T and $\rho:\mathcal{L}\to\mathcal{O}_T$ is a morphism of line bundles on T.

In what follows, for a line bundle $\mathcal{L}$ with a morphism $\rho:\mathcal{L}\to\mathcal{O}_T$ we write $\mathcal{S}_H(\alpha)_{(\mathcal{L},\rho)}$ (or just $\mathcal{S}_{(\mathcal{L},\rho)}$ if the reference to H and α is clear) for the fiber product of the diagram

$$\begin{array}{ccc} & & \overline{\mathcal{S}}_H(\alpha) \\ & & \downarrow \\ T & \xrightarrow{(\mathcal{L},\rho)} & [\mathbb{A}^1/\mathbb{G}_m]. \end{array} \tag{6.1.11.1}$$

For a ring R and an element $f\in R$, we also sometimes write $\mathcal{S}_H(\alpha)_{R,(f)}$ or $\mathcal{S}_H(\alpha)_{R,f}$ for the pullback of $\overline{\mathcal{S}}_H(\alpha)$ via the map $\mathrm{Spec}(R)\to[\mathbb{A}^1/\mathbb{G}_m]$ induced by the morphism of line bundles $\times f:\mathcal{O}_R\to\mathcal{O}_R$.

6.1.12. — The stack $\overline{\mathcal{S}}_H(\alpha)$ can also be described as follows.

Denote by $\mathcal{T}_H$ the stack over $\mathbb{Z}$ associated to the prestack $\mathcal{T}_H^{\mathrm{ps}}$ defined as follows. The objects of $\mathcal{T}_H^{\mathrm{ps}}$ over some scheme T are collections $(x_1,\dots,x_r)$ of elements in $\Gamma(T,\mathcal{O}_T)$, and a morphism $(x_1,\dots,x_r)\to(x_1',\dots,x_r')$ is a bijection $h:E(x)\to E(x')$ which is the restriction of an element of H and a collection of units $\{u_i\}_{i\in E(x)}$ such that $x'_{h(i)} = u_i x_i$ for each $i \in E(x)$. Then just as in [**63**] the stack $\mathcal{T}_H$ associated to the prestack $\mathcal{T}_H^{\mathrm{ps}}$ is algebraic, and there is a natural étale surjection

$$[\mathrm{Spec}(\mathbb{Z}[X_1, \dots, X_r])/\mathbb{G}_m^r] \longrightarrow \mathcal{T}_H. \tag{6.1.12.1}$$

There is a natural map

$$q : \overline{\mathcal{S}}_H(\alpha) \longrightarrow \mathcal{T}_H, \tag{6.1.12.2}$$

defined on the level of prestack by sending $(x_1, \dots, x_r, v)$ to $(x_1, \dots, x_r)$.

Proposition 6.1.13. — *The map q is an isomorphism.*

Proof. — First note that there is a natural map $\epsilon : \mathcal{T}_H \to [\mathbb{A}^1/\mathbb{G}_m]$. To define such a map it suffices to define a map $\epsilon^{\mathrm{ps}} : \mathcal{T}_H^{\mathrm{ps}} \to [\mathbb{A}^1/\mathbb{G}_m]^{\mathrm{ps}}$, where $[\mathbb{A}^1/\mathbb{G}_m]^{\mathrm{ps}}$ denotes the prestack which to any scheme T associates the groupoid with objects elements $t \in \mathcal{O}_T$ and morphisms $t \to t'$ an object $u \in \mathcal{O}_T^*$ such that $t' = ut$. The map ϵ^{ps} is defined by associating to $(x_1, \dots, x_r) \in \mathcal{T}_H^{\mathrm{ps}}(T)$ the element $x_1^{\alpha_1} \cdots x_r^{\alpha_r} \in \mathbb{A}^1(T)$, and to a morphism $(h, \{u_i\}_{i \in E(x)})$ the element $(\prod_{i \notin E(x)} x_i^{-\alpha_i})(\prod_{i \notin E(x')} x_i'^{\alpha_i})(\prod_{i \in E(x)} u_i^{\alpha_i})$.

It follows from the definition of the map q that it extends to a morphism over $[\mathbb{A}^1/\mathbb{G}_m]$. Thus to verify that it is an isomorphism, it suffices to show that it becomes an isomorphism after base change to $\mathbb{A}^1$. On the other hand, the fiber product $\mathcal{T}_H \times_{[\mathbb{A}^1/\mathbb{G}_m]} \mathbb{A}^1$ is the stack associated to the prestack over $\mathbb{A}^1$ which to any scheme $f : T \to \mathbb{A}^1$ associates the groupoid of pairs $((x_1, \dots, x_r) \in \mathcal{T}_H^{\mathrm{ps}}(T), v \in \mathbb{G}_m(T))$ such that

$$f^*(t) = v x_1^{\alpha_1} \cdots x_r^{\alpha_r} \tag{6.1.13.1}$$

and morphisms as in 6.1.6. In other words, $\mathcal{T}_H \times_{[\mathbb{A}^1/\mathbb{G}_m]} \mathbb{A}^1 \simeq \mathcal{S}_H(\alpha)$, and the map $\mathcal{S}_H(\alpha) \to \mathcal{S}_H(\alpha)$ obtained from q by base change is the identity. □

Remark 6.1.14. — Even though the stacks $\overline{\mathcal{S}}_H(\alpha)$ and $\mathcal{T}_H$ are isomorphic, we use both notations in what follows. Usually we write $\overline{\mathcal{S}}_H(\alpha)$ when we wish to view this stack as a stack over $[\mathbb{A}^1/\mathbb{G}_m]$, and $\mathcal{T}_H$ when we view this stack as a stack over $\mathbb{Z}$ with the modular interpretation given in 6.1.12.

6.1.15. — For the remainder of this section assume that m is an integer that divides each α_i. Then there is a canonical μ_m-torsor

$$\mathcal{S}_H(\alpha)[m] \longrightarrow \mathcal{S}_H(\alpha). \tag{6.1.15.1}$$

To define this torsor it suffices to specify the following data:

1. For every object $(x, v) = (x_1, \dots, x_r, v) \in \mathcal{S}_H^{\mathrm{ps}}(\alpha)$ over some scheme T a μ_m-torsor

$$P_{(x,v)} \longrightarrow T. \tag{6.1.15.2}$$

2. Fix a morphism of schemes $g : T' \to T$, objects $(x, v) \in \mathcal{S}_H(\alpha)(T)$, $(x', v') \in \mathcal{S}_H^{\mathrm{ps}}(\alpha)(T')$, and collection $((u_i)_{i \in g^*x}, h)$ as in 6.1.6 defining an isomorphism $g^*(x, v) \to (x, v)$ in $\mathcal{S}_H^{\mathrm{ps}}(T')$. Then we need an isomorphism

$$g^* P_{(x,v)} \longrightarrow P_{(x',v')}. \tag{6.1.15.3}$$

Furthermore these isomorphisms have to satisfy the usual cocycle compatibility with compositions.

For this define $P_{(x,v)} \to T$ to be the μ_m-torsor which to any $h : T' \to T$ associates the set $\{w \in \Gamma(T', \mathcal{O}_{T'}^*) | w^m = h^* v\}$. Given data as in (2) we define the isomorphism (6.1.15.3) to be the map which to any $h : T'' \to T'$ associates the map

$$\{w \in \Gamma(T'', \mathcal{O}_{T''}^*) | w^m = h^* g^* v\} \longrightarrow \{w \in \Gamma(T'', \mathcal{O}_{T''}^*) | w^m = h^* v'\} \tag{6.1.15.4}$$

sending

$$w \longmapsto \Big(\prod_{i \notin E(x')} x_i'^{\beta_i} \Big)^{-1} \Big(\prod_{i \notin E(g^*x)} x_i^{\beta_i} \Big) \Big(\prod_{i \in E(g^*x)} u_i^{-\beta_i} \Big) w, \tag{6.1.15.5}$$

where β_i is defined to be α_i / m. This map is well-defined by 6.1.6 (iii). We leave to the reader the verification of the cocycle condition.

6.1.16. — Let $\pi : [U/G] \to \mathcal{S}_H(\alpha)$ be as in 6.1.4. The pullback

$$[U/G] \times_{\mathcal{S}_H(\alpha)} \mathcal{S}_H(\alpha)[m] \longrightarrow [U/G] \tag{6.1.16.1}$$

can be described as follows.

Let $G' \subset G$ denote the kernel of the homomorphism

$$G - \mathbb{G}_m^r \rtimes II \longrightarrow \mathbb{G}_m, \quad (u, h) \longmapsto \prod_i u_i^{\alpha_i}, \tag{6.1.16.2}$$

let $\mathcal{G}' \subset G$ denote the kernel of the homomorphism

$$G = \mathbb{G}_m^r \rtimes H \longrightarrow \mathbb{G}_m, \quad (u, h) \longmapsto \prod_i u_i^{\beta_i}, \tag{6.1.16.3}$$

and let $Z \subset U$ be the closed subscheme

$$\begin{array}{c} \mathrm{Spec}(\mathbb{Z}[t][X_1, \dots, X_r]/(X_1^{\alpha_1} \cdots X_r^{\alpha_r} = t)) \\ \Big\downarrow {\scriptstyle V=1} \\ \mathrm{Spec}(\mathbb{Z}[t][X_1, \dots, X_r, V^{\pm}]/(X_1^{\alpha_1} \cdots X_r^{\alpha_r} V = t)). \end{array} \tag{6.1.16.4}$$

The action of G' on U restrict to an action on Z, and so we obtain a diagram

$$[Z/\mathcal{G}] \xrightarrow{\ a\ } [Z/G'] \xrightarrow{\ b\ } [U/G]. \tag{6.1.16.5}$$

Lemma 6.1.17

(i) *The map b is an isomorphism.*

(ii) *The stack $[Z/\mathcal{G}]$ over $[U/G]$ is isomorphic to $[U/G] \times_{\mathcal{S}_H(\alpha)} \mathcal{S}_H(\alpha)[m]$.*

(iii) *The map $Z \to \mathcal{S}_H(\alpha)[m]$ is smooth.*

Proof. — For (i) let $[Z/G']^{\mathrm{ps}}$ (resp. $[U/G]^{\mathrm{ps}}$) be the prestack which to any $\mathbb{Z}[t]$-scheme T associates the groupoid with objects $Z(T)$ (resp. $U(T)$) and for which a morphism $v \to v'$ (resp. $u \to u'$) between elements of $Z(T)$ (resp. $U(T)$) is an element $g' \in G'(T)$ (resp. $g \in G(T)$) such that $v = g'v'$ (resp. $u = gu'$). Then $[Z/G']$ (resp. $[U/G]$) is the stack associated to the prestack $[Z/G']^{\mathrm{ps}}$ (resp. $[U/G]^{\mathrm{ps}}$) and the map b is induced by the natural morphism

$$\hat{b} : [Z/G']^{\mathrm{ps}} \longrightarrow [U/G]^{\mathrm{ps}}. \tag{6.1.17.1}$$

It therefore suffices to show that $\hat{b}$ is fully faithful and that every object of $[U/G]^{\mathrm{ps}}$ is fppf locally in the image of $\hat{b}$.

The full faithfulness is immediate from the definitions of G' and the action of G on U. To see that every object is locally in the image, let T be a $\mathbb{Z}[t]$-scheme and $(x_1, \dots, x_r, v) \in U(T)$. After possibly replacing T be an fppf cover, we may assume there exists an element $u \in \mathcal{O}_T^*$ with $u^{\alpha_1} = v$. The element $((u, 1, \dots, 1), \mathrm{id}) \in \mathbb{G}_m^r \rtimes H = G$ then defines an isomorphism

$$(ux_1, \dots, x_r, 1) \longrightarrow (x_1, \dots, x_r, v) \tag{6.1.17.2}$$

in $[U/G]^{\mathrm{ps}}$. This completes the proof of (i).

For (ii) let $P \to [Z/G']$ denote the μ_m-torsor $[Z/G'] \times_{\mathcal{S}_H(\alpha)} \mathcal{S}_H(\alpha)[m]$. The stack P is the stack associated to the prestack P^{ps} which to any scheme T associates the groupoid with objects the set

$$\{(x, w) | x \in Z(T), \quad w \in \Gamma(T, \mathcal{O}_T^*) \quad \text{such that} \quad w^m = 1\}, \tag{6.1.17.3}$$

and for which a morphism

$$(x, w) \longrightarrow (x', w') \tag{6.1.17.4}$$

is an element $g = (u_1, \dots, u_r, h) \in G'(T)$ such that $gx' = x$ and $w = \prod_i u_i^{-\beta_i} w'$. There is a map of prestacks

$$\rho : [Z/\mathcal{G}]^{\mathrm{ps}} \longrightarrow P^{\mathrm{ps}} \tag{6.1.17.5}$$

sending $x \in Z(T)$ to the element $(x, 1) \in P^{\mathrm{ps}}$. Note that by the definition of $\mathcal{G}$ this is compatible with morphisms, and extends to a fully faithful functor. Also, for every object $(x, w) \in P^{\mathrm{ps}}(T)$ over some scheme T there exists an fppf-covering of T such that (x, w) is isomorphic to an object in the essential image of ρ. Indeed after replacing T by an fppf-cover we may assume that there exists an element $g \in G'$ mapping to w under (6.1.16.3). Then $(gx, 1)$ is isomorphic to (x, w) in $P^{\mathrm{ps}}(T)$. It follows that ρ induces an isomorphism $[Z/\mathcal{G}] \simeq P$.

Finally for (iii), note that $\mathcal{G}$ is a smooth group scheme over $\mathbb{Z}$ and therefore the map $Z \to [Z/\mathcal{G}]$ is smooth. Since $[Z/\mathcal{G}] \to \mathcal{S}_H(\alpha)[m]$ is étale the morphism $Z \to \mathcal{S}_H(\alpha)[m]$ is also smooth. □

6.2. Maps to $\mathcal{S}_H(\alpha)$

6.2.1. — Let V be a complete discrete valuation ring with uniformizer π and mixed characteristic $(0,p)$, and let X/V be a flat regular scheme with smooth generic fiber for which the reduced closed fiber $X_{0,\mathrm{red}} \subset X$ is a divisor with normal crossings. Fix a sequence of natural numbers $\{\alpha_1,\dots,\alpha_r\}$ of the form $\{a_1,\dots,a_1,a_2,\dots,a_2,\cdots\}$ with the a_i's distinct and each a_i occurring r_i times, such that for any geometric point $\bar{x} \to X$ mapping to the closed fiber one can in an étale neighborhood of $\bar{x}$ in X write the set of branches D_i of the closed fiber at the image of $\bar{x}$ as $\{D_1,\dots,D_{s_1},D_{s_1+1},\dots,D_{s_1+s_2},\cdots\}$ with $D_1,\dots,D_{s_1}$ branches of multiplicity a_1, $D_{s_1+1},\dots,D_{s_1+s_2}$ branches of multiplicity a_2 etc., and $s_i \leq r_i$ for all i. Let H be the group of elements $\sigma \in \mathrm{S}_r$ for which $\alpha_{\sigma(i)} = \alpha_i$ for all i. Let $\mathcal{S}_{V,\pi}$ denote the stack $\mathcal{S}_H(\alpha)_{(\pi)\subset V}$ obtained by base change from $\overline{\mathcal{S}}_H(\alpha)$ from the map $\mathrm{Spec}(V) \to [\mathbb{A}^1/\mathbb{G}_m]$ defined by the invertible sheaf (π) with the inclusion into $\mathcal{O}_V$ (6.1.11). Then there is a canonical map $X \to \mathcal{S}_{V,\pi}$ defined as follows.

First note that the choice of the generator $\pi \in (\pi)$ identifies $\mathcal{S}_{V,\pi}$ with the base change of $\mathcal{S}_H(\alpha)$ to $\mathrm{Spec}(V)$ via the map $\mathbb{Z}[t] \to V$ sending t to π. We write U_V and $\tilde{R}_V$ for the base changes $U \times_{\mathrm{Spec}(\mathbb{Z}[t]),t\mapsto\pi} \mathrm{Spec}(V)$ and $U \times_{\mathrm{Spec}(\mathbb{Z}[t]),t\mapsto\pi} \mathrm{Spec}(V)$ respectively.

Lemma 6.2.2

(i) *For any geometric point $\bar{z} \to X$ with image $z \in X_0$ in the closed fiber, there exist an étale neighborhood W of $\bar{z}$ and a W-valued point $(x,v) = (x_1,\dots,x_r,v) \in U_V(W)$ such that the following condition holds: the set of irreducible components D_i of the closed fiber at the image of $\bar{z}$ in W is equal to the set of divisors $\{(x_i)\}_{i\in E(x)}$ (where the essential set $E(x)$ of x is defined in 6.1.5).*

(ii) *Let $\bar{z} \to X$ be a point geometric point with image in the closed fiber, and let $(x,v) = (x_1,\dots,x_r,v)$ and $(x',v') = (x'_1,\dots,x'_r,v')$ be two W-valued points of U_V satisfying the condition in* (i), *for some étale neighborhood W of $\bar{z}$. Let*

$$\rho_x, \rho_{x'} : W \longrightarrow \mathcal{S}_H^{\mathrm{ps}}(\alpha) \tag{6.2.2.1}$$

be the two morphisms obtained by composing (x,v) and (x,v') with the projection $U_V \to \mathcal{S}_H^{\mathrm{ps}}(\alpha)$. Then there exists a unique isomorphism between $\rho_{(x,v)}$ and $\rho_{(x',v')}$ in $\mathcal{S}_H^{\mathrm{ps}}(\alpha)(W)$.

Proof. — For (i), it suffices by a standard limit argument to find a point $(x,v) \in U_V(\mathcal{O}_{X,\bar{z}})$ such that the set of branches D_i of the closed fiber in $\mathrm{Spec}(\mathcal{O}_{X,\bar{z}})$ is equal to the set of divisors $\{(x_i) | x_i \notin \mathcal{O}^*_{X,\bar{z}}\}$.

With notation as in 6.2.1, let $\{D_1,\dots,D_{s_1},D_{s_1+1},\dots,D_{s_1+s_2},\cdots\}$ be the set of branches of the closed fiber in $\mathrm{Spec}(\mathcal{O}_{X,\bar{z}})$. To prove the lemma it suffices to consider the case when $s_i = r_i$ (set the remaining x_i's equal to 1). Let $x_i \in \mathcal{O}_{X,\bar{z}}$ be an element defining D_i. If $\eta_i \in D_i$ denotes the generic point, then the image of π in $\mathcal{O}_{\mathrm{Spec}(\mathcal{O}_{X,\bar{z}}),\eta_i}$ is equal to a unit times $x_i^{\alpha_i}$. Therefore the two elements $\pi, \prod_{i\in I} x_i^{\alpha_i} \in \mathcal{O}_{X,\bar{z}}$ differ by a unit at every codimension 1 point. Since $\mathcal{O}_{X,\bar{z}}$ is regular it follows that the element

$\pi/(\prod_{i\in I} x_i^{\alpha_i})$ of the fraction field of $\mathcal{O}_{X,\bar{z}}$ extends uniquely to a unit $v \in \mathcal{O}^*_{X,\bar{z}}$ with $\pi = v\prod_i x_i^{\alpha_i}$.

For (ii), let

$$h : E(x) \longrightarrow E(x') \tag{6.2.2.2}$$

be the map sending i to the unique integer $j \in [1, r]$ such that $D_i = (x'_j)$. With h so defined, there exists for every $i \in E(x)$ a unique unit u_i such that $x'_{h(i)} = u_i x_i$ since $D_i = (x_i) = (x'_{h(i)})$. The formula 6.1.6 (iii) holds because X is integral and the two sides both become equal to $\pi(\prod_{i\in E(x)} u_i^{-\alpha_i})$ when multiplied by $\prod_{i\in E(x)} x_i^{\alpha_i}$. Moreover, if $(u', h') \in F((x, v), (x', v'))$ is a second element, then the condition $D_i = (x'_{h'(i)})$ ensures that $(u, h) = (u', h')$. This implies the existence and uniqueness. □

6.2.3. — It follows from 6.2.2 that we obtain a globally defined map $X \to \mathcal{S}_{V,\pi}$.

If $\pi' \in (\pi)$ is a second choice of uniformizer, then there is a canonical isomorphism between the resulting two maps $\rho, \rho' : X \to \mathcal{S}_H(\alpha)$. For this write $\pi' = u_0 \cdot \pi$ with $u_0 \in V^*$. If locally on X we write $\pi = v\prod_i x_i^{\alpha_i}$ as above, then the map ρ to $\mathcal{S}_H(\alpha)$ is induced by the map to $U = \mathrm{Spec}(\mathbb{Z}[t][X_1, \ldots, X_r, V^{\pm}]/(X_1 \cdots X_r V - t))$ sending X_i to x_i, V to v, and t to π. The map ρ' is induced by the map sending X_i to x_i, V to u_0V, and t to π'. It follows that the unit u_0 defines an isomorphism in $\overline{\mathcal{S}}_H(\alpha)$ between the two maps $X \to \overline{\mathcal{S}}_H(\alpha)$ compatible with the isomorphism in $[\mathbb{A}^1/\mathbb{G}_m]$ induced by u_0. Moreover, by construction these locally defined isomorphisms glue to a globally defined isomorphism as desired.

Remark 6.2.4. — The choice of the α_i is not serious. As explained in Chapter 9, one can glue the $\mathcal{S}(\alpha)_{V,\pi}$ in a natural way to a stack $\mathcal{S}_{V,\pi}$. Then the above shows that any scheme X/V as above admits a canonical map $X \to \mathcal{S}_{V,\pi}$.

Remark 6.2.5. — Note that since the map $X \to \mathcal{S}_{V,\pi}$ is obtained by gluing together morphisms defined locally in the étale topology, we could also let X in the above be a Deligne-Mumford stack.

The following explains when $X \to \mathcal{S}_{V,\pi}$ is smooth.

Proposition 6.2.6. — *Let* $\mathrm{Spec}(A) \to U_{V,\pi}$ *be a morphism induced by elements* $x_1, \ldots, x_r \in A$ *and* $v \in A^*$. *Then the induced map* $\mathrm{Spec}(A) \to \mathcal{S}_{V,\pi}$ *is smooth if and only if for each* $h \in H$ *the map*

$$V[X_1, \ldots, X_r, V^{\pm}]/(X_1^{\alpha_1} \cdots X_r^{\alpha_r} V = \pi) \longrightarrow A[U_1^{\pm}, \ldots, U_r^{\pm}] \tag{6.2.6.1}$$

sending X_i *to* $x_{h^{-1}(i)}U_{h^{-1}(i)}$ *and* V *to* $v(\prod_i U_i^{-\alpha_i})$ *is smooth.*

Proof. — Let $R_A = \mathrm{Spec}(A) \times_{\mathcal{S}} U$ and let $\tilde{R}_A = \mathrm{Spec}(A) \times_{\tilde{\mathcal{S}}} U$. Then $\mathrm{Spec}(A)$ is smooth over $\mathcal{S}_{V,\pi}$ if and only if R_A is smooth over U. Now the natural map $\tilde{R}_A \to R_A$ is étale and surjective, and hence R_A is smooth over U if and only if $\tilde{R}_A$ is smooth over U. But $\tilde{R}_A$ is naturally isomorphic to the disjoint union over $h \in H$ of the schemes described in the proposition. □

Corollary 6.2.7. — *If $A = V[Y_1, \dots, Y_r]/(Y_1^{\alpha_1} \cdots Y_r^{\alpha_r} = \pi)$, then the natural map*

$$\operatorname{Spec}(A) \longrightarrow \mathcal{S}_{\pi,V} \tag{6.2.7.1}$$

is smooth if one of the $\alpha_i = 1$.

Proof. — Without loss of generality assume that $\alpha_1 = 1$. Denote by $\mathcal{O}$ the ring

$$V[X_1, \dots, X_r, V^{\pm 1}]/(X_1 X_2^{\alpha_2} \cdots X_r^{\alpha_r} V = \pi). \tag{6.2.7.2}$$

The map (6.2.6.1) induces an isomorphism

$$A[U_1^{\pm}, \dots, U_r^{\pm}] \simeq \mathcal{O}[U_1, \dots, U_r]/(V = \prod U_i^{-\alpha_i}), \tag{6.2.7.3}$$

and since

$$\mathcal{O}[U_1, \dots, U_r]/(V = \prod U_i^{-\alpha_i}) \simeq \mathcal{O}[U_2^{\pm}, \dots, U_r^{\pm}]. \tag{6.2.7.4}$$

It follows that (6.2.6.1) is smooth. □

6.3. The map Λ_e

6.3.1. — Throughout this section we work with a fixed $\mathcal{S}_H(\alpha)$ and write simply $\mathcal{S}$ for this stack. We also write $\tilde{\mathcal{S}}$ for the stack $[U/G]$.

If R is a ring and $f \in R$ is an element, we write $\mathcal{S}_{R,f}$, $U_{R,f}$ etc., for the objects over R obtained by base change via the map $\mathbb{Z}[t] \to R$ sending t to f. If the reference to R is clear we sometimes also just write $\mathcal{S}_f$, U_f etc. For example, for each positive integer e, we write $\mathcal{S}_{t^e}$ (resp. U_{t^e} etc.) for the fiber product $\mathcal{S} \otimes_{\mathbb{Z}[t], t \mapsto t^e} \mathbb{Z}[t]$ (resp. $U \times_{\mathbb{Z}[t], t \mapsto t^e} \mathbb{Z}[t]$).

6.3.2. — Define

$$\Lambda_e^{\mathrm{ps}} : \mathcal{S}_t^{\mathrm{ps}} \longrightarrow \mathcal{S}_{t^e}^{\mathrm{ps}} \tag{6.3.2.1}$$

to be the map over the map $\mathbb{Z}[t] \to \mathbb{Z}[t]$, $\mapsto t^e$, which sends a pair (x, v) to the pair (x^e, v^e), where x^e denotes the set $(x_1^e, \dots, x_r^e)$, and whose map on morphisms is that induced by

(6.3.2.2)

$$F((x,v),(x',v')) \longrightarrow F((x^e,v^e),(x'^e,v'^e)), \quad ((u_i)_{i \in E(x)}, h) \longmapsto ((u_i^e)_{i \in E(x^e)} . h),$$

The map Λ_e^{ps} induces a morphism of stacks $\Lambda_e : \mathcal{S}_t \to \mathcal{S}_{t^e}$. Let μ denote the diagonalizable group scheme μ_e^r over $\mathbb{Z}$.

Proposition 6.3.3. — *The fiber product $U_{t^e} \times_{\mathcal{S}_{t^e}, \Lambda_e} \mathcal{S}_t$ is naturally isomorphic to the stack-theoretic quotient of*

$$U_t = \operatorname{Spec}\big(\mathbb{Z}[t][X_1, \dots, X_r, V^{\pm}]/(X_1^{\alpha_1} \cdots X_r^{\alpha_r} V - t)\big) \tag{6.3.3.1}$$

by the action of μ given by

$$(\zeta_1, \dots, \zeta_r) \cdot X_i = \zeta_i X_i, \quad V \longmapsto \Big(\prod \zeta_i^{-\alpha_i}\Big) V. \tag{6.3.3.2}$$

Proof. — Let $U^{(e)}$ denote the fiber product $U_{t^e} \times_{\mathcal{S}_{t^e},\Lambda_e} \mathcal{S}_t$ and let

$$\pi : U^{(e)} \longrightarrow U_{t^e} \tag{6.3.3.3}$$

be the projection. The map

$$\tilde{\Lambda}_e : U_t \longrightarrow U_{t^e}, \quad X_i \longmapsto X_i^e, \quad V \longmapsto V^e \tag{6.3.3.4}$$

fits naturally into a 2-commutative diagram

$$\begin{array}{ccc} U_t & \xrightarrow{\tilde{\Lambda}_e} & U_{t^e} \\ \downarrow & & \downarrow \\ \mathcal{S}_t & \xrightarrow{\Lambda_e} & \mathcal{S}_{t^e}, \end{array} \tag{6.3.3.5}$$

and hence we obtain a map $z : U_t \to U^{(e)}$ with $\tilde{\Lambda}_e$ equal to the composition $\pi \circ z$. It follows from the definition of the action of μ on U_t that the corresponding map $U_t \to U^{(e)}$ descends to a map

$$[U_t/\mu] \longrightarrow U^{(e)} \tag{6.3.3.6}$$

which we claim is an isomorphism.

It is clear that (6.3.3.6) is fully faithful. To see that it is essentially surjective, let $\iota : T \to U^{(e)}$ be a map corresponding to a morphism $g : T \to U_{t^e}$ and elements $(x'_1, \ldots, x'_r, v')$ in $\Gamma(T, \mathcal{O}_T)$ for which

$$\Big(\prod_i x_i'^{\alpha_i}\Big)v = t, \tag{6.3.3.7}$$

together with $((u_i), h) \in G(T)$ such that $x'^e_{h(i)} = u_i g^*(X_i)$, $v'^e = (\prod_i u_i^{-\alpha_i})g^*(V)$. Every morphism to $U^{(e)}$ is étale locally obtained from such data. Now locally in the flat topology on T, we can find units $\tilde{u}_i$ such that $\tilde{u}_i^e = u_i$. Define a map $\gamma : T \to U$ by sending X_i to $\tilde{u}_i^{-1}x'_{h(i)}$ and V to $\prod \tilde{u}_i^{-\alpha_i}v'$. Then ι and $z \circ \gamma$ are naturally isomorphic, and hence (6.3.3.6) is an isomorphism. □

Corollary 6.3.4. — *The map Λ_e is quasi-compact and quasi-separated.*

Proof. — This can be verified on the flat cover U_{t^e} of $\mathcal{S}_{t^e}$. □

Corollary 6.3.5. — *The formation of $R^i\Lambda_{e*}\mathcal{O}_{\mathcal{S}_t}$ is compatible with arbitrary base change $\mathbb{Z}[t] \to R$, and these groups are zero for $i > 0$.*

Proof. — Again we can work locally in the flat topology on $\mathcal{S}_{t^e}$ and hence may base change to U_{t^e}.

Since μ is a diagonalizable group scheme, it follows from [**14**, I.5.3.3] that

$$R^i\pi_*\mathcal{O}_{U^{(e)}\otimes_{\mathbb{Z}[t]}R} = 0 \tag{6.3.5.1}$$

for all $i > 0$ and all base changes $\mathbb{Z}[t] \to R$ (here π is the projection 6.3.3.3).

From this we can also describe the algebra $\pi_*\mathcal{O}_{U^{(e)}\otimes_{\mathbb{Z}[t]}R}$ explicitly. Indeed the isomorphism (where we abusively write also t for the image of t in R)

$$U^{(e)} \otimes_{\mathbb{Z}[t]} R \simeq \big[\operatorname{Spec}\big(R[X_1, \dots, X_r, V^{\pm}]/(X_1^{\alpha_1}\cdots X_r^{\alpha_r}V = t)\big)/\mu\big] \tag{6.3.5.2}$$

identifies the global sections of the quasi-coherent sheaf $\pi_*\mathcal{O}_{U^{(e)}\otimes_{\mathbb{Z}[t]}R}$ on U_{t^e} with the sub-R-module of μ-invariants in

$$R[X_1, \dots, X_r, V^{\pm}]/(X_1^{\alpha_1}\cdots X_r^{\alpha_r}V = t). \tag{6.3.5.3}$$

Since the action of μ is given by the action on monomials, this submodule of invariants is isomorphic to the free module on the monomials $X_1^{b_1}\cdots X_r^{b_r}V^c$ with the property that the element

$$(b_i - \alpha_i c)_{i=1}^r \in (\mathbb{Z}/(e))^r \tag{6.3.5.4}$$

is zero. Evidently formation of this submodule is compatible with arbitrary base change $R \to R'$. □

6.3.6. — Let R be an $\mathbb{F}_p$-algebra and $f \in R$ an element. Consider the stack $\mathcal{S}_{R,f}$ and let $X \to \mathcal{S}_{R,f}$ be a smooth morphism with X a scheme. Let $F_{\mathcal{S}_{R,f}} : \mathcal{S}_{R,f} \to \mathcal{S}_{R,f}$ be the Frobenius morphism, and let X' denote the fiber product of the diagram

$$\begin{array}{ccc} & & X \\ & & \downarrow \\ \mathcal{S}_{R,f} & \xrightarrow{F_{\mathcal{S}_{R,f}}} & \mathcal{S}_{R,f}. \end{array} \tag{6.3.6.1}$$

Denote by

$$X' \xrightarrow{P} \overline{X}' \longrightarrow X \tag{6.3.6.2}$$

the factorization of the projection $X' \to X$ defined in 3.3.1 (the Stein factorization of $X' \to X$).

Corollary 6.3.7. — *For any quasi-coherent sheaf $\mathcal{E}$ on $\overline{X}'$ the natural map $\mathcal{E} \to RP_*P^*\mathcal{E}$ is an isomorphism. In particular, $\mathcal{S}_{R,f}$ is Frobenius acyclic in the sense of 3.2.1.*

Proof. — The Frobenius morphism on $\mathcal{S}_{R,f}$ is simply the map Λ_p. After replacing X by an étale cover, we can assume that the map $X \to \mathcal{S}_{R,f}$ factors through an affine morphism $\tau : X \to U_{R,f}$. In this case, 6.3.3 shows that X' is the quotient of an affine X-scheme by an action of a diagonalizable group scheme μ. The statement that $R^iP_*P^*\mathcal{E} = 0$ for $i > 0$ then follows from [**14**, I.5.3.3], as in the proof of 6.3.5. Moreover, to see that $\mathcal{E} = P_*P^*\mathcal{E}$, it suffices to show that $\tau_*\mathcal{E} = P_{U_{R,f}*}P^*_{U_{R,f}}\tau_*\mathcal{E}$, and hence it suffices to consider the case when $X = U_{R,f}$.

Now as an $\mathcal{O}_{\overline{U}'_{R,f}}$-module, the ring (6.3.5.3) is isomorphic to a direct sum $\oplus_{a\in A}M_a$, where A denotes the group $(\mathbb{Z}/(e))^r$ and M_a is the free rank 1 submodule generated by monomials $X_1^{b_1}\cdots X_r^{b_r}V^c$ for which the class is equal to a. From this the result follows. □

6.3.8. — The map Λ_e extends naturally to a map on the stack $\mathcal{T}_H$ defined in 6.1.12. More precisely, let $\Lambda_{e,\mathcal{T}} : \mathcal{T}_H \to \mathcal{T}_H$ denote the map induced by the map on prestacks which sends $(x_1, \ldots, x_r)$ to $(x_1^e, \ldots, x_r^e)$ and a morphism $(h, \{u_i\})$ to $(h, \{u_i^e\})$. Then there is a natural commutative diagram

$$\begin{array}{ccc} \mathcal{S}_H(\alpha)_t & \xrightarrow{\Lambda_e} & \mathcal{S}_H(\alpha)_{t^e} \\ \downarrow & & \downarrow \\ \mathcal{T}_H \otimes_{\mathbb{Z}} \mathbb{Z}[t] & \xrightarrow{\Lambda_{e,\mathcal{T}}} & \mathcal{T}_H \otimes_{\mathbb{Z}} \mathbb{Z}[t], \end{array} \tag{6.3.8.1}$$

where the vertical arrows are as in (6.1.12.2).

Observe also that if W denotes the Witt ring of a perfect field of characteristic $p > 0$ and if $\mathcal{T}_{H,W}$ denotes the base change of $\mathcal{T}_H$ to W, then there is a natural isomorphism $\mathcal{T}_{H,W} \simeq \mathcal{T}_{H,W} \otimes_{W,\sigma} W$, where σ denotes the canonical lifting of Frobenius to W. This is simply because both sides are obtained by base change from $\mathcal{T}_H$ (which is defined over $\mathbb{Z}$). Moreover, the induced map

$$\mathcal{T}_{H,W} \xrightarrow{\Lambda_p} \mathcal{T}_{H,W} \simeq \mathcal{T}_{H,W} \otimes_{W,\sigma} W \xrightarrow{\mathrm{pr}_1} \mathcal{T}_{H,W} \tag{6.3.8.2}$$

is a lifting of the Frobenius endomorphism of the reduction of $\mathcal{T}_{H,W}$.

6.3.9. — We now make a calculation whose corollaries 6.3.19–6.3.26 below play an important role in the construction of the (φ, N, G)-module structure on de Rham cohomology (specifically the corollaries of this calculation are used in 6.4.3 (ii), and the proofs of 7.1.6 and 7.1.9).

From now until 6.3.27, assume that there are integers $a_1 \geq a_2 \geq \cdots \geq a_r$ such that $\alpha_i = p^{a_i}$ and assume $e = p^n$ for some $n > a_1$.

In the key technical result 6.3.18 below, we give an explicit description of the $\mathcal{O}_{U_{t^e}}$-algebra, $\Lambda_{e*}\mathcal{O}_{\mathcal{S}_t}(U_{t^e})$ by generators and relations.

In preparation for this result, note first by 6.3.3 we can view $\Lambda_{e*}\mathcal{O}_{\mathcal{S}_t}(U_{t^e})$ as the global sections of the structure sheaf on the stack $[U_t/\mu]$. Equivalently $\Lambda_{e*}\mathcal{O}_{\mathcal{S}_t}(U_{t^e})$ is the subalgebra of μ-invariant elements of

$$\mathbb{Z}[t][X_1, \ldots, X_r, V^{\pm}]/(X_1^{p^{a_1}} \cdots X_r^{p^{a_r}} V = t). \tag{6.3.9.1}$$

As mentioned in the proof of 6.3.5, this subalgebra admits as a basis over $\mathbb{Z}[t]$ the monomials $(\prod_i X_i^{m_i})V^l$, where

$$m_i \equiv p^{a_i} l \pmod{p^n}. \tag{6.3.9.2}$$

For integers $\epsilon \in (a_r, a_1]$ and $\gamma \in (0, p^\epsilon)$ with $(\gamma, p) = 1$, set

$$S(\epsilon) := \{i | \epsilon > a_i\}, \tag{6.3.9.3}$$

and

$$M_{(\epsilon,\gamma)} := \Big(\prod_{i \in S(\epsilon)} X_i^{p^{n-(\epsilon-a_i)}\beta_i}\Big) V^{p^{n-\epsilon}\gamma} \in \mathcal{O}_{U_t}, \tag{6.3.9.4}$$

where $\beta_i \in (0, p^{\epsilon - a_i})$ is a representative for γ modulo $p^{\epsilon - a_i}$. Note that for every i we have by the definition of β_i

$$p^{n-(\epsilon - a_i)}\beta_i \equiv p^{a_i} \cdot p^{n-\epsilon}\gamma \pmod{p^n} \tag{6.3.9.5}$$

so $M_{(\epsilon,\gamma)}$ is an element of $\Lambda_{e*}\mathcal{O}_{\mathcal{S}_t}(U_{t^e}) \subset \mathcal{O}_{U_t}$. Also let $Z \in \Lambda_{e*}\mathcal{O}_{\mathcal{S}_t}(U_{t^e})$ denote the element defined by $V^{p^{n-a_r}} \in \mathcal{O}_{U_t}$.

Lemma 6.3.10. — *As a $\mathcal{O}_{U_{t^e}}$-subalgebra of $\mathcal{O}_{U_t}$, the algebra $\Lambda_{e*}\mathcal{O}_{\mathcal{S}_t}(U_{t^e})$ is generated by Z and the $M_{(\epsilon,\gamma)}$.*

Proof. — To prove the lemma it suffices to show that any monomial $(\prod_i X_i^{m_i})V^l$, with $m_i \equiv p^{a_i}l \pmod{p^n}$ for all i, is in the $\mathcal{O}_{U_{t^e}}$-subalgebra generated by Z and the $M_{(\epsilon,\gamma)}$. For this we may without loss of generality assume that m_i and l are all nonnegative and smaller than p^n. In addition we can assume that at least one m_i is strictly less than p^{a_i}. Write $l = p^{n-\epsilon}\gamma$ with $(p,\gamma) = 1$, so the condition on the m_i can be written as $m_i \equiv p^{n-\epsilon+a_i}\gamma \pmod{p^n}$.

We consider two cases. If $\epsilon \leq a_r$, then $p^{n-\epsilon+a_i}$ is divisible by p^n for all i so in this case all the m_i are zero. Therefore, our monomial is of the form $V^{p^{n-a_r}p^{a_r-\epsilon}\gamma}$ and hence is in the subalgebra generated by Z. If $\epsilon > a_r$, then $S(\epsilon)$ is non-empty, but we still have $m_i = 0$ unless $i \in S(\epsilon)$. As for $i \in S(\epsilon)$, let β_i be the unique representative for γ modulo $p^{\epsilon - a_i}$. Then we have

$$m_i \equiv p^{n-\epsilon+a_i}\beta_i \pmod{p^n}, \tag{6.3.10.1}$$

and since $m_i < p^n$ we in fact have $m_i = p^{n-\epsilon+a_i}\beta_i$. This also implies that $\epsilon \leq a_1$. Indeed if $\epsilon > a_1$, then $S(\epsilon) = \{1, \ldots, r\}$ and hence each m_i is non-zero and greater than or equal to p^{a_i} contradicting our assumptions. Thus our monomial is equal to $M_{(\epsilon,\gamma)}$. □

Lemma 6.3.11. — *As a $\mathbb{Z}[t]$-module $\Lambda_{e*}\mathcal{O}_{\mathcal{S}_t}(U_{t^e})$ is free on generators*

$$\Big(\prod_i X_i^{m_i}\Big)V^l M_{(\epsilon,\gamma)}, \quad \Big(\prod_i X_i^{m_i}\Big)V^l Z^a, \quad \prod_i X_i^{m_i}V^l \tag{6.3.11.1}$$

where $(\prod_i X_i^{m_i})V^l \in \mathcal{O}_{U_{t^e}}$ (acting on $M_{(\epsilon,\gamma)}$ and Z through the $\mathcal{O}_{U_{t^e}}$-module structure on $\Lambda_{e*}\mathcal{O}_{\mathcal{S}_t}(U_{t^e})$), $l \in \mathbb{Z}$, and in the first case there exists an $i \in S(\epsilon)$ such that $m_i = 0$, and the second and third case at least one m_i is equal to zero and $a \in (0, p^{a_r})$.

Proof. — Under the inclusion of $\mathbb{Z}[t]$-modules $\Lambda_{e*}\mathcal{O}_{\mathcal{S}_t}(U_{t^e}) \subset \mathcal{O}_{U_t}$ the generators (6.3.11.1) map respectively to

$$A_{(\epsilon,\gamma)}(\underline{m}, l) := \Big(\prod_i X_i^{m_i p^n}\Big)V^{p^n l} \prod_{i \in S(\epsilon)} X_i^{p^{n-(\epsilon-a_i)}\beta_i}V^{p^{n-\epsilon}\gamma}, \tag{6.3.11.2}$$

$$B(\underline{m}, a) := \Big(\prod_i X_i^{p^n m_i}\Big)V^{p^n l + p^{n-a_r}a}, \tag{6.3.11.3}$$

and

$$C(\underline{m}, l) := \prod_i X_i^{p^n m_i} V^{p^n l}. \tag{6.3.11.4}$$

Suppose given a linear relationship
(6.3.11.5)

$$\Big(\sum_k \alpha_k A_{(\epsilon_k, \gamma_k)}(\underline{m}^{(k)}, l^{(k)})\Big) + \Big(\sum_s \beta_s B(\underline{m}^{(s)}, a^{(s)})\Big) + \Big(\sum_t \gamma_t C(\underline{m}^{(t)}, l^{(t)})\Big) = 0.$$

By definition for every $A_{(\epsilon,\gamma)}(\underline{m}, l)$ there exists some i such that the exponent of X_i in $A_{(\epsilon,\gamma)}(\underline{m}, l)$ is not divisible by p^n. Therefore by looking at the monomials with each exponent of X_i divisible by p^n we see that (6.3.11.5) induces relations

$$\Big(\sum_s \beta_s B(\underline{m}^{(s)}, a^{(s)})\Big) + \Big(\sum_t \gamma_t C(\underline{m}^{(t)}, l^{(t)})\Big) = 0 \tag{6.3.11.6}$$

and

$$\sum_k \alpha_k A_{(\epsilon_k, \gamma_k)}(\underline{m}^{(k)}, l^{(k)}) = 0. \tag{6.3.11.7}$$

Furthermore, note that the exponent of V in $B(\underline{m}, a)$ is not divisible by p^n so the relation (6.3.11.6) is in fact obtained from relations

$$\sum_s \beta_s B(\underline{m}^{(s)}, a^{(s)}) = 0 \tag{6.3.11.8}$$

and

$$\sum_t \gamma_t C(\underline{m}^{(t)}, l^{(t)}) = 0. \tag{6.3.11.9}$$

For any two $A_{(\epsilon,\gamma)}(\underline{m}, l)$ and $A_{(\epsilon',\gamma')}(\underline{m}', l')$ the exponents of V in these monomials are distinct unless $(\epsilon, \gamma) = (\epsilon', \gamma')$ and $l = l'$, and in this case the exponents of X_i are distinct unless we have $m_i = m_i'$ for all i. It follows that in (6.3.11.7) we must have all $\alpha_k = 0$.

Similarly for any two $B(\underline{m}, a)$ and $B(\underline{m}', a')$ the exponents of V are distinct unless $a = a'$ and the exponents of X_i are distinct unless $m_i = m_i'$. It follows that we also have $\beta_s = 0$ for all s.

Finally for two monomials $C(\underline{m}, l)$ and $C(\underline{m}', l')$ the exponents of V are distinct unless $l = l'$ and the exponents of X_i are distinct unless $m_i = m_i'$. It follows that $\gamma_t = 0$ for all t as well.

We conclude that the elements (6.3.11.1) are linearly independent.

That these elements generate $\Lambda_{e*}\mathcal{O}_{\mathbb{S}_t}(U_{t^e})$ as a $\mathbb{Z}[t]$-submodule of $\mathcal{O}_{U_t}$ follows from 6.3.10. □

6.3.12. — To completely characterize the $\mathcal{O}_{U_{t^e}}$-algebra $\Lambda_{e*}\mathcal{O}_{\mathcal{S}_t}(U_{t^e})$, it remains to determine the multiplicative relations among the generators $M_{(\epsilon,\gamma)}$ and Z. This is a rather complicated calculation, so we break it into several pieces. First let us introduce some notation.

Let (ϵ,γ) and (ϵ',γ') be two pairs as in the definition of the $M_{(\epsilon,\gamma)}$'s (6.3.9), and assume $\epsilon \geq \epsilon'$.

For $j \in S(\epsilon')$ define $e_j((\epsilon,\gamma),(\epsilon',\gamma')) \in \mathbb{Z}$ to be the unique integer such that

$$\beta_j + p^{\epsilon-\epsilon'}\beta'_j = p^{\epsilon-a_j}e_j((\epsilon,\gamma),(\epsilon',\gamma')) + r, \quad r \in \{0,\dots,p^{\epsilon-a_j}-1\}, \tag{6.3.12.1}$$

where as before β_j (resp. β'_j) denotes the unique representative in $(0,p^{\epsilon-a_j})$ (resp. $(0,p^{\epsilon'-a_j})$) of $\gamma \pmod{p^{\epsilon-a_j}}$ (resp. $\gamma' \pmod{p^{\epsilon'-a_j}}$). Also define $e((\epsilon,\gamma),(\epsilon',\gamma')) \in \mathbb{Z}$ by the equation

$$\gamma + p^{\epsilon-\epsilon'}\gamma' = p^{\epsilon}e((\epsilon,\gamma),(\epsilon',\gamma')) + r, \quad r \in \{0,\dots,p^{\epsilon}-1\}, \tag{6.3.12.2}$$

and set

$$E((\epsilon,\gamma),(\epsilon',\gamma')) := \Big(\prod_{j\in S(\epsilon')} X_j^{e_j((\epsilon,\gamma),(\epsilon',\gamma'))}\Big)\cdot V^{e((\epsilon,\gamma),(\epsilon',\gamma'))} \in \mathcal{O}_{U_{t^e}}. \tag{6.3.12.3}$$

Now assume further that $p^{\epsilon-a_r}$ does not divide $\gamma + p^{\epsilon-\epsilon'}\gamma'$. Define $\epsilon'' \in \mathbb{N}$ by the equation

$$n-\epsilon'' := \mathrm{ord}_p(p^{n-\epsilon}\gamma + p^{n-\epsilon'}\gamma'), \tag{6.3.12.4}$$

which since $\epsilon \geq \epsilon'$ can also be written as

$$\mathrm{ord}_p(p^{n-\epsilon}(\gamma + p^{\epsilon-\epsilon'}\gamma')). \tag{6.3.12.5}$$

Remark 6.3.13. — Note that if $\epsilon \neq \epsilon'$ then $\gamma + p^{\epsilon-\epsilon'}\gamma'$ is prime to p so $\epsilon'' = \epsilon$.

Since $p^{\epsilon-a_r}$ does not divide $\gamma + p^{\epsilon-\epsilon'}\gamma'$ we find

$$n-\epsilon \leq n-\epsilon'' < n-\epsilon+(\epsilon-a_r) = n-a_r \tag{6.3.13.1}$$

which yields

$$\epsilon \geq \epsilon'' > a_r. \tag{6.3.13.2}$$

In particular we have $\epsilon'' \in (a_r, a_1]$. Also define $\gamma'' \in (0,p^{\epsilon''})$ to be the unique element prime to p such that

$$p^{n-\epsilon}\gamma + p^{n-\epsilon'}\gamma' \equiv p^{n-\epsilon''}\gamma'' \pmod{p^n}. \tag{6.3.13.3}$$

We define

$$(\epsilon,\gamma) * (\epsilon',\gamma') := (\epsilon'',\gamma''). \tag{6.3.13.4}$$

Note that the equation (6.3.13.3) characterizes $(\epsilon,\gamma)*(\epsilon',\gamma')$. Also, by multiplying the equation (6.3.12.2) by $p^{n-\epsilon}$ we obtain

$$p^n e((\epsilon,\gamma),(\epsilon',\gamma')) + p^{n-\epsilon''}\gamma'' = p^{n-\epsilon}\gamma + p^{n-\epsilon'}\gamma'. \tag{6.3.13.5}$$

Lemma 6.3.14. — *Let (ϵ,γ) and (ϵ',γ') be two pairs as in 6.3.9, and assume $\epsilon \geq \epsilon'$.*

(i) *If $p^{\epsilon-a_r}$ does not divide $\gamma + p^{\epsilon-\epsilon'}\gamma'$ then*

$$M_{(\epsilon,\gamma)} \cdot M_{(\epsilon',\gamma')} = E((\epsilon,\gamma),(\epsilon',\gamma')) \cdot M_{(\epsilon,\gamma)*(\epsilon',\gamma')}. \tag{6.3.14.1}$$

(ii) *If $p^{\epsilon-a_r}$ divides $\gamma + p^{\epsilon-\epsilon'}\gamma'$ then*

$$M_{(\epsilon,\gamma)} \cdot M_{(\epsilon',\gamma')} = E((\epsilon,\gamma),(\epsilon',\gamma')) \cdot Z^{\rho((\epsilon,\gamma),(\epsilon',\gamma'))}, \tag{6.3.14.2}$$

where $\rho((\epsilon,\gamma),(\epsilon',\gamma'))$ is the natural number characterized by the condition that the remainder of $\gamma + p^{\epsilon-\epsilon'}\gamma'$ upon division by p^ϵ is $p^{\epsilon-a_r}\rho((\epsilon,\gamma),(\epsilon',\gamma'))$.

Remark 6.3.15. — Note that in case (ii), we must have $\epsilon = \epsilon'$ for if $\epsilon > \epsilon'$ then $\gamma + p^{\epsilon-\epsilon'}\gamma' \equiv \gamma \pmod p$ and γ is prime to p.

Proof of 6.3.14. — Viewing the $M_{(\epsilon,\gamma)}$'s as μ-invariant elements of the ring $\mathcal{O}_{U_t}$ we need to compute the product

$$\Big(\prod_{i\in S(\epsilon)} X_i^{p^{n-(\epsilon-a_i)}\beta_i} V^{p^{n-\epsilon}\gamma}\Big)\Big(\prod_{i\in S(\epsilon')} X_i^{p^{n-(\epsilon'-a_i)}\beta'_i} V^{p^{n-\epsilon'}\gamma'}\Big). \tag{6.3.15.1}$$

Note that since $\epsilon \geq \epsilon'$ we have $S(\epsilon') \subset S(\epsilon)$. For $i \in S(\epsilon)$ let c_i denote the exponent of X_i in (6.3.15.1) (so $c_i = p^{n-(\epsilon-a_i)}\beta_i + p^{n-(\epsilon'-a_i)}\beta'_i$ if $i \in S(\epsilon')$ and $c_i = p^{n-(\epsilon-a_i)}\beta_i$ otherwise), and let c denote the exponent of V (so $c = p^{n-\epsilon}\gamma + p^{n-\epsilon'}\gamma'$).

By (6.3.13.5), we have in case (i)

$$c = p^n e((\epsilon,\gamma),(\epsilon',\gamma')) + p^{n-\epsilon''}\gamma'' \tag{6.3.15.2}$$

so the exponents of V on either side of (6.3.14.1) agree. In case (ii) we have by the definition of $\rho((\epsilon,\gamma),(\epsilon',\gamma'))$ and $e((\epsilon,\gamma),(\epsilon',\gamma'))$ an equality

$$p^{n-\epsilon}\gamma + p^{n-\epsilon'}\gamma' = p^n e((\epsilon,\gamma),(\epsilon',\gamma')) + p^{n-a_r}\rho((\epsilon,\gamma),(\epsilon',\gamma')). \tag{6.3.15.3}$$

Therefore the exponents of V in both sides of (6.3.14.2) agree.

To verify that the exponents of the X_i agree we consider two cases.

Case 1: $\epsilon \neq \epsilon'$. In this case $\epsilon = \epsilon''$ by 6.3.13 and we just need to show (i) by 6.3.15. We therefore need to show that

$$p^{n-(\epsilon-a_i)}\beta''_i = p^{n-(\epsilon-a_i)}\beta_i \quad \text{if } i \in S(\epsilon) - S(\epsilon'), \tag{6.3.15.4}$$

and

$$p^n e_i((\epsilon,\gamma),(\epsilon',\gamma')) + p^{n-(\epsilon-a_i)}\beta''_i = p^{n-(\epsilon-a_i)}\beta_i + p^{n-(\epsilon'-a_i)}\beta'_i \quad \text{if } i \in S(\epsilon'). \tag{6.3.15.5}$$

To verify (6.3.15.4), note that if $i \in S(\epsilon) - S(\epsilon')$, then $a_i \geq \epsilon'$ which implies that $p^{\epsilon-a_i} | p^{\epsilon-\epsilon'}$. Therefore

$$\gamma + p^{\epsilon-\epsilon'}\gamma' \equiv \gamma \pmod{p^{\epsilon-a_i}}. \tag{6.3.15.6}$$

On the other hand, by definition of γ'' we also have

$$\gamma + p^{\epsilon-\epsilon'}\gamma' \equiv \gamma'' \pmod{p^{\epsilon-a_i}}. \tag{6.3.15.7}$$

It follows that $\beta_i'' = \beta_i$ and hence (6.3.15.4) holds.

To verify (6.3.15.5) note that by definition of β_i'' we have

$$\beta_i'' \equiv \gamma'' \equiv \gamma + p^{\epsilon-\epsilon'}\gamma' \equiv \beta_i + p^{\epsilon-\epsilon'}\beta_i' \pmod{p^{\epsilon-a_i}}. \tag{6.3.15.8}$$

Therefore $p^{n-(\epsilon-a_i)}\beta_i''$ is equal to the remainder of $p^{n-(\epsilon-a_i)}\beta_i + p^{n-(\epsilon'-a_i)}\beta_i'$ upon division by p^n. From this and the definition of $e_i((\epsilon,\gamma),(\epsilon',\gamma'))$ equation (6.3.15.5) follows. This proves (i) for $\epsilon \neq \epsilon'$.

Case 2: $\epsilon = \epsilon'$. In this case we need to show that for $i \in S(\epsilon)$.

(6.3.15.9)

$$p^{n-(\epsilon-a_i)}(\beta_i+\beta_i') = p^n e_i((\epsilon,\gamma),(\epsilon',\gamma')) + \begin{cases} 0 & \text{if } p^{\epsilon-a_r} \nmid (\gamma+\gamma') \\ & \quad \text{and } i \notin S(\epsilon''), \\ 0 & \text{if } p^{\epsilon-a_r} \mid (\gamma+\gamma'), \\ p^{n-(\epsilon''-a_i)}\beta_i'' & \text{if } p^{\epsilon-a_r} \nmid (\gamma+\gamma') \\ & \quad \text{and } i \in S(\epsilon''). \end{cases}$$

If $i \notin S(\epsilon'')$, then we have $a_i \geq \epsilon''$ which by the definition of ϵ'' implies that p^{n-a_i} divides $p^{n-\epsilon}(\gamma+\gamma')$. Since $p^{n-a_i} = p^{n-\epsilon}p^{\epsilon-a_i}$ this implies that $p^{\epsilon-a_i}$ divides $\gamma+\gamma'$. Therefore in both of the first two cases in (6.3.15.9) we have $p^{\epsilon-a_i} \mid (\gamma+\gamma')$ (note that $a_i \geq a_r$ for all i). In those two cases we therefore have by the definition of $e_i((\epsilon,\gamma),(\epsilon',\gamma'))$ an equality

$$\beta_i + \beta_i' = p^{\epsilon-a_i}e_i((\epsilon,\gamma),(\epsilon',\gamma')). \tag{6.3.15.10}$$

Multiplying this equation by $p^{n-(\epsilon-a_i)}$ we obtain the first two cases of (6.3.15.9). In the last case note that by the definitions of γ'' and β_i'' we have

$$p^{n-\epsilon''}\gamma'' \equiv p^{n-\epsilon}(\gamma+\gamma') \pmod{p^n} \tag{6.3.15.11}$$

and $p^{n-(\epsilon''-a_i)}\beta_i''$ is the remainder of $p^{n-(\epsilon''-a_i)}\gamma''$ divided by p^n. Since β_i (resp. β_i') is congruent to γ (resp. γ') modulo $p^{\epsilon-a_i}$, this gives

$$p^{n-(\epsilon-a_i)}(\beta_i+\beta_i') \equiv p^{n-(\epsilon-a_i)}(\gamma+\gamma') \equiv p^{n-(\epsilon''-a_i)}\gamma'' \pmod{p^n}. \tag{6.3.15.12}$$

It follows that $p^{n-(\epsilon''-a_i)}\beta_i''$ is the remainder of $p^{n-(\epsilon-a_i)}(\beta_i+\beta_i')$ divided by p^n. This together with the definition of $e_i((\epsilon,\gamma),(\epsilon',\gamma'))$ proves the last equality in (6.3.15.9). □

Lemma 6.3.16. — *Let (ϵ,γ) be a pair with $\epsilon \in (a_r, a_1]$ and $\gamma \in (0, p^\epsilon)$ with $(\gamma, p) = 1$.*

(i) *If $\epsilon \neq a_1$ then*

$$\Big(\prod_{i \notin S(\epsilon)} X_i\Big) \cdot M_{(\epsilon,\gamma)} = t^{p^{n-a_1}} M_{(a_1, p^{a_1-\epsilon}\gamma-1)}. \tag{6.3.16.1}$$

(ii) *If* $\epsilon = a_1$ *and* $\gamma = p^s\iota + 1$ *for some* $0 < s < a_1 - a_r$ *and* ι *prime to* p, *then*

$$\Big(\prod_{i\notin S(\epsilon)} X_i\Big) \cdot M_{(\epsilon,\gamma)} = t^{p^{n-a_1}} M_{(a_1-s,\iota)}. \tag{6.3.16.2}$$

(iii) *If* $(\epsilon, \gamma) = (a_1, p^{a_1-a_r}\iota + 1)$ *for some natural number* ι *then*

$$\Big(\prod_{i\notin S(\epsilon)} X_i\Big) \cdot M_{(\epsilon,\gamma)} = t^{p^{n-a_1}} Z^{\iota}. \tag{6.3.16.3}$$

Proof. — We compute the product

$$\Big(\prod_{i\notin S(\epsilon)} X_i^{p^n}\Big)\Big(\prod_{i\in S(\epsilon)} X_i^{p^{n-(\epsilon-a_i)}\beta_i}\Big)V^{p^{n-\epsilon}\gamma}, \tag{6.3.16.4}$$

inside the ring $\mathcal{O}_{U_t}$.

To prove (i), let $\tilde{\beta}_i \in (0, p^{a_1-a_i})$ $(i \in S(a_1))$ denote the representative for $p^{a_1-\epsilon}\gamma - 1$. To prove that (6.3.16.1) holds it suffices by comparing exponents of both sides and using the equality

$$t^{p^{n-a_1}} = \Big(\Big(\prod_i X_i^{p^{a_i}}\Big)V\Big)^{p^{n-a_1}} \tag{6.3.16.5}$$

to show that

(6.3.16.6)

$$(\text{Exponent of } V): \quad p^{n-\epsilon}\gamma = p^{n-a_1} + p^{n-a_1}(p^{a_1-\epsilon}\gamma - 1)$$

(6.3.16.7)

$$(\text{Exponent of } X_i): \quad \begin{cases} p^n = p^{a_1} \cdot p^{n-a_1} & i \notin S(a_1) \\ p^n = p^{a_i}p^{n-a_1} + p^{n-(a_1-a_i)}\tilde{\beta}_i & i \in S(a_1) - S(\epsilon) \\ p^{n-(\epsilon-a_i)}\beta_i = p^{n-(a_1-a_i)} + p^{n-(a_1-a_i)}\tilde{\beta}_i & i \in S(\epsilon) \end{cases}$$

The equality (6.3.16.6) is immediate as is the first equality in (6.3.16.7). The second equality in (6.3.16.7) follows from noting that if $a_i \geq \epsilon$ then $p^{a_1-a_i}$ divides $p^{a_1-\epsilon}\gamma$, and therefore $\tilde{\beta}_i = p^{a_1-a_i} - 1$. The last equality in (6.3.16.7) follows from noting that for $i \in S(\epsilon)$ the element $p^{a_1-\epsilon}\beta_i \in (0, p^{a_1-a_i})$ is a representative for $p^{a_1-\epsilon}\gamma$ and therefore $p^{a_1-\epsilon}\beta_i = \tilde{\beta}_i + 1$.

For case (iii) note first that we have $\gamma \equiv 1 \pmod{p^{a_1-a_i}}$ for all i so $\beta_i = 1$ for all i. Therefore for every i (including $i \notin S(\epsilon)$) the exponent of X_i in the left side of (6.3.16.4) is equal to

$$p^{n-(a_1-a_i)} = p^{a_i}p^{n-a_1}. \tag{6.3.16.8}$$

Also the exponent of V in the left side of (6.3.16.4) is equal to

$$p^{n-a_1}(p^{a_1-a_r}\iota + 1) = p^{n-a_1} + p^{n-a_r}\iota. \tag{6.3.16.9}$$

Statement (iii) therefore follows from (6.3.16.5).

For case (ii), let $\tilde{\beta}_i \in (0, p^{a_1-s-a_i})$ ($i \in S(a_1-s)$) denote the representative of ι modulo $p^{a_1-s-a_i}$. Using (6.3.16.5) and comparing exponents of V and the X_i, it suffices to show that the following equations hold:

(6.3.16.10) (Exponent of V): $p^{n-a_1}(p^s\iota+1) = p^{n-a_1} + p^{n-(a_1-s)}\iota,$

(6.3.16.11) (Exponent of X_i for $i \notin S(a_1-s)$): $p^{n-(a_1-a_i)} = p^{n-(a_1-a_i)}\beta_i,$

and

(6.3.16.12)

(Exponent of X_i for $i \in S(a_1-s)$): $p^{n-(a_1-a_i)}\beta_i = p^{n-(a_1-a_i)} + p^{n-(a_1-s-a_i)}\tilde{\beta}_i.$

The validity of (6.3.16.10) is immediate. Equation (6.3.16.11) holds because the condition $i \notin S(a_1-s)$ is equivalent to the condition that $a_1 - a_i \leq s$ which as in case (iii) treated above implies that $\beta_i = 1$. Equation (6.3.16.12) follows by noting that $p^s\tilde{\beta}_i \in (0, p^{a_1-a_i})$ is a representative for $p^s\iota$ modulo $p^{a_1-a_i}$, and therefore $\beta_i = p^s\tilde{\beta}_i + 1$ for $i \in S(a_1-s)$. □

Lemma 6.3.17

(i) *$Z^{p^{a_r}} = V$ (here $V \in \mathcal{O}_{U_{te}}$ which maps to V^{p^n} in $\mathcal{O}_{U_t}$).*

(ii) *$Z \cdot M_{(\epsilon,\gamma)} = V^a M_{(\epsilon,b)}$, where $a \in \mathbb{N}$ and $b \in (0, p^\epsilon)$ are characterized by the equation $\gamma + p^{\epsilon-a_r} = p^\epsilon a + b$.*

(iii) $$X_1 \cdots X_r \cdot V = t^{p^{n-a_1}} \cdot M_{(a_1, p^{a_1}-1)}$$

if $a_1 > a_r$, and otherwise $X_1 \cdots X_r \cdot V = t^{p^{n-a_1}} Z^{p^{a_1}-1}$ (in this case we have $a_1 = \cdots = a_r$).

Proof. — This follows immediately from the definitions. □

Proposition 6.3.18. — *Let $\mathcal{A}$ denote the quotient of the polynomial algebra*

(6.3.18.1) $$\mathcal{O}_{U_{te}}[T_{(\epsilon,\gamma)}, W^{\pm}]_{\{(\epsilon,\gamma)|\epsilon\in(a_r,a_1],\gamma\in(0,p^\epsilon),(\gamma,p)=1\}}$$

by the following relations (i)–(v).

(i) *Let (ϵ,γ) and (ϵ',γ') be pairs of integers with $\epsilon,\epsilon' \in (a_r, a_1]$, $\gamma \in (0,p^\epsilon)$, $\gamma' \in (0, p^{\epsilon'})$, γ and γ' prime to p, and with $\epsilon \geq \epsilon'$.*

(a) *If $p^{\epsilon-a_r}$ does not divide $\gamma + p^{\epsilon-\epsilon'}\gamma'$ then*

(6.3.18.2) $$T_{(\epsilon,\gamma)} \cdot T_{(\epsilon',\gamma')} = E((\epsilon,\gamma),(\epsilon',\gamma')) \cdot T_{(\epsilon,\gamma)*(\epsilon',\gamma')}.$$

(b) *If $p^{\epsilon-a_r}$ divides $\gamma + p^{\epsilon-\epsilon'}\gamma'$, then*

(6.3.18.3) $$T_{(\epsilon,\gamma)} \cdot T_{(\epsilon',\gamma')} = E((\epsilon,\gamma),(\epsilon',\gamma')) \cdot W^{\rho((\epsilon,\gamma)*(\epsilon',\gamma'))},$$

where $\rho((\epsilon,\gamma)*(\epsilon',\gamma'))$ is defined as in 6.3.14.

(ii) *Let (ϵ,γ) be a pair defining a generator $T_{(\epsilon,\gamma)}$.*

(a) *If $\epsilon \neq a_1$ then*

(6.3.18.4) $$\Big(\prod_{i\notin S(\epsilon)} X_i\Big) \cdot T_{(\epsilon,\gamma)} = t^{p^{n-a_1}} T_{(\epsilon,\gamma)}.$$

(b) *If* $\epsilon = a_1$ *and* $\gamma = p^s\iota + 1$ *with* $0 < s < a_1 - a_r$ *and* ι *prime to* p, *then*

$$\Big(\prod_{i \notin S(\epsilon)} X_i\Big) \cdot T_{(\epsilon,\gamma)} = t^{p^{n-a_1}} T_{(a_1 - s, \iota)}. \tag{6.3.18.5}$$

(c) *If* $(\epsilon, \gamma) = (a_1, p^{a_1 - a_r}\iota + 1)$ *for some natural number* ι, *then*

$$\Big(\prod_{i \notin S(\epsilon)} X_i\Big) \cdot T_{(\epsilon,\gamma)} = t^{p^{n-a_1}} W^{\iota}. \tag{6.3.18.6}$$

(iii) $W^{p^{a_r}} = V \in \mathcal{O}_{U_{t^e}}$.

(iv) $W \cdot T_{(\epsilon,\gamma)} = V^a T_{(\epsilon,b)}$, *where* $a \in \mathbb{N}$ *and* $b \in (0, p^{\epsilon})$ *are characterized by the equation* $\gamma + p^{\epsilon - a_r} = p^{\epsilon} a + b$.

(v) $X_1 \cdots X_r \cdot V = t^{p^{n-a_1}} \cdot T_{(a_1, p^{a_1} - 1)}$ *if* $a_1 > a_r$ *and otherwise* $X_1 \cdots X_r \cdot V = t^{p^{n-a_1}} W^{p^{a_1} - 1}$.

Then the map

$$\mathcal{O}_{U_{t^e}}[T_{(\epsilon,\gamma)}, W^{\pm}]_{\{(\epsilon,\gamma) | \epsilon \in (a_r, a_1], \gamma \in (0, p^{\epsilon}), (\gamma, p) = 1\}} \longrightarrow \Lambda_{e*}\mathcal{O}_{U_t}(U_{t^e}) \tag{6.3.18.7}$$

$$T_{(\epsilon,\gamma)} \longmapsto M_{(\epsilon,\gamma)}, \quad W \longmapsto Z \tag{6.3.18.8}$$

factors through an isomorphism

$$\Phi : \mathcal{A} \longrightarrow \Lambda_{e*}\mathcal{O}_{U_t}(U_{t^e}). \tag{6.3.18.9}$$

Proof. — That (6.3.18.7) factors through a map Φ follows from 6.3.14, 6.3.16, and 6.3.17. The surjectivity of Φ follows from 6.3.10.

For the injectivity of Φ note that relations (i), (iii), and (iv) imply that there is a surjection of $\mathbb{Z}[t]$-modules

$$\pi : (\oplus_{(\epsilon,\gamma,\underline{m},l)} \mathbb{Z}[t] \cdot f_{(\epsilon,\gamma,\underline{m},l)}) \oplus (\oplus_{(j,\underline{m},l)} \mathbb{Z}[t] \cdot g_{(j,\underline{m},l)}) \twoheadrightarrow \mathcal{A}, \tag{6.3.18.10}$$

where the left side is the free $\mathbb{Z}[t]$-module on the generators $f_{(\epsilon,\gamma,\underline{m},l)}$ and $g_{(j,\underline{m},l)}$ (here (ϵ, γ) is a pair of integers with $\epsilon \in (a_r, a_1]$ and $\gamma \in (0, p^{\epsilon})$, $\underline{m} \in \mathbb{N}^r$, $l \in \mathbb{Z}$, and $0 \leq j < p^{a_r}$). The map π sends

$$f_{(\epsilon,\gamma,\underline{m},l)} \longmapsto \Big(\prod_i X_i^{m_i}\Big) V^l T_{(\epsilon,\gamma)} \tag{6.3.18.11}$$

and

$$g_{(j,\underline{m},l)} \longmapsto \Big(\prod_i X_i^{m_i}\Big) V^l W^j. \tag{6.3.18.12}$$

The kernel of π is generated by the relations (ii) and (v). From this it follows that $\mathcal{A}$ is free as a $\mathbb{Z}[t]$-module with basis the monomials

$$\Big(\prod_i X_i^{m_i}\Big) V^l T_{(\epsilon,\gamma)}, \quad \Big(\prod_i X_i^{m_i}\Big) V^l W^a, \quad \prod_i X_i^{m_i} V^l \tag{6.3.18.13}$$

where $l \in \mathbb{Z}$ and in the first case there exists an $i \in S(\epsilon)$ such that $m_i = 0$, and the second and third case at least one m_i is equal to zero and $a \in (0, p^{a_r})$. The injectivity therefore follows 6.3.11. □

Corollary 6.3.19

(i) *If* $a_1 = \cdots = a_r$, *then*

$$\mathcal{A} \simeq \mathcal{O}_{U_{t^e}}[W^{\pm}]/(W^{p^{a_1}} = V, X_1 \cdots X_r V = t^{p^{n-a_1}} W^{p^{a_1}-1}). \tag{6.3.19.1}$$

In particular, if $a_1 = \cdots a_r = 0$, *then the map*

$$\mathcal{O}_{U_{t^e}} \longrightarrow \mathcal{A} \tag{6.3.19.2}$$

is an isomorphism.

(ii) *If* $a_1 = \cdots = a_r = 0$, *then for any morphism* $Y \to \mathcal{S}_{t^e}$ *from a stack* Y, *the map*

$$\mathcal{O}_Y \longrightarrow RP_{e*}\mathcal{O}_{\widetilde{Y}} \tag{6.3.19.3}$$

is also an isomorphism, where $\widetilde{Y}$ *denote* $Y \times_{\mathcal{S}_{t^e}, \Lambda_e} \mathcal{S}_t$ *and* $P_e : \widetilde{Y} \to Y$ *is the projection.*

Proof. — Statement (i) follows from 6.3.18.

For (ii), note that by 6.3.5 it suffices to consider the case when $Y \to \mathcal{S}_{t^e}$ is a faithfully flat morphism. In particular, it suffices to consider the case when $Y = U_{t^e}$ so (ii) follows from (i) and the vanishing of higher cohomology groups in 6.3.5. □

Remark 6.3.20. — As the referee suggests, an alternate proof of 6.3.19 (i) can be obtained in the following steps. Fix throughout this remark an integer $r \geq 1$.

(1) For an integer $f \geq 1$ let M_f denote the quotient of the free monoid on generators $e_1, \ldots, e_r, \pm w$, and z modulo the relation

$$e_1 + \cdots + e_r + w = fz. \tag{6.3.20.1}$$

Note that there is a natural map

$$\rho : M_f \longrightarrow \mathbb{Z}^r \oplus \mathbb{Z} \tag{6.3.20.2}$$

sending e_i to the i-th standard generator of $\mathbb{Z}^r$, z to the generator of $\mathbb{Z}$, and w to

$$f\rho(z) - \sum_{i=1}^{r} \rho(x_i). \tag{6.3.20.3}$$

Then ρ is an inclusion and identifies M_f^{gp} with $\mathbb{Z}^r \oplus \mathbb{Z}$. One verifies immediately that the map ρ identifies M_f with the submonoid of elements

$$((a_1, \ldots, a_r), b) \in \mathbb{Z}^r \oplus \mathbb{Z} \tag{6.3.20.4}$$

for which

$$a_i + \frac{b}{f} \geq 0 \tag{6.3.20.5}$$

for all i. In particular, M_f is a saturated monoid.

(2) Fix integers e and α with $\alpha | e$. Let

$$\pi : \mathbb{Z}^r \oplus \mathbb{Z} \longrightarrow (\mathbb{Z}/(e))^r \tag{6.3.20.6}$$

be the map sending the i-th standard generator $\mathbb{Z}^r$ to the i-th standard generator of $(\mathbb{Z}/(e))^r$, and which sends $(\underline{0}, 1)$ to the element

$$(-\alpha, \dots, -\alpha) \in (\mathbb{Z}/(e))^r. \tag{6.3.20.7}$$

There is a map

$$M_{e/\alpha} \longrightarrow \mathrm{Ker}(\pi) \cap (\mathbb{N}^r \oplus \mathbb{Z}) \subset \mathbb{Z}^r \oplus \mathbb{Z} \tag{6.3.20.8}$$

which sends $e_i \in M_{f,e/\alpha}$ to e times the i-th standard generator in $\mathbb{Z}^r$, w to the element

$$((0, \dots, 0), e/\alpha) \in \mathbb{Z}^r \oplus \mathbb{Z}, \tag{6.3.20.9}$$

and z to the element

$$((\alpha, \dots, \alpha), 1) \in \mathbb{Z}^r \oplus \mathbb{Z}. \tag{6.3.20.10}$$

This map (6.3.20.8) is in fact an isomorphism. Indeed suppose

$$((a_1, \dots, a_r), n) \in \mathrm{Ker}(\pi) \cap (\mathbb{N}^r \oplus \mathbb{Z}), \tag{6.3.20.11}$$

and let $0 \leq m < e/\alpha$ be a representative for n modulo e/α. We then have

$$a_i \equiv \alpha \cdot n \pmod{e} \tag{6.3.20.12}$$

for all i, which implies that we can write

$$a_i = e\gamma_i + \alpha m \tag{6.3.20.13}$$

for some $\gamma_i \geq 0$. It follows that we can write

$$((a_1, \dots, a_r), n) = ((e\gamma_1, \dots, e\gamma_r), 0) + m((\alpha, \dots, \alpha), 1) + \kappa((0, \dots, 0), e/\alpha), \tag{6.3.20.14}$$

where κ is characterized by the equation

$$\kappa(e/\alpha) + m = n. \tag{6.3.20.15}$$

This shows that (6.3.20.8) is surjective, and it is injective as the induced map on groups

$$M_{e/\alpha}^{\mathrm{gp}} \simeq \mathbb{Z}^r \oplus \mathbb{Z} \longrightarrow \mathbb{Z}^r \oplus \mathbb{Z} \tag{6.3.20.16}$$

is clearly injective.

(3) The geometric interpretation of the calculation in (2) is the following. Consider the scheme

$$U_t = \mathrm{Spec}(\mathbb{Z}[t][X_1, \dots, X_r, V^{\pm}]/(X_1^{\alpha} \cdots X_r^{\alpha} V = t)) \simeq \mathrm{Spec}(\mathbb{Z}[\mathbb{N}^r \oplus \mathbb{Z}]). \tag{6.3.20.17}$$

The map π in (6.3.20.6) defines an inclusion of diagonalizable group schemes

$$\mu = D((\mathbb{Z}/(e))^r) \subset D(\mathbb{Z}^r \oplus \mathbb{Z}) \tag{6.3.20.18}$$

and hence an action of μ on the affine toric variety $\mathrm{Spec}(\mathbb{Z}[\mathbb{N}^r \oplus \mathbb{Z}])$ (note that this is the same as the action in 6.3.3). The isomorphism (6.3.20.8) then gives an identification

$$\big(\mathbb{Z}[t][X_1, \ldots, X_r, V^{\pm}]/(X_1^{\alpha} \cdots X_r^{\alpha} V = t)\big)^{\mu} \simeq \mathbb{Z}[M_{e/\alpha}]. \tag{6.3.20.19}$$

(4) Let Q denote the pushout of the diagram

$$\begin{array}{ccc} \mathbb{N} & \xrightarrow{\cdot e} & \mathbb{N} \\ {\scriptstyle (\alpha,1)}\downarrow & & \\ \mathbb{N}^r \oplus \mathbb{Z}, & & \end{array} \tag{6.3.20.20}$$

where the map $(\alpha, 1)$ sends 1 to the element

$$((\alpha, \ldots, \alpha), 1) \in \mathbb{N}^r \oplus \mathbb{Z}. \tag{6.3.20.21}$$

The commutative diagram

$$\begin{array}{ccc} \mathbb{N}^r \oplus \mathbb{Z} & \xrightarrow{\times e} & \mathbb{N}^r \oplus \mathbb{Z} \\ {\scriptstyle (\alpha,1)}\uparrow & & \uparrow{\scriptstyle (\alpha,1)} \\ \mathbb{N} & \xrightarrow{\cdot e} & \mathbb{N}, \end{array} \tag{6.3.20.22}$$

induces a map

$$Q \to \mathrm{Ker}(\pi) \cap (\mathbb{N}^r \oplus \mathbb{Z}) \simeq M_{e/\alpha}. \tag{6.3.20.23}$$

Let $e_1, \ldots, e_r, \pm w, z$ be the generators of $M_{e/\alpha}$ considered in (1). Recall that under the embedding $M_{e/\alpha} \subset \mathbb{N}^r \oplus \mathbb{Z}$ the element e_i (resp. w, z) maps to e times the i-th standard generator of $\mathbb{N}^r$ (resp. (6.3.20.9), (6.3.20.10)). It follows that Q is the submonoid of $M_{e/\alpha}$ generated by the e_i, $v := \alpha w$, and z. In particular, as a monoid we can describe $M_{e/\alpha}$ as

$$Q \oplus \mathbb{N}/\Big((v, 0) = (0, \alpha), \sum_{i=1}^{r}(e_i, 0) + (0, 1) = (e/\alpha)\cdot(z, 0)\Big). \tag{6.3.20.24}$$

Geometrically we have

$$\mathrm{Spec}(\mathbb{Z}[Q]) \simeq U_{t^e}, \tag{6.3.20.25}$$

and hence this discussion shows that we have an isomorphism of $\mathcal{O}_{U_{t^e}}$-algebras

$$\begin{array}{c} \big(\mathbb{Z}[t][X_1, \ldots, X_r, V^{\pm}]/(X_1^{\alpha} \cdots X_r^{\alpha} V = t)\big)^{\mu} \\ \downarrow \simeq \\ \mathcal{O}_{U_{t^e}}[W^{\pm}]/(W^{\alpha} = V, X_1 \cdots X_r V = t^{e/\alpha} W^{\alpha-1}). \end{array} \tag{6.3.20.26}$$

This recovers the isomorphism (6.3.19.1) (and in fact a stronger result as we do not need the assumption that α and e are powers of a prime in the above argument).

Corollary 6.3.21. — *Assume all $\alpha_i = 1$, and let R be a perfect ring of characteristic $p > 0$. Then the stack $\mathcal{S}_{R,0}$ is a perfect stack in the sense of 4.4.1.*

Proof. — Take $Y = \mathcal{S}_{R,0}$ in 6.3.19 (ii). Then $\widetilde{Y}$ is also isomorphic to $\mathcal{S}_{R,0}$ and the map P_e is the relative Frobenius morphism of $\mathcal{S}_{R,0}$. The result therefore follows from 6.3.19 (ii). □

Corollary 6.3.22. — *Let R be a ring. Then for $l > a_1 + 1$ the map*

$$L : \Lambda_{p^{l-1}*}\mathcal{O}_{\mathcal{S}_{R,0}} \longrightarrow \Lambda_{p^l*}\mathcal{O}_{\mathcal{S}_{R,0}} = \Lambda_{p^{l-1}*}\Lambda_{p*}\mathcal{O}_{\mathcal{S}_{R,0}} \tag{6.3.22.1}$$

induced by the natural map $\mathcal{O}_{\mathcal{S}_{R,0}} \to \Lambda_{p*}\mathcal{O}_{\mathcal{S}_{R,0}}$ is an isomorphism.

Proof. — It suffices to show that the map L becomes an isomorphism when restricted to the smooth cover $U_{R,0}$ of $\mathcal{S}_{R,0}$. Let $\mathcal{A}$ (resp. $\mathcal{B}$) denote the ring $\Lambda_{p^{l-1}*}\mathcal{O}_{\mathcal{S}_{R,0}}(U_{R,0})$ (resp. $\Lambda_{p^l*}\mathcal{O}_{\mathcal{S}_{R,0}}(U_{R,0})$). Let $\mathcal{E}$ denote the ring

$$\mathcal{E} := R[X_1, \ldots, X_r, V]/(X_1^{p_1^{a_1}} \cdots X_r^{p^{a_r}} V), \tag{6.3.22.2}$$

and denote by $\tilde{\Lambda}_{p^j} : \mathcal{E} \to \mathcal{E}$ the map sending $X_i \mapsto X_i^{p^j}$ and $V \mapsto V^{p^j}$.

Then, as in 6.3.10, $\mathcal{A}$ (resp. $\mathcal{B}$) is identified with the subalgebra of $\mathcal{E}$ generated by $Z := V^{p^{(l-1)-a_r}}$ (resp. $Z' := V^{p^{l-a_r}}$) and the monomials

$$\begin{aligned} M^{\mathcal{A}}_{(\epsilon,\gamma)} &= \prod_{i \in S(\epsilon)} X_i^{p^{(l-1)-(\epsilon-a_i)}\beta_i} V^{p^{(l-1)-\epsilon}\gamma} \\ \text{(resp. } M^{\mathcal{B}}_{(\epsilon,\gamma)} &= \prod_{i \in S(\epsilon)} X_i^{p^{l-(\epsilon-a_i)}\beta_i} V^{p^{l-\epsilon}\gamma}), \end{aligned} \tag{6.3.22.3}$$

where $\epsilon \in (a_r, a_1]$, $\gamma \in (0, p^\epsilon)$, $(\gamma, p) = 1$, and $\beta_i \in (0, p^{\epsilon - a_i})$ is the unique representative for $\gamma \pmod{p^{\epsilon - a_i}}$. Since the diagram

$$\begin{array}{ccccc} U_{R,0} & \xrightarrow{\tilde{\Lambda}_p} & U_{R,0} & \xrightarrow{\tilde{\Lambda}_{p^{l-1}}} & U_{R,0} \\ \downarrow & & \downarrow & & \downarrow \\ \mathcal{S}_{R,0} & \xrightarrow{\Lambda_p} & \mathcal{S}_{R,0} & \xrightarrow{\Lambda_{p^{l-1}}} & \mathcal{S}_{R,0} \end{array} \tag{6.3.22.4}$$

commutes, the diagram

$$\begin{array}{ccc} \mathcal{A} & \xrightarrow{L_{U_{R,0}}} & \mathcal{B} \\ \downarrow & & \downarrow \\ \mathcal{E} & \xrightarrow{\tilde{\Lambda}_p} & \mathcal{E} \end{array} \tag{6.3.22.5}$$

also commutes, where $L_{U_{R,0}}$ denotes the restriction of L to $U_{R,0}$. Since $\tilde{\Lambda}_p$ is injective, it follows that $L_{U_{R,0}}$ is also injective. For the surjectivity it suffices to show that

$$L_{U_{R,0}}(Z) = Z', \quad L_{U_{R,0}}(M^{\mathcal{A}}_{(\epsilon,\gamma)}) = M^{\mathcal{B}}_{(\epsilon,\gamma)}, \tag{6.3.22.6}$$

which is immediate from the definition of $\tilde{\Lambda}_p$. □

6.3.23. — To completely characterize the sheaf $\Lambda_{e*}\mathcal{O}_{\mathcal{S}_t}$ on $\mathcal{S}_{t^e}$, it suffices to describe the action of the group scheme G defined in 6.1.1 on the algebra $\mathcal{A}$. Indeed if $s, b : R_{t^e} \to U_{t^e}$ denote the source and target morphisms respectively, then we know there is an isomorphism $s^*\mathcal{A} \to b^*\mathcal{A}$ over R_{t^e} giving the descent of $\mathcal{A}$ to the stack $\mathcal{S}_{t^e}$, and since the natural map $\pi : \tilde{R} \to R$ is étale and surjective, it suffices to understand what isomorphism we get over $\tilde{R}_{t^e}$ by pulling back.

Lemma 6.3.24. — *An element $g = (u_1, \ldots, u_r, h) \in G$ acts on $\mathcal{A}$ by*

$$T_{(\epsilon,\gamma)} \longmapsto u_1^{\lambda_1} \cdots u_r^{\lambda_r} T_{(\epsilon,\gamma)}, \quad W \longmapsto \Big(\prod_i u_i^{-p^{a_i - a_r}}\Big) W, \tag{6.3.24.1}$$

where $\lambda_i := (\beta_i - \gamma)/p^{\epsilon - a_i}$ for $i \in S(\epsilon)$ and $\lambda_i := -p^{a_i - \epsilon}\gamma$ for $i \notin S(\epsilon)$.

Proof. — Under the natural map $\Lambda_{e*}(\mathcal{O}_{\mathcal{S}_t})(U_{t^e}) \to \mathcal{O}_{U_t}$ the element $M_{(\epsilon,\gamma)}$ maps to

$$\Big(\prod_{i \in S(\epsilon)} X_i^{p^{n-(\epsilon - a_i)}\beta_i}\Big) V^{p^{n-\epsilon}\gamma}. \tag{6.3.24.2}$$

From this it is clear that an element of the form $(1, \ldots, 1, h) \in G$ fixes $T_{(\epsilon,\gamma)}$ and W.

To determine the action of $\mathbb{G}_m^r \subset G$, it suffices to determine the corresponding $\mathbb{Z}^r$-grading on $\mathcal{A}$. Write $\mathcal{A} = \oplus_{\underline{d} \subset \mathbb{Z}^r} \mathcal{A}_{\underline{d}}$, and let $\mathcal{O}_{U_t} = \oplus_{\underline{f} \subset \mathbb{Z}^r} K_{\underline{f}}$ denote the $\mathbb{Z}^r$-grading on $\mathcal{O}_{U_t}$ defined by the $\mathbb{G}^r$-action on $\mathcal{O}_{U_t}$. Then the inclusion $\mathcal{A} \subset \mathcal{O}_{U_t}$ sends $\mathcal{A}_{\underline{d}}$ to $K_{p^n \underline{d}}$, where if $\underline{d} = (d_1, \ldots, d_r)$ we define $p^n\underline{d}$ to be $(p^n d_1, \ldots, p^n d_r)$. It follows that to prove the lemma it suffices to show that (6.3.24.2) lies in the $(p^n\lambda_1, \ldots, p^n\lambda_r)$-graded piece of $\mathcal{O}_{U_t}$, and that $V^{p^{n-a_r}}$ lies in the $(\cdots, -p^{n+a_i-a_r}, \cdots)$-graded piece of $\mathcal{O}_{U_t}$. This is immediate from the definition of the action of G on $\mathcal{O}_{U_t}$. □

6.3.25. — With notation as in 6.3.9, let R be a ring with an element $\pi \in R$ with the property that $\pi^{p^{n-a_1}} = 0$. The base change $\mathbb{Z}[t] \to R$ given by $1 \mapsto \pi$ induces a map

$$\Lambda_e : \mathcal{S}_{R,\pi} \longrightarrow \mathcal{S}_{R,0} \tag{6.3.25.1}$$

and the map $t \mapsto 0$ induces a map

$$\Lambda'_e : \mathcal{S}_{R,0} \longrightarrow \mathcal{S}_{R,0}. \tag{6.3.25.2}$$

Theorem 6.3.26. — *There is a natural isomorphism of $\mathcal{O}_{\mathcal{S}_{R,0}}$-algebras*

$$\Lambda_{e*}\mathcal{O}_{\mathcal{S}_{R,\pi}} \simeq \Lambda'_{e*}\mathcal{O}_{\mathcal{S}_{R,0}}. \tag{6.3.26.1}$$

Proof. — Let $\mathcal{A}_{R,\pi}$ (resp. $\mathcal{A}_{R,0}$) denote $\Lambda_{e*}\mathcal{O}_{\mathcal{S}_{R,\pi}}(U_{R,0})$ (resp. $\Lambda'_{e*}\mathcal{O}_{\mathcal{S}_{R,0}}(U_{R,0})$). To prove the theorem it suffices to construct an isomorphism of $\mathcal{O}_{U_{R,0}}$-algebras

$$\sigma : \mathcal{A}_{R,\pi} \longrightarrow \mathcal{A}_{R,0} \tag{6.3.26.2}$$

compatible with the descent data to $\mathcal{S}_{R,0}$. As in 6.3.23, the descent data on $\mathcal{A}_{R,\pi}$ and $\mathcal{A}_{R,0}$ is determined by the actions of G. Therefore it suffices to construct a G-equivariant isomorphism (6.3.26.2).

For this note that since the formation of $\Lambda_{e*}\mathcal{O}_{\mathcal{S}_t}$ commutes with arbitrary base change on $\mathrm{Spec}(\mathbb{Z}[t])$ by 6.3.5, we have by 6.3.18 surjections

$$\rho : \mathcal{O}_{U_{R,0}}[T_{(\epsilon,\gamma)}, W^{\pm}]_{\epsilon\in(a_r,a_1],\gamma\in(0,p^{\epsilon}),(\gamma,p)=1} \longrightarrow \mathcal{A}_{R,\pi} \tag{6.3.26.3}$$

and

$$\rho' : \mathcal{O}_{U_{R,0}}[T_{(\epsilon,\gamma)}, W^{\pm}]_{\epsilon\in(a_r,a_1],\gamma\in(0,p^{\epsilon}),(\gamma,p)=1} \longrightarrow \mathcal{A}_{R,0}. \tag{6.3.26.4}$$

The kernel of ρ (resp. ρ') is the ideal generated by replacing t by π (resp. 0) in the relations (i)–(iv) in 6.3.18. Since $\pi^{p^{n-a_1}} = 0$ in R this implies that $\mathrm{Ker}(\rho) = \mathrm{Ker}(\rho')$. Therefore there exists a unique isomorphism (6.3.26.2) such that $\rho' = \sigma \circ \rho$. To check that this isomorphism σ is compatible with the G-actions, it suffices to show that for any element $g \in G$ we have

$$\sigma(g^*\rho(T_{(\epsilon,\gamma)})) = g^*\rho'(T_{(\epsilon,\gamma)}), \quad \sigma(g^*\rho(W)) = g^*\rho'(W) \tag{6.3.26.5}$$

which follows from 6.3.24. □

6.3.27. — There is a variant of the map Λ_e which we will use in example 7.2.13. Returning now to the situation of a general sequence $\{\alpha_1, \ldots, \alpha_r\}$, suppose m is an integer dividing every α_i (say $\alpha_i = m \cdot \beta_i$ and write $\beta := (\beta_1, \ldots, \beta_r)$), and let $\mathcal{S}_H(\alpha)[m] \to \mathcal{S}_H(\alpha)$ be the μ_m-torsor defined in 6.1.15. As above, let $\mathcal{S}_H(\alpha)[m]_{t^m}$ (resp. $\mathcal{S}_H(\alpha)_{t^m}$) denote the stack
(6.3.27.1)

$$\mathcal{S}_H(\alpha)[m] \times_{\mathrm{Spec}(\mathbb{Z}[t]),t\mapsto t^m} \mathrm{Spec}(\mathbb{Z}[t]) \ \ (\text{resp. } \mathcal{S}_H(\alpha) \times_{\mathrm{Spec}(\mathbb{Z}[t]),t\mapsto t^m} \mathrm{Spec}(\mathbb{Z}[t])).$$

Then there is a morphism

$$\Psi_m : \mu_m \times \mathcal{S}_H(\beta)_t \longrightarrow \mathcal{S}_H(\alpha)[m]_{t^m} \tag{6.3.27.2}$$

defined as follows. Let $\mathcal{S}_H(\beta)_t^{\mathrm{ps}}$ be the prestack defined in 6.1.8. The stack $\mathcal{S}_H(\alpha)[m]$ is the stack associated to the prestack $\mathcal{S}_H(\alpha)[m]_{t^m}^{\mathrm{ps}}$ which to any scheme T associates the following groupoid:

Objects: Collections of data $((x,v),w)$, where $(x,v) \in \mathcal{S}_H(\alpha)_{t^m}^{\mathrm{ps}}(T)$ and $w \in \Gamma(T, \mathcal{O}_T^*)$ such that $w^m = v$.

Morphisms: For two objects $((x,v),w), ((x',v'),w') \in \mathcal{S}_H(\alpha)[m]_{t^m}^{\mathrm{ps}}(T)$ the sheaf of isomorphisms between them is the sheaf over T associated to the presheaf which to any $g : T' \to T$ associates the set of pairs $((u_i)_{i\in E(g^*x)}, h)$ as in 6.1.6 such that

$$w' = \Big(\prod_{i\notin E(g^*x')} x_i'^{\beta_i}\Big)^{-1}\Big(\prod_{i\in E(g^*x)} u_i^{-\beta_i}\Big)\Big(\prod_{i\notin E(g^*x)} x_i^{\beta_i}\Big)w. \tag{6.3.27.3}$$

The map Ψ_m is induced by the map on prestacks

$$\Psi_m^{\mathrm{ps}} : \mu_m \times \mathcal{S}_H(\beta)_t^{\mathrm{ps}} \longrightarrow \mathcal{S}_H(\alpha)[m]_{t^m}^{\mathrm{ps}} \tag{6.3.27.4}$$

which over a $\mathbb{Z}[t]$-scheme $h : T \to \mathrm{Spec}(\mathbb{Z}[t])$ is the morphism of groupoids defined as follows:

1. If $\zeta \in \mu_m(T)$, $y_1, \ldots, y_r \in \Gamma(T, \mathcal{O}_T)$, and $s \in \Gamma(T, \mathcal{O}_T^*)$ such that

$$y_1^{\beta_1} \cdots y_r^{\beta_r} s = h^*(t), \tag{6.3.27.5}$$

then Ψ_m^{ps} sends the object $(\zeta, (y, s)) \in \mu_m \times \mathcal{S}_H(\beta)_t(T)$ to the object

$$((y_1, \ldots, y_r, s^m), \zeta s) \in \mathcal{S}_H(\alpha)_{t^m}[m]^{\mathrm{ps}}(T). \tag{6.3.27.6}$$

2. Given two objects $(\zeta, (y, s)), (\zeta', (y', s')) \in \mu_m \times \mathcal{S}_H^{\mathrm{ps}}(\beta)_t(T)$ the sheaf

$$\underline{\mathrm{Isom}}_{\mu_m \times \mathcal{S}_H(\beta)}((\zeta, y, s), (\zeta', y', s')) \tag{6.3.27.7}$$

is the sheaf associated to the presheaf which to any $g : T' \to T$ associates

$$\begin{cases} \varnothing & \text{if } g^*\zeta \neq g^*\zeta' \\ \underline{\mathrm{Isom}}_{\mathcal{S}_H(\beta)}((y, s), (y', s')) & \text{otherwise.} \end{cases} \tag{6.3.27.8}$$

The map on isomorphism sheaves

$$\underline{\mathrm{Isom}}_{\mu_m \times \mathcal{S}_H(\beta)}((\zeta, y, s), (\zeta', y', s')) \longrightarrow \underline{\mathrm{Isom}}_{\mathcal{S}_H(\beta)[m]_{t^m}}(((y, s^m), \zeta s), ((y', s'^m), \zeta' s')) \tag{6.3.27.9}$$

is the map associated to the map of presheaves which sends a collection of data $((u_i)_{i \in E(g^*y)}, h)$ as in 6.1.6 over $g : T' \to T$ such that $g^*\zeta = g^*\zeta'$ to the same collection $((u_i)_{i \in E(g^*y)}, h)$ which defines an isomorphism $((y, s^m), \zeta s) \to ((y', s'^m), \zeta' s')$ in $\mathcal{S}_H(\beta)[m]_{t^m}(T')$.

Proposition 6.3.28. — *The map Ψ_m in (6.3.27.2) is finite and restricts to an isomorphism over* $\mathrm{Spec}(\mathbb{Z}[t^{\pm}]) \subset \mathrm{Spec}(\mathbb{Z}[t])$.

Proof. — Let $h : T \to \mathrm{Spec}(\mathbb{Z}[t])$ be a $\mathbb{Z}[t]$-scheme, and let $(x_1, \ldots, x_r, v, w) \in \mathcal{S}_H(\alpha)[m]_{t^m}^{\mathrm{ps}}$ be an object. Then the fiber product of the diagram

$$\begin{array}{ccc} & & \mu_m \times \mathcal{S}_H(\beta)_t \\ & & \downarrow \Psi_m \\ T & \xrightarrow{(x,v,w)} & \mathcal{S}_H(\alpha)[m]_{t^m} \end{array} \tag{6.3.28.1}$$

is by the above description of the prestacks $\mathcal{S}_H(\alpha)[m]_{t^m}^{\mathrm{ps}}$ and $\mu_m \times \mathcal{S}_H(\beta)^{\mathrm{ps}}$ equal to the functor which to any morphism $g : T' \to T$ associates the set of elements $\zeta \in \mu_m(T')$ such that

$$x_1^{\beta_1} \cdots x_r^{\beta_r}(\zeta^{-1} w) = (hg)^* t. \tag{6.3.28.2}$$

This is represented by the affine T-scheme

$$\underline{\mathrm{Spec}}_T(\mathcal{O}_T[z]/(z^m - 1, x_1^{\beta_1} \cdots x_r^{\beta_r} w z^{-1} - (hg)^* t)). \tag{6.3.28.3}$$

From this the proposition follows. □

6.4. The Hyodo-Kato isomorphism: case of semistable reduction and trivial coefficients

6.4.1. — Let k be a perfect field of positive characteristic p, W the ring of Witt vectors of k, and let $\sigma : W \to W$ be the canonical lift of Frobenius. Denote by $W\langle t\rangle$ the p-adic completion of the divided power envelope of the surjection $W[t] \to k$ sending t to 0. Let $F : W\langle t\rangle \to W\langle t\rangle$ be the lifting of Frobenius induced by the map σ and $t \mapsto t^p$.

Fix an integer r, and let $\mathcal{S}_{W[t]}(r)$ denote the stack $\mathcal{S}_H(\alpha)$ over $\mathbb{A}^1_W$ obtained by taking $\alpha = (1, \cdots, 1)$ (r-copies) and H the full symmetric group on r letters in 6.1.1, and let $\mathcal{S}_{W\langle t\rangle}(r)$ be the base change via $W[t] \to W\langle t\rangle$ of $\mathcal{S}_{W[t]}(r)$ to $W\langle t\rangle$. In what follows r will be fixed, so to ease the notation we usually write just $\mathcal{S}_{W[t]}$ (resp. $\mathcal{S}_{W\langle t\rangle}$ etc.) for $\mathcal{S}_{W[t]}(r)$ (resp. $\mathcal{S}_{W\langle t\rangle}(r)$ etc.). Let $\mathcal{S}_W$ (resp. $\mathcal{S}_k$) denote the reduction of $\mathcal{S}_{W\langle t\rangle}$, and let $F_{\mathcal{S}_{W\langle t\rangle}} : \mathcal{S}_{W\langle t\rangle} \to \mathcal{S}_{W\langle t\rangle}$ be the lifting of Frobenius obtained from the map Λ_p (6.3.2). Denote by $F_{\mathcal{S}_W} : \mathcal{S}_W \to \mathcal{S}_W$ the lifting of Frobenius obtained by reduction.

Let $\mathcal{Y} \to \mathcal{S}_k$ be a smooth representable morphism of algebraic stacks with $\mathcal{Y}$ a tame Deligne-Mumford stack proper over k (see 2.5.14 for the definition of a tame Deligne-Mumford stack).

6.4.2. — Define graded projective systems

$$D. := \{H^*((\mathcal{Y}_{\mathrm{et}}/\mathcal{S}_{W_n})_{\mathrm{cris}}, \mathcal{O}_{\mathcal{Y}_{\mathrm{et}}/\mathcal{S}_{W_n}})\} \in \mathrm{ps}(W), \tag{6.4.2.1}$$

$$E. := \{H^*((\mathcal{Y}_{\mathrm{et}}/\mathcal{S}_{W_n\langle t\rangle})_{\mathrm{cris}}, \mathcal{O}_{\mathcal{Y}_{\mathrm{et}}/\mathcal{S}_{W_n\langle t\rangle}})\} \in \mathrm{ps}(W\langle t\rangle). \tag{6.4.2.2}$$

By 2.6.8 and 5.1.20 the projective system $D.$ is free of finite type mod $\mathcal{T}$ in $\mathrm{ps}(W)$.

By functoriality the lifting of Frobenius $F_{\mathcal{S}_{W\langle t\rangle}}$ induces maps

$$\varphi_E : F^*E. \longrightarrow E., \quad \varphi_D : \sigma^* D. \longrightarrow D.. \tag{6.4.2.3}$$

These maps extend to semi-linear endomorphisms $\varphi_E^{\cdot\cdot}$ and $\varphi_D^{\cdot\cdot}$ of the projective system of Leray spectral sequences

(6.4.2.4)
$$E_1^{pq} = H^q(\mathcal{Y}_{\mathrm{et}}, R^p u_{\mathcal{Y}_{\mathrm{et}}/\mathcal{S}_{W_n}*}\mathcal{O}_{\mathcal{Y}_{\mathrm{et}}/\mathcal{S}_{W_n}}) \implies H^{p+q}((\mathcal{Y}_{\mathrm{et}}/\mathcal{S}_{W_n})_{\mathrm{cris}}, \mathcal{O}_{\mathcal{Y}_{\mathrm{et}}/\mathcal{S}_{W_n}}),$$

and

(6.4.2.5)
$$E_1^{pq} = H^q(\mathcal{Y}_{\mathrm{et}}, R^p u_{\mathcal{Y}_{\mathrm{et}}/\mathcal{S}_{W_n\langle t\rangle}*}\mathcal{O}_{\mathcal{Y}_{\mathrm{et}}/\mathcal{S}_{W_n\langle t\rangle}}) \implies H^{p+q}((\mathcal{Y}_{\mathrm{et}}/\mathcal{S}_{W_n})_{\mathrm{cris}}, \mathcal{O}_{\mathcal{Y}_{\mathrm{et}}/\mathcal{S}_{W_n\langle t\rangle}}).$$

Reduction also defines a map of projective systems

$$\rho : E. \longrightarrow D. \tag{6.4.2.6}$$

compatible with the maps φ_E and φ_D, and this map ρ extends canonically to a morphism of projective systems of Leray spectral sequences

(6.4.2.7)
$$\begin{array}{ccc} E_1^{pq} = H^q(\mathcal{Y}_{\text{et}}, R^p u_{\mathcal{Y}_{\text{et}}/\mathcal{S}_{W_n\langle t\rangle}*}\mathcal{O}_{\mathcal{Y}_{\text{et}}/\mathcal{S}_{W_n\langle t\rangle}}) & \Longrightarrow & H^{p+q}((\mathcal{Y}_{\text{et}}/\mathcal{S}_{W_n\langle t\rangle})_{\text{cris}}, \mathcal{O}_{\mathcal{Y}_{\text{et}}/\mathcal{S}_{W_n\langle t\rangle}}) \\ & & \downarrow \rho^{\cdot\cdot} \\ E_1^{pq} = H^q(\mathcal{Y}_{\text{et}}, R^p u_{\mathcal{Y}_{\text{et}}/\mathcal{S}_{W_n}*}\mathcal{O}_{\mathcal{Y}_{\text{et}}/\mathcal{S}_{W_n}}) & \Longrightarrow & H^{p+q}((\mathcal{Y}_{\text{et}}/\mathcal{S}_{W_n})_{\text{cris}}, \mathcal{O}_{\mathcal{Y}_{\text{et}}/\mathcal{S}_{W_n}}) \end{array}$$

compatible with the endomorphisms $\varphi_D^{\cdot\cdot}$ and $\varphi_E^{\cdot\cdot}$.

6.4.3. — The stack $\mathcal{S}_{W\langle t\rangle}$ satisfies all of the assumptions of 4.6.8. In the present situation, this amounts to the following.

(i) The stack $\mathcal{S}_W$ is flat over W. This follows from the fact that $\mathcal{S}_H(\alpha)$ is flat over $\mathbb{Z}[t]$ by 6.1.3.

(ii) Let $\mathcal{S}_W^{(n)}$ denote $\mathcal{S}_W \times_{\text{Spec}(W),\sigma^n} \text{Spec}(W)$, and let $F^n_{\mathcal{S}_W/W} : \mathcal{S}_W \to \mathcal{S}_W^{(n)}$ denote the map induced by $F^n_{\mathcal{S}_W}$. For a morphism $Q \to \mathcal{S}_W$, let $Q^{(n)}$ denote $Q \times_{\mathcal{S}_W} \mathcal{S}_W^{(n)} = Q \times_{\text{Spec}(W),\sigma^n} \text{Spec}(W)$, let $\widetilde{Q}^{(n)}$ denote the fiber product $Q \times_{\mathcal{S}_W, F^n_{\mathcal{S}_W}} \mathcal{S}_W$, and let $P_n : \widetilde{Q}^{(n)} \to Q^{(n)}$ be the projection. Then the second condition is that for any morphism $Q \to \mathcal{S}_W$ from a scheme Q the canonical map

$$\mathcal{O}_{Q^{(n)}} \longrightarrow RP_{n*}\mathcal{O}_{\widetilde{Q}^{(n)}} \tag{6.4.3.1}$$

is an isomorphism. This can be seen as follows. Since there is a commutative diagram

$$\begin{array}{ccc} \text{Spec}(W) & \xrightarrow{\sigma} & \text{Spec}(W) \\ & {}_{t=0}\searrow \quad \swarrow_{t=0} & \\ & \text{Spec}(\mathbb{Z}[t]), & \end{array} \tag{6.4.3.2}$$

there is a canonical isomorphism $\mathcal{S}_W^{(n)} \simeq \mathcal{S}_W$. Under this isomorphism, the map $F^n_{\mathcal{S}_W/W}$ becomes identified with the base change of Λ_{p^n} along the morphism $\text{Spec}(W) \to \text{Spec}(\mathbb{Z}[t])$ defined by sending t to 0. That (6.4.3.1) is an isomorphism therefore follows from 6.3.19.

Theorem 4.6.9 therefore gives a canonical isomorphism of sheaves on $\mathcal{Y}_{\text{et}}$

$$\iota : R^p u_{\mathcal{Y}_{\text{et}}/\mathcal{S}_{W_n}*}\mathcal{O}_{\mathcal{Y}_{\text{et}}/\mathcal{S}_{W_n}} \hat{\otimes}_W W\langle t\rangle \simeq R^p u_{\mathcal{Y}_{\text{et}}/\mathcal{S}_{W_n\langle t\rangle}*}\mathcal{O}_{\mathcal{Y}_{\text{et}}/\mathcal{S}_{W_n\langle t\rangle}} \tag{6.4.3.3}$$

compatible with the Frobenius endomorphisms, such that the composition of ι with the reduction map

$$R^p u_{\mathcal{Y}_{\text{et}}/\mathcal{S}_{W_n\langle t\rangle}*}\mathcal{O}_{\mathcal{Y}_{\text{et}}/\mathcal{S}_{W_n\langle t\rangle}} \longrightarrow R^p u_{\mathcal{Y}_{\text{et}}/\mathcal{S}_{W_n}*}\mathcal{O}_{\mathcal{Y}_{\text{et}}/\mathcal{S}_{W_n}} \tag{6.4.3.4}$$

is equal to the map

$$R^p u_{\mathcal{Y}_{\text{et}}/\mathcal{S}_{W_n}*}\mathcal{O}_{\mathcal{Y}_{\text{et}}/\mathcal{S}_{W_n}} \otimes_W W\langle t\rangle \longrightarrow R^p u_{\mathcal{Y}_{\text{et}}/\mathcal{S}_{W_n}*}\mathcal{O}_{\mathcal{Y}_{\text{et}}/\mathcal{S}_{W_n}} \tag{6.4.3.5}$$

defined by the surjection $W\langle t\rangle \to W$ sending t to 0.

We now apply 5.3.33 with

$$E_1^{pq} = A_\cdot^{pq} \implies M_\cdot \quad (6.4.3.6)$$

the first spectral sequence in (6.4.2.7),

$$E_1^{pq} = B_\cdot^{pq} \implies N_\cdot \quad (6.4.3.7)$$

the second spectral sequence in (6.4.2.7), and $\gamma^{\cdot\cdot}$ the map $\rho^{\cdot\cdot}$. Then with notation as in 5.3.33 the map $\overline{A}_\cdot^{pq} \to B_\cdot^{pq}$ is an isomorphism, since by the above it is equal to the inverse of the isomorphism

$$\bar{\iota} : R^p u_{\mathcal{Y}_{\mathrm{et}}/\mathcal{S}_{W_\cdot} *} \mathcal{O}_{\mathcal{Y}_{\mathrm{et}}/\mathcal{S}_{W_\cdot}} \simeq R^p u_{\mathcal{Y}_{\mathrm{et}}/\mathcal{S}_{W_\cdot \langle t \rangle} *} \mathcal{O}_{\mathcal{Y}_{\mathrm{et}}/\mathcal{S}_{W_\cdot \langle t \rangle}} \otimes_{W\langle t \rangle} W \quad (6.4.3.8)$$

obtained by reduction from ι. By 5.3.33 we then obtain the following.

Theorem 6.4.4. — *There is a canonical isomorphism in* $\mathrm{ps}(W\langle t \rangle)_{\mathbb{Q}}$

$$D_\cdot \otimes_W W\langle t \rangle \simeq E_\cdot \quad (6.4.4.1)$$

compatible with φ_D *and* φ_E. *In particular,* $E_\cdot \in \mathrm{ps}(W\langle t \rangle)_{\mathbb{Q}}$ *is free of finite type mod* $\mathcal{T}$.

6.4.5. — Following the method of Berthelot and Ogus [**9**] and Hyodo and Kato [**31**, proof of 5.2], the above also gives information about cohomology over ramified extensions of W.

Let V be a complete discrete valuating ring of mixed characteristic and residue field k, and let $\pi \in V$ be a uniformizer. Let R_n denote the divided power envelope of the surjection $W_n[t] \to V_n$ sending t to π, and let R denote the inverse limit $\varprojlim R_n$. Denote by $\mathcal{S}_R$ the base change over $W_n[t]$ of $\mathcal{S}_{W_n[t]}$ to R, and by $\mathcal{S}_V$ the stack $\mathcal{S}_{W[t]} \otimes_{W[t], t \mapsto \pi} V$.

Let $\mathcal{X} \to \mathcal{S}_V$ be a smooth representable morphism of algebraic stacks with $\mathcal{X}$ a tame Deligne-Mumford stack proper over $\mathrm{Spec}(V)$. Denote by $\mathcal{Y}/\mathcal{S}_k$ the reduction of $\mathcal{X}$ modulo π. Let $\mathcal{X}_0$ denote $\mathcal{X} \otimes (V/pV)$, and define $D_\cdot$ and $E_\cdot$ as in 6.4.2. Also define $C_\cdot \in \mathrm{ps}(V)$ to be the projective system

$$C_\cdot := \{H^*((\mathcal{X}_{0,\mathrm{et}}/\mathcal{S}_{R_n})_{\mathrm{cris}}, \mathcal{O}_{\mathcal{X}_{0,\mathrm{et}}/\mathcal{S}_{R_n}})\}. \quad (6.4.5.1)$$

Theorem 6.4.6. — *There is a canonical isomorphism in* $\mathrm{ps}(R)_{\mathbb{Q}}$

$$D_\cdot \otimes_W R \simeq C_\cdot. \quad (6.4.6.1)$$

In particular, the projective system $C_\cdot \in \mathrm{ps}(R)$ *is free of finite type mod* $\mathcal{T}$.

The proof is in several steps 6.4.7–6.4.8.

6.4.7. — For every integer r, let $q_r : W[t] \to R$ denote the W-linear map sending t to the image of t^{p^r} under the canonical map $W[t] \to R$, and let $g_r : W\langle t \rangle \to R$ be the map which is equal to σ^r on W and sends t to t^{p^r}.

The diagram

$$\begin{array}{ccc} R_0 & \xrightarrow{F^r_{R_0}} & R_0 \\ \uparrow & & \uparrow \\ k[t] & \xrightarrow{F^r_{k[t]}} & k[t] \end{array} \tag{6.4.7.1}$$

commutes. It follows that $\mathcal{S}_{R_0} \otimes_{R_0, F^r_{R_0}} R_0$ is canonically isomorphic to the stack $\mathcal{S}_{W[t]} \otimes_{W[t], q_r} R_0 \simeq \mathcal{S}_{W\langle t\rangle} \otimes_{W\langle t\rangle, g_r} R$. Let $\mathcal{S}^{(i)}_R$ denote the stack $\mathcal{S}_{W[t]} \otimes_{W[t], q_i} R$. The map Λ_p defined in 6.3.2 induces a map

$$\tilde{F}_{\mathcal{S}^{(i)}/R} : \mathcal{S}^{(i-1)}_R \longrightarrow \mathcal{S}^{(i)}_R \tag{6.4.7.2}$$

as in 3.4.44. Let $\overline{\mathcal{X}}^{(p^r)}_0$ denote the stack $\mathcal{X}_0 \otimes_{R_0, F^r_{R_0}} R_0$.

Let w be an integer such that $(\pi^{p^w}) \subset pV$. Then the map $F^w_{V_0} : V_0 \to V_0$ factors through the surjection $V_0 \to k$. It follows that there is a commutative diagram of cartesian squares

$$\begin{array}{ccc} \overline{\mathcal{X}}^{(p^w)} & \longrightarrow & \mathcal{Y} \\ \downarrow & & \downarrow \\ \mathcal{S}^{(w)}_R & \longrightarrow & \mathcal{S}_{W\langle t\rangle} \\ \downarrow & & \downarrow \\ \mathrm{Spec}(R) & \xrightarrow{g_w} & \mathrm{Spec}(W\langle t\rangle). \end{array} \tag{6.4.7.3}$$

6.4.8. — Let $C^{(w)}_\cdot$ denote the projective system

$$C^{(w)}_\cdot := \{H^*((\overline{\mathcal{X}}^{(p^w)}_{\mathrm{et}} / \mathcal{S}^{(w)}_{R_n})_{\mathrm{cris}}, \mathcal{O}_{\overline{\mathcal{X}}^{(p^w)}_{\mathrm{et}} / \mathcal{S}^{(w)}_{R_n}})\} \in \mathrm{ps}(R). \tag{6.4.8.1}$$

By the base change theorem 2.6.2 we obtain from the diagram (6.4.7.3) an isomorphism of projective systems

$$C^{(w)}_\cdot \simeq E_\cdot \otimes_{W\langle t\rangle, g_w} R. \tag{6.4.8.2}$$

On the other hand, by functoriality of crystalline cohomology we obtain a map of projective systems

$$C^{(w)}_\cdot \longrightarrow C_\cdot \tag{6.4.8.3}$$

from the commutative diagram

$$\begin{array}{ccc} \mathcal{X}_0 & \xrightarrow{F^n_{\mathcal{X}_0/R}} & \overline{\mathcal{X}}^{(p^w)}_0 \\ \downarrow & & \downarrow \\ \mathcal{S}_R & \xrightarrow{H} & \mathcal{S}^{(w)}_R, \end{array} \tag{6.4.8.4}$$

where $F^n_{\mathfrak{X}_0/R}$ is the relative Frobenius morphism defined by the n-th power of Frobenius on $\mathfrak{X}_0$ and

$$H := \tilde{F}_{\mathcal{S}^{(1)}/R} \circ \tilde{F}_{\mathcal{S}^{(2)}/R} \circ \cdots \circ \tilde{F}_{\mathcal{S}^{(i)}/R}. \tag{6.4.8.5}$$

By 3.4.45 there is also a map of projective systems

$$V_. : C_. \longrightarrow C_.^{(w)} \tag{6.4.8.6}$$

such that the composites

$$H \circ (6.4.8.3) \text{ and } (6.4.8.3) \circ H \tag{6.4.8.7}$$

are both equal to multiplication by p^κ for some integer κ. In particular the map (6.4.8.3) induces an isomorphism in $\mathrm{ps}(R)_{\mathbb{Q}}$. We therefore obtain a sequence of isomorphisms in $\mathrm{ps}(R)_{\mathbb{Q}}$

$$\begin{aligned} C_. &\overset{(6.4.8.3)}{\simeq} C_.^{(w)} \\ &\overset{(6.4.8.2)}{\simeq} E_. \otimes_{W\langle t\rangle, g_w} R \\ &\overset{(6.4.4.1)}{\simeq} D_. \otimes_{W,\varphi^w} R \\ &\overset{\varphi^w\otimes 1}{\simeq} D_. \otimes_W R. \end{aligned} \tag{6.4.8.8}$$

This completes the proof of 6.4.6. □

Proposition 6.4.9. — *Let B be a ring and $C^\bullet$ a bounded complex of B-modules such that there exists an integer r such that for every i there exists a morphism $P \to H^i(C^\bullet)$ with kernel and cokernel annihilated by p^r and P a free B-module of finite rank. Then there exists a morphism of complexes $P^\bullet \to C^\bullet$ with each P^i a free module of finite rank and all the maps $P^i \to P^{i+1}$ equal to zero, such that the kernels and cokernels of the maps $P^i \simeq H^i(P^\bullet) \to H^i(C^\bullet)$ are annihilated by p^r.*

Proof. — Without loss of generality we may assume that $C^i = 0$ for $i \notin [0, m]$ for some m. We then prove the result by induction on m. The case $m = 0$ is trivial, so we prove the result for $m+1$ given the result for m. Let $P^{m+1} \to H^{m+1}(C^\bullet) \simeq C^{m+1}/d(C^m)$ be a map with kernel and cokernel annihilated by p^r and P^{m+1} free of finite rank. Denote by K^m the kernel of $C^m \to C^{m+1}$, and choose a section $s : P^{m+1} \to C^{m+1}$. There is then a morphism of complexes

$$\begin{array}{ccccccccc} (C^0 & \longrightarrow & \cdots & \longrightarrow & C^{m-1} & \longrightarrow & K^m & \xrightarrow{0} & P^{m+1}) \\ \downarrow{\scriptstyle \mathrm{id}} & & & & \downarrow{\scriptstyle \mathrm{id}} & & \downarrow & & \downarrow{\scriptstyle s} \\ (C^0 & \longrightarrow & \cdots & \longrightarrow & C^{m-1} & \longrightarrow & C^m & \longrightarrow & C^{m+1}) \end{array} \tag{6.4.9.1}$$

which induces an isomorphism on cohomology for $i < m+1$ and the map $P^{m+1} \to H^{m+1}(C^\bullet)$ in degree $m+1$. By induction we can find a morphism of complexes

$$(P^0 \to \cdots \to P^m) \longrightarrow (C^0 \to \cdots \to C^{m-1} \to K^m) \tag{6.4.9.2}$$

as in the proposition. The induced morphism of complexes

$$(P^0 \to \cdots \to P^m \xrightarrow{0} P^{m+1}) \longrightarrow C^\bullet \tag{6.4.9.3}$$

then works for $C^\bullet$. □

Corollary 6.4.10. — *There exists an integer ℓ such that for any integer n and R_n-module M the kernels and cokernels of the natural map*

$$C_n \otimes_{R_n} M \longrightarrow H^*(R\Gamma((\mathfrak{X}_{0,\text{et}}/\mathcal{S}_{R_n})_{\text{cris}}, \mathcal{O}_{\mathfrak{X}_{0,\text{et}}/\mathcal{S}_{R_n}}) \otimes^{\mathbb{L}}_{R_n} M) \tag{6.4.10.1}$$

are annihilated by p^ℓ.

Proof. — By 6.4.6, there exists an integer r such that

$$C^\bullet := R\Gamma((\mathfrak{X}_{0,\text{et}}/\mathcal{S}_{R_n})_{\text{cris}}, \mathcal{O}_{\mathfrak{X}_{0,\text{et}}/\mathcal{S}_{R_n}}) \tag{6.4.10.2}$$

satisfies the assumptions of 6.4.9 with $B = R_n$ (that is, the integer r can be chosen independently of n). Thus if $K^\bullet$ represents the complex $R\Gamma((\mathfrak{X}_{0,\text{et}}/\mathcal{S}_{R_n})_{\text{cris}}, \mathcal{O}_{\mathfrak{X}_{0,\text{et}}/\mathcal{S}_{R_n}})$, there exists a morphism of complexes $P^\bullet \to K^\bullet$ as in 6.4.9. Let $Q^\bullet$ denote the cone of this morphism of complexes. Since the kernel and cokernels of the maps $H^i(P^\bullet) \to H^i(K^\bullet)$ are all annihilated by p^r, the map $p^{2r} : Q^\bullet \to Q^\bullet$ is quasi-isomorphic to the zero map. Similarly, for any R_n-module M we have a distinguished triangle

$$P^\bullet \otimes^{\mathbb{L}} M \longrightarrow C^\bullet \otimes^{\mathbb{L}} M \longrightarrow Q^\bullet \otimes^{\mathbb{L}} M \longrightarrow P^\bullet \otimes^{\mathbb{L}} M[1] \tag{6.4.10.3}$$

which shows that the map

$$H^i(P^\bullet \otimes^{\mathbb{L}} M) \simeq P^i \otimes M \longrightarrow H^i(C^\bullet \otimes^{\mathbb{L}} M) \tag{6.4.10.4}$$

has kernel and cokernel annihilated by p^{2r}. Since the cokernel of (6.4.10.1) is a quotient of the cokernel of (6.4.10.4), this implies that the cokernel of (6.4.10.1) is annihilated by p^{2r}. Furthermore, for any element $m \in H^i(C^\bullet) \otimes M$, the element $p^r m$ can be lifted to an element $\tilde{m} \in P^i \otimes M$ since the cokernel of $P^i \to H^i(C^\bullet)$ is annihilated by p^r. If m maps to zero under (6.4.10.1), it follows that $\tilde{m}$ is in the kernel of (6.4.10.4) and hence is annihilated by p^{2r}. It follows that m is annihilated by p^{3r}. Consequently taking $\ell = 3r$ we obtain the corollary. □

Corollary 6.4.11. — *The natural map*

$$C_\cdot \otimes_R V \longrightarrow \{H^*((\mathfrak{X}_{0,\text{et}}/\mathcal{S}_{V_n})_{\text{cris}}, \mathcal{O}_{\mathfrak{X}_{0,\text{et}}/\mathcal{S}_{V_n}})\} \tag{6.4.11.1}$$

of projective systems induces an isomorphism in $\text{ps}(V)_{\mathbb{Q}}$.

Proof. — By the base change theorem 2.6.2 we have

$$R\Gamma((\mathfrak{X}_{0,\text{et}}/\mathcal{S}_{R_n})_{\text{cris}}, \mathcal{O}_{\mathfrak{X}_{0,\text{et}}/\mathcal{S}_{R_n}}) \otimes^{\mathbb{L}}_{R_n} V_n \simeq R\Gamma((\mathfrak{X}_{0,\text{et}}/\mathcal{S}_{V_n})_{\text{cris}}, \mathcal{O}_{\mathfrak{X}_{0,\text{et}}/\mathcal{S}_{V_n}}) \tag{6.4.11.2}$$

so the result follows from 6.4.10 □

Corollary 6.4.12. — *There is a natural isomorphism*

(6.4.12.1)
$$D. \otimes_W V \simeq \{H^*_{\mathrm{dR}}(\mathcal{X}_n/\mathcal{S}_{V_n})\}$$

in $\mathrm{ps}(V)_{\mathbb{Q}}$.

Proof. — This follows from the preceding corollary and the Comparison theorem 2.5.4. □

Remark 6.4.13. — In Chapter 7 we give a proof of 6.4.12 which also works with coefficients and more general types of reduction.

Remark 6.4.14. — Let $D \in \mathrm{Mod}(W)$ denote $\varprojlim D.$. The isomorphism

(6.4.14.1)
$$D \otimes_W K \simeq H^*((\mathcal{X}_{0,\mathrm{et}}/\mathcal{S}_V)_{\mathrm{cris}}, \mathcal{O}_{\mathcal{X}_{0,\mathrm{et}}/\mathcal{S}_V}) \otimes \mathbb{Q}$$

obtained from (6.4.11.1) can be described more explicitly as follows. The sequence of isomorphisms in (6.4.8.8) induce after passing to the limit and applying $\otimes_R K$ a sequence of isomorphisms

(6.4.14.2)
$$\begin{aligned} H^*((\mathcal{X}_{0,\mathrm{et}}/\mathcal{S}_V)_{\mathrm{cris}}, \mathcal{O}_{\mathcal{X}_{0,\mathrm{et}}/\mathcal{S}_V}) \otimes \mathbb{Q} &\simeq (\varprojlim C^{(w)}_{\cdot}) \otimes \mathbb{Q} \\ &\simeq (\varprojlim H^*((\mathcal{Y}^{(p^w)} \otimes_k V_0/\mathcal{S}^{(w)}_{V_n})_{\mathrm{cris}}, \mathcal{O}_{\mathcal{Y}^{(p^w)} \otimes_k V_0/\mathcal{S}^{(w)}_{V_n}})) \\ &\simeq (\varprojlim H^*((\mathcal{Y}^{(p^w)}/\mathcal{S}^{(w)}_{W_n})_{\mathrm{cris}}, \mathcal{O}_{\mathcal{Y}^{(p^w)}/\mathcal{S}^{(w)}_{W_n}})) \otimes_W K \\ &\overset{\varphi^w \otimes 1}{\simeq} D \otimes_W K. \end{aligned}$$

Let H denote

(6.4.14.3)
$$\mathrm{Hom}_V(D \otimes_W V, H^*((\mathcal{X}_{0,\mathrm{et}}/\mathcal{S}_V)_{\mathrm{cris}}, \mathcal{O}_{\mathcal{X}_{0,\mathrm{et}}/\mathcal{S}_V})),$$

Define a sequence $\nu_n := \mathrm{ord}_p(p^n!) = (p^n - 1)/(p - 1)$ as in 5.3.9, let m denote the relative dimension of $\mathcal{Y}$ over k, and define $\Omega_\nu(H)$ as in 5.2.2. By 5.2.4 there is a natural isomorphism $\xi : H \to \Omega_\nu(H)$. Therefore, the isomorphism (6.4.14.2) can be described by a collection of maps

(6.4.14.4)
$$h_n : H^*((\mathcal{Y}/\mathcal{S}_{W_{\nu_n}})_{\mathrm{cris}}, \mathcal{O}_{\mathcal{Y}/\mathcal{S}_{W_{\nu_n}}}) \longrightarrow H^*((\mathcal{X}_{0,\mathrm{et}}/\mathcal{S}_{V_{\nu_n}})_{\mathrm{cris}}, \mathcal{O}_{\mathcal{X}_{0,\mathrm{et}}/\mathcal{S}_{V_{\nu_n}}})$$

such that the reduction modulo p^{ν_n} of h_{n+1} is equal to $p^m h_n$. It follows from the construction of the isomorphism (6.4.4.1) that such a collection of maps is given by the composites

$$\begin{aligned} H^*(\mathcal{Y}/\mathcal{S}_{W_{\nu_n}}) \otimes V &\xrightarrow{\varphi^{w+n}} H^*(\mathcal{Y}^{(p^{w+n})}/\mathcal{S}^{(w+n)}_{W_{\nu_n}}) \otimes V \\ &\simeq H^*(\mathcal{X}_0^{(p^{w+n})}/\mathcal{S}^{(w+n)}_{V_{\nu_n}}) \xrightarrow{\psi^{n+w}} H^*(\mathcal{X}_0/\mathcal{S}_{V_{\nu_n}}), \end{aligned}$$

where to ease the notation we omit the structure sheaves and ψ denotes the canonical map for which $\varphi \circ \psi = p^d$ and $\psi \circ \varphi = p^d$ (3.4.42). This remark is the foundation for the construction in the next chapter.

6.5. The monodromy operator

6.5.1. — We continue with the notation and hypotheses of 6.4.1.

Let $\overline{\mathcal{S}}_{W[t]}$ be the stack over $[\mathbb{A}^1_W/\mathbb{G}_m]$ defined in 6.1.9 whose base change to $\mathbb{A}^1$ is $\mathcal{S}_{W[t]}$, and let $\overline{\mathcal{S}}_{W\langle t\rangle}$ denote the base change of $\overline{\mathcal{S}}_{W[t]}$ to the stack theoretic quotient

$$[\mathrm{Spec}(W\langle t\rangle)/\mathbb{G}_m], \tag{6.5.1.1}$$

where $u \in \mathbb{G}_m$ acts by multiplication by u^i on $t^{[i]}$. There is then a natural cartesian diagram

$$\begin{array}{ccc} \mathcal{S}_{W\langle t\rangle} \times \mathbb{G}_m & \xrightarrow{\mathrm{pr}_1} & \mathcal{S}_{W\langle t\rangle} \\ \chi\downarrow & & \downarrow \\ \mathcal{S}_{W\langle t\rangle} & \longrightarrow & \overline{\mathcal{S}}_{W\langle t\rangle}, \end{array} \tag{6.5.1.2}$$

such that the diagram

$$\begin{array}{ccc} \mathcal{S}_{W\langle t\rangle} \times \mathbb{G}_m & \xrightarrow{\chi} & \mathcal{S}_{W\langle t\rangle} \\ \downarrow & & \downarrow \\ \mathrm{Spec}(W\langle t\rangle) \times \mathbb{G}_m & \xrightarrow{(t,u)\mapsto ut} & \mathrm{Spec}(W\langle t\rangle) \end{array} \tag{6.5.1.3}$$

commutes. Under the identification $\mathcal{S}_{W\langle t\rangle} \times_{\overline{\mathcal{S}}_{W\langle t\rangle}} \mathcal{S}_{W\langle t\rangle} \simeq \mathcal{S}_{W\langle t\rangle} \times \mathbb{G}_m$ provided by (6.5.1.2) the diagonal map

$$\Delta : \mathcal{S}_{W\langle t\rangle} \longrightarrow \mathcal{S}_{W\langle t\rangle} \times_{\overline{\mathcal{S}}_{W\langle t\rangle}} \mathcal{S}_{W\langle t\rangle} \tag{6.5.1.4}$$

becomes identified with the closed immersion

$$\mathrm{id} \times e : \mathcal{S}_{W\langle t\rangle} \longrightarrow \mathcal{S}_{W\langle t\rangle} \times \mathbb{G}_m, \tag{6.5.1.5}$$

which we (abusively) also denote by Δ in what follows. Looking at the first infinitesimal neighborhood of the diagonal we then obtain for every $n \geq 0$ a commutative diagram

$$\begin{array}{ccccc} \mathcal{Y} & & & & \\ \downarrow & & & & \\ \mathcal{S}_{W_n\langle t\rangle} & \xrightarrow{\Delta} & \mathcal{S}_{W_n\langle t\rangle} \otimes_W W[(u-1)]/(u-1)^2 & \xrightarrow{\mathrm{pr}_1} & \mathcal{S}_{W_n\langle t\rangle} \\ & & \chi\downarrow & & \downarrow \\ & & \mathcal{S}_{W_n\langle t\rangle} & \longrightarrow & \overline{\mathcal{S}}_{W_n\langle t\rangle}. \end{array} \tag{6.5.1.6}$$

Since $(u-1) \subset W_n\langle t\rangle[(u-1)]/(u-1)^2$ is a square-zero ideal, there exists by [**8**, 3.2 (4)] a canonical divided power structure on the ideal $(p, u-1) + \langle t\rangle \subset W_n\langle t\rangle[(u-1)]/(u-1)^2$. We view $W_n\langle t\rangle[(u-1)]/(u-1)^2$ and $\mathcal{S}_{W_n\langle t\rangle[(u-1)]/(u-1)^2}$ as PD-stacks with this divided power structure.

Define

$$(6.5.1.7)\quad K_n := R\Gamma((\mathcal{Y}_{\mathrm{et}}/\mathcal{S}_{W_n\langle t\rangle})_{\mathrm{cris}}, \mathcal{O}_{\mathcal{Y}_{\mathrm{et}}/\mathcal{S}_{W_n\langle t\rangle}}),$$

$$(6.5.1.8)\quad K_n' := R\Gamma((\mathcal{Y}_{\mathrm{et}}/\mathcal{S}_{W_n\langle t\rangle\otimes W[(u-1)]/(u-1)^2})_{\mathrm{cris}}, \mathcal{O}_{\mathcal{Y}_{\mathrm{et}}/\mathcal{S}_{W_n\langle t\rangle\otimes W[(u-1)]/(u-1)^2}}).$$

By the base change theorem 2.6.2, there is a canonical isomorphism between $K_n\otimes_{W_n} W_n\langle t\rangle$ and
(6.5.1.9)

$$R\Gamma((\mathcal{Y}_{\mathrm{et}}\otimes_{W_n} W_n[(u-1)]/(u-1)^2/\mathcal{S}_{W_n\langle t\rangle\otimes W[(u-1)]/(u-1)^2})_{\mathrm{cris}}, \mathcal{O}_{\mathcal{Y}_{\mathrm{et}}/\mathcal{S}_{W_n\langle t\rangle\otimes W[(u-1)]/(u-1)^2}}).$$

On the other hand, since $\mathcal{Y}\subset \mathcal{Y}\times_W W[(u-1)]/(u-1)^2$ is defined by a sub-PD-ideal there is also a canonical isomorphism between K_n' and
(6.5.1.10)

$$R\Gamma((\mathcal{Y}_{\mathrm{et}}\otimes_{W_n} W_n[(u-1)]/(u-1)^2/\mathcal{S}_{W_n\langle t\rangle\otimes W[(u-1)]/(u-1)^2})_{\mathrm{cris}}, \mathcal{O}_{\mathcal{Y}_{\mathrm{et}}/\mathcal{S}_{W_n\langle t\rangle\otimes W[(u-1)]/(u-1)^2}}.$$

It follows that the projection in (6.5.1.6) induces an isomorphism

$$(6.5.1.11)\qquad \mathrm{pr}_1^*: K_n\otimes_W W[(u-1)]/(u-1)^2 \to K_n'.$$

Let $\lambda: W[(u-1)]/(u-1)^2 \simeq W\oplus W\cdot(u-1)\to W\cdot(u-1)\simeq W$ be the projection onto the second factor, and define $N: K_n\to K_n$ to be the composite map

$$(6.5.1.12)\qquad K_n \xrightarrow{\chi^*} K_n' \xrightarrow{\mathrm{pr}_1^{*-1}} K_n\otimes_W W[(u-1)]/(u-1)^2 \xrightarrow{\lambda} K_n.$$

Since $\chi^*(t^{[i]}) = \mathrm{pr}_1^*(t^{[i]}) + i\mathrm{pr}_1^* t^{[i]}\cdot(u-1)$ in $W_n\langle t\rangle[(u-1)]/(u-1)^2$ it follows that there is an equality of endomorphisms of K_n

$$(6.5.1.13)\qquad N(t^{[i]}(\cdot)) = it^{[i]}\cdot(-) + t^{[i]}N.$$

In particular, passing to cohomology we obtain an endomorphism $N: E_\cdot\to E_\cdot$ of the projective system $E_\cdot\in \mathrm{ps}(W\langle t\rangle)$ defined in (6.4.2.2).

Proposition 6.5.2. — *The operators $N\varphi_{E_\cdot}$ and $p\varphi_{E_\cdot}N$ are equal, where $\varphi_{E_\cdot}$ denotes the Frobenius endomorphism.*

Proof. — Let $F_{\mathbb{G}_m}:\mathbb{G}_m\to\mathbb{G}_m$ be the map $u\mapsto u^p$. The map Λ_p in 6.3.2 defines by making the base change $\mathbb{Z}[t]\to\mathbb{Z}$ sending t to 0 a map $\Lambda_{p,\mathbb{Z}}:\mathcal{S}_{\mathbb{Z},0}\to\mathcal{S}_{\mathbb{Z},0}$. The composite map

$$(6.5.2.1)\qquad \mathcal{S}_{W,0}\simeq \mathcal{S}_{\mathbb{Z},0}\otimes_{\mathbb{Z}} W \xrightarrow{\Lambda_{p,0}\otimes\sigma} \mathcal{S}_{\mathbb{Z},0}\otimes_{\mathbb{Z}} W\simeq \mathcal{S}_{W,0}$$

is then a lifting of Frobenius to $\mathcal{S}_{W,0}$ which we (abusively) denote also by Λ_p in what follows.

This lifting of Frobenius Λ_p is "$F_{\mathbb{G}_m}$-linear". That is, if $\chi:\mathcal{S}_{W\langle t\rangle}\times\mathbb{G}_m\to\mathcal{S}_{W\langle t\rangle}$ is as above, then the diagram

$$(6.5.2.2)\qquad \begin{array}{ccc} \mathcal{S}_{W\langle t\rangle}\times\mathbb{G}_m & \xrightarrow{\chi} & \mathcal{S}_{W\langle t\rangle} \\ {\scriptstyle \Lambda_p\times F_{\mathbb{G}_m}}\downarrow & & \downarrow{\scriptstyle \Lambda_p} \\ \mathcal{S}_{W\langle t\rangle}\times\mathbb{G}_m & \xrightarrow{\chi} & \mathcal{S}_{W\langle t\rangle} \end{array}$$

commutes. Thus for any $n \in \mathbb{N}$, we can extend the action of Frobenius on $\mathcal{S}_{W\langle t\rangle}$ to an action on the whole diagram (6.5.1.6), where the action on $W_n[(u-1)]/(u-1)^2$ is given by multiplication by p on $(u-1)$. It follows that the diagram

$$\begin{array}{ccc} K_n & \xrightarrow{N} & K_n \cdot (u-1) \\ {\scriptstyle \varphi_{K_n}} \downarrow & & \downarrow {\scriptstyle \varphi_{K_n} \cdot p} \\ K_n & \xrightarrow{N} & K_n \cdot (u-1) \end{array} \tag{6.5.2.3}$$

commutes, where φ_{K_n} denotes the Frobenius endomorphism of K_n. □

6.5.3. — If we replace $W\langle t\rangle$ by W and $\mathcal{S}_{W\langle t\rangle}$ by $\mathcal{S}_W$ in the above, the same arguments give an endomorphism

$$N_{D.} : D. \longrightarrow D. \tag{6.5.3.1}$$

such that if $\varphi_{D.}$ denotes the Frobenius endomorphism of $D.$ then $N\varphi_{D.} = p\varphi_{D.}N$. Furthermore, the reduction map $E. \to D.$ defined in (6.4.2.6) is compatible with these endomorphisms.

Corollary 6.5.4. — *The endomorphism N of $E.$ agrees in* $\mathrm{ps}(W\langle t\rangle)_{\mathbb{Q}}$ *with the endomorphism obtained from the map*

$$D. \otimes_W W\langle t\rangle \longrightarrow D. \otimes_W W\langle t\rangle, \quad d \otimes t^{[i]} \longmapsto N_{D.}(d) \otimes t^{[i]} + id \otimes t^{[i]} \tag{6.5.4.1}$$

and the isomorphism $D. \otimes_W W\langle t\rangle \simeq E.$ (6.4.4).

Proof. — This follows from 5.3.24. □

6.5.5. — The result 6.4.6 can also be strengthened as follows. The commutative diagram

$$\begin{array}{ccc} \mathcal{Y} & \longrightarrow & \mathrm{Spec}(k) \\ \downarrow & & \downarrow \\ \overline{\mathcal{S}}_{W\langle t\rangle} & \longrightarrow & [\mathrm{Spec}(W\langle t\rangle)/\mathbb{G}_m] \end{array} \tag{6.5.5.1}$$

induces for every n a morphism of topoi

$$g : (\mathcal{Y}_{\mathrm{et}}/\overline{\mathcal{S}}_{W_n\langle t\rangle})_{\mathrm{cris}} \longrightarrow (\mathrm{Spec}(k)/[\mathrm{Spec}(W_n\langle t\rangle)/\mathbb{G}_m])_{\mathrm{cris}} \tag{6.5.5.2}$$

such that the $W_n\langle t\rangle$-module E_n is obtained by evaluating $R^*g_*\mathcal{O}_{\mathcal{Y}_{\mathrm{et}}/\overline{\mathcal{S}}_{W_n\langle t\rangle}}$ on the object

$$(\mathrm{Spec}(k) \hookrightarrow \mathrm{Spec}(W_n\langle t\rangle)) \in \mathrm{Cris}(\mathrm{Spec}(k)/[\mathrm{Spec}(W_n\langle t\rangle)/\mathbb{G}_m])_{\mathrm{cris}}. \tag{6.5.5.3}$$

For every integer m, the m-fold fiber product of $\mathbb{A}^1_{W_n}$ over $[\mathbb{A}^1/\mathbb{G}_m]_{W_n}$ is isomorphic to

$$\mathrm{Spec}(W_n[t, u_1^{\pm}, \ldots, u_{m-1}^{\pm}]). \tag{6.5.5.4}$$

In particular, there is an isomorphism

$$D_{\mathrm{Spec}(k),\gamma}((\mathbb{A}^1)^{(m)}) \simeq \mathrm{Spec}(W_n\langle t\rangle\langle u_1 - 1, \dots, u_{m-1} - 1\rangle). \tag{6.5.5.5}$$

Since the projection maps to $\mathrm{Spec}(W_n\langle t\rangle)$ are flat, the base change theorem 2.6.2 implies that the $W_n\langle t\rangle$-module E_n has a canonical HPD-stratification ϵ_{E_n}. The monodromy operator N is simply the connection induced by this HPD-stratification. Thus the projective system $E.$ is a projective system in the category of modules with HPD-stratification. This HPD-stratification also induces a HPD-stratification on $D.$ relative to the map

$$\mathrm{Spec}(W) \longrightarrow (B\mathbb{G}_m)_W. \tag{6.5.5.6}$$

6.5.6. — Let $\Delta_n^{(r)}$ denote the divided power envelope of the closed immersion

$$[\mathrm{Spec}(W_n[t]/(t^r))/\mathbb{G}_m] \subset [\mathrm{Spec}(W_n[t])/\mathbb{G}_m], \tag{6.5.6.1}$$

and let $\Delta^{(r)}$ denote the divided power envelope (with compatibility with the divided power structure on $(p) \subset W$ of

$$[\mathrm{Spec}(W[t]/(t^r))/\mathbb{G}_m] \subset [\mathrm{Spec}(W[t])/\mathbb{G}_m]. \tag{6.5.6.2}$$

Define $\overline{\mathcal{L}}_{r,W_n}$ to be the pullback of $\overline{\mathcal{S}}_{W[t]}$ via the map $\Delta_n^{(r)} \to [\mathbb{A}^1/\mathbb{G}_m]$, and let $\overline{\mathcal{L}}_{r,W}$ be the pullback to $\Delta^{(r)}$.

Observe that the map $W[t] \to V$ sending t to a uniformizer π induces a cartesian square

$$\begin{array}{ccc} \mathrm{Spec}(V_0) & \longrightarrow & \mathrm{Spec}(R_n) \\ \downarrow & & \downarrow \\ [\mathrm{Spec}(W_n[t]/(t^e))/\mathbb{G}_m] & \longrightarrow & \Delta_n^{(e)}, \end{array} \tag{6.5.6.3}$$

where e denotes the ramification index of V/W. Also there is a commutative diagram

$$\begin{array}{ccc} \mathcal{X}_0 & \longrightarrow & \mathrm{Spec}(V) \\ \downarrow & & \downarrow \\ \overline{\mathcal{L}}_{e,W} & \longrightarrow & \Delta^{(e)} \end{array} \tag{6.5.6.4}$$

which induces a morphism of topoi

$$h : (\mathcal{X}_0/\overline{\mathcal{L}}_{e,W_n})_{\mathrm{cris}} \longrightarrow (\mathrm{Spec}(V_0)/\Delta_n^{(e)})_{\mathrm{cris}}. \tag{6.5.6.5}$$

The R_n-module C_n is equal to the evaluation of $R^*g_*\mathcal{O}_{\mathcal{X}_0/\overline{\mathcal{L}}_{e,W_n}}$ on the object

$$(\mathrm{Spec}(V_0) \hookrightarrow \mathrm{Spec}(R_n)) \in \mathrm{Cris}(\mathrm{Spec}(V_0)/\Delta_n^{(e)}). \tag{6.5.6.6}$$

It follows that the module C_n has a natural HPD-stratification ϵ_{C_n} and the projective system $C.$ can be viewed as a projective system of HPD-stratified modules.

6.5.7. — Let $\mathrm{ps}^{\mathrm{str}}(R)$ (resp. $\mathrm{ps}^{\mathrm{str}}(W)$) denote the category of projective systems in the category of R-modules (resp. W-modules) with HPD-stratification relative to the map $\mathrm{Spec}(R) \to \Delta^{(e)}$ (resp. $\mathrm{Spec}(W) \to B\mathbb{G}_m$). The commutative diagram

$$\begin{array}{ccc} \mathrm{Spec}(R) & \longrightarrow & \Delta^{(e)} \\ \downarrow & & \downarrow \\ \mathrm{Spec}(W) & \longrightarrow & B\mathbb{G}_{m,W} \end{array} \tag{6.5.7.1}$$

shows that the pullback functor $\mathrm{ps}(W) \to \mathrm{ps}(R)$ induces a functor

$$\mathrm{ps}^{\mathrm{str}}(W) \longrightarrow \mathrm{ps}^{\mathrm{str}}(R), \quad M_{\cdot} \longmapsto M_{\cdot} \otimes_W R. \tag{6.5.7.2}$$

Theorem 6.4.6 can now be strengthened as follows:

Theorem 6.5.8. — *The isomorphism in 6.4.6 induces an isomorphism in* $\mathrm{ps}^{\mathrm{str}}(R)_{\mathbb{Q}}$.

Proof. — Let $C_{\cdot}^{(w)}$ be as in 6.4.8, and let $\theta_w : [\mathbb{A}^1/\mathbb{G}_m] \to [\mathbb{A}^1/\mathbb{G}_m]$ be the map induced by the maps

$$\mathbb{A}^1 \longrightarrow \mathbb{A}^1, \quad a \longmapsto a^{p^w}, \tag{6.5.8.1}$$

$$\mathbb{G}_m \longrightarrow \mathbb{G}_m, \quad u \longmapsto u^{p^w}. \tag{6.5.8.2}$$

For any integer r, this map θ_w induces a map

$$\Delta^{(r+p^w)} \longrightarrow \Delta^{(r)}, \tag{6.5.8.3}$$

which we denote by the same letter θ_w. Denote by $\overline{\mathcal{L}}_W^{(w)}$ the stack over $\Delta^{(e)}$ obtained as the pullback of $\overline{\mathcal{S}}_{W[t]}$ via the composite

$$\Delta^{(e)} \longrightarrow \Delta^{(p^w)} \xrightarrow{\ \theta_w\ } \Delta^{(1)} \longrightarrow [\mathbb{A}^1/\mathbb{G}_m]. \tag{6.5.8.4}$$

The map $\Lambda_{p^w} : \overline{\mathcal{S}} \to \overline{\mathcal{S}}$ induces a natural map $\vartheta_{p^w} : \overline{\mathcal{L}}_{e,W} \to \overline{\mathcal{L}}_W^{(w)}$.

There is a commutative square

$$\begin{array}{ccc} \overline{\mathcal{X}}^{(p^w)} & \longrightarrow & \mathrm{Spec}(V_0) \\ \downarrow & & \downarrow \\ \overline{\mathcal{L}}_W^{(w)} & \longrightarrow & \Delta^{(e)} \end{array} \tag{6.5.8.5}$$

inducing a morphism of topoi

$$h^{(w)} : (\overline{\mathcal{X}}^{(p^w)}/\overline{\mathcal{S}}_W^{(w)})_{\mathrm{cris}} \longrightarrow (\mathrm{Spec}(V_0)/\Delta^{(e)})_{\mathrm{cris}}. \tag{6.5.8.6}$$

The module $C_n^{(w)}$ is then obtained by evaluating $R^*h_*^{(w)}\mathcal{O}_{\overline{\mathcal{X}}^{(p^w)}/\overline{\mathcal{L}}_{W_n}^{(w)}}$ on

$$(\mathrm{Spec}(V_0) \hookrightarrow \mathrm{Spec}(R_n)) \in \mathrm{Cris}(\mathrm{Spec}(V_0)/\Delta_n^{(e)}). \tag{6.5.8.7}$$

By the same reasoning as above, the projective system $C_{\cdot}^{(w)}$ has a natural structure of an object in $\mathrm{ps}^{\mathrm{str}}(R)$.

We verify that all the isomorphisms in (6.4.8.8) are compatible with the stratifications.

That the isomorphism (6.4.8.3) is compatible with the stratifications can be seen by observing that there is a commutative diagram

$$\begin{array}{ccc} \mathfrak{X}_0 & \longrightarrow & \overline{\mathfrak{X}}_0^{(p^w)} \\ \downarrow & & \downarrow \\ \overline{\mathfrak{L}}_{e,W} & \xrightarrow{\vartheta_{p^w}} & \overline{\mathfrak{L}}_W^{(w)} \\ \downarrow & & \downarrow \\ \Delta^{(e)} & \xrightarrow{\mathrm{id}} & \Delta^{(e)}. \end{array} \tag{6.5.8.8}$$

It follows that the map $C_n^{(w)} \to C_n$ is induced by a map of sheaves

$$R^* h_*^{(w)} \mathcal{O}_{\overline{\mathfrak{X}}^{(p^w)}/\overline{\mathfrak{L}}_W^{(w)}} \longrightarrow R^* h_* \mathcal{O}_{\mathfrak{X}/\overline{\mathfrak{L}}_{e,W}}. \tag{6.5.8.9}$$

In particular it is compatible with the stratifications.

For (6.4.8.2), consider the following commutative diagram

$$\begin{array}{ccc} \overline{\mathfrak{X}}_0^{(p^w)} & \longrightarrow & \mathcal{Y} \\ \downarrow & & \downarrow \\ \overline{\mathfrak{L}}_W^{(w)} & \longrightarrow & \overline{\mathcal{S}}_{W\langle t\rangle} \\ \downarrow & & \downarrow \\ \Delta^{(e)} & \xrightarrow{\theta_{p^w}\otimes\sigma^w} & [\mathrm{Spec}(W\langle t\rangle)/\mathbb{G}], \end{array} \tag{6.5.8.10}$$

and the $\theta_{p^w} \otimes \sigma^w$-PD-morphism

$$\begin{array}{ccc} (\mathrm{Spec}(V_0) & \longrightarrow & \mathrm{Spec}(R_n)) \\ {\scriptstyle F_{V_0}^w}\downarrow & & \downarrow{\scriptstyle g_w} \\ (\mathrm{Spec}(k) & \longrightarrow & \mathrm{Spec}(W_n\langle t\rangle)). \end{array} \tag{6.5.8.11}$$

From this it follows that the map (6.4.8.2) is obtained from a morphism of sheaves

$$R^* g_* \mathcal{O}_{\mathcal{Y}/\overline{\mathcal{S}}_{W\langle t\rangle}} \longrightarrow g_{w*} R^* h_*^{(w)} \mathcal{O}_{\overline{\mathfrak{X}}^{(p^w)}/\overline{\mathfrak{L}}_W^{(w)}}. \tag{6.5.8.12}$$

Therefore (6.4.8.2) is compatible with the stratifications.

That (6.4.4.1) is compatible with the stratifications follows from 5.3.24 which shows that the isomorphism is compatible with the monodromy operators. Since the map

$$\mathrm{Spec}(W\langle t\rangle) \longrightarrow [\mathrm{Spec}(W\langle t\rangle)/\mathbb{G}_m] \tag{6.5.8.13}$$

is smooth, it follows from 2.3.28 that the stratifications are determined by the monodromy operators so (6.4.4.1) is compatible with the stratifications.

Finally $\varphi^w \otimes 1$ is compatible with the stratifications since there is a commutative diagram

$$\begin{array}{ccc} \mathcal{Y} & \xrightarrow{F^w_{\mathcal{Y}}} & \mathcal{Y} \\ \downarrow & & \downarrow \\ \overline{\mathcal{S}}_{W\langle t\rangle} \times_{[\mathrm{Spec}(W\langle t\rangle)/\mathbb{G}_m]} B\mathbb{G}_{m,W} & \xrightarrow{\Lambda_{p^w}\otimes\sigma^w} & \overline{\mathcal{S}}_{W\langle t\rangle} \times_{[\mathrm{Spec}(W\langle t\rangle)/\mathbb{G}_m]} B\mathbb{G}_{m,W} \\ \downarrow & & \downarrow \\ B\mathbb{G}_{m,W} & \xrightarrow{\theta_{p^w}\otimes\sigma^w} & B\mathbb{G}_{m,W} \end{array} \tag{6.5.8.14}$$

and a $\theta_{p^w} \otimes \sigma^w$-PD-morphism

$$\sigma^w : \mathrm{Spec}(W) \longrightarrow \mathrm{Spec}(W). \tag{6.5.8.15}$$

□

Finally we note that the operator N is automatically nilpotent. Let K denote the field of fractions of W.

Proposition 6.5.9. — *Let M be a finite dimensional K-vector space with a semi-linear automorphism $\varphi_M : M \to M$, and $N : M \to M$ an endomorphism such that $N\varphi_M = p\varphi_M N$. Then N is nilpotent.*

Proof. — Let $M = \oplus_{\lambda\in\mathbb{Q}} M_\lambda$ be the slope decomposition of (M, φ_M) (see for example [**34**, II.3.4]). Let $\pi_\lambda : M \to M_\lambda$ be the projection and $j_\lambda : M_\lambda \hookrightarrow M$ the inclusion. For any λ and λ' the diagram

$$\begin{array}{ccccccc} M_\lambda & \xrightarrow{j_\lambda} & M & \xrightarrow{N} & M & \xrightarrow{\pi_{\lambda'}} & M_{\lambda'} \\ \downarrow{\scriptstyle \varphi_M|_{M_\lambda}} & & \downarrow{\scriptstyle \varphi_M} & & \downarrow{\scriptstyle p\varphi_M} & & \downarrow{\scriptstyle p\varphi_M|_{M_{\lambda'}}} \\ M_\lambda & \xrightarrow{j_\lambda} & M & \xrightarrow{N} & M & \xrightarrow{\pi_{\lambda'}} & M_{\lambda'} \end{array} \tag{6.5.9.1}$$

commutes. If the composite $\pi_{\lambda'} \circ N \circ j_\lambda$ is non-zero, then it follows from this that $\lambda' = \lambda - 1$. Since the set of λ's with $M_\lambda \neq 0$ is finite, this implies that N is nilpotent. □

CHAPTER 7

A VARIANT CONSTRUCTION OF THE (φ, N, G)-STRUCTURE

Let K be a complete discrete valuation field with ring of integers V and perfect residue field k of characteristic $p > 0$. Let W be the ring of Witt vectors of k, and let K_0 be the field of fractions of W. Fix a collection $\alpha = \{\alpha_1, \dots, \alpha_r\}$ and a group H as in 6.1.1, and let

$$f : \mathcal{X} \longrightarrow \mathcal{S}_H(\alpha)_{R,\mathfrak{m}_R} \tag{7.0.9.2}$$

be a smooth morphism from a proper tame Deligne-Mumford stack $\mathcal{X}/V$. Here $\mathcal{S}_H(\alpha)_{R,\mathfrak{m}_R}$ denotes the pullback of the stack $\overline{\mathcal{S}}_H(\alpha)$ defined in 6.1.9 by the map $\mathrm{Spec}(R) \to [\mathbb{A}^1/\mathbb{G}_m]$ defined by the maximal ideal of R. We denote by $\mathcal{Y}_0$ the reduction of $\mathcal{X}$ to k, and by $\mathcal{Y}$ the base change of $\mathcal{X}$ to $V_0 := V/pV$.

In this chapter we give a direct construction of the (φ, N, G)-structure on the de Rham cohomology of the generic fiber of $\mathcal{X}$ and also prove 0.1.8. We also prove results for coefficients. The main result is 7.1.3.

7.1. The case when the multiplicities are powers of p

Throughout this section $\alpha = \{\alpha_1, \dots, \alpha_r\}$ and H are fixed so we write simply $\mathcal{S}$ (resp. $\overline{\mathcal{S}}$, etc.) for $\mathcal{S}_H(\alpha)$ (resp. $\overline{\mathcal{S}}_H(\alpha)$, etc.). For a ring R we write $\overline{\mathcal{S}}_R$ for the fiber product $\overline{\mathcal{S}} \times_{\mathrm{Spec}(\mathbb{Z})} \mathrm{Spec}(R)$. Let $F_{\overline{\mathcal{S}}_W} : \overline{\mathcal{S}}_W \to \overline{\mathcal{S}}_W$ be the lifting of Frobenius obtained from the canonical lifting σ of Frobenius to W and the map $\Lambda_p : \overline{\mathcal{S}} \to \overline{\mathcal{S}}$ defined in 6.3.8.

Definition 7.1.1. — An *F-crystal of width b* on $\mathcal{Y}_0/\overline{\mathcal{S}}_W$ is a pair (E, φ), where E is a locally free crystal in $(\mathcal{Y}_0/\overline{\mathcal{S}}_W)_{\mathrm{cris}}$ and $\varphi : F^*E \to E$ is a morphism of crystals in $(\mathcal{Y}_0/\overline{\mathcal{S}}_W)_{\mathrm{cris}}$ such that there exists a map $\psi : E \to F^*E$ for which the composites $\varphi \circ \psi$ and $\psi \circ \varphi$ are both equal to multiplication by b.

7.1.2. — Let (E, φ) be an F-crystal of width b on $\mathcal{Y}_0/\overline{\mathcal{S}}_W$.

Let $\widehat{\mathcal{X}}$ be the p-adic completion of $\mathcal{X}$, and let $(\widehat{\mathcal{E}}, \nabla)$ be the module with integrable connection on the formal stack $\widehat{\mathcal{X}}/\mathcal{S}_{V,\mathfrak{m}_V}$ obtained from the restriction of E

to $(\mathcal{Y}_0/\mathcal{S}_{V,\mathfrak{m}_V})_{\mathrm{cris}}$ and 2.7.7. Since $\widehat{\mathcal{X}}$ is proper over V the pair $(\mathcal{E}, \nabla)$ is by the Grothendieck Existence theorem [**15**, III.5.1.4] obtained from a module with integrable connection $(\mathcal{E}, \nabla)$ on $\mathcal{X}/\mathcal{S}_{V,\mathfrak{m}_V}$. If X/K denotes the generic fiber of $\mathcal{X}$, we obtain a module with integrable connection $(\mathcal{E}_K, \nabla)$ on X/K. The following is the main result of this and the next section:

Theorem 7.1.3. — *There is a natural (φ, N, G)-structure (D^m, φ, N) on the de Rham cohomology groups $H^m_{\mathrm{dR}}(X/K, (\mathcal{E}_K, \nabla))$. The action of G on D^m factors through a tame quotient.*

Remark 7.1.4. — Recall (0.1.1) that by our conventions, a (φ, N, G)-module does *not* include a filtration.

Remark 7.1.5. — We show in 9.6 that if (E, φ) is equal to $\mathcal{O}_{\mathcal{X}/\overline{\mathcal{S}}_W}$ with the canonical map $\varphi : F^*E \to E$ induced by functoriality, then this (φ, N, G)-structure on $H^m_{\mathrm{dR}}(X/K)$ agrees with the one constructed in 8.5.

In this section we prove the theorem for the special case when each α_i is a power of p, say $\alpha_i = p^{a_i}$. We furthermore order the α_i so that $a_1 \geq a_2 \geq \cdots \geq a_r$. In the following section we then deduce the general case from this and also explain how to deduce 0.1.8.

The comparison isomorphism. — Fix a uniformizer $\pi \in V$. The following theorem furnishes the isomorphism ρ (which depends on the choice of π).

Theorem 7.1.6. — *There are natural isomorphisms*

(7.1.6.1)

$$H^*(\mathcal{X}_{K,\mathrm{et}}, \mathcal{E}_K \otimes \Omega^{\bullet}_{\mathcal{X}_K/K}) \simeq H^*((\mathcal{Y}/\mathcal{S}_{V,\mathfrak{m}_V})_{\mathrm{cris}}, E) \otimes \mathbb{Q} \simeq H^*((\mathcal{Y}_0/\mathcal{S}_{W,(0)})_{\mathrm{cris}}, E) \otimes_W K.$$

Proof. — The isomorphism $H^*(\mathcal{X}_{K,\mathrm{et}}, \mathcal{E}_K \otimes \Omega^{\bullet}_{\mathcal{X}_K/K}) \simeq H^*((\mathcal{Y}/\mathcal{S}_{V,\mathfrak{m}_V})_{\mathrm{cris}}, E) \otimes \mathbb{Q}$ is provided by 2.7.7 and [**15**, III.5.1.2].

For the second isomorphism the main difficulty is that the natural maps $\mathcal{O}_{\mathcal{S}_{V,(\pi^p)}} \to R\Lambda_{p*}\mathcal{O}_{\mathcal{S}_{V,(\pi)}}$ and $\mathcal{O}_{\mathcal{S}_{W,(0)}} \to R\Lambda_{p*}\mathcal{O}_{\mathcal{S}_{W,(0)}}$ are not isomorphisms. We overcome this using 6.3.26.

For each integer e, define

$$\mathcal{Y}^{(e)} := \mathcal{Y} \otimes_{V_0, F^e_V} V_0, \quad \mathcal{Y}_0^{(e)} := \mathcal{Y}_0 \otimes_{k, F^e_k} k \tag{7.1.6.2}$$

where F_{V_0} and F_k denote the Frobenius morphisms on V_0 and k respectively. Note that V_0 is an artinian k-algebra, and let $\mathrm{rk}(V_0)$ denote its rank as a k-module. If e is greater than $\log_p(\mathrm{rk}(V_0/k))$, then there is a natural isomorphism $\mathcal{Y}^{(e)} \simeq \mathcal{Y}_0^{(e)} \otimes_k V_0$ since $F^e_{V_0} : V_0 \to V_0$ factors through k.

Fix an integer $r > \log_p(\mathrm{rk}(V_0/k)) + a_1$, so that $\Lambda_{p^{n+r}}$ is a map $\mathcal{S}_{V/p^n!,(\pi)} \to \mathcal{S}_{V/p^n!,(0)}$ (since $\mathrm{ord}_p(\pi^{p^{n+r}}) > \mathrm{ord}_p(p^{p^n}) > \mathrm{ord}_p(p^n!)$). Let

(7.1.6.3)

$$\overline{\mathcal{S}}^{(n+r)}_{V/p^n!,(\pi)} := \mathrm{Spec}(\Lambda_{p^{n+r}*}\mathcal{O}_{\mathcal{S}_{V/p^n!,(\pi)}}), \quad \overline{\mathcal{S}}^{(n+r)}_{W/p^n!,(0)} := \mathrm{Spec}(\Lambda_{p^{n+r}*}\mathcal{O}_{\mathcal{S}_{W/p^n!,(0)}})$$

be the resulting affine stacks over $\mathcal{S}_{V/p^n!,(0)}$ and $\mathcal{S}_{W/p^n!,(0)}$ respectively. We also define

$$\overline{\mathcal{S}}^{(n+r)}_{V/p^n!,(0)} := \mathrm{Spec}(\Lambda_{p^{n+r}*}\mathcal{O}_{\mathcal{S}_{V/p^n!,(0)}}) \simeq \overline{\mathcal{S}}^{(n+r)}_{W/p^n!,(0)} \otimes_W V. \tag{7.1.6.4}$$

We denote by $\overline{\mathcal{Y}}^{(n+r)}$ (resp. $\overline{\mathcal{Y}}^{(n+r)}_0$) the pullback of $\mathcal{Y}^{(n+r)}$ (resp. $\mathcal{Y}^{(n+r)}_0$) to $\overline{\mathcal{S}}^{(n+r)}_{V/p^n!,(\pi)}$ (resp. $\overline{\mathcal{S}}^{(n+r)}_{W/p^n!,(0)}$).

Let $E_{\overline{\mathcal{Y}}^{(n+r)}_0/\overline{\mathcal{S}}^{(n+r)}_{W/p^n!,(0)}}$ (resp. $E_{\overline{\mathcal{Y}}^{(n+r)}/\overline{\mathcal{S}}_{V/p^n!,(\pi)}}$, etc.) denote the pullback of E to the topos $(\overline{\mathcal{Y}}^{(n+r)}_0/\overline{\mathcal{S}}^{(n+r)}_{W/p^n!,(0)})_{\mathrm{cris}}$ (resp. $(\overline{\mathcal{Y}}^{(n+r)}/\overline{\mathcal{S}}_{V/p^n!,(\pi)})_{\mathrm{cris}}$, etc.). By the base change theorem 2.6.2 the canonical map

$$Ru_*E_{\overline{\mathcal{Y}}^{(n+r)}_0/\overline{\mathcal{S}}^{(n+r)}_{W/p^n!,(0)}} \otimes_W V \longrightarrow Ru_*E_{(\overline{\mathcal{Y}}^{(n+r)}_0\otimes_k V_0)/(\overline{\mathcal{S}}^{(n+r)}_{W/p^n!,(0)}\otimes_W V)} \tag{7.1.6.5}$$

is an isomorphism. On the other hand, by 6.3.26 there is a canonical isomorphism

$$\overline{\mathcal{S}}^{(n+r)}_{W/p^n!,(0)} \otimes_W V \simeq \overline{\mathcal{S}}^{(n+r)}_{V/p^n!,(\pi)} \tag{7.1.6.6}$$

of affine stacks over $\mathcal{S}_{V/p^n!,(0)}$. From the isomorphism $\mathcal{Y}^{(n+r)} \simeq \mathcal{Y}^{(n+r)}_0 \otimes_k V_0$ over $\mathcal{S}_{V/p^n!,(0)}$ we therefore obtain a commutative diagram

$$\begin{array}{ccc} \overline{\mathcal{Y}}^{(n+r)}_0 \otimes_k V_0 & \longrightarrow & \overline{\mathcal{Y}}^{(n+r)} \\ \downarrow & & \downarrow \\ \overline{\mathcal{S}}^{(n+r)}_{W/p^n!,(0)} \otimes_W V & \longrightarrow & \overline{\mathcal{S}}^{(n+r)}_{V/p^n!,(\pi)}, \end{array} \tag{7.1.6.7}$$

where the horizontal arrows are isomorphisms. This diagram induces an isomorphism

$$Ru_*E_{\overline{\mathcal{Y}}^{(n+r)}_0\otimes_k V_0/\overline{\mathcal{S}}^{(n+r)}_{W/p^n!,(0)}\otimes_W V} \simeq Ru_*E_{\overline{\mathcal{Y}}^{(n+r)}/\overline{\mathcal{S}}^{(n+r)}_{V/p^n!,(\pi)}}. \tag{7.1.6.8}$$

Composing (7.1.6.5) and (7.1.6.8) we obtain an isomorphism

$$Ru_*E_{\overline{\mathcal{Y}}^{(n+r)}_0/\overline{\mathcal{S}}^{(n+r)}_{W/p^n!,(0)}} \otimes_W V \simeq Ru_*E_{\overline{\mathcal{Y}}^{(n+r)}/\overline{\mathcal{S}}^{(n+r)}_{V/p^n!,(\pi)}}. \tag{7.1.6.9}$$

For each i, the triple

$$(E_{\overline{\mathcal{Y}}^{(i)}/\overline{\mathcal{S}}^{(i)}_{V,(\pi)}}, E_{\overline{\mathcal{Y}}^{(i+1)}/\overline{\mathcal{S}}^{(i+1)}_{V,(\pi)}}, \Phi) \quad (\text{resp. } (E_{\overline{\mathcal{Y}}^{(i)}_0/\overline{\mathcal{S}}^{(i)}_{W,(0)}}, E_{\overline{\mathcal{Y}}^{(i+1)}_0/\overline{\mathcal{S}}^{(i+1)}_{W,(0)}}, \Phi)), \tag{7.1.6.10}$$

where Φ denotes the map induced by the F-crystal structure on E, forms an F-span of width b in the sense of 3.4.41. Therefore, by 3.4.42 we obtain maps
(7.1.6.11)

$$Ru_*E_{\mathcal{Y}/\mathcal{S}_{V,(\pi)}} \underset{\psi^{n+r}}{\overset{\nu^{n+r}}{\rightleftarrows}} Ru_*E_{\overline{\mathcal{Y}}^{(n+r)}/\overline{\mathcal{S}}^{(n+r)}_{V,(\pi)}}, \quad Ru_*E_{\mathcal{Y}_0/\mathcal{S}_{W,(0)}} \underset{\psi^{n+r}}{\overset{\nu^{n+r}}{\rightleftarrows}} Ru_*E_{\overline{\mathcal{Y}}^{(n+r)}_0/\overline{\mathcal{S}}^{(n+r)}_{W,(0)}}.$$

Combining this with (7.1.6.9), we obtain maps

(7.1.6.12)
$$\rho_n := (\psi^{n+1} \otimes V) \circ (7.1.6.9) \circ \nu^{n+r} : Ru_* E_{\mathcal{Y}/\mathcal{S}_{V/p^n!,(\pi)}} \longrightarrow Ru_* E_{\mathcal{Y}_0/\mathcal{S}_{W/p^n!,(0)}} \otimes_W V,$$

(7.1.6.13)
$$\epsilon_n := \psi^{n+r} \circ (7.1.6.9) \circ (\nu^{n+r} \otimes V) : Ru_* E_{\mathcal{Y}_0/\mathcal{S}_{W/p^n!,(0)}} \otimes_W V \longrightarrow Ru_* E_{\mathcal{Y}/\mathcal{S}_{V/p^n!,(\pi)}}$$

such that $\rho_n \circ \epsilon_n$ and $\epsilon_n \circ \rho_n$ are both equal to multiplication by $p^{2b(n+r)d}$ where d is the relative dimension of $\mathcal{X}/V$ Moreover, it follows from the construction that the reduction of ρ_{n+1} modulo $p^n!$ is equal to $p^d \rho_n$.

Define

$$H := \mathrm{Hom}_V(H^*((\mathcal{Y}/\mathcal{S}_{V,\mathfrak{m}_V})_{\mathrm{cris}}, E), H^*((\mathcal{Y}_0/\mathcal{S}_{W,(0)})_{\mathrm{cris}}, E) \otimes_W V), \quad (7.1.6.14)$$

$$m := 2b(r+r)d, \quad \nu_n := (p^n - 1)/(p-1). \quad (7.1.6.15)$$

Then in the notation of 5.2.2 the maps ρ_n define an element of $\Omega(H)$. By 5.2.4 this element then induces a map

$$\xi : H^*((\mathcal{Y}/\mathcal{S}_{V,\mathfrak{m}_V})_{\mathrm{cris}}, E) \longrightarrow H^*((\mathcal{Y}_0/\mathcal{S}_{W,(0)})_{\mathrm{cris}}, E) \otimes_W V. \quad (7.1.6.16)$$

That this map is an isomorphism follows from the same argument used in 5.3.14. □

The monodromy operator. — The construction of N is essentially the same as that given in 6.5.

7.1.7. — The map $\mathrm{Spec}(W) \to [\mathbb{A}^1/\mathbb{G}_m]$ defined by the map of free modules $W \to W$ sending 1 to 0 factors as

$$\mathrm{Spec}(W) \longrightarrow B\mathbb{G}_m \subset [\mathbb{A}^1/\mathbb{G}_m]. \quad (7.1.7.1)$$

Let $[\mathcal{S}_{W,(0)}/\mathbb{G}_m]$ denote the pullback of $\overline{\mathcal{S}}$ to $B\mathbb{G}_{m,W}$. This notation is justified by the observation that there is a natural cartesian diagram

$$\begin{array}{ccc} [\mathcal{S}_{W,(0)}/\mathbb{G}_m] & \longleftarrow & \mathcal{S}_{W,(0)} \\ \downarrow & & \downarrow \\ B\mathbb{G}_{m,W} & \longleftarrow & \mathrm{Spec}(W) \end{array} \quad (7.1.7.2)$$

which makes $\mathcal{S}_{W,0}$ a $\mathbb{G}_m$-torsor over $[\mathcal{S}_{W,(0)}/\mathbb{G}_m]$.

In particular, (6.1.12.2) factors through a representable smooth morphism

$$\mathcal{S}_{W,(0)} \longrightarrow [\mathcal{S}_{W,(0)}/\mathbb{G}_m]. \quad (7.1.7.3)$$

Let $\mathcal{P}^1$ denote the first infinitesimal neighborhood of the diagonal

$$\Delta : \mathcal{S}_{W,(0)} \longrightarrow \mathcal{S}_{W,(0)} \times_{[\mathcal{S}_{W,(0)}/\mathbb{G}_m]} \mathcal{S}_{W,(0)} \simeq \mathcal{S}_{W,(0)} \times_{\mathrm{Spec}(W)} \mathbb{G}_{m,W}. \quad (7.1.7.4)$$

There is a canonical isomorphism

$$\mathcal{P}^1 \simeq \mathcal{S}_{W,(0)} \otimes_W W[(u-1)]/(u-1)^2, \quad (7.1.7.5)$$

and $\mathcal{P}^1$ becomes a PD-stack by giving $(p,(u-1))$ the canonical PD-structure. We thus have a 2-commutative diagram

$$\begin{array}{ccccc} & & & & \mathcal{Y}_0 \\ & & & & \downarrow \\ \mathcal{S}_{W,(0)} & \xleftarrow{p_1} & \mathcal{S}_{W,(0)} \otimes_W W[(u-1)]/(u-1)^2 & \xleftarrow{\Delta} & \mathcal{S}_{W,(0)} \\ \downarrow & & \downarrow \rho & & \\ [\mathcal{S}_{W,(0)}/\mathbb{G}_m] & \longleftarrow & \mathcal{S}_{W,(0)} & & \end{array} \tag{7.1.7.6}$$

where ρ is the map obtained from the action.

Define

$$K := R\Gamma((\mathcal{Y}_0/\mathcal{S}_{W,0})_{\mathrm{cris}}, E_{\mathcal{Y}_0/\mathcal{S}_{W,0}}), \tag{7.1.7.7}$$

$$K' := R\Gamma((\mathcal{Y}_0/\mathcal{S}_{W[u-1]/(u-1)^2,0})_{\mathrm{cris}}, E_{\mathcal{Y}_0/\mathcal{S}_{W[u-1]/(u-1)^2,0}}). \tag{7.1.7.8}$$

By the base change theorem 2.6.2, pullback defines isomorphisms ρ^* and p_1^* between $W[u-1]/(u-1)^2 \otimes_W K$ and

(7.1.7.9)
$$R\Gamma((\mathcal{Y}_0 \times_{\mathcal{S}_{k,(0)},\rho} \mathcal{S}_{k[u-1]/(u-1)^2,(0)}/\mathcal{S}_{W[u-1]/(u-1)^2},(0))_{\mathrm{cris}}, E_{\mathcal{Y}_0/\mathcal{S}_{W[u-1]/(u-1)^2,(0)}})$$

and

(7.1.7.10)
$$R\Gamma((\mathcal{Y}_0 \times_{\mathcal{S}_{k,(0)},p_1} \mathcal{S}_{k[u-1]/(u-1)^2,(0)}/\mathcal{S}_{W[u-1]/(u-1)^2,(0)})_{\mathrm{cris}}, E_{\mathcal{Y}_0/\mathcal{S}_{W[u-1]/(u-1)^2,(0)}}).$$

On the other hand, both of these complexes are naturally isomorphic to K' since $(u-1)$ defines a sub-PD ideal of $W[u-1]/(u-1)^2$.

Let $\lambda : W[u-1]/(u-1)^2 \simeq W \oplus W \to W$ be the projection onto the second factor. We define the monodromy operator N on D^m to be the map induced by the composite

$$K \xrightarrow{\rho^*} K' \xrightarrow{p_1^{*-1}} W[u-1]/(u-1)^2 \otimes_W K \xrightarrow{\lambda} K. \tag{7.1.7.11}$$

The Frobenius morphism

7.1.8. — Define

$$\Phi : Ru_* E_{\mathcal{Y}_0/\mathcal{S}_{W,(0)}} \longrightarrow Ru_* E_{\mathcal{Y}_0/\mathcal{S}_{W,(0)}} \tag{7.1.8.1}$$

to be the semi-linear map induced by the F-crystal structure on E.

Proposition 7.1.9. — *The induced map $\varphi : D^m \otimes K_0 \to D^m \otimes K_0$ is an isomorphism.*

Proof. — Let $F_{\mathcal{S}_{W,0}} : \mathcal{S}_{W,0} \to \mathcal{S}_{W,0}$ denote the lifting of Frobenius defined as the map

(7.1.9.1)
$$\mathcal{S}_{W,0} \simeq \mathcal{S}_{\mathbb{Z},0} \times \mathrm{Spec}(W) \xrightarrow{\Lambda_p \times \sigma} \mathcal{S}_{\mathbb{Z},0} \times \mathrm{Spec}(W) \simeq \mathcal{S}_{W,0}.$$

For $l \geq 0$ let $F_{\mathcal{Y}}^{l*}E$ denote the crystal obtained from E by pullback along the morphism of topoi defined by the commutative diagram

(7.1.9.2)
$$\begin{array}{ccc} \mathcal{Y} & \xrightarrow{F_{\mathcal{Y}}^l} & \mathcal{Y} \\ \downarrow & & \downarrow \\ \mathcal{S}_{W,0} & \xrightarrow{F_{\mathcal{S}}^l} & \mathcal{S}_{W,0}, \end{array}$$

where $F_{\mathcal{Y}}$ denotes the absolute Frobenius morphism on $\mathcal{Y}$. For $l \geq 0$ let

(7.1.9.3)
$$\delta_l : R\Gamma((\mathcal{Y}/\mathcal{S}_{W,0})_{\mathrm{cris}}, F_{\mathcal{Y}}^{l-1*}E_{\mathcal{Y}/\mathcal{S}_{W,0}}) \to R\Gamma((\mathcal{Y}/\mathcal{S}_{W,0})_{\mathrm{cris}}, F_{\mathcal{Y}}^{l*}E_{\mathcal{Y}/\mathcal{S}_{W,0}})$$

be the map induced by the commutative diagram

(7.1.9.4)
$$\begin{array}{ccc} \mathcal{Y} & \xrightarrow{F_{\mathcal{Y}}} & \mathcal{Y} \\ \downarrow & & \downarrow \\ \mathcal{S}_{W,0} & \xrightarrow{F_{\mathcal{S}_{W,0}}} & \mathcal{S}_{W,0} \end{array}$$

and functoriality of the crystalline topos. The map $\varphi : F^*E \to E$ induces by restriction a map $F_{\mathcal{Y}}^*E_{\mathcal{Y}/\mathcal{S}_{W,0}} \to E_{\mathcal{Y}/\mathcal{S}_{W,0}}$ (which we again denote by φ), and the induced maps

(7.1.9.5)
$$H^m(\varphi^l) : H^m((\mathcal{Y}/\mathcal{S}_{W,0})_{\mathrm{cris}}, F_{\mathcal{Y}}^{l*}E_{\mathcal{Y}/\mathcal{S}_{W,0}})\otimes_W K_0 \longrightarrow H^m((\mathcal{Y}/\mathcal{S}_{W,0})_{\mathrm{cris}}, E_{\mathcal{Y}/\mathcal{S}_{W,0}})\otimes_W K_0$$

are all isomorphisms, as by assumption there exists a map $\psi : E \to F_{\mathcal{Y}}^*E$ such that the induced maps

(7.1.9.6)
$$H^m(\psi^l) : H^m((\mathcal{Y}/\mathcal{S}_{W,0})_{\mathrm{cris}}, E_{\mathcal{Y}/\mathcal{S}_{W,0}})\otimes_W K_0 \longrightarrow H^m((\mathcal{Y}/\mathcal{S}_{W,0})_{\mathrm{cris}}, F_{\mathcal{Y}}^{l*}E_{\mathcal{Y}/\mathcal{S}_{W,0}})\otimes_W K_0$$

satisfies $H^m(\varphi^l) \circ H^m(\psi^l) = b^l$ and $H^m(\psi^l) \circ H^m(\varphi^l) = b^l$. It follows that it suffices to show that the map

(7.1.9.7)
$$H^m(\delta_1) : H^m((\mathcal{Y}/\mathcal{S}_{W,0})_{\mathrm{cris}}, E_{\mathcal{Y}/\mathcal{S}_{W,0}}) \otimes_W K_0 \longrightarrow H^m((\mathcal{Y}/\mathcal{S}_{W,0})_{\mathrm{cris}}, F_{\mathcal{Y}}^*E_{\mathcal{Y}/\mathcal{S}_{W,0}}) \otimes_W K_0$$

is an isomorphism. From the commutativity of the diagrams

(7.1.9.8)
$$\begin{array}{ccc} H^m((\mathcal{Y}/\mathcal{S}_{W,0})_{\mathrm{cris}}, F_{\mathcal{Y}}^{l-1*}E_{\mathcal{Y}/\mathcal{S}_{W,0}}) \otimes_W K_0 & \xrightarrow{H^m(\delta_l)} & H^m((\mathcal{Y}/\mathcal{S}_{W,0})_{\mathrm{cris}}, F_{\mathcal{Y}}^{l*}E_{\mathcal{Y}/\mathcal{S}_{W,0}}) \otimes_W K_0 \\ \downarrow{\scriptstyle H^m(\varphi^{l-1})} & & \downarrow{\scriptstyle H^m(\varphi^{l-1})} \\ H^m((\mathcal{Y}/\mathcal{S}_{W,0})_{\mathrm{cris}}, E_{\mathcal{Y}/\mathcal{S}_{W,0}}) \otimes_W K_0 & \xrightarrow{H^m(\delta_1)} & H^m((\mathcal{Y}/\mathcal{S}_{W,0})_{\mathrm{cris}}, F_{\mathcal{Y}}^*E_{\mathcal{Y}/\mathcal{S}_{W,0}}) \otimes_W K_0 \end{array}$$

we conclude that to prove 7.1.9 it suffices to show that for l sufficiently big the map (7.1.9.9)

$$H^m(\delta_l) : H^m((\mathcal{Y}/\mathcal{S}_{W,0})_{\mathrm{cris}}, F_{\mathcal{Y}}^{l-1*}E_{\mathcal{Y}/\mathcal{S}_{W,0}})\otimes K_0 \longrightarrow H^m((\mathcal{Y}/\mathcal{S}_{W,0})_{\mathrm{cris}}, F_{\mathcal{Y}}^{l*}E_{\mathcal{Y}/\mathcal{S}_{W,0}})\otimes K_0$$

is an isomorphism. Since this is a semi-linear map between vector spaces of the same dimension, it suffices to show that (7.1.9.9) is surjective for l sufficiently big.

We claim that (7.1.9.9) is surjective for

$$l > \max\{\mathrm{ord}_p(\alpha_i) + 1\}. \tag{7.1.9.10}$$

For any $j \geq 1$ let

$$\overline{\Lambda}_{p^j} : \mathcal{S}_{W,(0)} \longrightarrow \overline{\mathcal{S}}^{(j)}_{W,(0)} \tag{7.1.9.11}$$

be the canonical map, so we have a commutative diagram of solid arrows

(7.1.9.12)

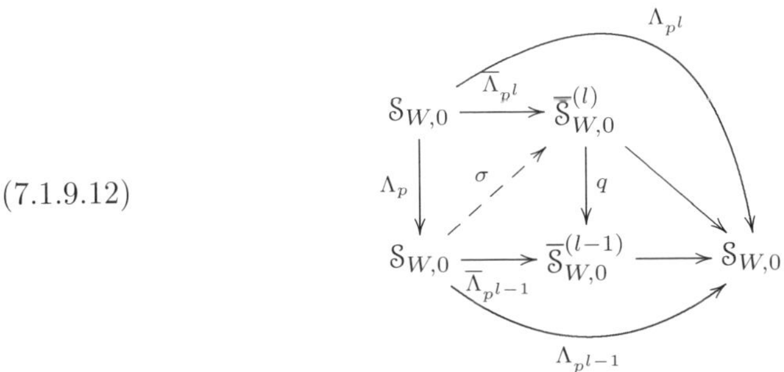

where q is the map induced by the natural map $L : \Lambda_{p^{l-1}*}\mathcal{O}_{\mathcal{S}_{W,0}} \to \Lambda_{p^l*}\mathcal{O}_{\mathcal{S}_{W,0}}$ considered in 6.3.22. By *loc. cit.* the map q is an isomorphism, and therefore there exists a unique dotted arrow σ filling in (7.1.9.12). From this and the definition of the $\overline{\mathcal{Y}}^{(e)}$'s, we obtain a commutative diagram

$$\begin{array}{ccccc} \mathcal{Y} & \xrightarrow{F_{\mathcal{Y}/k}} & \mathcal{Y}^{(1)} & \xrightarrow{F^{l-1}_{\mathcal{Y}^{(1)}/k}} & \overline{\mathcal{Y}}^{(l)} \\ \downarrow & & \downarrow & & \downarrow \\ \mathcal{S}_{W,0} & \xrightarrow{\Lambda_p} & \mathcal{S}_{W,0} & \xrightarrow{\sigma} & \overline{\mathcal{S}}^{(l)}_{W,0}, \end{array} \tag{7.1.9.13}$$

(with the top curved arrow $F^l_{\mathcal{Y}/k} : \mathcal{Y} \to \overline{\mathcal{Y}}^{(l)}$ and the bottom curved arrow $\overline{\Lambda}_{p^l} : \mathcal{S}_{W,0} \to \overline{\mathcal{S}}^{(l)}_{W,0}$)

where the top horizontal arrows are the morphisms induced by absolute Frobenius on $\mathcal{Y}$. Let

$$\begin{gathered} \pi_{\mathcal{Y}} : \mathcal{Y}^{(1)} = \mathcal{Y} \times_{\mathrm{Spec}(k), F_k} \mathrm{Spec}(k) \to \mathcal{Y} \\ (\text{resp. } \pi_{\mathcal{S}} : \mathcal{S}_{W,0} \simeq \mathcal{S}_{W,0} \times_{\mathrm{Spec}(W),\sigma} \mathrm{Spec}(W) \to \mathcal{S}_{W,0}) \end{gathered} \tag{7.1.9.14}$$

denote the projection onto the first factor. Then (7.1.9.13) fits into a larger commutative diagram

(7.1.9.15)

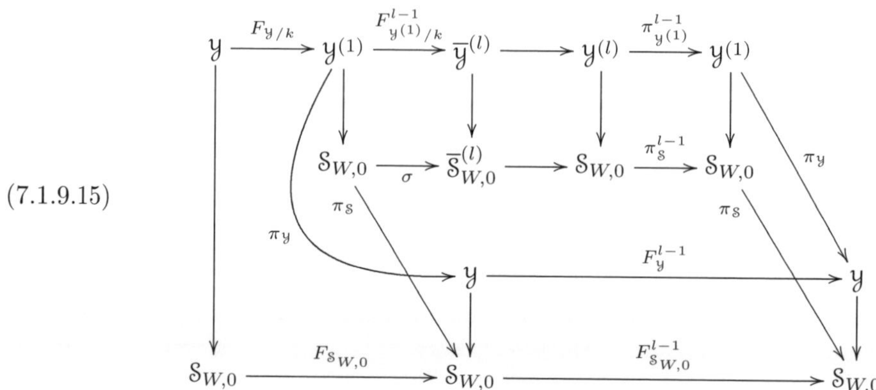

From this diagram we obtain isomorphisms

(7.1.9.16)
$$F^{l*}_{\mathcal{Y}/k} E_{\overline{\mathcal{Y}}^{(l)}/\overline{\mathcal{S}}^{(l)}_{W,0}} \simeq F^{l*}_{\mathcal{Y}} E_{\mathcal{Y}/\mathcal{S}_{W,0}}$$

and

(7.1.9.17)
$$F^{l-1*}_{\mathcal{Y}^{(1)}/k} E_{\overline{\mathcal{Y}}^{(l)}/\overline{\mathcal{S}}^{(l)}_{W,0}} \simeq \pi^*_{\mathcal{Y}} F^{l-1*}_{\mathcal{Y}} E_{\mathcal{Y}/\mathcal{S}_{W,0}}.$$

We therefore obtain a commutative diagram

(7.1.9.18)
$$\begin{array}{ccc}
R\Gamma((\overline{\mathcal{Y}}^{(l)}/\overline{\mathcal{S}}^{(l)}_{W,0})_{\text{cris}}, E_{\overline{\mathcal{Y}}^{(l)}/\overline{\mathcal{S}}^{(l)}_{W,0}}) & \xrightarrow{\kappa_l} & R\Gamma((\mathcal{Y}/\mathcal{S}_{W,0})_{\text{cris}}, F^{l*}_{\mathcal{Y}/k} E_{\overline{\mathcal{Y}}^{(l)}/\overline{\mathcal{S}}^{(l)}_{W,0}}) \\
\downarrow \sigma^* & & \downarrow \simeq \\
R\Gamma((\mathcal{Y}^{(1)}/\mathcal{S}_{W,0})_{\text{cris}}, \pi^*_{\mathcal{Y}} F^{l-1*}_{\mathcal{Y}} E_{\mathcal{Y}/\mathcal{S}_{W,0}}) & \xrightarrow{F^*_{\mathcal{Y}/k}} & R\Gamma((\mathcal{Y}/\mathcal{S}_{W,0})_{\text{cris}}, F^{l*}_{\mathcal{Y}} E_{\mathcal{Y}/\mathcal{S}_{W,0}}) \\
\uparrow \pi^*_{\mathcal{Y}} & \nearrow \delta_l & \\
R\Gamma((\mathcal{Y}/\mathcal{S}_{W,0})_{\text{cris}}, F^{l-1*}_{\mathcal{Y}} E_{\mathcal{Y}/\mathcal{S}_{W,0}}), & &
\end{array}$$

where $\pi^*_{\mathcal{Y}}$ is an isomorphism and κ_l denotes the map defined by $F^{l*}_{\mathcal{Y}/k}$. From this it follows that to prove that (7.1.9.9) is surjective, it suffices to show that

(7.1.9.19)
$$H^m(\kappa_l) : H^m((\overline{\mathcal{Y}}^{(l)}/\overline{\mathcal{S}}^{(l)}_{W,0})_{\text{cris}}, E_{\overline{\mathcal{Y}}^{(l)}/\overline{\mathcal{S}}^{(l)}_{W,0}}) \longrightarrow H^m((\mathcal{Y}/\mathcal{S}_{W,0})_{\text{cris}}, F^{l*}_{\mathcal{Y}/k} E_{\overline{\mathcal{Y}}^{(l)}/\overline{\mathcal{S}}^{(l)}_{W,0}})$$

is an isomorphism (in fact it suffices to show that $H^m(\kappa_l)$ is surjective). This follows from 3.4.45. □

The relation $N\varphi = p\varphi N$. — This is essentially the same as the proof of 6.5.2.

7.1.10. — Let $F_{\mathbb{G}_m} : \mathbb{G}_m \to \mathbb{G}_m$ be the map $u \mapsto u^p$. Define a lifting of Frobenius

$$\Lambda_p : \mathcal{S}_{W,0} \to \mathcal{S}_{W,0} \tag{7.1.10.1}$$

as in the proof of 6.5.2, and note that as in *loc. cit.* this lifting is "$F_{\mathbb{G}_m}$-linear". That is, if ρ denotes the action of $\mathbb{G}_m$ (7.1.7), then the diagram

$$\begin{array}{ccc} \mathcal{S}_{W,(0)} \times \mathbb{G}_m & \xrightarrow{\rho} & \mathcal{S}_{W,(0)} \\ \downarrow{\scriptstyle \Lambda_p \times F_{\mathbb{G}_m}} & & \downarrow{\scriptstyle \Lambda_p} \\ \mathcal{S}_{W,(0)} \times \mathbb{G}_m & \xrightarrow{\rho} & \mathcal{S}_{W,(0)} \end{array} \tag{7.1.10.2}$$

commutes. Thus we can extend the action of Frobenius on $\mathcal{S}_{W,(0)}$ to an action on the whole diagram (7.1.7.6), where the action on $W_n[(u-1)]/(u-1)^2$ is given by multiplication by p on $(u-1)$. It follows that the diagram

$$\begin{array}{ccc} K & \xrightarrow{(7.1.7.11)} & K \cdot (u-1) \\ \downarrow{\scriptstyle \Phi} & & \downarrow{\scriptstyle \Phi \cdot p} \\ K & \xrightarrow{(7.1.7.11)} & K \cdot (u-1) \end{array} \tag{7.1.10.3}$$

commutes. Passing to cohomology we find that the operators φ and N on D^m satisfy the relation $N\varphi = p\varphi N$.

Remark 7.1.11. — In the case when all α_i are equal to 1, it follows from 6.4.14 that the above constructed (φ, N, G)-structure agrees with the one constructed in 6.4.

7.2. Lowering the exponents

7.2.1. — Write each $\alpha_i = p^{a_i}\beta_i$ with $(\beta_i, p) = 1$, set $\beta := \prod_i \beta_i$, and write H for the group of elements $\sigma \in S_r$ for which $\alpha_{\sigma(i)} = \alpha_i$. Let $\mathcal{S}(\alpha)$ and $\overline{\mathcal{S}}(\alpha)$ denote the stacks obtained from $\alpha = (\alpha_1, \ldots, \alpha_r)$ and the group H. Denote by K' the Galois closure of $K[T]/(T^\beta - \pi)$ and let V' be the ring of integers of K'. Denote by $\mathcal{X}'$ the scheme obtained by base change to V'. We denote by π' the uniformizer of V' given by "T". Let $G_{K'/K}$ be the Galois group of K'/K. There is a natural homomorphism $G_{K'/K} \to \mu_\beta$, denoted $g \mapsto \zeta_g$, characterized by $(\pi')^g = \zeta_g \pi'$. Let k' be the residue field, W' the ring of Witt vectors of k', $\sigma : W' \to W'$ the canonical lift of Frobenius, and K'_0 the field of fractions of W'.

7.2.2. — There is a canonical map

$$B : \mathcal{S}_H(p^{\mathrm{a}})_{V',(\pi')} \longrightarrow \mathcal{S}(\alpha)_{V,(\pi)}, \tag{7.2.2.1}$$

where p^{a} denotes the set $\{p^{a_1}, \ldots, p^{a_r}\}$. It is the map associated to the map of prestacks

$$\mathcal{S}_H(p^{\mathrm{a}})^{\mathrm{ps}}_{V',(\pi')} \longrightarrow \mathcal{S}(\alpha)^{\mathrm{ps}}_{V,(\pi)} \tag{7.2.2.2}$$

which sends a collection $(x_1, \ldots, x_r, v)$ satisfying

$$x_1^{p^{a_1}} \cdots x_r^{p^{a_r}} v = \pi' \tag{7.2.2.3}$$

to the collection

$$(x_1^{\prod_{j\neq 1}\beta_j}, \ldots, x_r^{\prod_{j\neq r}\beta_i}, v^{\beta}). \tag{7.2.2.4}$$

Note the equality (7.2.2.3) implies that

$$(x_1^{\prod_{j\neq 1}\beta_j})^{\alpha_1} \cdots (x_r^{\prod_{j\neq r}\beta_i})^{\alpha_r} v^{\beta} = (x_1^{p^{a_1}} \cdots x_r^{p^{a_r}} v)^{\beta} = (\pi')^{\beta} = \pi. \tag{7.2.2.5}$$

Similarly, there are natural maps

$$B_0 : \mathcal{S}_H(p^{\mathrm{a}})_{W',(0)} \longrightarrow \mathcal{S}(\alpha)_{W,(0)}, \tag{7.2.2.6}$$

and

$$B_{\overline{\mathcal{S}}} : \overline{\mathcal{S}}_H(p^{\mathrm{a}}) \longrightarrow \overline{\mathcal{S}}_H(p^{\mathrm{a}}), \tag{7.2.2.7}$$

and B and B_0 induce the same map on the closed fiber.

Lemma 7.2.3. — *The map B (resp. B_0) is finite, relatively Deligne-Mumford, and is an isomorphism over K (resp. K_0).*

Proof. — This is proven as in [**63**, 4.3]. □

7.2.4. — Define

$$\widetilde{\mathcal{X}} := \mathcal{X} \times_{\mathcal{S}(\alpha)_{V,(\pi)}} \mathcal{S}_H(p^{\mathrm{a}})_{V',(\pi')}. \tag{7.2.4.1}$$

The stack $\widetilde{\mathcal{X}}$ is a proper tame Deligne-Mumford stack whose coarse moduli space equals $\mathcal{X}_{V'}$. Moreover, the map

$$\widetilde{\mathcal{X}} \longrightarrow \mathcal{S}_H(p^{\mathrm{a}})_{V',(\pi')} \tag{7.2.4.2}$$

is smooth. Hence we can apply the results of the preceding section to $\widetilde{\mathcal{X}}$.

7.2.5. — Set $\mathcal{Y} := \mathcal{X} \otimes V/pV$ and $\mathcal{Y}_0 := \mathcal{X} \otimes_V k$. Define $\widetilde{\mathcal{Y}} := \widetilde{\mathcal{X}} \otimes V'/pV'$ and $\widetilde{\mathcal{Y}}_0 := \widetilde{\mathcal{X}} \otimes_{V'} k'$.

Let (E, φ) be a F-crystal of width b on $\mathcal{Y}_0/\overline{\mathcal{S}}(\alpha)_V$. There is a natural commutative diagram

$$\begin{array}{ccc} \mathcal{S}_H(p^{\mathrm{a}})_{V',(\pi')} & \xrightarrow{B} & \mathcal{S}(\alpha)_{V,(\pi)} \\ \downarrow & & \downarrow \\ \overline{\mathcal{S}}_H(p^{\mathrm{a}})_{V'} & \xrightarrow{B_{\overline{\mathcal{S}}}} & \overline{\mathcal{S}}(\alpha)_V, \end{array} \tag{7.2.5.1}$$

where the vertical maps are those given by (6.1.12.2). It follows that we can pullback (E, φ) to an F-crystal of width b, denoted E', on $\widetilde{\mathcal{Y}}_0/\overline{\mathcal{S}}_H(p^{\mathrm{a}})_{V'}$. We define

$$D^m := H^m((\widetilde{\mathcal{Y}}_0/\mathcal{S}_H(p^{\mathrm{a}})_{W',(0)})_{\mathrm{cris}}, E') \otimes_{W'} K_0^{ur}, \tag{7.2.5.2}$$

where as in the introduction $K_0^{ur} \subset \overline{K}$ denotes the maximal unramified extension of K_0 in $\overline{K}$. By 7.1.6 there is a natural isomorphism

$$D^m \otimes_{K_0^{ur}} \overline{K} \simeq H^*((\widetilde{\mathcal{Y}}/\mathcal{S}_H(p^{\mathrm{a}})_{V',\pi'})_{\mathrm{cris}}, E') \otimes_{V'} \overline{K}. \tag{7.2.5.3}$$

As in 7.1.2, the Grothendieck Existence theorem for stacks [**68**, 8.1] implies that the right hand side of (7.2.5.3) computes the de Rham cohomology of the module with connection $(\mathcal{E}_{K'}, \nabla)$ on $\widetilde{\mathcal{X}}_{K'}/K'$ obtained from E. By 7.2.3 the natural map $\widetilde{\mathcal{X}}_{K'} \to \mathcal{X}_{K'}$ is an isomorphism. It follows that there is a natural isomorphism

$$D^m \otimes_{K_0^{ur}} \overline{K} \simeq H^m_{\mathrm{dR}}(\mathcal{X}_K/K, (\mathcal{E}_K, \nabla)) \otimes_K \overline{K}, \tag{7.2.5.4}$$

where $(\mathcal{E}_K, \nabla)$ denotes the module with integrable connection on $\mathcal{X}/K$ obtained from E.

Moreover, by the constructions of the preceding section the module D^m comes equipped with a monodromy operator $N_{K'}$ and Frobenius operator φ satisfying 0.1.1 2). We define the operator N to be $\beta \cdot N_{K'}$.

7.2.6. — It remains to construct an action of $G := \mathrm{Gal}(K'/K)$ on D^m.

Define an action of G^{op} on $\mathcal{S}_H(p^{\mathrm{a}})_{W',(0)}$ by letting $\tau_g : \mathcal{S}_H(p^{\mathrm{a}})_{W',(0)} \to \mathcal{S}_H(p^{\mathrm{a}})_{W',(0)}$ be the map associated to the map on prestacks which sends

$$(x_1, \ldots, x_r, v) \longmapsto (x_1, \ldots, x_r, \zeta_g^{-1} v). \tag{7.2.6.1}$$

Note that by the definition of (6.1.12.2), the two maps (6.1.12.2), (6.1.12.2) $\circ\, \tau_g$: $\mathcal{S}_H(p^{\mathrm{a}})_{W',(0)} \to \overline{\mathcal{S}}_H(p^{\mathrm{a}})_{W'}$ are equal. It follows that we have an action of G^{op} on the square

$$\begin{array}{ccc} \widetilde{\mathcal{Y}}_0 & \xrightarrow{F_{\widetilde{\mathcal{Y}}_0}} & \widetilde{\mathcal{Y}}_0 \\ \downarrow & & \downarrow \\ \mathcal{S}_H(p^{\mathrm{a}})_{W',(0)} & \xrightarrow{(1\otimes\sigma)\circ\Lambda_p} & \mathcal{S}_H(p^{\mathrm{a}})_{W',(0)}, \end{array} \tag{7.2.6.2}$$

and for every $g \in G$ an isomorphism of F-crystals $g^*(E, \varphi) \simeq (E, \varphi)$. Moreover, these isomorphisms are compatible with the group structure on G. We therefore obtain an action of G on $H^m((\widetilde{\mathcal{Y}}_0/\mathcal{S}_H(p^{\mathrm{a}})_{W',(0)})_{\mathrm{cris}}, E')$ compatible with the action on W'. The action of G on D^m is defined using this action and the natural action on K_0^{ur}.

We now verify that condition 0.1.1 3) holds.

Lemma 7.2.7. — *The isomorphism*

$$H^*((\widetilde{\mathcal{Y}}_0/\mathcal{S}_H(p^{\mathrm{a}})_{W',(0)})_{\mathrm{cris}}, E') \otimes_{W'} K' \simeq H^*(\mathcal{X}_{K,\mathrm{et}}, \mathcal{E}_K \otimes \Omega^\bullet_{\mathcal{X}_K/K}) \otimes_K K' \tag{7.2.7.1}$$

provided by 7.1.6 is G-equivariant.

Proof. — First note that the isomorphism

$$H^*((\widetilde{\mathcal{Y}}/\mathcal{S}_H(p^{\mathrm{a}})_{V',\pi'})_{\mathrm{cris}}, E') \otimes_{V'} \overline{K} \simeq H^*(\mathcal{X}_{K,\mathrm{et}}, \mathcal{E}_K \otimes \Omega^{\bullet}_{\mathcal{X}_K/K}) \otimes_K K' \tag{7.2.7.2}$$

is G-equivariant, where the action of G on $H^*((\widetilde{\mathcal{Y}}/\mathcal{S}_H(p^{\mathrm{a}})_{V',(\pi')})_{\mathrm{cris}}, E')$ is obtained from the action of G on $\mathcal{S}_H(p^{\mathrm{a}})_{V',\pi'}$ over $\mathcal{S}_H(\alpha)_{V,\pi}$ induced by the action on prestacks for which $g \in G$ acts by

$$(x_1, \dots, x_r, v) \longmapsto (x_1, \dots, x_r, \zeta_g^{-1} v). \tag{7.2.7.3}$$

It follows that G also acts on $Ru_* E_{\widetilde{\mathcal{Y}}/_H(p^{\mathrm{a}})_{V'/p^n!,(\pi')}}$, and to prove that (7.2.7.1) is G-equivariant it suffices to show that each of the maps

$$\epsilon_n : Ru_* E'_{\widetilde{\mathcal{Y}}_0/\mathcal{S}_H(p^{\mathrm{a}})_{W'/p^n!,(0)}} \longrightarrow Ru_* E'_{\widetilde{\mathcal{Y}}/\mathcal{S}_H(p^{\mathrm{a}})_{V'/p^n!,(\pi')}} \tag{7.2.7.4}$$

used in the proof of 7.1.6 are G-equivariant.

Write $\mathcal{S}_H(p^{\mathrm{a}})^{(n+r)}_{W',(0)}$ for the stack $\mathcal{S}_H(p^{\mathrm{a}})_{W',(0)}$ with G-action given by

$$(x_1, \dots, x_r, v) \longmapsto (x_1, \dots, x_r, \zeta_g^{-p^{n+r}} v). \tag{7.2.7.5}$$

Then the maps

$$\Lambda_{p^{r+n}} : \mathcal{S}_H(p^{\mathrm{a}})_{W'/p^n!,(0)} \longrightarrow \mathcal{S}_H(p^{\mathrm{a}})^{(n+r)}_{W'/p^n!,(0)}, \tag{7.2.7.6}$$

$$\Lambda_{p^{r+n}} : \mathcal{S}_H(p^{\mathrm{a}})_{V'/p^n!,(\pi')} \longrightarrow \mathcal{S}_H(p^{\mathrm{a}})_{V'/p^n!,(0)} \simeq \mathcal{S}_H(p^{\mathrm{a}})^{(n+r)}_{W'/p^n!,(0)} \otimes_{W'} V' \tag{7.2.7.7}$$

are G-equivariant. We thus obtain algebraic stacks with G-action

$$\overline{\mathcal{S}}^{(n+r)}_{V'/p^n!,(\pi')} := \mathrm{Spec}(\Lambda_{p^{r+n},*}(\mathcal{O}_{\mathcal{S}_H(p^{\mathrm{a}})_{V'/p^n!,(\pi')}})), \tag{7.2.7.8}$$

and

$$\overline{\mathcal{S}}^{(n+r)}_{W'/p^n!,(0)} := \mathrm{Spec}(\Lambda_{p^{r+n},*}(\mathcal{O}_{\mathcal{S}_H(p^{\mathrm{a}})_{W'/p^n!,(0)}})), \tag{7.2.7.9}$$

and it follows from the construction that the isomorphism $\overline{\mathcal{S}}^{(n+r)}_{V'/p^n!,(0)} \simeq \overline{\mathcal{S}}^{(n+r)}_{W'/p^n!,(0)} \otimes_{W'} V'$ is G-equivariant.

Writing $\overline{\overline{\mathcal{Y}}}^{(n+r)}$ (resp. $\overline{\overline{\mathcal{Y}}}_0^{(n+r)}$) as in the proof of 7.1.6, the map ϵ_n is the composite

$$\begin{aligned} Ru_* E'_{\widetilde{\mathcal{Y}}_0/\mathcal{S}_H(p^{\mathrm{a}})'_{W'/p^n!,(0)}} &\xrightarrow{\nu} Ru_* E'_{\overline{\overline{\mathcal{Y}}}_0^{(n+r)}/\overline{\mathcal{S}}^{(n+r)}_{W'/p^n!,(0)}} \\ &\longrightarrow Ru_* E'_{\overline{\overline{\mathcal{Y}}}^{(n+r)}/\overline{\mathcal{S}}^{(n+r)}_{V'/p^n!,(\pi')}} \\ &\xrightarrow{\psi} Ru_* E'_{\widetilde{\mathcal{Y}}/\mathcal{S}_H(p^{\mathrm{a}})_{V'/p^n!,(\pi')}}, \end{aligned}$$

where ν and ψ are as in 3.4.42. By the above discussion the middle arrow is G-equivariant, and by the naturality of the maps ν and ψ they are also G-equivariant. It follows that ϵ_n is also equivariant. □

Lemma 7.2.8. — *The action of G on D^m commutes with the action of Frobenius.*

Proof. — The Frobenius action is induced by the diagram

$$(7.2.8.1)\qquad \begin{array}{ccc} \widetilde{\mathcal{Y}}_0 & \xrightarrow{F_{\widetilde{\mathcal{Y}}_0}} & \widetilde{\mathcal{Y}}_0 \\ \downarrow & & \downarrow \\ \mathcal{S}_{W',(0)} & \longrightarrow & \mathcal{S}_{W',(0)}, \end{array}$$

where the bottom arrow is the composite

$$(7.2.8.2)\qquad \mathcal{S}_{W',(0)} \xrightarrow{\Lambda_p} \mathcal{S}_{W',(0)} \xrightarrow{1\otimes\sigma} \mathcal{S}_{W',(0)}.$$

Since the action of G commutes with $(1 \otimes \sigma) \circ \Lambda_p$ the result follows. □

Lemma 7.2.9. — *The monodromy operator N commutes with the G-action.*

Proof. — Recall that N is induced by the $\mathbb{G}_m$-action on $\mathcal{S}_{W',(0)}$ given by

$$(7.2.9.1)\qquad u \cdot (x_1, \dots, x_r, v) \longmapsto (x_1, \dots, x_r, uv).$$

Evidently this commutes with the G-action. We thus obtain an action of G^{op} on the whole diagram

$$(7.2.9.2)\qquad \begin{array}{ccccc} & & & & \widetilde{\mathcal{Y}}_0 \\ & & & & \downarrow \\ \mathcal{S}_{W',(0)} & \longleftarrow & \mathcal{P}^1 & \longleftarrow & \widetilde{\mathcal{Y}}_0 \\ \downarrow & & \downarrow & & \\ [\mathcal{S}_{W',(0)}/\mathbb{G}_m] & \longleftarrow & \mathcal{Z}_{W',(0)}, & & \end{array}$$

where $\mathcal{P}^1$ is as in 7.1.7. From this the result follows. □

Finally we observe:

Proof of 0.1.8. — Assuming that the (φ, N, G)-structure on $H^m_{\mathrm{dR}}(\mathcal{X})$ constructed in 7.1.3 agrees with the one constructed in 8.5 below (which is shown in 9.6), the theorem follows immediately from the above construction. Indeed the above shows that the action of Galois on D^m factors through the Galois group of the Galois closure of $K(\pi^{1/p})$ (notation as in 7.2.1). □

***Example 7.2.10* (Compare with [31, 3.3]).** — As above, let k be a perfect field of positive characteristic p, let W denote the ring of Witt vectors of k, and let K_0 denote the field of fractions of W. Let s be an integer prime to p and set

$$(7.2.10.1)\qquad X := \mathrm{Spec}(W[t]/(t^s - p)).$$

In this case we can work through the construction of the (φ, N, G)-structure on D^m explicitly.

First of all, the collection of α_i's reduces to the set $\{s\}$. Let K' be the Galois closure of $K_0[T]/(T^s - p)$, let $V' \subset K'$ be the ring of integers of K', and let $\pi' \in V'$ be the uniformizer defined by "T". The stack $\mathcal{S}_H(\alpha)_{W,(p)}$ is equal to

$$\big[\mathrm{Spec}(W[x, v^{\pm}]/(x^s v - p))/\mathbb{G}_m\big], \tag{7.2.10.2}$$

where $u \in \mathbb{G}_m$ acts by $x \mapsto ux$ and $v \mapsto u^{-s}v$. Since the map $\mathbb{G}_m \to \mathbb{G}_m$ sending u to u^{-s} is surjective, this stack can also be written as (this is a special case of 6.1.17 (i))

$$\big[\mathrm{Spec}(W[x]/(x^s - p))/\mu_s\big] = [X/\mu_s]. \tag{7.2.10.3}$$

With this identification the map $X \to \mathcal{S}_H(\alpha)_{W,(p)}$ is simply the quotient map.

We also have (recall that $r = 1$, $a = a_1 = 0$, and $H = \{1\}$)

$$\mathcal{S}_H(p^{\mathrm{a}})_{V',(\pi')} \simeq \big[\mathrm{Spec}(V'[z, w^{\pm}]/zw = \pi')/\mathbb{G}_m\big], \tag{7.2.10.4}$$

where $u \in \mathbb{G}_m$ acts by $z \mapsto uz$ and $w \mapsto u^{-1}w$. Making the change of variables $z' := zv$ we see that

$$\mathcal{S}_H(p^{\mathrm{a}})_{V',(\pi')} \simeq \big(\mathrm{Spec}(V'[z']/(z' - p))\big) \times [\mathbb{G}_m/\mathbb{G}_m] \simeq \mathrm{Spec}(V'). \tag{7.2.10.5}$$

The map $B : \mathrm{Spec}(V') \to [X/\mu_s]$ is simply the composite map

$$\mathrm{Spec}(V') \longrightarrow X \xrightarrow{\text{projection}} [X/\mu_s]. \tag{7.2.10.6}$$

To compute D^m, observe that there is a cartesian diagram

$$\begin{array}{ccc} X \times \mu_s & \xrightarrow{\mathrm{pr}_1} & X \\ \downarrow{\scriptstyle \rho} & & \downarrow \\ X & \longrightarrow & [X/\mu_s], \end{array} \tag{7.2.10.7}$$

where ρ denotes the action of μ_s. It follows that we have a commutative diagram with cartesian squares

$$\begin{array}{ccc} \mathrm{Spec}(V') \times \mu_s & \xrightarrow{\mathrm{pr}_1} & \mathrm{Spec}(V') \simeq \overline{\mathcal{S}}_H(p^{\mathrm{a}})_{V',(\pi')} \\ \downarrow & & \downarrow \\ X \times \mu_s & \xrightarrow{\mathrm{pr}_1} & X \\ \downarrow{\scriptstyle \rho} & & \downarrow \\ X & \longrightarrow & \mathcal{S}_H(\alpha)_{W,(p)}. \end{array} \tag{7.2.10.8}$$

We conclude that $\widetilde{\mathcal{X}}$ in (7.2.4.1) is isomorphic to $\mathrm{Spec}(V') \times \mu_s$.

Let G denote the Galois group of K'/K_0 and let $\zeta : G \to \mu_s$ be the homomorphism defined in 7.2.1. To determine the action of $g \in G$ on $\widetilde{\mathfrak{X}} = \mathrm{Spec}(V') \times \mu_s$, let $\chi_g : \mu_s \to \mu_s$ be the map such that the map of rings

$$g^* : \prod_{\eta \in \mu_s} V' \longrightarrow \prod_{\eta \in \mu_s} V' \tag{7.2.10.9}$$

sends the η-component of $\prod_{\eta\in\mu_s} V'$ to the $\chi_g(\eta)$-component. Then $g^*((x_\eta)_{\eta\in\mu_s})$ is equal to the element of $\prod_{\eta\in\mu_s} V'$ whose $\chi_g(\eta)$-component is $g(x_\eta)$. On the other hand, for every $g \in G$ the diagram

$$\begin{array}{ccc} \mathrm{Spec}(V') \times \mu_s & \xrightarrow{\;g\;} & \mathrm{Spec}(V') \times \mu_s \\ & {\scriptstyle\rho}\searrow \quad \swarrow{\scriptstyle\rho} & \\ & X & \end{array} \tag{7.2.10.10}$$

commutes, which implies that the diagram of rings

$$\begin{array}{ccc} \prod_{\eta\in\mu_s} V' & \xleftarrow{\;g^*\;} & \prod_{\eta\in\mu_s} V' \\ {\scriptstyle x\mapsto(\eta(x))_\eta}\nwarrow & & \nearrow{\scriptstyle x\mapsto(\eta(x))_\eta} \\ & V & \end{array} \tag{7.2.10.11}$$

commutes. It follows that

$$g(\eta(x)) = \chi_g(\eta)(x). \tag{7.2.10.12}$$

Therefore $\chi_g(\eta) = \zeta_g \cdot \eta$. In other words, G acts on $\mathrm{Spec}(V') \times \mu_s$ through the product of the natural action on $\mathrm{Spec}(V')$ and the action on μ_s given by the character ζ.

Let k' denote the residue field of V', let W' denote the ring of Witt vectors of k', and let K_0' denote the field of fractions of W'. Let $\varphi : K_0' \to K_0'$ denote the endomorphism defined by the canonical lifting of Frobenius to W'. Then by the definition of D^m in (7.2.5.2), we have

$$D^m = H^m((\widetilde{\mathfrak{X}} \otimes_{V'} k'/W')_{\mathrm{cris}}, \mathcal{O}) \otimes_{W'} K_0^{ur} \tag{7.2.10.13}$$

which is zero for $m > 0$ and

$$D^0 = \prod_{\eta\in\mu_s} K_0^{ur} \tag{7.2.10.14}$$

The action of Frobenius is given by the component-wise action on K_0^{ur}, the monodromy operator is trivial, and the action of an element $g \in G_{\overline{K}/K_0}$ is given by sending an element $(x_\eta) \in \prod_{\eta\in\mu_s} K_0^{ur}$ to the element of $\prod_{\eta\in\mu_s} K_0^{ur}$ with $\zeta_g \cdot \eta$-component $g(x_\eta)$, where $\overline{K}$ is an algebraic closure of K_0 and we abusively write also ζ for the character

$$G_{\overline{K}/K_0} \xrightarrow{\text{restriction}} G_{K'/K_0} \xrightarrow{\;\zeta\;} \mu_s. \tag{7.2.10.15}$$

Note that if K_0 contains the s-th roots of unity, then $\zeta : G_{K'/K_0} \to \mu_s$ is an isomorphism, and the action G_{K'/K_0} on D^0 is faithful (*i.e.*, the bound $K_0(p^{1/s})$ on the Galois action on D^0 is sharp).

Example 7.2.11. — Let p be a prime, and let $M_{\Gamma_1(p),\mathbb{Q}_p}$ denote the stack over $\mathbb{Q}_p$ whose fiber over a scheme S is the groupoid of pairs (E, P), where E/S is a generalized elliptic curve in the sense of [**12**, II.1.12] and $P \in E^{\mathrm{sm}}(S)$ is a section in the smooth locus of $E \to S$, such that for every geometric point $\bar{s} \to S$ the subgroup $\langle P_{\bar{s}} \rangle \subset E^{\mathrm{sm}}_{\bar{s}}$ generated by $P_{\bar{s}}$ meets every irreducible component of $E_{\bar{s}}$. By [**12**, IV.3.4] the stack $M_{\Gamma_1(p),\mathbb{Q}_p}$ is a smooth and proper Deligne-Mumford stack over $\mathbb{Q}_p$. Let (D, φ, N) denote the (φ, N, G)-module associated to the p-adic étale cohomology $H^m(M_{\Gamma_1(p),\overline{\mathbb{Q}}_p}, \mathbb{Q}_p)$, where G denotes $\mathrm{Gal}_{\overline{\mathbb{Q}}_p/\mathbb{Q}_p}$. Define

$$K' = \mathbb{Q}_p(p^{1/(p-1)}), \tag{7.2.11.1}$$

and let $G_{K'} \subset G$ denote $\mathrm{Gal}(\overline{\mathbb{Q}}_p/K')$.

Theorem 7.2.12. — *The natural map*

$$D^{G_{K'}} \otimes_{K'_0} K_0^{ur} \longrightarrow D \tag{7.2.12.1}$$

*is an isomorphism (*i.e., *the potentially semistable representation $H^m(M_{\Gamma_1(p),\overline{\mathbb{Q}}_p}, \mathbb{Q}_p)$ becomes semistable over K').*

Proof. — Let $M_{\Gamma_1(p),\mathbb{Z}_p}$ denote the normalization of $M_{1,1,\mathbb{Z}_p}$ in $M_{\Gamma_1(p),\mathbb{Q}_p}$, where $M_{1,1,\mathbb{Z}_p}$ denotes the proper smooth $\mathbb{Z}_p$-stack classifying generalized elliptic curves with no level structure, and let $\bar{s} : \mathrm{Spec}(\overline{\mathbb{F}}_p) \to M_{\Gamma_1(p),\mathbb{Z}_p}$ be a geometric point. By the same argument proving [**12**, V.2.8] using the description of finite flat group schemes of rank p in [**12**, V.2.4] one obtains that the local ring $\widehat{\mathcal{O}}_{M_{\Gamma_1(p),\mathbb{Z}_p},\bar{s}}$ can be described as follows:

$$\widehat{\mathcal{O}}_{M_{\Gamma_1(p),\mathbb{Z}_p},\bar{s}} \simeq \begin{cases} \begin{matrix} W(\overline{\mathbb{F}}_p)[[s]] \\ \text{or} \\ W(\overline{\mathbb{F}}_p)[[z]][s]/(s^{p-1} - p) \end{matrix} & \text{if } \bar{s} \text{ is ordinary} \\ W(\overline{\mathbb{F}}_p)[[x, y]]/(x^{p-1}y - p) & \text{if } \bar{s} \text{ is supersingular.} \end{cases} \tag{7.2.12.2}$$

In case $p = 2$ this shows that $M_{\Gamma_1(2),\mathbb{Z}_p}$ has semistable reduction, and if $p \neq 2$ this shows that the (reduced) irreducible components of the closed fiber are smooth. Theorem 7.2.12 therefore follows from 0.1.8. □

Example 7.2.13. — Let X/V and $\{\alpha_1, \ldots, \alpha_r\}$ be as in 6.2.1. Let f be the maximal integer such that $p^f | \alpha_i$ for all i, write $\alpha_i = p^f \gamma_i$, and set $\gamma := (\gamma_1, \ldots, \gamma_r)$. Assume that the following two conditions hold:

1. The group $\mu_{p^f}(K)$ has order p^f (*i.e.*, K contains all p^f-th roots of unity).

2. For any geometric point $\bar{x} \to X$ with image in the closed fiber, there exist an étale neighborhood U of $\bar{x}$ and an étale morphism

$$U \longrightarrow \operatorname{Spec}\big(V[X_1, \dots, X_n]/(X_1^{\alpha_1} \cdots X_r^{\alpha_r} = \pi)\big), \tag{7.2.13.1}$$

for some $n \geq r$.

Fix an algebraic closure $K \hookrightarrow \overline{K}$ with Galois group G_K, and let H denote the G_K-representation $H^*(X_{\overline{K},\mathrm{et}}, \mathbb{Q}_p)$. Let D denote $D_{\mathrm{pst}}(H)$ (0.1.1.6).

Let $\pi' \subset \overline{K}$ be an element with $(\pi')^{p^f} = \pi$, and let K' denote $K(\pi') \subset \overline{K}$. Let $V' \subset K'$ be the ring of integers. Note that the map $V \to V'$ induces an isomorphism on residue fields. Let $K_0 \subset K$ denote the field of fractions of the ring of Witt vectors of the residue field of K, and let $K_0^{ur} \subset \overline{K}$ denote the maximal unramified extension of K_0 in $\overline{K}$. Denote by $G_{K'}$ the Galois group $\mathrm{Gal}(\overline{K}/K')$.

Theorem 7.2.14. — *The action of $G_{K'}$ on D (obtained by restricting the G_K-action) factors through a tame quotient of $G_{K'}$.*

Proof. — Let $\mathcal{S}(\alpha)$ be as in 7.2.1, and let $\mathcal{S}(\alpha)[p^f] \to \mathcal{S}(\alpha)$ denote the μ_{p^f}-torsor defined in 6.1.15. Let $Y \to X$ denote the fiber product of the diagram

$$\begin{array}{ccc} & & \mathcal{S}(\alpha)[p^f] \\ & & \downarrow \\ X & \longrightarrow & \mathcal{S}(\alpha), \end{array} \tag{7.2.14.1}$$

where the horizontal arrow is the map defined in 6.2.1. Then Y is a μ_{p^f}-torsor over X, and in particular Y_K is a $\mu_{p^f}(K)$-torsor over X_K. It follows that H is a direct summand (in the category of G_K-representations) of the G_K-representation $W := H^*(Y_{\overline{K},\mathrm{et}}, \mathbb{Q}_p)$. In fact, a retraction $W \to H$ is given by the operator

$$\frac{1}{p^f} \sum_{\zeta \in \mu_e(K)} \zeta : W \longrightarrow W \tag{7.2.14.2}$$

defined by the $\mu_{p^f}(K)$-action on W. It follows that if E denotes $D_{\mathrm{pst}}(W)$ then it suffices to show that the action of $G_{K'}$ on E factors through a tame quotient.

If $\bar{x} \to X$ is a geometric point in the closed fiber, and $U \to X$ is an étale neighborhood of $\bar{x}$ with a morphism (7.2.13.1), then as discussed in 6.1.15, the torsor $Y \times_X U \to U$ is the μ_{p^f}-torsor of p^f-th roots of $1 \in \mathcal{O}_U$. In particular, the μ_{p^f}-torsor $Y \to X$ is étale locally on X trivial.

Let P_Y denote the fppf-sheaf on X corresponding to Y. Then if $\epsilon : X_{\mathrm{fppf}} \to X_{\mathrm{et}}$ is the natural morphism of topoi, we deduce that $\epsilon_* P_Y$ is a $\epsilon_* \mu_{p^f} = \mu_{p^f}(K)$ torsor. Let $Y' \to X$ denote the corresponding scheme with $\mu_{p^f}(K)$-action. The map of sheaves $\epsilon^* \epsilon_* P_Y \to P_Y$ induces a morphism of schemes $Y' \to Y$ over X, which is an isomorphism over K.

Lemma 7.2.15. — *The composite map $Y' \to Y \to \mathcal{S}(\alpha)[p^f]_{V,\pi}$ is smooth.*

Proof. — The assertion is étale local on Y, so we may assume that there exists a morphism (7.2.13.1). Then as in the proof of 6.1.17 we have

$$Y = \operatorname{Spec}\big(V[X_1, \ldots, X_n][z^{\pm}]/(X_1^{\alpha_1} \cdots X_r^{\alpha_r} - \pi, z^{p^f} - 1)\big), \tag{7.2.15.1}$$

from which it follows that

$$Y' = \coprod_{\zeta \in \mu_{p^f}(K)} \operatorname{Spec}(V[X_1, \ldots, X_n]/(X_1^{\alpha_1} \cdots X_r^{\alpha_r} - \pi)). \tag{7.2.15.2}$$

The result therefore follows from 6.1.17. □

Let $\delta : \mu_{p^f}(K) \to \mu_{p^f}$ be the natural map (where $\mu_{p^f}(K)$ is viewed as a constant sheaf), and let

$$\Psi_{p^f} : \mu_{p^f} \times \mathcal{S}(\gamma)_{V', \pi'} \to \mathcal{S}(\alpha)[p^f]_{V', \pi} \tag{7.2.15.3}$$

be the map defined in 6.3.27. Let $Y'' \to Y'$ be the fiber product of the diagram

$$\begin{array}{ccc} & & \mu_{p^f}(K) \times \mathcal{S}(\gamma)_{V', \pi'} \\ & & \downarrow \delta \times \mathrm{id} \\ & & \mu_{p^f} \times \mathcal{S}(\gamma)_{V', \pi'} \\ & & \downarrow \Psi_{p^f} \\ Y' & \longrightarrow & \mathcal{S}(\alpha)[p^f]_{V', \pi}. \end{array} \tag{7.2.15.4}$$

Then Y''/V' is proper with generic fiber $Y_{K'}$, and Y'' admits a smooth morphism to $\mathcal{S}(\gamma)_{V', \pi'}$. This implies that $Y_{K'}$ has log smooth reduction, so 7.2.14 now follows from 0.1.7. □

Example 7.2.16. — A special case of 7.2.13 is obtained by taking

$$X = \operatorname{Spec}(V[T]/(T^p - \pi)). \tag{7.2.16.1}$$

Let K_X denote the field of fractions of $V[T]/(T^p - \pi)$. The group $H^*(X_{\overline{K}}, \mathbb{Q}_p)$ is then canonically isomorphic to

$$\prod_{\sigma : K_X \hookrightarrow \overline{K}} \mathbb{Q}_p \simeq \operatorname{Hom}(S_{K_X}, \mathbb{Q}_p), \tag{7.2.16.2}$$

where S_{K_X} denotes the set of embeddings $K_X \hookrightarrow \overline{K}$ compatible with the given embedding $K \hookrightarrow \overline{K}$. The action of $g \in G_K$ is given by sending a function $F : S_{K_X} \to \mathbb{Q}_p$ to the function

$$(\sigma : K_X \hookrightarrow \overline{K}) \longmapsto F(g\sigma). \tag{7.2.16.3}$$

Thus the associated (φ, N, G_K)-module is equal to

$$\operatorname{Hom}(S_{K_X}, K_0^{ur}), \tag{7.2.16.4}$$

with G_K-action given also by the formula (7.2.16.3), replacing $\mathbb{Q}_p$ by K_0^{ur}.

This description of D coincides with that obtained from the construction in the proof of 7.2.13. Indeed in this case the μ_p-torsor $Y \to X$ is trivial, and therefore the $\mu_p(K)$-torsor $Y' \to X$ is also trivial. It follows that the G_K-representation W arising in the proof of 7.2.13 is simply

$$\operatorname{Hom}(S_{K_X}, \mathbb{Q}_p) \otimes_{\mathbb{Q}_p} \operatorname{Hom}(\mu_p(K), \mathbb{Q}_p), \tag{7.2.16.5}$$

with G_K-action given by (7.2.16.3) on the first factor and the trivial action on the second factor, and $\mu_p(K)$-action (induced by the torsor structure) given by the trivial action on the first factor and the standard action on the second factor.

CHAPTER 8

COMPARISON WITH SYNTOMIC COHOMOLOGY

For the convenience of the reader who wishes to compare this chapter to [**41**], we deviate in this chapter from the notation in the rest of the text and use the letter A to denote a p-adically complete discrete valuation ring instead of V.

8.1. Syntomic morphisms of algebraic stacks

Recall [**25**, II.1.1] that a morphism $f : X \to Y$ of schemes is *syntomic* if it is flat and locally a complete intersection (for the definition of a local complete intersection morphism in the non-noetherian setting see [**6**, VIII.1.1], where the terminology "complete intersection morphism" is used).

Lemma 8.1.1

(i) *If $f : X \to Y$ and $g : Y \to Z$ are syntomic morphisms, then $g \circ f : X \to Z$ is syntomic.*

(ii) *Let $f : X \to Y$ be a morphism of schemes, and $Y' \to Y$ a quasi-compact, flat, and surjective morphism. Then f is syntomic if and only if the base change $f' : X \times_Y Y' \to Y'$ is syntomic.*

Proof. — Statement (i) follows from [**6**, VIII.1.5] and (ii) follows from [**6**, VIII.1.6]. □

Lemma 8.1.2. — *A locally of finite type morphism $f : X \to Y$ between locally noetherian schemes is syntomic if and only if f is flat and the cotangent complex $L_{X/Y}$ has perfect amplitude in* $[-1, 0]$ [**6**, I 4.7 and 4.8].

Proof. — This follows from [**32**, III.3.2.6]. □

Remark 8.1.3. — Let S be a locally noetherian scheme and $a \leq b$ integers. By [**6**, I.5.8], if $E \in D^b_{\mathrm{coh}}(S)$ is a bounded complex with coherent cohomology sheaves, then E has perfect amplitude in $[a, b]$ if and only if for every $\mathcal{O}_S$-module M the sheaf $\mathcal{H}^i(M \otimes^{\mathbb{L}}_{\mathcal{O}_S} E)$ is zero for $i \notin [a, b]$. From this it follows that the condition that E has perfect amplitude in $[a, b]$ is local for the flat topology on S.

Lemma 8.1.4. — *Consider a commutative diagram of schemes*

(8.1.4.1)

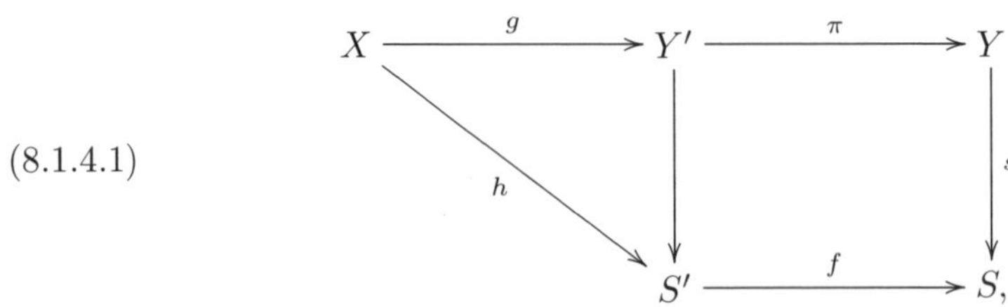

where f and g are smooth and surjective, and the square is cartesian. Then the morphism s is syntomic if and only if the morphism h is syntomic.

Proof. — If s is syntomic, then by 8.1.1 (ii) the morphism $Y' \to S'$ is syntomic, and hence by 8.1.1 (i) the composite $X \to Y' \to S'$ is also syntomic.

Conversely, assume the morphism h is syntomic. To prove that s is syntomic, it suffices by 8.1.1 (ii) to prove that the morphism $Y' \to S'$ is syntomic. For this we show that $L_{Y'/S'}$ has perfect amplitude in $[-1, 0]$. For this note that it suffices by 8.1.3 to show that $g^*L_{Y'/S'}$ has perfect amplitude in $[-1, 0]$. Consider the distinguished triangle

$$g^*L_{Y'/S'} \longrightarrow L_{X/S'} \longrightarrow L_{X/Y'} \longrightarrow g^*L_{Y'/S'}[1]. \tag{8.1.4.2}$$

Since $L_{X/Y'} \simeq \Omega^1_{X/Y'}$ (since $X \to Y'$ is smooth) and $L_{X/S'}$ has perfect amplitude in $[0, -1]$ because h is syntomic, this implies the lemma. □

8.1.5. — By [**49**, p. 33], it follows that there is a well-defined notion of a syntomic morphism locally of finite type between locally noetherian algebraic stacks. A locally of finite type morphism $f : \mathcal{X} \to \mathcal{Y}$ between locally noetherian algebraic stacks is syntomic if there exists a commutative diagram

(8.1.5.1)

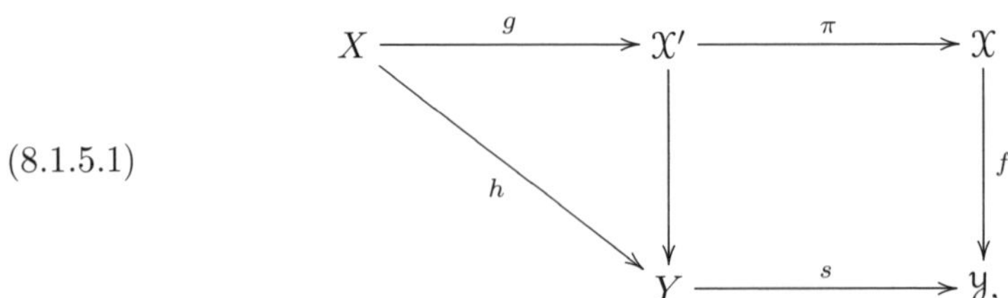

with s, f, and g smooth and surjective, X and Y schemes, and h a syntomic morphism of schemes.

Proposition 8.1.6. — *Let $f : \mathcal{X} \to \mathcal{Y}$ be a morphism locally of finite type between locally noetherian algebraic stacks. Assume further that $\mathcal{X}$ is an algebraic space. Then f is syntomic if and only if f is flat and there exists étale locally on $\mathcal{X}$ a regular embedding $j : \mathcal{X} \hookrightarrow P$ over $\mathcal{Y}$ with $P \to \mathcal{Y}$ smooth.*

Proof. — For the "if" direction assume that étale locally there exists such an immersion j. Since the property of f being syntomic is étale local on $\mathcal{X}$, we may assume that $\mathcal{X}$ is a scheme and that such an embedding $j : \mathcal{X} \hookrightarrow P$ is defined globally. Then for any smooth surjection $Y \to \mathcal{Y}$ with Y a scheme, we obtain by base change a regular embedding $\mathcal{X} \times_{\mathcal{Y}} Y \hookrightarrow P \times_{\mathcal{Y}} Y$ over Y. It follows that $\mathcal{X} \times_{\mathcal{Y}} Y \to Y$ is syntomic, and hence f is syntomic as well (by the definition of a syntomic morphism of algebraic stacks).

For the "only if" direction, assume f is syntomic and fix first any smooth cover $Y \to \mathcal{Y}$. By the existence of quasi-sections for smooth morphisms [**15**, IV.17.16.3 (ii)], there exists after replacing $\mathcal{X}$ by an étale cover a section $s : \mathcal{X} \to Y \times_{\mathcal{Y}} \mathcal{X}$. Hence after shrinking on $\mathcal{X}$ some more, we may assume that there exists an immersion $\mathcal{X} \hookrightarrow \mathbb{A}^n_Y$ over Y. This immersion is in fact a regular immersion since this can be verified after making a smooth base change $V \to \mathcal{Y}$ in which case it follows from [**6**, VIII.1.2]. □

The key property of syntomic morphisms that we use is the following:

Proposition 8.1.7. — *Let $\mathcal{S}_0 \hookrightarrow \mathcal{S}$ be a closed immersion of algebraic stacks defined by a PD-ideal, and let $\mathcal{X} \to \mathcal{S}_0$ be a syntomic morphism of algebraic stacks. Then for any closed immersion $j : \mathcal{X} \hookrightarrow \mathcal{Y}$ into a smooth $\mathcal{S}$-stack $\mathcal{Y}$, the divided power envelope D of $\mathcal{X}$ in $\mathcal{Y}$ is flat over $\mathcal{S}$.*

Proof. — It suffices to verify this after making a smooth base change $S \to \mathcal{S}$ with S a scheme. Hence we may assume that $\mathcal{S}$ is a scheme. Furthermore, we can replace $\mathcal{Y}$ by a smooth cover $Y \to \mathcal{Y}$ with Y a scheme. Hence we may also assume that $\mathcal{X}$ and $\mathcal{Y}$ are schemes in which case the result follows from [**7**, I.3.4.4]. □

The base change theorem 2.6.2 can be generalized to a result for syntomic morphisms as follows. As in 2.6.1, let $u : (B', I', \gamma') \to (B, I, \gamma)$ be a morphism of PD-algebraic spaces, and let $\mathcal{S}/B$ be an algebraic stack which we assume flat over B. Set $\mathcal{S}' := \mathcal{S} \times_B B'$, and note that γ (resp. γ') extends to $\mathcal{S}$ (resp. $\mathcal{S}'$) since $\mathcal{S}/B$ and $\mathcal{S}'/B'$ are flat (1.1.12). Let $B'_0 \subset B'$ and $B_0 \subset B$ denote closed subspaces defined by sub-PD-ideals such that the composite

$$B'_0 \longrightarrow B' \longrightarrow B \tag{8.1.7.1}$$

factors through B_0. Set $\mathcal{S}_0$ denote $\mathcal{S} \times_B B_0$ and let $\mathcal{S}'_0$ denote $\mathcal{S}' \times_{B'} B'_0$.

Theorem 8.1.8. — *Let $f : \mathcal{X} \to \mathcal{S}_0$ be a syntomic morphism of algebraic stacks with $\mathcal{X}$ a tame noetherian Deligne-Mumford stack (2.5.14), and let $f' : \mathcal{X}' \to \mathcal{S}'_0$ be the base change to $\mathcal{S}'_0$. Then there is a natural isomorphism in the derived category of sheaves of $\mathcal{O}_{B'_{\mathrm{et}}}$-modules*

$$Lu^* Rh_{\mathcal{X}/B*}\mathcal{O}_{\mathcal{X}_{\mathrm{et}}/\mathcal{S}} \simeq Rh_{\mathcal{X}'/B'*}\mathcal{O}_{\mathcal{X}'_{\mathrm{et}}/\mathcal{S}'}. \tag{8.1.8.1}$$

Proof. — By the same arguments used in the proof of 2.6.2, one is reduced to the case when $\mathcal{X}$, B, and B' are affine schemes, and there exists a closed immersion $\mathcal{X} \hookrightarrow \mathcal{Y}$ over $\mathcal{S}$ with $\mathcal{Y}/\mathcal{S}$ smooth. Denote by $\mathcal{X}' \hookrightarrow \mathcal{Y}'$ the closed immersion over $\mathcal{S}'$ obtained by base change, and let D (resp. D') denote the divided power envelope of $\mathcal{X}$ in $\mathcal{Y}$ (resp. $\mathcal{X}'$ in $\mathcal{Y}'$) and let $\mathcal{D}$ (resp. $\mathcal{D}'$) denote the coordinate ring of D (resp. D').

By 8.1.7, the scheme D is flat over $\mathcal{S}$. In particular, there is an isomorphism

$$Lu^* Rh_{\mathcal{X}/B*}\mathcal{O}_{\mathcal{X}_{\mathrm{et}}/\mathcal{S}} \simeq \mathcal{O}_{B'} \otimes_{\mathcal{O}_B} (\mathcal{D} \otimes \Omega^\bullet_{\mathcal{Y}/\mathcal{S}}), \tag{8.1.8.2}$$

and the arrow in question becomes identified with the natural map

$$\mathcal{O}_{B'} \otimes_{\mathcal{O}_B} (\mathcal{D} \otimes \Omega^\bullet_{\mathcal{Y}/\mathcal{S}}) \longrightarrow \mathcal{D}' \otimes \Omega^\bullet_{\mathcal{Y}'/\mathcal{S}'}. \tag{8.1.8.3}$$

Thus to prove the theorem it suffices to show that the natural map $\mathcal{O}_{B'} \otimes_{\mathcal{O}_B} \mathcal{D} \to \mathcal{D}'$ is an isomorphism. This follows from [**7**, I.2.8.2]. □

8.2. The rings B_{cris}, B_{dR}, and B_{st} of Fontaine

8.2.1. — Let A be a complete discrete valuation ring of mixed characteristic $(0,p)$, K the field of fractions of A, and k the residue field which is assumed perfect. Let W be the ring of Witt vectors of k and set $W_n := W \otimes \mathbb{Z}/p^n$. Fix an algebraic closure $K \subset \overline{K}$, and let $\overline{A}$ denote the integral closure of A in $\overline{K}$. Define

$$A_n := A \otimes \mathbb{Z}/p^n, \quad \overline{A}_n := \overline{A} \otimes \mathbb{Z}/p^n, \tag{8.2.1.1}$$

$$S := \mathrm{Spec}(A), \quad S_n := \mathrm{Spec}(A_n), \quad \overline{S} := \mathrm{Spec}(\overline{A}), \quad \overline{S}_n := \mathrm{Spec}(\overline{A}_n). \tag{8.2.1.2}$$

8.2.2. — Define

(8.2.2.1)
$$B_n := \Gamma((\overline{S}_n/W_n)_{\mathrm{cris}}, \mathcal{O}_{\overline{S}_n/W_n}) = \varinjlim_{A\subset A'\subset\overline{A}} \Gamma((\mathrm{Spec}(A'_n)/W_n)_{\mathrm{cris}}, \mathcal{O}_{\mathrm{Spec}(A'_n)/W_n}),$$

where the limit is taken over subalgebras $A' \subset \overline{A}$ finitely generated over A.

As explained in [**73**, A.1.1] the canonical map $B_n \to \overline{A}_n$ is surjective and the kernel J_n is a PD-ideal.

8.2.3. — For a sequence $s = (s_n)_{n\geq 0}$ of elements in $\overline{A}_n$ with $s^p_{n+1} = s_n$ for all n, define

$$\epsilon(s) := ((\tilde{s}_n)^{p^n})_n \in \varprojlim_n B_n, \tag{8.2.3.1}$$

where $\tilde{s}_n \in B_n$ is any lifting of s_n. If $\tilde{s}'_n = \tilde{s}_n + h$ is a second choice of lifting of s_n then

$$\tilde{s}'^{p^n}_n = \sum_{i=0}^n \binom{p^n}{i} \tilde{s}^{p^n-i}_n h^i = \tilde{s}^{p^n} + \sum_{i=1}^n \frac{p^n!}{(p^n-i)!} \tilde{s}^{p^n-i}_n h^{[i]} = \tilde{s}^{p^n}, \tag{8.2.3.2}$$

so $\epsilon(s)$ is independent of the choices.

For a sequence $s = (s_n) \in \varprojlim \mu_{p^n}(\overline{K}) = \mathbb{Z}_p(1)$ the image of $\epsilon(s)$ in $\varprojlim \overline{A}_n$ is zero, and hence we obtain a map

$$\log \epsilon : \mathbb{Z}_p(1) \longrightarrow \varprojlim_n B_n, \quad \log(\epsilon(s)) = \sum_{n\geq 1} (-1)^{n+1}(\epsilon(s)-1)^n/n. \tag{8.2.3.3}$$

which is injective [**23**, 1.5.4].

Set

$$B^+_{\mathrm{cris}} := \mathbb{Q} \otimes \varprojlim_n B_n, \quad B_{\mathrm{cris}} := B^+_{\mathrm{cris}}[t^{-1}], \tag{8.2.3.4}$$

$$B^+_{\mathrm{dR}} := \varprojlim_r (\mathbb{Q} \otimes \varprojlim_n B_n/J^{[r]}), \quad B_{\mathrm{dR}} := B^+_{\mathrm{dR}}[t^{-1}], \tag{8.2.3.5}$$

where t is any element generating $\mathbb{Q}_p(1) \subset B^+_{\mathrm{cris}}$.

8.2.4. — Fix a uniformizer $\pi \in A$. For $s = (s_n)$ a sequence of elements of $\overline{A}$ such that $s_0 = \pi$, $s^p_{n+1} = s_n$ we have

$$\epsilon(s)\pi^{-1} \in \mathrm{Ker}\big((B^+_{\mathrm{dR}})^* \longrightarrow \mathbb{C}^*_p = (\mathbb{Q} \otimes \varprojlim_n \overline{A}_n)^*\big), \tag{8.2.4.1}$$

and so $u_s := \log(\epsilon(s)\pi^{-1}) \in B^+_{\mathrm{dR}}$ is defined.

Define

$$B^+_{\mathrm{st}} := B^+_{\mathrm{cris}}[u_s] \subset B_{\mathrm{dR}}, \quad B_{\mathrm{st}} := B_{\mathrm{cris}}[u_s] \subset B_{\mathrm{dR}}. \tag{8.2.4.2}$$

As explained in [**73**, 4.1] this ring does not depend on the choice of the sequence $\{s_n\}$, and the element u_s is transcendental over B_{cris}.

In particular, if $\varphi : B_{\mathrm{cris}} \to B_{\mathrm{cris}}$ denotes the Frobenius endomorphism induced by the natural Frobenius endomorphisms of the B_n, we can extend φ to B_{st} by declaring $\varphi(u_s) := pu_s$. Furthermore, we define a B_{cris}-linear operator $N : B_{\mathrm{st}} \to B_{\mathrm{st}}$ by $N(u^i_s) := iu^{i-1}_s$. The resulting data $(B_{\mathrm{st}}, \varphi, N)$ depends only on the uniformizer π and not on the choice of the sequence s used in the construction.

Remark 8.2.5. — As discussed in [**22**, 2.8], the ring B_{dR} is a complete discrete valuation field with valuation ring B^+_{dR} and uniformizer t. This discrete valuation ν defines a filtration $\mathrm{Fil}_{B_{\mathrm{dR}}}$ on B_{dR} by setting

$$\mathrm{Fil}^i_{B_{\mathrm{dR}}} := \{x \in B_{\mathrm{dR}} | \nu(x) \geq i\}, \tag{8.2.5.1}$$

where ν is normalized by the condition that $\nu(B^*_{\mathrm{dR}}) = \mathbb{Z}$. This filtration induces a filtration $\mathrm{Fil}_{B_{\mathrm{st}}}$ on B_{st} by setting

$$\mathrm{Fil}^i_{B_{\mathrm{st}}} := B_{\mathrm{st}} \cap \mathrm{Fil}^i_{B_{\mathrm{dR}}}. \tag{8.2.5.2}$$

8.2.6. — Following [**41**] we now give a crystalline interpretation of the triple $(B_{\mathrm{st}}, \varphi, N)$. The main result is 8.2.30 below, but in preparation for this theorem we need several auxiliary results.

For any integer $e \geq 1$, let $\theta_e : [\mathbb{A}^1/\mathbb{G}_m] \to [\mathbb{A}^1/\mathbb{G}_m]$ be the map induced by the maps

$$\mathbb{A}^1 \xrightarrow{f \mapsto f^e} \mathbb{A}^1 \quad \mathbb{G}_m \xrightarrow{u \mapsto u^e} \mathbb{G}_m. \tag{8.2.6.1}$$

Observe that this map θ_e is flat. For an integer r, let $\Delta^{(r)}$ denote the divided power envelope (1.2.3) of the closed immersion

$$[\mathrm{Spec}(W[t]/t^r)/\mathbb{G}_m] \hookrightarrow [\mathrm{Spec}(W[t])/\mathbb{G}_m], \tag{8.2.6.2}$$

and let $\Delta_n^{(r)}$ denote the divided power envelope of

$$[\mathrm{Spec}(W_n[t]/t^r)/\mathbb{G}_m] \hookrightarrow [\mathrm{Spec}(W_n[t])/\mathbb{G}_m]. \tag{8.2.6.3}$$

The maps θ_e induce maps, denoted by the same letter,

$$\theta_e : \Delta^{(re)} \longrightarrow \Delta^{(r)}, \quad \theta_e : \Delta_n^{(re)} \longrightarrow \Delta_n^{(r)}. \tag{8.2.6.4}$$

Observe that the diagram

$$\begin{array}{ccc} [\mathrm{Spec}(W_n[t]/t^{re})/\mathbb{G}_m] & \longrightarrow & [\mathrm{Spec}(W_n[t])/\mathbb{G}_m] \\ {\scriptstyle\theta_e}\downarrow & & \downarrow{\scriptstyle\theta_e} \\ [\mathrm{Spec}(W_n[t]/t^{r})/\mathbb{G}_m] & \longrightarrow & [\mathrm{Spec}(W_n[t])/\mathbb{G}_m] \end{array} \tag{8.2.6.5}$$

is cartesian. Since the formation of divided power envelopes commutes with flat base change (1.2.3), it follows that the diagram

$$\begin{array}{ccc} \Delta_n^{(re)} & \longrightarrow & [\mathbb{A}^1/\mathbb{G}_m] \\ {\scriptstyle\theta_e}\downarrow & & \downarrow{\scriptstyle\theta_e} \\ \Delta_n^{(r)} & \longrightarrow & [\mathbb{A}^1/\mathbb{G}_m] \end{array} \tag{8.2.6.6}$$

is also cartesian.

For a finite algebraic extension $K \subset K' \subset \overline{K}$, let $e_{K'}$ denote the absolute ramification index of K'.

If $K \subset K' \subset \overline{K}$ is a finite algebraic extension with ring of integers A' and ramification index $e_{K'/K}$ (also sometimes written $e_{A'/A}$), then there is a canonical 2-commutative diagram

$$\begin{array}{ccc} \mathrm{Spec}(A') & \xrightarrow{\mathfrak{m}_{A'}} & [\mathbb{A}^1/\mathbb{G}_m] \\ \downarrow & & \downarrow{\scriptstyle\theta_{e_{K'/K}}} \\ \mathrm{Spec}(A) & \xrightarrow{\mathfrak{m}_A} & [\mathbb{A}^1/\mathbb{G}_m], \end{array} \tag{8.2.6.7}$$

where the horizontal arrows are defined by the inclusions $\mathfrak{m}_A \subset A$ and $\mathfrak{m}_{A'} \subset A'$ and the interpretation of $[\mathbb{A}^1/\mathbb{G}_m]$ given in 6.1.11. For every $n \geq 1$ the induced map $\mathfrak{m}_A : \mathrm{Spec}(A_1) \to [\mathbb{A}^1/\mathbb{G}_m]$ factors through $\Delta_n^{(e_K)}$, and therefore from (8.2.6.6) we obtain a commutative diagram

$$\begin{array}{ccc} \mathrm{Spec}(A'_1) & \longrightarrow & \Delta_n^{(e_{K'})} \\ \downarrow & & \downarrow{\scriptstyle\theta_{e_{K'/K}}} \\ \mathrm{Spec}(A_1) & \longrightarrow & \Delta^{(e_K)} \end{array} \tag{8.2.6.8}$$

which induces a morphism of topoi

$$\delta_{A'/A} : (\mathrm{Spec}(A'_1)_{\mathrm{et}}/\Delta_n^{(e_{K'})})_{\mathrm{cris}} \longrightarrow (\mathrm{Spec}(A_1)_{\mathrm{et}}/\Delta_n^{(e_K)})_{\mathrm{cris}}. \tag{8.2.6.9}$$

Furthermore, if $K' \subset K'' \subset \overline{K}$ is a second finite extension then it follows from the construction that $\delta_{A''/A} \simeq \delta_{A''/A'} \circ \delta_{A'/A}$.

Let $(\mathrm{Spec}(\overline{A}_1)/\widetilde{\Delta}_n)_{\mathrm{cris}}$ denote the inverse limit topos [**5**, VI.8.1.1]

$$(8.2.6.10)\qquad (\mathrm{Spec}(\overline{A}_1)/\widetilde{\Delta}_n)_{\mathrm{cris}} := \varprojlim_{K\subset K'\subset \overline{K}} (\mathrm{Spec}(A'_1)_{\mathrm{et}}/\Delta_n^{(e_{K'})})_{\mathrm{cris}}$$

with transition maps the morphisms of topoi $\delta_{A''/A'}$ defined above. Since the morphisms $\delta_{A''/A'}$ are morphisms of ringed topoi there is a natural structure sheaf

$$(8.2.6.11)\qquad \mathscr{O}_{\mathrm{Spec}(\overline{A}_1)/\widetilde{\Delta}_n} \in (\mathrm{Spec}(\overline{A}_1)/\widetilde{\Delta}_n)_{\mathrm{cris}}.$$

The projections $\delta_{A'/A}$ induce a morphism of ringed topoi

$$(8.2.6.12)\qquad h_{\mathrm{cris}} : (\mathrm{Spec}(\overline{A}_1)/\widetilde{\Delta}_n)_{\mathrm{cris}} \longrightarrow (\mathrm{Spec}(A_1)/\Delta_n^{(e_K)})_{\mathrm{cris}}.$$

Proposition 8.2.7. — *The sheaf $h_{\mathrm{cris}*}\mathscr{O}_{\mathrm{Spec}(\overline{A}_1)/\widetilde{\Delta}_n}$ is a quasi-coherent flat crystal in the topos $(\mathrm{Spec}(A_1)/\Delta_n^{(e_K)})_{\mathrm{cris}}$ and $R^q h_{\mathrm{cris}*}\mathscr{O}_{\mathrm{Spec}(\overline{A}_1)/\widetilde{\Delta}_n} = 0$ for $q > 0$.*

The proof is in several steps 8.2.8–8.2.18.

8.2.8. — Let $U \hookrightarrow F$ be any object of $\mathrm{Cris}(\mathrm{Spec}(A_1)/\Delta_n^{(e_K)})$, and for any integer $e \geq 1$ let $F^{(e)}$ denote $F \times_{[\mathbb{A}_1/\mathbb{G}_m],\theta_e} [\mathbb{A}^1/\mathbb{G}_m]$. We view $F^{(e)}$ is a PD-stack by pulling back the divided power ideal on F (this is possible because the map θ_e is flat). The closed substack $U^{(e)} \subset F^{(e)}$ defined by this divided power ideal is equal to $U \times_{[\mathbb{A}_1/\mathbb{G}_m],\theta_e} [\mathbb{A}^1/\mathbb{G}_m]$.

If $K \subset K' \subset \overline{K}$ is a finite algebraic extension with ramification index $e_{K'/K}$, then there is a canonical map

$$(8.2.8.1)\qquad U \times_{\mathrm{Spec}(A)} \mathrm{Spec}(A') \longrightarrow F^{(e_{K'/K})}$$

induced by the map $U \to F$ and the map $\mathfrak{m}_{A'} : \mathrm{Spec}(A') \to [\mathbb{A}^1/\mathbb{G}_m]$. From this and [**5**, V.5.1] it follows that for any $q \geq 0$ the sheaf $R^q h_{\mathrm{cris}*}\mathscr{O}_{\mathrm{Spec}(\overline{A}_1)/\widetilde{\Delta}_n}$ is equal to the sheaf associated to the presheaf

$$(8.2.8.2)\qquad (U \hookrightarrow F) \longmapsto \varinjlim_{K\subset K'\subset \overline{K}} H^q\big((U \otimes_A A'/F^{(e_{A'/A})})_{\mathrm{cris}}, \mathscr{O}_{U\otimes_A A'/F^{(e_{A'/A})}}\big).$$

Hence to prove the proposition it suffices to show that for $q > 0$

$$(8.2.8.3)\qquad \varinjlim H^q\big((U \otimes_A A'/F^{(e_{A'/A})})_{\mathrm{cris}}, \mathscr{O}_{U\otimes_A A'/F^{(e_{A'/A})}}\big) = 0,$$

and that the formation of $\varinjlim H^0((U \otimes_A A'/F^{(e_{A'/A})})_{\mathrm{cris}}, \mathscr{O}_{U\otimes_A A'/F^{(e_{A'/A})}})$ is compatible with base changes by morphisms $(U' \hookrightarrow T') \to (U \hookrightarrow T)$ in $\mathrm{Cris}(\mathrm{Spec}(A_1)/\Delta_n^{(e_K)})$.

Lemma 8.2.9. — *The morphism $U \otimes_{A_1} A'_1 \to U^{(e_{K'/K})}$ is syntomic.*

Proof. — Since $U^{(e_{K'/K})}$ is étale over $\mathrm{Spec}(A_1)\times_{[\mathbb{A}^1/\mathbb{G}_m],\theta_{e_{K'/K}}}[\mathbb{A}^1/\mathbb{G}_m]$, it suffices to consider the case when $U=\mathrm{Spec}(A_1)$. In this case the stack $U^{(e_{K'/K})}$ is isomorphic to the stack

$$\big[\mathrm{Spec}(A_1[t]/t^{e_{K'/K}}=\pi)/\mu_{e_{K'/K}}\big], \tag{8.2.9.1}$$

where a scheme-valued point $\zeta\in\mu_{e_{K'/K}}$ acts by multiplication on t. To verify that the map $\mathrm{Spec}(A_1')\to U^{(e_{K'/K})}$ is syntomic, it suffices to verify that the morphism obtained by base change

$$\mathrm{Spec}(A_1')\times_{U^{(e_{K'/K})}}\mathrm{Spec}(A_1[t]/t^{e_{K'/K}}=\pi)\longrightarrow\mathrm{Spec}(A_1[t]/t^{e_{K'/K}}=\pi) \tag{8.2.9.2}$$

is syntomic. This map is isomorphic to the map of schemes

$$\mathrm{Spec}(A_1'[t,u]/(u^{e_{K'/K}}=1,t^{e_{K'/K}}=\pi,\pi'=tu))\longrightarrow\mathrm{Spec}(A_1[t]/t^{e_{K'/K}}=\pi). \tag{8.2.9.3}$$

After base changing to a finite extension of the residue field k so that there exists a unit $v\in k$ so that $(v\pi')^{e_{K'/K}}=\pi$, this map is isomorphic to the map of schemes

$$\mathrm{Spec}(A_1[t,u]/(t^{e_{K'/K}}=\pi,u^{e_{K'/K}}=1))\longrightarrow\mathrm{Spec}(A_1[t]/t^{e_{K'/K}}=\pi) \tag{8.2.9.4}$$

which is clearly syntomic. □

Lemma 8.2.10. — *Let $(U_1\hookrightarrow F_1)\to(U_2\hookrightarrow F_2)$ be a morphism in the category* $\mathrm{Cris}(\mathrm{Spec}(A_1)/\Delta_n^{(e_K)})$. *Then the induced map*

$$\begin{array}{c}\mathcal{O}_{F_2}\otimes^{\mathbb{L}}_{\mathcal{O}_{F_1}}R\Gamma((U_1\otimes_{A_1}A_1'/F_1^{(e_{K'/K})})_{\mathrm{cris}},\mathcal{O}_{U_1\otimes_{A_1}A_1'/F_1^{(e_{K'/K})}})\\ \downarrow\\ R\Gamma((U_2\otimes_{A_2}A_2'/F_2^{(e_{K'/K})})_{\mathrm{cris}},\mathcal{O}_{U_2\otimes_{A_2}A_2'/F_2^{(e_{K'/K})}})\end{array} \tag{8.2.10.1}$$

is an isomorphism.

Proof. — Let Q denote the fiber product over $F_1^{(e_{K'/K})}$ of $U_1\otimes_{A_1}A_1'$ and $F_2^{(e_{K'/K})}$. By 8.2.9 and the base change theorem 8.1.8 there is a natural isomorphism

$$\begin{array}{c}\mathcal{O}_{F_2}\otimes^{\mathbb{L}}_{\mathcal{O}_{F_1}}R\Gamma((U_1\otimes_{A_1}A_1'/F_1^{(e_{K'/K})})_{\mathrm{cris}},\mathcal{O}_{U_1\otimes_{A_1}A_1'/F_1^{(e_{K'/K})}})\\ \simeq\downarrow\\ R\Gamma((Q/F_2^{(e_{K'/K})})_{\mathrm{cris}},\mathcal{O}_{\mathrm{Spec}(A_n')/F_2^{(e_{K'/K})}}).\end{array} \tag{8.2.10.2}$$

On the other hand, the closed immersion $U_2\otimes_{A_1}A_1'\subset Q$ is defined by a PD-ideal in Q, and hence as in [**8**, 5.17] there is an isomorphism

$$\begin{aligned}&R\Gamma((Q/F_2^{(e_{K'/K})})_{\mathrm{cris}},\mathcal{O}_{Q/F_2^{(e_{K'/K})}})\\&\simeq R\Gamma((U_2\otimes_{A_1}A_1'/F_2^{(e_{K'/K})})_{\mathrm{cris}},\mathcal{O}_{U_2\otimes_{A_1}A_1'/F_2^{(e_{K'/K})}}).\end{aligned}$$
□

8.2.11. — We first show that $\varinjlim H^q((U\otimes_A A'/F^{(e_{A'/A})})_{\mathrm{cris}}, \mathcal{O}_{U\otimes_A A'/F^{(e_{A'/A})}})$ is zero for $q>0$. For this note first that we may without loss of generality assume that U is connected. Furthermore, we can without loss of generality replace K by a finite unramified extension $K\subset K'$ (since then $F^{(e_{A'/A})}\simeq F$). It follows that we may assume that $U=\mathrm{Spec}(A_1)$.

8.2.12. — Let $K\subset K'\subset \overline{K}$ be a finite extension, and let $\pi\in A$ and $\pi'\in A'$ be uniformizing elements with $(\pi')^{e_{K'/K}}a=\pi$ for some $a\in A'^*$. Then if $\tilde{\pi}\in\mathcal{O}_F$ is a section lifting π so that $F\to[\mathbb{A}^1/\mathbb{G}_m]$ is induced by $\tilde{\pi}:\mathcal{O}_F\to\mathcal{O}_F$, there is an isomorphism of stacks over F

$$F^{(e)}\simeq\big[\mathrm{Spec}(\mathcal{O}_F[t]/t^{e_{K'/K}}=\tilde{\pi})/\mu_{e_{K'/K}}\big], \tag{8.2.12.1}$$

where the group $\mu_{e_{K'/K}}$ acts by multiplication on t. This stack can also be described as

$$F^{(e)}\simeq\big[\mathrm{Spec}(\mathcal{O}_F[t,w^{\pm}]/t^{e_{K'/K}}w=\pi)/\mathbb{G}_m\big] \tag{8.2.12.2}$$

where $u\in\mathbb{G}_m$ acts by $t\mapsto ut$ and $w\mapsto u^{-e_{K'/K}}w$.

The map $\mathrm{Spec}(A_1')\to F^{(e)}$ is equal to the composite of the closed immersion

$$\mathrm{Spec}(A_1')\hookrightarrow\mathrm{Spec}(\mathcal{O}_F[t,w^{\pm}]/t^{e_{K'/K}}w=\pi),\quad \pi'\longleftarrow t,\quad a\longleftarrow w, \tag{8.2.12.3}$$

with the natural smooth surjection

$$\mathrm{Spec}\big(\mathcal{O}_F[t,w^{\pm}]/t^{e_{K'/K}}w=\pi\big)\longrightarrow\big[\mathrm{Spec}(\mathcal{O}_F[t,w^{\pm}]/t^{e_{K'/K}}w=\pi)/\mathbb{G}_m\big]. \tag{8.2.12.4}$$

It follows that the group $H^q((\mathrm{Spec}(A_1')/F^{(e_{K'/K})})_{\mathrm{cris}}, \mathcal{O}_{\mathrm{Spec}(A_1')/F^{(e_{K'/K})}})$ is equal to the q-th cohomology group of the de Rham complex of the divided power envelope of (8.2.12.3) over $F^{(e)}$.

Lemma 8.2.13

(i) *The module* $\Omega^1_{\mathrm{Spec}(\mathcal{O}_F[t,w^{\pm}]/t^{e_{K'/K}}w=\pi)/F^{(e_{K'/K})}}$ *is free on one generator* $\mathrm{dlog}(t)$, *and the differential sends* t *to* $t\mathrm{dlog}(t)$ *and* w *to* $-e_{K'/K}w\mathrm{dlog}(t)$.

(ii) *If* $K'\subset K''\subset\overline{K}$ *is a second finite extension, then the natural map*

$$A''\otimes_{A'}\Omega^1_{\mathrm{Spec}(\mathcal{O}_F[t,w^{\pm}]/t^{e_{K'/K}}w=\pi)/F^{(e_{K'/K})}}\longrightarrow\Omega^1_{\mathrm{Spec}(\mathcal{O}_F[t,w^{\pm}]/t^{e_{K''/K}}w=\pi)/F^{(e_{K''/K})}} \tag{8.2.13.1}$$

sends $\mathrm{dlog}(t)$ *to* $e_{K''/K'}\cdot\mathrm{dlog}(t)$.

Proof. — To see (i), consider the cartesian square

$$\begin{array}{ccc}\mathrm{Spec}(\mathcal{O}_F[t,w^{\pm}]/t^{e_{K'/K}}w=\pi) & \xleftarrow{\mathrm{pr}_1} & \mathrm{Spec}(\mathcal{O}_F[t,w^{\pm}]/t^{e_{K'/K}}w=\pi)\times\mathbb{G}_m\\ \downarrow & & \downarrow\rho\\ F^{(e)} & \longleftarrow & \mathrm{Spec}(\mathcal{O}_F[t,w^{\pm}]/t^{e_{K'/K}}w=\pi),\end{array} \tag{8.2.13.2}$$

where pr_1 denotes projection onto the first factor and ρ denotes the action. If we write $\mathbb{G}_m = \mathrm{Spec}(W[u^{\pm}])$, then the pullback to $\mathrm{Spec}(\mathcal{O}_F[t, w^{\pm}]/t^{e_{K'/K}}w = \pi)$ of the ideal of the diagonal is generated by $(u-1)$. Let $\mathrm{dlog}(t)$ denote this basis for $\Omega^1_{\mathrm{Spec}(\mathcal{O}_F[t,w^{\pm}]/t^{e_{K'/K}}w=\pi)/F^{(e_{K'/K})}}$. Then the statements in (i) amount to the observation that

(8.2.13.3)
$$\rho^*(t) - \mathrm{pr}_1^* t = \mathrm{pr}_1^*(t)(u-1), \quad \rho^*(w) - \mathrm{pr}_1^*(w) \equiv -e_{K'/K}w(u-1) \pmod{(u-1)^2}.$$

Statement (ii) can be seen by noting that the transition map

$$F^{(e_{K''/K})} \longrightarrow F^{(e_{K'/K})} \tag{8.2.13.4}$$

is induced by the maps

(8.2.13.5)
$$(\mathcal{O}_F[t, w^{\pm}]/t^{e_{K'/K}}w = \pi) \longrightarrow (\mathcal{O}_F[\tilde{t}, \tilde{w}^{\pm}]/\tilde{t}^{e_{K''/K}}\tilde{w} = \pi), \ t \longmapsto \tilde{t}^{e_{K''/K'}}, \ w \mapsto \tilde{w}^{e_{K''/K'}},$$

and

$$\mathbb{G}_m \longrightarrow \mathbb{G}_m, \quad u \longmapsto u^{e_{K''/K'}}. \tag{8.2.13.6}$$

This shows that the generator $(u-1)$ of $\Omega^1_{\mathrm{Spec}(\mathcal{O}_F[t,w^{\pm}]/t^{e_{K'/K}}w=\pi)/F^{(e_{K'/K})}}$ is sent to the class of $(u^{e_{K''/K'}} - 1)$. Since

$$u^{e_{K''/K'}} - 1 \equiv e_{K''/K'}(u-1) \pmod{(u-1)^2} \tag{8.2.13.7}$$

this implies (ii). □

8.2.14. — From the lemma it follows that if $K' \subset K'' \subset \overline{K}$ is a second extension with $p^n | e_{K''/K'}$ (where n is the integer in 8.2.7), then since $p^n\mathcal{O}_F = 0$ (since F is a scheme over $\Delta_n^{(e_K)}$) the map

$$\begin{array}{c} H^q((\mathrm{Spec}(A_1')/F^{(e_{K'/K})})_{\mathrm{cris}}, \mathcal{O}_{\mathrm{Spec}(A_1')/F^{(e_{K'/K})}}) \\ \downarrow \\ H^q((\mathrm{Spec}(A_1')/F^{(e_{K'/K})})_{\mathrm{cris}}, \mathcal{O}_{\mathrm{Spec}(A_1'')/F^{(e_{K''/K})}}) \end{array} \tag{8.2.14.1}$$

is zero for $q > 0$. This therefore proves that $R^q h_{\mathrm{cris}*}\mathcal{O}_{\mathrm{Spec}(\overline{A}_1)/\widetilde{\Delta}_n} = 0$ for $q > 0$.

8.2.15. — That $h_{\mathrm{cris}*}\mathcal{O}_{\mathrm{Spec}(\overline{A}_1)/\widetilde{\Delta}_n}$ is a quasi-coherent crystal can be seen as follows.

By the same reasoning used in 2.5.12, for any object

$$(U \hookrightarrow F) \in \mathrm{Cris}(\mathrm{Spec}(A_1)/\Delta_n^{(e_K)}) \tag{8.2.15.1}$$

the restriction of $h_{\mathrm{cris}*}\mathcal{O}_{\mathrm{Spec}(\overline{A}_1)/\widetilde{\Delta}_n}$ to F_{et} is a quasi-coherent sheaf. Furthermore, for any morphism $(U' \hookrightarrow F') \to (U \hookrightarrow F)$ in $\mathrm{Cris}(\mathrm{Spec}(A_1')/\Delta_n^{(e_{K'})})$ with F' and F affine, 8.2.10 implies that the natural map

$$\mathcal{O}_{F'} \otimes^{\mathbb{L}}_{\mathcal{O}_F} h_{\mathrm{cris}*}\mathcal{O}_{\mathrm{Spec}(\overline{A}_1)/\widetilde{\Delta}_n}(F) \longrightarrow h_{\mathrm{cris}*}\mathcal{O}_{\mathrm{Spec}(\overline{A}_1)/\widetilde{\Delta}_n}(F') \tag{8.2.15.2}$$

is an isomorphism. In particular, $h_{\mathrm{cris}*}\mathcal{O}_{\mathrm{Spec}(\overline{A}_1)/\widetilde{\Delta}_n}$ is a quasi-coherent crystal.

8.2.16. — Finally for the flatness of $h_{\mathrm{cris}*}\mathcal{O}_{\mathrm{Spec}(\overline{A}_1)/\widetilde{\Delta}_n}$, we use an argument we learned from [**73**, 4.1.5].

Let $(U \hookrightarrow F) \in \mathrm{Cris}(\mathrm{Spec}(A_1)/\Delta_n^{(e_K)})$ be an object with F affine, and let M be a $\mathcal{O}_F$-module.

For any $K \subset K' \subset \overline{K}$ define a crystal $\mathcal{M}_{A'}$ in $(\mathrm{Spec}(A'_n)/F^{(e_{K'/K})})_{\mathrm{cris}}$ by associating to any object T the global sections of the pullback of M to T. Then these crystals are compatible with the morphisms of topoi associated to further extensions $K \subset K' \subset K'' \subset \overline{K}$. The same argument used in the proof that $R^q h_{\mathrm{cris}*}\mathcal{O}_{\mathrm{Spec}(\overline{A}_1)/\widetilde{\Delta}_n} = 0$ for $q > 0$ shows that

$$\varinjlim H^q((\mathrm{Spec}(A'_1)/F^{(e_{K'/K})})_{\mathrm{cris}}, \mathcal{M}_{A'}) = 0 \tag{8.2.16.1}$$

for $q > 0$.

Lemma 8.2.17. — *For any finite extension $K \subset K' \subset \overline{K}$, the natural map*

$$\begin{array}{c} M \otimes^{\mathbb{L}}_{\mathcal{O}_F} R\Gamma\big((\mathrm{Spec}(A'_1)/F^{(e_{K'/K})})_{\mathrm{cris}}, \mathcal{O}_{\mathrm{Spec}(A'_1)/F^{(e_{K'/K})}}\big) \\ \downarrow \\ R\Gamma\big((\mathrm{Spec}(A'_1)/F^{(e_{K'/K})})_{\mathrm{cris}}, \mathcal{M}_{A'}\big) \end{array} \tag{8.2.17.1}$$

is an isomorphism.

Proof. — Consider the closed immersion $\mathrm{Spec}(A'_1) \hookrightarrow Z$ defined in (8.2.12.3) over $F^{(e_{K'/K})}$, and let $\mathcal{D}$ be the divided power envelope. Since $\mathcal{D}$ is flat over $\mathcal{O}_F$ by 8.1.7

$$M \otimes^{\mathbb{L}}_{\mathcal{O}_F} R\Gamma\big((\mathrm{Spec}(A'_1)/F^{(e_{K'/K})})_{\mathrm{cris}}, \mathcal{O}_{\mathrm{Spec}(A'_1)/F^{(e_{K'/K})}}\big) \simeq M \otimes_{\mathcal{O}_F} \big(\mathcal{D} \otimes \Omega^{\bullet}_{Z/F^{(e_{K'/K})}}\big), \tag{8.2.17.2}$$

which by the definition of $\mathcal{M}_{A'}$ is equal to the right hand side of (8.2.17.1). □

8.2.18. — From this lemma it follows that there is a spectral sequence

$$\begin{array}{c} E_2^{pq} = \mathrm{Tor}_{-p,\mathcal{O}_F}\big(M, H^q((\mathrm{Spec}(A'_1)/F^{(e_{K'/K})})_{\mathrm{cris}}, \mathcal{O}_{\mathrm{Spec}(A'_1)/F^{(e_{K'/K})}})\big) \\ \Downarrow \\ H^{p+q}(\mathrm{Spec}(A'_1)/F^{(e_{K'/K})}, \mathcal{M}_{A'}). \end{array} \tag{8.2.18.1}$$

From the vanishing (8.2.16.1) it then follows that

$$\mathrm{Tor}_{-p,\mathcal{O}_F}\big(M, H^0((\mathrm{Spec}(A'_1)/F^{(e_{K'/K})})_{\mathrm{cris}}, \mathcal{O}_{\mathrm{Spec}(A'_1)/F^{(e_{K'/K})}})\big) = 0 \tag{8.2.18.2}$$

for $p > 0$. This completes the proof of 8.2.7. □

8.2.19. — The quasi-coherent crystal $h_{\mathrm{cris}*}\mathcal{O}_{\mathrm{Spec}(\overline{A}_1)/\widetilde{\Delta}_n}$ can be described explicitly as follows. Fix a uniformizer $\pi \in A$, let R_n denote the divided power envelope of the surjection $W_n[t] \to A_n$ sending t to π, let $E_n = \mathrm{Spec}(R_n)$, and define

$$P_n := h_{\mathrm{cris}*}\mathcal{O}_{\mathrm{Spec}(\overline{A}_1)/\widetilde{\Delta}_n}(E_n). \tag{8.2.19.1}$$

Observe that the diagram

$$\begin{array}{ccc} E_n & \longrightarrow & \mathrm{Spec}(W_n[t]) \\ \downarrow & & \downarrow \\ \Delta_n^{(e_K)} & \longrightarrow & [\mathbb{A}_1/\mathbb{G}_m]_{W_n} \end{array} \tag{8.2.19.2}$$

is cartesian. In particular, since the map $\mathrm{Spec}(W_n[t]) \to [\mathbb{A}^1/\mathbb{G}_m]_{W_n}$ is smooth, the crystal $h_{\mathrm{cris}*}\mathcal{O}_{\mathrm{Spec}(\overline{A}_1)/\widetilde{\Delta}_n}$ is determined by P_n and the canonical connection

$$\nabla : P_n \longrightarrow P_n \cdot \mathrm{dlog}(t), \tag{8.2.19.3}$$

where we write $\mathrm{dlog}(t)$ for the canonical generator of $\Omega^1_{W_n[t]/[\mathbb{A}^1/\mathbb{G}_m]_{W_n}}$ corresponding to the element $(u-1)$ in the ideal defining the closed immersion

$$\mathrm{id} \times e : \mathbb{A}^1 \hookrightarrow \mathbb{A}^1 \times \mathbb{G}_m. \tag{8.2.19.4}$$

The connection ∇ satisfies $\nabla(t^{[i]} \cdot m) = t^{[i]}\nabla(m) + it^{[i]} \cdot m$. Note also that there is a natural map

$$B_n = \Gamma((\overline{S}_n/W_n)_{\mathrm{cris}}, \mathcal{O}_{\overline{S}_n/W_n}) \longrightarrow P_n \tag{8.2.19.5}$$

whose image is horizontal for ∇.

Proposition 8.2.20 **(Stack version of [41, 3.3])**

(i) *For each p^n-th root β of π in $\overline{A}$ there exists a canonical element $\nu_\beta \in \mathrm{Ker}(P_n^* \to \overline{A}_n^*)$ such that the map*

$$B_n\langle V\rangle \longrightarrow P_n, \quad V \longmapsto \nu_\beta - 1 \tag{8.2.20.1}$$

is an isomorphism. If $\zeta \in \overline{A}$ is a root of unity with $\zeta^{p^n} = 1$ then $\nu_{\zeta\beta} = \tilde{\zeta}^{p^n}\nu_\beta$, where $\tilde{\zeta} \in B_n$ is any lifting of ζ.

(ii) *The map ∇ is the unique B_n-linear map satisfying*

$$\nabla((\nu_\beta - 1)^{[i]}) = (\nu_\beta - 1)^{[i-1]}\nu_\beta \mathrm{dlog}(t). \tag{8.2.20.2}$$

(iii) *The Frobenius on P_n is given by $\varphi(\nu_\beta) = \nu_\beta^p$ and the Frobenius on B_n.*

(iv) *The action of $\mathrm{Gal}(\overline{K}/K)$ is characterized by the condition that it extends the action on B_n, it preserves the divided power structure, and $\sigma(\nu_\beta) = \nu_{\sigma(\beta)}$ for any $\sigma \in \mathrm{Gal}(\overline{K}/K)$.*

The proof is in steps 8.2.21–8.2.26.

8.2.21. — Fix $\beta \in \overline{A}$ such that $\beta^{p^n} = \pi$. For any finite extension $K \subset K' \subset \overline{K}$ with $\beta \in K'$ and $p^{2n} | e_{K'/K}$, let $B_n(A')$ denote $\Gamma((\mathrm{Spec}(A'_1)/W_n)_{\mathrm{cris}}, \mathcal{O}_{A'_1/W_n})$.

There is a canonical map $\rho_{K'} : \mathrm{Spec}(B_n(A')\langle V\rangle) \to [\mathbb{A}^1/\mathbb{G}_m]$ defined as follows. Let $\pi' \in A'$ be a uniformizer, and let $\tilde{\pi}' \in B_n(A')$ be any lifting of π'. Then the map $\rho_{K'}$ is given by the map of free rank 1-modules $\times \tilde{\pi}'^{p^n} : B_n(A'_1)\langle V\rangle \to B_n(A'_1)\langle V\rangle$ and the interpretation (6.1.11) of $[\mathbb{A}^1/\mathbb{G}_m]$. By an argument as in (8.2.3.2) this map is independent of the choices, and there is a commutative diagram

$$\begin{array}{ccc} \mathrm{Spec}(A'_1) & \longrightarrow & S_n \\ \downarrow & & \downarrow \\ \mathrm{Spec}(B_n(A')\langle V\rangle) & \xrightarrow{\vartheta} & E_n \\ {\scriptstyle \rho_{K'}}\downarrow & & \downarrow \\ [\mathbb{A}^1/\mathbb{G}_m]_{W_n} & \xrightarrow{\theta_{d_{K'/K}}} & [\mathbb{A}^1/\mathbb{G}_m]_{W_n}, \end{array} \tag{8.2.21.1}$$

where $d_{K'/K} := e_{K'/K}/p^n$ and ϑ is the map induced by $t \mapsto \tilde{\beta}^{p^n}(1+V)^{-1}$. Let $\mathcal{L}_{B_n(A')\langle V\rangle} \to B_n(A')\langle V\rangle$ denote the line bundle corresponding to $\rho_{K'}$, and let $\gamma : \mathcal{L}_{B_n(A')\langle V\rangle}^{d_{K'/K}} \to \vartheta^*(\mathcal{O}_{E_n} \cdot t)$ be the canonical isomorphism making the bottom square commutative (the isomorphism γ is obtained from $(\tilde{\beta}^{p^n})^{\otimes d_{K'/K}} \mapsto (1+V)\cdot t$).

8.2.22. — For any object $(\mathrm{Spec}(A'_1) \hookrightarrow T) \in \mathrm{Cris}(\mathrm{Spec}(A'_1)/[\mathbb{A}^1/\mathbb{G}_m] \times_{\theta_{e_{K'/K}}, [\mathbb{A}^1/\mathbb{G}_m]} E_n)$, we claim that there exists a unique map $\lambda : T \to \mathrm{Spec}(B_n(A')\langle V\rangle)$ such that the diagram

$$\begin{array}{ccccc} T & \xrightarrow{\lambda} & \mathrm{Spec}(B_n(A')\langle V\rangle) & \longrightarrow & E_n \\ \downarrow & & \downarrow{\scriptstyle \rho_{K'}} & & \downarrow \\ [\mathbb{A}^1/\mathbb{G}_m]_{W_n} & \xrightarrow{\theta_{p^n}} & [\mathbb{A}^1/\mathbb{G}_m]_{W_n} & \xrightarrow{\theta_{d_{K'/K}}} & [\mathbb{A}^1/\mathbb{G}_m]_{W_n} \end{array} \tag{8.2.22.1}$$

commutes and the composite of the top row is the given map to E_n.

For this we use the interpretation of the stack $[\mathbb{A}^1/\mathbb{G}_m]$ given in 6.1.11. Let $\mathcal{L}_T \to \mathcal{O}_T$ be the morphism of line bundles corresponding to the projection $T \to [\mathbb{A}^1/\mathbb{G}_m]_{W_n}$. The given morphism

$$T \to [\mathbb{A}^1/\mathbb{G}_m] \times_{\theta_{e_{K'/K}}, [\mathbb{A}^1/\mathbb{G}_m]} E_n \tag{8.2.22.2}$$

corresponds to a map $T \to E_n$ and an isomorphism

$$\iota : \mathcal{L}_T^{e_{K'/K}} \simeq \mathcal{O}_T \cdot t \tag{8.2.22.3}$$

compatible with the maps to $\mathcal{O}_T$, where we write t also for the image of the coordinate t on $\mathbb{A}^1$ under the composite $T \to E_n \to \mathbb{A}^1_{W_n}$.

There is also a second isomorphism $\iota' : \mathcal{L}_T^{e_{K'/K}} \simeq \mathcal{O}_T \cdot t$ defined as follows. The pullback $\mathcal{L}_T|_{\mathrm{Spec}(A'_1)}$ comes equipped with an isomorphism with $\mathfrak{m}_{A'}$. The element β

therefore defines a generator of $\mathcal{L}_T^{d_{K'/K}}|_{\mathrm{Spec}(A'_1)}$. Let $\tilde{\beta} \in \mathcal{L}_T^{d_{K'/K}}$ be a lifting of this generator. Then the generator $\tilde{\beta}^{p^n} \in \mathcal{L}_T^{e_{K'/K}}$ is independent of the choice of lifting $\tilde{\beta}$ (again because the kernel of $\mathcal{O}_T \to A'_1$ has divided powers). The second isomorphism ι' is defined by sending $\tilde{\beta}^{p^n}$ to t.

Since ι and ι' are isomorphisms of trivial line bundles, there exists a unique element $\nu_\beta \in \mathcal{O}_T^*$ such that $\iota' = \nu_\beta \iota$. Furthermore, by construction the element ν_β maps to 1 in A'_1. We define the map λ by sending V to $\nu_\beta - 1$.

Let π' be a uniformizer in A'. This uniformizer π' defines a trivialization of the line bundle $\mathcal{L}_{B_n(A')\langle V\rangle}$ since by construction of $\rho_{K'}$ it is defined as the pullback of the map of line bundles $(t) \to \mathcal{O}_{\mathbb{A}^1}$ by the map $W_n[t] \to B_n(A')\langle V\rangle$ sending t to $\tilde{\pi}'^{p^n}$, for any lifting $\tilde{\pi}'$ of π' to $B_n(A')$. Similarly, π' defines a trivialization of the line bundle $\mathcal{L}_T^{p^n}$. Let

$$\mathcal{L}_T^{p^n} \longrightarrow \lambda^* \mathcal{L}_{B_n(A')\langle V\rangle} \tag{8.2.22.4}$$

be the isomorphism defined by these trivializations. If $u \in A'$ is a unit, then the isomorphism defined by $u\pi'$ is equal to that defined by c so in fact this isomorphism is independent of the choice of π'.

8.2.23. — The uniqueness of the map λ is seen as follows. Let $\mathcal{L}_{B_n(A')\langle V\rangle} \to \mathcal{O}_{B_n(A')\langle V\rangle}$ be the data corresponding to the map $\rho_{K'}$.

Let $\lambda' : T \to \mathrm{Spec}(B_n(A')\langle V\rangle)$ be a second map with an isomorphism

$$\tau : \mathcal{L}_T^{p^n} \longrightarrow \lambda'^* \mathcal{L}_{B_n(A')\langle V\rangle} \tag{8.2.23.1}$$

compatible with the maps to $\mathcal{O}_T$ such that the composite

$$\mathcal{L}_T^{e_{K'/K}} \xrightarrow{\ \tau\ } \lambda'^* \mathcal{L}_{B_n(A')\langle V\rangle}^{d_{K'/K}} \xrightarrow{\tilde{\beta}^{p^n} \mapsto (1+\lambda'(V))t} \mathcal{O}_T \cdot t \tag{8.2.23.2}$$

is equal to the map induced by ι. By the universal property of $B_n(A')$ we have $\lambda = \lambda'$ when they are restricted to $B_n(A')$.

As in 8.2.22, there is a canonical isomorphism $c : \mathcal{L}_T^{p^n} \to \lambda'^* \mathcal{L}_{B_n(A')\langle V\rangle}$ defined as follows. Let π' be a uniformizer in A'. This uniformizer π' defines a trivialization of the line bundle $\mathcal{L}_{B_n(A')\langle V\rangle}$ since by construction of $\rho_{K'}$ it is defined as the pullback of the map of line bundles $(t) \to \mathcal{O}_{\mathbb{A}^1}$ by the map $W_n[t] \to B_n(A')\langle V\rangle$ sending t to $\tilde{\pi}'^{p^n}$, for any lifting $\tilde{\pi}'$ of π' to $B_n(A')$. Similarly, π' defines a trivialization of the line bundle $\mathcal{L}_T^{p^n}$. Let

$$c : \mathcal{L}_T^{p^n} \longrightarrow \lambda'^* \mathcal{L}_{B_n(A')\langle V\rangle} \tag{8.2.23.3}$$

denote the isomorphism defined by these trivializations. Then $\tau = uc$ for some element $u \in \mathcal{O}_T^*$ reducing to 1 in A'_1.

The isomorphism $\mathcal{L}_T^{e_{K'/K}} \to \lambda'^* \mathcal{L}_{B_n(A')\langle V\rangle}^{d_{K'/K}}$ is then equal to $u^{d_{K'/K}} \cdot c^{\otimes d_{K'/K}}$. Since $p^n | d_{K'/K}$ by assumption and $u = 1 + h$ for some h in the divided power ideal of T, the element $u^{d_{K'/K}}$ is equal to

$$(1+h)^{p^n} = \sum_{i=0}^{p^n} \frac{p^n!}{(p^n-i)!} h^{[i]} = 1. \tag{8.2.23.4}$$

Let $\pi'^{d_{K'/K}} = a\beta$ for some $a \in A'^*$, and let $\tilde{a} \in B_n(A')$ be a lifting of a. Then the composite (8.2.23.2) sends $\tilde{\pi}'^{e_{K'/K}}$ to $\tilde{a}^{p^n}(1+\lambda'(V))\cdot t$. On the other hand, the map ι sends $\tilde{\pi}'^{e_{K'/K}}$ to $a(1+\lambda(V))\cdot t$ by the definition of the map λ. It follows that $\lambda(V) = \lambda'(V)$ and hence $\lambda = \lambda'$.

8.2.24. — It follows that the choice of β defines a map

$$B_n\langle V\rangle = \varinjlim_{K\subset K'\subset \overline{K}} B_n(A')\langle V\rangle \longrightarrow P_n \tag{8.2.24.1}$$

which we claim is an isomorphism.

For this note that if $K \subset K' \subset \overline{K}$ is any finite extension with $p^{2n}|e_{K'/K}$, then there exists an extension $K' \subset K'' \subset \overline{K}$ with $p^n|e_{K''/K'}$. There is then a commutative diagram

$$\begin{array}{ccccc}
\mathrm{Spec}(A_1'') & \xrightarrow{\lambda} & \mathrm{Spec}(B_n(A'')\langle V\rangle) & \longrightarrow & E_n \\
\downarrow{\scriptstyle \mathrm{id}} & & \downarrow{\scriptstyle \rho_{K''}} & & \downarrow{\scriptstyle \mathrm{id}} \\
\mathrm{Spec}(A_1'') & \longrightarrow & [\mathbb{A}^1/\mathbb{G}_m]_{W_n} & & E_n \\
\downarrow & & \downarrow{\scriptstyle \theta_{d_{K''/K}/e_{K'/K}}} & & \downarrow \\
\mathrm{Spec}(A_1') & \longrightarrow & [\mathbb{A}^1/\mathbb{G}_m]_{W_n} & \xrightarrow{\theta_{e_{K'/K}}} & [\mathbb{A}^1/\mathbb{G}_m]_{W_n}.
\end{array} \tag{8.2.24.2}$$

It follows that the image in P_n of any global section

$$s \in \Gamma\big((\mathrm{Spec}(A_1')_{\mathrm{et}}/[\mathbb{A}^1/\mathbb{G}_m]_{W_n})_{\mathrm{cris}}, \mathcal{O}_{\mathrm{Spec}(A_1')_{\mathrm{et}}/[\mathbb{A}^1/\mathbb{G}_m]_{W_n}}\big) \tag{8.2.24.3}$$

is in the image of (8.2.24.1). In particular (8.2.24.1) is surjective.

To see that the map (8.2.24.1) is injective, let $K \subset K' \subset \overline{K}$ be an extension with $p^{2n}|e_{K'/K}$, and let $m \in B_n(A')\langle V\rangle$ be a section that maps to zero in P_n. After replacing K' by an extension we may assume that the image of m in $\Gamma((\mathrm{Spec}(A_1')_{\mathrm{et}}/\Delta_n^{(e_{K'})})_{\mathrm{cris}}, \mathcal{O}_{\mathrm{Spec}(A_1')_{\mathrm{et}}/\Delta_n^{(e_{K'})}})$ is zero. Let $K' \subset K'' \subset \overline{K}$ be an extension with $p^n|e_{K''/K'}$ as above. Then the diagram (8.2.24.2) shows that the image of m in $B_n(A_1'')\langle V\rangle$ is zero. Consequently the kernel of (8.2.24.1) is zero.

This completes the construction of the element ν_β and the proof that the resulting map (8.2.20.1) is an isomorphism. The statement that $\nu_{\zeta\beta} = \tilde{\zeta}^{p^n}\nu_\beta$ follows immediately from the construction. From this property (iv) also follows.

8.2.25. — To prove property (iii), for any $K \subset K' \subset \overline{K}$ as in 8.2.21, let

$$f : \mathrm{Spec}(B_n(A')\langle V\rangle) \longrightarrow \mathrm{Spec}(B_n(A')\langle V\rangle) \tag{8.2.25.1}$$

be the map induced by the canonical lifting φ of Frobenius to $B_n(A')$ and $(1+V) \mapsto (1+V)^p$. For any lifting $\tilde{\beta} \in B_n(A')$ of β, $\varphi(\tilde{\beta}) = \tilde{\beta}^p + h$, where h is in the divided power ideal of $B_n(A')$. In particular, $\varphi(\tilde{\beta})^{p^n} = \tilde{\beta}^{p^{n+1}}$. It follows that f and the canonical liftings of Frobenius to E_n and $[\mathbb{A}^1/\mathbb{G}_m]_{W_n}$ induce a semi-linear endomorphism of the commutative square

$$\begin{array}{ccc} \mathrm{Spec}(B_n(A')\langle V\rangle) & \longrightarrow & E_n \\ \downarrow{\scriptstyle \rho_{K'}} & & \downarrow \\ [\mathbb{A}^1/\mathbb{G}_m]_{W_n} & \xrightarrow{\theta_{d_{K'/K}}} & [\mathbb{A}^1/\mathbb{G}_m]_{W_n}. \end{array} \tag{8.2.25.2}$$

From this and the universal property described in 8.2.22 statement (iii) follows.

8.2.26. — Finally to prove property (ii), note that since ∇ is a derivation it suffices to show that $\nabla(V) = (1+V)\mathrm{dlog}(t)$. To see this, let $K \subset K' \subset \overline{K}$ be any extension as in 8.2.21, and define an action of $\mathbb{G}_m$ on $\mathrm{Spec}(B_n(A')\langle V\rangle)$ for which a scheme valued point $u \in \mathbb{G}_m$ acts by $(1+V) \mapsto u(1+V)$. With this definition the group $\mathbb{G}_m$ acts on the entire diagram (8.2.25.2). In particular, looking at the first infinitesimal neighborhood of the diagonal there is a commutative diagram

$$\begin{array}{ccc} \mathrm{Spec}(B_n(A')) & \longrightarrow & E_n \\ \downarrow{\scriptstyle \Delta} & & \downarrow{\scriptstyle \Delta} \\ \mathrm{Spec}(B_n(A') \otimes W_n[(u-1)]/(u-1)^2) & \longrightarrow & E_n \otimes W_n[(u-1)]/(u-1)^2 \\ \downarrow\downarrow & & \downarrow\downarrow \\ \mathrm{Spec}(B_n(A')) & \longrightarrow & E_n \\ \downarrow & & \downarrow \\ [\mathbb{A}^1/\mathbb{G}_m]_{W_n} & \xrightarrow{\theta_{d_{K'/K}}} & [\mathbb{A}^1/\mathbb{G}_m]_{W_n}. \end{array} \tag{8.2.26.1}$$

It follows from 8.2.22 that for any object

$$(\mathrm{Spec}(A'_1) \hookrightarrow T) \in \mathrm{Cris}(\mathrm{Spec}(A'_1)/[\mathbb{A}^1/\mathbb{G}_m] \times_{\theta_{e_{K'/K}}, [\mathbb{A}^1/\mathbb{G}_m]} E_n \otimes W_n[(u-1)]/(u-1)^2), \tag{8.2.26.2}$$

there exists a unique map $\lambda : T \to \mathrm{Spec}(B_n(A')\langle V\rangle \otimes W_n[(u-1)]/(u-1)^2)$ such that the diagram

$$\begin{array}{ccc} T & \longrightarrow & [\mathbb{A}^1/\mathbb{G}_m]_{W_n} \\ {\scriptstyle \lambda}\downarrow & & \downarrow{\scriptstyle \theta_{p^n}} \\ \mathrm{Spec}(B_n(A')\langle V\rangle \otimes W_n[(u-1)]/(u-1)^2) & \xrightarrow{\rho_{K'}} & [\mathbb{A}^1/\mathbb{G}_m]_{W_n} \\ \downarrow & & \downarrow{\scriptstyle \theta_{d_{K'/K}}} \\ E_n \otimes W_n[(u-1)]/(u-1)^2 & \longrightarrow & [\mathbb{A}^1/\mathbb{G}_m]_{W_n}. \end{array} \tag{8.2.26.3}$$

commutes and the composite of the left column is the given map to $E_n \otimes W_n[(u-1)]/(u-1)^2$. Using this one sees as in 8.2.22 that there is an isomorphism

(8.2.26.4)
$$B_n\langle V\rangle \otimes W_n[(u-1)]/(u-1)^2 \longrightarrow h_{\mathrm{cris}*}\mathcal{O}_{\mathrm{Spec}(\overline{A}_1)/\widetilde{\Delta}_n}(E_n \otimes W_n[(u-1)]/(u-1)^2)$$

such that for $i=1,2$ the diagram

(8.2.26.5)
$$\begin{array}{ccc} B_n\langle V\rangle \otimes W_n[(u-1)]/(u-1)^2 & \xleftarrow{\mathrm{pr}_i^*} & B_n\langle V\rangle \\ \downarrow{\scriptstyle (8.2.26.4)} & & \downarrow{\scriptstyle (8.2.20.1)} \\ h_{\mathrm{cris}*}\mathcal{O}_{\mathrm{Spec}(\overline{A}_1)/\widetilde{\Delta}_n}(E_n \otimes W_n[(u-1)]/(u-1)^2) & \xleftarrow{\mathrm{pr}_i^*} & P_n \end{array}$$

commutes. In particular, $\nabla(1+V)$ is given by the class of

(8.2.26.6)
$$\mathrm{pr}_1^*(1+V) - \mathrm{pr}_2^*(1+V) = (1+V)\cdot(u-1) = (1+V)\mathrm{dlog}(t)$$

so the connection must satisfy

(8.2.26.7)
$$\nabla(V) = \nabla(1+V) = (1+V)\mathrm{dlog}(t).$$

This completes the proof of 8.2.20. □

Remark 8.2.27. — It follows from the construction that the image of ν_β in P_{n-1} is equal to ν_{β^p}. In particular the map $P_n \to P_{n-1}$ is surjective.

8.2.28. — Let $\mathcal{N} : P_n \to P_n$ denote the endomorphism characterized by $\nabla(m) = \mathcal{N}(m)\mathrm{dlog}(t)$ for all $m \in P_n$.

For a p^n-th root β of π in $\overline{A}$, define

(8.2.28.1)
$$u_\beta := \log(\nu_\beta) := \sum_{n>0}(-1)^{n+1}(n-1)!(\nu_\beta - 1)^{[n]},$$

where we use the fact that ν_β maps to 1 in $\overline{A}_n$.

Corollary 8.2.29 ([**41**, 3.6]). — *The map $\mathcal{N} : P_n \to P_n$ is surjective,*

(8.2.29.1)
$$\{a \in P_n | \mathcal{N}^i(a) = 0\} = \oplus_{0\leq j<i} B_n \cdot u_\beta^{[j]},$$

(8.2.29.2)
$$\{a \in P_n | \mathcal{N}^i(a) = 0 \textit{ for some } i\} = B_n\langle u_\beta\rangle,$$

and u_β is transcendental over B_n.

Proof. — For any element $\alpha = \sum_{n\geq 0}\alpha_n(\nu_\beta - 1)^{[n]} \in P_n$ we have

(8.2.29.3)
$$\mathcal{N}(\alpha) = \sum_{n\geq 0}(n\alpha_n + \alpha_{n+1})(\nu_\beta - 1)^{[n]}.$$

If $\gamma = \sum_{n\geq 0}\gamma_n(\nu_\beta - 1)^{[n]}$ is any element then defining α_n inductively by

(8.2.29.4)
$$\alpha_1 = \gamma_0, \quad \alpha_{n+1} = \gamma_n - n\alpha_n$$

we obtain an element with image under $\mathcal{N}$ equal to γ so $\mathcal{N}$ is surjective.

The formula (8.2.29.3) also shows that if $\mathcal{N}(a) = 0$ then $a \in B_n$. Furthermore,

$$\mathcal{N}(u_\beta) = 1 + \sum_{n\geq 1}(n(-1)^{n+1}(n-1)! + (-1)^{n+2}n!)(\nu_\beta - 1)^{[n]} = 1. \tag{8.2.29.5}$$

It follows that for any $j \geq 0$ we have $\mathcal{N}(u_\beta^{[j]}) = u_\beta^{[j-1]}$. From this the identifications (8.2.29.1) and (8.2.29.2) follow. Indeed if $\mathcal{N}^i(\alpha) = 0$, then by induction $\mathcal{N}(\alpha) = \sum_{0\leq j<i-1} a_j u_\beta^{[j]}$. Consequently $\alpha - \sum_{0\leq j<i-1} a_j u_\beta^{[j+1]}$ is in the kernel of $\mathcal{N}$ and hence lies in B_n. This implies (8.2.29.1) and (8.2.29.2) follows from (8.2.29.1). The statement that u_β is transcendental over B_n follows from the equation $\mathcal{N}(u_\beta^{[j]}) = u_\beta^{[j-1]}$. □

Theorem 8.2.30 (**[41**, 3.7]). — *For any choice of π, there is a canonical B^+_{cris}-linear isomorphism between the ring B^+_{st} (defined using the chosen π) and*

$$\{a \in \mathbb{Q} \otimes \varprojlim_n P_n | \mathcal{N}^i(a) = 0 \textit{ for some } i \geq 0\} \tag{8.2.30.1}$$

which preserves the Frobenius endomorphism φ, $\mathcal{N}$, and the action of $\mathrm{Gal}(\overline{K}/K)$.

Proof. — Let $s = (s_n)$ be a sequence in $\overline{A}$ with $s_0 = \pi$ and $s^p_{n+1} = s_n$, and let $u_s \in B^+_{\mathrm{st}}$ be the element obtained as in 8.2.4. The isomorphism in the theorem is obtained by sending u_s to $(u_{s_n})_n \in \varprojlim P_n$ (this is well-defined by 8.2.27). The inverse map is given by

$$((\nu_{s_n} - 1)^{[i]})_n \longmapsto (i!)^{-1}(\epsilon(s)\pi^{-1} - 1)^i, \tag{8.2.30.2}$$

where $\epsilon(s)$ is defined as in 8.2.3. □

8.3. Crystalline interpretation of $(B_{\mathrm{st}} \otimes D)^{N=0}$

8.3.1. — Let A, k, K, etc. be as in 8.2.1, and let $\mathcal{X}/A$ be a tame (see 2.5.14), proper, regular Deligne-Mumford stack whose generic fiber $\mathcal{X}_K \to \mathrm{Spec}(K)$ is proper and smooth and whose closed fiber is a divisor with normal crossings. By definition, this means that for any étale morphism $U \to \mathcal{X}$, with U a scheme, the scheme U is regular and the closed fiber of U is a divisor with normal crossings in U. Let $\mathcal{Y}/k$ denote the closed fiber of $\mathcal{X}$, and let r be an integer such that for any geometric point $\bar{x} \to \mathcal{Y}$ the scheme $\mathrm{Spec}(\mathcal{O}_{\mathcal{Y},\bar{x}})$ has less than equal to r irreducible components.

Let $\mathcal{S}_{W[t]}$ denote the stack $\mathcal{S}_H(\alpha)$ over $\mathbb{A}^1$ obtained as in 6.4.1 by taking $\alpha = (1, \ldots, 1)$ (r-times) and H the full symmetric group on r letters. Let $\overline{\mathcal{S}}$ denote the stack over $[\mathbb{A}^1/\mathbb{G}_m]$ described in 6.1.9, and let $\mathcal{S}_{A,\mathfrak{m}_A}$ denote the stack over $\mathrm{Spec}(A)$ obtained by base change from the map $\mathrm{Spec}(A) \to [\mathbb{A}^1/\mathbb{G}_m]$ defined by the maximal ideal of A.

Recall (6.1.9) that $\overline{\mathcal{S}}$ is the stack associated to the prestack $\overline{\mathcal{S}}^{\mathrm{ps}}$ which to any scheme T associates the following groupoid $\overline{\mathcal{S}}^{\mathrm{ps}}(T)$:

Objects: Collections $(x_1, \ldots, x_r, v)$ of elements of $\Gamma(T, \mathcal{O}_T)$ with $v \in \Gamma(T, \mathcal{O}^*_T)$. For such an object we write (as before) $E(x)$ for $\{i | x_i \notin \Gamma(T, \mathcal{O}^*_T)\}$.

Morphisms: A morphism $(x_1, \dots, x_r, v) \to (x'_1, \dots, x'_r, v')$ is a collection of data $((u, h), \lambda)$, where $u = \{u_i\}_{i \in E(x)}$ is a set of elements $u_i \in \Gamma(T, \mathcal{O}_T^*)$, $h : E(x) \to E(x')$ is a bijection, and $\lambda \in \Gamma(T, \mathcal{O}_T^*)$. Composition is defined by the formula

$$((u', h'), \lambda') \circ ((u, h), \lambda) = ((\{u'_{h(i)} \cdot u_i\}_{i \in E(x)}, h' \circ h), \lambda \cdot \lambda'). \tag{8.3.1.1}$$

If $\pi \in \mathfrak{m}_A$ denotes a uniformizer, then by 6.1.8 the stack $\mathcal{S}_{A,\mathfrak{m}_A}$ can be described as the stack associated to the prestack which to any A-scheme T associates the following groupoid:

Objects: Objects $(x_1, \dots, x_r, v) \in \overline{\mathcal{S}}^{\mathrm{ps}}(T)$ with $x_1 \cdots x_r v = \pi$ (where we abusively write also π for its image in $\mathcal{O}_T$).

Morphisms: A morphism $(x_1, \dots, x_r, v) \to (x'_1, \dots, x'_r, v')$ is a pair (u, h), where $h : E(x) \to E(x')$ is a bijection and $u = \{u_i\}_{i \in E(x)}$ is a set of elements of $\Gamma(T, \mathcal{O}_T^*)$ such that

$$\Big(\prod_{i \notin E(x')} x'_i\Big) v' = \Big(\prod_{i \in E(x)} u_i^{-1}\Big)\Big(\prod_{i \notin E(x)} x_i\Big) v. \tag{8.3.1.2}$$

Composition is defined by the formula

$$(u', h') \circ (u, h) = (\{u'_{h(i)} \cdot u_i\}_{i \in E(x)}, h' \circ h). \tag{8.3.1.3}$$

We also consider the stacks $\mathcal{S}_{W\langle t \rangle}$, $\mathcal{S}_W$, and $\mathcal{S}_k$ defined as in 6.4.1. Recall that these stacks are defined as follows.

1. $\mathcal{S}_{W\langle t \rangle}$ is the base change $\overline{\mathcal{S}} \times_{[\mathbb{A}^1/\mathbb{G}_m]} \mathrm{Spec}(W\langle t \rangle)$, where $\mathrm{Spec}(W\langle t \rangle) \to [\mathbb{A}^1/\mathbb{G}_m]$ is the composite map

$$\mathrm{Spec}(W\langle t \rangle) \xrightarrow{t \mapsto t} \mathrm{Spec}(\mathbb{Z}[t]) = \mathbb{A}^1 \xrightarrow{\text{projection}} [\mathbb{A}^1/\mathbb{G}_m]. \tag{8.3.1.4}$$

2. $\mathcal{S}_W = \mathcal{S}_{W\langle t \rangle} \times_{\mathrm{Spec}(W\langle t \rangle), t \mapsto 0} \mathrm{Spec}(W)$.
3. $\mathcal{S}_k = \mathcal{S}_W \times_{\mathrm{Spec}(W)} \mathrm{Spec}(k)$.

By the construction in 6.2 (see also 6.2.5) there is a canonical smooth map $\mathcal{X} \to \mathcal{S}_{A,\mathfrak{m}_A}$.

Warning 8.3.2. — Contrary to what the notation may suggest, it is not true that the stack $\mathcal{S}_{A,\mathfrak{m}_A}$ is equal to the base change of $\mathcal{S}_W$ to $\mathrm{Spec}(A)$ since the stack $\mathcal{S}_W$ is defined by pulling back $\overline{\mathcal{S}}$ via the map $\mathrm{Spec}(W) \to [\mathbb{A}^1/\mathbb{G}_m]$ defined by the zero map $\mathcal{O}_W \to \mathcal{O}_W$. It is true, however, that the reductions $\mathcal{S}_{A,\mathfrak{m}_A} \otimes_A k$ is isomorphic to $\mathcal{S}_W \otimes_W k \simeq \mathcal{S}_k$. Such an isomorphism is determined by the choice of a generator for $\mathfrak{m}_A/\mathfrak{m}_A^2$ which induces an isomorphism between the two arrows $\mathrm{Spec}(k) \to [\mathbb{A}^1/\mathbb{G}_m]$ obtained by reduction.

For an interpretation of $\mathcal{S}_{A,\mathfrak{m}_A}$ (resp. $\overline{\mathcal{S}}$) in terms of logarithmic structures see 9.1.26 (resp. 9.1.21).

8.3.3. — For any integer e define $\Delta^{(e_K e)}$ as in 8.2.6, and let $\mathcal{R}^{(e)} \to \Delta^{(e_K e)}$ denote the pullback of the stack $\overline{\mathcal{S}}$ via the map

(8.3.3.1) $$\Delta^{(e_K e)} \longrightarrow [\mathbb{A}^1/\mathbb{G}_m] \xrightarrow{\theta_e} [\mathbb{A}^1/\mathbb{G}_m].$$

Also define $\mathcal{R}_n^{(e)}$ to be the pullback of $\mathcal{R}^{(e)}$ to $\Delta_n^{(e_K e)}$.

For any finite extension $K \subset K' \subset \overline{K}$ with ring of integers A', there is a canonical map

(8.3.3.2) $$\mathcal{X}_{A'} \longrightarrow \mathcal{R}^{(e_{K'/K})}$$

defined by the map $\mathcal{X} \to \mathcal{S}_{A,\mathfrak{m}_A}$ and the composite

(8.3.3.3) $$\mathcal{X}_{A'} \longrightarrow \mathrm{Spec}(A') \xrightarrow{\mathfrak{m}_{A'}} [\mathbb{A}^1/\mathbb{G}_m].$$

Let $\mathcal{E}^{(e_{K'/K})} \subset \mathcal{R}^{(e_{K'/K})}$ be the inverse image of

(8.3.3.4) $$[\mathrm{Spec}(k[t]/t^{e_{K'}})/\mathbb{G}_m] \subset [\mathrm{Spec}(W[t])/\mathbb{G}_m].$$

Observe that by the definition of $\Delta^{(e_{K'})}$ there is a natural divided power structure on the ideal of $\mathcal{E}^{(e_{K'/K})}$ in $\mathcal{R}^{(e_{K'/K})}$. We view $\mathcal{R}^{(e_{K'/K})}$ as a divided power stack with this PD-ideal.

For $K = K'$ we write simply $\mathcal{E} \subset \mathcal{R}^{(1)}$. Observe that $\mathcal{E}^{(e_{K'/K})} \simeq \mathcal{E} \times_{[\mathbb{A}^1/\mathbb{G}_m],\theta_{e_{K'/K}}} [\mathbb{A}^1/\mathbb{G}_m]$ and that the pullback of $\mathcal{E}$ via the map $\mathrm{Spec}(A_1) \to \Delta^{(e_K)}$ is equal to $\mathcal{S}_H(\alpha)_{A_1,\pi}$.

Lemma 8.3.4. — *The map (8.3.3.2) induces a syntomic morphism $\mathcal{X}_{A'_1} \to \mathcal{E}^{(e_{K'/K})}$.*

Proof. — That the morphism (8.3.3.2) induces a map $\mathcal{X}_{A'_1} \to \mathcal{E}^{(e_{K'/K})}$ is clear because the map $\mathrm{Spec}(A'_1) \to [\mathbb{A}^1/\mathbb{G}_m]$ factors through $[\mathrm{Spec}(\mathbb{Z}[t]/t^{e_{K'}})/\mathbb{G}_m]$.

Let $\overline{\Delta}^{(e)} \subset \Delta^{(e)}$ denote the closed substack defined by the divided power ideal. Recall that the stack $\overline{\Delta}^{(e)}$ is isomorphic to $[\mathrm{Spec}(k[t]/t^e)/\mathbb{G}_m]$. Recall also that by 8.2.9, for any finite extension $K \subset K'$ the natural map

(8.3.4.1) $$\mathrm{Spec}(A'_1) \longrightarrow \overline{\Delta}^{(e_{K'})} \times_{\overline{\Delta}^{(e_K)}} \mathrm{Spec}(A_1)$$

is syntomic.

The map $\mathcal{X}_{A'_1} \to \mathcal{E}^{(e_{K'/K})}$ can be factored as the composite of the following maps: the map

(8.3.4.2)
$$\mathcal{X}_{A_1} \times_{\mathrm{Spec}(A_1)} \mathrm{Spec}(A'_1) \longrightarrow (\mathcal{E} \times_{\overline{\Delta}^{(e_K)}} \mathrm{Spec}(A_1)) \times_{\mathrm{Spec}(A_1)} (\overline{\Delta}^{(e_{K'})} \times_{\overline{\Delta}^{(e_K)}} \mathrm{Spec}(A_1))$$

which is syntomic being the product of two syntomic morphisms (note that $\mathcal{E} \times_{\overline{\Delta}^{(e_K)}} \mathrm{Spec}(A_1) \simeq \mathcal{S}_{A,\mathfrak{m}_A} \times_{\mathrm{Spec}(A)} \mathrm{Spec}(A_1)$), the isomorphism

(8.3.4.3)
$$\begin{array}{c} (\mathcal{E} \times_{\overline{\Delta}^{(e_K)}} \mathrm{Spec}(A_1)) \times_{\mathrm{Spec}(A_1)} (\overline{\Delta}^{(e_{K'})} \times_{\overline{\Delta}^{(e_K)}} \mathrm{Spec}(A_1)) \\ \downarrow \simeq \\ (\mathcal{E} \times_{\overline{\Delta}^{(e_K)}} \mathrm{Spec}(A_1)) \times_{\overline{\Delta}^{(e_K)}} \overline{\Delta}^{(e_{K'})}, \end{array}$$

and the projection

$$(\mathcal{E} \times_{\overline{\Delta}^{(e_K)}} \mathrm{Spec}(A_1)) \times_{\overline{\Delta}^{(e_K)}} \overline{\Delta}^{(e_{K'})} \longrightarrow \mathcal{E} \times_{\overline{\Delta}^{(e_K)}} \overline{\Delta}^{(e_{K'})} \simeq \mathcal{E}^{(e_{K'/K})} \tag{8.3.4.4}$$

which is obtained by base change from the syntomic morphism $\mathrm{Spec}(A_1) \to \overline{\Delta}^{(e_K)}$ and hence is syntomic. Since a composite of syntomic morphisms is again syntomic the lemma follows. □

8.3.5. — If $K' \subset K'' \subset \overline{K}$ is a second finite extension then there is a canonical commutative diagram

$$\begin{array}{ccccc} \mathcal{X}_{A''} & \longrightarrow & \mathcal{R}^{(e_{K''/K})} & \longrightarrow & \Delta^{(e_{K''})} \\ \downarrow & & \downarrow & & \downarrow \theta_{e_{K''/K'}} \\ \mathcal{X}_{A'} & \longrightarrow & \mathcal{R}^{(e_{K'/K})} & \longrightarrow & \Delta^{(e_{K'})} \\ \downarrow & & \downarrow & & \downarrow \theta_{e_{K'/K}} \\ \mathcal{X} & \longrightarrow & \mathcal{R}^{(1)} & \longrightarrow & \Delta^{(e_K)}. \end{array} \tag{8.3.5.1}$$

Consider the inverse limit topos [**5**, VI.8.1.1]

$$(\mathcal{X}_{\overline{A}_1}/\mathcal{R}_n)_{\mathrm{cris}} := \varprojlim_{K \subset K' \subset \overline{K}} (\mathcal{X}_{A'_1}/\mathcal{R}_n^{(e_{K'/K})})_{\mathrm{cris}}, \tag{8.3.5.2}$$

where as in 8.2 $\overline{A}$ denotes the integral closure of A in $\overline{K}$ and $\overline{A}_1$ denotes $\overline{A}/p\overline{A}$. The structure sheaves in $(\mathcal{X}_{A'_1}/\mathcal{R}_n^{(e_{K'/K})})_{\mathrm{cris}}$ define a structure sheaf $\mathcal{O}_{\mathcal{X}_{\overline{A}_1}/\mathcal{R}_n}$, and there is a natural commutative diagram of ringed topoi

$$\begin{array}{ccc} (\mathcal{X}_{\overline{A}_1}/\mathcal{R}_n)_{\mathrm{cris}} & \xrightarrow{j} & (\mathrm{Spec}(\overline{A}_1)/\widetilde{\Delta}_n)_{\mathrm{cris}} \\ f \downarrow & & \downarrow h \\ (\mathcal{X}_{A_1}/\mathcal{R}_n^{(1)})_{\mathrm{cris}} & \xrightarrow{g} & (\mathrm{Spec}(A_1)/\Delta_n^{(e_K)})_{\mathrm{cris}}, \end{array} \tag{8.3.5.3}$$

where $(\mathrm{Spec}(\overline{A}_1)/\widetilde{\Delta}_n)_{\mathrm{cris}}$ denotes the inverse limit topos in (8.2.6.10) and h is the morphism (8.2.6.12).

Define

$$D_n := H^*((\mathcal{Y}/\Delta_n^{(e_K)})_{\mathrm{cris}}, \mathcal{O}_{\mathcal{Y}/\Delta_n^{(e_K)}}), \tag{8.3.5.4}$$

and let $D = \mathbb{Q} \otimes (\varprojlim D_n)$. The group D is a module over the field of fractions K_0 of W. By the constructions of 6.5 the W-module D comes equipped with a monodromy operator $N : D \to D$ and a semi-linear Frobenius endomorphism $\varphi : D \to D$. Also define

$$H^*(\mathcal{X}_{\overline{A}}/\mathcal{R}) := \varprojlim_n H^*((\mathcal{X}_{\overline{A}_1}/\mathcal{R}_n)_{\mathrm{cris}}, \mathcal{O}_{\mathcal{X}_{\overline{A}_1}/\mathcal{R}_n}). \tag{8.3.5.5}$$

The main result of this section is the following:

Theorem 8.3.6 **([41, 4.1])**. — *The kernel of the map*

$$\mathcal{N} := \mathcal{N} \otimes 1 + 1 \otimes N : B_{\mathrm{st}}^+ \otimes_{K_0} D \to B_{\mathrm{st}}^+ \otimes_{K_0} D \tag{8.3.6.1}$$

is canonically isomorphic to $\mathbb{Q} \otimes H^*(\mathfrak{X}_{\overline{A}}/\mathcal{R})$.

The proof is in steps 8.3.7-8.3.16.

Let $\kappa : (\mathfrak{X}_{\overline{A}_1}/\mathcal{R}_n)_{\mathrm{cris}} \to (\mathrm{Spec}(A_1)/\Delta_n^{(e_K)})_{\mathrm{cris}}$ be the morphism of topoi $g \circ f = h \circ j$ (where the notation is as in (8.3.5.3)).

Proposition 8.3.7. — *Let $K \subset K' \subset \overline{K}$ be a finite extension, and write also h and κ for the morphisms of topoi*

$$(\mathrm{Spec}(A_1')/\Delta_n^{(e_{K'})})_{\mathrm{cris}} \longrightarrow (\mathrm{Spec}(A_1)/\Delta_n^{(e_K)})_{\mathrm{cris}}, \tag{8.3.7.1}$$

$$(\mathfrak{X}_{A_1'}/\mathcal{R}_n^{(e_{K'/K})})_{\mathrm{cris}} \longrightarrow (\mathrm{Spec}(A_1)/\Delta_n^{(e_K)})_{\mathrm{cris}}. \tag{8.3.7.2}$$

Then the natural map

$$(Rg_*\mathcal{O}_{\mathfrak{X}_{A_1}/\mathcal{R}_n^{(1)}}) \otimes^{\mathbb{L}} Rh_*\mathcal{O}_{\mathrm{Spec}(A_1')/\Delta_n^{(e_{K'})}} \longrightarrow R\kappa_*\mathcal{O}_{\mathfrak{X}_{A_1'}/\mathcal{R}_n^{(e_{K'/K})}} \tag{8.3.7.3}$$

is an isomorphism.

Proof. — Let $(U \hookrightarrow T) \in \mathrm{Cris}(\mathrm{Spec}(A_1)/\Delta_n^{(e_K)})$ be an object with T affine, set $U' := \mathrm{Spec}(A_1') \times_{\mathrm{Spec}(A)} U$, and let $\mathcal{R}_{n,T}^{(1)}$, $\Delta_{n,T}^{(e_{K'})}$, and $\mathcal{R}_{n,T}^{(e_{K'/K})}$ be the stacks obtained by base change $T \to \Delta_n^{(e_K)}$. It then suffices to show that the natural map

$$R\Gamma(\mathfrak{X}_{A_1} \times_{\mathrm{Spec}(A_1)} U/\mathcal{R}_{n,T}^{(1)}) \otimes_{\mathcal{O}_T}^{\mathbb{L}} R\Gamma(U'/\Delta_{n,T}^{(e_{K'})}) \longrightarrow R\Gamma((\mathfrak{X}_{A_1} \times_{\mathrm{Spec}(A_1)} U')/\mathcal{R}_{n,T}^{(e_{K'/K})}) \tag{8.3.7.4}$$

is an isomorphism, where we omit the structure sheaves from the notation.

Using the same cohomological descent argument as in the proof of [7, V.4.2.1], one sees furthermore that it suffices to prove that (8.3.7.4) is an isomorphism after replacing $\mathfrak{X}_{A_1}$ and U' by étale covers. By 8.1.6, we may therefore assume that $\mathfrak{X}_{A_1}$ and U' are affine and that there exist regular embeddings $\mathfrak{X}_{A_1} \times_{A_1} U \hookrightarrow Y$ over $\mathcal{R}_{n,T}^{(1)}$ and $U' \hookrightarrow Z$ over $\Delta_{n,T}^{(e_{K'})}$ with $Y \to \mathcal{R}_{n,T}^{(1)}$ and $Z \to \Delta_{n,T}^{(e_{K'})}$ smooth. Let $\mathcal{D}_1$ (resp. $\mathcal{D}_2$) denote the divided power envelope of $\mathfrak{X}_{A_1} \times_{A_1} U \subset Y$ (resp. $U' \subset Z$), and observe that by 8.1.7 the rings $\mathcal{D}_1$ and $\mathcal{D}_2$ are flat over T. Since

$$Y \times_T Z \simeq (Y \times_{\mathcal{R}_{n,T}^{(1)}} \mathcal{R}_{n,T}^{(e_{K'/K})}) \times_{\mathcal{R}_{n,T}^{(e_{K'/K})}} (Z \times_{\Delta_{n,T}^{(e_{K'})}} \mathcal{R}_{n,T}^{(e_{K'/K})}) \tag{8.3.7.5}$$

and the divided power envelope of $\mathfrak{X}_{A_1} \times_{\mathrm{Spec}(A_1)} U'$ in $Y \times_T Z$ is equal to $\mathrm{Spec}(\mathcal{D}_1 \otimes_{\mathcal{O}_T} \mathcal{D}_2)$, it follows that the arrow (8.3.7.4) can in this local situation be identified with the natural isomorphism

$$(\mathcal{D}_1 \otimes \Omega^\bullet_{Y/\mathcal{R}_{n,T}^{(1)}}) \otimes_{\mathcal{O}_T} (\mathcal{D}_2 \otimes \Omega^\bullet_{Z/\Delta_{n,T}^{(e_{K'})}}) \longrightarrow (\mathcal{D}_1 \otimes_{\mathcal{O}_T} \mathcal{D}_2) \otimes \Omega^\bullet_{Y \times_T Z/\mathcal{R}_{n,T}^{(e_{K'/K})}}. \quad \square \tag{8.3.7.6}$$

Corollary 8.3.8. — *The natural map*

$$(Rg_*\mathcal{O}_{\mathcal{X}_{A_1}/\mathcal{R}_n^{(1)}}) \otimes^{\mathbb{L}} Rh_*\mathcal{O}_{\mathrm{Spec}(\overline{A}_1)/\widetilde{\Delta}_n} \longrightarrow R\kappa_*\mathcal{O}_{\mathcal{X}_{\overline{A}_1}/\mathcal{R}_n} \tag{8.3.8.1}$$

is an isomorphism in the derived category of sheaves in $(\mathrm{Spec}(A_1)/\Delta_n^{(e_K)})_{\mathrm{cris}}$.

Proof. — This follows from 8.3.7 by passage to the limit and [**5**, VI.8.7.3]. □

Corollary 8.3.9. — *There is a natural isomorphism*

$$(Rg_*\mathcal{O}_{\mathcal{X}_{A_1}/\mathcal{R}_n^{(1)}}) \otimes^{\mathbb{L}} h_*\mathcal{O}_{\mathrm{Spec}(\overline{A}_1)/\widetilde{\Delta}_n} \simeq R\kappa_*\mathcal{O}_{\mathcal{X}_{\overline{A}_1}/\mathcal{R}_n}. \tag{8.3.9.1}$$

Proof. — Combine 8.3.8 with 8.2.7. □

For ease of notation define

$$M_n^{pq} := H^q((\mathrm{Spec}(A_1)/\Delta_n^{(e_K)})_{\mathrm{cris}}, (R^q g_*\mathcal{O}_{\mathcal{X}_{A_1}/\mathcal{R}_n^{(1)}}) \otimes^{\mathbb{L}} h_*\mathcal{O}_{\mathrm{Spec}(\overline{A}_1)/\widetilde{\Delta}_n}). \tag{8.3.9.2}$$

Corollary 8.3.10. — *There is a canonical spectral sequence in the category of projective systems of abelian groups*

$$E_1^{pq} = M_n^{pq} \Longrightarrow H^{p+q}((\mathcal{X}_{\overline{A}_1}/\mathcal{R}_n)_{\mathrm{cris}}, \mathcal{O}_{\mathcal{X}_{\overline{A}_1}/\mathcal{R}_n}). \tag{8.3.10.1}$$

8.3.11. — Let $\mathrm{Spec}(A_1) \hookrightarrow E_n = \mathrm{Spec}(R_n)$ be the object of $\mathrm{Cris}(\mathrm{Spec}(A_1)/\Delta_n^{(e_K)})$ defined in 8.2.19, and let $K_n^{\cdot}$ denote the complex of R_n-modules obtained by evaluating $Rg_*\mathcal{O}_{\mathcal{X}_{A_1}/\mathcal{R}_n^{(1)}}$ on E_n. Observe that we have a canonical isomorphism

$$K_n^{\cdot} \simeq R\Gamma((\mathcal{X}_{A_1}/\mathcal{S}_{R_n})_{\mathrm{cris}}, \mathcal{O}_{\mathcal{X}_{A_1}/\mathcal{S}_{R_n}}). \tag{8.3.11.1}$$

By 6.4.6 there exists an integer s independent of n such that $\mathcal{H}^m(K_n^{\cdot})$ admits a morphism to a free R_n-module whose kernel and cokernel is annihilated by p^s.

For any object $(U \hookrightarrow T) \in \mathrm{Cris}(\mathrm{Spec}(A_1)/\Delta_n^{(e_K)})$ with T affine and a retraction $r : T \to E_n$, the base change theorem 8.1.8 implies that the natural map

$$K_n^{\cdot} \otimes^{\mathbb{L}} \mathcal{O}_T \longrightarrow Rg_*\mathcal{O}_{\mathcal{X}_{A_1}/\mathcal{R}_n^{(1)}}(T) \tag{8.3.11.2}$$

is an isomorphism. By the same argument used in the proof of 6.4.10 this implies that the kernels and cokernels of the natural map

$$\mathcal{H}^m(K^\bullet) \otimes \mathcal{O}_T \longrightarrow R^m g_*\mathcal{O}_{\mathcal{X}_{A_1}/\mathcal{R}_n^{(1)}}(T) \tag{8.3.11.3}$$

are annihilated by some p^s (for some possibly bigger s).

Lemma 8.3.12. — *Let n and m be integers, and let $\mathcal{M}$ be the crystal on the site* $\mathrm{Cris}(\mathrm{Spec}(A_1)/\Delta_n^{(1)})$ *defined by the module $D_n^m \otimes_W R$ with the stratification defined in 6.5.5 (where D_n^m is defined as in 6.4.2). Then there exist an integer s independent of n and a map $\mathcal{M} \to R^m g_*\mathcal{O}_{\mathcal{X}_{A_1}/\mathcal{R}_n^{(1)}}$ whose kernel and cokernel is annihilated by p^s.*

Proof. — Let C_n^m denote the evaluation of $R^m g_* \mathcal{O}_{\mathfrak{X}_{A_1}/\mathcal{R}_n^{(1)}}$ on the object

$$(\mathrm{Spec}(A_1) \hookrightarrow E_n) \in \mathrm{Cris}(\mathrm{Spec}(A_1)/\Delta_n^{(e_K)}). \tag{8.3.12.1}$$

As explained in 6.5.5 the module C_n^m also has a natural stratification, and

$$\mathrm{Hom}(\mathcal{M}, R^m g_* \mathcal{O}_{\mathfrak{X}_{A_1}/\mathcal{R}_n^{(1)}}) \simeq (\mathrm{Hom}(D_n^m, C_n^m))^{\epsilon=0}, \tag{8.3.12.2}$$

where the right hand side denotes the set of morphisms horizontal for the stratifications. By 6.5.8 there exists such a horizontal morphism inducing a map $\psi : \mathcal{M} \to R^m g_* \mathcal{O}_{\mathfrak{X}_{A_1}/\mathcal{R}_n^{(1)}}$ such that the induced map $D_n^m \otimes_W R \to C_n$ has kernel and cokernel annihilated by p^s for some integer s independent of n. From 8.3.11 it then follows that this map ψ has the desired properties. □

Proposition 8.3.13. — *There exists an integer r (independent of n) such that M_n^{pq} is annihilated by p^r for all $q > 0$.*

Proof. — By the preceding lemma it suffices to show that

$$H^q((\mathrm{Spec}(A_1)/\Delta_n^{(1)})_{\mathrm{cris}}, \mathcal{M} \otimes h_* \mathcal{O}_{\mathrm{Spec}(\overline{A}_1)/\tilde{\Delta}_n}) \tag{8.3.13.1}$$

is annihilated by p^r for $q > 0$ and some r independent of n. As in 8.2.19, let P_n denote the value of $h_* \mathcal{O}_{\mathrm{Spec}(\overline{A}_1)/\tilde{\Delta}_n}$ on $\mathrm{Spec}(A_n) \hookrightarrow \mathrm{Spec}(R_n)$, and by definition the value of $\mathcal{M}$ on $\mathrm{Spec}(A_n) \hookrightarrow \mathrm{Spec}(R_n)$ is canonically isomorphic to $D_n \otimes_W R$. Using the descriptions of the connections on P_n and $D \otimes_W R$ in 8.2.28 and 6.5.5 respectively, we obtain that the cohomology groups (8.3.13.1) are computed by the de Rham complex

$$(D_n \otimes_W R) \otimes_R P_n \xrightarrow{N \otimes 1 + 1 \otimes \mathcal{N}} (D_n \otimes_W R) \otimes_R P_n. \tag{8.3.13.2}$$

Therefore the following lemma completes the proof of 8.3.13. □

Lemma 8.3.14. — *The map*

$$N \otimes 1 + 1 \otimes \mathcal{N} : (D_n \otimes_W R) \otimes_R P_n \longrightarrow (D_n \otimes_W R) \otimes_R P_n \tag{8.3.14.1}$$

has cokernel annihilated by p^s for some integer s independent of n.

Proof. — Let $D_\infty = \varprojlim D_n$. By 2.6.8 and 5.1.20 the projective system $D_\cdot \in \mathrm{ps}(W)$ is free of finite type mod $\mathcal{T}$ and therefore the natural map $D_\infty / p^n D_\infty \to D_n$ has kernel and cokernel annihilated by some integer independent of n. Replacing D_n by $D_\infty/(T + p^n D_\infty)$, where $T \subset D_\infty$ denotes the torsion subgroup, we may therefore assume that D_∞ is torsion free and that the projection $D_\infty \otimes \mathbb{Z}/p^n \to D_n$ is an isomorphism for all n.

By 6.5.9 the operator N is nilpotent on D_∞. Let $D_\infty^{N=0}$ denote the elements annihilated by N, and let $\overline{D}_\infty$ denote the cokernel. If $N(p^e d) = 0$ then $N(d) = 0$ so $\overline{D}_\infty$ is also p-torsion free. In particular, writing $(D_\infty^{N=0})_n := (D_\infty^{N=0} \otimes \mathbb{Z}/p^n)$ and $\overline{D}_n := \overline{D}_\infty \otimes \mathbb{Z}/p^n$ there is an exact sequence

$$0 \longrightarrow (D_\infty^{N=0})_n \longrightarrow D_n \longrightarrow \overline{D}_n \longrightarrow 0. \tag{8.3.14.2}$$

Using the flatness of P_n, there is a commutative diagram with exact rows
(8.3.14.3)

$$\begin{array}{ccccccccc} 0 & \longrightarrow & (D_\infty^{N=0})_n \otimes_W P_n & \longrightarrow & D_n \otimes_W P_n & \longrightarrow & \overline{D}_n \otimes_W P_n & \longrightarrow & 0 \\ & & \Big\downarrow{\scriptstyle 1\otimes\mathcal{N}} & & \Big\downarrow{\scriptstyle N\otimes 1+1\otimes\mathcal{N}} & & \Big\downarrow{\scriptstyle \overline{N}\otimes 1+1\otimes\mathcal{N}} & & \\ 0 & \longrightarrow & (D_\infty^{N=0})_n \otimes_W P_n & \longrightarrow & D_n \otimes_W P_n & \longrightarrow & \overline{D}_n \otimes_W P_n & \longrightarrow & 0, \end{array}$$

where $\overline{N} : \overline{D}_n \to \overline{D}_n$ denotes the endomorphism defined by N. By 8.2.29, the map $1 \otimes \mathcal{N} : (D_\infty^{N=0})_n \otimes_W P_n \to (D_\infty^{N=0})_n \otimes_W P_n$ is surjective, so by the snake lemma it suffices to show that there exists an integer s such that p^s annihilates

$$\operatorname{Coker}(\overline{N} \otimes 1 + 1 \otimes \mathcal{N} : \overline{D}_n \otimes_W P_n \to \overline{D}_n \otimes_W P_n) \tag{8.3.14.4}$$

for all n. Since N induces a nilpotent operator on $\overline{D}_\infty$ this follows by induction on the rank of D_∞. □

Remark 8.3.15. — In what follows we write just $\mathcal{N}$ for the endomorphism $N\otimes 1+1\otimes\mathcal{N}$ in (8.3.14.1).

8.3.16. — From 8.3.13 and (8.3.10.1) it follows that there is an isomorphism in $\mathrm{ps}(W)_\mathbb{Q}$

$$\{H^p((\mathcal{X}_{\overline{A}_1}/\mathcal{R}_n)_{\mathrm{cris}}, \mathcal{O}_{\mathcal{X}_{\overline{A}_1}/\mathcal{R}_n})\} \simeq \{M_n^{p0}\}. \tag{8.3.16.1}$$

By the proof of 8.3.13 there is a canonical isomorphism

$$\{M_n^{p0}\} \simeq \operatorname{Ker}(\{D_n \otimes_W P_n\} \xrightarrow{\mathcal{N}} \{D_n \otimes_W P_n\}) \tag{8.3.16.2}$$

in $\mathrm{ps}(W)_\mathbb{Q}$. By 8.2.27, the maps $P_n \to P_{n-1}$ are surjective so the system $\{D_n \otimes_W P_n\}$ satisfies the Mittag-Leffler condition. Passing to the inverse limit we obtain an exact sequence

$$0 \longrightarrow \mathbb{Q} \otimes H^*(\mathcal{X}_{\overline{A}}/\mathcal{R}) \longrightarrow (\varprojlim P_n) \otimes D \xrightarrow{\mathcal{N}} (\varprojlim P_n) \otimes D \longrightarrow 0, \tag{8.3.16.3}$$

where $D = \mathbb{Q} \otimes \varprojlim D_n$ as in 8.3.5. Since the monodromy operator on D is nilpotent, we can replace $\varprojlim P_n$ by the part on which the monodromy operator is nilpotent which by 8.2.30 is equal to B_{st}^+. This completes the proof of 8.3.6. □

8.4. Syntomic complexes

We continue with the notation of 8.3.1.

8.4.1. — First we need some general facts about hypercovers of $\mathcal{X}$.

We consider collections of data $(X_\bullet, X_{A',\bullet} \hookrightarrow Z_{A',\bullet}, F_{Z_{A',\bullet}}, \tau_{K''/K'})$ as follows. The simplicial space $X_\bullet$ is an étale hypercover of $\mathcal{X}$, for every $K \subset K' \subset \overline{K}$ the simplicial space $X_{A',\bullet}$ is the base change $X_\bullet \otimes_A A'$ and $X_{A',\bullet} \hookrightarrow Z_{A',\bullet}$ is an immersion of simplicial spaces over the stack $\mathcal{R}^{(e_{K'/K})}$ such that each morphism $Z_{A',n} \to \mathcal{R}^{(e_{K'/K})}$ is smooth, $F_{Z_{A',\bullet}} : Z_{A',\bullet} \to Z_{A',\bullet}$ is a lifting of Frobenius compatible with the canonical lifting of Frobenius to $\mathcal{R}^{(e_{K'/K})}$, and for every $K' \subset K'' \subset \overline{K}$ the map $\tau_{K''/K'}$: $Z_{A'',\bullet} \to Z_{A',\bullet}$ is a morphism of simplicial spaces compatible with the liftings of Frobenius such that the diagram

$$\begin{array}{ccccccc} X_{A'',\bullet} & \longrightarrow & Z_{A'',\bullet} & \longrightarrow & \mathcal{R}^{(e_{K''/K})} & \longrightarrow & \Delta^{(e_{K''})} \\ \downarrow & & \downarrow{\scriptstyle \tau_{K''/K'}} & & \downarrow & & \downarrow{\scriptstyle \theta_{e_{K''/K'}}} \\ X_{A',\bullet} & \longrightarrow & Z_{A',\bullet} & \longrightarrow & \mathcal{R}^{(e_{K'/K})} & \longrightarrow & \Delta^{(e_{K'})} \end{array} \tag{8.4.1.1}$$

commutes. We further require that for a third extension $K' \subset K'' \subset K''' \subset \overline{K}$

$$\tau_{K'''/K'} = \tau_{K''/K'} \circ \tau_{K'''/K''}. \tag{8.4.1.2}$$

Lemma 8.4.2. — *There exist such data* $(X_\bullet, X_{A',\bullet} \hookrightarrow Z_{A',\bullet}, F_{Z_{A',\bullet}}, \tau_{K''/K'})$.

Proof. — By Galois theory, for any given $K \subset K' \subset \overline{K}$ there exist only finitely many extensions $K \subset L \subset K'$. Choose for each K' an embedding $\mathrm{Spec}(A') \hookrightarrow P_{A'}$ into a smooth scheme $P_{A'} \to \Delta^{(e_{K'})}$ with a lifting of Frobenius $F_{P_{A'}}$ compatible with the lifting of Frobenius to $\Delta^{(e_{K'})}$, and define

$$Q_{A'} := \prod_{K \subset L \subset K'} (P_L \times_{\Delta^{(e_L)}, \theta_{K'/L}} \Delta^{(e_{K'})}), \tag{8.4.2.1}$$

where the product is taken over $\Delta^{(e_{K'})}$. The liftings of Frobenius $F_{P_{A'}}$ define a lifting of Frobenius $F_{Q_{A'}}$ on $Q_{A'}$. Since each of the projections

$$P_L \times_{\Delta^{(e_L)}, \theta_{K'/L}} \Delta^{(e_{K'})} \to \Delta^{(e_{K'})} \tag{8.4.2.2}$$

are smooth and representable, and the scheme $P_{A'}$ appears in the product, $Q_{A'}$ is an algebraic space smooth over $\Delta^{(e_{K'})}$ and the natural map $\mathrm{Spec}(A') \to Q_{A'}$ is an immersion since the composite with the projection $Q_{A'} \to P_{A'}$ is an immersion.

For a second finite extension $K \subset K' \subset K'' \subset \overline{K}$ there is a natural projection $Q_{A''} \to Q_{A'}$ obtained by projection onto the components corresponding to extensions $L \subset K''$ contained in K'. By construction, this map is compatible with the liftings of Frobenius.

To construct the data $(X_\bullet, X_{A',\bullet} \hookrightarrow Z_{A',\bullet}, F_{Z_{A',\bullet}}, \tau_{K''/K'})$, choose first a hypercover $X_\bullet$ and an embedding $X_\bullet \hookrightarrow Z_\bullet$ into a simplicial space over $\mathcal{R}^{(1)}$ such that each $Z_n \to \mathcal{R}^{(1)}$ is smooth. Furthermore, choose a lifting of Frobenius $F_{Z_\bullet} : Z_\bullet \to Z_\bullet$.

Define

$$Z_{A',\bullet} := Z_\bullet \times_{\Delta^{(e_K)}, \vartheta} Q_{A'}, \tag{8.4.2.3}$$

where $\vartheta : Q_{A'} \to \Delta^{(e_K)}$ denotes the composite

$$Q_{A'} \longrightarrow \Delta^{(e_{K'})} \xrightarrow{\theta_{e_{K'/K}}} \Delta^{(e_K)}. \tag{8.4.2.4}$$

Then there is a natural smooth map $Z_{A',\bullet} \to \mathcal{R}^{(e_{K'/K})}$ induced by the maps $Z_\bullet \to \mathcal{R}^{(1)}$ and $Q_{A'} \to \Delta^{(e_{K'})}$ and an immersion $X_{A',\bullet} \hookrightarrow Z_{A',\bullet}$. The liftings of Frobenius to $Z_\bullet$

and $Q_{A'}$ induce a lifting of Frobenius $F_{Z_{A',\bullet}}$ on $Z_{A',\bullet}$, and for any finite extension $K' \subset K'' \subset \overline{K}$ the map $Q_{A''} \to Q_{A'}$ induces a map

$$\tau_{K''/K'} : Z_{A'',\bullet} \longrightarrow Z_{A',\bullet} \tag{8.4.2.5}$$

with the desired properties. □

8.4.3. — With the natural notion of morphism, the collections of data $(X_\bullet, X_{A',\bullet} \hookrightarrow Z_{A',\bullet}, F_{Z_{A',\bullet}}, \tau_{K''/K'})$ form a category which we denote by $HC(\mathfrak{X})$. The category $HC(\mathfrak{X})$ has products given for two objects $(X_\bullet^{(i)}, X_{A',\bullet}^{(i)} \hookrightarrow Z_{A',\bullet}^{(i)}, F_{Z_{A',\bullet}^{(i)}}, \tau_{K''/K'}^{(i)})$ $(i=1,2)$ by
(8.4.3.1)

$$(X_\bullet^{(1)} \times_{\mathfrak{X}} X_\bullet^{(2)}, X_\bullet^{(1)} \times_{\mathfrak{X}} X_{A',\bullet}^{(2)} \hookrightarrow Z_{A',\bullet}^{(1)} \times_{\overline{\mathcal{S}}_W^{(e_{K'/K})}} Z_{A',\bullet}^{(2)}, F_{Z_{A',\bullet}^{(1)}} \times F_{Z_{A',\bullet}^{(2)}}, \tau_{K''/K'}^{(1)} \times \tau_{K''/K'}^{(2)}).$$

In particular the category $HC(\mathfrak{X})$ is connected.

8.4.4. — Let $(X_\bullet, X_{A',\bullet} \hookrightarrow Z_{A',\bullet}, F_{Z_{A',\bullet}}, \tau_{K''/K'}) \in HC(\mathfrak{X})$ be an object, and for every n denote by $X_{A'_n,\bullet}$ and $Z_{A'_n,\bullet}$ the reductions modulo p^{n+1}. Let $D_{A'_n,\bullet}$ be the divided power envelope of the immersion $X_{A'_n,\bullet} \hookrightarrow Z_{A'_n,\bullet}$, and let $J_{D_{A'_n,\bullet}}$ denote the divided power ideal

$$J_{D_{A'_n,\bullet}} := \mathrm{Ker}(\mathcal{O}_{D_{A'_n,\bullet}} \to \mathcal{O}_{X_{A'_n,\bullet}}). \tag{8.4.4.1}$$

We view $J_{D_{A'_n,\bullet}}$ as a sheaf on $X_{A'_n,\bullet\mathrm{et}} \simeq D_{A'_n,\bullet\mathrm{et}}$. We also consider the p-adic completion $D_{A',\bullet} := \varinjlim D_{A'_n,\bullet}$ and the ideal $J_{D_{A',\bullet}} := \varprojlim J_{A'_n,\bullet}$. The lifting of Frobenius $F_{Z_{A',\bullet}}$ induces a lifting of Frobenius φ to $D_{A',\bullet}$.

Lemma 8.4.5. — *The sheaf of rings $\mathcal{O}_{D_{A',\bullet}}$ is p-torsion free. For any integer r in the interval $[0,p-1]$ we have $\varphi(J_{D_{A',\bullet}}^{[r]}) \subset p^r\mathcal{O}_{D_{A',\bullet}}$.*

Proof. — That $\mathcal{O}_{D_{A',\bullet}}$ is p-torsion free follows from the fact that each of the morphisms $X_{A'_1,n} \to \mathcal{E}^{(e_{K'/K})}$ are syntomic by 8.3.4 and 8.1.7.

For the second assertion, it suffices to show that for any element $x \in J_{D_{A',\bullet}}^{[r]}$ and integer $r' \geq r$ the element $\varphi(x^{[r']})$ is in $p^r\mathcal{O}_{D_{A',\bullet}}$. For this write $\varphi(x) = x^p + py$ for some y. Then

$$\varphi(x^{[r']}) = (x^p + py)^{[r']} = p^{[r']}((p-1)!x^{[p]} + y)^{r'}, \tag{8.4.5.1}$$

and $p^r | p^{[r']}$ if $r \leq p-1$. □

8.4.6. — For $r \in [0,p-1]$, define a map

$$p^{-r}\varphi : J_{D_{A'_n,\bullet}}^{[r]} \to \mathcal{O}_{D_{A'_n,\bullet}} \tag{8.4.6.1}$$

by sending a local section x to the class of an element $a \in \mathcal{O}_{D_{A',\bullet}}$ for which $p^r a = \varphi(\tilde{x})$ for some lifting $\tilde{x} \in \mathcal{O}_{D_{A',\bullet}}$ of x. The above lemma implies that this is well-defined.

Define $j^{A'}_{n,X_\bullet/\mathcal{R}}(r)$ to be the complex

(8.4.6.2)
$$J^{[r]}_{D_{A'_n,\bullet}} \xrightarrow{d} J^{[r-1]}_{D_{A'_n,\bullet}} \otimes \mathcal{O}_{Z_{A',\bullet}} \Omega^1_{Z_{A',\bullet}/\mathcal{R}^{(e_{K'/K})}} \xrightarrow{d} J^{[r-2]}_{D_{A'_n,\bullet}} \otimes \mathcal{O}_{Z_{A',\bullet}} \Omega^2_{Z_{A',\bullet}/\mathcal{R}^{(e_{K'/K})}} \longrightarrow \cdots .$$

For $r \in [0, p-1]$ there is a map

$$p^{-r}\varphi : j^{A'}_{n,X_\bullet/\mathcal{R}}(r) \longrightarrow j^{A'}_{n,X_\bullet/\mathcal{R}}(0) \tag{8.4.6.3}$$

given in degree q by the map $p^{q-r}\varphi \otimes p^{-q}\varphi$ on $J^{[r-q]}_{D_{A'_n,\bullet}} \otimes \mathcal{O}_{Z_{A',\bullet}} \Omega^q_{Z_{A',\bullet}/\mathcal{R}^{(e_{K'/K})}}$. Define the complex $s^{A'}_{n,X_\bullet/\mathcal{R}}(r)$ to be the mapping fiber of the map $1 - p^{-r}\varphi : j^{A'}_{n,X_\bullet/\mathcal{R}}(r) \to j^{A'}_{n,X_\bullet/\mathcal{R}}(0)$.

8.4.7. — As explained in [**39**, 2.1], the product structure on $j^{A'}_{n,X_\bullet/\mathcal{R}}(r)$ induces a product structure on $s^{A'}_{n,X_\bullet/\mathcal{R}}(r)$. Let $r, r' \in [0, p-1]$ be integers with $r + r' \leq p-1$, and consider two local sections in degrees q and q'

(8.4.7.1)
$$(x, y) \in \big(J^{[r-q]}_{D_{A'_n,\bullet}} \otimes \mathcal{O}_{Z_{A',\bullet}} \Omega^q_{Z_{A',\bullet}/\mathcal{R}^{(e_{K'/K})}}\big) \oplus \big(\Omega^{q-1}_{Z_{A',\bullet}/\mathcal{R}^{(e_{K'/K})}}\big) = s^{A'}_{n,X_\bullet/\mathcal{R}}(r)^q,$$

(8.4.7.2)
$$(x', y') \in \big(J^{[r'-q']}_{D_{A'_n,\bullet}} \otimes \mathcal{O}_{Z_{A',\bullet}} \Omega^{q'}_{Z_{A',\bullet}/\mathcal{R}^{(e_{K'/K})}}\big) \oplus \big(\Omega^{q'-1}_{Z_{A',\bullet}/\mathcal{R}^{(e_{K'/K})}}\big) = s^{A'}_{n,X_\bullet/\mathcal{R}}(r')^{q'}.$$

Then their product is given by

$$(x, y) \cdot (x', y') := (xx', (-1)^q p^r xy' + y\varphi(x')) \tag{8.4.7.3}$$

in

(8.4.7.4)
$$\big(J^{[r+r'-q'-q]}_{D_{A'_n,\bullet}} \otimes \mathcal{O}_{Z_{A',\bullet}} \Omega^{q'+q}_{Z_{A',\bullet}/\mathcal{R}^{(e_{K'/K})}}\big) \oplus \big(\Omega^{q+q'-1}_{Z_{A',\bullet}/\mathcal{R}^{(e_{K'/K})}}\big) = s^{A'}_{n,X_\bullet/\mathcal{R}}(r + r')^{q+q'}.$$

By construction the forgetful map

$$s^{A'}_{n,X_\bullet/\mathcal{R}}(r) \longrightarrow j^{A'}_{n,X_\bullet/\mathcal{R}}(0), \quad (x, y) \longmapsto x \tag{8.4.7.5}$$

is compatible with the product structures.

8.4.8. — Set

$$\mathcal{Y} := \mathcal{X} \otimes_A k, \quad \overline{\mathcal{Y}} := \mathcal{Y} \otimes_k \bar{k}, \quad Y_\bullet := X_\bullet \otimes_A k, \quad \overline{Y}_\bullet := Y_\bullet \otimes_k \bar{k}, \tag{8.4.8.1}$$

and let $\theta : \overline{Y}_{\bullet,\mathrm{et}} \to \overline{\mathcal{Y}}_{\mathrm{et}}$ be the natural morphism of topoi, where the étale topos of $\overline{Y}_\bullet$ is defined as in 0.2.6.

For every finite extension $K \subset K' \subset \overline{K}$ let $t_{A'} : \overline{Y}_{\bullet,\mathrm{et}} \to X_{A'_n,\bullet}$ be the natural morphism of topoi, and define complex $s_{n,X_\bullet/\mathcal{R}}(r)$ and $j_{n,X_\bullet/\mathcal{R}}(r)$ in $\overline{Y}_{\bullet,\mathrm{et}}$ by

$$s_{n,X_\bullet/\mathcal{R}}(r) := \varinjlim_{K \subset K' \subset \overline{K}} t^{-1}_{A'} s^{A'}_{n,X_\bullet/\mathcal{R}}(r), \tag{8.4.8.2}$$

and

$$j_{n,X_\bullet/\mathcal{R}}(r) := \varinjlim_{K\subset K'\subset \overline{K}} t_{A'}^{-1} j^{A'}_{n,X_\bullet/\mathcal{R}}(r). \tag{8.4.8.3}$$

Also define

$$s_{n,\mathcal{X}/\mathcal{R}}(r) := R\theta_* s_{n,X_\bullet/\mathcal{R}}(r) \in D(\overline{\mathcal{Y}}_{\mathrm{et}}, \mathbb{Z}/p^n). \tag{8.4.8.4}$$

The product structure on the $s^{A'}_{n,X_\bullet/\mathcal{R}}(r)$ induce a map

$$s_{n,\mathcal{X}/\mathcal{R}}(r) \otimes^{\mathbb{L}} s_{n,\mathcal{X}/\mathcal{R}}(r') \longrightarrow s_{n,\mathcal{X}/\mathcal{R}}(r+r'), \tag{8.4.8.5}$$

for $r, r' \in [0, p-1]$ with $r + r' \leq p-1$.

Of course the above construction of $s_{n,\mathcal{X}/\mathcal{R}}(r)$ depends on the choice of an object in $HC(\mathcal{X})$, but the following lemma shows that the ambiguous notation is justified:

Lemma 8.4.9. — *If we perform the above construction with another object*

$$(\widetilde{X}_\bullet, \widetilde{X}_{A',\bullet} \hookrightarrow \widetilde{Z}_{A',\bullet}, F_{\widetilde{Z}_{A',\bullet}}, \widetilde{\tau}_{K''/K'}) \in HC(\mathcal{X}) \tag{8.4.9.1}$$

to obtain a second complex $\tilde{s}_{n,\mathcal{X}/\mathcal{R}}(r)$ then there is a canonical isomorphism $s_{n,\mathcal{X}/\mathcal{R}}(r) \simeq \tilde{s}_{n,\mathcal{X}/\mathcal{R}}(r)$ in $D(\overline{\mathcal{Y}}_{\mathrm{et}}, \mathbb{Z}/p^n)$ compatible with the product structure.

Proof. — Consider first a morphism
(8.4.9.2)

$$f : (\widetilde{X}_\bullet, \widetilde{X}_{A',\bullet} \hookrightarrow \widetilde{Z}_{A',\bullet}, F_{\widetilde{Z}_{A',\bullet}}, \widetilde{\tau}_{K''/K'}) \longrightarrow (X_\bullet, X_{A',\bullet} \hookrightarrow Z_{A',\bullet}, F_{Z_{A',\bullet}}, \tau_{K''/K'})$$

in $HC(\mathcal{X})$. For any $K \subset K' \subset \overline{K}$, there is a canonical map

$$j^{A'}_{n,X_\bullet/\mathcal{R}}(r)|_{\widetilde{X}_{A'_n,\bullet}} \longrightarrow j^{A'}_{n,\widetilde{X}_\bullet/\mathcal{R}}(r) \tag{8.4.9.3}$$

compatible with the Frobenius endomorphisms. These morphisms induce a morphism

$$s^{A'}_{n,X_\bullet/\mathcal{R}}(r)|_{\widetilde{X}_{A',\bullet}} \longrightarrow s^{A'}_{n,\widetilde{X}_\bullet/\mathcal{R}}(r) \tag{8.4.9.4}$$

compatible with the product structures, and by passage to the limit a morphism $s_{n,X_\bullet/\mathcal{R}}(r)|_{\overline{Y}^\sim_\bullet} \to s_{n,\widetilde{X}_\bullet/\mathcal{R}}(r)$ which induces a map

$$s_{n,\mathcal{X}/\mathcal{R}}(r) = R\tilde{\theta}_* s_{n,X_\bullet/\mathcal{R}}(r)|_{\overline{Y}^\sim_\bullet} \longrightarrow R\tilde{\theta}_* s_{n,\widetilde{X}_\bullet/\mathcal{R}}(r) = \tilde{s}_{n,\mathcal{X}/\mathcal{R}}(r), \tag{8.4.9.5}$$

where $\tilde{\theta} : \overline{Y}^\sim_{\bullet,\mathrm{et}} \to \overline{\mathcal{Y}}_{\mathrm{et}}$ denotes the projection. By the construction of $s_{n,\mathcal{X}/\mathcal{R}}(r)$ this map extends to a map of distinguished triangles
(8.4.9.6)

$$\begin{array}{ccccccc}
s_{n,\mathcal{X}/\mathcal{R}}(r) & \longrightarrow & R\theta_* \varinjlim_{K'} j^{A'}_{n,X_\bullet/\mathcal{R}}(r) & \xrightarrow{1-p^{-r}\varphi} & R\theta_* \varinjlim_{K'} j^{A'}_{n,X_\bullet/\mathcal{R}}(0) & \xrightarrow{+1} & \\
\alpha\big\downarrow & & \beta\big\downarrow & & \beta\big\downarrow & & \\
\tilde{s}_{n,\mathcal{X}/\mathcal{R}}(r) & \longrightarrow & R\tilde{\theta}_* \varinjlim_{K'} j^{A'}_{n,\widetilde{X}_\bullet/\mathcal{R}}(r) & \xrightarrow{1-p^{-r}\varphi} & R\tilde{\theta}_* \varinjlim_{K'} j^{A'}_{n,\widetilde{X}_\bullet/\mathcal{R}}(0) & \xrightarrow{+1} & .
\end{array}$$

By 2.5.5 the complex $R\theta_* j^{A'}_{n,X_\bullet/\mathcal{R}}(r)$ is quasi-isomorphic to the restriction to $\overline{\mathcal{Y}}_{\mathrm{et}}$ of the complex $Ru_{\mathcal{X}_{\mathrm{et}}/\mathcal{R}_n^{(e_{K'/K})}*} I^{[r]}_{\mathcal{X}_{\mathrm{et}}/\mathcal{R}_n^{(e_{K'/K})}}$, and similarly for $R\tilde{\theta}_* j^{A'}_{n,\tilde{X}_\bullet/\mathcal{R}}(r)$. It follows that the map β is an isomorphism, and hence α is also an isomorphism.

For a general second object $(\widetilde{X}_\bullet, \widetilde{X}_{A',\bullet} \hookrightarrow \widetilde{Z}_{A',\bullet}, F_{\widetilde{Z}_{A',\bullet}}, \widetilde{\tau}_{K''/K'}) \in HC(\mathcal{X})$, let $(X^\clubsuit_\bullet, X^\clubsuit_{A',\bullet} \hookrightarrow Z^\clubsuit_{A',\bullet}, F_{Z^\clubsuit_{A',\bullet}}, \tau^\clubsuit_{K''/K'})$ denote the product in $HC(\mathcal{X})$ with $(X_\bullet, X_{A',\bullet} \hookrightarrow Z_{A',\bullet}, F_{Z_{A',\bullet}}, \tau_{K''/K'})$, and let $s^\clubsuit_{n,\mathcal{X}/\mathcal{R}}(r)$ denote the complex obtained using the product. Then there are isomorphisms

$$(8.4.9.7) \qquad s_{n,\mathcal{X}/\mathcal{R}}(r) \xrightarrow{\mathrm{pr}_1^*} s^\clubsuit_{n,\mathcal{X}/\mathcal{R}}(r) \xleftarrow{\mathrm{pr}_2^*} \tilde{s}_{n,\mathcal{X}/\mathcal{R}}(r),$$

which gives an isomorphism $s_{n,\mathcal{X}/\mathcal{R}}(r) \simeq \tilde{s}_{n,\mathcal{X}/\mathcal{R}}(r)$. If there exists a morphism f as in (8.4.9.2) then in fact this isomorphism agrees with the one defined by f as in the start of the proof. Indeed, let Γ denote the graph of f in $HC(\mathcal{X})$. Then $f^* : s_{n,\mathcal{X}/\mathcal{R}}(r) \to \tilde{s}_{n,\mathcal{X}/\mathcal{R}}(r)$ is equal to the composite

$$(8.4.9.8) \qquad s_{n,\mathcal{X}/\mathcal{R}}(r) \xrightarrow{\mathrm{pr}_1^*} s^\clubsuit_{n,\mathcal{X}/\mathcal{R}}(r) \xrightarrow{\Gamma^*} \tilde{s}_{n,\mathcal{X}/\mathcal{R}}(r).$$

On the other hand, since $\mathrm{pr}_2 \circ \Gamma$ is the identity we have $\Gamma = \mathrm{pr}_2^{*-1}$ so $f^* = \mathrm{pr}_2^{*-1} \circ \mathrm{pr}_1^*$. □

8.4.10. — The above lemma also implies (modulo the verification of the appropriate transitivity relations which we leave to the reader) that there is a natural action of $G := \mathrm{Gal}(\overline{K}/K)$ on $s_{n,\mathcal{X}/\mathcal{R}}(r)$ compatible with the natural action on $\overline{\mathcal{Y}}$. For any element $\sigma \in G$, this action $\sigma^* s_{n,\mathcal{X}/\mathcal{R}}(r) \to s_{n,\mathcal{X}/\mathcal{R}}(r)$ can be described as follows.

Fix an object $(X_\bullet, X_{A',\bullet} \hookrightarrow Z_{A',\bullet}, F_{Z_{A',\bullet}}, \tau_{K''/K'}) \in HC(\mathcal{X})$, and let

$$(8.4.10.1) \qquad (X_\bullet, X_{A',\bullet} \hookrightarrow Z^\sigma_{A',\bullet}, F_{Z^\sigma_{A',\bullet}}, \tau^\sigma_{K''/K'}) \in HC(\mathcal{X})$$

be the object with $Z^\sigma_{A',\bullet} = Z_{A',\bullet}$, $F_{Z^\sigma_{A',\bullet}} = F_{Z_{A',\bullet}}$, and $\tau^\sigma_{K''/K} = \tau_{K''/K'}$, but the closed immersions $X_{A',\bullet} \hookrightarrow Z^\sigma_{A',\bullet}$ given by the composites

$$(8.4.10.2) \qquad X_{A'} \xrightarrow{\sigma} X_{A'} \longrightarrow Z_{A',\bullet}.$$

From the construction there is a natural map of complexes $\sigma^* s_{n,X_\bullet/\mathcal{R}}(r) \to \tilde{s}_{n,X_\bullet/\mathcal{R}}(r)$ on $\overline{Y}_{\mathrm{et}}$, where $s_{n,X_\bullet/\mathcal{R}}(r)$ is constructed using $(X_\bullet, X_{A',\bullet} \hookrightarrow Z_{A',\bullet}, F_{Z_{A',\bullet}}, \tau_{K''/K'})$ and $\tilde{s}_{n,X_\bullet/\mathcal{R}}(r)$ is constructed using $(X_\bullet, X_{A',\bullet} \hookrightarrow Z^\sigma_{A',\bullet}, F_{Z^\sigma_{A',\bullet}}, \tau^\sigma_{K''/K'})$. Applying $R\theta_*$ we obtain the map $\sigma^* s_{n,\mathcal{X}/\mathcal{R}}(r) \to s_{n,\mathcal{X}/\mathcal{R}}(r)$ which by the proof of the lemma is independent of all the choices and compatible with multiplication in G. This action is also compatible with the product structure (8.4.8.5).

Proposition 8.4.11. — *There is a natural isomorphism*

$$(8.4.11.1) \qquad H^*(\overline{\mathcal{Y}}_{\mathrm{et}}, j_{n,X_\bullet/\mathcal{R}}(0)) \simeq H^*((\mathcal{X}_{\overline{A}_n}/\mathcal{R}_n)_{\mathrm{cris}}, \mathcal{O}_{\mathcal{X}_{\overline{A}_n}/\mathcal{R}_n}),$$

where the right hand side is defined as in 8.3.5.

Proof. — By 2.5.4, there is a natural isomorphism

$$j^{A'}_{n,X_\bullet/\mathcal{R}}(0) \simeq Ru_{\mathcal{X}_{A'_1,\mathrm{et}}/\mathcal{R}_n^{(e_{K'/K})}*}\mathcal{O}_{\mathcal{X}_{A'_1,\mathrm{et}}/\mathcal{R}_n^{(e_{K'/K})}} \tag{8.4.11.2}$$

which is functorial. It follows from this functoriality and [**5**, VI.8.7.7] that

$$\begin{aligned} H^*(\overline{\mathcal{Y}}_{\mathrm{et}}, j_{n,X_\bullet/\mathcal{R}}(0)) &\simeq \varinjlim H^*(\mathcal{X}_{A'_n,\bullet}, j^{A'}_{n,X_\bullet/\mathcal{R}}(0)) \\ &\simeq \varinjlim H^*(\mathcal{X}_{A'_n,\bullet}, Ru_{\mathcal{X}_{A'_n,\mathrm{et}}/\mathcal{R}_n^{(e_{K'/K})}*}\mathcal{O}_{\mathcal{X}_{A'_n,\mathrm{et}}/\mathcal{R}_n^{(e_{K'/K})}}), \end{aligned}$$

and

$$\begin{aligned} &\varinjlim H^*(\mathcal{X}_{A'_n,\bullet}, Ru_{\mathcal{X}_{A'_n,\mathrm{et}}/\mathcal{R}_n^{(e_{K'/K})}*}\mathcal{O}_{\mathcal{X}_{A'_n,\mathrm{et}}/\mathcal{R}_n^{(e_{K'/K})}}) \\ &\qquad \simeq H^*((\mathcal{X}_{\overline{A}_n}/\mathcal{R}_n)_{\mathrm{cris}}, \mathcal{O}_{\mathcal{X}_{\overline{A}_n}/\mathcal{R}_n}). \quad \square \end{aligned} \tag{8.4.11.3}$$

Corollary 8.4.12. — *There is a natural map*

$$H^*(\overline{\mathcal{Y}}_{\mathrm{et}}, s_{n,\mathcal{X}/\mathcal{R}}(r)) \longrightarrow H^*((\mathcal{X}_{\overline{A}_n}/\mathcal{R}_n)_{\mathrm{cris}}, \mathcal{O}_{\mathcal{X}_{\overline{A}_n}/\mathcal{R}_n})^{\varphi=p^r}, \tag{8.4.12.1}$$

where the right hand side denotes the submodule on which Frobenius acts as multiplication by p^r.

Proof. — The natural map $s_{n,X_\bullet/\mathcal{R}}(r) \to j_{n,X_\bullet/\mathcal{R}}(0)$ induced by the maps (8.4.7.5) induce a map

$$H^*(\overline{\mathcal{Y}}_{\mathrm{et}}, s_{n,\mathcal{X}/\mathcal{R}}(r)) \longrightarrow H^*((\mathcal{X}_{\overline{A}_n}/\mathcal{R}_n)_{\mathrm{cris}}, \mathcal{O}_{\mathcal{X}_{\overline{A}_n}/\mathcal{R}_n}), \tag{8.4.12.2}$$

which by the definition of $s_{n,\mathcal{X}/\mathcal{R}}(r)$ as the mapping fiber of $1 - p^{-r}\varphi$ lands in the part on which Frobenius acts by multiplication by p^r. $\square$

8.4.13. — Following [**41**], we now briefly indicate for the convenience of the reader the remaining pieces needed to complete the proof of the C_{st}-conjecture in the case when $p > 2\dim(\mathcal{X} \otimes_A K) + 1$. The reader interested in more details should consult *loc. cit.* and [**73**]. Also, it hopefully is clear from the preceding that stack-theoretic techniques can also be used in Tsuji's arguments in [**73**] which remove the hypothesis on the dimension. We do not discuss these things here since we do not have anything to contribute which is not already in the above references.

Theorem 8.4.14 ([**41**, 5.4]). — *Let $\mathcal{X}_K$ denote the generic fiber of $\mathcal{X}$ and let $\bar{i} : \overline{\mathcal{Y}} \to \mathcal{X} \otimes_A \overline{A}$ and $\bar{j} : \mathcal{X}_{\overline{K}} \to \mathcal{X} \otimes_A \overline{A}$ be the natural maps. Then for $0 \leq r \leq p-1$ there is a natural isomorphism*

$$\tau_{\leq r}\bar{i}_*\bar{i}^*R\bar{j}_*(\mathbb{Z}/p^n\mathbb{Z}(r)) \simeq \bar{i}_*s_{n,\mathcal{X}/\mathcal{R}}(r) \tag{8.4.14.1}$$

in $D((\mathcal{X} \otimes_A \overline{A})_{\mathrm{et}}, \mathbb{Z}/p^n)$ compatible with the Galois actions. In particular, if $m \leq r \leq p-1$ or $\dim(\mathcal{X}_K) \leq r < p-1$ then by the proper base change theorem for étale cohomology [**5**, XII.5.1] *there is a natural isomorphism*

$$H^*(\overline{\mathcal{Y}}_{\mathrm{et}}, s_{n,\mathcal{X}/\mathcal{R}}(r)) \longrightarrow H^*(\mathcal{X}_{\overline{K}}, \mathbb{Z}/p^n(r)) \tag{8.4.14.2}$$

compatible with the Galois action.

8.4.15. — From this and 8.4.12 we obtain a map

$$H^m(\mathfrak{X}_{\overline{K}}, \mathbb{Z}/p^n(r)) \longrightarrow H^m((\mathfrak{X}_{\overline{A}_n}/\mathcal{R}_n)_{\text{cris}}, \mathcal{O}_{\mathfrak{X}_{\overline{A}_n}/\mathcal{R}_n})^{\varphi=p^r} \tag{8.4.15.1}$$

for $m \leq r < p-1$. By 8.3.6, this induces by applying $\varprojlim_n$ and tensoring with $\mathbb{Q}$ a natural map (using the notation of 8.3.6

$$H^m(\mathfrak{X}_{\overline{K}}, \mathbb{Q}_p(r)) \longrightarrow (B^+_{\text{st}} \otimes_{K_0} D^m)^{\mathcal{N}=0, \varphi=p^r}. \tag{8.4.15.2}$$

Recall (8.2.3) that there is a canonical map $\mathbb{Q}_p(1) \hookrightarrow B^+_{\text{st}}$ such that B_{st} is obtained by inverting the image of a generator of $\mathbb{Q}_p(1)$. There is thus a canonical map $\mathbb{Q}_p(-r) \hookrightarrow B_{\text{st}}$. Tensoring (8.4.15.2) with $\mathbb{Q}_p(-r)$ we obtain a map

$$V^m := H^m(\mathfrak{X}_{\overline{K}}, \mathbb{Q}_p) \longrightarrow (B_{\text{st}} \otimes_{K_0} D^m)^{\mathcal{N}=0, \varphi=1}. \tag{8.4.15.3}$$

Extending scalars we obtain a map

$$B_{\text{st}} \otimes_{\mathbb{Q}_p} V^m \longrightarrow B_{\text{st}} \otimes_{K_0} D^m \tag{8.4.15.4}$$

compatible with the monodromy operators, Frobenii, Galois actions, and product structure when it makes sense. Note that since $K \otimes_{K_0} D^m \simeq H^m_{\text{dR}}(\mathfrak{X}_K/K)$ this map (8.4.15.4) can also be written as a map

$$B_{\text{st}} \otimes_{\mathbb{Q}_p} V^m \longrightarrow B_{\text{st}} \otimes_K H^m_{\text{dR}}(\mathfrak{X}_K/K). \tag{8.4.15.5}$$

8.4.16. — To prove that (8.4.15.4) is an isomorphism when the dimension d of $\mathfrak{X}_K$ satisfies $2d+1 < p$, one proceeds as follows.

First observe that it suffices by replacing K by a finite extension to consider the case when $\mathfrak{X}_K$ is geometrically connected, and also we may assume that K/K_0 is Galois. Define a trace map

$$\text{tr} : D^{2d} \longrightarrow K_0 \tag{8.4.16.1}$$

by noting that the usual trace map on de Rham-cohomology

$$H^{2d}_{\text{dR}}(\mathfrak{X}_K) \simeq K \otimes_{K_0} D^{2d} \longrightarrow K \tag{8.4.16.2}$$

is Gal(K/K_0)-invariant. We also have the trace map $V^{2d} \to \mathbb{Q}_p(-d)$. The canonical inclusion $\mathbb{Q}_p(-d) \subset B_{\text{st}}$ induces an isomorphism $B_{\text{st}}(-d) \simeq B_{\text{st}}$. Hence the trace map on étale cohomology induces a map

$$\text{tr} : B_{\text{st}} \otimes_{\mathbb{Q}_p} V^{2d} \longrightarrow B_{\text{st}}. \tag{8.4.16.3}$$

The following proposition is proven by showing that the isomorphism in 8.4.14 is compatible with certain Chern classes of line bundles.

Proposition 8.4.17 ([**73**, proof of 4.10.3]). — *The diagram*

$$\begin{array}{ccc} B_{\mathrm{st}} \otimes_{\mathbb{Q}_p} V^{2d} & \xrightarrow{\mathrm{tr}} & B_{\mathrm{st}} \\ {\scriptstyle (8.4.15.4)}\downarrow & & \downarrow{\scriptstyle (8.4.15.4)} \\ B_{\mathrm{st}} \otimes_{K_0} D^{2d} & \xrightarrow{\mathrm{tr}} & B_{\mathrm{st}} \end{array} \tag{8.4.17.1}$$

commutes.

8.4.18. — To see that (8.4.15.4) is an isomorphism, note that it suffices to consider the case when $m < 2d$. Then there is a commutative diagram

$$\begin{array}{ccc} B_{\mathrm{st}} \otimes_{\mathbb{Q}_p} V^m \times B_{\mathrm{st}} \otimes_{\mathbb{Q}_p} V^{2d-m} & \longrightarrow & B_{\mathrm{st}} \otimes_{\mathbb{Q}_p} V^{2d} \simeq B_{\mathrm{st}} \\ \downarrow & & \downarrow{\scriptstyle (8.4.15.4)} \\ B_{\mathrm{st}} \otimes_{K_0} D^m \times B_{\mathrm{st}} \otimes_{K_0} D^{2d-m} & \longrightarrow & B_{\mathrm{st}} \otimes_{K_0} D^{2d} \simeq B_{\mathrm{st}}, \end{array} \tag{8.4.18.1}$$

where the horizontal arrows are given by cup-product. By Poincaré duality, the horizontal arrows are perfect pairings between finitely generated free B_{st}-modules. In particular, the map (8.4.15.4) is injective. In fact it is canonically split since a complement is given by the set of elements $m \in B_{\mathrm{st}} \otimes_{K_0} D^m$ annihilating $B_{\mathrm{st}} \otimes_{\mathbb{Q}_p} V^{2d-m} \subset B_{\mathrm{st}} \otimes_{K_0} D^{2d-m}$. Since V^m and D^m have the same rank (for example after tensoring with $\mathbb{C}$ they are both isomorphic to Betti cohomology), it follows that (8.4.15.4) is an isomorphism.

The only thing that remains for the proof of the C_{st}-conjecture is the following result. Recall (8.2.5) that there is a filtration $\mathrm{Fil}^{\cdot}_{B_{\mathrm{st}}}$ on the ring B_{st}. This defines a filtration $\mathrm{Fil}^{\cdot}_{B_{\mathrm{st}}} \otimes V^m$ on $B_{\mathrm{st}} \otimes_{\mathbb{Q}_p} V^m$, and a filtration on $B_{\mathrm{st}} \otimes_K H^m_{\mathrm{dR}}(\mathfrak{X}_K/K)$ by taking the tensor product of $\mathrm{Fil}^{\cdot}_{B_{\mathrm{st}}}$ and the Hodge filtration on $H^m_{\mathrm{dR}}(\mathfrak{X}_K/K)$

Proposition 8.4.19 ([**73**, 4.10.3]). — *The isomorphism (8.4.15.4) is compatible with these filtrations.*

Finally let us mention two results of Tsuji concerning functoriality which are needed in the next section.

Theorem 8.4.20 ([**73**, 4.10.4]). — *Let $K \subset L \subset \overline{K}$ be a finite extension, $\mathfrak{X}'/\mathcal{O}_L$ a semistable proper scheme, and $g : \mathfrak{X}' \to \mathfrak{X}$ a morphism over* $\mathrm{Spec}(\mathcal{O}_L) \to \mathrm{Spec}(V)$. *Then the diagram*

$$\begin{array}{ccc} B_{\mathrm{dR}} \otimes_{\mathbb{Q}_p} H^*(\overline{\mathfrak{X}}'_L, \mathbb{Q}_p) & \xrightarrow{(8.4.15.5)} & B_{\mathrm{dR}} \otimes_L H^*_{\mathrm{dR}}(\mathfrak{X}'_L/L) \\ {\scriptstyle g^*}\uparrow & & \uparrow{\scriptstyle g^*} \\ B_{\mathrm{dR}} \otimes_{\mathbb{Q}_p} H^*(\overline{\mathfrak{X}}_K, \mathbb{Q}_p) & \xrightarrow{(8.4.15.5)} & B_{\mathrm{dR}} \otimes_K H^*_{\mathrm{dR}}(\mathfrak{X}'_K/K) \end{array} \tag{8.4.20.1}$$

commutes.

Theorem 8.4.21 ([**74**, A2.7]). — *Let $K \subset L \subset \overline{K}$ be a finite extension, and $\mathfrak{X}/\mathcal{O}_L$ a semistable scheme with general fiber $\mathfrak{X}_L$. Let $\sigma \in \mathrm{Gal}(\overline{K}/K)$ be an automorphism, and let $\mathfrak{X}^\sigma$ denote the semistable scheme $\mathfrak{X}^\sigma := \mathfrak{X} \otimes_{\mathcal{O}_L,\sigma} \mathcal{O}_L$. Then the diagram*

$$(8.4.21.1)\qquad \begin{array}{ccc} B_{\mathrm{dR}} \otimes_{\mathbb{Q}_p} H^*(\overline{\mathfrak{X}}^\sigma_L, \mathbb{Q}_p) & \xrightarrow{c_{\mathfrak{X}^\sigma}} & B_{\mathrm{dR}} \otimes_L H^*_{\mathrm{dR}}(\mathfrak{X}^\sigma_L/L) \\ {\scriptstyle \sigma^*\otimes\sigma^*}\uparrow & & \uparrow{\scriptstyle \sigma^*\otimes\sigma^*} \\ B_{\mathrm{dR}} \otimes_{\mathbb{Q}_p} H^*(\overline{\mathfrak{X}}_L, \mathbb{Q}_p) & \xrightarrow{c_{\mathfrak{X}}} & B_{\mathrm{dR}} \otimes_L H^*_{\mathrm{dR}}(\mathfrak{X}_L/L), \end{array}$$

commutes, where $c_{\mathfrak{X}}$ (resp. $c_{\mathfrak{X}^\sigma}$) denotes the map (8.4.15.5) for $\mathfrak{X}/\mathcal{O}_L$ (resp. $\mathfrak{X}^\sigma/\mathcal{O}_L$).

8.5. Construction of the (φ, N, G)-structure in general

Following [**74**, Appendix], we explain in this section how to associated to any smooth proper scheme X/K a canonical (φ, N, G)-module structure on $H^*_{\mathrm{dR}}(X/K)$. For more details the reader should consult *loc. cit.*.

8.5.1. — First we recall some general facts about correspondences. Let $H^*(-)$ be a Weil cohomology theory defined on the category of varieties over $\overline{K}$ and taking values in the category of graded C-vector spaces for a field C of characteristic 0 (cf. [**45**, 1.2], but note that Tate twists are neglected in this reference). The case of interest for us is $H^*(-)$ equal to p-adic étale cohomology or de Rham cohomology. For an integer s and a smooth scheme X we write $H^*(X)(s)$ for the s-th Tate twist of the cohomology ring $H^*(X)$.

For technical reasons it will be useful to also consider smooth proper $\overline{K}$-schemes, which are possibly not connected. Let us briefly discuss how to extend some of the results of [**45**] to this slightly more general setting.

(i) If X is a smooth proper $\overline{K}$ scheme with connected components $\{X_i\}_{i\in I}$ then we define $H^*(X) := \prod_{i\in I} H^*(X_i)$.

(ii) If X and Y are smooth proper $\overline{K}$-schemes, then for every integer n the pullback map

$$(8.5.1.1)\qquad \mathrm{pr}_1^* \times \mathrm{pr}_2^* : \oplus_{p+q=n} H^p(X) \otimes H^q(Y) \longrightarrow H^n(X \times Y)$$

is an isomorphism (the resulting decomposition of $H^n(X \times Y)$ is called the *Kunneth decomposition*). This follows from noting that if $\{X_i\}_{i\in I}$ and $\{Y_j\}_{j\in J}$ are the connected components of X and Y respectively, then there is a commutative diagram

$$(8.5.1.2)\qquad \begin{array}{ccc} \oplus_{p+q=n} H^p(X) \otimes H^q(Y) & \xrightarrow{\mathrm{pr}_1^*\times\mathrm{pr}_2^*} & H^n(X \times Y) \\ \downarrow{\scriptstyle \mathrm{def.}} & & \\ \oplus_{p+q=n}(\prod_{i\in I} H^p(X_i)) \otimes (\prod_{j\in J} H^q(Y_j)) & & \downarrow{\scriptstyle \mathrm{def.}} \\ \downarrow{\scriptstyle \simeq} & & \\ \prod_{i,j}(\oplus_{p+q=n} H^p(X_i) \otimes H^q(Y_j)) & \xrightarrow{\mathrm{pr}_1^*\times\mathrm{pr}_2^*} & \prod_{i,j} H^n(X_i \times Y_j) \end{array}$$

where the vertical arrows are isomorphisms, and the bottom horizontal arrow is an isomorphism by [**45**, 1.2 B].

(iii) If X is a smooth proper $\overline{K}$-scheme of pure dimension d, then there is a trace map $\mathrm{tr} : H^2d(X)(d) \to C$ which is an isomorphism if X is connected. If $\{X_i\}_{i\in I}$ then $H^{2d}(X)(d) = \prod_{i\in I} H^{2d}(X_i)(d)$ and the trace map for X is defined to be the sum of the trace maps of the X_i [**45**, 1.2 A].

(iv) Let X and Y be smooth proper connected $\overline{K}$-schemes of pure dimension d and d' respectively. A class $\alpha \in H^{2d'}(X \times Y)(d')$ defines a map $\alpha^* : H^*(Y) \to H^*(X)$ as follows (we call such a class α a *correspondence*). First note that the Kunneth decomposition

$$H^n(X \times Y) \simeq \oplus_{p+q=n} H^p(X) \otimes H^q(Y) \tag{8.5.1.3}$$

defines a canonical map $H^{2d'+*}(X \times Y)(d') \to H^*(X) \otimes H^{2d'}(Y)(d')$ whose composition with $1 \otimes \mathrm{tr} : H^*(X) \otimes H^{2d'}(Y)(d') \to H^*(X)$ we denote by q. Define α^* to be the composite morphism

$$H^*(Y) \xrightarrow{\mathrm{pr}_2^*} H^*(X \times Y) \xrightarrow{\alpha\cdot(-)} H^{2d'+*}(X \times Y)(d') \xrightarrow{q} H^*(X). \tag{8.5.1.4}$$

Observe that if $\{X_i\}_{i\in I}$ and $\{Y_j\}_{j\in J}$ are the connected components of X and Y respectively, and if α_{ij} denotes the (i,j)-component of α in $H^{2d'}(X \times Y)(d') = \prod_{i,j} H^{2d'}(X_i \times Y_j)(d')$ then the diagram

$$\begin{array}{ccc} H^*(Y) & \xrightarrow{\alpha^*} & H^*(X) \\ \| & & \| \\ \prod_j H^*(Y_j) & \xrightarrow{\sum_i \alpha^*_{i,j}} & \prod_i H^*(X_i) \end{array} \tag{8.5.1.5}$$

commutes, where $\sum \alpha^*_{i,j}$ denotes the map which sends $(v_j)_{j\in J} \in \prod_j H^*(Y_j)$ to the element of $\prod_i H^*(X_i)$ with i-th component $\sum_{j\in J} \alpha^*_{ij}(v_j)$. Note also that for two classes $\alpha, \beta \in H^{2d'}(X \times Y)(d')$ we have $(\alpha + \beta)^* = \alpha^* + \beta^*$.

Proposition 8.5.2

(i) *Let $f : X \to Y$ be a morphism of smooth, and proper $\overline{K}$-schemes of pure dimension d and d' respectively, and let $\alpha \in H^{2d'}(X \times Y)(d')$ be the class of the graph of f. Then α^* is equal to the map $f^* : H^*(Y) \to H^*(X)$.*

(ii) *Assume X and Y connected, and let Z be a third smooth proper connected $\overline{K}$-scheme of dimension d'' and let $\beta \subset H^{2d''}(Y \times Z)(d'')$ be a class. Then for any $\alpha \in H^{2d'}(X \times Y)(d')$, the composite $\alpha^* \circ \beta^* : H^*(Z) \to H^*(X)$ is equal to the map defined by*

$$\mathrm{pr}_{13*}(\mathrm{pr}_{12}^*(\alpha) \cdot \mathrm{pr}_{23}^*(\beta)) \in H^{2d''}(X \times Z)(d''). \tag{8.5.2.1}$$

Proof. — For (i) note first that when X and Y are connected this is [**45**, 1.3.7 (iii)]. For the general case, let $\{X_i\}_{i\in I}$ and $\{Y_j\}_{j\in J}$ be the connected components of X and Y respectively and observe that if we write α_{ij} for the (i,j)-component of α as above, then $\alpha_{ij}=0$ unless $X_i\subset f^{-1}(Y_j)$. Therefore if $\alpha_j\subset H^{2d'}(f^{-1}(Y_j)\times Y_j)(d')\subset H^{2d'}(X\times Y)(d')$ denotes the class of the graph of the morphism $f^{-1}(Y_j)\to Y_j$ we have $\alpha^*=\sum_{j\in J}\alpha_j^*$. From the commutativity of (8.5.1.5) it follows that it suffices to consider the case when Y is connected. In this case, if $\alpha_i\in H^{2d'}(X_i\times Y)(d')\subset H^{2d'}(X\times Y)(d')$ denotes the class of the graph of $X_i\to Y$ we have $\alpha=\sum_{i\in I}\alpha_i$, and $\alpha^*=\sum_i\alpha_i^*$. Again using the commutativity of (8.5.1.5) this reduces the proof to the connected case.

Statement (ii) follows from [**45**, 1.3.3]. □

Corollary 8.5.3. — *Let $f:Y\to X$ be a generically étale morphism of proper smooth $\overline{K}$-schemes of pure dimension d, and let $\alpha\in H^{2d}(X\times Y)(d)$ denotes the class of the cycle $f\times\mathrm{id}:Y\hookrightarrow X\times Y$.*

(i) *If X and Y are connected, then $\alpha^*\circ f^*:H^*(X)\to H^*(X)$ is equal to multiplication by the degree of the field extension $k(X)\to k(Y)$.*

(ii) *In general the map f^* is injective, and α^* gives a splitting of the inclusion.*

Proof. — Let $\Gamma_f^t\subset X\times Y$ denote the cycle $f\times\mathrm{id}:Y\subset X\times Y$.

To prove (i) it suffices by 8.5.2 (i) and (ii) to show that if $\Gamma_f\subset Y\times X$ denotes the graph of f, then

(8.5.3.1)
$$\mathrm{pr}_{13*}(\mathrm{pr}_{12}^*(\Gamma_f)\cdot\mathrm{pr}_{23}^*(\Gamma_f^t))=[k(Y):k(X)]\cdot\Delta_X$$

in the Chow ring of $X\times X$ which is clear.

To prove (ii), let $\{X_i\}$ (resp. $\{Y_j\}$) denote the connected components of X (resp. Y), and for $Y_j\subset f^{-1}(X_i)$ let $f_{ij}:Y_j\to X_i$ be the restriction of f, and let $\alpha_{ij}\in H^{2d}(X_i\times Y_j)(d)\subset H^{2d}(X\times Y)(d)$ be the class of $\Gamma_{f_{ij}}^t\subset X_i\times Y_j$. Using the commutativity of (8.5.1.5) once more, it follows that the map

(8.5.3.2)
$$\alpha^*\circ f^*:H^*(X)\longrightarrow H^*(X)$$

is equal to product of the maps

(8.5.3.3)
$$\sum_{Y_j\subset f^{-1}(X_i)}\alpha_{ij}^*\circ f_{ij}^*:H^*(X_i)\longrightarrow H^*(X_i).$$

If d_{ij} denotes the generic degree of the map f_{ij}, we then get from (i) that (8.5.3.3) is equal to multiplication by the positive integer $\sum_{Y_j\subset f^{-1}(X_i)}d_{ij}$. Therefore $\alpha^*\circ f^*$ is an isomorphism, which implies (ii). □

If $f:X\to Y$ is a morphism of proper smooth K-schemes of pure dimension d and d' respectively, then the preceding discussion can also be applied to the de Rham cohomology groups $H_{\mathrm{dR}}^*(X)$ and $H_{\mathrm{dR}}^*(Y)$ over K. In the case when $H^*(-)=H_{\mathrm{dR}}^*(-)$, the above argument also gives the following:

Corollary 8.5.4. — *Let $f : Y \to X$ be a generically étale morphism between proper smooth K-schemes. Then the map $f^* : H^*_{\mathrm{dR}}(X) \to H^*_{\mathrm{dR}}(Y)$ is strictly compatible with the Hodge filtrations.*

Proof. — Let $H^*_{\mathrm{Hod}}(X)$ (resp. $H^*_{\mathrm{Hod}}(Y)$) denote the graded (with respect to the Hodge filtration) algebra associated to $H^*_{\mathrm{dR}}(X)$ (resp. $H^*_{\mathrm{dR}}(Y)$). It suffices to show that the map $f^* : H^*_{\mathrm{Hod}}(X) \to H^*_{\mathrm{Hod}}(Y)$ is injective.

For this it suffices to show that the map $\alpha^* : H^*_{\mathrm{dR}}(Y) \to H^*_{\mathrm{dR}}(X)$ defined in 8.5.3 respects the filtrations. For then there is a map $\alpha^* : H^*_{\mathrm{Hod}}(Y) \to H^*_{\mathrm{Hod}}(X)$ such that $\alpha^* \circ f^* : H^*_{\mathrm{Hod}}(X) \to H^*_{\mathrm{Hod}}(X)$ is an isomorphism.

To see that α^* respects the Hodge filtrations, let Fil denote the Hodge filtration on $H^*_{\mathrm{dR}}(X \times Y)$. The class α of Γ^t_f lies in $\mathrm{Fil}^d H^{2\mathrm{d}}_{\mathrm{dR}}(X \times Y)$. Hence if $\beta \in H^p_{\mathrm{dR}}(Y)$ lies in the j-th step of the Hodge filtration on $H^p_{\mathrm{dR}}(Y)$, we have $\alpha \cdot \mathrm{pr}_2^* \beta \in \mathrm{Fil}^{d+j} H^{2d+p}_{\mathrm{dR}}(X \times Y)$. Since the Kunneth decomposition $H^{2d+p}_{\mathrm{dR}}(X \times Y) \simeq \oplus_{p+q=2d+p} H^p_{\mathrm{dR}}(X) \otimes H^q_{\mathrm{dR}}(Y)$ is strictly compatible with the filtrations, it follows that the component of $\alpha \cdot \mathrm{pr}_2^*(\beta)$ lying in $H^p(X) \otimes H^{2d}(Y)(d)$ in fact lies in $\mathrm{Fil}^j \otimes \mathrm{Fil}^{2d}$. It follows that the composite

$$H^*_{\mathrm{dR}}(X \times Y) \xrightarrow{\alpha \cdot (-)} H^{2d+*}_{\mathrm{dR}}(X \times Y)(d) \xrightarrow{\quad q \quad} H^*_{\mathrm{dR}}(X) \tag{8.5.4.1}$$

is compatible with the filtrations. □

Theorem 8.5.5 **([74, A2]).** — *Let X/K be a smooth proper scheme with semistable reduction, and let $\mathfrak{X}/V$ be a semistable model. Then the isomorphism*

$$B_{\mathrm{dR}} \otimes_{\mathbb{Q}_p} H^*(\overline{X}, \mathbb{Q}_p) \simeq B_{\mathrm{dR}} \otimes_K H^*_{\mathrm{dR}}(X/K) \tag{8.5.5.1}$$

induced by (8.4.15.4) is compatible with the endomorphisms defined by correspondences $\alpha \in CH^d(X \times X)$.

8.5.6. — We use this result in conjunction with de Jong's alterations theorem.

Recall [37, 2.20] that if S is a noetherian integral algebraic space, then an *alteration* S' of S is a dominant proper morphism $\phi : S' \to S$ of noetherian integral schemes such that over some non-empty open set $U \subset S$ the morphism $\phi^{-1}(U) \to U$ is finite. Observe that if S is a $\mathbb{Q}$-scheme then this last condition implies that ϕ is generically étale. This notion generalizes to reduced noetherian algebraic spaces. If S is such a space over $\mathbb{Q}$, then an alteration is a dominant proper morphism $\phi : S' \to S$ of reduced noetherian algebraic spaces such that for every point $s' \in S'$ of codimension 0 the morphism ϕ is étale in a neighborhood of s' (and in particular $\phi(s')$ is a codimension 0 point of S). In what follows we will use this slightly more general definition of alteration.

The main result about alterations that we need is the following:

Theorem 8.5.7 **([37, 6.5]).** — *Let $\mathfrak{X}/V$ be a flat, proper, and reduced V-scheme. Then there exist a finite extension $K \subset L \subset \overline{K}$ and an alteration $\mathfrak{X}' \to \mathfrak{X}$ over* $\mathrm{Spec}(\mathcal{O}_L) \to \mathrm{Spec}(V)$ *with $\mathfrak{X}'/\mathcal{O}_L$ a semistable scheme.*

Proof. — In the case when $\mathcal{X}$ is also integral this is [**37**, 6.5]. For the slightly more general case of the theorem, note that by replacing $\mathcal{X}$ by the disjoint union of its irreducible components, we may assume that $\mathcal{X}$ is a disjoint union $\mathcal{X} = \coprod_i \mathcal{X}_i$, with each $\mathcal{X}_i$ a flat, proper, and integral V-scheme. An examination of the proof of [**37**, 6.5] shows that in this case there exists a single extension $K \subset L \subset \overline{K}$ such that for every i there exists an alteration $\mathcal{X}'_i \to \mathcal{X}_i$ over $\mathrm{Spec}(\mathcal{O}_L) \to \mathrm{Spec}(V)$. We then obtain the theorem by setting $\mathcal{X}' := \coprod_i \mathcal{X}'_i$. □

Corollary 8.5.8. — *Let X/K be a smooth proper algebraic space. Then there exist a finite extension $K \subset L \subset \overline{K}$ and a semistable proper scheme $\mathcal{Y}/\mathcal{O}_L$ and an alteration $\phi : \mathcal{Y}_L \to X$ over* $\mathrm{Spec}(L) \to \mathrm{Spec}(K)$.

Proof. — By [**46**, IV.3.1], there exists a birational map $X' \to X$ with X' an integral projective scheme. Replacing X by X' we may therefore assume that X is a projective scheme. Choose an embedding $X \subset \mathbb{P}_K$ into some projective space, and let $\mathcal{X} \subset \mathbb{P}_V$ be the scheme-theoretic closure. Then $\mathcal{X}$ is a flat, proper, and integral V-scheme so the result follows from de Jong's theorem. □

8.5.9. — Let X/K be a proper smooth algebraic space, and let $\mathrm{Alt}(X)$ denote the category of pairs $(L, \mathcal{Y}, \phi)$, where $K \subset L \subset \overline{K}$ is a finite extension, $\mathcal{Y}/\mathcal{O}_L$ is a proper semistable scheme, and $\phi : \mathcal{Y}_L \to X$ is an alteration. The set of morphisms $(L', \mathcal{Y}', \phi') \to (L, \mathcal{Y}, \phi)$ is the empty set unless $L \subset L'$ in which case a morphism is an X-map $g : \mathcal{Y}' \to \mathcal{Y}$ over $\mathrm{Spec}(\mathcal{O}_{L'}) \to \mathrm{Spec}(\mathcal{O}_L)$.

Proposition 8.5.10. — *The category* $\mathrm{Alt}(X)$ *is non-empty and connected.*

Proof. — The preceding corollary states that the category is nonempty.

To see that it is connected, let $(L_i, \mathcal{Y}_i, \phi_i)$ $(i = 1, 2)$ be two objects. We construct an object $(L, \mathcal{Y}, \psi) \in \mathrm{Alt}(X)$ mapping to both $(L_i, \mathcal{Y}_i, \phi_i)$. Choose first a finite extension $L' \subset \overline{K}$ containing both L_i, and let $\mathcal{Y}'_i$ denote the base change of $\mathcal{Y}_i$ to $\mathrm{Spec}(\mathcal{O}_{L'})$. Let $\mathcal{Y}'$ denote $\mathcal{Y}'_1 \times_{\mathrm{Spec}(\mathcal{O}_{L'})} \mathcal{Y}'_2$. Since the maps $\mathcal{Y}_{i,L_i} \to X_K$ are generically étale, the projection $\mathcal{Y}'_{1,L'} \times_{X_{L'}} \mathcal{Y}'_{2,L'} \to X_{L'}$ is generically étale. In particular, $\mathcal{Y}'_{1,L'} \times_{X_{L'}} \mathcal{Y}'_{2,L'}$ is generically reduced. Let $\mathcal{Z}'$ denote the closure of the natural immersion $\mathcal{Y}'_{1,L'} \times_{X_{L'}} \mathcal{Y}'_{2,L'} \hookrightarrow \mathcal{Y}'$ with the reduced structure. Note that $\mathcal{Z}'$ is equal to the scheme-theoretic closure of $(\mathcal{Y}'_{1,L'} \times_{X_{L'}} \mathcal{Y}'_{2,L'})_{\mathrm{red}}$ and hence $\mathcal{Z}'$ is proper, reduced, and flat over $\mathcal{O}_{L'}$. Using 8.5.7, choose a finite extension $L' \subset L \subset \overline{K}$, and an alteration $\mathcal{Y} \to \mathcal{Z}$ over $\mathrm{Spec}(\mathcal{O}_L) \to \mathrm{Spec}(\mathcal{O}_{L'})$ with $\mathcal{Y}/\mathcal{O}_L$ semistable. Then $\mathcal{Y}$ with the natural projection $\phi : \mathcal{Y}_L \to X$ is an object of $\mathrm{Alt}(X)$ mapping to both $(L_i, \mathcal{Y}_i, \phi_i)$. □

8.5.11. — Let X/K be a smooth proper scheme, and let m be an integer. We now construct the (φ, N, G)-structure (D^m, φ, N) on $H^m_{\mathrm{dR}}(X/K)$.

Choose a finite extension $K \subset L \subset \overline{K}$ and a proper semistable scheme $\mathcal{Y}/\mathcal{O}_L$ with an alteration $\phi : \mathcal{Y}_L \to X$ over $\mathrm{Spec}(L) \to \mathrm{Spec}(K)$. The C_{st}-conjecture for $\mathcal{Y}/\mathcal{O}_L$ provides an isomorphism

$$B_{\mathrm{dR}} \otimes_{\mathbb{Q}_p} H^m(\overline{\mathcal{Y}}_L, \mathbb{Q}_p) \simeq B_{\mathrm{dR}} \otimes_L H^m_{\mathrm{dR}}(\mathcal{Y}_L/L) \quad (8.5.11.1)$$

which by 8.5.4 and 8.5.5 induces an isomorphism

$$B_{\mathrm{dR}} \otimes_{\mathbb{Q}_p} H^m(\overline{X}, \mathbb{Q}_p) \simeq B_{\mathrm{dR}} \otimes_L H^m_{\mathrm{dR}}(X_L/L) \tag{8.5.11.2}$$

compatible with the filtrations and action of $\mathrm{Gal}(\overline{K}/L)$. Define

$$D^m := (B_{\mathrm{st}} \otimes_{\mathbb{Q}_p} H^m(\overline{X}, \mathbb{Q}_p))^{\mathrm{Gal}(\overline{K}/L)} \otimes_{L_0} K_0^{ur}, \tag{8.5.11.3}$$

where L_0 denotes the maximal unramified extension of K_0 in L. The space D^m inherits operators φ and N from B_{st} satisfying 0.1.1 1) and 0.1.1 2).

Note that if $\alpha^* : H^m(\overline{\mathcal{Y}}_L, \mathbb{Q}_p) \to H^m(\overline{\mathcal{Y}}_L, \mathbb{Q}_p)$ denotes the endomorphism giving the projection to $H^m(\overline{X}, \mathbb{Q}_p)$ as in 8.5.3, then by 8.5.5 the map α^* induces a $\mathrm{Gal}(\overline{K}/L)$-equivariant endomorphism of $B_{\mathrm{dR}} \otimes_{\mathbb{Q}_p} H^m(\overline{\mathcal{Y}}_L, \mathbb{Q}_p)$. It follows that there is an isomorphism

$$D^m \otimes_{K_0^{ur}} \overline{K} \simeq H^m_{\mathrm{dR}}(X_L/L) \otimes_L \overline{K} \simeq H^m_{\mathrm{dR}}(X/K) \otimes_K \overline{K}. \tag{8.5.11.4}$$

Lemma 8.5.12. — *The data (D^m, φ, N) and the isomorphism (8.5.11.4) is independent of the choices.*

Proof. — To see that the data (D^m, φ, N) is independent of the choices, note that if $L \subset L' \subset \overline{K}$ is another finite extension then there is a natural map
(8.5.12.1)

$$(B_{\mathrm{st}} \otimes_{\mathbb{Q}_p} H^m(\overline{X}, \mathbb{Q}_p))^{\mathrm{Gal}(\overline{K}/L)} \otimes_{L_0} K_0^{ur} \longrightarrow (B_{\mathrm{st}} \otimes_{\mathbb{Q}_p} H^m(\overline{X}, \mathbb{Q}_p))^{\mathrm{Gal}(\overline{K}/L')} \otimes_{L'_0} K_0^{ur}$$

which we claim is an isomorphism. Since both spaces have the same dimension (the dimension of $H^m_{\mathrm{dR}}(X_K/K)$), it suffices to show that it is an injection. Since B_{st} is flat over K_0^{ur}, it suffices to verify this after base changing to B_{st}. For this consider the commutative diagram
(8.5.12.2)

$$\begin{array}{ccc}
(B_{\mathrm{st}} \otimes_{\mathbb{Q}_p} H^m(X, \mathbb{Q}_p))^{\mathrm{Gal}(\overline{K}/L)} \otimes_{L_0} B_{\mathrm{st}} & \xrightarrow{(8.5.12.1)} & (B_{\mathrm{st}} \otimes_{\mathbb{Q}_p} H^m(\overline{X}, \mathbb{Q}_p))^{\mathrm{Gal}(\overline{K}/L')} \otimes_{L'_0} B_{\mathrm{st}} \\
\downarrow a & & \downarrow b \\
B_{\mathrm{st}} \otimes_{\mathbb{Q}_p} V & \xrightarrow{\mathrm{id}} & B_{\mathrm{st}} \otimes_{\mathbb{Q}_p} V.
\end{array}$$

By [**24**, 5.1.2 (ii)] the maps a and b are injective, and therefore the map (8.5.12.1) is also injective.

To see that the isomorphism (8.5.11.4) is independent of the choices, it suffices by 8.5.10 to show that if $(L', \mathcal{Y}', \phi') \to (L, \mathcal{Y}, \phi)$ is a morphism in $\mathrm{Alt}(X)$, then the isomorphism (8.5.11.4) obtained from $(L', \mathcal{Y}', \phi')$ and $(L, \mathcal{Y}, \phi)$ agree. For this in turn it suffices to show that the diagram

$$\begin{array}{ccc}
B_{\mathrm{dR}} \otimes_{\mathbb{Q}_p} H^m(\overline{\mathcal{Y}}'_{L'}, \mathbb{Q}_p) & \xrightarrow{\simeq} & B_{\mathrm{dR}} \otimes_{L'} H^m_{\mathrm{dR}}(\mathcal{Y}'_{L'}/L') \\
\uparrow & & \uparrow \\
B_{\mathrm{dR}} \otimes_{\mathbb{Q}_p} H^m(\overline{\mathcal{Y}}_L, \mathbb{Q}_p) & \xrightarrow{\simeq} & B_{\mathrm{dR}} \otimes_L H^m_{\mathrm{dR}}(\mathcal{Y}_L/L)
\end{array} \tag{8.5.12.3}$$

commutes which follows from 8.4.20. □

8.5.13. — To complete the construction of the (φ, N, G)-structure, it remains to construct an action of G on D^m such that the isomorphism (8.5.11.4) is G-equivariant.

For this fix an object $(L, \mathcal{Y}, \phi) \in \mathrm{Alt}(X)$. For any $\sigma \in \mathrm{Gal}(\overline{K}/K)$, we obtain a second object $(\sigma(L), \mathcal{Y}^\sigma, \phi^\sigma) \in \mathrm{Alt}(X)$ by defining $\mathcal{Y}^\sigma := \mathcal{Y} \times_{\mathrm{Spec}(\mathcal{O}_L),\sigma} \mathrm{Spec}(\mathcal{O}_L)$. There is then a commutative diagram

$$\begin{array}{ccc} \mathcal{Y}^\sigma & \longrightarrow & \mathcal{Y} \\ \downarrow & & \downarrow \\ \mathrm{Spec}(\mathcal{O}_L) & \xrightarrow{\sigma} & \mathrm{Spec}(\mathcal{O}_L). \end{array} \tag{8.5.13.1}$$

By 8.4.21 the resulting diagram

$$\begin{array}{ccc} B_{\mathrm{dR}} \otimes_{\mathbb{Q}_p} H^m(\overline{\mathcal{Y}}^\sigma, \mathbb{Q}_p) & \xrightarrow{c_{\mathcal{Y}^\sigma}} & B_{\mathrm{dR}} \otimes_L H^m_{\mathrm{dR}}(\mathcal{Y}^\sigma_L/L) \\ \uparrow{\scriptstyle \sigma^*\otimes\sigma^*} & & \uparrow{\scriptstyle \sigma^*\otimes\sigma^*} \\ B_{\mathrm{dR}} \otimes_{\mathbb{Q}_p} H^m(\overline{\mathcal{Y}}, \mathbb{Q}_p) & \xrightarrow{c_{\mathcal{Y}}} & B_{\mathrm{dR}} \otimes_L H^m_{\mathrm{dR}}(\mathcal{Y}_L/L) \end{array} \tag{8.5.13.2}$$

commutes. From this and the independence on the chosen model in $\mathrm{Alt}(X)$ it follows that σ^* induces an automorphism of D^m compatible with the automorphism on K_0^{ur} induced by σ and also compatible with the isomorphism (8.5.11.4). This therefore completes the construction of the (φ, N, G)-structure on $H^m_{\mathrm{dR}}(X/K)$.

CHAPTER 9

COMPARISON WITH LOG GEOMETRY IN THE SENSE OF FONTAINE AND ILLUSIE

Throughout this chapter we assume the reader is familiar with the basic notions of log geometry [**40**].

9.1. The stacks $\mathcal{L}og_{(S,M_S)}$

9.1.1. — Let (S, M_S) be a fine log scheme, and let

$$\mathcal{L}og_{(S,M_S)} \longrightarrow (S\text{-schemes}) \tag{9.1.1.1}$$

be the fibered category whose fiber over an S-scheme $f : X \to S$ is the groupoid of pairs (M_X, f^b), where M_X is a fine log structure on X and $f^b : f^*M_S \to M_X$ is a morphism of fine log structures on X.

Thus the data of a morphism of fine log schemes $(X, M_X) \to (S, M_S)$ is by definition equivalent to the data of an "ordinary" morphism $X \to \mathcal{L}og_{(S,M_S)}$.

Theorem 9.1.2 ([**62**, 1.1]). — *The fibered category $\mathcal{L}og_{(S,M_S)}$ is an algebraic stack locally of finite presentation over S.*

9.1.3. — This result enables one to translate many of the basic notions in log geometry into the stack theoretic language. For example it is shown in [**62**, 4.6] that a morphism of fine log schemes $(X, M_X) \to (S, M_S)$ is log smooth (resp. log étale, log flat) in the sense of [**40**] if and only if the corresponding morphism $X \to \mathcal{L}og_{(S,M_S)}$ is a smooth (resp. étale, flat) morphism of algebraic stacks. Furthermore, as explained in [**64**, 3.8] there is a canonical isomorphism

$$\Omega^1_{(X,M_X)/(S,M_S)} \simeq \Omega^1_{X/\mathcal{L}og_{(S,M_S)}}, \tag{9.1.3.1}$$

where the left hand side is the sheaf of logarithmic differentials defined in [**40**, 1.7]. If $(X, M_X) \to (S, M_S)$ is log smooth, then $\Omega^1_{(X,M_X)/(S,M_S)}$ is locally free of finite type and (9.1.3.1) also induces an isomorphism on tangent sheaves

$$T_{(X,M_X)/(S,M_S)} \simeq T_{X/\mathcal{L}og_{(S,M_S)}}. \tag{9.1.3.2}$$

9.1.4. — An important aspect of the stacks $\mathcal{L}og_{(S,M_S)}$ discussed in [**62**, §5] is their relationship with toric stacks. Let P be a finitely generated integral monoid, and let $\mathcal{S}_P$ denote the stack-theoretic quotient of $\mathrm{Spec}(\mathbb{Z}[P])$ by the natural action of the diagonalizable group scheme $D(P^{\mathrm{gp}})$ with underlying scheme $\mathrm{Spec}(\mathbb{Z}[P^{\mathrm{gp}}])$. Any morphism of finitely generated integral monoids $l : Q \to P$ induces a morphism of algebraic stacks $\mathcal{S}(l) : \mathcal{S}_P \to \mathcal{S}_Q$.

Let A be a ring, and let $\beta : \mathbb{Z}[Q] \to A$ be a ring homomorphism defining a morphism $\mathrm{Spec}(A) \to \mathcal{S}_Q$. Define maps

$$m : Q \longrightarrow P \oplus Q^{\mathrm{gp}}, \quad q \longmapsto (l(q), q) \tag{9.1.4.1}$$

$$\pi : P \oplus Q^{\mathrm{gp}} \longrightarrow P^{\mathrm{gp}}, \quad (p, q) \longmapsto p - l(q), \tag{9.1.4.2}$$

and

$$\kappa : P \longrightarrow P \oplus Q^{\mathrm{gp}}, \quad p \longmapsto (p, 0). \tag{9.1.4.3}$$

We view $\mathrm{Spec}(\mathbb{Z}[P \oplus Q^{\mathrm{gp}}])$ as a $\mathbb{Z}[Q]$-scheme using the map m.

The map π defines a homomorphism $D(P^{\mathrm{gp}}) \to D(P^{\mathrm{gp}} \oplus Q^{\mathrm{gp}})$ and hence an action of $D(P^{\mathrm{gp}})$ on $\mathrm{Spec}(\mathbb{Z}[P \oplus Q^{\mathrm{gp}}])$ over $\mathrm{Spec}(\mathbb{Z}[Q])$ (since the composite $\pi \circ m$ is the zero map). We therefore also obtain an action of $D(P^{\mathrm{gp}})$ on $\mathrm{Spec}(A \otimes_{\mathbb{Z}[Q]} \mathbb{Z}[P \oplus Q^{\mathrm{gp}}])$. The commutative diagram

$$\begin{array}{ccc} \mathrm{Spec}(A \otimes_{\mathbb{Z}[Q]} \mathbb{Z}[P \oplus Q^{\mathrm{gp}}]) & \xrightarrow{\kappa} & \mathrm{Spec}(\mathbb{Z}[P]) \\ \downarrow{\scriptstyle m} & & \downarrow{\scriptstyle l} \\ \mathrm{Spec}(A) & \xrightarrow{\beta} & \mathrm{Spec}(\mathbb{Z}[Q]) \end{array} \tag{9.1.4.4}$$

is compatible with the action of $D(P^{\mathrm{gp}})$ and $D(Q^{\mathrm{gp}})$, so by passing to the stack-theoretic quotients we obtain a commutative diagram

$$\begin{array}{ccc} [\mathrm{Spec}(A \otimes_{\mathbb{Z}[Q]} \mathbb{Z}[P \oplus Q^{\mathrm{gp}}])/D(P^{\mathrm{gp}})] & \longrightarrow & \mathcal{S}_P \\ \downarrow & & \downarrow \\ \mathrm{Spec}(A) & \xrightarrow{\beta} & \mathcal{S}_Q. \end{array} \tag{9.1.4.5}$$

Proposition 9.1.5. — *The diagram (9.1.4.5) is cartesian.*

Proof. — The diagram (9.1.4.5) is functorial in A in the sense that if $g : A \to A'$ is a ring homomorphism and $\beta' : \mathbb{Z}[Q] \to A'$ denotes the composite map

$$\mathbb{Z}[Q] \xrightarrow{\beta} A \longrightarrow A', \tag{9.1.5.1}$$

then there is a commutative diagram

$$
\begin{array}{ccc}
[\mathrm{Spec}(A' \otimes_{\mathbb{Z}[Q]} \mathbb{Z}[P \oplus Q^{\mathrm{gp}}])/D(P^{\mathrm{gp}})] & \longrightarrow & \mathrm{Spec}(A') \\
\downarrow & & \downarrow \\
[\mathrm{Spec}(A \otimes_{\mathbb{Z}[Q]} \mathbb{Z}[P \oplus Q^{\mathrm{gp}}])/D(P^{\mathrm{gp}})] & \longrightarrow & \mathrm{Spec}(A) \\
\downarrow & & \downarrow \\
\mathcal{S}_P & \longrightarrow & \mathcal{S}_Q,
\end{array}
\tag{9.1.5.2}
$$

where the inside bottom square is (9.1.4.5) for $\beta : \mathbb{Z}[Q] \to A$, the big outside square is (9.1.4.5) for $\beta' : \mathbb{Z}[Q] \to A'$, and the top inside square is cartesian. Therefore if the proposition holds for $\beta : \mathbb{Z}[Q] \to A$ then it also holds for $\beta' : \mathbb{Z}[Q] \to A'$. It follows that it suffices to prove the proposition for $A = \mathbb{Z}[Q]$ and β the identity map.

Let $\mathcal{P}$ denote the fiber product $\mathrm{Spec}(\mathbb{Z}[Q]) \times_{\mathcal{S}_Q} \mathcal{S}_P$. The stack $\mathcal{P}$ associates to any $\mathbb{Z}[Q]$-scheme $g : T \to \mathrm{Spec}(\mathbb{Z}[Q])$ the groupoid of triples (Z, f, s), where $Z \to T$ is a $D(P^{\mathrm{gp}})$-torsor, $f : Z \to \mathrm{Spec}(\mathbb{Z}[P])$ is a $D(P^{\mathrm{gp}})$-equivariant map, and s is a trivialization of the $D(Q^{\mathrm{gp}})$-torsor $Z \times^{D(P^{\mathrm{gp}})} D(Q^{\mathrm{gp}})$ such that the induced diagram

$$
\begin{array}{ccc}
Z & \xrightarrow{f} & \mathrm{Spec}(\mathbb{Z}[P]) \\
\downarrow y & & \downarrow l \\
T \times D(Q^{\mathrm{gp}}) & \xrightarrow{g \times \mathrm{action}} & \mathrm{Spec}(\mathbb{Z}[Q]) \\
\downarrow \mathrm{pr}_1 & & \\
T & &
\end{array}
\tag{9.1.5.3}
$$

is commutative. Here the map y is the composite of the canonical projection

$$Z \longrightarrow Z \times^{D(P^{\mathrm{gp}})} D(Q^{\mathrm{gp}}) \tag{9.1.5.4}$$

and the isomorphism

$$Z \times^{D(P^{\mathrm{gp}})} D(Q^{\mathrm{gp}}) \simeq T \times D(Q^{\mathrm{gp}}) \tag{9.1.5.5}$$

defined by s. Note that giving the trivialization s is equivalent to giving this $D(P^{\mathrm{gp}})$-equivariant map y.

Now given the $D(P^{\mathrm{gp}})$-torsor $Z \to T$, specifying the commutative diagram (9.1.5.3) is equivalent to giving a $D(P^{\mathrm{gp}})$-equivariant map

$$\tilde{f} : Z \to (T \times D(Q^{\mathrm{gp}})) \times_{\mathrm{Spec}(\mathbb{Z}[Q])} \mathrm{Spec}(\mathbb{Z}[P]) \simeq T \times_{\mathrm{Spec}(\mathbb{Z}[Q]),m} \mathrm{Spec}(\mathbb{Z}[P \oplus Q^{\mathrm{gp}}]), \tag{9.1.5.6}$$

where the action of $D(P^{\mathrm{gp}})$ on the right side is given by the map π. Now the category of pairs $(Z, \tilde{f})$ consisting of a $D(P^{\mathrm{gp}})$-torsor $Z \to T$ and a $D(P^{\mathrm{gp}})$-equivariant morphism (9.1.5.6) is by definition the groupoid

$$[\mathrm{Spec}(\mathbb{Z}[P \oplus Q^{\mathrm{gp}}])/D(P^{\mathrm{gp}})](T). \tag{9.1.5.7}$$

In this way we obtain an isomorphism

$$[\mathrm{Spec}(\mathbb{Z}[P \oplus Q^{\mathrm{gp}}])/D(P^{\mathrm{gp}})] \simeq \mathcal{P}. \tag{9.1.5.8}$$

It follows from the construction that this isomorphism agrees with the map defined by the commutative diagram (9.1.4.5). □

Remark 9.1.6

(i) If $l : Q \to P$ is injective, then the stack

$$[\mathrm{Spec}(A \otimes_{\mathbb{Z}[Q]} \mathbb{Z}[P \oplus Q^{\mathrm{gp}}])/D(P^{\mathrm{gp}})] \tag{9.1.6.1}$$

can also be described as follows. Let G denote $\mathrm{Coker}(Q^{\mathrm{gp}} \to P^{\mathrm{gp}})$ so we have an inclusion $D(G) \hookrightarrow D(P^{\mathrm{gp}})$. There is an inclusion

$$\mathrm{Spec}(A \otimes_{\mathbb{Z}[Q],l} \mathbb{Z}[P]) \hookrightarrow \mathrm{Spec}(A \otimes_{\mathbb{Z}[Q]} \mathbb{Z}[P \oplus Q^{\mathrm{gp}}]) \tag{9.1.6.2}$$

obtained from the morphism of monoids

$$P \oplus Q^{\mathrm{gp}} \longrightarrow P, \quad (p, q) \longmapsto p. \tag{9.1.6.3}$$

The inclusion (9.1.6.2) is $D(G)$-equivariant, where $D(G)$ acts on $\mathrm{Spec}(A \otimes_{\mathbb{Z}[Q]} \mathbb{Z}[P])$ by the action induced by the natural action on $\mathrm{Spec}(\mathbb{Z}[P])$. We therefore obtain a morphism of stacks

$$[\mathrm{Spec}(A \otimes_{\mathbb{Z}[Q],l} \mathbb{Z}[P])/D(G)] \longrightarrow [\mathrm{Spec}(A \otimes_{\mathbb{Z}[Q]} \mathbb{Z}[P \oplus Q^{\mathrm{gp}}])/D(P^{\mathrm{gp}})] \tag{9.1.6.4}$$

which we claim is an isomorphism. For this it suffices, as in the proof of 9.1.5, to consider the universal case $A = \mathbb{Z}[Q]$. The map (9.1.6.2) induces a $D(P^{gp})$-equivariant map

$$\mathrm{Spec}(\mathbb{Z}[P]) \times {}^{D(G)} D(P^{\mathrm{gp}}) \longrightarrow \mathrm{Spec}(\mathbb{Z}[P \oplus Q^{\mathrm{gp}}]), \tag{9.1.6.5}$$

and since we a factorization of (9.1.6.4) as

$$\begin{array}{c} [\mathrm{Spec}(\mathbb{Z}[P])/D(G)] \\ \downarrow \simeq \\ [\mathrm{Spec}(\mathbb{Z}[P]) \times {}^{D(G)} D(P^{\mathrm{gp}})/D(P^{\mathrm{gp}})] \\ \downarrow \\ [\mathrm{Spec}(\mathbb{Z}[P \oplus Q^{\mathrm{gp}}])/D(P^{\mathrm{gp}})] \end{array} \tag{9.1.6.6}$$

it suffices to show that the map (9.1.6.5) is an isomorphism.

Let $\sigma : P^{\mathrm{gp}} \to G$ be the projection and define

$$\delta : P \oplus P^{\mathrm{gp}} \longrightarrow G, \quad (p, p') \longmapsto \sigma(p) + \sigma(p'). \tag{9.1.6.7}$$

Then δ induces an action of $D(G)$ on the product

$$\mathrm{Spec}(\mathbb{Z}[P \oplus P^{\mathrm{gp}}]) \simeq \mathrm{Spec}(\mathbb{Z}[P]) \times D(P^{\mathrm{gp}}) \tag{9.1.6.8}$$

and we have

$$\mathrm{Spec}(\mathbb{Z}[P]) \times^{D(G)} D(P^{\mathrm{gp}}) \simeq \mathrm{Spec}(\mathbb{Z}[P \oplus P^{\mathrm{gp}}]^{D(G)}), \tag{9.1.6.9}$$

where the right side is the spectrum of the ring of $D(G)$-invariants. With this identification the map (9.1.6.5) is the map induced by the morphism of monoids

$$\tau : P \oplus Q^{\mathrm{gp}} \longrightarrow P \oplus P^{\mathrm{gp}}, \quad (p, q) \longmapsto (p, -p + l(q)). \tag{9.1.6.10}$$

Note that τ is injective as l is injective and the composition $\delta \circ \tau$ is the zero map $P \oplus Q^{\mathrm{gp}} \to G$. We therefore obtain an inclusion

$$P \oplus Q^{\mathrm{gp}} \hookrightarrow \delta^{-1}(0). \tag{9.1.6.11}$$

This map is surjective, for if $(p, p') \in \delta^{-1}(0)$ then $p+p'$ is in Q^{gp} and therefore $(p, p') = \tau(p, p + p')$. Now as a $D(P^{\mathrm{gp}})$ representation the ring $\mathbb{Z}[P \oplus P^{\mathrm{gp}}]$ is equal to a direct sum of rank 1 subrepresentations, and therefore the ring of invariants $\mathbb{Z}[P \oplus P^{\mathrm{gp}}]^{D(G)}$ is equal to the monoid algebra $\mathbb{Z}[\delta^{-1}(0)]$. It follows that the map (9.1.6.5), and hence also the map (9.1.6.4), is an isomorphism. In summary, when l is injective then for any morphism $\beta : \mathbb{Z}[Q] \to A$ we have a cartesian diagram

$$\begin{array}{ccc} [\mathrm{Spec}(A \otimes_{\mathbb{Z}[Q]} \mathbb{Z}[P])/D(G)] & \longrightarrow & \mathcal{S}_P \\ \downarrow & & \downarrow \\ \mathrm{Spec}(A) & \xrightarrow{\beta} & \mathcal{S}_Q. \end{array} \tag{9.1.6.12}$$

(ii) Suppose $l : Q \to P$ is surjective and furthermore that l is exact, which means that the diagram

$$\begin{array}{ccc} Q & \twoheadrightarrow & P \\ \downarrow & & \downarrow \\ Q^{\mathrm{gp}} & \twoheadrightarrow & P^{\mathrm{gp}} \end{array} \tag{9.1.6.13}$$

is cartesian. In this case we claim that the projection induced by the map m (9.1.4.1)

$$[\mathrm{Spec}(\mathbb{Z}[P \oplus Q^{\mathrm{gp}}])/D(P^{\mathrm{gp}})] \longrightarrow \mathrm{Spec}(\mathbb{Z}[Q]) \tag{9.1.6.14}$$

is an isomorphism, and therefore the map $\mathcal{S}(l) : \mathcal{S}_Q \to \mathcal{S}_P$ is also an isomorphism.

For this note that since l is surjective, the map $D(P^{\mathrm{gp}}) \to D(Q^{\mathrm{gp}})$ induced by $-l$ is an inclusion and therefore the action of $D(P^{\mathrm{gp}})$ on

$$\mathrm{Spec}(\mathbb{Z}[P \oplus Q^{\mathrm{gp}}]) \simeq \mathrm{Spec}(\mathbb{Z}[P]) \times D(Q^{\mathrm{gp}}) \tag{9.1.6.15}$$

is a free action. We therefore have

$$[\mathrm{Spec}(\mathbb{Z}[P \oplus Q^{\mathrm{gp}}])/D(P^{\mathrm{gp}})] \simeq \mathrm{Spec}(\mathbb{Z}[P \oplus Q^{\mathrm{gp}}]^{D(P^{\mathrm{gp}})}). \tag{9.1.6.16}$$

It therefore suffices to show that the map $m : Q \to P \oplus Q^{\mathrm{gp}}$ identifies Q with the submonoid

(9.1.6.17)
$$\pi^{-1}(0) \subset P \oplus Q^{\mathrm{gp}}.$$

This is clear, for if $(p, q) \in P \oplus Q^{\mathrm{gp}}$ maps to zero under π, then $p + q = 0$ in P^{gp} which implies that $(p, -q)$ defines an element of the fiber product of the diagram

(9.1.6.18)
$$\begin{array}{ccc} & & P \\ & & \downarrow \\ Q^{\mathrm{gp}} & \longrightarrow & P^{\mathrm{gp}}, \end{array}$$

which since l is exact is equal to Q.

Remark 9.1.7. — Note that the definition of the stack $\mathcal{S}_P$ makes sense also for finitely generated, but not necessarily integral, monoids P, and 9.1.5 and 9.1.6 remain valid in this more general setting. One place where such monoids arise naturally is the following. Consider a diagram of fine monoids

(9.1.7.1)
$$\begin{array}{ccc} Q & \xrightarrow{a} & Q' \\ {\scriptstyle b}\downarrow & & \\ P, & & \end{array}$$

and let P' denote the pushout of this diagram in the category of monoids. Even though Q, Q', and P are integral the monoid P' need not be integral. Nonetheless we obtain a commutative diagram of stacks

(9.1.7.2)
$$\begin{array}{ccc} \mathcal{S}_{P'} & \longrightarrow & \mathcal{S}_P \\ \downarrow & & \downarrow \\ \mathcal{S}_{Q'} & \longrightarrow & \mathcal{S}_Q. \end{array}$$

If b is injective the proposition 9.1.5 implies that this diagram is in fact cartesian.

To see this let $X = \mathrm{Spec}(\mathbb{Z}[Q'])$ and let $\pi : X \to \mathcal{S}_{Q'}$ be the projection. Since π is faithfully flat, to verify that (9.1.7.2) is cartesian it suffices to show that the natural map

(9.1.7.3)
$$X \times_{\mathcal{S}_{Q'}} \mathcal{S}_{P'} \longrightarrow X \times_{\mathcal{S}_Q} \mathcal{S}_P$$

is an isomorphism. Let G' (resp. G) denote the cokernel of the map $Q'^{\mathrm{gp}} \to P'^{\mathrm{gp}}$ (resp. $Q^{\mathrm{gp}} \to P^{\mathrm{gp}}$). Then by 9.1.6 (i) we have

(9.1.7.4)
$$X \times_{\mathcal{S}_{Q'}} \mathcal{S}_{P'} \simeq [\mathrm{Spec}(\mathbb{Z}[P'])/D(G')],$$

and

(9.1.7.5)
$$X \times_{\mathcal{S}_Q} \mathcal{S}_P \simeq [\mathrm{Spec}(\mathbb{Z}[Q'] \otimes_{\mathbb{Z}[Q]} \mathbb{Z}[P])/D(G)].$$

The map (9.1.7.3) is the map induced by the map (which is compatible with the group actions)

$$\mathbb{Z}[Q'] \otimes_{\mathbb{Z}[Q]} \mathbb{Z}[P] \longrightarrow \mathbb{Z}[P']. \tag{9.1.7.6}$$

Therefore the statement that (9.1.7.2) is cartesian follows from the observation that the map (9.1.7.6) and the canonical map $G \to G'$ are isomorphisms.

Remark 9.1.8. — One can reinterpret the notions of integral and saturated monoids as follows. Let P be a finitely generated monoid and let $\gamma : P \to P^{\mathrm{gp}}$ be the natural map. We claim that P is integral (resp. saturated) if and only if the stack $\mathcal{S}_P$ is reduced and irreducible (resp. reduced and irreducible and normal).

For this choose first a finitely generated free group G and a surjection

$$\delta : G \longrightarrow\!\!\!\!\!\rightarrow P^{\mathrm{gp}} \tag{9.1.8.1}$$

and let F denote the fiber product $G \times_{P^{\mathrm{gp}}} P$ so we have a cartesian diagram

$$\begin{array}{ccc} F & \xrightarrow{\alpha} & P \\ \downarrow{\scriptstyle\beta} & & \downarrow{\scriptstyle\gamma} \\ G & \xrightarrow{\delta} & P^{\mathrm{gp}}. \end{array} \tag{9.1.8.2}$$

Then F is integral (resp. saturated) if and only if P is integral (resp. saturated). On the other hand the induced map of stacks $\mathcal{S}_P \to \mathcal{S}_F$ is an isomorphism by 9.1.7 (ii). Replacing P by F we may therefore assume that P^{gp} is torsion free.

In this case the projection

$$\mathrm{Spec}(\mathbb{Z}[P]) \longrightarrow \mathcal{S}_P \tag{9.1.8.3}$$

is a smooth surjection. If P is integral then $\mathbb{Z}[P]$ embeds into the ring $\mathbb{Z}[P^{\mathrm{gp}}]$ which is an integral domain (since P^{gp} is assumed torsion free). It follows that if P is integral then $\mathcal{S}_P$ is reduced and irreducible. Conversely if P is not integral then $\mathbb{Z}[P]$ cannot be integral. For if $x, y \in P$ are two distinct elements with same image in P^{gp} then x and y define two morphisms

$$f_x, f_y : \mathrm{Spec}(\mathbb{Z}[P]) \longrightarrow \mathbb{A}^1 \tag{9.1.8.4}$$

whose restriction to $\mathrm{Spec}(\mathbb{Z}[P^{\mathrm{gp}}])$ are equal. Here f_x (resp. f_y) is the map induced by the map of rings $\mathbb{Z}[t] \to \mathbb{Z}[P]$ sending t to x (resp. y). It follows that if P is not integral then one of the following hold:

(i) $\mathbb{Z}[P]$ is not reduced in which case the stack $\mathcal{S}_P$ is not reduced either.

(ii) $\mathrm{Spec}(\mathbb{Z}[P])$ is not irreducible. In this case each irreducible component of $\mathrm{Spec}(\mathbb{Z}[P])$ is $D(P^{\mathrm{gp}})$-invariant as $D(P^{\mathrm{gp}})$ is a connected group. It follows that each irreducible component of $\mathrm{Spec}(\mathbb{Z}[P])$ defines a proper closed substack of $\mathcal{S}_P$ thereby showing that $\mathcal{S}_P$ is not irreducible.

We deduce that P is integral if and only if $\mathcal{S}_P$ is reduced and irreducible.

Similarly P is saturated if and only if the ring $\mathbb{Z}[P]$ is normal [**56**, 3.3.1], and since the map (9.1.8.3) is a smooth surjection $\mathbb{Z}[P]$ is normal if and only if $\mathcal{S}_P$ is normal.

Using this and 9.1.7 we also obtain an interpretation of the notion of integral (resp. saturated) morphism of monoids. Namely an injective morphism of fine monoids $Q \to P$ is integral (resp. saturated) if and only if for any morphism of fine monoids $Q \to Q'$ the fiber product of the diagram of stacks

$$\begin{array}{ccc} & & \mathcal{S}_P \\ & & \downarrow \\ \mathcal{S}_{Q'} & \longrightarrow & \mathcal{S}_Q \end{array} \tag{9.1.8.5}$$

is reduced and irreducible (resp. normal).

Remark 9.1.9. — In what follows we will only consider the stacks $\mathcal{S}_P$ for fine monoids P unless explicitly stated otherwise (the only place we use non-fine monoids is in the proof of 9.3.5 below).

9.1.10. — For a log structure M on a scheme T, let $\overline{M}$ denote the sheaf of monoids $M/\mathcal{O}_T^*$. By [**62**, 5.20] the stack $\mathcal{S}_P$ is naturally viewed as the stack which to any scheme T associates the groupoid of pairs $(M_T, \beta : P \to \overline{M}_T)$, where M_T is a fine log structure on T and β is a morphism of sheaves of monoids which fppf-locally lifts to a chart for M_T.

In particular, if (S, M_S) is a fine log scheme and $\beta : Q \to M_S$ is a chart with induced map $\bar{\beta} : Q \to \overline{M}_S$, then there is a natural map $S \to \mathcal{S}_Q$. For any morphism of fine monoids $l : Q \to P$ the stack $\mathcal{S}_P \times_{\mathcal{S}_Q} S$ is by [**62**, 5.20] isomorphic to the stack which to any S-scheme $f : T \to S$ associates the groupoid of triples (N, η, γ), where N is a fine log structure on T, $\eta : f^*M_S \to N$ is a morphism of log structures on T, and $\gamma : P \to \overline{N}$ is a morphism of monoids which fppf-locally lifts to a chart such that the diagram

$$\begin{array}{ccc} Q & \xrightarrow{l} & P \\ {\scriptstyle\bar{\beta}}\downarrow & & \downarrow{\scriptstyle\gamma} \\ f^{-1}\overline{M}_S & \xrightarrow{\eta} & \overline{N} \end{array} \tag{9.1.10.1}$$

commutes. In particular, there is a natural map

$$\mathcal{S}_P \times_{\mathcal{S}_Q} S \longrightarrow \mathcal{L}og_{(S,M_S)}. \tag{9.1.10.2}$$

Theorem 9.1.11 ([**62**, 5.24]). — *The map (9.1.10.2) is representable and étale.*

9.1.12. — This theorem is useful because it often enables one to replace the stack $\mathcal{L}og_{(S,M_S)}$ by the simpler stack $\mathcal{S}_P \times_{\mathcal{S}_Q} S$. For example, let $f : (X, M_X) \to (S, M_S)$

be a log smooth morphism. Then étale locally on S and X there exists by [**40**, 3.5] a chart for f:

$$\begin{array}{ccc} X & \longrightarrow & \operatorname{Spec}(\mathbb{Z}[P]) \\ \downarrow & & \downarrow \\ S & \longrightarrow & \operatorname{Spec}(\mathbb{Z}[Q]) \end{array} \tag{9.1.12.1}$$

for some morphism of monoids $l : Q \to P$. It follows that the map $X \to \mathcal{L}og_{(S,M_S)}$ factors through $\mathcal{S}_P \times_{\mathcal{S}_Q} S$. Since the map (9.1.10.2) is étale this implies for example that the de Rham complex of X over $\mathcal{L}og_{(S,M_S)}$ is equal to the de Rham complex of X over the stack $\mathcal{S}_P \times_{\mathcal{S}_Q} S$.

Proposition 9.1.13. — *Let $(X, M_X) \to (S, M_S)$ be a log smooth morphism between fine log schemes.*

(i) *The isomorphism (9.1.3.2) is an isomorphism of sheaves of Lie algebras, where the left hand side is given a Lie algebra structure as in* [**58**, 1.1.7] *and the right hand side is viewed as a sheaf of Lie algebras using the definition in 2.2.6.*

(ii) *If S is of characteristic $p > 0$, then (9.1.3.2) is an isomorphism of sheaves of restricted p-Lie algebras where the left hand side is given the restricted p-Lie algebra structure defined in* [**58**, 1.2.1] *and the right hand side is given the restricted p-Lie algebra structure defined in 3.1.6.*

Proof. — The issue is étale local on S and X, so by [**40**, 3.4] we may assume that

$$X = \operatorname{Spec}(\mathbb{Z}[N]) \times_{\operatorname{Spec}(\mathbb{Z}[Q])} S \tag{9.1.13.1}$$

for some injective morphism of fine monoids $\theta : Q \to N$ such that the torsion part of $G := \operatorname{Coker}(Q^{\mathrm{gp}} \to N^{\mathrm{gp}})$ is invertible on S. Let $R := X \times_{\operatorname{Spec}(\mathbb{Z})} \operatorname{Spec}(\mathbb{Z}[G]) = X \times_{\mathcal{S}_N \times_{\mathcal{S}_Q} S} X$, and let pr (resp. ρ) be the projection $R \to X$ (resp. the map induced by $\mathbb{Z}[N] \to \mathbb{Z}[N] \otimes_{\mathbb{Z}} \mathbb{Z}[G]$ sending $n \in N$ to $n \otimes g_n$, where g_n denotes the image of n in G). Let $\gamma : G \to \mathcal{O}_R$ denote the map

$$G \xrightarrow{g \mapsto 1 \otimes g} \mathcal{O}_X \otimes_{\mathbb{Z}} \mathbb{Z}[G] = \mathcal{O}_R, \tag{9.1.13.2}$$

and let $\beta : N \to \mathcal{O}_R$ be the map

$$N \xrightarrow{n \mapsto n \otimes 1 \otimes 1} (\mathbb{Z}[N] \otimes_{\mathbb{Z}[Q]} \mathcal{O}_S) \otimes_{\mathbb{Z}} \mathbb{Z}[G] \simeq \mathcal{O}_R. \tag{9.1.13.3}$$

Then $\rho^* M_X$ (resp. $\mathrm{pr}^* M_X$) is the log structure associated to the prelog structure

$$N \longrightarrow \mathcal{O}_R, \quad n \longmapsto \gamma(g_n) \cdot \beta(n) \quad (\text{resp.} \quad n \longmapsto \beta(n)). \tag{9.1.13.4}$$

The map

$$N \longrightarrow \mathcal{O}_R^* \oplus N, \quad n \longmapsto (\gamma(g_n), n) \tag{9.1.13.5}$$

therefore induces an isomorphism $\iota : \mathrm{pr}^* M_X \to \rho^* M_X$. Using the identification

$$X \times_{\mathcal{L}og_{(S,M_S)}} X \simeq \underline{\mathrm{Isom}}_{X \times X}(\mathrm{pr}_1^* M_X, \mathrm{pr}_2^* M_X), \tag{9.1.13.6}$$

the isomorphism ι defines a morphism $\tau : R \to X \times_{\mathcal{L}og_{(S,M_S)}} X$ over the map $\mathrm{pr} \times \rho : R \to X \times X$. By 9.1.11 the map τ is étale. Let J denote the ideal of the diagonal map $\Delta : X \to R$ induced by the map of rings $\mathbb{Z}[G] \to \mathcal{O}_X$ sending $g \in G$ to 1, so that $\Omega^1_{X/\mathcal{L}og_{(S,M_S)}}$ is isomorphic to J/J^2. The map $g \mapsto 1 - g$ induces an isomorphism $J/J^2 \simeq \mathcal{O}_X \otimes_{\mathbb{Z}} G$, and $dn = n(1 - g_n)$. Furthermore, under the isomorphism $\Omega^1_{X/\mathcal{L}og_{(S,M_S)}} \simeq \Omega^1_{(X,M_X)/(S,M_S)}$ of [**62**, 5.24] the element $(1 - g_n)$ maps to $\mathrm{dlog}(n)$ (notation as in [**40**, 1.7]).

Let $g_1, \ldots, g_r \in G$ be elements which form a basis for the free part of G, so that $\{\mathrm{dlog}(g_i)\}$ form a basis for $\Omega^1_{(X,M_X)/(S,M_S)}$, and let $\{\partial_{g_i}\}$ denote the dual basis. Then the above discussion implies that the Lie bracket on $T_{(X,M_X)/(S,M_S)}$ induced by the one in 2.2.6 satisfies $[\partial_g, \partial_{g'}] = 0$. Moreover, this condition determines the Lie bracket structure completely, and hence it agrees with that defined in [**58**, 1.1.7].

Furthermore, if S is a scheme in characteristic $p > 0$, then the definition of the p-Lie algebra structure in 3.1.6 shows that $\partial_g^{(p)} = \partial_g$ which by [**58**, 1.2.2] agrees with the one given there. □

Corollary 9.1.14. — *Let $(X, M_X) \to (S, M_S)$ be a log smooth morphism between fine log schemes. Then the logarithmic de Rham complex $\Omega^\bullet_{(X,M_X)/(S,M_S)}$ is canonically isomorphic to the de Rham complex $\Omega^\bullet_{X/\mathcal{L}og_{(S,M_S)}}$ defined in 2.2.16.*

Remark 9.1.15. — In [**43**, 4.1.3 (2)] a *framed log scheme* is defined to be a triple $(X, M_X, \beta : P \to \overline{M}_X)$ consisting of a fine saturated log scheme (X, M_X) and a map $\beta : P \to \overline{M}_X$ from a fine saturated monoid P which étale locally on X lifts to a chart. The map β is called a *frame* for (X, M_X). By [**62**, 5.20] giving a frame β for a fine saturated log scheme (X, M_X) is equivalent to giving a dotted arrow

$$\begin{array}{ccc} & & \mathcal{S}_P \\ & \overset{\beta}{\nearrow} & \downarrow \\ X & \xrightarrow[M_X]{} & \mathcal{L}og_{(\mathrm{Spec}(\mathbb{Z}), \mathcal{O}_{\mathrm{Spec}(\mathbb{Z})})} \end{array} \tag{9.1.15.1}$$

making the diagram commute.

9.1.16. — Let $\alpha = (\alpha_1, \ldots, \alpha_r)$ be a sequence of positive integers, and let $H \subset S_r$ be the subgroup of the symmetric group consisting of permutations $\sigma \in S_r$ such that $\alpha_{\sigma(i)} = \alpha_i$ for all i. The stack $\mathcal{S}_H(\alpha)$ defined in 6.1.1 can then be interpreted as follows.

Let $M_{\mathbb{A}^1}$ denote the log structure on $\mathbb{A}^1 = \mathrm{Spec}(\mathbb{Z}[t])$ defined by the map $\mathbb{N} \to \mathbb{Z}[t]$ sending 1 to t.

Proposition 9.1.17. — *The stack $\mathcal{S}_H(\alpha)$ is canonically isomorphic to the open substack of $\mathcal{L}og_{(\mathbb{A}^1, M_{\mathbb{A}^1})}$ which to any $f : X \to \mathbb{A}^1$ associates the groupoid of pairs (M_X, f^b), where M_X is a fine log structure on X and $f^b : f^* M_{\mathbb{A}^1} \to M_X$ is a morphism such* *that for every geometric point $\bar{x} \to X$ there exist an integer ℓ and an isomorphism $\sigma : \overline{M}_{X,\bar{x}} \oplus \mathbb{N}^\ell \simeq \mathbb{N}^r$ such that the diagram*

$$
\begin{array}{ccc}
\overline{M}_{\mathbb{A}^1,f(\bar{x})} & \xrightarrow{f^b} & \overline{M}_{X,\bar{x}} \\
\uparrow & & \uparrow{\scriptstyle\tau} \\
\mathbb{N} & \xrightarrow{k} & \mathbb{N}^r
\end{array}
\tag{9.1.17.1}
$$

commutes, where τ is the projection obtained from σ and k is the map induced by $1 \mapsto (\alpha_1, \ldots, \alpha_r)$.

Remark 9.1.18. — If $(f, f^b) : (X, M_X) \to (\mathbb{A}^1, M_{\mathbb{A}^1})$ is a morphism of fine log scheme and $\tau : \mathbb{N}^r \to \overline{M}_X$ is a surjection which étale locally on X lifts to a chart and such that the diagram

$$
\begin{array}{ccc}
f^{-1}\overline{M}_{\mathbb{A}^1} & \xrightarrow{f^b} & \overline{M}_X \\
\uparrow & & \uparrow{\scriptstyle\tau} \\
\mathbb{N} & \xrightarrow{k} & \mathbb{N}^r
\end{array}
\tag{9.1.18.1}
$$

commutes, then for every geometric point $\bar{x} \to X$ there exists an isomorphism $\sigma : \overline{M}_{X,\bar{x}} \oplus \mathbb{N}^\ell \simeq \mathbb{N}^r$ identifying $\tau_{\bar{x}} : \mathbb{N}^r \to \overline{M}_{X,\bar{x}}$ with the projection onto the first factor.

To see this we may work étale locally on X, and may therefore assume that τ lifts to a chart $\alpha : \mathbb{N}^r \to M_X$. In this case, for every geometric point $\bar{x} \to X$ the projection $\mathbb{N}^r \to \overline{M}_{X,\bar{x}}$ identifies $\overline{M}_{X,\bar{x}}$ with the quotient of $\mathbb{N}^r$ by the face $F_{\bar{x}} \subset \mathbb{N}^r$ generated by the standard generators $e_i \in \mathbb{N}^r$ for which $\alpha(e_i)$ maps to a unit in $M_{X,\bar{x}}$. Then $F_{\bar{x}} \simeq \mathbb{N}^\ell$ for some ℓ (the number of generators mapping to units in $M_{X,\bar{x}}$) and if $N \subset \mathbb{N}^r$ denotes the submonoid generated by those generators e_i which map to nonzero elements in $\overline{M}_{X,\bar{x}}$ then the natural map $N \oplus F_{\bar{x}} \to \mathbb{N}^r$ is an isomorphism and the composite $N \hookrightarrow \mathbb{N}^r \to \overline{M}_{X,\bar{x}}$ is an isomorphism defining a splitting of the surjection $\mathbb{N}^r \to \overline{M}_{X,\bar{x}}$.

Conversely, suppose $(f, f^b) : (X, M_X) \to (\mathbb{A}^1, M_{\mathbb{A}^1})$ is a morphism of fine log schemes such that for some geometric point $\bar{x} \to X$ there exists an isomorphism $\sigma : \overline{M}_{X,\bar{x}} \oplus \mathbb{N}^\ell \simeq \mathbb{N}^r$ as in 9.1.17. Then in some étale neighborhood of $\bar{x}$ the morphism τ lifts to a chart for $\mathbb{N}^r$. Indeed the existence of σ implies that $\overline{M}_{X,\bar{x}}$ is a free monoid, and therefore there exists a section $s : \overline{M}_{X,\bar{x}} \to M_{X,\bar{x}}$ of the projection $M_{X,\bar{x}} \to \overline{M}_{X,\bar{x}}$. The composite map $s \circ \tau : \mathbb{N}^r \to M_{X,\bar{x}}$ then extends to a chart in some étale neighborhood of $\bar{x}$ by [**56**, 2.2.4].

In summary, if $(f, f^b) : (X, M_X) \to (\mathbb{A}^1, M_{\mathbb{A}^1})$ is a morphism of fine log schemes and $\bar{x} \to X$ is a geometric point, then the condition that there exists an isomorphism $\sigma : \overline{M}_{X,\bar{x}} \oplus \mathbb{N}^\ell \simeq \mathbb{N}^r$ as in 9.1.17 is equivalent to the condition that in some étale neighborhood of $\bar{x}$ there exists a morphism $\tau : \mathbb{N}^r \to \overline{M}_X$ which étale locally lifts to a chart such that the diagram (9.1.18.1) commutes.

In particular, the set of points $x \in X$ for which there exists an isomorphism $\sigma : \overline{M}_{X,\bar{x}} \oplus \mathbb{N}^\ell \simeq \mathbb{N}^r$ as in 9.1.17 is an open subset of X.

Proof of 9.1.17. — Denote temporarily by $\mathcal{F} \subset \mathcal{L}og_{(\mathbb{A}^1, M_{\mathbb{A}^1})}$ the substack described in the theorem.

Lemma 9.1.19. — *The substack $\mathcal{F} \subset \mathcal{L}og_{(\mathbb{A}^1, M_{\mathbb{A}^1})}$ is an open substack.*

Proof. — Let $(f, f^b) : (X, M_X) \to (\mathbb{A}^1, M_{\mathbb{A}^1})$ be a morphism of fine log schemes corresponding to a morphism $X \to \mathcal{L}og_{(\mathbb{A}^1, M_{\mathbb{A}^1})}$. Let $x \in X$ be a geometric point such that there exists an isomorphism $\sigma : \overline{M}_{X,\bar{x}} \oplus \mathbb{N}^\ell \simeq \mathbb{N}^r$ for some ℓ such that the diagram (9.1.17.1) commutes. To prove the lemma we must show that there exists an open neighborhood $x \in U \subset X$ of x such that for every geometric point $\bar{x}' \to U$ there exists an isomorphism $\sigma' : \overline{M}_{X,\bar{x}'} \oplus \mathbb{N}^{\ell'} \simeq \mathbb{N}^r$ such that the diagram (9.1.17.1) for $\bar{x}'$ commutes. This follows from 9.1.18. □

Define $\mathcal{S}_H(\alpha)^{\mathrm{ps}}$ as in 6.1.8. Recall that an object of $\mathcal{S}_H(\alpha)^{\mathrm{ps}}$ over some scheme $f : X \to \mathbb{A}^1$ is given by elements $(x_1, \ldots, x_r) \in \Gamma(X, \mathcal{O}_X)$ and an element $v \in \Gamma(X, \mathcal{O}_X^*)$ such that

(9.1.19.1) $$f^{-1}(t) = x_1^{\alpha_1} \cdots x_r^{\alpha_r} v.$$

For such an object let $M_{(x,v)}$ denote the log structure on X associated to the map $\mathbb{N}^r \to \mathcal{O}_X$ sending the i-th standard generator e_i to x_i. The map

(9.1.19.2) $$\mathbb{N} \longrightarrow \mathcal{O}_X^* \oplus \mathbb{N}^r, \quad 1 \longmapsto (v, \alpha_1, \ldots, \alpha_r)$$

defines a morphism $f^b_{(x,v)} : f^* M_{\mathbb{A}^1} \to M_{(x,v)}$ making $(M_{(x,v)}, f^b_{(x,v)})$ an object of $\mathcal{F}$.

If $(x', v') \in \mathcal{S}_H(\alpha)^{\mathrm{ps}}(X)$ is a second object, then an isomorphism

(9.1.19.3) $$\iota : (M_{(x,v)}, f^b_{(x,v)}) \longrightarrow (M_{(x',v')}, f^b_{(x',v')})$$

in $\mathcal{F}$ is by the universal property of the log structure associated to a prelog structure obtained from a morphism $\mathbb{N}^r \to M_{(x',v')}$ such that the composite

(9.1.19.4) $$\mathbb{N} \xrightarrow{1 \mapsto (v, \alpha_1, \ldots, \alpha_r)} \mathcal{O}_X^* \oplus \mathbb{N}^r \longrightarrow M_{(x',v')}$$

is equal to the map induced by $f^b_{(x',v')}$. Let $\bar{t} \to X$ be a geometric point. The commutativity of the diagram

(9.1.19.5) $$\begin{array}{ccc} \overline{M}_{\mathbb{A}^1, f(\bar{t})} & \xrightarrow{\mathrm{id}} & \overline{M}_{\mathbb{A}^1, f(\bar{t})} \\ {\scriptstyle f^b_{(x,v)}} \downarrow & & \downarrow {\scriptstyle f^b_{(x',v')}} \\ \overline{M}_{(x,v),\bar{t}} & \xrightarrow{\iota} & \overline{M}_{(x',v'),\bar{t}} \end{array}$$

shows that every standard generator e_i of $\mathbb{N}^r$ which does not map to 0 in $\overline{M}_{(x,v),\bar{t}}$ is mapped under ι to an element $e_{h(i)}$ with $\alpha_i = \alpha_{h(i)}$. From this it follows that an isomorphism ι as above is given locally by a section of the presheaf $F((x,v),(x',v'))$ (notation as in 6.1.8). It follows that the prestack $\mathcal{S}_H(\alpha)^{\mathrm{ps}}$ is equivalent to the prestack whose objects are objects (M_X, f^b) of $\mathcal{F}$ together with a chart $\mathbb{N}^r \to M_X$. Passing to the associated stack it follows that $\mathcal{S}_H(\alpha) \simeq \mathcal{F}$. □

Remark 9.1.20. — If in 9.1.21 we instead take H to be the trivial group $\{e\} \subset S_r$, then an argument similar to the one used in the proof of 9.1.21 shows that the stack $\mathcal{S}_{\{e\}}(\alpha)$ is canonically isomorphic to the stack over $\mathbb{A}^1$ which to any $f : X \to \mathbb{A}^1$ associates the groupoid of triples $(M_X, \tau : \mathbb{N}^r \to \overline{M}_X, f^b)$, where (M_X, β) is a framed fine saturated log structure on X (in the sense of [**43**, 4.1.3]), and $f^b : f^*M_{\mathbb{A}^1} \to M_X$ is a morphism log structures on X such that the following diagram commutes

$$
\begin{array}{ccc}
f^{-1}\overline{M}_{\mathbb{A}^1} & \xrightarrow{f^b} & \overline{M}_X \\
\uparrow & & \uparrow \beta \\
\mathbb{N} & \xrightarrow{k} & \mathbb{N}^r,
\end{array}
\tag{9.1.20.1}
$$

where as in 9.1.21 k is the map induced by $1 \mapsto (\alpha_1, \ldots, \alpha_r)$. Equivalently, with the terminology and notation of [**43**, 4.1.3] the stack $\mathcal{S}_{\{e\}}(\alpha)$ is isomorphic to the stack over $\mathbb{A}^1$ (which we view as the underlying scheme of the framed log scheme $((\mathbb{A}^1, M_{\mathbb{A}^1}), [\mathbb{N}]))$ which to any scheme X associates the groupoid of triples $(M_X, [\mathbb{N}^r], g)$, where $(M_X, [\mathbb{N}^r])$ is a framed log structure on X and

$$
g : ((X, M_X), [\mathbb{N}^r]) \to ((\mathbb{A}^1, M_{\mathbb{A}^1}), [\mathbb{N}])
\tag{9.1.20.2}
$$

is a morphism of framed log schemes with underlying morphism of schemes f over the morphism $[\mathbb{N}^r] \to [\mathbb{N}]$ defined by the map k.

Proposition 9.1.21. — *Assume all $\alpha_i = 1$ and that H is equal to the full symmetric group on r letters. Then the stack $\overline{\mathcal{S}}_H(\alpha)$ defined in 6.1.9 is naturally isomorphic to the open substack of $\mathcal{L}og_{(\mathrm{Spec}(\mathbb{Z}), \mathcal{O}^*_{\mathrm{Spec}(\mathbb{Z})})}$ which to any scheme T associates the groupoid of fine log structures M on T such that for every geometric point $\bar{t} \to T$ the monoid $\overline{M}_{\bar{t}}$ is a free monoid of rank $\leq r$ The map $\mathcal{S}_H(\alpha) \to \overline{\mathcal{S}}_H(\alpha)$ is the map which sends a pair (M_X, f^b) to M_X.*

Proof. — Let $\mathcal{F} \subset \mathcal{L}og_{(\mathrm{Spec}(\mathbb{Z}), \mathcal{O}^*_{\mathrm{Spec}(\mathbb{Z})})}$ be the substack classifying log structures as in the proposition. It is an open substack since the sheaves $\overline{M}$ are constructible sheaves (see for example [**62**, 3.5 (ii)]), and for any specialization $\bar{\zeta} \to \bar{t}$ the cospecialization map $\overline{M}_{\bar{t}} \to \overline{M}_{\bar{\eta}}$ is the quotient by a face. In other words, if $\overline{M}_{\bar{t}} \simeq \mathbb{N}^s$ then $\overline{M}_{\bar{\eta}}$ is the quotient of $\mathbb{N}^s$ by the irreducible elements e_i which lift to an element in $M_{\bar{t}}$ whose image in $\mathcal{O}_{T,\bar{t}}$ maps to a unit in $\mathcal{O}_{T,\bar{\eta}}$.

Let $\mathcal{T}_H^{\mathrm{ps}}$ be as in 6.1.12. There is a natural map $\mathcal{T}_H^{\mathrm{ps}} \to \mathcal{T}$ which associates to a collection $(x_1, \ldots, x_r) \in \mathcal{T}_H^{\mathrm{ps}}(T)$ the log structure M_x associated to the map $\mathbb{N}^r \to \mathcal{O}_T$ sending e_i to x_i. Note that the log structure M_x is canonically isomorphic to the log structure associated to $\oplus_{i \in E(x)} \mathbb{N} \to \mathcal{O}_T$ which associates to e_i ($i \in E(x)$) the element x_i. Thus for a second object $(x'_1, \ldots, x'_r) \in \mathcal{T}_H^{\mathrm{ps}}(T)$ an isomorphism $M_x \to M_{x'}$ is given by units $\{u_i\}_{i \in E(x)}$ and a bijection $h : E(x) \to E(x')$ such that $x'_{h(i)} = u_i x_i$. In other words, by an isomorphism in $\mathcal{T}_H^{\mathrm{ps}}(T)$. It follows that there is a fully faithful functor $\mathcal{T}_H \to \mathcal{F}$. To show that it is an equivalence it suffices to show that any object of $\mathcal{F}$ is étale locally in the image which follows from [**62**, 2.1].

Finally the description of the map $\mathcal{S}_H(\alpha) \to \overline{\mathcal{S}}_H(\alpha)$ follows from the preceding proof. □

9.1.22. — Since a free monoid of rank 1 has no automorphisms, if M is a fine log structure on a scheme T such that for every geometric point $\bar{t} \to T$ the stalk $\overline{M}_{\bar{t}}$ is a free monoid of rank ≤ 1, then there is a unique map $\beta : \mathbb{N} \to \overline{\mathcal{M}}$ which étale locally lifts to a chart. It follows from this and 6.1.11 that the stack $[\mathbb{A}^1/\mathbb{G}_m]$ can be viewed as the stack classifying fine log structures M whose stalks are free monoids of rank ≤ 1.

This is related to the interpretation of $[\mathbb{A}^1/\mathbb{G}_m]$ given in 6.1.11 as follows. For a log structure M on T as above, let P be the $\mathcal{O}_T^*$-torsor of liftings of the map β to a chart $\mathbb{N} \to M$. The map $M \to \mathcal{O}_T$ induces a map of sheaves with $\mathcal{O}_T^*$-action $P \to \mathcal{O}_T$. If $\mathcal{L}$ denotes the invertible sheaf corresponding to P, then this map defines a map of invertible sheaves $\mathcal{L} \to \mathcal{O}_T$. This functor is in fact an equivalence (this follows for example from 6.1.11, but can also be seen directly).

More generally, for a free monoid $N \simeq \mathbb{N}^s$ the diagonal map $\mathbb{N} \to N$ sending 1 to $(1, 1, \ldots, 1)$ is invariant under all automorphisms of N. It follows that for any fine log structure M on a scheme T such that the stalks $\overline{M}_{\bar{t}}$ are free monoids, there is a canonical map $\beta : \mathbb{N} \to \overline{M}$. The inverse image of $\beta(\mathbb{N})$ in M is then a fine log structure on T whose stalks are free monoids of rank ≤ 1. It follows from this that with notation and assumptions as in 6.1.11 there is a canonical map $\overline{\mathcal{S}}_H(\alpha) \to [\mathbb{A}^1/\mathbb{G}_m]$. This is simply the map described in 6.1.9.

9.1.23. — For an integer $e \geq 1$, let $\theta_e : [\mathbb{A}^1/\mathbb{G}_m] \to [\mathbb{A}^1/\mathbb{G}_m]$ be the map defined in 8.2.6, and let $\mathcal{P}^{(e)}$ denote the fiber product $\overline{\mathcal{S}}_H(\alpha) \times_{[\mathbb{A}^1/\mathbb{G}_m],\theta_e} [\mathbb{A}^1/\mathbb{G}_m]$ defined in 8.3.3, where $\alpha = (1, \ldots, 1)$ (r times) and H is the symmetric group on r letters. From the above interpretations of $[\mathbb{A}^1/\mathbb{G}_m]$ and $\overline{\mathcal{S}}_H(\alpha)$, for any scheme T the groupoid $\mathcal{P}^{(e)}(T)$ is equivalent to the groupoid of diagrams of fine log structures

$$\begin{array}{ccc} M_1 & \xrightarrow{a} & M_2 \\ {\scriptstyle b}\downarrow & & \\ M_3, & & \end{array} \tag{9.1.23.1}$$

where for every geometric point $\bar{t} \to T$ the stalks $\overline{M}_{1,\bar{t}}$ and $\overline{M}_{3,\bar{t}}$ are free monoids of rank ≤ 1, $\overline{M}_{2,\bar{t}}$ is a free monoid of rank $\leq r$, the map $\overline{M}_{1,\bar{t}} \to \overline{M}_{2,\bar{t}}$ is the diagonal map, and $\overline{M}_{1,\bar{t}} \to \overline{M}_{3,\bar{t}}$ is multiplication by e.

Let M denote the pushout in the category of fine log structures of the diagram (9.1.23.1). The diagram (9.1.23.1) can be recovered from M as follows. For every geometric point $\bar{t} \to T$, the stalk $\overline{M}_{\bar{t}}$ is either trivial or isomorphic to $P := \mathbb{N}^s \oplus_{\Delta,\mathbb{N},e} \mathbb{N}$ for some s. It follows that there exists a unique irreducible element p_0 in P such that ep_0 is equal to a sum of distinct irreducible elements in P, and that there are s other irreducible elements in P. It follows that that there is a canonical isomorphism $P \simeq N \oplus_{\mathbb{N},e} \mathbb{N}$ with $N \simeq \mathbb{N}^s$. The stalk $\overline{M}_{3,\bar{t}}$ is recovered as the submonoid of P generated by p_0 and $\overline{M}_{2,\bar{t}}$ is equal to N. Thus we recover the log structure M_3 as the subsheaf of M consisting of elements whose image in $\overline{M}_{\bar{t}}$ lands in $\mathbb{N} \cdot p_0$ for every geometric point $\bar{t}$, and M_2 is equal to the subsheaf of sections whose image in $\overline{M}_{\bar{t}}$ lands in N. Finally M_1 is equal to the subsheaf of sections whose image in $\overline{M}_{\bar{t}}$ is in $\mathbb{N} \cdot ep_0$.

Corollary 9.1.24. — *The stack $\mathcal{P}^{(e)}$ is isomorphic to the stack which to any scheme T associates the groupoid of fine log structures M on T which are isomorphic to a pushout of a diagram as in (9.1.23.1).*

Remark 9.1.25. — By an argument similar to the one used in 9.1.18, if X is a scheme and M is a fine log structure on X then M defines an object of $\mathcal{P}^{(e)}(X)$ if and only if for every geometric point $\bar{x} \to X$ the stalk $\overline{M}_{\bar{x}}$ is isomorphic to $\mathbb{N}^s \oplus_{\Delta,\mathbb{N},e} \mathbb{N}$ for some $s \leq r$. Furthermore, the condition that there exists such an isomorphism $\mathbb{N}^s \oplus_{\Delta,\mathbb{N},e} \mathbb{N} \simeq \overline{M}_{\bar{x}}$ is equivalent to the condition that in some étale neighborhood of $\bar{x}$ there exists a chart $\mathbb{N}^s \oplus_{\Delta,\mathbb{N},e} \mathbb{N} \to M$.

As in 9.1.19 this implies that $\mathcal{P}^{(e)}$ is an open substack of $\mathcal{L}og_{(\mathrm{Spec}(\mathbb{Z}),\mathcal{O}^*_{\mathrm{Spec}(\mathbb{Z})})}$.

9.1.26. — As in 6.2.1, let V be a complete discrete valuation ring with uniformizer π and mixed characteristic $(0,p)$, and let X/V be a flat regular scheme with smooth generic fiber for which the reduced closed fiber $X_{0,\mathrm{red}} \subset X$ is a divisor with normal crossings. Let $\{\alpha_1, \dots, \alpha_r\}$ be as in 6.2.1, and let H be the group of elements $\sigma \in \mathrm{S}_r$ for which $\alpha_{\sigma(i)} = \alpha_i$ for all i. Let $\mathcal{S}_{V,\pi}$ denote the stack $\mathcal{S}_H(\alpha)_{(\pi)\subset V}$ obtained by base change from $\overline{\mathcal{S}}_H(\alpha)$ from the map $\mathrm{Spec}(V) \to [\mathbb{A}^1/\mathbb{G}_m]$ defined by the invertible sheaf (π) with the inclusion into $\mathcal{O}_V$.

Let M_V denote the log structure on $\mathrm{Spec}(V)$ defined by the maximal ideal $(\pi) \subset V$. From the above discussion, the stack $\mathcal{S}_H(\alpha)_{V,\pi}$ can also be described as the substack of $\mathcal{L}og_{(\mathrm{Spec}(V),M_V)}$ which to any $f : T \to \mathrm{Spec}(V)$ associates the groupoid of pairs (M_T, f^b), where M_T is a fine log structure on T and $f^b : f^*M_V \to M_T$ is a morphism of fine log structures such that for every geometric point $\bar{t}$ there exists a diagram as in (9.1.17.1).

In particular, the canonical log structure M_X on X [**40**, 1.5 (1)] with the natural map $f^b : f^*M_V \to M_X$ defines a morphism $X \to \mathcal{S}_H(\alpha)_{V,\pi}$. It follows from the construction in 6.2 that this map agrees with the one constructed there.

9.1.27. — There is a generalization of the stacks $\mathcal{L}og_{(S,M_S)}$ which is sometimes useful. Let Γ be a category with finitely many objects and finitely many morphisms, and let $\mathcal{L}og^\Gamma$ be the stack over $\mathbb{Z}$ which to any scheme T associates the category of functors from Γ to the category of log structures on T. If $f : \Gamma' \to \Gamma$ is a functor then there is a natural functor

$$f^* : \mathcal{L}og^\Gamma \longrightarrow \mathcal{L}og^{\Gamma'}. \tag{9.1.27.1}$$

It is shown in [**64**, 2.4] that the stack $\mathcal{L}og^\Gamma$ is an algebraic stack locally of finite presentation over $\mathbb{Z}$. The following examples are especially useful.

Example 9.1.28. — For any integer $n \geq 0$ let $\mathcal{L}og^{[n]}$ denote the stack $\mathcal{L}og^\Gamma$ with Γ equal to the category corresponding to the ordered set $[n] = \{0, \dots, n\}$. The stack associates to any scheme T the groupoid of diagrams of fine log structures on T

(9.1.28.1) $$M_0 \longrightarrow M_1 \longrightarrow \cdots \longrightarrow M_n.$$

Any order preserving map $[n'] \to [n]$ defines a functor so the stacks $\mathcal{L}og^{[n]}$ in fact form a simplicial algebraic stack

(9.1.28.2) $$[n] \longmapsto \mathcal{L}og^{[n]}.$$

If (S, M_S) is a fine log scheme, then the stack $\mathcal{L}og_{(S,M_S)}$ is canonically isomorphic to the fiber product

(9.1.28.3) $$\mathcal{L}og^{[1]} \times_{\delta_1^*, \mathcal{L}og^{[0]}, M_S} S,$$

where $\delta_1 : [0] \to [1]$ is the map sending 0 to 0.

Proposition 9.1.29 (**[64**, 2.11 (i)]). — *Let $j \leq n$ be an integer and $\delta_j : [n] \to [n+1]$ the unique injective order preserving map whose image does not contain j. Then the induced map $\delta_j^* : \mathcal{L}og^{[n+1]} \to \mathcal{L}og^{[n]}$ is relatively Deligne-Mumford and étale.*

Example 9.1.30. — Let $\square$ denote the category with four objects $\{0, 1, 2, 3\}$ and morphisms given by

(9.1.30.1) $$\operatorname{Hom}(j, j') = \begin{cases} \{*\} & \text{if } j = j' \text{ or } (j, j') \in \{(0,1), (0,2), (0,3), (1,3), (2,3)\} \\ \varnothing & \text{otherwise.} \end{cases}$$

The stack $\mathcal{L}og^{\square}$ associates to any scheme T the groupoid of commutative diagrams of fine log structures on T

(9.1.30.2) $$\begin{array}{ccc} \mathcal{M}_0 & \longrightarrow & \mathcal{M}_1 \\ \downarrow & & \downarrow \\ \mathcal{M}_2 & \longrightarrow & \mathcal{M}_3. \end{array}$$

In addition to the maps described in the following proposition, we will also sometimes consider the maps $\kappa_i^* : \mathcal{L}og^{\square} \to \mathcal{L}og^{[1]}$ $(i = 1, 2)$ obtained from the functor $\kappa_i : [1] \to \square$ sending 0 to 0 and 1 to i.

Proposition 9.1.31 (**[64**, 2.11 (ii)])

(i) *Let $q_1 : [1] \to \square$ (resp. $q_2 : [1] \to \square$) be the functor sending 0 to 1 (resp. 2) and 1 to 3. Then the induced morphism $q_1^* : \mathcal{L}og^{\square} \to \mathcal{L}og^{[1]}$ (resp. $q_2^* : \mathcal{L}og^{\square} \to \mathcal{L}og^{[1]}$) is relatively Deligne-Mumford and étale.*

(ii) *For $i = 1, 2$, let $\tau_i : [2] \to \square$ be the functor sending 0 to 0, 1 to i, and 2 to 3. Then the induced morphism $\tau_i^* : \mathcal{L}og^{\square} \to \mathcal{L}og^{[2]}$ is relatively Deligne-Mumford and étale.*

Example 9.1.32. — To illustrate the utility of the above stacks, let us construct using the stack theoretic approach the functoriality morphism for log differentials. Consider a commutative diagram of fine log schemes

$$\begin{array}{ccc} (W, M_W) & \xrightarrow{a} & (X, M_X) \\ {\scriptstyle f}\downarrow & & \downarrow{\scriptstyle g} \\ (Z, M_Z) & \xrightarrow{b} & (Y, M_Y). \end{array} \tag{9.1.32.1}$$

We construct a morphism $a^*\Omega^1_{(X,M_X)/(Y,M_Y)} \to \Omega^1_{(W,M_W)/(Z,M_Z)}$ as follows. Let $\mathcal{L}$ be the fiber product of the diagram

$$\begin{array}{ccc} & & \mathcal{L}og^{\square} \\ & & \downarrow{\scriptstyle \kappa_2^*} \\ Z & \xrightarrow{b^*M_Y \to M_Z} & \mathcal{L}og^{[1]}. \end{array} \tag{9.1.32.2}$$

There is a natural commutative diagram

$$\begin{array}{ccc} \mathcal{L} & \longrightarrow & \mathcal{L}og^{\square} \\ \downarrow & & \downarrow{\scriptstyle \tau_2} \\ \mathcal{L}og_{(Z,M_Z)} & \xrightarrow{c} & \mathcal{L}og^{[2]} \\ \downarrow & & \downarrow \\ Z & \xrightarrow{b^*M_Y \to M_Z} & \mathcal{L}og^{[1]}, \end{array} \tag{9.1.32.3}$$

where the map c sends a morphism of fine log schemes $(f, f^b) : (T, M_T) \to (Z, M_Z)$ to the diagram $f^*b^*M_Y \to f^*M_Z \to M_T$. It follows from the definitions that the squares in (9.1.32.3) are all cartesian, and from 9.1.31 (ii) that the map $\mathcal{L} \to \mathcal{L}og_{(Z,M_Z)}$ is étale. Let $\rho : W \to \mathcal{L}$ be the morphism defined by the square

$$\begin{array}{ccc} M_Y|_W & \longrightarrow & a^*M_X \\ \downarrow & & \downarrow \\ f^*M_Z & \longrightarrow & M_W. \end{array} \tag{9.1.32.4}$$

Then it follows that there is a natural isomorphism

$$\Omega^1_{(W,M_W)/(Z,M_Z)} \simeq \Omega^1_{W/\mathcal{L}}. \tag{9.1.32.5}$$

The map $\kappa_1 : \mathcal{L}og^{\square} \to \mathcal{L}og^{[1]}$ induces a map $\kappa : \mathcal{L} \to \mathcal{L}og_{(Y,M_Y)}$ such that the diagram

$$\begin{array}{ccc} W & \xrightarrow{a} & X \\ {\scriptstyle \rho}\downarrow & & \downarrow \\ \mathcal{L} & \xrightarrow{\kappa} & \mathcal{L}og_{(Y,M_Y)} \end{array} \tag{9.1.32.6}$$

commutes. There is therefore a natural map

$$a^*\Omega^1_{(X,M_X)/(Y,M_Y)} \simeq a^*\Omega^1_{X/\mathcal{L}og_{(Y,M_Y)}} \longrightarrow \Omega^1_{W/\mathcal{L}} \simeq \Omega^1_{(W,M_W)/(Z,M_Z)} \tag{9.1.32.7}$$

which is the desired functoriality morphism.

Example 9.1.33. — Fiber products in the logarithmic category can also be described using the above techniques. Consider a diagram of fine log schemes

$$\begin{array}{ccc} & & (X, M_X) \\ & & \downarrow \\ (Z, M_Z) & \longrightarrow & (Y, M_Y). \end{array} \tag{9.1.33.1}$$

Let $\mathcal{P}$ denote the fiber product of the diagram

$$\begin{array}{ccc} & & X \times_Y Z \\ & & \downarrow{\scriptstyle (M_Y|X \to M_X)\times(M_Y|_Z \to M_Z)} \\ \mathcal{L}og^{\square} & \xrightarrow{\kappa_2\times\kappa_1} & \mathcal{L}og^{[1]} \times \mathcal{L}og^{[1]}. \end{array} \tag{9.1.33.2}$$

The stack $\mathcal{P}$ is the stack which to any scheme $T \to X \times_Y Z$ associates the groupoid of diagrams of fine log structures

$$\begin{array}{ccc} M_Y|_T & \longrightarrow & M_X|_T \\ \downarrow & & \downarrow \\ M_Z|_T & \longrightarrow & M. \end{array} \tag{9.1.33.3}$$

Let $X' \subset \mathcal{P}$ be the open substack classifying diagrams (9.1.33.3) which are cocartesian. Since such a diagram admits no nontrivial automorphisms the stack X' is in fact an algebraic space, and there is a tautological commutative diagram of log structures on X'

$$\begin{array}{ccc} M_Y|_T & \longrightarrow & M_X|_T \\ \downarrow & & \downarrow \\ M_Z|_T & \longrightarrow & M_{X'}. \end{array} \tag{9.1.33.4}$$

Finally by the universal property of pushout there is a canonical isomorphism $\mathcal{P} \simeq \mathcal{L}og_{(X',M_{X'})}$. In particular, $(X', M_{X'})$ is the fiber product in the category of fine log schemes.

Remark 9.1.34. — Let LOG be the category fibered (not in groupoids) over the category of schemes which to any scheme T associates the category of fine log structures on T. Any morphism of fibered categories $\Lambda : \mathrm{LOG} \to \mathrm{LOG}$ induces for any finite category Γ a functor $\Lambda^\Gamma : \mathcal{L}og^\Gamma \to \mathcal{L}og^\Gamma$. For any scheme T and functor $(f : \Gamma \to \mathrm{LOG}(T)) \in \mathcal{L}og^\Gamma(T)$ the image $\Lambda^\Gamma(f)$ is the composite functor

$$\Gamma \xrightarrow{f} \mathrm{LOG}(T) \xrightarrow{\Lambda} \mathrm{LOG}(T). \tag{9.1.34.1}$$

For example, let $e \geq 1$ be an integer and take Λ to be the functor which associates to a fine log structure M on T the log structure associated to the pre-log structure

$$M \xrightarrow{\times e} M \longrightarrow \mathcal{O}_T. \tag{9.1.34.2}$$

If $e = p$ is a prime, then the induced morphism $\Lambda^p : \mathcal{L}og^\Gamma \otimes \mathbb{F}_p \to \mathcal{L}og^\Gamma \otimes \mathbb{F}_p$ is simply the Frobenius morphism. It follows that if R is any ring over $\mathbb{Z}_p$ with a lifting of Frobenius $F_R : R \to R$, then there is a lift of Frobenius to $\mathcal{L}og^\Gamma \otimes_{\mathbb{Z}} R$ given by $\Lambda_p^\Gamma \otimes F_R$. Furthermore, if $\Gamma' \to \Gamma$ is a functor then the induced morphism $\mathcal{L}og^\Gamma \otimes R \to \mathcal{L}og^{\Gamma'} \otimes R$ is compatible with these lifts of Frobenius.

Finally, it follows from the proof of 9.1.17 that with the modular interpretation given in the *loc. cit.* of $\mathcal{S}_H(\alpha)$, the liftings of Frobenius used in the preceding chapters defined using the map Λ_p in 6.3 agrees with the liftings of Frobenius defined on $\mathcal{L}og^\Gamma$ above.

9.2. Comparison of crystalline topoi

9.2.1. — Let (S, M_S, I, γ) be a fine log scheme with a divided power ideal (I, γ), and let $f : (X, M_X) \to (S, M_S)$ be a morphism of fine log schemes such that γ extends to X. Recall [**40**, 5.2] that the log crystalline site, denoted $\mathrm{Cris}((X, M_X)/(S, M_S))$, is the site whose underlying category is the category of strict closed immersions $(U, M_U) \hookrightarrow (T, M_T)$ over f, with $U \to X$ étale, $M_X|_U = M_U$, and $U \hookrightarrow T$ defined by a divided power ideal compatible with γ. A family of morphisms $\{((U_i, M_{U_i}) \hookrightarrow (T_i, M_{T_i})) \to ((U, M_U) \hookrightarrow (T, M_T))\}$ is a covering if the map $\coprod T_i \to T$ is étale and surjective. We denote by $((X, M_X)/(S, M_S))_{\mathrm{cris}}$ the associated ringed topos.

Assume that there exists an open substack $\mathcal{U} \subset \mathcal{L}og_{(S,M_S)}$ such that (I, γ) extends to $\mathcal{U}$ and the morphism $X \to \mathcal{L}og_{(S,M_S)}$ factors through $\mathcal{U}$.

Proposition 9.2.2. — *There is a natural equivalence of sites*

$$\mathrm{Cris}((X, M_X)/(S, M_S)) \simeq \mathrm{Cris}(X_{\mathrm{et}}/\mathcal{U}). \tag{9.2.2.1}$$

Proof. — By definition of $\mathcal{L}og_{(S,M_S)}$, to give an exact closed immersion $(U, M_U) \hookrightarrow (T, M_T)$ as in the definition of $\mathrm{Cris}((X, M_X)/(S, M_S))$ is equivalent to giving a 2-commutative diagram

$$\begin{array}{ccc} U & \xrightarrow{j} & T \\ \downarrow & & \downarrow \\ X & \longrightarrow & \mathcal{L}og_{(S,M_S)}, \end{array} \tag{9.2.2.2}$$

with j a closed immersion defined by a PD-ideal compatible with (I, γ). Since j is defined by a nil-ideal the map $T \to \mathcal{L}og_{(S,M_S)}$ factors through $\mathcal{U}$, and hence giving $(U, M_U) \hookrightarrow (T, M_T)$ in $\mathrm{Cris}((X, M_X)/(S, M_S))$ is equivalent to giving an object $(U \hookrightarrow T, \delta) \in \mathrm{Cris}(X/\mathcal{U})$. □

Corollary 9.2.3. — *There is an equivalence of categories between the category of quasi-coherent crystals in* $((X, M_X)/(S, M_S))_{\mathrm{cris}}$ *and the category of quasi-coherent crystals in* $(X/\mathcal{U})_{\mathrm{cris}}$.

9.2.4. — The equivalence in 9.2.2 is compatible with the projections to the étale topos

$$u_{X_{\mathrm{et}}/\mathcal{U}} : (X_{\mathrm{et}}/\mathcal{U})_{\mathrm{cris}} \longrightarrow X_{\mathrm{et}}, \tag{9.2.4.1}$$

and

$$u_{(X,M_X)/(S,M_S)} : ((X, M_X)/(S, M_S))_{\mathrm{cris}} \longrightarrow X_{\mathrm{et}}, \tag{9.2.4.2}$$

where the morphism $u_{(X,M_X)/(S,M_S)}$ is defined as in [**40**, 5.9]. This follows from observing that the inverse image functors $u^*_{X_{\mathrm{et}}/\mathcal{U}}$ and $u^*_{(X,M_X)/(S,M_S)}$ are equal by definition.

9.2.5. — Consider an exact closed immersion $(X, M_X) \hookrightarrow (Y, M_Y)$ over (S, M_S) inducing a closed immersion $X \hookrightarrow Y$ over $\mathcal{L}og_{(S,M_S)}$. After replacing Y by an open subset we may assume that the morphism $Y \to \mathcal{L}og_{(S,M_S)}$ also factors through $\mathcal{U}$.

Proposition 9.2.6. — *The logarithmic divided power envelope* $D_{(X,M_X),\gamma}(Y, M_Y)$ *defined in* [**40**, 5.4] *is canonically isomorphic to the divided power envelope defined as in 1.2.3 of the closed immersion* $X \hookrightarrow Y$ *over the PD-stack* $(\mathcal{U}, I_{\mathcal{U}}, \gamma)$ *with log structure equal to the pullback of* M_Y.

Proof. — To give an exact closed immersion $(X, M_X) \hookrightarrow (T, M_T)$ of (S, M_S)-log schemes with a divided power structure on the ideal of X in T compatible with γ is equivalence to giving a closed immersion $X \hookrightarrow T$ over $\mathcal{U}$ with a divided power structure on the ideal of X in T compatible with $(I_{\mathcal{U}}, \gamma)$. Hence the result follows from the universal properties of the PD-envelopes in 1.2.3 and [**40**, 5.4]. □

9.2.7. — Let (S, M_S, I, γ) be a fine log scheme with a PD-ideal (I, γ), let $(X, M_X) \to (S, M_S)$ be a morphism of fine log schemes such that γ extends to X, and let $(X, M_X) \hookrightarrow (Y, M_Y)$ be an exact closed immersion over (S, M_S) with $(Y, M_Y) \to (S, M_S)$ log smooth. Assume that the morphism $X \to \mathcal{L}og_{(S,M_S)}$ factors through an open substack $\mathcal{U} \subset \mathcal{L}og_{(S,M_S)}$ to which (I, γ) extends, and let (D, M_D) be the divided power envelope of $(X, M_X) \hookrightarrow (Y, M_Y)$. Let $\mathcal{D}$ denote the coordinate ring of D viewed as a sheaf on Y_{et}. Let $d : \mathcal{D} \to \mathcal{D} \otimes \Omega^1_{(Y,M_Y)/(S,M_S)}$ denote the canonical connection.

From 2.4.7, 2.3.28, 2.5.4, and 9.1.14 we now obtain the following:

Theorem 9.2.8 **([40, 6.2]).** — *There is a natural equivalence of categories between the category of quasi-coherent crystals in* $((X, M_X)/(S, M_S))_{\mathrm{cris}}$ *and the category of quasi-coherent* $\mathcal{D}$*-modules* $\mathcal{E}$ *with a map*

$$\nabla : \mathcal{E} \longrightarrow \mathcal{E} \otimes_{\mathcal{O}_Y} \Omega^1_{(Y,M_Y)/(S,M_S)} \tag{9.2.8.1}$$

such that for an local sections $e \in \mathcal{E}$ *and* $f \in \mathcal{D}$ *we have*

$$\nabla(fe) = f\nabla(e) + e \otimes df, \tag{9.2.8.2}$$

and the composite

$$\mathcal{E} \xrightarrow{\ \nabla\ } \mathcal{E} \otimes_{\mathcal{O}_Y} \Omega^1_{(Y,M_Y)/(S,M_S)} \xrightarrow{\ \nabla^2\ } \mathcal{E} \otimes_{\mathcal{O}_Y} \Omega^2_{(Y,M_Y)/(S,M_S)} \tag{9.2.8.3}$$

is zero, where ∇^2 is the map defined by $e \otimes \omega \mapsto \nabla(e) \wedge \omega + e \otimes d\omega$.

Theorem 9.2.9 **([40, 6.4]).** — *With notation as in 9.2.8, let E be a quasi-coherent crystal in $((X, M_X)/(S, M_S))_{\mathrm{cris}}$ and let $(\mathcal{E}, \nabla)$ be the corresponding quasi-coherent $\mathcal{D}$-module with integrable connection. Then $Ru_{(X,M_X)/(S,M_S)*}E$ is canonically isomorphic to the de Rham complex $\mathcal{E} \otimes \Omega^{\bullet}_{(D,M_D)/(S,M_S)}$ of $(\mathcal{E}, \nabla)$.*

9.2.10. — The functoriality of the log crystalline site can be understood using the method of 9.1.32. Let (Z, M_Z, J, δ) and (Y, M_Y, I, γ) be fine log schemes with divided power ideals, and let $b : (Z, M_Z) \to (Y, M_Y)$ be a PD-morphism. Consider a commutative square as in (9.1.32.1) such that (I, γ) (resp. (J, δ)) extends to X (resp. W). We define a canonical morphism of ringed topoi

$$((W, M_W)/(Z, M_Z))_{\mathrm{cris}} \longrightarrow ((X, M_X)/(Y, M_Y))_{\mathrm{cris}} \tag{9.2.10.1}$$

as follows. Let $\mathcal{L}$ be as in 9.1.32. Since the morphism $\mathcal{L} \to \mathcal{L}og_{(Z,M_Z)}$ is étale, the infinitesimal lifting property for étale morphisms implies that any object $(U \hookrightarrow T) \in \mathrm{Cris}(W/\mathcal{L}og_{(Z,M_Z)})$ admits a unique structure of an object in $\mathrm{Cris}(W/\mathcal{L})$. It follows that there is a canonical isomorphism of topoi

$$(W/\mathcal{L}og_{(Z,M_Z)})_{\mathrm{cris}} \simeq (W/\mathcal{L})_{\mathrm{cris}}. \tag{9.2.10.2}$$

From the commutative diagram

$$\begin{array}{ccc} W & \longrightarrow & X \\ \big\downarrow & & \big\downarrow \\ \mathcal{L} & \longrightarrow & \mathcal{L}og_{(Y,M_Y)} \end{array} \tag{9.2.10.3}$$

and 1.4.14 we therefore obtain a morphism of ringed topoi

(9.2.10.4)

$$((W, M_W)/(Z, M_Z))_{\mathrm{cris}} \simeq (W/\mathcal{L})_{\mathrm{cris}} \longrightarrow (X/\mathcal{L}og_{(Y,M_Y)})_{\mathrm{cris}} \simeq ((X, M_X)/(Y, M_Y))_{\mathrm{cris}}.$$

9.3. The Cartier type property

9.3.1. — Let $(X, M_X) \to (S, M_S)$ be a log smooth morphism between fine log schemes over $\mathbb{F}_p$ for some prime p. Let $F_{(S,M_S)} : (S, M_S) \to (S, M_S)$ be the Frobenius morphism, and let $(X', M_{X'})$ denote the fiber product

$$(X', M_{X'}) := (X, M_X) \times_{(S,M_S),F_{(S,M_S)}} (S, M_S) \tag{9.3.1.1}$$

in the category of fine log schemes. There is a canonical map $F : (X, M_X) \to (X', M'_X)$ defined by the Frobenius morphism $F_{(X,M_X)} : (X, M_X) \to (X, M_X)$. By [40, 4.10 (2)] there is a canonical factorization of F

$$(X, M_X) \xrightarrow{\ g\ } (X'', M_{X''}) \xrightarrow{\ h\ } (X', M_{X'}), \tag{9.3.1.2}$$

where the morphism g is exact and h is purely inseparable [**40**, 4.9]. The morphism $g : (X, M_X) \to (X'', M_{X''})$ is called the *exact relative Frobenius morphism.*

9.3.2. — We can also consider the diagram

(9.3.2.1)

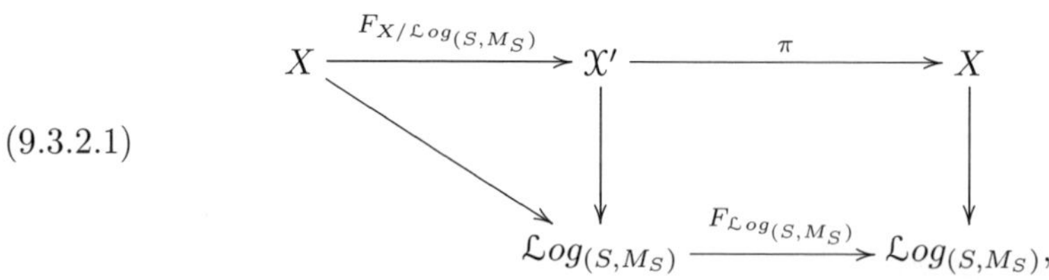

where the square is cartesian. Recall that $\mathcal{X}'$ is only an algebraic stack, but by 3.3.2 there exists a universal map $\epsilon : \mathcal{X}' \to \overline{\mathcal{X}}'$ to a scheme. Let $\tilde{g} : X \to \overline{\mathcal{X}}'$ denote the composite map

$$X \xrightarrow{F_{X/\mathcal{L}og_{(S,M_S)}}} \mathcal{X}' \xrightarrow{\epsilon} \overline{\mathcal{X}}'. \tag{9.3.2.2}$$

The stack $\mathcal{X}'$ is by the definition of the fiber product of stacks isomorphic to the stack which to any S-scheme $f : T \to S$ associates the groupoid of data (M_T, f^b, γ), where M_T is a fine log structure on T, $f^b : f^*M_S \to M_T$ is a morphism of fine log structures on T, and $\gamma : (T, F_T^*M_T) \to (X, M_X)$ is a strict morphism of log schemes such that the diagram

$$\begin{array}{ccc} (T, F_T^*M_T) & \xrightarrow{\gamma} & (X, M_X) \\ & \searrow^{(F_S\circ f, F_T^*(f^b))} & \downarrow \\ & & (S, M_S) \end{array} \tag{9.3.2.3}$$

commutes, where $F_T^*(f^b)$ denotes the composite map

$$f^*F_S^*M_S \xrightarrow{\simeq} F_T^*f^*M_S \xrightarrow{F_T^*(f^b)} F_T^*M_T. \tag{9.3.2.4}$$

Here F_T and F_S denote the Frobenius morphisms of T and S respectively. In particular we obtain a commutative diagram

$$\begin{array}{ccccc} (T, M_T) & \xrightarrow{c} & (T, F_T^*M_T) & \xrightarrow{\gamma} & (X, M_X) \\ \downarrow{\scriptstyle (f,f^b)} & & & & \downarrow \\ (S, M_S) & & \xrightarrow{F_{(S,M_S)}} & & (S, M_S), \end{array} \tag{9.3.2.5}$$

where c is the morphism of log schemes which is the identity on T and the natural map $F_T^*M_T \to M_T$ on log structures. By the universal property of $(X', M_{X'})$ we therefore obtain a morphism $d : (T, M_T) \to (X', M_{X'})$. The functor sending (M_X, f^b, γ) to the underlying morphism $T \to X'$ of d defines a morphism of stacks

(9.3.2.6) $$h' : \mathfrak{X}' \longrightarrow X',$$

which in turn induces a morphism $\tilde{h} : \overline{\mathfrak{X}}' \to X'$.

The composite map

(9.3.2.7) $$X \xrightarrow{\ \tilde{g}\ } \overline{\mathfrak{X}}' \xrightarrow{\ \tilde{h}\ } X' \xrightarrow{\ e\ } X,$$

where e denotes the projection, is equal to the Frobenius morphism of X.

Proposition 9.3.3. — *There exists a unique isomorphism $\iota : \overline{\mathfrak{X}}' \simeq X''$ such that $\iota \circ \tilde{g} = g$ and $\tilde{h} = h \circ \iota$ (here we consider only the underlying morphisms of schemes of h and g).*

Proof. — Since $\mathfrak{X}' \to \overline{\mathfrak{X}}'$ is universal for maps to schemes, the map $\iota : \overline{\mathfrak{X}}' \to X''$ is determined by the induced map $\tilde{\iota} : \mathfrak{X}' \to X''$. Since the map $F_{X/\mathcal{L}og_{(S,M_S)}} : X \to \mathfrak{X}'$ is faithfully flat (3.1.4), the map $\tilde{\iota}$ is in turn determined by the map

(9.3.3.1) $$\tilde{\iota} \circ F_{X/\mathcal{L}og_{(S,M_S)}} = \iota \circ \tilde{g}.$$

The uniqueness of ι therefore follows from the condition $\iota \circ \tilde{g} = g$.

By the uniqueness, it suffices to prove the proposition after replacing S and X by étale covers. By [**40**, 3.5 and 3.6], we may therefore assume given a chart $\beta : Q \to M_S$ for M_S and an injective map $\theta : Q \to P$ such that $X = \mathrm{Spec}(\mathbb{Z}[P]) \times_{\mathrm{Spec}(\mathbb{Z}[Q])} S$. Furthermore, we may assume that the cokernel $G := \mathrm{Coker}(Q^{\mathrm{gp}} \to P^{\mathrm{gp}})$ is p-torsion free, and that $S = \mathrm{Spec}(A)$ for some ring A.

Let $P' \subset Q^{\mathrm{gp}} \oplus_{\times p, Q^{\mathrm{gp}}} P^{\mathrm{gp}}$ denote the image of the map $Q \oplus_{\times p, Q} P \to Q^{\mathrm{gp}} \oplus_{\times p, Q^{\mathrm{gp}}} P^{\mathrm{gp}}$ (so the natural map $Q \oplus_{\times p, Q} P \to P'$ is the universal map from $Q \oplus_{\times p, Q} P$ to an integral monoid). Let $w : P' \to P$ denote the map induced by the map

(9.3.3.2) $$Q \oplus P \longrightarrow P, \quad (q, m) \longmapsto \theta(q) + p \cdot m,$$

let $\theta' : Q \to P'$ be the composite map

(9.3.3.3) $$Q \xrightarrow{q \mapsto (q,0)} Q \oplus P \twoheadrightarrow P',$$

and let $j : P \to P'$ be the composite map

(9.3.3.4) $$P \xrightarrow{m \mapsto (1,m)} Q \oplus P \twoheadrightarrow P'.$$

By the construction of fiber products in the category of integral log schemes [**40**, 2.7] we have in the present situation

(9.3.3.5) $$X' = \mathrm{Spec}(A \otimes_{\mathbb{Z}[Q]} \mathbb{Z}[P']),$$

and the canonical factorization

(9.3.3.6)
$$X \longrightarrow X' \longrightarrow X \quad (\text{composite } F_X : X \to X)$$

of the absolute Frobenius of X is identified with the composite

(9.3.3.7) $$\mathrm{Spec}(A \otimes_{\mathbb{Z}[Q]} \mathbb{Z}[P]) \xrightarrow{a} \mathrm{Spec}(A \otimes_{\mathbb{Z}[Q]} \mathbb{Z}[P']) \xrightarrow{b} \mathrm{Spec}(A \otimes_{\mathbb{Z}[Q]} \mathbb{Z}[P]),$$

where a is the map defined by the map of rings

(9.3.3.8) $$\mathrm{id}_A \otimes w : A \otimes_{\mathbb{Z}[Q]} \mathbb{Z}[P'] \longrightarrow A \otimes_{\mathbb{Z}[Q]} \mathbb{Z}[P],$$

and b is the map defined by the map of rings

(9.3.3.9) $$F_A \otimes j : A \otimes_{\mathbb{Z}[Q]} \mathbb{Z}[P] \longrightarrow A \otimes_{\mathbb{Z}[Q]} \mathbb{Z}[P'].$$

Lemma 9.3.4. — *Let* G_p *denote* $\mathrm{Coker}(\times p : G \to G)$. *Then the projection*

(9.3.4.1) $$P^{\mathrm{gp}} \longrightarrow \mathrm{Coker}(w^{gp} : P'^{\mathrm{gp}} \to P^{\mathrm{gp}})$$

factors through an isomorphism $G_p \simeq \mathrm{Coker}(P'^{\mathrm{gp}} \to P^{\mathrm{gp}})$.

Proof. — By construction of P', j, and w there is a commutative diagram with exact rows

(9.3.4.2)
$$\begin{array}{ccccccccc}
0 & \longrightarrow & Q^{\mathrm{gp}} & \longrightarrow & P^{\mathrm{gp}} & \longrightarrow & G & \longrightarrow & 0 \\
 & & \downarrow{\scriptstyle \times p} & & \downarrow{\scriptstyle j} & & \| & & \\
0 & \longrightarrow & Q^{\mathrm{gp}} & \longrightarrow & P'^{\mathrm{gp}} & \longrightarrow & G & \longrightarrow & 0 \\
 & & \| & & \downarrow{\scriptstyle w} & & \downarrow{\scriptstyle \times p} & & \\
0 & \longrightarrow & Q^{\mathrm{gp}} & \longrightarrow & P^{\mathrm{gp}} & \longrightarrow & G & \longrightarrow & 0.
\end{array}$$

The result therefore follows from the snake lemma. □

In particular the map $\mathrm{Spec}(\mathbb{Z}[P]) \to \mathrm{Spec}(\mathbb{Z}[P'])$ defined by w descends to a morphism of stacks

(9.3.4.3) $$\bar{w} : [\mathrm{Spec}(\mathbb{Z}[P])/D(G_p)] \longrightarrow \mathrm{Spec}(\mathbb{Z}[P']).$$

The commutative diagram of monoids

(9.3.4.4)
$$\begin{array}{ccccc}
P & \xleftarrow{w} & P' & \xleftarrow{j} & P \\
 & \nwarrow{\scriptstyle \theta} & \uparrow{\scriptstyle \theta'} & & \uparrow{\scriptstyle \theta} \\
 & & Q & \xleftarrow{\times p} & Q
\end{array} \quad (\text{composite } \times p : P \to P)$$

induces a commutative diagram of stacks

$$(9.3.4.5)\qquad \begin{array}{ccccc} [\mathrm{Spec}(\mathbb{Z}[P])/D(G_p)] & \xrightarrow{\bar{w}} & \mathrm{Spec}(\mathbb{Z}[P']) & \xrightarrow{j} & \mathrm{Spec}(\mathbb{Z}[P]) \\ \downarrow{\scriptstyle f''} & & \downarrow{\scriptstyle f'} & & \downarrow{\scriptstyle f} \\ \mathcal{S}_P & \xrightarrow{\mathcal{S}(w)} & \mathcal{S}_{P'} & \xrightarrow{\mathcal{S}(j)} & \mathcal{S}_P \\ & \searrow^{\mathcal{S}(\theta)} & \downarrow{\scriptstyle \mathcal{S}(\theta')} & & \downarrow{\scriptstyle \mathcal{S}(\theta)} \\ & & \mathcal{S}_Q & \xrightarrow{\mathcal{S}(\times p)} & \mathcal{S}_Q. \end{array}$$

The map $\mathcal{S}(\times p) : \mathcal{S}_Q \to \mathcal{S}_Q$ is a lifting of Frobenius, so we also have a commutative diagram

$$(9.3.4.6)\qquad \begin{array}{ccc} S & \xrightarrow{F_S} & S \\ \downarrow & & \downarrow \\ \mathcal{S}_Q & \xrightarrow{\mathcal{S}(\times p)} & \mathcal{S}_Q. \end{array}$$

Base changing along $S \to \mathcal{S}_Q$ we therefore obtain a commutative diagram

$$(9.3.4.7)\qquad \begin{array}{ccccc} [X/D(G_p)] & \longrightarrow & X' & \longrightarrow & X \\ \downarrow & & \downarrow & & \downarrow \\ \mathcal{S}_P \times_{\mathcal{S}_Q} S & \xrightarrow{\mathcal{S}(w)\times \mathrm{id}} & \mathcal{S}_{P'} \times_{\mathcal{S}_Q} S & \xrightarrow{\mathcal{S}(j)\times F_S} & \mathcal{S}_P \times_{\mathcal{S}_Q} S. \end{array}$$

(with the composite of the bottom row equal to $F_{\mathcal{S}_P \times_{\mathcal{S}_Q} S}$)

Lemma 9.3.5

(i) *The squares*

$$(9.3.5.1)\qquad \begin{array}{ccc} [X/D(G_p)] & \longrightarrow & X' \\ \downarrow & & \downarrow \\ \mathcal{S}_P \times_{\mathcal{S}_Q} S & \xrightarrow{\mathcal{S}(w)\times \mathrm{id}} & \mathcal{S}_{P'} \times_{\mathcal{S}_Q} S \end{array}$$

and

$$(9.3.5.2)\qquad \begin{array}{ccc} X' & \longrightarrow & X \\ \downarrow & & \downarrow \\ \mathcal{S}_{P'} \times_{\mathcal{S}_Q} S & \xrightarrow{\mathcal{S}(j)\times F_S} & \mathcal{S}_P \times_{\mathcal{S}_Q} S \end{array}$$

are cartesian.

(ii) *The square*

$$(9.3.5.3)\qquad \begin{array}{ccc} [X/D(G_p)] & \longrightarrow & X \\ \downarrow & & \downarrow \\ \mathcal{S}_P \times_{\mathcal{S}_Q} S & \xrightarrow{F_{\mathcal{S}_P \times_{\mathcal{S}_Q} S}} & \mathcal{S}_P \times_{\mathcal{S}_Q} S \end{array}$$

is cartesian.

Proof. — Note that (ii) follows from (i).

To see that (9.3.5.1) is cartesian note that it suffices to show that the diagram

$$(9.3.5.4)\qquad \begin{array}{ccc} [X/D(G_p)] & \longrightarrow & X' \\ \downarrow & & \downarrow \\ \mathcal{S}_P & \xrightarrow{\mathcal{S}(w)} & \mathcal{S}_{P'} \end{array}$$

is cartesian. This follows from 9.1.5, 9.3.4, and the observation that the canonical map

$$(9.3.5.5)\qquad X \longrightarrow X' \times_{\mathrm{Spec}(\mathbb{Z}[P'])} \mathrm{Spec}(\mathbb{Z}[P])$$

is an isomorphism.

That (9.3.5.2) is cartesian can be seen as follows. Let P'' denote the (not necessarily integral) monoid $Q \oplus_{\times p, Q} P$ so we have a commutative diagram

$$(9.3.5.6)\qquad \begin{array}{ccccc} P' & \xleftarrow{\alpha} & P'' & \xleftarrow{\gamma} & P \\ & \nwarrow^{\theta'} & \uparrow{\scriptstyle \theta''} & & \uparrow{\scriptstyle \theta} \\ & & Q & \xleftarrow[\times p]{} & Q, \end{array}$$

(with a curved arrow $j: P \to P'$ along the top)

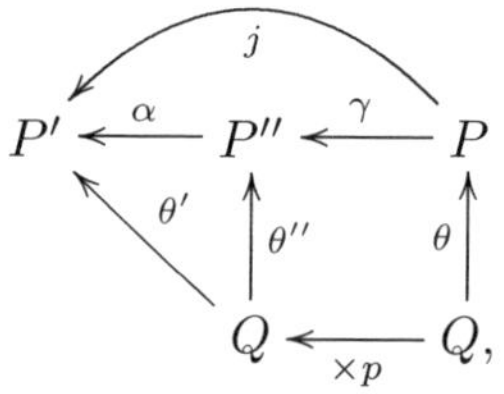

where θ'' is the composite map

$$(9.3.5.7)\qquad Q \xrightarrow{q \mapsto (q,0)} Q \oplus P \xrightarrow{\text{projection}} P'',$$

the map γ is the composite map

$$(9.3.5.8)\qquad P \xrightarrow{m \mapsto (0,m)} Q \oplus P \xrightarrow{\text{projection}} P'',$$

and α is the surjection from P'' to its image in $P''^{\mathrm{gp}} = Q^{\mathrm{gp}} \oplus_{\times p, Q^{\mathrm{gp}}} P^{\mathrm{gp}}$.

Set

$$(9.3.5.9)\qquad Y = \mathrm{Spec}(A \otimes_{\mathbb{Z}[Q], \theta''} \mathbb{Z}[P'']),$$

and let

$$(9.3.5.10)\qquad X' \xrightarrow{\epsilon} Y \xrightarrow{\delta} X$$

be the natural factorization of the map $b : X' \to X$ (9.3.3.7). We then obtain a commutative diagram

$$\begin{array}{ccccc} X' & \xrightarrow{\delta} & Y & \xrightarrow{\epsilon} & X \\ \downarrow & & \downarrow & & \downarrow \\ \mathcal{S}_{P'} \times_{\mathcal{S}_Q} S & \xrightarrow{\mathcal{S}(\alpha)\times \mathrm{id}} & \mathcal{S}_{P''} \times_{\mathcal{S}_Q} S & \xrightarrow{\mathcal{S}(\gamma)\times F_S} & \mathcal{S}_P \times_{\mathcal{S}_Q} S \\ & & \downarrow & & \downarrow \\ & & S & \xrightarrow{F_S} & S, \end{array} \tag{9.3.5.11}$$

where $\mathcal{S}_{P''}$ for the not necessarily integral monoid P'' is as in 9.1.7.

Since the natural maps

$$X' \longrightarrow Y \times_{\mathrm{Spec}(\mathbb{Z}[P'])} \mathrm{Spec}(\mathbb{Z}[P]), \quad Y \longrightarrow X \times_{S,F_S} S \tag{9.3.5.12}$$

are isomorphisms, to prove that (9.3.5.2) is cartesian it suffices to show that the square

$$\begin{array}{ccc} \mathcal{S}_{P''} \times_{\mathcal{S}_Q} S & \xrightarrow{\mathcal{S}(\gamma)\times F_S} & \mathcal{S}_P \times_{\mathcal{S}_Q} S \\ \downarrow & & \downarrow \\ S & \xrightarrow{F_S} & S \end{array} \tag{9.3.5.13}$$

is cartesian. This follows from 9.1.7 which shows that the natural map

$$\mathcal{S}_{P''} \longrightarrow \mathcal{S}_P \times_{\mathcal{S}_Q, \mathcal{S}(\times p)} \mathcal{S}_Q \tag{9.3.5.14}$$

is an isomorphism. □

Lemma 9.3.6. — *Let S be an $\mathbb{F}_p$-scheme and $f : \mathcal{X} \to \mathcal{Y}$ be an étale representable morphism of algebraic stacks over S. Then the diagram*

$$\begin{array}{ccc} \mathcal{X} & \xrightarrow{F_{\mathcal{X}}} & \mathcal{X} \\ \downarrow f & & \downarrow f \\ \mathcal{Y} & \xrightarrow{F_{\mathcal{Y}}} & \mathcal{Y} \end{array} \tag{9.3.6.1}$$

is cartesian.

Proof. — Let $\mathcal{P}$ denote the fiber product $\mathcal{Y} \times_{F_{\mathcal{Y}}, \mathcal{Y}} \mathcal{X}$, and let $q : \mathcal{X} \to \mathcal{P}$ be the map defined by the square (9.3.6.1). We need to show that q is an isomorphism.

In the case when $\mathcal{X}$ and $\mathcal{Y}$ are schemes the result is standard [**28**, 5.1], as the morphism q is étale (being a morphism between two étale $\mathcal{Y}$-schemes), surjective, and radicial (since the Frobenius morphism $F_{\mathcal{X}} : \mathcal{X} \to \mathcal{X}$ is radicial).

For the general case, let $\pi : Y \to \mathcal{Y}$ be a smooth surjection. To verify that q is an isomorphism, it suffices to show that the map

$$\tilde{q} : \mathcal{X} \times_{\mathcal{Y}} Y \longrightarrow \mathcal{P} \times_{\mathcal{Y}} Y \tag{9.3.6.2}$$

is an isomorphism. On the other hand, we have

$$\mathcal{P} \times_{\mathcal{Y}} Y \simeq \mathcal{X} \times_{f, \mathcal{Y}, \pi \circ F_Y} Y \simeq (\mathcal{X} \times_{\mathcal{Y}} Y) \times_{Y, F_Y} Y. \tag{9.3.6.3}$$

Under this isomorphism the map $\tilde{q}$ becomes identified with the map

$$(\mathcal{X} \times_{\mathcal{Y}} Y) \longrightarrow (\mathcal{X} \times_{\mathcal{Y}} Y) \times_{Y, F_Y} Y \tag{9.3.6.4}$$

defined by the commutative square

$$\begin{array}{ccc} \mathcal{X} \times_{\mathcal{Y}} Y & \xrightarrow{F_{\mathcal{X} \times_{\mathcal{Y}} Y}} & \mathcal{X} \times_{\mathcal{Y}} Y \\ \downarrow & & \downarrow \\ Y & \xrightarrow{F_Y} & Y. \end{array} \tag{9.3.6.5}$$

This therefore reduces the proof to the case when Y is a scheme. Combining this with the case of schemes discussed at the beginning of the proof, this therefore proves the case when f is schematic (*i.e.*, for every morphism $Y \to \mathcal{Y}$ with Y a scheme the fiber product $\mathcal{X} \times_{\mathcal{Y}} Y$ is a scheme).

It remains to consider the case when $\mathcal{Y}$ is a scheme and $\mathcal{X}$ is an algebraic space (note that if f is separated then $\mathcal{X}$ is automatically a scheme by [**46**, II.6.17]). In this case let $U \to \mathcal{X}$ be an étale surjection. We then obtain a commutative diagram

$$\begin{array}{ccc} U & \xrightarrow{F_U} & U \\ \downarrow & & \downarrow \\ \mathcal{X} & \xrightarrow{F_{\mathcal{X}}} & \mathcal{X} \\ \downarrow f & & \downarrow f \\ \mathcal{Y} & \xrightarrow{F_{\mathcal{Y}}} & \mathcal{Y}, \end{array} \tag{9.3.6.6}$$

where the top square is cartesian by the case when f is schematic (applied to $U \to \mathcal{X}$), and the big outside square is cartesian by the case of schemes. Since $U \to \mathcal{X}$ is étale surjective this implies that the bottom inside square is also cartesian. □

By 9.1.11, the natural map $\mathcal{S}_P \times_{\mathcal{S}_Q} S \to \mathcal{L}og_{(S,M_S)}$ is étale and representable, and therefore by 9.3.6 the square

$$\begin{array}{ccc} \mathcal{S}_P \times_{\mathcal{S}_Q} S & \xrightarrow{F_{\mathcal{S}_P \times_{\mathcal{S}_Q} S}} & \mathcal{S}_P \times_{\mathcal{S}_Q} S \\ \downarrow & & \downarrow \\ \mathcal{L}og_{(S,M_S)} & \xrightarrow{F_{\mathcal{L}og_{(S,M_S)}}} & \mathcal{L}og_{(S,M_S)} \end{array} \tag{9.3.6.7}$$

is cartesian. Using 9.3.5 (ii), we therefore get an isomorphism
(9.3.6.8)
$$\mathcal{X}' = X \times_{\mathcal{L}og_{(S,M_S)}, F_{\mathcal{L}og_{(S,M_S)}}} \mathcal{L}og_{(S,M_S)} \simeq X \times_{\mathcal{S}_P \times_{\mathcal{S}_Q} S, F_{\mathcal{S}_P \times_{\mathcal{S}_Q} S}} \mathcal{S}_P \times_{\mathcal{S}_Q} S \simeq [X/D(G_p)].$$

This isomorphism identifies the diagram

(9.3.6.9) $$X \xrightarrow{F_{X/\mathcal{L}og_{(S,M_S)}}} \mathfrak{X}' \xrightarrow{h'} X',$$

where h' is defined as in (9.3.2.6), with the diagram

(9.3.6.10) $$X \xrightarrow{\text{quotient}} [X/D(G_p)] \xrightarrow{z} X',$$

where z is the map induced by the map $\bar{w}$ in (9.3.4.3). It follows that in this case the scheme $\overline{\mathfrak{X}}'$ is the spectrum of the ring

(9.3.6.11) $$(A \otimes_{\mathbb{Z}[Q]} \mathbb{Z}[P])^{D(G_p)}$$

of $D(G_p)$-invariants in $A \otimes_{\mathbb{Z}[Q]} \mathbb{Z}[P]$.

Define a submonoid $H \subset P$ by

(9.3.6.12)
$$H := \{a \in P| \text{ image of } a \text{ in } P^{\text{gp}} \text{ equals } b^p c, \text{ for some } b \in P^{\text{gp}}, \text{ and } c \in Q^{\text{gp}}\}.$$

By the explicit description of X'' given in [**40**, proof of 4.12] we have

(9.3.6.13) $$X'' = \mathrm{Spec}(A \otimes_{\mathbb{Z}[Q]} \mathbb{Z}[H]).$$

To complete the proof of 9.3.3, it therefore suffices to prove the following lemma. □

Lemma 9.3.7. — *For any $\mathbb{Z}[Q]$-algebra A, the map $A \otimes_{\mathbb{Z}[Q]} \mathbb{Z}[H] \to A \otimes_{\mathbb{Z}[Q]} \mathbb{Z}[P]$ induced by the inclusion $H \subset P$ is injective, and identifies $A \otimes_{\mathbb{Z}[Q]} \mathbb{Z}[H]$ with the ring of invariants $(A \otimes_{\mathbb{Z}[Q]} \mathbb{Z}[P])^{D(G_p)}$.*

Proof. — Since $D(G_p)$ is diagonalizable, for any $\mathbb{Z}[Q]$-algebra A the natural map

(9.3.7.1) $$A \otimes_{\mathbb{Z}[Q]} (\mathbb{Z}[P])^{D(G_p)} \longrightarrow (A \otimes_{\mathbb{Z}[Q]} \mathbb{Z}[P])^{D(G_p)}$$

is an isomorphism. Therefore it suffices to consider the case when $A = \mathbb{Z}[Q]$ (note that the injectivity for general A also follows, for if the result holds for $A = \mathbb{Z}[Q]$ then $\mathbb{Z}[H]$ is a direct summand of the $\mathbb{Z}[Q]$-module $\mathbb{Z}[P]$).

As a $D(G_p)$-representation, $\mathbb{Z}[P] = \oplus_{g \in G} M_g$, where $M_g \subset \mathbb{Z}[P]$ is the free subgroup

(9.3.7.2) $$\oplus_{m \in P, m \mapsto g} \mathbb{Z} \cdot e_m \subset \mathbb{Z}[P],$$

where for $m \in P$ we write $e_m \in \mathbb{Z}[P]$ for the image of m under the natural map $P \to \mathbb{Z}[P]$ and the sum is taken over the set of elements $m \in P$ with image g in G_p. To prove the lemma it therefore suffices to show that the natural map

(9.3.7.3) $$H \to P \cap \mathrm{Ker}(P^{\text{gp}} \longrightarrow G_p) \subset P^{\text{gp}}$$

is an isomorphism, which is immediate from the definition. □

Corollary 9.3.8. — *The Frobenius morphism $F_{\mathcal{L}og_{(S,M_S)}} : \mathcal{L}og_{(S,M_S)} \to \mathcal{L}og_{(S,M_S)}$ is Frobenius acyclic in the sense of 3.2.1.*

Proof. — It suffices to show that for any smooth morphism $X \to \mathcal{L}og_{(S,M_S)}$ with X affine, and any quasi-coherent sheaf $\mathcal{E}$ on $\mathfrak{X}'$ we have $H^i(\mathfrak{X}', \mathcal{E}) = 0$ for $i > 0$. Furthermore, it suffices to prove this after replacing X by an étale cover, and hence it is enough to consider the local situation as in the proof of 9.3.3 where $X = \mathrm{Spec}(A \otimes_{\mathbb{Z}[Q]} \mathbb{Z}[P])$. In this case, we have the isomorphism $\mathfrak{X}' \simeq [X/D(G_p)]$ given by (9.3.6.8). Therefore a quasi-coherent sheaf $\mathcal{E}$ on $\mathfrak{X}'$ is equivalent to the data of a quasi-coherent $\mathcal{O}_X$-module E with an action of $D(G_p)$ compatible with the action on X. Since X is affine, the cohomology $H^i(\mathfrak{X}, \mathcal{E})$ is canonically isomorphic to the group cohomology $H^i(D(G_p), E)$. The result therefore follows from [**14**, I.5.5]. □

Corollary 9.3.9. — *The log smooth morphism $(X, M_X) \to (S, M_S)$ is of Cartier type in the sense of* [**40**, 4.8] *if and only if the morphism $X \to \mathcal{L}og_{(S,M_S)}$ factors through the open substack of $\mathcal{L}og_{(S,M_S)}$ over which the natural map $\mathcal{O}_{\mathcal{L}og_{(S,M_S)} \times_{S,F_S} S} \to F_{\mathcal{L}og_{(S,M_S)}*}\mathcal{O}_{\mathcal{L}og_{(S,M_S)}}$ is an isomorphism.*

From the above discussion and 3.3.21 we also obtain a canonical isomorphism of $\mathcal{O}_{X''}$-modules

$$C^{-1} : \Omega^q_{(X'', M_{X''})/(S,M_S)} \simeq \mathcal{H}^q(\Omega^\bullet_{(X,M_X)/(S,M_S)}). \tag{9.3.9.1}$$

Proposition 9.3.10. — *The isomorphism (9.3.9.1) agrees with the one defined in* [**40**, 4.12 (1)].

Proof. — Denote temporarily by

$$\widetilde{C}^{-1} : \Omega^q_{(X'', M_{X''})/(S,M_S)} \simeq \mathcal{H}^q(\Omega^\bullet_{(X,M_X)/(S,M_S)}) \tag{9.3.10.1}$$

the isomorphism defined in [**40**, 4.12 (1)].

Let $\pi : (X'', M_{X''}) \to (X, M_X)$ be the projection, and let

$$D : \Omega^q_{(X,M_X)/(S,M_S)} \longrightarrow \mathcal{H}^q(\Omega^\bullet_{(X,M_X)/(S,M_S)}) \tag{9.3.10.2}$$

denote the composite map

$$\Omega^q_{(X,M_X)/(S,M_S)} \xrightarrow{\pi^*} \Omega^q_{(X'', M_{X''})} \xrightarrow{C^{-1}} \mathcal{H}^q(\Omega^\bullet_{(X,M_X)/(S,M_S)}). \tag{9.3.10.3}$$

Similarly define

$$\widetilde{D} : \Omega^q_{(X,M_X)/(S,M_S)} \to \mathcal{H}^q(\Omega^\bullet_{(X,M_X)/(S,M_S)}) \tag{9.3.10.4}$$

to be the map $\widetilde{C}^{-1} \circ \pi^*$. Since the natural map $\pi^*\Omega^q_{(X,M_X)/(S,M_S)} \to \Omega^q_{(X'', M_{X''})/(S,M_S)}$ is an isomorphism, the maps C^{-1} and $\widetilde{C}^{-1}$ are determined by the maps D and $\widetilde{D}$. It therefore suffices to show that $D = \widetilde{D}$.

To prove that $D^{-1} = \widetilde{D}^{-1}$ we may work étale locally on S and X, and may therefore assume that $X = \mathrm{Spec}(A \otimes_{\mathbb{Z}[Q]} \mathbb{Z}[P])$ as in the proof of 9.3.3. Let G denote the cokernel of $Q^{\mathrm{gp}} \to P^{\mathrm{gp}}$. Then by [**40**, 1.8] we have

$$\Omega^1_{(X,M_X)/(S,M_S)} \simeq \mathcal{O}_X \otimes_{\mathbb{Z}} G, \tag{9.3.10.5}$$

and. For $g_1, \dots, g_q \in G$ we write

$$\mathrm{dlog}(g_1) \wedge \cdots \wedge \mathrm{dlog}(g_q) \in \Omega^q_{(X,M_X)/(S,M_S)} \tag{9.3.10.6}$$

for the element

$$(1 \otimes g_1) \wedge \cdots \wedge (1 \otimes g_q) \in \bigwedge^q \mathcal{O}_X \otimes_{\mathbb{Z}} G. \tag{9.3.10.7}$$

The maps D^{-1} and $\widetilde{D}^{-1}$ are characterized by their values on the elements (9.3.10.6), and by [**40**, 4.12 (1)] we have

$$\widetilde{D}^{-1}(\mathrm{dlog}(g_1) \wedge \cdots \wedge \mathrm{dlog}(g_q)) = [\mathrm{dlog}(g_1) \wedge \cdots \wedge \mathrm{dlog}(g_q)], \tag{9.3.10.8}$$

where the right side denotes the class of the form in $\mathcal{H}^q(\Omega^\bullet_{(X,M_X)/(S,M_S)})$.

It therefore suffices to show that we also have

$$D^{-1}(\mathrm{dlog}(g_1) \wedge \cdots \wedge \mathrm{dlog}(g_q)) = [\mathrm{dlog}(g_1) \wedge \cdots \wedge \mathrm{dlog}(g_q)]. \tag{9.3.10.9}$$

To ease the notation write just $\mathcal{S}$ for the stack $\mathcal{S}_P \times_{\mathcal{S}_Q} S$, and note that by 9.1.6 (i) we have an isomorphism

$$[X/D(G)] \simeq \mathcal{S}. \tag{9.3.10.10}$$

Let $X[G]$ denote the product $X \times D(G)$, let $\mathrm{pr} : X[G] \to X$ be the projection, and let $\rho : X[G] \to X$ be the action. Let $\Omega^1_{X[G]/X}$ denote the relative differentials of $\mathrm{pr} : X[G] \to X$..

For $g \in G$ write $f_g \in \mathcal{O}_{X[G]}$ for the image of g under the map

$$G \xrightarrow{g \mapsto 1 \otimes g} \mathcal{O}_X \otimes_{\mathbb{Z}} \mathbb{Z}[G] = \mathcal{O}_{X[G]} \tag{9.3.10.11}$$

and note that the induced map

$$\mathcal{O}_{X[G]} \otimes_{\mathbb{Z}} G \longrightarrow \Omega^1_{X[G]/X}, \quad a \otimes g \longmapsto a \cdot df_g \tag{9.3.10.12}$$

is an isomorphism.

There is a cartesian diagram

$$\begin{array}{ccc} X & \xleftarrow{\rho} & X[G] \\ \downarrow & & \downarrow \mathrm{pr} \\ \mathcal{S} & \longleftarrow & X \end{array} \tag{9.3.10.13}$$

which induces a map

$$\rho^* : \Omega^1_{(X,M_X)/(S,M_S)} \simeq \Omega^1_{X/\mathcal{S}} \to \Omega^1_{X[G]/X}. \tag{9.3.10.14}$$

Lemma 9.3.11. — *For $g \in G$ the image of* $\mathrm{dlog}(g)$ *under the composite map*

$$\Omega^1_{(X,M_X)/(S,M_S)} \xrightarrow{\iota} \Omega^1_{X/S} \xrightarrow{\rho^*} \Omega^1_{X[G]/X} \tag{9.3.11.1}$$

is equal to $f_g^{-1} \cdot df_g$.

Proof. — Note that for $g, h \in G$ we have $f_g \cdot f_h = f_{g+h}$ and

$$f_g^{-1} \cdot df_g + f_h^{-1} \cdot df_h = (f_h \cdot df_g + f_g \cdot df_h)/(f_g f_h) = f_{g+h}^{-1} df_{g+h}. \tag{9.3.11.2}$$

Using this additivity and the fact that the image of $P \to G$ generates G as a group, we conclude that it suffices to prove the result for elements $g \in G$ which are in the image of P.

Also it suffices to consider the universal case of $A = \mathbb{Z}[Q]$.

Let $m \in P$ be an element, let $g \in G$ be the image of m, and let $e_m \in \mathcal{O}_X = A \otimes_{\mathbb{Z}[Q]} \mathbb{Z}[P]$ be the image of m under the natural map $P \to \mathcal{O}_X$. Since the isomorphism (9.1.3.1) is compatible with the derivations from $\mathcal{O}_X$ (by construction of the isomorphism in [**64**, 3.8]), we have

$$\rho^*(de_m) = d\rho^*(e_m), \tag{9.3.11.3}$$

and under the map

$$\rho^* : \mathcal{O}_X = A \otimes_{\mathbb{Z}[Q]} \mathbb{Z}[P] \to (A \otimes_{\mathbb{Z}[Q} \mathbb{Z}[P]) \otimes_{\mathbb{Z}} \mathbb{Z}[G] = \mathcal{O}_{X[G]} \tag{9.3.11.4}$$

the element e_m maps to $\mathrm{pr}^*(e_m) \cdot f_g$. Therefore

$$\rho^*(de_m) = d(\mathrm{pr}^*(e_m) \cdot f_g) = \mathrm{pr}^*(e_m) df_g. \tag{9.3.11.5}$$

Since

$$e_m \mathrm{dlog}(g) = de_m \tag{9.3.11.6}$$

in $\Omega^1_{(X,M_X)/(S,M_S)}$ we obtain

$$\mathrm{pr}^*(e_m) \cdot f_g \rho^*(\mathrm{dlog}(g)) = \rho^*(e_m)\rho^* \mathrm{dlog}(g) = \mathrm{pr}^*(e_m) df_g. \tag{9.3.11.7}$$

Since we are considering the universal case $A = \mathbb{Z}[Q]$, the element $\mathrm{pr}^*(e_m)$ is not a zero-divisor and we obtain $\rho^*(\mathrm{dlog}(g)) = f_g^{-1} df_g$. □

Consider now the commutative diagram

(9.3.11.8)

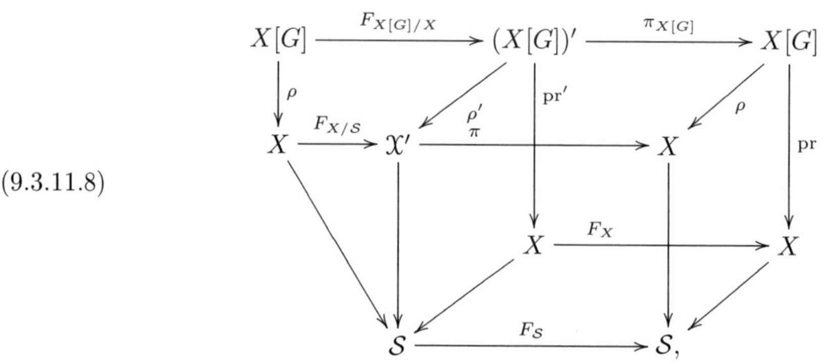

where the square

$$(9.3.11.9)\qquad \begin{array}{ccc} (X[G])' & \xrightarrow{\pi_{X[G]}} & X[G] \\ \downarrow{\scriptstyle \mathrm{pr}'} & & \downarrow{\scriptstyle \mathrm{pr}} \\ X & \xrightarrow{F_X} & X \end{array}$$

is cartesian and

$$(9.3.11.10)\qquad X[G] \xrightarrow{F_{X[G]/X}} (X[G])' \xrightarrow{\pi_{X[G]}} X[G]$$

and

$$(9.3.11.11)\qquad X \xrightarrow{F_{X/S}} \mathcal{X}' \xrightarrow{\pi} X$$

are the canonical factorizations of Frobenius. By the construction of the map C^{-1} in 3.3.21, we then have a commutative diagram

$$(9.3.11.12)\qquad \begin{array}{ccc} \Omega^q_{X/\mathcal{S}} & \xrightarrow{\rho^*} & \Omega^q_{X[G]/X} \\ \downarrow{\scriptstyle D} & & \downarrow{\scriptstyle D_{X[G]}} \\ \mathcal{H}^q(\Omega^\bullet_{X/\mathcal{S}}) & \xrightarrow{\rho^*} & \mathcal{H}^q(\Omega^\bullet_{X[G]/X}), \end{array}$$

where if $C^{-1}_{X[G]} : \Omega^q_{(X[G])'/X} \to \mathcal{H}^q(\Omega^\bullet_{X[G]/X})$ is the Cartier isomorphism for $\mathrm{pr} : X[G] \to X$ the map $D_{X[G]}$ is defined to be the composite

$$(9.3.11.13)\qquad \Omega^q_{X[G]/X} \xrightarrow{\pi^*_{X[G]}} \Omega^q_{(X[G])'/X} \xrightarrow{C^{-1}_{X[G]}} \mathcal{H}^q(\Omega^\bullet_{X[G]/X}).$$

Moreover, by the proof of 3.3.7 the horizontal arrows in (9.3.11.12) are injective. By 9.3.11 it therefore suffices to show that for $g_1, \dots, g_q \in G$ we have

$$(9.3.11.14)\qquad D_{X[G]}((f_{g_1}^{-1}df_{g_1}) \wedge \cdots \wedge (f_{g_q}^{-1}df_{g_q})) = [(f_{g_1}^{-1}df_{g_1}) \wedge \cdots \wedge (f_{g_q}^{-1}df_{g_q})]$$

in $\mathcal{H}^q(\Omega^\bullet_{X[G]/X})$. This follows from the classical construction of the Cartier isomorphism [**44**, 7.2.3]. □

Remark 9.3.12. — From the stack-theoretic point of view and 3.1.8 the difficulty in generalizing the theory of Cartier descent to the logarithmic setting can be reinterpreted as the problem of understanding the relationship between the category of quasi-coherent sheaves on a stack and the category of quasi-coherent sheaves on the coarse moduli space. Indeed 3.1.8 shows that the category of quasi-coherent sheaves with logarithmic integrable connection and p-curvature 0 on $(X, M_X)/(S, M_S)$ is canonically equivalent to the category of quasi-coherent sheaves on the stack $\mathcal{X}'$. We would like to relate this category to the category of quasi-coherent sheaves on X'' which by 3.1.8 is the coarse moduli space of $\mathcal{X}'$. It would be interesting to study the work of Lorenzon [**50**] and Ogus [**58**, 1.3] from this point of view.

9.4. Comparison of de Rham-Witt complexes

9.4.1. — Let k be a perfect field of characteristic $p > 0$, and let L be a fine saturated log structure on $\mathrm{Spec}(k)$. Let $f : (Y, M) \to (\mathrm{Spec}(k), L)$ be a smooth morphism of Cartier type between fine saturated log schemes. For $n \geq 1$ let $(W_n, W_n(L))$ denote the hollow log scheme defined in [**31**, 3.1], whose underlying scheme is the spectrum of the truncated ring of Witt vectors $W_n(k)$.

Remark 9.4.2. — For the results of this section, one does not need the log structures to be saturated, but we make this assumption because it simplifies the theory of charts, and is satisfied in all the applications we have in mind.

9.4.3. — In this setup, Hyodo and Kato define the *logarithmic de Rham-Witt complex* $W_n^{HK}\omega_Y^\bullet$ [**31**, 4.1].

By definition,

$$W_n^{HK}\omega_Y^q = R^q u_{(Y,M)/(W_n,W_n(L))*}(\mathcal{O}_{(Y,M)/(W_n,W_n(L))}). \tag{9.4.3.1}$$

This graded $W_n(k)$-module comes equipped with operators d, F, V, and π_n as in 4.3.7 [**31**, 4.1 and 4.2].

9.4.4. — Let W denote the spectrum of the ring of Witt vector $W(k)$ of k, and let $\sigma : W \to W$ denote the canonical lifting of Frobenius. The log structures $W_n(L)$ on the reductions W_n define a compatible system of fine log structures $\{W_n(L)\}$ on the reductions of W. In other words, a compatible family of objects in $\mathcal{L}og_{(\mathrm{Spec}(\mathbb{Z}),\mathcal{O}^*)}(W_n)$. Since the stack $\mathcal{L}og_{(\mathrm{Spec}(\mathbb{Z}),\mathcal{O}^*)}$ is algebraic this compatible system of log structures is uniquely algebraizable to a fine log structure $W(L)$ on W (this follows for example from the converse to [**4**, 5.3] alluded to in *loc. cit.*, p. 182. Note that in the case when L is equal to the log structure obtained from a map $Q \to k$ from a fine sharp monoid Q sending all nonzero elements to zero, then $W(L)$ is simply the log structure obtained from the map $Q \to W$ sending all nonzero elements to zero. In particular, the log structure $W(L)$ is a hollow log structure on W, as this can be verified after replacing W by a finite flat covering where it is given by this construction.

Similarly, the canonical liftings of Frobenius to $(W_n, W_n(L))$ defined in [**31**, 3.1] are obtained from a unique morphism of log schemes

$$F_{(W,W(L))} : (W, W(L)) \longrightarrow (W, W(L)), \tag{9.4.4.1}$$

which in turn induces a lifting of Frobenius

$$F_{\mathcal{L}og_{(W,W(L))}} : \mathcal{L}og_{(W,W(L))} \longrightarrow \mathcal{L}og_{(W,W(L))}. \tag{9.4.4.2}$$

The morphism $F_{\mathcal{L}og_{(W,W(L))}}$ is defined as follows. As in 9.1.34, for a log structure M on a scheme T let $M^{(p)}$ denote the log structure associated to the prelog structure

$$M \xrightarrow{\times p} M \longrightarrow \mathcal{O}_T. \tag{9.4.4.3}$$

Note that there is a canonical morphism $\gamma_M : M^{(p)} \to M$ of log structures on T. For a morphism $\delta : M \to N$ of log structures on T, let $\delta^{(p)} : M^{(p)} \to N^{(p)}$ denote the morphism induced by δ.

By the construction of $W(L)$ in [**31**, 3.1], there is a canonical isomorphism of log structures on W

(9.4.4.4) $$W(L)^{(p)} \simeq \sigma^* W(L),$$

which identifies the map $\sigma^* W(L) \to W(L)$ giving the lifting of Frobenius $F_{(W,W(L))}$ with the map $\gamma_{W(L)}$.

Since the map $F_{\mathcal{L}og_{(W,W(L))}}$ covers the map $\sigma : W \to W$, it is specified by a morphism

(9.4.4.5) $$F'_{\mathcal{L}og_{(W,W(L))}} : \mathcal{L}og_{(W,W(L))} \longrightarrow \mathcal{L}og_{(W,\sigma^* W(L))}$$

over W. We define this map to be the functor sending

(9.4.4.6) $$(T, M_T) \xrightarrow{(g,g^b)} (W, W(L))$$

to the morphism

(9.4.4.7) $$(T, M_T^{(p)}) \xrightarrow{(g,g^{b,(p)})} (W, W(L)^{(p)}) \simeq (W, \sigma^* W(L)).$$

9.4.5. — Let T/W be a flat W-scheme with a lifting of Frobenius $F_T : T \to T$, and let M_T be a fine log structure on T. Let

(9.4.5.1) $$j : (W, W(L)) \hookrightarrow (T, M_T)$$

be an exact closed immersion defined by a divided power ideal. This map induces a closed immersion

(9.4.5.2) $$\mathcal{L}og(j) : \mathcal{L}og_{(W,W(L))} \hookrightarrow \mathcal{L}og_{(T,M_T)}.$$

For every integer n, let (T_n, M_{T_n}) denote the reduction of (T, M_T) modulo p^n. Then (T_n, M_{T_n}) is an object of $\mathrm{Cris}((\mathrm{Spec}(k), L)/W_n)$, where W_n is endowed with the trivial log structure.

In this situation, Hyodo and Kato construct in [**31**, 4.8] an isomorphism of graded $\mathcal{O}_T$-algebras

(9.4.5.3) $$\iota^{HK} : \oplus_{q\geq 0} \mathcal{O}_T \otimes_{W_n(k)} W_n^{HK}\omega_Y^q \longrightarrow \oplus_{q\geq 0} R^q u_{(Y,M)/(T_n,M_{T_n})*} \mathcal{O}_{(Y,M)/(T_n,M_{T_n})}.$$

We now explain how we recover this isomorphism using our stack-theoretic methods, in the case when there exists a chart $\beta_T : Q \to M_T$ such that the induced map $Q \to \overline{L}$ is an isomorphism, and such that the diagram

(9.4.5.4) $$\begin{array}{ccc} Q & \xrightarrow{\times p} & Q \\ \downarrow & & \downarrow \\ W(k) & \xrightarrow{\sigma} & W(k) \end{array}$$

defines a chart for the lifting of Frobenius $F_{(W,W(L))}$.

Note that we may without loss of generality assume that T is the spectrum of a p-adically complete local ring (replace T by the p-adic completion of the local ring of the point $\mathrm{Spec}(k) \to T$ defined by j).

Remark 9.4.6. — The most important example for the purposes of this text is when $T = \mathrm{Spec}(W\langle t\rangle)$ (where $W\langle t\rangle$ is defined as in 5.3.1), and the log structure is defined by the map $\mathbb{N} \to W\langle t\rangle$ sending 1 to t.

Proposition 9.4.7. — *There exists an open substack $\mathscr{U}_T \subset \mathcal{L}og_{(T,M_T)}$ flat over T such that the open substack $\mathscr{U}_W := \mathcal{L}og(j)^{-1}(\mathscr{U}_T)$ of $\mathcal{L}og_{(W,W(L))}$ is stable under the lifting of Frobenius $F_{\mathcal{L}og_{(W,W(L))}}$ (let $F_{\mathscr{U}_W}$ be the induced lifting of Frobenius to $\mathscr{U}_W$), the pair $(\mathscr{U}_W, F_{\mathscr{U}_W})$ satisfies the assumptions of 4.4.5, and the morphism $Y \to \mathcal{L}og_{(\mathrm{Spec}(k),L)}$ factors through the reduction $\mathscr{U}_k$ of $\mathscr{U}_T$.*

Proof. — Define $\mathscr{U}_T \subset \mathcal{L}og_{(T,M_T)}$ to be the full substack whose objects are morphisms of fine log schemes

$$f : (X, M_X) \longrightarrow (T, M_T) \tag{9.4.7.1}$$

such that for any geometric point $\bar{x} \to X$ there exist an étale neighborhood U of $\bar{x}$ and a chart for the restriction f_U of f to U

$$\begin{array}{ccc} (U, M_X|_U) & \longrightarrow & \mathrm{Spec}(P \to \mathbb{Z}[P]) \\ {\scriptstyle f_U}\downarrow & & \downarrow{\scriptstyle \gamma^*} \\ (T, M_T) & \xrightarrow{\beta_T} & \mathrm{Spec}(Q \to \mathbb{Z}[Q]), \end{array} \tag{9.4.7.2}$$

where γ^* is induced by a morphism of monoids $\gamma : Q \to P$ satisfying the following conditions:

(i) γ is integral and injective;

(ii) The quotient $P^{\mathrm{gp}}/Q^{\mathrm{gp}}$ is p-torsion free;

(iii) If H is the submonoid of P given by (cf. (9.3.6.12))

$$H := \{a \in P | a = b^p c \text{ for some } b \in P^{gp}, c \in Q^{gp}\} \tag{9.4.7.3}$$

then the natural map

$$P \oplus_{Q,\times p} Q \longrightarrow H, \quad (b, c) \mapsto b^p c \tag{9.4.7.4}$$

is an isomorphism.

Note that $\mathscr{U}_T$ is clearly an open substack of $\mathcal{L}og_{(T,M_T)}$. We claim that $\mathscr{U}_T$ has the desired properties.

To verify this, fix a morphism $\gamma : Q \to P$ as above, and let $\mathcal{S}_{P/Q,T}$ denote the stack $\mathcal{S}_P \times_{\mathcal{S}_Q,\beta_T} T$ defined in 9.1.4. If G denotes the quotient $P^{\mathrm{gp}}/Q^{\mathrm{gp}}$ and $D(G)$ is the associated diagonalizable group scheme, then $\mathcal{S}_{P/Q,T}$ is the stack-theoretic quotient of

$$\mathrm{Spec}(\mathbb{Z}[P]) \times_{\mathrm{Spec}(\mathbb{Z}[Q]),\beta_T} T \tag{9.4.7.5}$$

by the natural action of $D(G)$ induced by the action of the torus $D(P^{\mathrm{gp}})$ on the toric variety $\mathrm{Spec}(\mathbb{Z}[P])$. By 9.1.11 there is a natural projection $\mathcal{S}_{P/Q,T} \to \mathscr{U}_T$ which is representable and étale, and the images of these morphisms cover $\mathscr{U}_T$.

To verify the flatness of $\mathscr{U}_T$ over T, it therefore suffices to show that the schemes (9.4.7.5) are flat over T, and this in turn follows from the fact that $\mathbb{Z}[Q] \to \mathbb{Z}[P]$ is flat since γ is integral and injective [**40**, 4.1].

The reduction $\mathcal{S}_{P/Q,W}$ of $\mathcal{S}_{P/Q,T}$ to W also comes equipped with a lifting of Frobenius $F_{\mathcal{S}_{P/Q,W}}$ induced by the map

$$\mathbb{Z}[P] \otimes_{\mathbb{Z}[Q]} W \longrightarrow \mathbb{Z}[P] \otimes_{\mathbb{Z}[Q]} W \tag{9.4.7.6}$$

which is equal to σ on W and multiplication by p on P, and the map $D(G) \to D(G)$ induced by multiplication by p on G. It follows from the modular interpretation of $\mathcal{S}_{P/Q,W}$ given in [**62**, 5.20] that the projection $\mathcal{S}_{P/Q,W} \to \mathcal{L}og_{(W,W(L))}$ is compatible with the liftings of Frobenius. Since $\mathscr{U}_W$ is the union of the images of these maps it follows that $\mathscr{U}_W$ is stable under the lifting of Frobenius $F_{\mathcal{L}og_{(W,W(L))}}$.

To verify the assumptions of 4.4.5, recall that we need to show that for any morphism $X \to \mathscr{U}$ the natural map

$$\mathcal{O}_{X^{(n)}} \longrightarrow RP_{n*}\mathcal{O}_{\widetilde{X}^{(n)}} \tag{9.4.7.7}$$

is an isomorphism, where $X^{(n)}$ denotes $X \times_{W,\sigma^n} W$, $\widetilde{X}^{(n)}$ denotes $X \times_{\mathscr{U}_W, F^n_{\mathscr{U}_W}} \mathscr{U}_W$, and

$$P_n : X \times_{\mathscr{U}_W, F^n_{\mathscr{U}_W}} \mathscr{U}_W \longrightarrow X \times_{W,\sigma^n} W \tag{9.4.7.8}$$

is the projection (in 4.3.2 the "test scheme" X is denoted W, but we change the notation so as not to conflict with our notation for Witt vectors). Clearly to verify this we can work étale locally on X, and therefore it suffices to consider morphisms $X \to \mathscr{U}$ which factor through some $\mathcal{S}_{P/Q,W}$. It follows that it suffices to verify the above condition for $\mathcal{S}_{P/Q,W}$ instead of $\mathscr{U}_W$, since the diagram

$$\begin{array}{ccc} \mathcal{S}_{P/Q,W} & \xrightarrow{F^n_{\mathcal{S}_{P/Q,W}}} & \mathcal{S}_{P/Q,W} \\ \downarrow & & \downarrow \\ \mathscr{U} & \xrightarrow{F^n_{\mathscr{U}}} & \mathscr{U} \end{array} \tag{9.4.7.9}$$

is cartesian for all $n \geq 0$ since the vertical arrows are representable and étale.

Lemma 9.4.8. — *For every $n \geq 1$ the map*
(9.4.8.1)

$$P \oplus_{Q,\times p^n} Q \longrightarrow \{a \in P | a = b^{p^n} c \text{ for some } b \in P^{gp} \text{ and } c \in Q^{gp}\}, \quad (a,c) \longmapsto a^{p^n} c$$

is an isomorphism.

Proof. — To see the injectivity of (9.4.8.1), note that since $Q \to P$ is integral, the monoid $P \oplus_{Q,\times p^n} Q$ is integral, so it suffices to verify that the induced map

$$(9.4.8.2) \qquad (P \oplus_{Q,\times p^n} Q)^{\mathrm{gp}} \simeq P^{\mathrm{gp}} \oplus_{Q^{\mathrm{gp}},\times p^n} Q^{\mathrm{gp}} \longrightarrow P^{\mathrm{gp}}, \quad (b,c) \longmapsto b^{p^n} c$$

is injective. This is clear because if $b \in P^{\mathrm{gp}}$ and $c \in Q^{\mathrm{gp}}$ are elements such that $b^{p^n} c = 0$ then $b \in Q^{\mathrm{gp}}$ because $P^{\mathrm{gp}}/Q^{\mathrm{gp}}$ is p-torsion free and hence (b,c) maps to 0 in $P^{\mathrm{gp}} \oplus_{Q^{\mathrm{gp}},\times p^n} Q^{\mathrm{gp}}$.

For the surjectivity of (9.4.8.1), we proceed by induction on n. For $n = 1$, we have the result by assumption.

So we assume the surjectivity holds for $n-1$ and prove it for n. Let $a \in P$ be an element such that $a = b^{p^n} c$ for some $b \in P^{\mathrm{gp}}$ and $c \in Q^{\mathrm{gp}}$. Then by the case $n = 1$, we can write $a = \alpha^p c'$, where $\alpha \in P$ and $c' \in Q$. Then $\gamma := b^{p^{n-1}} \alpha^{-1}$ is an element of P^{gp} whose p-th power is in Q^{gp}. Since $P^{\mathrm{gp}}/Q^{\mathrm{gp}}$ is p-torsion free it follows that $\gamma \in Q^{\mathrm{gp}}$. Therefore $\alpha = b^{p^{n-1}} \gamma^{-1}$, which implies by the induction hypothesis that we can write $\alpha = e^{p^{n-1}} q$, where $e \in P$ and $q \in Q$. This in turn gives

$$(9.4.8.3) \qquad a = e^{p^n} (q^p c')$$

which proves the surjectivity for n. □

Let $Z_{P/Q,W}$ denote the scheme

$$(9.4.8.4) \qquad \mathrm{Spec}(\mathbb{Z}[P]) \times_{\mathrm{Spec}(\mathbb{Z}[Q]),\beta_W} W$$

so that the projection $Z_{P/Q,W} \to \mathcal{S}_{P/Q,W}$ is a smooth surjection. Let G_{p^n} denote the cokernel of the map

$$(9.4.8.5) \qquad \times p^n : P^{\mathrm{gp}}/Q^{\mathrm{gp}} \longrightarrow P^{\mathrm{gp}}/Q^{\mathrm{gp}},$$

and let $D(G_{p^n})$ denote the corresponding diagonalizable group scheme. The action of the torus $D(P^{\mathrm{gp}})$ on the toric variety $\mathrm{Spec}(\mathbb{Z}[P])$ induces an action of $D(G_{p^n})$ on $Z_{P/Q,W}$. By the same argument proving 9.3.5 (ii) the fiber product of the diagram

$$(9.4.8.6) \qquad \begin{array}{ccc} & & Z_{P/Q,W} \\ & & \downarrow \\ \mathcal{S}_{P/Q,W} & \xrightarrow{F^n_{\mathcal{S}_{P/Q,W}}} & \mathcal{S}_{P/Q,W} \end{array}$$

is isomorphic to the stack-quotient of $Z_{P/Q,W}$ by this action of $D(G_{p^n})$. The projection

$$(9.4.8.7) \qquad \pi_n : [Z_{P/Q,W}/D(G_{p^n})] \longrightarrow Z_{P/Q,W}$$

is induced by the map

$$(9.4.8.8) \qquad F^n_Z : \mathbb{Z}[P] \otimes_{\mathbb{Z}[Q]} W \longrightarrow \mathbb{Z}[P] \otimes_{\mathbb{Z}[Q]} W$$

given by multiplication by p^n on P and σ^n on W.

Since any morphism to $\mathcal{S}_{P/Q,W}$ étale locally factors through $Z_{P/Q,W}$, this implies that for any morphism $X \to \mathcal{S}_{P/Q,W}$ the stack $\widetilde{X}^{(n)}$ is étale locally on X (with $\widetilde{X}^{(n)}$ viewed as an X-stack via the projection) isomorphic to the stack-theoretic quotient of a finite X-scheme by the action of a finite diagonalizable group scheme. This implies that the sheaves

(9.4.8.9) $$R^i P_{n*}\mathcal{O}_{\widetilde{X}^{(n)}} = 0$$

for $i > 0$. This also implies that to prove that the map $\mathcal{O}_X \to R^0 P_{n*}\mathcal{O}_{\widetilde{X}^{(n)}}$ is an isomorphism, it suffices to prove it for $X = Z_{P/Q,W}$. For if $\mathcal{F}$ is a quasi-coherent sheaf with action of $D(G_{p^n})$ on $Z_{P/Q,W}$ (a $D(G_{p^n}) - \mathcal{O}_{Z_{P/Q,W}}$-module in the sense of [**14**, I.4.7.1] then since $D(G_{p^n})$ is diagonalizable the formation of the subsheaf of invariants of $\mathcal{F}$ commutes with arbitrary base change on $Z_{P/Q,W}$ (this follows for example from [**14**, I.4.7.3]).

To prove it for $Z_{P/Q,W}$, note that the map (9.4.8.8) is obtained by base change along $\beta_W : \mathbb{Z}[Q] \to W$ from the map

(9.4.8.10) $$\mathbb{Z}[P \oplus_{Q,\times p^n} Q] \longrightarrow \mathbb{Z}[P],$$

induced by the morphism of monoids (9.4.8.1). The result therefore follows from 9.4.8. This completes the verification that the pair $(\mathscr{U}, F_{\mathscr{U}})$ satisfies the assumptions in 4.4.5.

Finally the statement that $Y \to \mathcal{L}og_{(\mathrm{Spec}(k),L)}$ factors through $\mathscr{U}_k$ follows from the argument used in the proof of [**40**, 4.12] which shows that locally on Y the map $Y \to \mathcal{L}og_{(\mathrm{Spec}(k),L)}$ factors through some $\mathcal{S}_{P/Q,k}$. □

9.4.9. — Let $\mathcal{A}^\bullet_{n,Y/\mathscr{U}_W}$ and $\mathcal{A}^\bullet_{n,Y/\mathscr{U}_T}$ be defined as in 4.3. Using the equivalence (9.2.2) we obtain canonical isomorphisms

(9.4.9.1) $$W_n^{HK}\omega^q_Y \simeq \mathcal{A}^q_{n,Y/\mathscr{U}_W}, \quad R^q u_{(Y,M)/(T_n,M_{T_n})*}\mathcal{O}_{(Y,M)/(T_n,M_{T_n})} \simeq \mathcal{A}^q_{n,Y/\mathscr{U}_{T_n}}.$$

Lemma 9.4.10. — *The first isomorphism in (9.4.9.1) is compatible with the operators $d, F, V,$ and π_n defined in* [**31**, 4.1 and 4.2] *for $W_\bullet^{HK}\omega^\bullet_Y$ and in 4.3.7 for $\mathcal{A}^\bullet_{\bullet,Y/\mathscr{U}_W}$.*

Proof. — It suffices to verify the lemma étale locally on Y, so we may assume that there exists a compatible system of liftings of (Y, M) to log schemes $(\widetilde{Y}_n, \widetilde{M}_n)$ smooth over $(W_n, W_n(L))$ as well as a compatible system of liftings of Frobenius to the log schemes $(\widetilde{Y}_n, \widetilde{M}_n)$. Let C_n denote the logarithmic de Rham-complex

(9.4.10.1) $$C^\bullet_n := \Omega^\bullet_{(\widetilde{Y}_m,\widetilde{M}_m)/(W_m,W_m(L))}.$$

By 9.1.14 the complex $C^\bullet_m$ is canonically identified with the de Rham-complex

(9.4.10.2) $$\Omega^\bullet_{\widetilde{Y}/\mathcal{L}og_{(W_m,W_m(L))}},$$

and this identification is compatible with the Frobenius endomorphisms. The lemma can now be verified as follows.

(i) The maps d at level n are in both cases obtained from the boundary morphism arising from the long exact sequence associated to the short exact sequences

$$0 \longrightarrow C_n^\bullet \xrightarrow{p^n} C_{2n}^\bullet \longrightarrow C_n^\bullet \longrightarrow 0. \tag{9.4.10.3}$$

(ii) The maps F are induced by the projection $C_{n+1}^\bullet \to C_n^\bullet$, and the maps V are induced by the maps $\times p : C_n^\bullet \to C_{n+1}^\bullet$.

(iii) The maps π_n are in both cases defined using the method described in 4.1.9. □

9.4.11. — From 4.6.9 and with T as in 9.4.5, we therefore obtain an isomorphism

$$\iota : \oplus_{q\geq 0}\mathcal{O}_T \otimes_W \mathcal{A}^q_{n,Y/\mathcal{U}_{W_n}} \longrightarrow \oplus_{q\geq 0}\mathcal{A}^q_{n,Y/\mathcal{U}_{T_n}}. \tag{9.4.11.1}$$

Proposition 9.4.12. — *The diagram*
(9.4.12.1)

$$\begin{array}{ccc} \oplus_{q\geq 0}\mathcal{O}_T \otimes_W W_n^{HK}\omega^q_Y & \xrightarrow{(9.4.9.1)} & \oplus_{q\geq 0}\mathcal{O}_T \otimes_W \mathcal{A}^q_{n,Y/\mathcal{U}_{W_n}} \\ \downarrow{\scriptstyle \iota^{HK}} & & \downarrow{\scriptstyle \iota} \\ \oplus_{q\geq 0}R^q u_{(Y,M)/(T_n,M_{T_n})*}\mathcal{O}_{(Y,M)/(T_n,M_{T_n})} & \xrightarrow{(9.4.9.1)} & \oplus_{q\geq 0}\mathcal{A}^q_{n,Y/\mathcal{U}_{T_n}} \end{array}$$

commutes.

Proof. — The assertion is étale local on Y, so it suffices to consider the case when Y is equal to the scheme $Z_{P/Q,k}$ associated to a morphism of monoids $Q \to P$ as in the proof of 9.4.7. Let $M_{Z_{P/Q,T_n}}$ be the log structure on $Z_{P/Q,T_n}$ defined by the projection to $\mathcal{S}_{P/Q,T_n}$, so that $(Z_{P/Q,T_n}, M_{Z_{P/Q,T_n}})$ is a log smooth lifting of $(Z_{P/Q,k}, M_{Z_{P/Q,k}})$ to (T_n, M_{T_n}).

In this case the square (9.4.12.1) is induced from the diagram of differential graded algebras

(9.4.12.2)

$$\begin{array}{ccc} \oplus_q\mathcal{H}^q(\Omega^\bullet_{(Z_{P/Q,W_n},M_{Z_{P/Q,W_n}})/(W_n,W_n(L))}) & \xrightarrow{\simeq} & \oplus_q\mathcal{H}^q(\Omega^\bullet_{Z_{P/Q,W_n}/\mathcal{S}_{P/Q,W_n}}) \\ \downarrow{\scriptstyle \iota^{HK}} & & \downarrow{\scriptstyle \iota} \\ \oplus_{q\geq 0}\mathcal{H}^q(\Omega^\bullet_{(Z_{P/Q,T_n},M_{Z_{P/Q,T_n}})/(T_n,M_{T_n})}) & \xrightarrow{\simeq} & \oplus_{q\geq 0}\mathcal{H}^q(\Omega^\bullet_{Z_{P/Q,T_n}/\mathcal{S}_{P/Q,T_n}}) \end{array}$$

by tensoring the top row with $\mathcal{O}_T$. It therefore suffices to show that this diagram (9.4.12.2) commutes.

Let G denote $P^{\mathrm{gp}}/Q^{\mathrm{gp}}$ so that

$$\Omega^1_{(Z_{P/Q,W_n},M_{Z_{P/Q,W_n}})/(W_n,W_n(L))} \simeq \mathcal{O}_{Z_{P/Q,W_n}} \otimes_{\mathbb{Z}} G, \tag{9.4.12.3}$$

and similarly

$$\Omega^1_{(Z_{P/Q,T_n},M_{Z_{P/Q,T_n}})/(T_n,M_{T_n})} \simeq \mathcal{O}_{Z_{P/Q,T_n}} \otimes_{\mathbb{Z}} G. \tag{9.4.12.4}$$

For $g \in G$ write $\mathrm{dlog}(g)$ for the element $1 \otimes g$ in either of these two modules. By the proof of [**31**, 4.8], the map ι^{HK} is determined by the map in degree $q = 0$ and by the condition that

$$\iota^{HK}[\mathrm{dlog}(g)] = [\mathrm{dlog}(g)] \tag{9.4.12.5}$$

for all $g \in G$. It therefore suffices to verify that (9.4.12.2) commutes for $q = 0$, and that $\iota[\mathrm{dlog}(g)] = [\mathrm{dlog}(g)]$ for all $g \in G$.

The commutativity for $q = 0$ follows from the construction. Indeed it follows from the definition of the map ρ_n in 4.2.2 and descent theory that the map ι in degree 0 is given by the map

$$W_n(\mathcal{O}_{Z_{P/Q,k}}) \longrightarrow W_n(\mathcal{O}_{Z_{P/Q,T_n}}) \tag{9.4.12.6}$$

sending a Witt vector $(a_0, \dots, a_{n-1})$ to

$$\sum_{i=0}^{n-1} p^i \tilde{a}_i^{p^{n-i}} \in \mathcal{H}^0(\Omega^\bullet_{Z_{P/Q,T_n}/\mathcal{S}_{P/Q,T_n}}), \tag{9.4.12.7}$$

where $\tilde{a}_i \in \mathcal{O}_{Z_{P/Q,T_n}}$ is any lifting of a_i. By the construction in [**31**, 4.9] this agrees with the map ι^{HK} in degree 0.

To verify that $\iota[\mathrm{dlog}(g)] = [\mathrm{dlog}(g)]$, let $Z_{P/Q,T_n}[G]$ denote the scheme

$$Z_{P/Q,T_n} \times D(G), \tag{9.4.12.8}$$

and note that $Z_{P/Q,T_n}[G]$ with the two projections to $Z_{P/Q,T_n}$ defined by the first projection and the action is isomorphic to the fiber product of the diagram

$$\begin{array}{ccc} & & Z_{P/Q,T_n} \\ & & \downarrow \\ Z_{P/Q,T_n} & \longrightarrow & \mathcal{S}_{P/Q,T_n}. \end{array} \tag{9.4.12.9}$$

We define $Z_{P/Q,W_n}[G]$ similarly. By 4.3.20 the pullback map

$$\mathcal{H}^1(\Omega^\bullet_{Z_{P/Q,T_n}/\mathcal{S}_{P/Q,T_n}}) \longrightarrow \mathcal{H}^1(\Omega^\bullet_{Z_{P/Q,T_n}[G]/Z_{P/Q,T_n}}) \tag{9.4.12.10}$$

is injective, and by the same argument used in the proof of 9.1.13 the image of $[\mathrm{dlog}(g)]$ in $\mathcal{H}^1(\Omega^\bullet_{Z_{P/Q,T_n}[G]/Z_{P/Q,T_n}})$ is the class of $g^{-1}dg$, where we view $g \in G$ as a unit in the ring

$$\mathcal{O}_{Z_{P/Q,T_n}[G]} = \mathcal{O}_{Z_{P/Q,T_n}} \otimes_{\mathbb{Z}} \mathbb{Z}[G]. \tag{9.4.12.11}$$

The liftings of Frobenius to $Z_{P/Q,W_n}$ and $\mathcal{S}_{P/Q,W_n}$ also define a lifting of Frobenius to $Z_{P/Q,W_n}[G]$, which in fact is the lifting of Frobenius defined by the lifting of Frobenius on each of the two factors of (9.4.12.8). In other words, the lifting of Frobenius on $Z_{P/Q,W_n}[G]$ is given by the map

$$\mathcal{O}_{Z_{P/Q,W_n}} \otimes_{\mathbb{Z}} \mathbb{Z}[G] \longrightarrow \mathcal{O}_{Z_{P/Q,W_n}} \otimes_{\mathbb{Z}} \mathbb{Z}[G] \tag{9.4.12.12}$$

which is $F_{Z_{P/Q,W_n}}$ on the first factor and multiplication by p on G. By [**34**, 0.1.3] there exists a canonical commutative diagram

(9.4.12.13)
$$\begin{array}{ccc} W_n(Z_{P/Q,k}[G]) & \xrightarrow{s} & Z_{P/Q,W_n}[G] \\ \downarrow & & \downarrow \\ W_n(Z_{P/Q,k}) & \xrightarrow{t} & Z_{P/Q,W_n}, \end{array}$$

where s and t are compatible with the Frobenius morphisms and reduce to the identities modulo p. By [**34**, 0.1.3.18], the map s sends $g \in \mathcal{O}_{Z_{P/Q,W_n}[G]}$ to the Teichmuller lifting $[g] \in W_n(\mathcal{O}_{Z_{P/Q,k}[G]})$.

Let $\alpha_g \in W_n^{\mathrm{LZ}}\Omega^1_{Z_{P/Q,k}[G]/Z_{P/Q,k}}$ be the image of $g^{-1}dg \in \Omega^1_{Z_{P/Q,W_n}[G]/Z_{P/Q,W_n}}$ under the composite

(9.4.12.14)
$$\Omega^1_{Z_{P/Q,W_n}[G]/Z_{P/Q,W_n}} \xrightarrow{s^*} \Omega^1_{W_n(Z_{P/Q,k}[G])/W_n(Z_{P/Q,k})} \xrightarrow{q_n} W_n^{\mathrm{LZ}}\Omega^1_{Z_{P/Q,k}[G]/Z_{P/Q,k}},$$

where the second map is the projection map used in the construction of the Langer-Zink de Rham-Witt complex (4.2.4–4.2.6).

The following two lemmas now complete the proof of 9.4.12, as they show that the images of $\iota[\mathrm{dlog}(g)]$ and $[\mathrm{dlog}(g)]$ under the injective map (9.4.12.10) are equal, and in fact equal to α_g.

Lemma 9.4.13. — *The image of α_g under the map*

(9.4.13.1)
$$\rho_n : W_n^{\mathrm{LZ}}\Omega^1_{Z_{P/Q,k}[G]/Z_{P/Q,k}} \longrightarrow \mathcal{A}^1_{n,Z_{P/Q,k}[G]/Z_{P/Q,T}} \simeq \mathcal{H}^1(\Omega^\bullet_{Z_{P/Q,T_n}[G]/Z_{P/Q,T_n}})$$

defined in 4.2.3 is equal to the class $[g^{-1}dg]$.

Proof. — Note first that the image of g under the composite map

(9.4.13.2)
$$\mathcal{O}_{Z_{P/Q,W_n}[G]/Z_{P/Q,W_n}} \xrightarrow{s^*} W_n(\mathcal{O}_{Z_{P/Q,k}[G]}) \xrightarrow{\rho_n} \mathcal{H}^0(\Omega^\bullet_{Z_{P/Q,T_n}[G]/Z_{P/Q,T_n}})$$

is equal to the class of g^{p^n} by the definition of the map ρ_n in 4.2.2. Since

(9.4.13.3)
$$\rho_n : W_n^{\mathrm{LZ}}\Omega^\bullet_{Z_{P/Q,k}[G]/Z_{P/Q,k}} \to \mathcal{A}^\bullet_{n,Z_{P/Q,k}[G]/Z_{P/Q,T}}$$

is a map of differential graded algebras, it follows that the image of α_g under ρ_n is equal to $[(g^{p^n})^{-1}]\cdot d[g^{p^n}]$. Now recall that $d[g^{p^n}]$ is obtained by choosing a lifting ℓ of g^{p^n} to $\mathcal{O}_{Z_{P/Q,T_{2n}}[G]}$, and defining $d[g^{p^n}]$ to be the class of an element $\omega \in \Omega^1_{Z_{P/Q,T_n}[G]/Z_{P/Q,T_n}}$ for which $p^n\omega = d\ell$ in $\Omega^1_{Z_{P/Q,T_{2n}}[G]/Z_{P/Q,T_{2n}}}$. Taking $\ell = g^{p^n}$ we see that

(9.4.13.4)
$$[(g^{p^n})^{-1}]\cdot d[g^{p^n}] = [(g^{p^n})^{-1}g^{p^n-1}dg] = [g^{-1}dg]$$

as desired. □

Lemma 9.4.14. — *The image of the class* $[\mathrm{dlog}(g)]$ *under the pullback map*

$$\mathcal{H}^1(\Omega^\bullet_{Z_{P/Q,W_n}/\mathcal{S}_{P/Q,W_n}}) \simeq \mathcal{A}^1_{n,Z_{P/Q,k}/\mathcal{S}_{P/Q,W_n}} \longrightarrow W_n^{\mathrm{LZ}}\Omega^1_{Z_{P/Q,k}[G]/Z_{P/Q,k}} \tag{9.4.14.1}$$

is equal to α_g.

Proof. — Let $R_{P/Q,W_n}$ denote the fiber product

$$R_{P/Q,W_n} := Z_{P/Q,W_n}[G] \times_{\mathrm{action}, Z_{P/Q,W_n}, \mathrm{action}} Z_{P/Q,W_n}[G] \tag{9.4.14.2}$$

so that we have a commutative diagram of schemes with liftings of Frobenius

$$\begin{array}{ccccc} Z_{P/Q,W_n} & \longleftarrow & Z_{P/Q,W_n}[G] & \leftleftarrows & R_{P/Q,W_n} \\ \downarrow & & \downarrow & & \downarrow \\ \mathcal{S}_{P/Q,W_n} & \longleftarrow & Z_{P/Q,W_n} & \leftleftarrows & Z_{P/Q,W_n}[G] \end{array} \tag{9.4.14.3}$$

This diagram in turn induces a commutative diagram of modules

(9.4.14.4)

$$\begin{array}{ccccc} \Omega^1_{R_{P/Q,W_n}/Z_{P/Q,W_n}[G]} & \xrightarrow{\tilde{s}^*} & \Omega^1_{W_n(R_{P/Q,k})/W_n(Z_{P/Q,k}[G])} & \xrightarrow{\tilde{q}_n} & W_n^{\mathrm{LZ}}\Omega^1_{R_{P/Q,k}/Z_{P/Q,k}[G]} \\ \upuparrows & & \upuparrows & & \upuparrows \\ \Omega^1_{Z_{P/Q,W_n}[G]/Z_{P/Q,W_n}} & \xrightarrow{s^*} & \Omega^1_{W_n(Z_{P/Q,k}[G])/W_n(Z_{P/Q,k})} & \xrightarrow{q_n} & W_n^{\mathrm{LZ}}\Omega^1_{Z_{P/Q,k}[G]/Z_{P/Q,k}} \\ \uparrow & & & & \\ \Omega^1_{Z_{P/Q,W_n}/\mathcal{S}_{P/Q,W_n}} & & & & \end{array}$$

where the left column is exact, and $\tilde{s}^*$ and $\tilde{q}_n^*$ are defined analogously to s^* and q_n^*. Since the element $g^{-1}dg \in \Omega^1_{Z_{P/Q,W_n}[G]/Z_{P/Q,W_n}}$ is in the image of $\Omega^1_{Z_{P/Q,W_n}/\mathcal{S}_{P/Q,W_n}}$ (in fact $g^{-1}dg$ is the image of the element $\mathrm{dlog}(g)$), it follows that α_g is in the equalizer of the two maps

$$W_n^{\mathrm{LZ}}\Omega^1_{Z_{P/Q,k}[G]/Z_{P/Q,k}} \rightrightarrows W_n^{\mathrm{LZ}}\Omega^1_{R_{P/Q,k}/Z_{P/Q,k}[G]}. \tag{9.4.14.5}$$

Since the morphism of diagrams

$$\begin{array}{ccc} W_n^{\mathrm{LZ}}\Omega^1_{R_{P/Q,k}/Z_{P/Q,k}[G]} & \xrightarrow{\rho_n} & \mathcal{A}^1_{n,R_{P/Q,k}/Z_{P/Q,W}[G]} \\ \upuparrows & & \upuparrows \\ W_n^{\mathrm{LZ}}\Omega^1_{Z_{P/Q,k}[G]/Z_{P/Q,k}} & \xrightarrow{\rho_n} & \mathcal{A}^1_{n,Z_{P/Q,k}[G]/Z_{P/Q,W}} \end{array} \tag{9.4.14.6}$$

induces an isomorphism between the equalizers of the vertical arrows by 4.6.7, it follows that to prove the lemma it suffices to show that the image of α_g in

$$\mathcal{A}^1_{n,Z_{P/Q,k}[G]/Z_{P/Q,W}} \simeq \mathcal{H}^1(\Omega^\bullet_{Z_{P/Q,W_n}[G]/Z_{P/Q,W_n}}) \tag{9.4.14.7}$$

is equal to the class of $[g^{-1}dg]$. This follows from the same argument used in the proof of 9.4.13. □

This completes the proof of 9.4.12. □

9.5. Equivalence of definitions of syntomic complexes

9.5.1. — Let A be a complete discrete valuation ring of mixed characteristic $(0,p)$ and perfect residue field k. Let K be the field of fractions of A, W the ring of Witt vectors of k, and $K_0 \subset K$ the field of fractions of W. Let $\mathfrak{X}/A$ be a proper scheme étale locally isomorphic to

$$\text{Spec}(A[x_1, \ldots, x_n]/(x_1 \cdots x_s - \pi)) \tag{9.5.1.1}$$

for some integers $s \leq n$. Let M_A (resp. $M_{\mathfrak{X}}$) be the log structure defined by the closed point (resp. closed fiber) so there is a natural log smooth morphism $(\mathfrak{X}, M_{\mathfrak{X}}) \to (\text{Spec}(A), M_A)$. For $r \in [0, p-1]$ let $s_{n,\mathfrak{X}/\mathfrak{R}}(r)$ be the complex constructed in (8.4.8.4), and let $s^{\log}_{n,\mathfrak{X}}(r)$ be the logarithmic syntomic complex defined in [**41**, §5].

Proposition 9.5.2. — *There is a natural isomorphism* $s_{n,\mathfrak{X}/\mathfrak{R}}(r) \simeq s^{\log}_{n,\mathfrak{X}}(r)$ *in* $D(\mathfrak{X}_{\text{et}}, \mathbb{Z}/p^n)$ *compatible with the product structure and with the action of* $\text{Gal}(\overline{K}/K)$.

Proof. — For each finite extension $K \subset K' \subset \overline{K}$ with ring of integers $A' \subset K'$, let $(\mathfrak{X}_{A'}, M_{\mathfrak{X}_{A'}})$ denote $(\mathfrak{X}, M_{\mathfrak{X}}) \times_{(\text{Spec}(A), M_A)} (\text{Spec}(A'), M_{A'})$. As in [**73**, p. 263] choose data as follows:

(9.5.2.1) An étale hypercover $X_\bullet \to \mathfrak{X}$.

(9.5.2.2) For each finite extension $K \subset K' \subset \overline{K}$ an exact closed immersion

$$(X_{A'\bullet}, M_{\mathfrak{X}_{A'\bullet}}) \hookrightarrow (Z_{A'\bullet}, M_{Z_{A'\bullet}})$$

over $(\text{Spec}(W), \mathcal{O}^*_{\text{Spec}(W)})$, where $M_{\mathfrak{X}_{A'\bullet}}$ denotes the pullback of the log structure $M_{\mathfrak{X}_{A'}}$ on $\mathfrak{X}_{A'}$.

(9.5.2.3) For every inclusion $K' \subset K'' \subset \overline{K}$ of finite extensions of K a morphism

$$\tau_{K'K''} : (Z_{A''\bullet}, M_{Z_{A''\bullet}}) \to (Z_{A'\bullet}, M_{Z_{A'\bullet}})$$

over $(\text{Spec}(W), \mathcal{O}^*_{\text{Spec}(W)})$ such that the diagram

$$\begin{array}{ccc} (X_{A''\bullet}, M_{\mathfrak{X}_{A''\bullet}}) & \longrightarrow & (Z_{A''\bullet}, M_{Z_{A''\bullet}}) \\ \downarrow & & \downarrow{\scriptstyle \tau_{K'K''}} \\ (X_{A'\bullet}, M_{\mathfrak{X}_{A'\bullet}}) & \longrightarrow & (Z_{A'\bullet}, M_{Z_{A'\bullet}}) \end{array}$$

commutes and for $K' \subset K'' \subset K''' \subset \overline{K}$ we have

$$\tau_{K'K'''} = \tau_{K'K''} \circ \tau_{K''K'''}.$$

(9.5.2.4) A compatible collection of liftings of Frobenius $F_{Z_{A'\bullet}}$ to the $(Z_{A'\bullet}, M_{Z_{A'\bullet}})$.

It follows from 9.1.24 that such a collection of data is equivalent to an object $(X_\bullet, X_{A',\bullet} \hookrightarrow Z_{A',\bullet}, F_{Z_{A',\bullet}}, \tau_{K''/K'})$ of the category $HC(\mathfrak{X})$ defined in 8.4.3.

Let $D_{A',\bullet}$ denote the divided power envelope of $X_{A',\bullet} \hookrightarrow Z_{A',\bullet}$, and let $j^{A',\log}_{n,X_\bullet,Z_\bullet}(r)$ denote the complex

(9.5.2.5)

$$J^{[r]}_{D_{A'_n,\bullet}} \xrightarrow{\ d\ } J^{[r-1]}_{D_{A'_n,\bullet}} \otimes_{\mathcal{O}_{Z_{A',\bullet}}} \omega^1_{Z_{A'\bullet}/W} \xrightarrow{\ d\ } J^{[r-2]}_{D_{A'_n,\bullet}} \otimes_{\mathcal{O}_{Z_{A',\bullet}}} \omega^2_{Z_{A'\bullet}/W} \longrightarrow \cdots,$$

where $\omega^1_{Z_{A'\bullet}/W} := \Omega^1_{(Z_{A',\bullet}, M_{Z_{A'\bullet}})/(\mathrm{Spec}(W), \mathcal{O}^*_{\mathrm{Spec}(W)})}$. As in [**41**, 5.1] for $0 \le r < p$ there is a well-defined map

$$p^{-r}\varphi : j^{A',\log}_{n,X_\bullet,Z_\bullet}(r) \longrightarrow j^{A',\log}_{n,X_\bullet,Z_\bullet}(0). \tag{9.5.2.6}$$

Let $s^{A',\log}_{n,X_\bullet,Z_\bullet}(r)$ be the mapping fiber of

$$1 - p^{-r}\varphi : j^{A',\log}_{n,X_\bullet,Z_\bullet}(r) \longrightarrow j^{A',\log}_{n,X_\bullet,Z_\bullet}(0). \tag{9.5.2.7}$$

Define $j^{A'}_{n,X_\bullet/\mathcal{R}}(r)$ and $s^{A'}_{n,X_\bullet/\mathcal{R}}(r)$ as in 8.4.6. Then it follows from (9.1.3.1) and 9.2.6 that there is a canonical isomorphism

$$j^{A'}_{n,X_\bullet/\mathcal{R}}(r) \simeq j^{A',\log}_{n,X_\bullet,Z_\bullet}(r) \tag{9.5.2.8}$$

compatible with the Frobenius endomorphisms and the product structure. From this it follows that there is a natural isomorphism compatible with the product structure

$$s^{A'}_{n,X_\bullet/\mathcal{R}}(r) \simeq s^{A',\log}_{n,X_\bullet,Z_\bullet}(r). \tag{9.5.2.9}$$

As in 8.4.8, set

$$\mathcal{Y} := \mathfrak{X} \otimes_A k, \ \ \overline{\mathcal{Y}} := \mathcal{Y} \otimes_k \bar{k}, \ \ Y_\bullet := X_\bullet \otimes_A k, \ \ \overline{Y}_\bullet := Y_\bullet \otimes_k \bar{k}, \tag{9.5.2.10}$$

and let $\theta : \overline{Y}_{\bullet,\mathrm{et}} \to \overline{\mathcal{Y}}_{\mathrm{et}}$ be the natural morphism of topoi, and for each $K \subset K' \subset \overline{K}$ let $\pi_{K'} : \overline{Y}_{\bullet,\mathrm{et}} \to \mathcal{Y}_{A',\bullet,\mathrm{et}}$ be the natural projection. Then by definition we have

$$s_{n,\mathfrak{X}/\mathcal{R}}(r) = R\theta_*(\varinjlim \pi^*_{K'} s^{A'}_{n,X_\bullet/\mathcal{R}}(r)), \tag{9.5.2.11}$$

and

$$s^{\log}_{n,\mathfrak{X}}(r) = R\theta_*(\varinjlim \pi^*_{K'} s^{A',\log}_{n,X_\bullet,Z_\bullet}(r)). \tag{9.5.2.12}$$

We thus obtain an isomorphism $s_{n,\mathfrak{X}/\mathcal{R}}(r) \simeq s^{\log}_{n,\mathfrak{X}}(r)$ from (9.5.2.9). That this isomorphism is compatible with the Galois action follows from the construction in 8.4.10. □

9.6. Equivalence of the different constructions of (φ, N, G)-structure

First we compare the construction of 6.4 and 6.5 with the logarithmic theory.

9.6.1. — Let $\mathcal{X}/A$ be as in 9.5.1, let $\mathcal{Y}/k$ be the reduction of $\mathcal{X}$ modulo the maximal ideal of A, and let $\mathcal{X}_0$ denote the reduction modulo p. Denote by $W_n\langle t\rangle$ the divided power envelope of $0 : \mathrm{Spec}(k) \hookrightarrow \mathrm{Spec}(W_n[t])$, let A_n denote A/p^n, let R_n denote the divided power envelope of the surjection $W_n[t] \to A_n$ sending t to a uniformizer $\pi \in A$, and let R denote the inverse limit $R := \varprojlim R_n$.

Let $M_{\mathcal{X}}$ denote the natural log structure on $\mathcal{X}$, and let $M_{\mathcal{Y}}$ and $M_{\mathcal{X}_0}$ be the log structures obtained by pullback. Also define the log structures M_{R_n} and $M_{W_n\langle t\rangle}$ on $\mathrm{Spec}(R_n)$ and $\mathrm{Spec}(W_n\langle t\rangle)$ respectively as the log structures obtained from the map $\mathbb{N} \to W_n[t]$ sending 1 to t. Also let M_W be the log structure on $\mathrm{Spec}(W)$ obtained from the map $\mathbb{N} \to W$ sending 1 to 0. Observe that the lifting of Frobenius to $W[t]$ induced by the canonical lifting σ to W and $t \mapsto t^p$ induces a lifting of Frobenius

$$F_{(\mathrm{Spec}(W_n\langle t\rangle), M_{W_n\langle t\rangle})} : (\mathrm{Spec}(W_n\langle t\rangle), M_{W_n\langle t\rangle}) \to (\mathrm{Spec}(W_n\langle t\rangle), M_{W_n\langle t\rangle}). \tag{9.6.1.1}$$

9.6.2. — By 9.1.21, we can write the projective systems $C.$, $D.$, and $E.$ of 6.4.6 as

$$C. = \{H^*((\mathcal{X}_{0,\mathrm{et}}/\mathcal{L}og_{(\mathrm{Spec}(R_n), M_{R_n})})_{\mathrm{cris}}, \mathcal{O}_{\mathcal{X}_{0,\mathrm{et}}/\mathcal{L}og_{(\mathrm{Spec}(R_n), M_{R_n})}})\}, \tag{9.6.2.1}$$

$$D. = \{H^*((\mathcal{Y}_{\mathrm{et}}/\mathcal{L}og_{(\mathrm{Spec}(W_n), M_{W_n})})_{\mathrm{cris}}, \mathcal{O}_{\mathcal{Y}_{\mathrm{et}}/\mathcal{L}og_{(\mathrm{Spec}(W_n), M_{W_n})}})\}, \tag{9.6.2.2}$$

$$E. = \{H^*((\mathcal{Y}_{\mathrm{et}}/\mathcal{L}og_{(\mathrm{Spec}(W_n\langle t\rangle), M_{W_n\langle t\rangle})})_{\mathrm{cris}}, \mathcal{O}_{\mathcal{Y}_{\mathrm{et}}/\mathcal{L}og_{(\mathrm{Spec}(W_n\langle t\rangle), M_{W_n\langle t\rangle})}})\}. \tag{9.6.2.3}$$

Also define projective systems

$$C.^{\log} = \{H^*(((\mathcal{X}_0, M_{\mathcal{X}_0})/(\mathrm{Spec}(R_n), M_{R_n}))_{\mathrm{cris}}, \mathcal{O}_{(\mathcal{X}_0, M_{\mathcal{X}_0})/(\mathrm{Spec}(R_n), M_{R_n})})\}, \tag{9.6.2.4}$$

$$D.^{\log} = \{H^*(((\mathcal{Y}, M_{\mathcal{Y}})/(\mathrm{Spec}(W_n), M_{W_n}))_{\mathrm{cris}}, \mathcal{O}_{(\mathcal{Y}, M_{\mathcal{Y}})/(\mathrm{Spec}(W_n), M_{W_n})})\}, \tag{9.6.2.5}$$

$$E.^{\log} = \{H^*(((\mathcal{Y}, M_{\mathcal{Y}})/(\mathrm{Spec}(W_n\langle t\rangle), M_{W_n\langle t\rangle}))_{\mathrm{cris}}, \mathcal{O}_{(\mathcal{Y}, M_{\mathcal{Y}})/(\mathrm{Spec}(W_n\langle t\rangle), M_{W_n\langle t\rangle})})\}. \tag{9.6.2.6}$$

Lemma 9.6.3. — *There are natural isomorphisms of projective systems $C. \simeq C.^{\log}$, $D. \simeq D.^{\log}$, and $E. \simeq E.^{\log}$. Furthermore, the isomorphism $D. \simeq D.^{\log}$ and $E. \simeq E.^{\log}$ are compatible with the Frobenius endomorphisms.*

Proof. — The isomorphisms are obtained from the equivalences of sites provided by 9.2.2. The statement that the isomorphisms $D. \simeq D.^{\log}$ and $E. \simeq E.^{\log}$ are compatible with Frobenius can be seen as follows. The arguments for both are the same, so we prove the result for $D.$ leaving the case of $E.$ to the reader. Let $\varphi^{\log}$ denote the endomorphism of $D.$ obtained from the logarithmic theory, and let φ denote the endomorphism obtained from the stack-theoretic approach. Let

$$\mathcal{L}og(F_{(\mathrm{Spec}(W_n\langle t\rangle), M_{W_n\langle t\rangle})}) : \mathcal{L}og_{(\mathrm{Spec}(W_n\langle t\rangle), M_{W_n\langle t\rangle})} \longrightarrow \mathcal{L}og_{(\mathrm{Spec}(W_n\langle t\rangle), M_{W_n\langle t\rangle})} \tag{9.6.3.1}$$

be the morphism sending a morphism of fine log schemes

$$(T, M_T) \longrightarrow (\mathrm{Spec}(W_n\langle t\rangle), M_{W_n\langle t\rangle}) \tag{9.6.3.2}$$

to the composite

$$(T, M_T) \longrightarrow (\mathrm{Spec}(W_n\langle t\rangle), M_{W_n\langle t\rangle}) \xrightarrow{F_{(\mathrm{Spec}(W_n\langle t\rangle), M_{W_n\langle t\rangle})}} (\mathrm{Spec}(W_n\langle t\rangle), M_{W_n\langle t\rangle}). \tag{9.6.3.3}$$

Then $\varphi^{\log}$ is obtained from functoriality and the commutative diagram

$$\begin{array}{ccc} \mathcal{Y} & \xrightarrow{F_{\mathcal{Y}}} & \mathcal{Y} \\ \downarrow & & \downarrow \\ \mathcal{L}og_{(\mathrm{Spec}(W_n\langle t\rangle), M_{W_n\langle t\rangle})} & \xrightarrow{\mathcal{L}og(F_{(\mathrm{Spec}(W_n\langle t\rangle), M_{W_n\langle t\rangle})})} & \mathcal{L}og_{(\mathrm{Spec}(W_n\langle t\rangle), M_{W_n\langle t\rangle})}. \end{array} \tag{9.6.3.4}$$

Thus to prove that $\varphi^{\log} = \varphi$ it suffices to show that the diagram

$$\begin{array}{ccc} \mathcal{S}_{W_n\langle t\rangle} & \xrightarrow{\Lambda_p \otimes F_{W_n\langle t\rangle}} & \mathcal{S}_{W_n\langle t\rangle} \\ {\scriptstyle 9.1.21}\downarrow & & \downarrow{\scriptstyle 9.1.21} \\ \mathcal{L}og_{(\mathrm{Spec}(W_n\langle t\rangle), M_{W_n\langle t\rangle})} & \xrightarrow{\mathcal{L}og(F_{(\mathrm{Spec}(W_n\langle t\rangle), M_{W_n\langle t\rangle})})} & \mathcal{L}og_{(\mathrm{Spec}(W_n\langle t\rangle), M_{W_n\langle t\rangle})} \end{array} \tag{9.6.3.5}$$

commutes which follows from 9.1.34. □

Corollary 9.6.4. — *The isomorphism $D_\cdot \otimes W\langle t\rangle \simeq E_\cdot$ in $\mathrm{ps}(W\langle t\rangle)_{\mathbb{Q}}$ constructed in 6.4.4 agrees under the isomorphisms in 9.6.3 with the isomorphism $D_\cdot^{\log} \otimes W\langle t\rangle \simeq E_\cdot^{\log}$ constructed in* [**31**, 4.13].

Proof. — We have a commutative diagram of topoi (see the discussion in 9.2.10)

$$\begin{array}{ccc} (\mathcal{Y}_{\mathrm{et}}/\mathcal{L}og_{(\mathrm{Spec}(W_n), M_{W_n})})_{\mathrm{cris}} & \longrightarrow & (\mathcal{Y}_{\mathrm{et}}/\mathcal{L}og_{(\mathrm{Spec}(W_n\langle t\rangle), M_{W_n\langle t\rangle})})_{\mathrm{cris}} \\ \downarrow{\scriptstyle 9.2.2} & & \downarrow{\scriptstyle 9.2.2} \\ ((\mathcal{Y}, M_{\mathcal{Y}})/(\mathrm{Spec}(W_n), M_{W_n}))_{\mathrm{cris}} & \longrightarrow & ((\mathcal{Y}, M_{\mathcal{Y}})/(\mathrm{Spec}(W_n\langle t\rangle), M_{W_n\langle t\rangle}))_{\mathrm{cris}} \end{array} \tag{9.6.4.1}$$

and therefore a commutative diagram in $\mathrm{ps}(W)_{\mathbb{Q}}$

$$\begin{array}{ccc} E_\cdot & \xrightarrow{\text{projection}} & D_\cdot \\ {\scriptstyle 9.6.3}\downarrow & & \downarrow{\scriptstyle 9.6.3} \\ E_\cdot^{\log} & \xrightarrow{\text{projection}} & D_\cdot^{\log}. \end{array} \tag{9.6.4.2}$$

The isomorphism $D_\cdot^{\log} \otimes W\langle t\rangle \to E_\cdot^{\log}$ therefore defines a section of $E_\cdot \to D_\cdot$ compatible with Frobenius. By the uniqueness statement in 5.3.17 (i) it follows that this section agrees with the one defined in 6.4.4. □

Proposition 9.6.5. — *The isomorphism $D_\cdot \otimes_{W_n} R \simeq C_\cdot$ in $\mathrm{ps}(R)_{\mathbb{Q}}$ constructed in 6.4.6 agrees under the isomorphisms in 9.6.3 with the one constructed in* [**31**, 5.2].

Proof. — As in 6.4.7, let $g_w : W\langle t\rangle \to R$ be the map which is equal to σ^w on W and sends t to t^{p^w}. Note that this extends naturally to a morphism of log schemes

$$(\mathrm{Spec}(R), M_R) \longrightarrow (\mathrm{Spec}(W\langle t\rangle), M_{W\langle t\rangle}) \tag{9.6.5.1}$$

which we again denote by g_w. Since $(\mathrm{Spec}(k), M_k) \hookrightarrow (\mathrm{Spec}(W\langle t\rangle), M_{W\langle t\rangle})$ is a PD-immersion, there exists a unique dotted arrow d filling in the following diagram

$$\begin{array}{ccc} \mathrm{Spec}(k) & \hookrightarrow & \mathrm{Spec}(A_1) \\ \downarrow & & \downarrow \\ \mathrm{Spec}(W\langle t\rangle) & \overset{d}{\dashrightarrow} & \mathrm{Spec}(R) \\ & \searrow^{\mathrm{id}} & \downarrow \\ & & \mathrm{Spec}(W\langle t\rangle). \end{array} \tag{9.6.5.2}$$

As in 6.4.7, choose w such that $(\pi^{p^w}) \subset pA$. Then the p^w-th power Frobenius morphism $F^w_{A_1} : \mathrm{Spec}(A_1) \to \mathrm{Spec}(A_1)$ factors through k, and therefore we obtain a commutative diagram of log schemes

$$\begin{array}{ccccc} (\mathcal{X}_0, M_{\mathcal{X}_0}) & \longrightarrow & (\mathcal{Y}, M_{\mathcal{Y}}) & \longrightarrow & (\mathcal{X}_0, M_{\mathcal{X}_0}) \\ \downarrow & & \downarrow & & \downarrow \\ (\mathrm{Spec}(A_1), M_{A_1}) & \longrightarrow & (\mathrm{Spec}(k), M_k) & \hookrightarrow & (\mathrm{Spec}(A_1), M_{A_1}) \\ \downarrow & & \downarrow & & \downarrow \\ (\mathrm{Spec}(R), M_R) & \xrightarrow{g_w} & (\mathrm{Spec}(W\langle t\rangle), M_{W\langle t\rangle}) & \xrightarrow{d} & \mathrm{Spec}(R). \end{array} \tag{9.6.5.3}$$

Let

$$\tau^{\log} : R \otimes_{g_w, W\langle t\rangle} E^{\log}_\cdot \longrightarrow C^{\log}_\cdot \tag{9.6.5.4}$$

be the map in $\mathrm{ps}(R)_{\mathbb{Q}}$ defined by the left side of the diagram (9.6.5.3). By [**31**, 5.3] the map $\tau^{\log}$ is an isomorphism. The isomorphism $D^{\log} \otimes_W R \simeq C^{\log}_\cdot$ defined in [**31**, 5.2] is by the construction in *loc. cit.* the composite morphism

$$\begin{aligned} C^{\log}_\cdot &\overset{(\tau^{\log})^{-1}}{\simeq} R \otimes_{g_w, W\langle t\rangle} E^{\log}_\cdot \\ &\overset{[\mathbf{31},\, 4.13]}{\simeq} R \otimes_{\varphi^r, W} D^{\log}_\cdot \\ &\overset{1\otimes\varphi^r}{\simeq} R \otimes_W D^{\log}_\cdot. \end{aligned} \tag{9.6.5.5}$$

On the other hand, the isomorphism (6.4.6) is constructed as follows. With notation as in 6.4.7, let

$$H : \mathcal{S}_R \longrightarrow \mathcal{S}^{(w)}_R \tag{9.6.5.6}$$

denote the map

$$H := \tilde{F}_{\mathcal{S}(1)/R} \circ \tilde{F}_{\mathcal{S}(2)/R} \circ \cdots \circ \tilde{F}_{\mathcal{S}(w)/R} \tag{9.6.5.7}$$

as in (6.4.8.5). We then have a commutative diagram (compare with (6.4.7.3))

$$\begin{array}{ccccc} \mathcal{X} & \xrightarrow{F^w_{\mathcal{X}/A_1}} & \overline{\mathcal{X}}^{(p^w)} & \longrightarrow & \mathcal{Y} \\ \downarrow & & \downarrow & & \downarrow \\ \mathcal{S}_R & \xrightarrow{H} & \mathcal{S}_R^{(w)} & \longrightarrow & \mathcal{S}_{W\langle t\rangle} \\ & \searrow & \downarrow & & \downarrow \\ & & \operatorname{Spec}(R) & \xrightarrow{g_w} & \operatorname{Spec}(W\langle t\rangle). \end{array} \tag{9.6.5.8}$$

Let

$$\tau : R \otimes_{g_w, W\langle t\rangle} E_{\cdot} \longrightarrow C_{\cdot} \tag{9.6.5.9}$$

be the resulting morphism in $\operatorname{ps}(R)_{\mathbb{Q}}$. By construction this map is equal to the composite

$$R \otimes_{g_w, W\langle t\rangle} E_{\cdot} \xrightarrow{(6.4.8.3)} C_{\cdot}^{(w)} \xrightarrow{(6.4.8.2)} C_{\cdot}. \tag{9.6.5.10}$$

By the construction in (6.4.8.3) the isomorphism $D_{\cdot} \otimes_W R \simeq C_{\cdot}$ is equal to the composite

$$\begin{aligned} C_{\cdot} &\overset{\tau^{-1}}{\simeq} R \otimes_{g_w, W\langle t\rangle} E_{\cdot} \\ &\overset{6.5}{\simeq} R \otimes_{\varphi^r, W} D_{\cdot} \\ &\overset{1\otimes\varphi^r}{\simeq} R \otimes_W D_{\cdot}. \end{aligned} \tag{9.6.5.11}$$

To prove the proposition it therefore suffices to show that the following three squares commute

$$\begin{array}{ccc} R \otimes_W D_{\cdot}^{\log} & \xrightarrow{1\otimes\varphi^r} & R \otimes_{\varphi^r, W} D_{\cdot}^{\log} \\ \downarrow{\scriptstyle 9.6.3} & & \downarrow{\scriptstyle 9.6.3} \\ R \otimes_W D_{\cdot} & \xrightarrow{1\otimes\varphi^r} & R \otimes_{\varphi^r, W} D_{\cdot}, \end{array} \tag{9.6.5.12}$$

$$\begin{array}{ccc} R \otimes_{\varphi^r, W} D_{\cdot}^{\log} & \xrightarrow{[\mathbf{31}, 4.13]} & R \otimes_{g_w, W\langle t\rangle} E_{\cdot}^{\log} \\ \downarrow{\scriptstyle 9.6.3} & & \downarrow{\scriptstyle 9.6.3} \\ R \otimes_{\varphi^r, W} D_{\cdot} & \xrightarrow{6.5} & R \otimes_{g_w, W\langle t\rangle} E_{\cdot}, \end{array} \tag{9.6.5.13}$$

and

$$(9.6.5.14)\qquad \begin{array}{ccc} R\otimes_{g_w,W\langle t\rangle} E^{\log}_{\cdot} & \xrightarrow{\tau^{\log}} & C^{\log}_{\cdot} \\ \big\downarrow{\scriptstyle 9.6.3} & & \big\downarrow{\scriptstyle 9.6.3} \\ R\otimes_{g_w,W\langle t\rangle} E_{\cdot} & \xrightarrow{\tau} & C_{\cdot}. \end{array}$$

The diagram (9.6.5.12) commutes by the compatibility of 9.6.3 with the Frobenius endomorphisms, and (9.6.5.13) commutes by 9.6.4. Finally the commutativity of (9.6.5.14) follows from an argument similar to the proof of 9.6.3 which we leave to the reader. □

Lemma 9.6.6. — *The monodromy operator N on $D_{\cdot}$ constructed in 6.5 agrees under the isomorphism in 9.6.3 with the monodromy operator $N^{\log}$ on $D^{\log}_{\cdot}$ constructed in* [**31**, §3].

Proof. — Let $D = \operatorname{Spec}(W_n\langle u-1\rangle)$, and let M_D be the log structure on D induced by the map $\mathbb{N}\to\mathcal{O}_D$ sending 1 to 0. There are natural projections

$$(9.6.6.1)\qquad \rho, \operatorname{pr} : (D, M_D) \longrightarrow (\operatorname{Spec}(W_n), M_{W_n})$$

over W induced by the morphisms of log structures $M_{W_n}\to M_D$ induced by the maps $\mathbb{N}\to\mathcal{O}_D^*\oplus\mathbb{N}$ sending 1 to $(u,1)$ and $(1,1)$ respectively. There is also a natural closed immersion $(\operatorname{Spec}(W_n), M_{W_n})\hookrightarrow(D, M_D)$ obtained from the map $W_n\langle u-1\rangle\to W_n$ sending u to 1. Let $(D^{(1)}, M_{D^{(1)}})$ be the first infinitesimal neighborhood of the diagonal, and set

$$(9.6.6.2)\qquad K_n^{\log} := R\Gamma\big(((\mathcal{Y}, M_{\mathcal{Y}})/\mathcal{L}og_{(\operatorname{Spec}(W_n),M_{W_n})})_{\mathrm{cris}}, \mathcal{O}_{(\mathcal{Y},M_{\mathcal{Y}})/\mathcal{L}og_{(\operatorname{Spec}(W_n),M_{W_n})}}\big),$$

$$(9.6.6.3)\qquad K_n'^{\log} := R\Gamma\big(((\mathcal{Y}, M_{\mathcal{Y}})/\mathcal{L}og_{(D^{(1)},M_{D^{(1)}})})_{\mathrm{cris}}, \mathcal{O}_{(\mathcal{Y},M_{\mathcal{Y}})/\mathcal{L}og_{(D^{(1)},M_{D^{(1)}})}}\big).$$

Note that there is a canonical isomorphism $\mathcal{O}_{D^{(1)}}\simeq W_n[(u-1)]/(u-1)^2$. Then the base change maps

$$(9.6.6.4)\qquad \rho^*, \operatorname{pr}^* : K_n^{\log}\otimes_{W_n} W_n[(u-1)]/(u-1)^2 \longrightarrow K'^{\log}$$

are isomorphisms. Let $\lambda : W_n[(u-1)]/(u-1)^2\to W_n\cdot(u-1)$ be the map sending $a+b\cdot(u-1)$ to b. Then by [**31**, 3.5] the monodromy operator $N^{\log}$ is induced by the composite
(9.6.6.5)

$$K_n^{\log} \xrightarrow{\rho^*} K_n'^{\log} \xrightarrow{(\operatorname{pr}^*)^{-1}} K_n^{\log}\otimes W_n[(u-1)]/(u-1)^2 \xrightarrow{\lambda} K_n^{\log}\cdot(u-1).$$

Propositions 9.1.17 and 9.1.21 provide a natural morphism of diagrams of algebraic stacks from
(9.6.6.6)

$$\begin{array}{ccccc} \mathcal{S}_H(\alpha)_{W_n\langle t\rangle} & \xrightarrow{\Delta} & \mathcal{S}_H(\alpha)_{W_n\langle t\rangle}\otimes_W W[(u-1)]/(u-1)^2 & \xrightarrow{\operatorname{pr}_1} & \mathcal{S}_H(\alpha)_{W_n\langle t\rangle} \\ & & \big\downarrow{\scriptstyle\rho} & & \big\downarrow \\ & & \mathcal{S}_H(\alpha)_{W_n\langle t\rangle} & \longrightarrow & \overline{\mathcal{S}}_H(\alpha)_{W_n\langle t\rangle} \end{array}$$

defined as in (6.5.1.6) to
(9.6.6.7)

$$\begin{array}{ccccc} \mathcal{L}og_{(\mathrm{Spec}(W_n), M_{W_n})} & \longrightarrow & \mathcal{L}og_{(D^{(1)}, M_{D^{(1)}})} & \xrightarrow{\mathrm{pr}} & \mathcal{L}og_{(\mathrm{Spec}(W_n), M_{W_n})} \\ & & \rho \big\downarrow & & \big\downarrow \\ & & \mathcal{L}og_{(\mathrm{Spec}(W_n), M_{W_n})} & \longrightarrow & \mathcal{L}og_{(\mathrm{Spec}(W_n), \mathcal{O}^*_{\mathrm{Spec}(W_n)})}. \end{array}$$

This morphism of diagrams identifies (9.6.6.5) with the composite (6.5.1.12). From this the lemma follows. □

Corollary 9.6.7. — *The (φ, N, G)-structure on $H^*_{\mathrm{dR}}(\mathcal{X}_K/K)$ constructed in 6.4 and 6.5 agrees with the one constructed in* [**31**].

9.6.8. — Next we consider the construction of Chapter 7.

Let $\mathcal{X}/A$ be a proper, tame Deligne-Mumford stack with a smooth morphism $\mathcal{X} \to \mathcal{S}_H(\alpha)_A$, for some α and H as in Chapter 7, and assume that the generic fiber of $\mathcal{X}$ is a scheme.

Theorem 9.6.9. — *The (φ, N, G)-structure on $H^*_{\mathrm{dR}}(\mathcal{X}_K/K)$ constructed in Chapter 7 agrees with the one constructed in 8.5.*

The proof is in several steps 9.6.10–9.6.17.

9.6.10. — Note first that as mentioned in 7.1.11, when $\mathcal{X}$ is semistable the structure constructed in Chapter 7 agrees with the one obtained in 6.4. Therefore, by the above comparison with the logarithmic approach in this situation, the theorem holds when $\mathcal{X}$ is semistable.

9.6.11. — For general $\mathcal{X}$ we use an argument using alterations as follows. Let $(\widetilde{D}^m, \tilde{\varphi}, \widetilde{N})$ denote the (φ, N, G)-structure on $H^m_{\mathrm{dR}}(\mathcal{X}_K/K)$ constructed in Chapter 7, and let (D^m, φ, N) denote the one constructed in 8.5. The choice of a uniformizer $\pi \in A$ gives isomorphisms

(9.6.11.1) $$\widetilde{D}^m \otimes_{K_0^{ur}} \overline{K} \simeq H^m_{\mathrm{dR}}(\mathcal{X}_K/K) \otimes_K \overline{K}, \quad D^m \otimes_{K_0^{ur}} \overline{K} \simeq H^m_{\mathrm{dR}}(\mathcal{X}_K/K) \otimes_K \overline{K},$$

and hence also a Galois equivariant isomorphism

(9.6.11.2) $$\iota : \widetilde{D}^m \otimes_{K_0^{ur}} \overline{K} \longrightarrow D^m \otimes_{K_0^{ur}} \overline{K}.$$

We must show that $\iota(\widetilde{D}^m) = D^m$ and that the resulting isomorphism $\iota_0 : \widetilde{D}^m \to D^m$ is compatible with the Frobenii and monodromy operators. Observe also that since $\widetilde{D}^m$ and D^m have the same dimension it suffices to show that $\iota(\widetilde{D}^m) \subset D^m$.

For this we can without loss of generality replace K by a finite extension, and hence by the construction of $(\widetilde{D}^m, \tilde{\varphi}, \widetilde{N})$ may also assume that $\mathcal{X}$ is a tame regular Deligne-Mumford stack whose reduced closed fiber is a divisor with normal crossings and all the multiplicities of the components of the closed fiber are powers of p.

9.6.12. — By [**49**, 16.6.1], there exists a generically finite proper surjective map $Y \to \mathfrak{X}$ with Y a reduced scheme. By 8.5.7, there exist therefore a finite extension $K \subset K' \subset \overline{K}$, a semistable scheme $\mathfrak{X}'/A'$ (where A' is the ring of integers of K'), and an alteration $\phi : \mathfrak{X}' \to \mathfrak{X}$ over $\mathrm{Spec}(A') \to \mathrm{Spec}(A)$. Let $(D^m_{\mathfrak{X}'}, \varphi_{\mathfrak{X}'}, N_{\mathfrak{X}'})$ denote the (φ, N, G)-structure on $H^m_{\mathrm{dR}}(\mathfrak{X}'_{K'}/K')$. By the construction in 8.5.11, the natural map $H^m_{\mathrm{dR}}(\mathfrak{X}_K/K) \to H^m_{\mathrm{dR}}(\mathfrak{X}'_{K'}/K')$ is injective and admits a retraction $r : H^m_{\mathrm{dR}}(\mathfrak{X}'_{K'}/K') \to H^m_{\mathrm{dR}}(\mathfrak{X}_K/K) \otimes_K K'$ compatible with the (φ, N, G)-structures (D^m, φ, N) and $(D^m_{\mathfrak{X}'}, \varphi_{\mathfrak{X}'}, N_{\mathfrak{X}'})$. In particular,

$$D^m = D^m_{\mathfrak{X}'} \cap (H^m_{\mathrm{dR}}(\mathfrak{X}_K/K) \otimes \overline{K}) \subset H^m_{\mathrm{dR}}(\mathfrak{X}'_{K'}/K') \otimes_{K'} \overline{K}. \quad (9.6.12.1)$$

Thus to prove that $\iota(\widetilde{D}^m) \subset D^m$ and that the resulting map $\iota_0 : \widetilde{D}^m \to D^m$ is compatible with φ and N, it suffices to show that the image of $\widetilde{D}^m$ in $H^m_{\mathrm{dR}}(\mathfrak{X}'_{K'}/K') \otimes_{K'} \overline{K}$ is contained in $D^m_{\mathfrak{X}'}$ and that the resulting map $j : \widetilde{D}^m \to D^m_{\mathfrak{X}'}$ is compatible with the Frobenei and monodromy operators.

9.6.13. — Let $M_{\mathfrak{X}}$ (resp. $M_{\mathfrak{X}'}$, $M_{A'}$, M_A) be the log structure on $\mathfrak{X}$ (resp. $\mathfrak{X}'$, $\mathrm{Spec}(A')$, $\mathrm{Spec}(A)$) defined by the closed fiber so that there is a commutative diagram of fine log schemes

$$\begin{array}{ccc} (\mathfrak{X}', M_{\mathfrak{X}'}) & \xrightarrow{\phi} & (\mathfrak{X}, M_{\mathfrak{X}}) \\ {\scriptstyle f}\downarrow & & \downarrow{\scriptstyle g} \\ (\mathrm{Spec}(A'), M_{A'}) & \longrightarrow & (\mathrm{Spec}(A), M_A), \end{array} \quad (9.6.13.1)$$

where f and g are log smooth.

9.6.14. — The key point to the comparison, is that in the construction in Chapter 7, we can replace the stack $\mathcal{S}_H(\alpha)$ by a variant constructed using the stack $\mathcal{L}og^{\square}$ defined in 9.1.30.

For a log structure M on a scheme W, let $M^{(r)}$ denote the log structure associated to the prelog structure

$$M \xrightarrow{\times r} M \longrightarrow \mathcal{O}_W. \quad (9.6.14.1)$$

Let $\pi' \in A'$ be a uniformizer. For an integer r let $\mathcal{L}_{A',(\pi'^r)}$ denote the fiber product of the diagram

$$\begin{array}{ccc} & & \mathcal{L}og^{\square} \\ & & \downarrow{\scriptstyle \tau_2} \\ \mathcal{L}og_{(\mathrm{Spec}(A'), M^{(r)}_{A'})} & \xrightarrow{c_r} & \mathcal{L}og^{[2]}, \end{array} \quad (9.6.14.2)$$

where τ_2 is as in 9.1.31 (ii) and the map c_r sends a morphism $M^{(r)}_{A'}|_T \to M$ over a A'-scheme T to the diagram $M^{(r)}_A|_T \to M^{(r)}_{A'}|_T \to M$. The projection maps $\mathcal{L}_{A',(\pi'^r)} \to \mathcal{L}og_{(\mathrm{Spec}(A'), M^{(r)}_{A'})}$ are all étale by 9.1.31 (ii).

Fix an integer w. Recall from 9.1.34 that for any finite category Γ the functor $M \mapsto M^{(p^w)}$ induces a functor

(9.6.14.3) $$\Lambda_{p^w} : \mathcal{L}og^{\Gamma} \longrightarrow \mathcal{L}og^{\Gamma}.$$

Moreover, if $f : \Gamma' \to \Gamma$ is a functor between finite categories then the induced diagram

(9.6.14.4) $$\begin{array}{ccc} \mathcal{L}og^{\Gamma} & \xrightarrow{\Lambda_{p^w}} & \mathcal{L}og^{\Gamma} \\ \downarrow{\scriptstyle f^*} & & \downarrow{\scriptstyle f^*} \\ \mathcal{L}og^{\Gamma'} & \xrightarrow{\Lambda_{p^w}} & \mathcal{L}og^{\Gamma'} \end{array}$$

commutes. Let $\delta_1 : [0] \to [1]$ denote the map sending 0 to 0 so that we have

(9.6.14.5) $$\mathcal{L}og_{(\mathrm{Spec}(A'), M_{A'}^{(r)})} = \mathrm{Spec}(A') \times_{M_{A'}^{(r)}, \mathcal{L}og^{[0]}, \delta_1^*} \mathcal{L}og^{[1]}.$$

From the commutative diagram

(9.6.14.6) $$\begin{array}{ccccc} & & & & \mathcal{L}og^{[1]} \\ & & & \swarrow{\scriptstyle \delta_1^*} & \downarrow{\scriptstyle \Lambda_{p^w}} \\ \mathrm{Spec}(A') & \xrightarrow{M_{A'}^{(r)}} & \mathcal{L}og^{[0]} & & \mathcal{L}og^{[1]} \\ & \searrow{\scriptstyle M_{A'}^{(rp^w)}} & \downarrow{\scriptstyle \Lambda_{p^w}} & \swarrow{\scriptstyle \delta_1^*} & \\ & & \mathcal{L}og^{[0]} & & \end{array}$$

we therefore obtain a morphism

(9.6.14.7) $$\mathcal{L}og_{(\mathrm{Spec}(A'), M_{A'}^{(r)})} \to \mathcal{L}og_{(\mathrm{Spec}(A'), M_{A'}^{(rp^w)})},$$

which we again denote by Λ_{p^w}. It follows immediately from the definition of c_r that the following diagram commutes

(9.6.14.8) $$\begin{array}{ccccc} & & & & \mathcal{L}og^{\square} \\ & & & \swarrow{\scriptstyle \tau_2} & \downarrow{\scriptstyle \Lambda_{p^w}} \\ \mathcal{L}og_{(\mathrm{Spec}(A'), M_{A'}^{(r)})} & \xrightarrow{c_r} & \mathcal{L}og^{[2]} & & \mathcal{L}og^{\square} \\ \downarrow{\scriptstyle \Lambda_{p^w}} & & \downarrow{\scriptstyle \Lambda_{p^w}} & \swarrow{\scriptstyle \tau_2} & \\ \mathcal{L}og_{(\mathrm{Spec}(A'), M_{A'}^{(rp^w)})} & \xrightarrow{c_{rp^w}} & \mathcal{L}og^{[2]}, & & \end{array}$$

and therefore we also obtain a map (which we again abusively denote by Λ_{p^w})

(9.6.14.9) $$\Lambda_{p^w} : \mathfrak{L}_{A', (\pi'^r)} \longrightarrow \mathfrak{L}_{A', (\pi'^{rp^w})}$$

such that the diagram

$$\begin{array}{ccc} \mathcal{L}_{A',(\pi'^r)} & \xrightarrow{\Lambda_{p^w}} & \mathcal{L}_{A',(\pi'^{rp^w})} \\ \downarrow & & \downarrow \\ \mathcal{L}og_{(\mathrm{Spec}(A'),M_{A'}^{(r)})} & \xrightarrow{\Lambda_{p^w}} & \mathcal{L}og_{(\mathrm{Spec}(A'),M_{A'}^{(rp^w)})} \end{array} \tag{9.6.14.10}$$

commutes.

Let W' denote the ring of Witt vectors of the residue field k' of A', and let e be the ramification induced of K'/K. Let $M_{W'}$ (resp. M_W) be the log structure on $\mathrm{Spec}(W')$ (resp. $\mathrm{Spec}(W)$) associated to the map $\mathbb{N} \to W'$ (resp. $\mathbb{N} \to W$) sending 1 to 0. Let $\mathcal{Y}'$ (resp. $\mathcal{Y}$) denote the reduction of $\mathcal{X}'$ (resp. $\mathcal{X}$) to k' (resp. k). There is a natural commutative diagram of log schemes

$$\begin{array}{ccc} (\mathcal{Y}', M_{\mathcal{Y}'}) & \xrightarrow{\phi} & (\mathcal{Y}, M_{\mathcal{Y}}) \\ \downarrow & & \downarrow \\ (\mathrm{Spec}(W'), M_{W'}) & \longrightarrow & (\mathrm{Spec}(W), M_W), \end{array} \tag{9.6.14.11}$$

where the map $M_W|_{W'} \to M_{W'}$ is induced by multiplication by e on $\mathbb{N}$.

Let $\mathcal{L}_{W',(0)}$ denote the fiber product of the diagram

$$\begin{array}{ccc} & & \mathcal{L}^{\square} \\ & & \downarrow \tau_2 \\ \mathcal{L}og_{(\mathrm{Spec}(W'),M_{W'})} & \xrightarrow{c} & \mathcal{L}^{[2]}, \end{array} \tag{9.6.14.12}$$

where c sends $M_{W'}|_T \to M$ over some T to $M_W|_T \to M_{W'}|_T \to M$. Again the map $\mathcal{L}_{W',(0)} \to \mathcal{L}og_{(\mathrm{Spec}(W'),M_{W'})}$ is étale.

The stack $\mathcal{L}_{W',(0)}$ is a $\mathbb{G}_m$-torsor over another stack $\overline{\mathcal{L}}_{W'}$ defined as follows. For any ring R, let $\mathcal{L}og_{(B\mathbb{G}_{m,R},M_{B\mathbb{G}_{m,R}})}$ be the closed substack of $\mathcal{L}^{[1]}$ which to any R-scheme T associates the groupoid of morphisms of fine log structures $M_1 \to M_2$ on T, such that the image of every non-zero section of M_1 in $\mathcal{O}_T$ is zero and the sheaf $\overline{M}_1$ is isomorphic to the constant sheaf $\mathbb{N}$. For any integer n there is a natural map

$$\theta_n : \mathcal{L}og_{(B\mathbb{G}_{m,R},M_{B\mathbb{G}_{m,R}})} \longrightarrow \mathcal{L}og_{(B\mathbb{G}_{m,R},M_{B\mathbb{G}_{m,R}})} \tag{9.6.14.13}$$

sending a morphism $M_1 \to M_2$ to the composite $M_1^{(n)} \to M_1 \to M_2$. The map c above then factors as

$$\mathcal{L}og_{(\mathrm{Spec}(W'),M_{W'})} \longrightarrow \mathcal{L}og_{(B\mathbb{G}_{m,W'},M_{B\mathbb{G}_{m,W'}})} \longrightarrow \mathcal{L}^{[2]}, \tag{9.6.14.14}$$

and $\overline{\mathcal{L}}_{W',(0)}$ is defined to be the cartesian product

$$\overline{\mathcal{L}}_{W',(0)} := \mathcal{L}og_{(B\mathbb{G}_{m,W'},M_{B\mathbb{G}_{m,W'}})} \times_{\mathcal{L}^{[2]}} \mathcal{L}^{\square}. \tag{9.6.14.15}$$

The commutative diagram of log structures

$$\begin{array}{ccc} M_A|_{\mathcal{X}'} & \longrightarrow & M_{\mathcal{X}}|_{\mathcal{X}'} \\ \downarrow & & \downarrow \\ M_{A'}|_{\mathcal{X}'} & \longrightarrow & M_{\mathcal{X}'} \end{array} \tag{9.6.14.16}$$

induces a morphism $\mathcal{X}' \to \mathcal{L}_{A',(\pi)}$. There is a commutative diagram

$$\begin{array}{ccc} \mathcal{Y}' & & \\ \downarrow & & \\ \mathcal{L}_{W',(0)} & \longrightarrow & \mathcal{L}og_{(\mathrm{Spec}(W'),M_{W'})} \\ \downarrow & & \downarrow \\ \overline{\mathcal{L}}_{W',(0)} & \longrightarrow & \mathcal{L}og_{(B\mathbb{G}_{m,W'},M_{B\mathbb{G}_{m,W'}})}. \end{array} \tag{9.6.14.17}$$

Remark 9.6.15. — As explained in [**62**, 5.1], one can define a notion of fine log structure on an algebraic stack. There is a natural such log structure $M_{B\mathbb{G}_m}$ on the stack $B\mathbb{G}_m$. To describe this log structure, let N be the log structure on $\mathrm{Spec}(\mathbb{Z})$ associated to the map $\mathbb{N} \to \mathbb{Z}$ sending 1 to 0. The trivial action of $\mathbb{G}_m$ on $\mathrm{Spec}(\mathbb{Z})$ extends to an action on the log scheme $(\mathrm{Spec}(\mathbb{Z}), N)$ by associating to a T-valued point $u \in \mathbb{G}_m(T)$ the automorphism of $N|_T$ induced by the map $\mathbb{N} \to \mathcal{O}_T^* \oplus \mathbb{N}$ sending 1 to $(u, 1)$. The stack $\mathcal{L}og_{(B\mathbb{G}_m,M_{\mathbb{G}_m})}$ is then equal to a generalization of the stacks $\mathcal{L}og_{(S,M_S)}$ obtained by replacing (S, M_S) by an algebraic stack with a log structure.

9.6.16. — In summary, the stacks $\mathcal{L}_{A',(\pi'^r)}$ and $\mathcal{L}_{W',(0)}$ with the maps

$$\Lambda_{p^w} : \mathcal{L}_{A',(\pi'^r)} \to \mathcal{L}_{A',(\pi'^{rp^w})} \tag{9.6.16.1}$$

enjoy all the same formal properties as the stacks $\mathcal{S}_{A',(\pi'^r)}$ and $\mathcal{S}_{W',(0)}$ used in 7.1. Since the natural maps

$$\mathcal{L}_{A',(\pi'^r)} \longrightarrow \mathcal{L}og_{(\mathrm{Spec}(A'),M_{A'}^{(r)})}, \quad \mathcal{L}_{W',(0)} \longrightarrow \mathcal{L}og_{(\mathrm{Spec}(W'),M_{W'})} \tag{9.6.16.2}$$

are étale and compatible with the maps Λ_{p^w} and the $\mathbb{G}_m$-action, it follows that the (φ, N, G)-structure on $H^*_{\mathrm{dR}}(\mathcal{X}'_{K'}/K')$ can be constructed using the stacks $\mathcal{L}_{A',(\pi'^r)}$ and $\mathcal{L}_{W',(0)}$ throughout 7.1 instead of the stacks $\mathcal{S}_{A',(\pi'^r)}$ and $\mathcal{S}_{W',(0)}$.

9.6.17. — The map $\kappa_1 : \mathcal{L}^{\square} \to \mathcal{L}^{[1]}$ induces morphisms

$$\mathcal{L}_{A',(\pi'^r)} \longrightarrow \mathcal{L}og_{(\mathrm{Spec}(A),M_A^{(r)})}, \quad \mathcal{L}_{W',(0)} \longrightarrow \mathcal{L}og_{(\mathrm{Spec}(W),M_W)} \tag{9.6.17.1}$$

such that the diagram

$$\begin{array}{ccc} \mathcal{X}' & \longrightarrow & \mathcal{X} \\ \downarrow & & \downarrow \\ \mathcal{L}_{A',(\pi')} & \longrightarrow & \mathcal{L}og_{(\mathrm{Spec}(A),M_A)} \end{array} \tag{9.6.17.2}$$

commutes. This map κ_1 also induces a morphism $\bar{\theta} : \overline{\mathcal{L}}_{W',(0)} \to \mathcal{L}og_{(B\mathbb{G}_{m,W}, M_{B\mathbb{G}_{m,W}})}$ over the map

$$B\mathbb{G}_{m,W'} \longrightarrow B\mathbb{G}_{m,W} \tag{9.6.17.3}$$

induced by multiplication by $e_{K'/K}$ (the ramification index of K'/K) on $\mathbb{G}_m$. There is then also a commutative diagram

$$\begin{array}{ccc} \mathcal{Y}' & \longrightarrow & \mathcal{Y} \\ \downarrow & & \downarrow \\ \mathcal{L}_{W',(0)} & \longrightarrow & \mathcal{L}og_{(\mathrm{Spec}(W), M_W)} \\ \downarrow & & \downarrow \\ \overline{\mathcal{L}}_{W',(0)} & \xrightarrow{\bar{\theta}} & \mathcal{L}og_{(B\mathbb{G}_{m,W}, M_{B\mathbb{G}_{m,W}})}. \end{array} \tag{9.6.17.4}$$

From this it follows that there is an induced map

$$\begin{aligned} D^m &\simeq H^m\big((\mathcal{Y}/\mathcal{L}og_{(\mathrm{Spec}(W),M_W)})_{\mathrm{cris}}, \mathcal{O}_{\mathcal{Y}/\mathcal{L}og_{(\mathrm{Spec}(W),M_W)}}\big) \\ &\to H^m\big((\mathcal{Y}'/\mathcal{L}_{W',(0)})_{\mathrm{cris}}, \mathcal{O}_{\mathcal{Y}'/\mathcal{L}_{W',(0)}}\big) \\ &\simeq D^m_{\mathcal{X}'} \end{aligned} \tag{9.6.17.5}$$

compatible with Frobenius and that the diagram

$$\begin{array}{ccc} D^m & \longrightarrow & D^m_{\mathcal{X}'} \\ {\scriptstyle N}\downarrow & & \downarrow{\scriptstyle eN_{\mathcal{X}'}} \\ D^m & \longrightarrow & D^m_{\mathcal{X}'} \end{array} \tag{9.6.17.6}$$

commutes.

Finally that the map (9.6.17.5) is compatible with the isomorphisms

$$D^m \otimes_{K_0} K \simeq H^m_{\mathrm{dR}}(\mathcal{X}_K/K), \quad D^m_{\mathcal{X}'} \otimes_{K'_0} K' \simeq H^m_{\mathrm{dR}}(\mathcal{X}_{K'}/K') \tag{9.6.17.7}$$

follows from the construction of these isomorphisms (7.1.6), the commutativity of (9.6.17.2), and the fact that the morphisms $\mathcal{L}_{A',(\pi'^r)} \to \mathcal{L}og_{(\mathrm{Spec}(A), M_A^{(r)})}$ are compatible with the morphisms Λ_{p^w}. This completes the proof of 9.6.9.

9.7. Theorem 0.1.8 implies 0.1.7

Let $(\mathcal{X}, \mathcal{U})/V$ be a log smooth model as in 0.1.7, and let M_V be the log structure on Spec(V) defined by the closed point.

Definition 9.7.1. — A *modification* of a scheme $\mathcal{Z}$ is a proper morphism of schemes $\mathcal{Y} \to \mathcal{Z}$ which is an isomorphism over some dense open subset in $\mathcal{Z}$.

By 8.5.3 (ii) if $f : \mathcal{Y} \to \mathcal{X}$ is a modification of $\mathcal{X}$ over V with induced morphism $Y \to X$ over K and if Y/K is smooth then the map on étale cohomology groups

$$f^* : H^*(\overline{X}, \mathbb{Q}_p) \longrightarrow H^*(\overline{Y}, \mathbb{Q}_p) \tag{9.7.1.1}$$

is injective, where $\overline{X} := X_{\overline{K}}$ and $\overline{Y} := Y_{\overline{K}}$.

To deduce 0.1.7 from 0.1.8 it therefore suffices to prove the following.

Theorem 9.7.2. — *There exists a modification $\mathcal{Y} \to \mathcal{X}$ over V such that the generic fiber Y of $\mathcal{Y}$ satisfies the assumptions of 0.1.8 (and in fact we can take $\mathcal{Y}$ to be the log smooth model in 0.1.8).*

The proof is in steps 9.7.3–9.7.7.

Lemma 9.7.3. — *Let $\mathcal{X}'$ be the normalization of $\mathcal{X}$, and let $\mathcal{U}' \subset \mathcal{X}'$ be the inverse image of $\mathcal{U} \subset \mathcal{X}$. Then $(\mathcal{X}', \mathcal{U}')$ is also a log smooth model for X, and étale locally on $\mathcal{X}'$ there exists a morphism as in (0.1.4.6) with the monoid P saturated (*i.e.*, if $p \in P^{\mathrm{gp}}$ is an element and $np \in P$ for some $n \geq 1$ then $p \in P$).*

Proof. — The assertion is étale local on $\mathcal{X}$ so we may assume that $\mathcal{X}$ is equal to $\mathrm{Spec}(V \otimes_{\mathbb{Z}[\mathbb{N}]} \mathbb{Z}[P])$ for a fine monoid P. Let $P' \subset P^{\mathrm{gp}}$ be the set of elements $p \in P^{\mathrm{gp}}$ for which there exists an integer n with $np \in P$. The scheme

$$\widetilde{\mathcal{X}} := \mathrm{Spec}(V \otimes_{\mathbb{Z}[\mathbb{N}]} \mathbb{Z}[P']) \tag{9.7.3.1}$$

is finite over $\mathcal{X}$ and normal by [**42**, 8.2 and 4.1]. It follows that there is a unique finite birational $\mathcal{X}$-map $\widetilde{\mathcal{X}} \to \mathcal{X}'$ which since both schemes are normal is an isomorphism [**15**, III.4.4.9]. □

Replacing $\mathcal{X}$ by its normalization, we may therefore assume that $\mathcal{X}$ is normal and that étale locally there exists a morphism as in (0.1.4.6) with the monoid P saturated.

The following result makes 0.1.5 more precise.

Proposition 9.7.4. — *Let $M_{\mathcal{X}}$ be the log structure on $\mathcal{X}$ which to any étale $W \to \mathcal{X}$ associates the set of elements $f \in \Gamma(W, \mathcal{O}_W)$ whose restriction to $W \times_{\mathcal{X}} \mathcal{U}$ is invertible. Then $M_{\mathcal{X}}$ is a fine saturated log structure on $\mathcal{X}$, and the natural morphism $(\mathcal{X}, M_{\mathcal{X}}) \to (\mathrm{Spec}(V), M_V)$ is log smooth.*

Proof. — The assertion is étale local on $\mathcal{X}$, and hence we may assume

$$\mathcal{X} = \mathrm{Spec}(V \otimes_{\mathbb{Z}[\mathbb{N}]} \mathbb{Z}[P]) \tag{9.7.4.1}$$

with $\mathbb{N} \to P$ injective, $\mathrm{Coker}(\mathbb{Z} \to P^{\mathrm{gp}})$ p-torsion free, and P saturated. Let M' denote the log structure on $\mathcal{X}$ defined by the natural map $P \to V \otimes_{\mathbb{Z}[\mathbb{N}]} \mathbb{Z}[P]$. By our assumptions, the natural map $(\mathcal{X}, M') \to (\mathrm{Spec}(V), M_V)$ is log smooth. Hence by [**42**, 8.2] the log scheme $(\mathcal{X}, M')$ is log regular in the sense of [**42**, 2.1]. From this and [**55**, 2.6] it follows that $M_{\mathcal{X}} = M'$. □

9.7.5. — By [**55**, 5.6] there exists a proper log étale morphism $(\mathcal{X}', M_{\mathcal{X}'}) \to (\mathcal{X}, M_{\mathcal{X}})$ which is an isomorphism over a dense open set and such that for every geometric point $\bar{x} \to \mathcal{X}'$ the stalk $\overline{M}_{\mathcal{X}', \bar{x}}$ is a free monoid.

Lemma 9.7.6. — *The scheme $\mathfrak{X}'$ is regular and the reduced closed fiber is a divisor with normal crossings on $\mathfrak{X}'$.*

Proof. — We can without loss of generality replace V by an unramified extension. Hence since the residue field k is assumed perfect we may without loss of generality assume that k is algebraically closed.

That $\mathfrak{X}'$ is regular follows from [**55**, 5.2].

To describe the closed fiber, let $\bar{x} \to \mathfrak{X}'$ be a geometric point in the closed fiber. Choose an isomorphism $\mathbb{N}^r \to \overline{M}_{\mathfrak{X}',\bar{x}}$, and a lifting $\mathbb{N}^r \to M_{\mathfrak{X}',\bar{x}}$. Let $x_i \in \mathcal{O}_{\mathfrak{X}',\bar{x}}$ be the image of the i-th standard generator of $\mathbb{N}^r$, and let $I \subset \widehat{\mathcal{O}}_{\mathfrak{X}',\bar{x}}$ be the ideal $(x_1, \ldots, x_r)$. By [**42**, 8.2 and 2.1 (i)] the quotient $A := \widehat{\mathcal{O}}_{\mathfrak{X}',\bar{x}}/I$ is a complete regular local k-algebra with residue field k. By the Cohen Structure theorem [**16**, 7.7], the ring A is isomorphic to $k[\![t_1, \ldots, t_n]\!]$ for some integer n. Let $\tilde{t}_i \in \widehat{\mathcal{O}}_{\mathfrak{X}',\bar{x}}$ be a lifting of t_i.

Let $(a_1, \ldots, a_r) \in \mathbb{N}^r$ be the image of $1 \in \mathbb{N}$ under the composite

$$\mathbb{N} \simeq H^0(\mathrm{Spec}(V), \overline{M}_V) \longrightarrow \overline{M}_{\mathfrak{X}',\bar{x}} \simeq \mathbb{N}^r. \tag{9.7.6.1}$$

By the definition of of $(a_1, \ldots, a_r)$ and $(x_1, \ldots, x_r)$ there exists a unit $u \in \mathcal{O}_{\mathfrak{X}',\bar{x}}$ such that

$$x_1^{a_1} \cdots x_r^{a_r} u = \pi. \tag{9.7.6.2}$$

Also since $\bar{x}$ maps to the closed fiber of $\mathfrak{X}'$ not all a_i are zero. Let

$$\rho : V[\![X_1, \ldots, X_r, T_1, \ldots, T_n]\!] \to \widehat{\mathcal{O}}_{\mathfrak{X}',\bar{x}} \tag{9.7.6.3}$$

be the surjection sending X_i to x_i and T_i to $\tilde{t}_i$. If $\tilde{u} \in V[\![X_1, \ldots, X_r, T_1, \ldots, T_n]\!]^*$ is a lifting of u, then $\theta := X_1^{a_1} \cdots X_r^{a_r}\tilde{u} - \pi$ maps to zero under ρ. We thus obtain a surjection

$$\bar{\rho} : V[\![X_1, \ldots, X_r, T_1, \ldots, T_n]\!]/(\theta) \longrightarrow \widehat{\mathcal{O}}_{\mathfrak{X}',\bar{x}} \tag{9.7.6.4}$$

which we claim is an isomorphism.

For this observe first that the ring $V[\![X_1, \ldots, X_r, T_1, \ldots, T_n]\!]/(\theta)$ is an integral domain by [**42**, 3.4]. The dimension of this ring is $r + n$. On the other hand, the dimension of the ring $\widehat{\mathcal{O}}_{\mathfrak{X}',\bar{x}}$ is by [**42**, 2.1 and 8.2] also equal to $r + n$. From this it follows that the kernel of $\bar{\rho}$ is zero and hence $\bar{\rho}$ is an isomorphism. □

9.7.7. — To complete the reduction of 0.1.7 to 0.1.8 it remains to show that there exists a modification $\mathfrak{X}'' \to \mathfrak{X}'$ with $\mathfrak{X}''$ a regular scheme with reduced closed fiber a *simple* normal crossing divisor. This is done for example in [**43**, 4.2.12]. This therefore completes the proof of 9.7.2 and 0.1.7. □

Remark 9.7.8. — Assume $\mathfrak{U} \subset \mathfrak{X}$ is equal to the generic fiber and that $\mathfrak{X}$ is normal. Let $M_{\mathfrak{X}}$ be the log structure on $\mathfrak{X}$ defined in 9.7.4 so there is a log smooth morphism

$$(\mathfrak{X}, M_{\mathfrak{X}}) \longrightarrow (\mathrm{Spec}(V), M_V). \tag{9.7.8.1}$$

By [**55**, 5.6] the morphism $(\mathcal{X}', M_{\mathcal{X}'}) \to (\mathcal{X}, M_{\mathcal{X}})$ can taken to be a so-called log-blowup along a coherent ideal in $M_{\mathcal{X}}$. By [**75**, 2.4.3.3] the natural map on log crystalline cohomology

$$H^*_{\mathrm{cris}}((\mathcal{X}, M_{\mathcal{X}})/(\mathrm{Spec}(V), M_V)) \longrightarrow H^*_{\mathrm{cris}}((\mathcal{X}', M_{\mathcal{X}'})/(\mathrm{Spec}(V), M_V)) \tag{9.7.8.2}$$

is an isomorphism. In particular, if the closed fiber of $\mathcal{X}'$ is reduced, then the module D^m associated to $\mathcal{X}_K$ is given by

$$H^*_{\mathrm{cris}}((\mathcal{X}, M_{\mathcal{X}})/(\mathrm{Spec}(V), M_V)) \otimes_{K_0} K_0^{ur} \tag{9.7.8.3}$$

with Galois action given by the action on K_0^{ur}.

BIBLIOGRAPHY

[1] D. Abramovich, M. C. Olsson & A. Vistoli – "Tame stacks in positive characteristic", to appear in *Ann. Inst. Fourier (Grenoble)* **58** (2008), no. 4.

[2] D. Abramovich & A. Vistoli – "Compactifying the space of stable maps", *J. Amer. Math. Soc.* **15** (2002), no. 1, p. 27–75.

[3] M. Artin – "Algebraic approximation of structures over complete local rings", *Publ. Math. Inst. Hautes Études Sci.* **36** (1969), p. 23–58.

[4] ______, "Versal deformations and algebraic stacks", *Invent. Math.* **27** (1974), p. 165–189.

[5] M. Artin, A. Grothendieck & J.-L. Verdier – *Théorie des topos et cohomologie étale des schémas*, Lecture Notes in Math., vol. 269, 270 & 305, Springer Verlag, Berlin, 1972.

[6] P. Berthelot, A. Grothendieck & L. Illusie – *Théorie des intersections et théorème de Riemann-Roch (SGA 6)*, Lecture Notes in Math., vol. 225, Springer-Verlag, Berlin, 1971.

[7] P. Berthelot – *Cohomologie cristalline des schémas de caractéristique $p > 0$*, Lecture Notes in Math., vol. 407, Springer-Verlag, Berlin, 1974.

[8] P. Berthelot & A. Ogus – *Notes on crystalline cohomology*, Princeton University Press, Princeton, N.J., 1978.

[9] ______, "F-isocrystals and de Rham cohomology. I", *Invent. Math.* **72** (1983), no. 2, p. 159–199.

[10] S. Bloch – "Algebraic K-theory and crystalline cohomology", *Publ. Math. Inst. Hautes Études Sci.* **47** (1977), p. 187–268.

[11] B. Conrad & A. J. de Jong – "Approximation of versal deformations", *J. Algebra* **255** (2002), no. 2, p. 489–515.

[12] P. Deligne & M. Rapoport – "Les schémas de modules de courbes elliptiques", Modular functions of one variable, II (Proc. Internat. Summer School, Univ. Antwerp, Antwerp, 1972), Lecture Notes in Math., vol. 349, Springer, Berlin, 1973, p. 143–316.

[13] P. Deligne – "Théorie de Hodge. III", *Publ. Math. Inst. Hautes Études Sci.* **44** (1974), p. 5–77.

[14] M. Demazure & A. Grothendieck – *Schémas en groupes*, Lecture Notes in Math., vol. 151–153, Springer-Verlag, Berlin, 1970.

[15] J. Dieudonné & A. Grothendieck – "Éléments de géométrie algébrique", *Publ. Math. Inst. Hautes Études Sci.* **4**, **8**, **11**, **17**, **20**, **24**, **28** & **32** (1961–1967).

[16] D. Eisenbud – *Commutative algebra*, Grad. Texts in Math., vol. 150, Springer-Verlag, New York, 1995.

[17] T. Ekedahl – "On the multiplicative properties of the de Rham-Witt complex. I", *Ark. Mat.* **22** (1984), no. 2, p. 185–239.

[18] ———, "On the multiplicative properties of the de Rham-Witt complex. II", *Ark. Mat.* **23** (1985), no. 1, p. 53–102.

[19] G. Faltings – "p-adic Hodge theory", *J. Amer. Math. Soc.* **1** (1988), no. 1, p. 255–299.

[20] ———, "Crystalline cohomology and p-adic Galois-representations", Algebraic analysis, geometry, and number theory (Baltimore, MD, 1988), Johns Hopkins Univ. Press, Baltimore, MD, 1989, p. 25–80.

[21] ———, "Almost étale extensions", Cohomologies p-adiques et applications arithmétiques (II) (J.-M. Berthelot, P.and Fontaine, L. Illusie, K. Kato & M. Rapoport, eds.), Astérisque, vol. 279, Soc. Math. France, Paris, 2002, p. 185–270.

[22] J.-M. Fontaine – "Sur certains types de représentations p-adiques du groupe de Galois d'un corps local; construction d'un anneau de Barsotti-Tate", *Annals of Math.* **115** (1982), p. 529–577.

[23] ———, "Le corps des périodes p-adiques", Périodes p-adiques, Astérisque, vol. 223, Soc. Math. France, Paris, 1994, p. 59–111.

[24] ———, *Représentations p-adiques semi-stables*, Astérisque, vol. 223, Soc. Math. France, Paris, 1994.

[25] J.-M. FONTAINE & W. MESSING – "p-adic periods and p-adic étale cohomology", Current trends in arithmetical algebraic geometry (Arcata, Calif., 1985), Contemp. Math., vol. 67, Amer. Math. Soc., Providence, RI, 1987, p. 179–207.

[26] P. GABRIEL – "Des catégories abéliennes", *Bull. Soc. Math. France* **90** (1962), p. 323–448.

[27] J. GIRAUD – *Cohomologie non abélienne*, Grundlehren Math. Wiss., vol. 179, Springer-Verlag, Berlin, 1971.

[28] A. GROTHENDIECK – *Revêtements étales et Groupe Fondamental*, Lecture Notes in Math., vol. 224, Springer-Verlag, Berlin, 1971.

[29] ______, *Groupes de monodromie en géométrie algébrique (SGA7)*, Lecture Notes in Math., vol. 288 & 340, Springer-Verlag, Berlin, 1972–1973.

[30] O. HYODO – "On the de Rham-Witt complex attached to a semi-stable family", *Compositio Math.* **78** (1991), no. 3, p. 241–260.

[31] O. HYODO & K. KATO – "Semi-stable reduction and crystalline cohomology with logarithmic poles", Périodes p-adiques (J.-M. Fontaine, ed.), Astérisque, vol. 223, Soc. Math. France, Paris, 1994, p. 221–268.

[32] L. ILLUSIE – *Complexe cotangent et déformations. I*, Lecture Notes in Math., vol. 239, Springer-Verlag, Berlin, 1971.

[33] ______, *Complexe cotangent et déformations. II*, Lecture Notes in Math., vol. 283, Springer-Verlag, Berlin, 1972.

[34] ______, "Complexe de de Rham-Witt et cohomologie cristalline", *Ann. Sci. École Norm. Sup. (4)* **12** (1979), no. 4, p. 501–661.

[35] ______, "Finiteness, duality, and Künneth theorems in the cohomology of the de Rham-Witt complex", Algebraic geometry (Tokyo/Kyoto, 1982), Lecture Notes in Math., vol. 1016, Springer, Berlin, 1983, p. 20–72.

[36] L. ILLUSIE & M. RAYNAUD – "Les suites spectrales associées au complexe de de Rham-Witt", *Publ. Math. Inst. Hautes Études Sci.* **57** (1983), p. 73–212.

[37] A. J. DE JONG – "Smoothness, semi-stability and alterations", *Publ. Math. Inst. Hautes Études Sci.* **83** (1996), p. 51–93.

[38] F. W. KAMBER & P. TONDEUR – "Invariant differential operators and cohomology of Lie algebra sheaves", *Mem. Amer. Math. Soc.* **113** (1971).

[39] K. KATO – "On p-adic vanishing cycles (application of ideas of Fontaine-Messing)", Algebraic geometry, Sendai, 1985, Adv. Stud. Pure Math., vol. 10, North-Holland, Amsterdam, 1987, p. 207–251.

[40] ———, "Logarithmic structures of Fontaine-Illusie", Algebraic analysis, geometry, and number theory (Baltimore, MD, 1988), Johns Hopkins Univ. Press, Baltimore, MD, 1989, p. 191–224.

[41] ———, "Semi-stable reduction and p-adic étale cohomology", Périodes p-adiques (J.-M. Fontaine, ed.), Astérisque, vol. 223, Soc. Math. France, Paris, 1994, p. 269–293.

[42] ———, "Toric singularities", *Amer. J. Math.* **116** (1994), no. 5, p. 1073–1099.

[43] K. KATO & T. SAITO – "On the conductor formula of Bloch", *Publ. Math. Inst. Hautes Études Sci.* **100** (2004), p. 5–151.

[44] N. KATZ – "Nilpotent connections and the monodromy theorem: applications of a result of Turrittin", *Publ. Math. Inst. Hautes Études Sci.* **39** (1971), p. 355–412.

[45] S. L. KLEIMAN – "Algebraic cycles and the Weil conjectures", Dix exposés sur la cohomologie des schémas, North-Holland, Amsterdam, 1968, p. 359–386.

[46] D. KNUTSON – *Algebraic spaces*, Lecture Notes in Math., vol. 203, Springer-Verlag, Berlin, 1971.

[47] A. LANGER & T. ZINK – "De Rham-Witt cohomology for a proper and smooth morphism", *J. Inst. Math. Jussieu* **3** (2004), no. 2, p. 231–314.

[48] Y. LASZLO & M. C. OLSSON – "The six operations for sheaves on Artin stacks I: Finite Coefficients", 2005, preprint, `http://arxiv.org/abs/math/0512097`.

[49] G. LAUMON & L. MORET-BAILLY – *Champs algébriques*, Ergeb. Math. Grenzgeb. (3), vol. 39, Springer-Verlag, Berlin, 2000.

[50] P. LORENZON – "Indexed algebras associated to a log structure and a theorem of p-descent on log schemes", *Manuscripta Math.* **101** (2000), no. 3, p. 271–299.

[51] ———, "Logarithmic Hodge-Witt forms and Hyodo-Kato cohomology", *J. Algebra* **249** (2002), no. 2, p. 247–265.

[52] C. NAKAYAMA – "Nearby cycles for log smooth families", *Compositio Math.* **112** (1998), no. 1, p. 45–75.

[53] W. NIZIOŁ – "Crystalline conjecture via K-theory", *Ann. Sci. École Norm. Sup. (4)* **31** (1998), no. 5, p. 659–681.

[54] ———, "Semistable conjecture via K-theory", 2005, preprint.

[55] ———, "Toric singularities: log-blow-ups and global resolutions", *J. Algebraic Geom.* **15** (2006), no. 1, p. 1–29.

[56] A. OGUS – "Lectures on logarithmic algebraic geometry", book in preparation; version of 09-15-2006.

[57] ———, "F-crystals and Griffiths transversality", *Proceedings of the International Symposium on Algebraic Geometry (Kyoto Univ., Kyoto, 1977)* (Tokyo), Kinokuniya Book Store, 1978, p. 15–44.

[58] ———, *F-crystals, Griffiths transversality, and the Hodge decomposition*, Astérisque, vol. 221, Soc. Math. France, Paris, 1994.

[59] ———, "F-crystals on schemes with constant log structure", *Compositio Math.* **97** (1995), no. 1–2, p. 187–225, special issue in honour of Frans Oort.

[60] ———, "Logarithmic de Rham cohomology", 2000, preprint.

[61] ———, "Higgs cohomology, p-curvature, and the Cartier isomorphism", *Compos. Math.* **140** (2004), no. 1, p. 145–164.

[62] M. C. OLSSON – "Logarithmic geometry and algebraic stacks", *Ann. Sci. École Norm. Sup. (4)* **36** (2003), no. 5, p. 747–791.

[63] ———, "A stacky semi-stable reduction theorem", *Int. Math. Res. Not.* **29** (2004), p. 1497–1509.

[64] ———, "The logarithmic cotangent complex", *Math. Ann.* **333** (2005), no. 4, p. 859–931.

[65] ———, "On proper coverings of Artin stacks", *Adv. Math.* **198** (2005), no. 1, p. 93–106.

[66] ———, "Deformation theory of representable morphisms of algebraic stacks", *Math. Z.* **253** (2006), no. 1, p. 25–62.

[67] ———, "$\underline{\mathrm{Hom}}$-stacks and restriction of scalars", *Duke Math. J.* **134** (2006), no. 1, p. 139–164.

[68] ———, "Sheaves on Artin stacks", *J. Reine Angew. Math.* **603** (2007), p. 55–112.

[69] M. C. OLSSON & J. STARR – "Quot functors for Deligne-Mumford stacks", *Comm. Algebra* **31** (2003), no. 8, p. 4069–4096.

[70] M. RAPOPORT & T. ZINK – "Über die lokale Zetafunktion von Shimuravarietäten. Monodromiefiltration und verschwindende Zyklen in ungleicher Charakteristik", *Invent. Math.* **68** (1982), no. 1, p. 21–101.

[71] J.-P. SERRE – *Local fields*, Grad. Texts in Math., vol. 67, Springer-Verlag, New York, 1979.

[72] N. SPALTENSTEIN – "Resolutions of unbounded complexes", *Compositio Math.* **65** (1988), no. 2, p. 121–154.

[73] T. TSUJI – "p-adic étale cohomology and crystalline cohomology in the semi-stable reduction case", *Invent. Math.* **137** (1999), no. 2, p. 233–411.

[74] ———, "Semi-stable conjecture of Fontaine-Jannsen: a survey", Cohomologies p-adiques et applications arithmétiques (II) (P. Berthelot, J.-M. Fontaine, L. Illusie, K. Kato & M. Rapoport, eds.), Astérisque, vol. 279, Soc. Math. France, Paris, 2002, p. 323–370.

[75] I. VIDAL – "Monodromie locale et fonctions Zêta des log schémas", Geometric Aspects of Dwork Theory, W. de Gruyter, 2004, p. 983–1038.

INDEX OF NOTATION

INDEX OF TERMINOLOGY

ASTÉRISQUE

2007

316. M. C. Olsson – *Crystalline cohomology of algebraic stacks and Hyodo-Kato cohomology*
315. J. AYOUB – *Les six opérations de Grothendieck et le formalisme des cycles évanescents dans le monde motivique (II)*
314. J. AYOUB – *Les six opérations de Grothendieck et le formalisme des cycles évanescents dans le monde motivique (I)*
313. T. NGO DAC – *Compactification des champs de chtoucas et théorie géométrique des invariants*
312. ARGOS seminar on intersections of modular correspondences
311. SÉMINAIRE BOURBAKI, volume 2005/2006, exposés 952-966

2006

310. J. NEKOVÁŘ – *Selmer Complexes*
309. T. MOCHIZUKI – *Kobayashi-Hitchin correspondence for tame harmonic bundles and an application*
308. D.-C. CISINSKI – *Les préfaisceaux comme modèles des types d'homotopie*
307. SÉMINAIRE BOURBAKI, volume 2004/2005, exposés 938-951
306. C. BONNAFÉ – *Sur les caractères des groupes réductifs finis à centre non connexe : applications aux groupes spéciaux linéaires et unitaires*
305. M. JUNGE, C. LE MERDY, Q. XU – H^∞ *functional calculus and square functions on noncommutative* L^p*-spaces*

2005

304. S. SHEFFIELD – *Random surfaces*
303. N. BERGERON, L. CLOZEL – *Spectre automorphe des variétés hyperboliques et applications topologiques*
302. Formes Automorphes (II), Le cas du groupe $\mathrm{GSp}(4)$, J. TILOUINE, H. CARAYOL, M. HARRIS, M.-F. VIGNÉRAS, éditeurs
301. G. MALTSINIOTIS – *La théorie de l'homotopie de Grothendieck*
300. C. SABBAH – *Polarizable twistor* $\mathcal{D}$*-modules*
299. SÉMINAIRE BOURBAKI, volume 2003/2004, exposés 924-937
298. Formes Automorphes (I), Actes du Semestre du Centre Émile Borel, printemps 2000, J. TILOUINE, H. CARAYOL, M. HARRIS, M.-F. VIGNÉRAS, éditeurs

2004

297. Analyse complexe, systèmes dynamiques, sommabilité des séries divergentes et théories galoisiennes (II), M. LODAY-RICHAUD éditeur
296. Analyse complexe, systèmes dynamiques, sommabilité des séries divergentes et théories galoisiennes (I), M. LODAY-RICHAUD éditeur
295. Cohomologies p-adiques et applications arithmétiques (III), P. BERTHELOT, J.-M. FONTAINE, L. ILLUSIE, K. KATO, M. RAPOPORT, éditeurs
294. SÉMINAIRE BOURBAKI, volume 2002/2003, exposés 909-923
293. M. EMERTON, M. KISIN – *The Riemann-Hilbert correspondence for unit* F*-crystals*
292. F. LE ROUX – *Homéomorphismes de surfaces. Théorèmes de la fleur de Leau-Fatou et de la variété stable*
291. L. FARGUES, E. MANTOVAN – *Variétés de Shimura, espaces de Rapoport-Zink et correspondances de Langlands locales*

2003

290. SÉMINAIRE BOURBAKI, volume 2001/2002, exposés 894-908
289. P. SCOTT, G.A. SWARUP – *Regular neighbourhoods and canonical decompositions for groups*
288. M. REES – *Views of Parameter Space : Topographer and Resident*
287. Geometric Methods in Dynamics (II), volume in honor of Jacob Palis, W. DE MELO, M. VIANA, J.-C. YOCCOZ, editors
286. Geometric Methods in Dynamics (I), volume in honor of Jacob Palis, W. DE MELO, M. VIANA, J.-C. YOCCOZ, editors
285. PO HU – *Duality for Smooth Families in Equivariant Stable Homotopy Theory*
284. Autour de l'analyse microlocale, volume en l'honneur de Jean-Michel Bony, G. LEBEAU, éditeur

2002

283. L. ROBBIANO, C. ZUILY – *Analytic theory for the quadratic scattering wave front set and application to the Schrödinger equation*
282. SÉMINAIRE BOURBAKI, volume 2000/2001, exposés 880-893
281. T. DUQUESNE, J.-F. LE GALL – *Random Trees, Lévy Processes and Spatial Branching Processes*
280. A. MOKRANE, P. POLO, J. TILOUINE – *Cohomology of Siegel varieties*
279. Cohomologies p-adiques et applications arithmétiques (II), P. BERTHELOT, J.-M. FONTAINE, L. ILLUSIE, K. KATO, M. RAPOPORT, éditeurs
278. Cohomologies p-adiques et applications arithmétiques (I), P. BERTHELOT, J.-M. FONTAINE, L. ILLUSIE, K. KATO, M. RAPOPORT, éditeurs
277. B. RÉMY – *Groupes de Kac-Moody déployés et presque déployés*
276. SÉMINAIRE BOURBAKI, volume 1999/2000, exposés 865-879

2001

275. J.-M. BISMUT, S. GOETTE – *Families torsion and Morse functions*
274. A. BONNET, G. DAVID – *Cracktip is a global Mumford-Shah minimizer*
273. K. NISHIYAMA, H. OCHIAI, K. TANIGUCHI, H. YAMASHITA, S. KATO – *Nilpotent orbits, associated cycles and Whittaker models for highest weight representations*
272. M. BOILEAU, J. PORTI – *Geometrization of 3-orbifolds of cyclic type (avec la collaboration de M. HEUSENER)*
271. M. KASHIWARA, P. SCHAPIRA – *Ind-Sheaves*
270. M. BONK, J. HEINONEN, P. KOSKELA – *Uniformizing Gromov hyperbolic spaces*
269. J.-L. WALDSPURGER – *Intégrales orbitales nilpotentes et endoscopie pour les groupes classiques non ramifiés*

2000

268. J. FRANCHETEAU, G. MÉTIVIER – *Existence de chocs faibles pour des systèmes quasi-linéaires hyperboliques multidimensionnels*
267. R. CERF – *Large Deviations for three Dimensional Supercritical Percolation*
266. SÉMINAIRE BOURBAKI, volume 1998/1999, exposés 850-864
265. O. BIQUARD – *Métriques d'Einstein asymptotiquement symétriques*
264. L. LIPSHITZ, Z. ROBINSON – *Rings of Separated Power Series and Quasi-Affinoid Geometry*
263. C. SABBAH – *Équations différentielles à points singuliers irréguliers et phénomène de Stokes en dimension 2*
262. A. VASY – *Propagation of singularities in three-body scattering*
261. Géométrie complexe et systèmes dynamiques, colloque en l'honneur d'Adrien Douady, Orsay 1995, M. FLEXOR, P. SENTENAC et J.-C. YOCCOZ, éditeurs